Handbook of Polymernanocomposites. Processing, Performance and Application

Volume A: Layered Silicates

Jitendra K. Pandey
Kummetha Raghunatha Reddy
Amar Kumar Mohanty
Manjusri Misra
Editors

Handbook of Polymernanocomposites. Processing, Performance and Application

Volume A: Layered Silicates

With 209 Figures and 32 Tables

Springer

Editors
Jitendra K. Pandey
University of Petroleum
and Energy Studies (UPES)
Dehradun, India

Kummetha Raghunatha Reddy
Department of Future Industry–Oriented
Basic Science and Materials
Toyota Technological Institute
Tempaku, Nagoya, Japan

and

School of Science and Technology
Kwansei Gakuin University
Sanda, Hyogo, Japan

Amar Kumar Mohanty
Bioproducts Discovery
and Development Centre
Department of Plant Agriculture
School of Engineering
University of Guelph
Guelph, ON, Canada

Manjusri Misra
Bioproducts Discovery
and Development Centre
Department of Plant Agriculture
School of Engineering
University of Guelph
Guelph, ON, Canada

ISBN 978-3-642-38648-0 ISBN 978-3-642-38649-7 (eBook)
DOI 10.1007/978-3-642-38649-7
Springer Heidelberg New York Dordrecht London

Library of Congress Control Number: 2013955726

Printed on acid-free paper

Springer is part of Springer Science+Business Media (www.springer.com)

Contents

Contributors

Abozar Akbari Enhanced Polymer Research Group (EnPRO), Department of Polymer Engineering, Faculty of Chemical Engineering, Universiti Teknologi Malaysia, Johor Bahru, Johor, Malaysia

S. Ananda Kumar Department of Chemistry, Anna University, Chennai, TN, India

Stefano Besco Department of Industrial Engineering, University of Padova, Padova, Italy

Stéphane Bruzaud Laboratoire d'Ingénierie des Matériaux de Bretagne (LIMATB), Université de Bretagne-Sud, Cedex, Lorient, France

Engin Burgaz Department of Materials Science and Engineering, Ondokuz Mayis University, Atakum Samsun, Turkey

Jin-Hae Chang School of Energy and Integrated Materials Engineering, Kumoh National Institute of Technology, Gumi, Gyeongbuk, South Korea

Hossien Cherghibidsorkhi Enhanced Polymer Research Group (EnPRO), Department of Polymer Engineering, Faculty of Chemical Engineering, Universiti Teknologi Malaysia, Johor Bahru, Johor, Malaysia

G. Choudalakis Department of Sciences, Technical University of Crete, Chania, Greece

Yi-Lin Chung Department of Agricultural Chemistry, National Taiwan University, Taipei, Taiwan

Marcia Maria Favaro Ferrarezi Institute of Chemistry, University of Campinas (UNICAMP), Campinas, SP, Brazil

Sara Filippi Department of Chemical Engineering, University of Pisa, Pisa, Italy

Maria do Carmo Gonçalves Institute of Chemistry, University of Campinas (UNICAMP), Campinas, SP, Brazil

A. D. Gotsis Department of Sciences, Technical University of Crete, Chania, Greece

E. Bhoje Gowd Materials Science and Technology Division, CSIR-National Institute for Interdisciplinary Science and Technology (CSIR-NIIST), Thiruvananthapuram, India

Mozhgan Hayasi Department of Chemistry, Faculty of Sciences, University of Semnan, Semnan, Semnan, Iran

Junichiro Iida Department of Orthodontics, Graduate School of Dental Medicine, Hokkaido University, Sapporo, Japan

Baiju John Sealy Centre for Structural Biology and Molecular Biophysics, University of Texas Medical Branch, Galveston, TX, USA

Advanced Polymeric Nanocomposites Materials Laboratory, Toyota Technological Institute, Nagoya, Japan

Graduate School of Science and Technology, Nagasaki University, Nagasaki, Japan

Nam Hoon Kim Department of Hydrogen and Fuel Cell Engineering, Chonbuk National University, Jeonju, Jeonbuk, Republic of Korea

Tapas Kuila Department of BIN Fusion Technology, Chonbuk National University, Jeonju, Jeonbuk, Republic of Korea

Hsi-Mei Lai Department of Agricultural Chemistry, National Taiwan University, Taipei, Taiwan

Joong Hee Lee Advanced Wind Power Research Center, Department of Polymer and Nano Science and Technology, Chonbuk National University, Jeonju, Jeonbuk, Republic of Korea

Department of BIN Fusion Technology, Chonbuk National University, Jeonju, Jeonbuk, Republic of Korea

Tianxi X. Liu State Key Laboratory of Molecular Engineering of Polymers, Department of Macromolecular Science, Fudan University, Shanghai, China

Alessandra Lorenzetti Department of Industrial Engineering, University of Padova, Padova, Italy

Yashwant Mahajan Centre for Knowledge Management of Nanoscience & Technology (CKMNT), Secunderabad, AP, India

Mainak Majumder Department of Mechanical and Aerospace Engineering, Monash University, Clayton, VIC, Australia

K. Shree Meenakshi Department of Chemistry, Anna University, Chennai, TN, India

Filippo Merusi Department of Civil and Environmental Engineering, University of Parma, Parma, Italy

Sławomir Michałowski Department of Chemistry and Technology of Polymers, Cracow University of Technology, Kraków, Poland

Ananta Kumar Mishra BIN Fusion Team, Department of Polymer and Nano Science and Technology, Chonbuk National University, Jeonju, Jeonbuk, Republic of Korea

Manjusri Misra Bioproducts Discovery and Development Centre, Department of Plant Agriculture, Crop Science Building, University of Guelph, Guelph, ON, Canada

School of Engineering, Thornbrough Building, University of Guelph, Guelph, ON, Canada

Michele Modesti Department of Industrial Engineering, University of Padova, Padova, Italy

Amar Kumar Mohanty Bioproducts Discovery and Development Centre, Department of Plant Agriculture, Crop Science Building, University of Guelph, Guelph, ON, Canada

School of Engineering, Thornbrough Building, University of Guelph, Guelph, ON, Canada

Mohsen Mosadegh Department of Nanochemistry, School of Chemistry, University College of Science, University of Tehran, Tehran, Tehran, Iran

Hassan Namazi Research Laboratory of Dendrimers and Nanopolymers, Faculty of Chemistry, University of Tabriz, Tabriz, EA, Iran

Research Center for Pharmaceutical Nanotechnology (RCPN), Tabriz University of Medical Sciences, Tabriz, Iran

James Njuguna Institute for Innovation, Design and Sustainability, School of Engineering, Robert Gordon University, Aberdeen, UK

Massimo Paci Department of Chemical Engineering, University of Pisa, Pisa, Italy

Vivek Patel Centre for Knowledge Management of Nanoscience & Technology (CKMNT), Secunderabad, AP, India

Krzysztof Pielichowski Department of Chemistry and Technology of Polymers, Cracow University of Technology, Kraków, Poland

Vinicius Pistor Laboratory of Polymers (LPOL), Center for Exact Sciences and Technology (CCET), Caxias do Sul University (UCS), Caxias do Sul, RS, Brazil

Giovanni Polacco Department of Chemical Engineering, University of Pisa, Pisa, Italy

C. Ramesh Polymer Science and Engineering Division, CSIR-National Chemical Laboratory (CSIR-NCL), Pune, India

Kummetha Raghunatha Reddy Department of Future Industry–Oriented Basic Science and Materials, Toyota Technological Institute, Tempaku, Nagoya, Japan

School of Science and Technology, Kwansei Gakuin University, Sanda, Hyogo, Japan

Makoto Schreiber Bioproducts Discovery and Development Centre, Department of Plant Agriculture, Crop Science Building, University of Guelph, Guelph, ON, Canada

Department of Physics, University of Guelph, Guelph, ON, Canada

Department of Chemistry, University of Guelph, Guelph, ON, Canada

E. Pradeep Jaya Sudhan Department of Chemistry, Anna University, Chennai, TN, India

Mahsa A. Tehrani School of Industrial Technology, Universiti Sains Malaysia, Penang, Malaysia

M. J. Umapathy Department of Chemistry, Anna University, Chennai, TN, India

Singaravelu Vivekanandhan Bioproducts Discovery and Development Centre, Department of Plant Agriculture, Crop Science Building, University of Guelph, Guelph, ON, Canada

School of Engineering, Thornbrough Building, University of Guelph, Guelph, ON, Canada

Fumio Watari Department of Biomedical Materials and Engineering, Graduate School of Dental Medicine, Hokkaido University, Sapporo, Japan

Shuichi Yamagata Department of Orthodontics, Graduate School of Dental Medicine, Hokkaido University, Sapporo, Japan

Ademir José Zattera Laboratory of Polymers (LPOL), Center for Exact Sciences and Technology (CCET), Caxias do Sul University (UCS), Caxias do Sul, RS, Brazil

Hong Zhu State Key Laboratory of Molecular Engineering of Polymers, Department of Macromolecular Science, Fudan University, Shanghai, China

Polyhydroxyalkanoates-Based Nanocomposites: An Efficient and Promising Way of Finely Controlling Functional Material Properties

1

Stéphane Bruzaud

Contents

Abstract

This chapter aims at highlighting on recent advances in preparation, characterization, and functional properties of polyhydroxyalkanoates (PHA) and their layered silicate nanocomposites. These materials have attracted considerable interest in material science research. PHA are microbial polyesters produced by numerous bacteria in nature as intracellular reserve of carbon or energy. They are also generally biodegradable, with good biocompatibility, making them attractive as biomaterials. Nevertheless, biodegradable polymers alone as PHA have limited physical and mechanical properties which, at present, do not allow them to fully replace the mainstream plastics. The preparation of bionanocomposites defined as a combination between PHA and inorganic nanofillers as layered silicates is a route to enhance some of the biodegradable PHA properties. Preparative techniques include essentially intercalation of PHA in solution and melt intercalation.

S. Bruzaud
Laboratoire d'Ingénierie des Matériaux de Bretagne (LIMATB), Université de Bretagne-Sud, Cedex, Lorient, France
e-mail: stephane.bruzaud@univ-ubs.fr

J.K. Pandey et al. (eds.), *Handbook of Polymernanocomposites. Processing, Performance and Application – Volume A: Layered Silicates*, DOI 10.1007/978-3-642-38649-7_7,

This critical review highlights the major developments in this area during the last decade and focuses on the control of the functional PHA properties using layered silicate as mechanical behavior, thermal stability, gas barrier properties, and biodegradability. This review also points out some contradictory trends, which result from combined and antagonist effects. This proves that the PHA-based nanocomposite morphologies should be comprehensively investigated for predicting and interpreting the complex phenomena which can take place in such systems.

Keywords
Functional properties • Layered silicates • Morphology • Nanocomposites • Polyhydroyxyalkanoates

1 Introduction

In the general context of sustainable development, the formulation of biodegradable plastics has given rise to increasing interest. Throughout the world today, the development of biodegradable materials with controlled properties has been a subject of great research challenge to the community of scientists. These materials tend to substitute synthetic plastics in many applications which cause huge amount of waste, as, for example, packaging. However, biodegradable polymers alone have limited physical and mechanical properties which, at present, do not allow them to fully replace the mainstream plastics. One approach for overcoming these drawbacks is the incorporation of silicate layers into biopolymers to form a bionanocomposite. The preparation of bionanocomposites defined as a combination between a biopolymer and an inorganic nanofiller is a route to enhance some of the biodegradable polymer properties [1–3]. Nanoclays such as montmorillonite (MMT), saponite, or hectorite are classically used to improve biodegradable polymer stress and stiffness, reduce their gas/water vapor barrier properties, increase their thermal stability, and modify their biodegradation rate [4–7]. The best results are generally obtained when the inorganic platelets are exfoliated within the polymer matrix, which can be obtained by different methods [8]. These three main processes are in situ intercalative polymerization, melt intercalation, and exfoliation-adsorption from polymer in solution. For PHA-based nanocomposites, only these two last processes were used as explained later in this chapter.

Among these biosourced and biodegradable polymers, the family of PHA is one of the most studied [9–13]. PHA are microbial polyesters produced by numerous bacteria in nature as intracellular reserve of carbon or energy [9, 14]. They are recognized as completely biosynthetic and biodegradable with zero toxic waste and recyclable into organic waste. They are also well known for presenting good biocompatibility, making them attractive as biomaterials [13, 15]. A wide variety of different monomer compositions of PHA that provide different properties and functionalities has been described, as well as their future prospects for applications where renewable character, biodegradability, or biocompatibility is required [16].

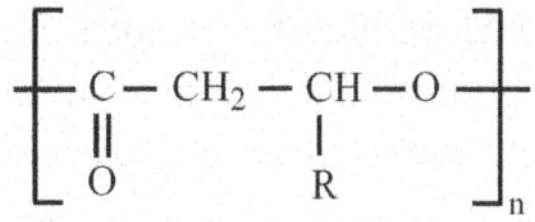

Scheme 1.1 Chemical structure of PHA

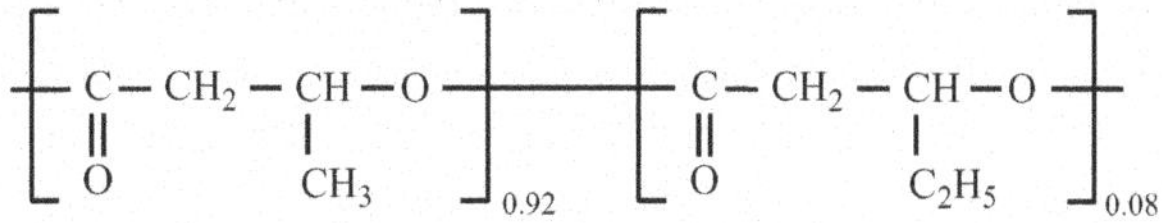

Scheme 1.2 Chemical structure of PHBV

PHA are efficiently degraded in the environment because many microorganisms in the soil are able to secrete PHA depolymerases, enzymes that hydrolyze the polymer ester bonds [17]. Microorganisms then metabolize these degradation products into water and carbon dioxide.

PHA are among the most fascinating and largest groups of biopolyesters, with over 150 different types of monomer compositions that yield polymers with tunable properties and potential applications [18, 19]. PHA are generally classified in two categories dependin g on the number of carbon atoms in their monomer units: small chain length (*scl*) PHA when the monomer units contain from 3 to 5 carbon atoms and medium chain length (*mcl*) PHA with monomer units possessing from 6 to 14 carbon atoms (Scheme 1.1).

Nevertheless, only few PHA are commercially available, mainly polyhydroxybutyrate (PHB), poly (3-hydroxybutyrate-*co*-3-hydroxyvalerate) (PHBV), and poly (3-hydroxybutyrate-*co*-3-hydroxyhexanoate) (PHBH) that are produced at a relatively large scale. Both these latter polymers are copolymers comprising 3-hydroxybutyrate units and a relatively small amount of other medium chain length 3-hydroxyalkanoate (*mcl*-3-HA) comonomers with side groups embedding at least three carbon units. The most promising PHA is PHBV in which monomer units are randomly sequenced with a majority proportion of 3-hydroxybutyrate units (more than 90 %) (Scheme 1.2). PHBV has attracted some scientific interest because it possesses characteristics similar to synthetic thermoplastics. PHBV is often compared to polypropylene (PP) with regard to its physical properties because they have similar melting points, degree of crystallinity, and glass transition temperatures. In general, PHB is stiffer and more brittle than PP. In addition, PHB exhibits much lower solvent resistance even if PHB has better natural resistance to ultraviolet radiation than PP.

There are several different grades of copolymers available, depending on the average molecular weight, average *mcl*-3-HA content within the copolymer, and side group chain length of the chosen *mcl*-3-HA unit. PHA copolymers with different *mcl*-3-HA types and contents can be synthesized either by bacterial fermentation or by chemical processes [20, 21]. The incorporation of *mcl*-3HA units into PHA effectively lowers both the crystallinity and the melting temperature (T_m) in a manner similar to the branching of R-olefins in linear low-density polyethylene [18]. T_m can be lowered far below the thermal decomposition temperature of PHA to make this material much easier to process.

The reduced crystallinity provides the ductility, toughness, and low shrinkage required for many practical applications. The *mcl*-3-HA content regulates T_m and crystallinity of copolymer almost independently of the branch size, as long as more than three carbons are present in the side group. On the other hand, the side group chain length of the *mcl*-3-HA has a significant effect on copolymer flexibility. Finally, the properties of PHA copolymers can be tuned by changing the type, content, and distribution of comonomer units comprising the polymer chains, as well as the average molecular weight and molecular weight distribution. These properties allow potential uses of PHA for a wide range of applications. Interest in the use of PHA for packaging, medical, agricultural, and fishery applications has recently increased [16]. There are also further applications of PHA within the medical and pharmaceutical industries, primarily due to their biodegradability and biocompatibility [13].

2 Preparative Techniques of PHA-Based Nanocomposites

Several usual strategies have been considered to prepare polymer-layered silicate nanocomposites. These different routes have been presented in detail several times elsewhere in this book. They include three main processes according to the starting materials and processing techniques [4]:

- *Exfoliation-adsorption*: The layered silicate is exfoliated into single layers using a solvent in which the polymer is soluble. It is well known that such layered silicates, owing to the weak forces that stack the layers together, can be easily dispersed in an adequate solvent.
- In situ *intercalative polymerization*: In this technique, the layered silicate is swollen within the liquid monomer (or a monomer solution) so as the polymer formation can occur in between the intercalated sheets. Polymerization can be initiated either by heat or radiation, by the diffusion of a suitable initiator, or by an organic initiator or catalyst fixed through cationic exchange inside the interlayer before the swelling step by the monomer.
- *Melt intercalation*: The layered silicate is mixed with the polymer matrix in the molten state. Under these conditions and if the layer surfaces are sufficiently compatible with the chosen polymer, the polymer can crawl into the interlayer space and form either an intercalated or an exfoliated nanocomposite. In this technique, no solvent is required. Obviously, the latter method is highly preferred in the context of sustainable development since it avoids the use of organic solvents, which are not eco-friendly and then alter the life cycle analysis.

While many processes for preparing PHA-based nanocomposites in solution or via melt intercalation have been investigated in detail during the last decade, the in situ intercalative polymerization has not been reported so far. The main reason is probably the difficulty to polymerize β-butyrolactones, substituted or not, in the experimental conditions for the ring-opening polymerization requiring specific metallic catalysts. This technique revolves around metal (essentially aluminum, zinc, tin, or rare earths)-containing complexes as initiating species for the ring-opening polymerization of β-butyrolactones [22]. In the literature, only exfoliation-adsorption in solution and melt intercalation were found as techniques allowing to the preparation of PHA-based nanocomposites [1, 3].

3 PHA-Based Nanocomposite Properties

3.1 Mechanical Properties

Some mechanical drawbacks of PHA, such as brittleness, may restrict their developments and uses. Therefore, bionanocomposites appear as an efficient way to overcome these problems and to improve mechanical behavior [23]. Biodegradable nanocomposites consisting of biodegradable polymers and layered silicates (organically modified or not) frequently exhibit remarkably improved mechanical properties when compared to those of virgin polymers [1]. Improvements generally include a higher modulus both in solid and melt state and an increased strength. The main reason for these improved properties in nanocomposites is the stronger interfacial interaction between the matrix and layered silicate, compared with conventional filler-reinforced systems.

PHB/layered silicate nanocomposites were successfully prepared through melt extrusion [24]. The nature and amount of clay and organic modifier present in the system markedly influence properties of the nanocomposites. Significant improvement of mechanical properties of nanocomposites as compared to pristine PHB has been observed. The storage modulus increases as the nanoclay content in the nanocomposite increases (Table 1.1). The increment in the storage modulus as compared to pure PHB at same temperature is 35 % when 3.6 wt% of organically modified nanoclays (montmorillonite ion exchanged with dimethyl-octadecylamine) is incorporated in PHB matrix.

The preparation of PHBV-based nanocomposites can be also achieved through a solution intercalation process in the presence of organically modified montmorillonite, i.e., Cloisite 15A [8]. Mechanical properties have been studied by several techniques such as tensile tests and nanoindentation. The Cloisite 15A incorporation into PHBV matrix leads to a large increase of the Young's moduli of nanocomposites, determined using usual tensile tests, at filler contents as low as a few weight percent (Table 1.2). It is drastically increased from 633 MPa for pure PHBV to 1,677 MPa for the nanocomposite containing 5 wt% of filler.

The reinforcement effect R, which corresponds to the ratio of the tensile modulus of the nanocomposite to the tensile modulus of the pure polymer, can be calculated. PHBV-based nanocomposites studied here yield R values of 1.6, 2.1 and 2.6 for 1, 2.5 and 5 wt%, respectively, which is significantly high. Concerning the evolution of the maximal stress at break which expresses the ultimate stress that the material can bear before break, the differences observed are sufficiently notable to draw some conclusions. The increase of the stress at break is probably explained by the presence of polar polymer (PHBV) leading to a strong interaction between the polymer and silicate layers. For instance, in the case of polypropylene-based nanocomposites, no

Table 1.1 Storage modulus at 20 °C of pure PHB and PHB-based nanocomposites [24]

MMT content (wt%)	Storage modulus (GPa)	% increment as compared to PHB (%)
0	3.2	–
1.2	3.6	13
2.3	4.0	25
3.6	4.5	35

Table 1.2 Tensile properties at 23 °C of pure PHBV and PHBV-based nanocomposites [8]

Cloisite 15A content (wt%)	Young's modulus (MPa)	Tensile stress (MPa)	Elongation at break (%)
0	633 ± 19	5.9 ± 2.3	3.3 ± 0.5
1	1,043 ± 58	11.8 ± 1.7	2.7 ± 0.4
2.5	1,311 ± 142	18.0 ± 3.2	1.8 ± 0.9
5	1,677 ± 121	28.9 ± 3.1	1.4 ± 0.3

or only very slight tensile stress enhancement are measured [25]. On the contrary, nanocomposites based on more polar polymers such as poly(methyl methacrylate)s or polyamides exhibit an increase in the stress break with the filler content, which is more pronounced for polyamides [26, 27]. Lastly, the elongation at break tends to decrease as expected for such materials when the interaction between the polymer and the filler becomes stronger. Nanoindentation is a powerful technique which can provide mechanical properties at the very first surface layers in comparison to classical tensile test. Mechanical moduli obtained using nanoindentation and their evolutions are very close to those measured in tensile tests (Fig. 1.1).

An improvement of modulus and hardness with the increase of clay loading is also observed. As regards the mechanical behavior of the different nanocomposites, it is shown a significant improvement in modulus, stress, and hardness with increase of clay loading, due to the addition of stiff clay nanofillers into the PHBV matrix. Finally, this work demonstrates the good agreement between the mechanical measurements carried out using tensile tests, which can be considered as a macroscopic characterization technique and those obtained by the nanoindentation technique which allows the determination of the mechanical properties at the nanometric scale. This good correlation of mechanical properties at the macro- and nanometric scales can be attributed to a high degree of dispersion of nanoplatelets within the polymer matrix. Indeed, similar moduli on the macroscopic scale and for localized points evidence that the tensile modulus is that of a homogeneous material and not an averaged value from different phases.

Chardron et al. reported a recent and original study concerning *mcl*-PHA which have been produced by fed-batch cultivation of *Pseudomonas oleovorans* [28]. The *mcl*-PHA obtained was composed of two different monomer units which are 3-hydroxyoctanoate and 3-hydroxyhexanoate with molar composition equal to 92.4 and 7.6 mol%, respectively. Then, *mcl*-PHA-based nanocomposites have been prepared using Cloisite 15A as nanofiller and characterized.

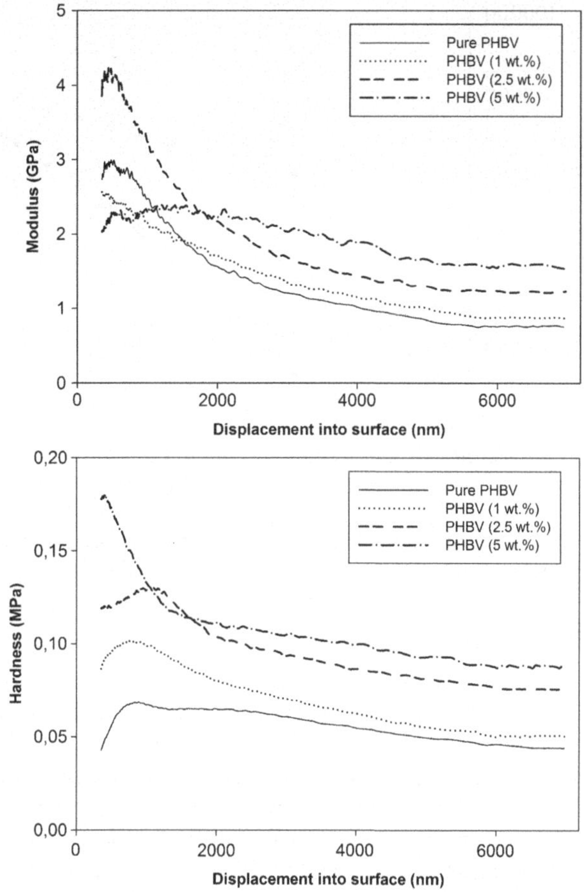

Fig. 1.1 Modulus (**a**) and hardness (**b**) profiles of pure PHBV and PHBV-based nanocomposites containing 1, 2.5, and 5 wt% of Cloisite 15A, determined using nanoindentation (Reprinted with permission from [8]. Copyright (2007) Elsevier)

The storage moduli G' measured using linear viscoelastic characterization for neat *mcl*-PHA and *mcl*-PHA nanocomposites with various clay loadings is compared in Fig. 1.2. The curves show that the storage modulus of nanocomposite samples filled with 1 and 2.5 wt% of Cloisite 15A is substantially higher than that of neat *mcl*-PHA in the low-frequencies domain. A more drastic increase of the storage modulus is observed for the sample containing 5 wt% of Cloisite 15A. The reasons evocated by the authors call for the confinement effect and the interparticle interactions, which result in the enhancement of low-frequencies G' in comparison with the polymer matrix.

Due to the small quantities of *mcl*-PHA produced via bacterial synthesis, the technique used for determining mechanical characteristics of *mcl*-PHA and their nanocomposites is nanoindentation. The storage modulus of nanocomposites is significantly higher compared to pure *mcl*-PHA. It is drastically increased from 100 MPa for neat *mcl*-PHA to 170, 490, and 940 MPa for nanocomposites that contain 1, 2.5, and 5 wt% of filler, respectively. The reinforcement effect R, which

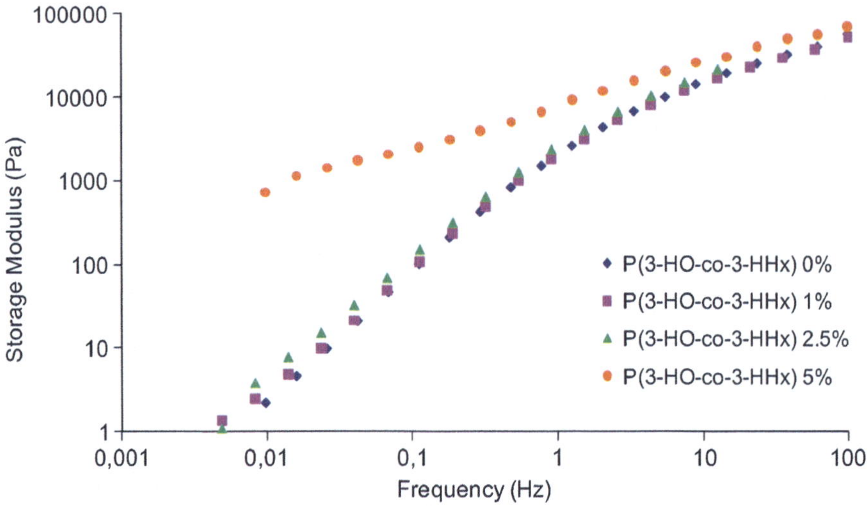

Fig. 1.2 Storage modulus G' as a function of frequency for pure *mcl*-PHA and *mcl*-PHA-based nanocomposites containing 1, 2.5, and 5 wt% of Cloisite 15A (Reprinted with permission from [28]. Copyright (2010) Elsevier)

corresponds to the ratio of the storage modulus of the nanocomposite to the storage modulus of the pure polymer, can be easily calculated. The *mcl*-PHA/Cloisite 15A nanocomposites studied here yield R values of 1.7, 4.9, and 9.4 for 1, 2.5, and 5 wt%, respectively, which corresponds to a significant increase in the modulus. Associated to the results previously obtained using rheology, this significant enhancement in modulus by the incorporation of a small amount of Cloisite 15A can be attributed to the portion of platelets exfoliated, resulting in a greater *mcl*-PHA matrix nanofiller interfacial area.

Chen et al. prepared a PHBV/clay composite by using polymer intercalation from solution [29]. Organically modified montmorillonite was prepared by cationic exchange between Na^+ in MMT galleries and hexadecyltrimethylammonium bromide in an aqueous solution. The influence of the clay content on the mechanical properties of the nanocomposites showed that with the incorporation of 3 wt% clay, the tensile strength of hybrid increased to 35.6 MPa, which is about 32 % higher than that of the original PHBV, and the tensile modulus was also increased. The PHBV/clay nanocomposites showed the best balance at 3 wt% clay loading among the samples prepared. When the clay content exceeds 3 wt%, both the tensile strength and the strain at break were decreased. These decreases may be caused by the aggregation of clay, which leads to the loss of the feature of a nanometer composite.

Tensile properties (i.e., Young's modulus, stress at break, and elongation at break) were also measured by Choi et al. for PHBV-Cloisite 30B nanocomposites prepared by melt intercalation [30]. The tensile strength is greatly improved with only a little increase in the stress at break for the PHBV-based nanocomposites compared with the PHBV copolymer alone. Even at low clay content (i.e., 3 wt% organoclay), the Young's modulus is significantly increased from 480 to 790 MPa. Small amounts of nanodispersed clay act as effective reinforcing filler to enhance the mechanical properties of the PHBV copolymer. This improvement is also attributed to the strong hydrogen bonding between PHBV copolymer and Cloisite 30B, indicating the importance of the strong interaction between polymer and organoclay for the formation of nanocomposites with fine dispersion, as seen in the results obtained from X-ray diffractometry and transmission electron microscopy.

3.2 Thermal Stability

The thermal stability of polymeric materials is usually studied by thermogravimetric analysis (TGA). The weight loss due to the formation of volatile products after degradation at high temperature is monitored as a function of temperature. When the heating occurs under an inert gas flow, a non-oxidative degradation occurs, while the use of air or oxygen allows oxidative degradation of the samples. Generally, the incorporation of clay into the polymer matrix was found to enhance thermal stability by acting as a superior insulator and mass transport barrier to the volatile products generated during decomposition [1]. Nevertheless, some exceptions were found in literature.

PHBV-organoclay nanocomposites have been prepared by melt intercalation using Cloisite 30B [30]. X-ray diffractometry and transmission electron microscopy analyses clearly confirm that an intercalated microstructure is formed and finely distributed in the PHBV copolymer matrix because PHBV has a strong hydrogen bond interaction with the hydroxyl group in the organic modifier of Cloisite 30B. The thermal stability was studied and indicated the 3 % weight loss of pure PHBV copolymer started at 252 °C, and that of the nanocomposites with 1 % and 2 % Cloisite 30B as organoclay increases to 259 °C. The nanocomposite with 3 % organoclay has a much higher weight loss temperature of 263 °C. The increase in thermal stability observed for nanocomposites may be related to the nanodispersion of the silicate layers. The well-dispersed and layered structure of clay in the polymer matrix is thought to be an effective barrier to the permeation of oxygen and combustion gas, which improves the thermal stability.

Bruzaud et al. showed that all the PHBV-based nanocomposites degrade at a higher temperature than the pure PHBV [8]. The thermal stability of the nanocomposites systematically increases with increasing Cloisite 15A, up to a loading of 5 wt% (Fig. 1.3).

For example, the temperature corresponding to 50 % degradation of pure PHBV is 270 °C as usually met in the literature [31], but that of the PHBV-based nanocomposite is 300 °C, indicating a 30 °C improvement with just 5 wt% of clay. This demonstrates that the thermal decomposition process of the material takes more time to start in the presence of a few percent organoclay. This shift to higher temperature for the matrix decomposition can be explained by a decrease in the diffusion of oxygen and volatile degradation products throughout the composite

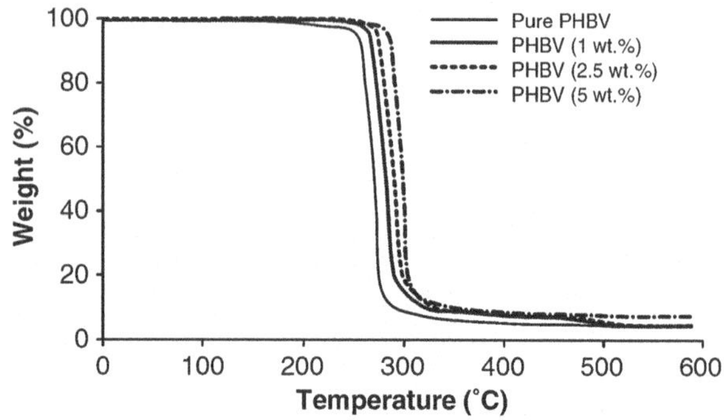

Fig. 1.3 TGA curves of pure PHBV and PHBV-based nanocomposites containing 1, 2.5, and 5 wt% of Cloisite 15A (Reprinted with permission from [8]. Copyright (2007) Elsevier)

material due to the homogeneous incorporation of clay sheets. The clay acts as a heat barrier, which enhances the overall thermal stability of the system.

Nevertheless, the opposite trend was also found in literature. Zhang et al. described the preparation and the thermal stability of layered silicate nanocomposites based on poly(3-hydroxybutyrate-*co*-3-hydroxyhexanoate) (PHBH, known as Nodax^{TM}) [32]. The dispersed phases were organically modified montmorillonites (clay 20A and clay 25A), and they were introduced by solution mixing. Wide-angle X-ray scattering results and transmission electron microscopy images confirmed that these two clays were intercalated and finely distributed in the Nodax^{TM} matrix. The results showed that the onset temperature was slightly decreased with increasing layered silicates concentration in these composites (Table 1.3).

Compared with that of neat Nodax^{TM}, the end set temperature was greatly improved at concentrations of fillers less than 5 wt%, for the two silicates. The interval temperature between degradation onset and end set showed a slight broadening for the Nodax^{TM} clay 20A series but remained nearly constant around a value of 26 °C for the other series with clay 25A. However, the interval temperatures were increased by about 10 °C compared with that for the neat Nodax^{TM}. These layered fillers, i.e., clay 20A and clay 25A, slightly increased the degradability of the Nodax^{TM} in agreement with studies on other reinforced biodegradable polymers in some cases [33].

Recently, Hablot et al. studied the influence of fermentation residues and quaternary ammonium salts on the thermal and thermomechanical degradation of a biodegradable bacterial PHB [34]. The obtained results reveal that ammonium cations greatly enhance the degradation leading to a dramatic decrease in PHB molecular weight, caused by the decomposition products of clay organomodifiers which have a catalyzing effect on the thermal or thermomechanical degradation. A possible mechanism of degradation has been proposed based on characterizations using nuclear magnetic resonance [35]. The ammonium surfactant turns into an amine through either nucleophilic attack of the ammonium counter-ion or Hofmann elimination. Then, the released acidic proton and/or the nucleophilic amine formed act as catalysts of the PHB random chain scission (Scheme 1.3). Besides, they showed that the presence of fermentation residues does not affect significantly the PHB thermal stability in comparison to the ammonium cations.

Table 1.3 Thermal stability parameters determined by TGA of pure PHBH and PHBH-based nanocomposites [32]

Filler	Filler content (wt%)	Onset temperature (°C)	End set temperature (°C)	ΔT (°C)
–		289.5	305.5	16.0
Cloisite 20A	1	286.2	316.7	20.5
Cloisite 20A	3	278.8	308.7	29.9
Cloisite 20A	5	277.0	307.4	29.6
Cloisite 20A	7	272.7	302.7	30.0
Cloisite 20A	10	247.9	292.5	44.6
Cloisite 20A	15	247.9	289.2	41.3
Cloisite 25A	1	285.2	311.4	26.2
Cloisite 25A	3	284.3	310.0	25.7
Cloisite 25A	5	286.4	312.6	26.2
Cloisite 25A	7	281.6	304.4	22.8
Cloisite 25A	10	280.7	303.8	23.1
Cloisite 25A	15	272.5	303.0	30.5

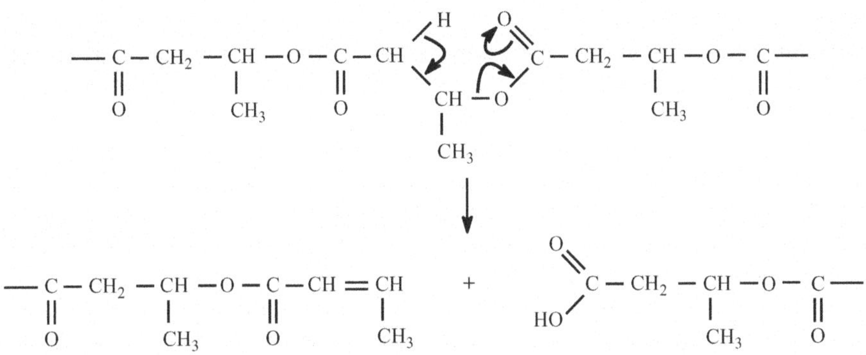

Scheme 1.3 PHB random chain scission

A similar study on PHB-based nanocomposites was carried out by Lim et al. [36]. They used solvent intercalation route to obtain PHB/Cloisite 25A with 3, 6, and 9 wt% of clay content. X-ray diffractometry data led to the conclusion of intercalated structures, the interlayer distance reaching 35 Å, but no dependence on clay content was observed. TGA results indicated an increase of the onset temperature of weight loss and a decrease of the degradation rate with 3 wt% of Cloisite 25A. This was attributed to the nanoscale layers dispersion decreasing the diffusion of volatile decomposition products. At higher clay contents (more than 6 wt%), although the onset of thermal degradation did not increase because of the organomodifier's thermal sensitivity, the nanocomposite degradation rates decreased due to restricted thermal motion of the polymer chains in the clay interlayer.

In fact, despite the general improvement of thermal stability, decreases in the thermal stability of polymers upon nanocomposite formation have also been reported, and various mechanisms have been put forward to explain the results. It has been argued, for example, that after the early stages of thermal decomposition the stacked silicate layers could hold accumulated heat, acting as a heat source to accelerate the decomposition process, in conjunction with the heat flow supplied by the outside heat source [5]. Also, the alkylammonium cations in the organoclay could suffer decomposition following the Hoffmann elimination reaction, and the product could catalyze the degradation of polymer matrices. Moreover, due to the presence of some residual impurities, the clay itself can also catalyze the degradation of polymer matrices [7].

Thus, it becomes obvious that the organoclay may have two opposing functions in thermal stability of nanocomposites: a barrier effect, which should improve the thermal stability, and a catalytic effect on the degradation of the polymer matrix, which should decrease the thermal stability.

3.3 Gas Barrier Properties

Generally, clays are believed to increase the barrier properties by creating a maze or "tortuous path" that retards the progress of the gas molecules through the matrix resin [37]. The presence of filler, spherical, plate, cylindrical, etc., introduces a tortuous path for a diffusing penetrant. The reduction of permeability arises from the longer diffusive path that the penetrants must travel in the presence of filler [38]. A sheet-like morphology is particularly efficient at maximizing the path length due to the large length-to-width ratio, when compared to other filler shapes such as spheres or cubes [39].

While numerous results have been published on the permeability of PLA-based or PCL-based nanocomposites, very few studies were found in the literature concerning the barrier properties of PHA and PHA-based nanocomposites [1].

Corrêa et al. recently measured the oxygen permeability values for different formulated PHBV-based nanocomposites with Cloisite 30B obtained by melt processing, in the presence, or not, of acetyltributylcitrate as plasticizer [40]. The nanofillers have a strong impact in the oxygen permeability reduction, corroborating the formation of intercalated/exfoliated plasticized bionanocomposites. The crystallinity content being equivalent, the authors compared each formulated systems. They showed that the plasticizer increases the permeability due to an increase of the matrix mobility; free volume is created, which increases gas diffusion through the polymer. The nanofiller reduces the permeability due to the increase of tortuosity, in correlation with an intercalated/exfoliated nanostructure. The values obtained for PHBV formulated with 3 wt% of Cloisite 30B and 10 wt% of plasticizer are an average compromise between two antagonist effects, a strong increase and a low decrease of the permeability, due to the plasticizer and the nanofiller addition, respectively. Interestingly, addition of nanoclays to the plasticized PHBV results in a significant decrease in oxygen permeability, attesting for the good nanostructure of the sample and for its promising properties.

Surprising and very interesting results have been published by Sanchez-Garcia et al. on permeability of PHB-nanocomposites [41]. A comprehensive study about the polymer morphology and its barrier properties is carried out at the light of the most commonly considered models for permeability reduction in nanocomposites. D-limonene and specially water molecules were, however, found to sorb in both hydrophobic and hydrophilic sites of the filler, respectively, hence diminishing the positive barrier effect of an enlarged tortuosity factor in the permeability. The barrier properties of the systems were not seen to fit the most widely applied models such as those of Nielsen and Fricke for oriented and random dispersion of the fillers, in that barrier enhancements were found to depend on the penetrant and did not clearly match morphological observations in terms of aspect ratio. The reason evocated by the authors for this disagreement must be attributed to limitations of the models to account for factors such as polymer morphology and crystallinity alterations, irregular morphology and orientation of the filler platelets, chemical alterations in the matrix, and solubility of the penetrants in the filler.

As generally reported in the literature, it is clear that nanocomposites containing layered silicates show enhanced barrier properties. However, from the above discussion, some contradictory results were found in this domain because the dependence of many factors on the nanocomposite permeability is not still well understood.

3.4 Biodegradability

Biodegradable polymers have great commercial potential for bioplastic. So, preparation to processing of biodegradable polymer-based nanocomposites, that is green nanocomposites, is the wave of the future and considered as the next generation materials [42]. The durability of PHA nanocomposites has been evaluated under different environments [43]. Degradation rate was checked in nanocomposites in the presence of clay particle by numerous authors. An interesting aspect of nanocomposite technology is the modification in biodegradability, often reported after nanocomposite formation.

Maiti et al. first reported the biodegradability of the PHB and its nanocomposites under compost [44]. The degradation started just after 1 week, and at the initial stage the weight loss was almost the same for both PHB and its nanocomposites. Deviation occurred after 3 weeks of exposure, but degradation tendency of nanocomposites was suppressed. They assumed that the retardation of biodegradation of PHB is because of the improvement of the barrier properties of the matrices after nanocomposite preparation with layered silicates, even though they did not report about the permeability [1]. However, according to Sinha Ray et al. in the case of PLA-based nanocomposites, they showed that there is no relation between the biodegradability and the gas barrier properties [43, 45]. Some nanocomposites were

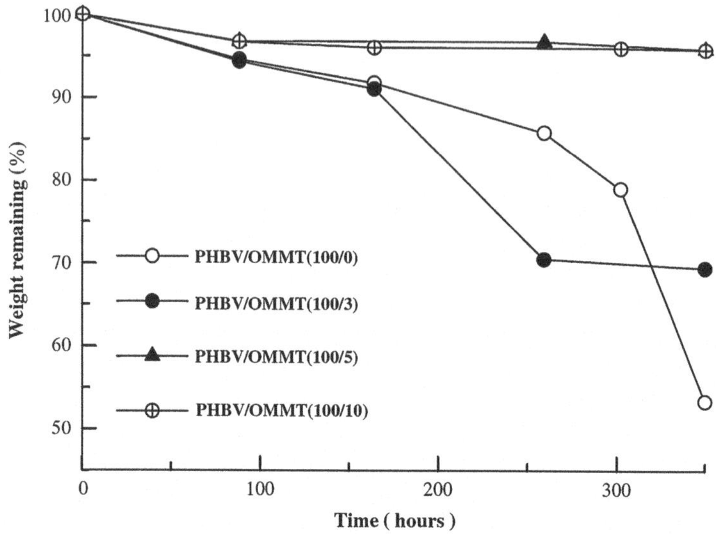

Fig. 1.4 Weight remaining of pure PHBV and PHBV-based nanocomposites containing 3, 5, and 10 wt% of OMMT, in soil suspension (Reprinted with permission from [46]. Copyright (2005) Elsevier)

degraded with higher rate having significantly improved barrier properties as compared with that of neat PLA [45].

PHBV/organophilic montmorillonite (OMMT) nanocomposites containing different loadings of OMMT (3, 5, and 10 wt%) were prepared by solution intercalation method [46]. Then, its biodegradability was investigated by a degrading cultivation method in soil suspension (Fig. 1.4).

As an example, after degradation for 88 h, the weight loss of PHBV was 5.35 %, weight losses of nanocomposites for 100/3, 100/5, and 100/10 were 5.61 %, 3.18 %, and 3.25 %, respectively. The weight loss of PHBV after degradation for 164 h was 8.19 %, and weight losses of nanocomposites for 100/3, 100/5, and 100/10 were 8.89 %, 3.22 %, and 3.93 %, respectively. These results indicated that the biodegradability of PHBV/OMMT in soil suspension decreased with increasing amount of OMMT, which is in agreement with the results of the biodegradability for aliphatic polyester/OMMT in activated soil [47]. For explaining these results, the authors suggest that the incorporation of clay in the polymer matrix affected the crystallization of the polymer in the composites in opposite ways. On the one hand, the filler platelets modified the nucleation, which increased the number of crystalline nuclei, causing a more rapid crystallization rate. On the other hand, the number of crystallizable PHBV chains was observed to decrease due to the restricted motion in the presence of the clay. Thus, two opposing factors of an increasing crystallization rate and decreasing crystallinity of the polymer as a function of amount of filler in the composites were observed. Similar restrictions to the segmental motion at the organic–inorganic interface were also suggested to be responsible for the decrease of the biodegradability in PHBV nanocomposites.

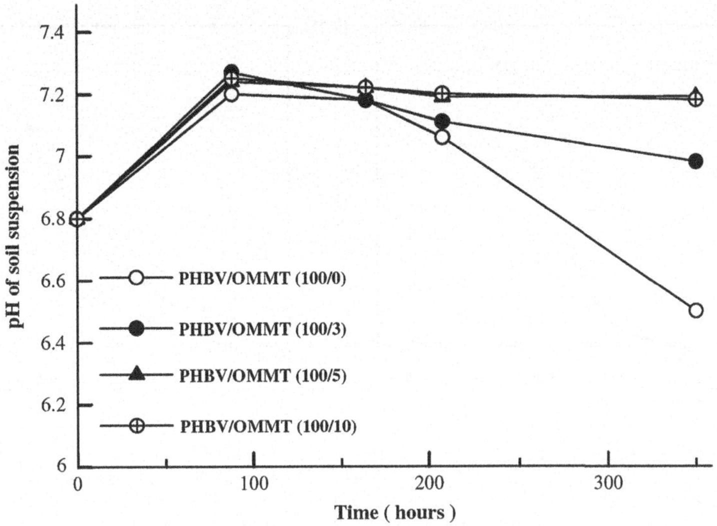

Fig. 1.5 Variations of pH in soil suspension during the biodegradation of pure PHBV and PHBV-based nanocomposites containing 3, 5, and 10 wt% of OMMT (Reprinted with permission from [46]. Copyright (2005) Elsevier)

These authors also showed that the pH of soil suspension decreased with the progress of biodegradation (Fig. 1.5).

In fact, the biodegradation of PHBV is a combination of hydrolysis and microbial metabolism. In the process, PHBV hydrolyzes under the action of depolymerase excreted by degraders; in the initial stage, PHBV with relatively high molecular weight is degraded to low molecular weight and terminal hydroxyl and carboxyl groups are gradually formed. Therefore, the pH of soil suspension decreased, and the weight losses increased with the progress of biodegradation. Figures 1.4 and 1.5 showed that the higher the weight loss of PHBV/organophilic montmorillonite nanocomposite, the lower the pH of soil suspension cultivation liquid.

Song et al. mainly discussed the effect of the structure of PHBV/OMMT nanocomposites on its biodegradability [48]. They assumed that there are at least four aspects affecting the biodegradation of material:

1. Essential properties of materials (composition, structure, molecular weight, etc.)
2. Processing of the material (type of processing, surface characteristics, etc.)
3. Physicochemical parameters of the ecosystem (temperature, pH, oxygen content, nutrient supply, etc.)
4. Microbial parameters of the ecosystem (population density, microbial diversity, microbial activity, etc.)

Ishida et al. also described the relationships between not only comonomer unit compositions but also their distributions and structures as well as properties for many bacterial copolyhydroxyalkanoates [49]. It was found that the physical properties and mainly biodegradabilities of copolyhydroxyalkanoates depend not only on the chemical structure and the comonomer unit composition but also on

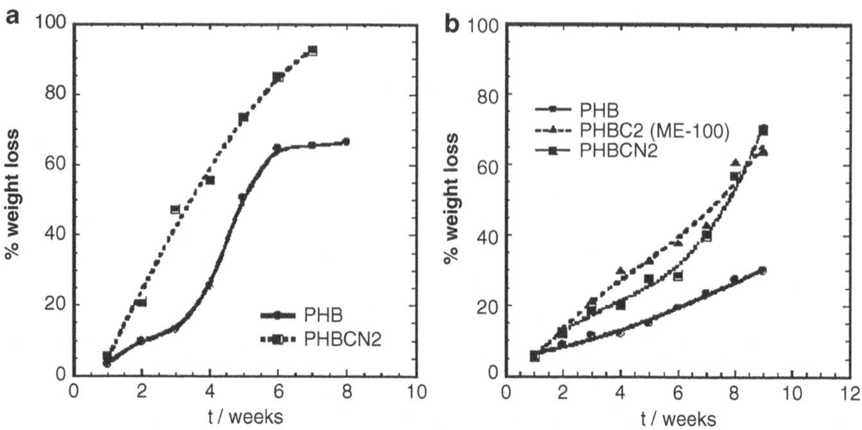

Fig. 1.6 Percentage weight loss during biodegradation, in the compost media, of pure PHB and PHB-based nanocomposites containing 2 wt% of OMMT (**a**) at room temperature and (**b**) at 60 °C (Reprinted with permission from [24]. Copyright (2007) American Chemical Society)

the comonomer unit compositional distribution. Moreover, the morphology of nanocomposites also plays an important role in determining the biodegradability [50]. Nanocomposites having exfoliated and intercalated structures were obtained by employing two different organically modified nanoclays. Nanocomposites with the exfoliated structure had better biodegradability than nanocomposites with the intercalated structure or pure polymer.

However, contradictory results concerning the effect of clay dispersion on polymer biodegradability are also found in the literature. These numerous parameters affecting the biodegradability of PHA-based nanocomposites may explain the opposite behaviors for others bionanocomposites based on PLA or PCL. Indeed, the biodegradability of reinforced PLA or PCL strongly depends on the nature of the layered silicate and the organic modifier, making it possible to tailor the material's biodegradability by adding an appropriate organically modified clay as nanofiller [51, 52]. Currently, it has been also stated that a faster hydrolytic degradation takes place for more hydrophilic fillers [53–56]. Moreover, the biodegradation rate increases in presence of MMT might be due to the presence of Al-lewis sites which catalyze the hydrolysis of the ester linkages. Even in the case of PHB-based nanocomposites, the rate of biodegradation of PHB can be enhanced dramatically in the nanohybrids [24]. Maiti et al. reported results on PHB biodegradability in compost media and showed remarkable enhancement of biodegradation rate in the presence of only 2 wt% of clay as a result of changes in crystallinity (Fig. 1.6).

Biodegradation is clearly enhanced with clay at 20 °C. For instance, near complete biodegradation is observed at 20 °C in about 7 weeks for PHB containing 2 wt% of clay. On the other hand, biodegradation drastically decreases for PHB and their nanocomposites at 60 °C. Biodegradation rate decreases 2–3 times by increasing

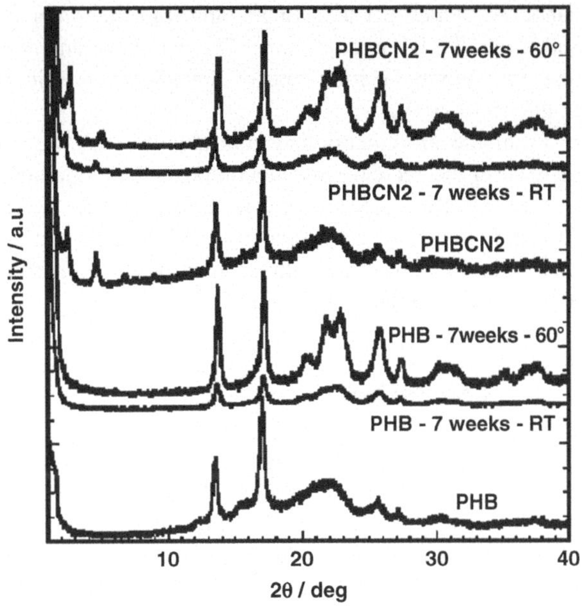

Fig. 1.7 X-ray diffractometry diagrams of pure PHB and PHB-based nanocomposites containing 2 wt% of OMMT, before and after 7 weeks of biodegradation at room temperature (RT) and at 60 °C (Reprinted with permission from [24]. Copyright (2007) American Chemical Society)

the media temperature from 20 °C to 60 °C. The reasons evocated by the authors are a lower concentration of microorganisms at 60 °C and a higher amount crystallinity of PHB at 60 °C. For samples studied at 20 °C, near glass transition temperature of PHB ($T_g \sim 16$ °C), there is not sufficient segmental motion in the polymer chains for PHB molecules to crystallize (Fig. 1.7). Hence, the crystallinity of PHB cannot increase further during the biodegradation studies at 20 °C.

Finally, the rate of biodegradation can be fine-tuned by either the addition of nanoparticles or other processing that affects the crystallinity of the samples.

A typical example of this complex phenomenon is the work very recently published by Corrêa et al. which have formulated different PHBV-Cloisite 30B nanocomposites, in the presence, or not, of acetyltributylcitrate as plasticizer [40]. In order to estimate the final biodegradability of the different systems, composting tests were performed. The time evolution of the samples weights was monitored and indicated the weight losses of the specimens after 30 and 90 days of biodegradation, respectively. Results showed that in the compost conditions, the weight losses are globally low (less than 10 %) even after 3 months of biodegradation. Since the crystallinity extents of the different formulated systems are similar, variations in the biodegradation behavior should only result from the formulations and the resulting nanostructures differences, between the samples. The authors showed that, on one hand, the addition of plasticizer causes a decrease in the PHBV biodegradation. This is likely due to the plasticizer which could hinder the biodegradation. On the other hand, the nanofiller is responsible for a significant enhancement in the biodegradation. This behavior is likely due to the nanoclays which act as sponges and trap the water molecules, promoting biotic and abiotic PHBV degradations. The combined antagonist effects of both additives lead to a biodegradation rate of the plasticized bionanocomposite quite similar to the neat PHBV.

From the aforementioned contradictory results, it becomes obvious that the increase or the decrease in nanocomposite biodegradability is still under discussion, and no definitive conclusion can be driven about mechanisms on the basis of the current literature [7, 43]. Numerous phenomena can take place, and the modification in the PHA biodegradation rate strongly depends on the predominance attributed to each phenomenon.

4 Conclusion

This paper provides a comprehensive and critical review on PHA-based nanocomposites. The formulation of layered silicates into PHA matrices, not only the preparation of PHA-based nanocomposites, but also the consequences on the PHA functional properties, has been discussed for a large number of PHA homo- or copolymers. Substantial progress has been made in the development, processing, and microstructural aspects of PHA reinforced with low-level loading of layered silicates over the recent years. As expected in comparison with other polymer matrices, the improvement in PHA properties generally includes high modulus, strength, and toughness; decrease gas permeability; increase and regulated biodegradability and thermal stability. Basic reason for those improvements in properties is interfacial interaction between polymer matrix and nanoparticle, attributed to the large aspect ratio. The above properties are attained at low silicate content (lower than 5 wt%) in comparison with conventionally filled systems. Therefore, polymer-layered silicate nanocomposites are far lighter in weight than a conventional composite, which makes them quite competitive for specific applications.

It has been clearly demonstrated that numerous parameters such as, for instance, elaboration route, polymer/clay affinity, clay choice, and clay content can affect the structure, the morphology, and then, the bionanocomposite properties. This explains why, in some cases, contradictory results were found in literature. The effect of these different parameters dramatically affects the material morphology, and all the reasons are not yet fully understood.

Understanding the synthesis-structure–property relationship of nanocomposites is vital for a better development of advanced PHA nanocomposites. Future research work should be focused on the development of techniques that provide precise control over the morphology of the PHA-based nanocomposites, which is the key factor determining the properties and applications of these promising materials. Despite some contradictory results reported in the literature and presented herein, concerning some aspects of polymer-layered silicate nanocomposite technology, these PHA-based materials are guaranteed to be technically competitive towards synthetic polymer-based nanocomposites or others nanocomposites containing a biobased and/or biodegradable matrices, opening a new dimension in this research field and for the plastic industry.

Acknowledgements The author express their sincere thanks to J.K. Pandey, K.R. Reddy, A.K. Mohanty, and M. Misra, the editors of this book, for their kind invitation for this contribution. Grateful appreciation is also extended to all my collaborators for their continuous help in this field (Dr. A. Bourmaud, Dr. S. Chardron, Dr. Y.M. Corre, Dr. I. Pillin and F. Peresse) and especially Dr. M. Castro for his help in the redaction of this review and Pr. Y. Grohens, director of LIMATB, for useful discussions.

References

1. Sinha Ray S, Bosmina M (2005) Prog Mater Sci 50:962
2. Darder M, Aranda P, Ruiz-Hitzk E (2007) Adv Mater 19:1309
3. Bordes P, Pollet E, Averous L (2009) Prog Polym Sci 34:125
4. Alexandre M, Dubois P (2000) Mater Sci Eng R 28:1
5. Sinha Ray S, Okamoto M (2003) Prog Polym Sci 28:1539
6. Paul DR, Robeson LM (2008) Polymer 49:3187
7. Pavlidou S, Papaspyrides CD (2008) Prog Polym Sci 33:1119
8. Bruzaud S, Bourmaud A (2007) Polym Test 26:652
9. Sudesh K, Abe H, Doi Y (2000) Prog Polym Sci 25:1503
10. Khanna S, Srivastava AK (2005) Proc Biochem 40:607
11. Vroman I, Tighzert L (2009) Materials 2:307
12. Chanprateep S (2010) J Biosci Bioeng 110:621
13. Rai R, Keshavarz T, Roether JA, Boccaccini AR, Roy I (2011) Mater Sci Eng 72:29
14. Bhubalan K, Lee WH, Loo CY, Yamamoto T, Tsuge T, Doi Y, Sudesh K (2008) Polym Degrad Stab 93:17
15. Chen GQ, Wu Q (2005) Biomaterials 26:6565
16. Akaraonye E, Keshavarz T, Roy I (2010) J Chem Technol Biotechnol 85:732
17. Philip S, Kershavarz T, Roy I (2007) J Chem Technol Biotechnol 82:233
18. Noda I, Green PR, Satkowski MM, Schechtman LA (2005) Biomacromolecules 6:580
19. Corre YM, Bruzaud S, Audic JL, Grohens Y (2012) Polym Test 31:226
20. Dufresne A, Samain E (1998) Macromolecules 31:6426
21. Mohanty AK, Misra M, Hinrichsen G (2000) Macromol Mater Eng 1:276–277
22. Amgoune A, Thomas CM, Ilinca S, Roisnel T, Carpentier JF (2006) Angew Chem Int Ed 118:2848
23. Pandey JK, Kumar AP, Misra M, Mohanty AK, Drzal LT, Singh RP (2005) J Nanosci Nanotechnol 5:497
24. Maiti P, Batt CA, Giannelis EP (2007) Biomacromolecules 8:3393
25. Hasegawa N, Kawasumi M, Kato M, Usuki A, Okada A (1998) J Appl Polym Sci 67:87
26. Kojima Y, Usuki A, Kawasumi M, Okada A, Kurauchi T, Kamigaito O (1993) J Polym Sci Part A Polym Chem 31:1755
27. Lee DC, Jang LW (1996) J Appl Polym Sci 61:1117
28. Chardron S, Bruzaud S, Lignot B, Elain A, Sire O (2010) Polym Test 29:966
29. Chen GX, Hao GJ, Guo TY, Song MD, Zhang BH (2002) J Mater Sci Lett 21:1587
30. Choi WM, Kim TW, Park OO, Chang YK, Lee JW (2003) J Appl Polym Sci 90:525
31. Carrasco F, Dionisi D, Martinelli M, Majone M (2006) J Appl Polym Sci 100:2111

32. Zhang X, Lin G, Abou-Hussein R, Hassan MK, Noda I, Mark JE (2007) Eur Polym J 43:3128
33. Sinha Ray S, Yamada K, Okamoto M, Ueda K (2002) Nano Lett 2:1093
34. Hablot E, Bordes P, Pollet E, Avérous L (2008) Polym Degrad Stab 93:413
35. Bordes P, Hablot E, Pollet E, Avérous L (2009) Polym Degrad Stab 94:789
36. Lim ST, Hyun YH, Lee CH, Choi HJ (2003) J Mater Sci Lett 22:299
37. Nielsen L (1967) J Macromol Sci Chem A1(5):929
38. Bharadwaj RK (2001) Macromolecules 34:9189
39. Gusev AA, Lusti HR (2001) Adv Mater 13:1641
40. Corrêa MCS, Branciforti MC, Pollet E, Agnelli JAM, Nascente PAP, Avérous L (2012) J Polym Env. doi:10.1007/s10924-011-0379-0
41. Sanchez-Garcia MD, Gimenez E, Lagaron JM (2008) J Appl Polym Sci 108:2787
42. Kumar AP, Depan D, Tomer NS, Singh RP (2009) Prog Polym Sci 34:479
43. Pandey JK, Reddy KR, Kumar AP, Singh RP (2005) Polym Degrad Stab 88:234
44. Maiti P, Batt CA, Giannelis EP (2003) Polym Mater Sci Eng 88:58
45. Sinha Ray S, Yamada K, Okamoto M, Fujimoto Y, Ogami A, Ueda K (2003) Polymer 44:6633
46. Wang S, Song C, Chen G, Guo T, Liu J, Zhang B, Takeuchi S (2005) Polym Degrad Stab 87:69
47. Lee SR, Park HM, Lim H, Kang T, Li X, Cho WJ (2002) Polymer 43:2495
48. Song CJ, Wang SF, Ono S, Shimasaki C, Inoue M (2001) Soil Sci Plant Nutr 48:159
49. Ishida K, Asakawa N, Inoue Y (2005) Macromol Symp 224:47
50. Lee SK, Seong DG, Youn JR (2005) Fib Polym 6:289
51. Sinha Ray S, Yamada K, Okamoto M, Ogami A, Ueda K (2003) Chem Mater 15:1456
52. Bruzaud S, Grohens Y (2008) Int J Nanotechnol 5:660
53. Chang JH, An YU, Cho D, Giannelis EP (2003) Polymer 44:3715
54. Sinha Ray S, Okamoto M (2003) Macromol Rapid Commun 24:815
55. Sinha Ray S, Yamada K, Okamoto M, Ogami A, Ueda K (2003) Comp Int 10:435
56. Sinha Ray S, Yamada K, Okamoto M, Ueda K (2003) J Nanosci Nanotechnol 3:503

New Developments in Polycaprolactone-Layered Silicate Nano-biocomposites: Fabrication and Properties

2

Hassan Namazi, Mohsen Mosadegh, and Mozhgan Hayasi

Contents

Abstract

In the family of synthetic biodegradable polymers, PCL, which is linear, hydrophobic, and partially crystalline polyester, is a biodegradable polymer. Its physical properties and commercial availability make it very attractive not only as a substitute for nonbiodegradable polymers of commodity applications but also as a specific plastic of medicine and agricultural areas. The main

H. Namazi (✉)
Research Laboratory of Dendrimers and Nanopolymers, Faculty of Chemistry, University of Tabriz, Tabriz, EA, Iran

Research Center for Pharmaceutical Nanotechnology (RCPN), Tabriz University of Medical Sciences, Tabriz, Iran
e-mail: namazi@tabrizu.ac.ir

M. Mosadegh
Department of Nanochemistry, School of Chemistry, University College of Science, University of Tehran, Tehran, Tehran, Iran
e-mail: mosadegh_mm@yahoo.com

M. Hayasi
Department of Chemistry, Faculty of Sciences, University of Semnan, Semnan, Semnan, Iran
e-mail: mozhgan_hayasi@yahoo.com

J.K. Pandey et al. (eds.), *Handbook of Polymernanocomposites. Processing, Performance and Application – Volume A: Layered Silicates*, DOI 10.1007/978-3-642-38649-7_28,

limitation of PCL is its low melting temperature (65 °C), which can be overcome by blending it with other polymers. In recent years, polymer/silicate hybrid nanocomposites have been of considerable interest as an effective method to improving polymer properties. This class of materials has improved the material properties due to the high aspect ratio and easy phase-to-phase energy transfer, even at very low filler concentration, if the filler is uniformly and completely dispersed in the host matrix. Polymer/layered silicate nanocomposites (PLSN) show a considerable enhancement of strength, modulus, gas barrier resistance, and heat distortion temperature compared to their pure polymer counterparts, even with silicate loadings as low as 3–5 wt%. Moreover, PLSN are also interesting from the fundamental point of view due to the nanoscale constraints of the filler to the polymer matrix and the ultra-large-specific interfacial area between the silicate and the polymer matrix.

One kind of nanometer-size reinforcement is the montmorillonite, which is a layered silicate whose interlayer ions can be exchanged by organ ions in order to produce an increment in the interlayer spacing and to improve the polymer/clay compatibility. These improvements allow the dispersion of clay platelets to be easier. Nanocomposites of poly(ε-caprolactone)/clay deserve interest because of the possible upgrading of polymer known for biocompatibility, biodegradability, and miscibility with a wide range of other polymers. There have been attempts to develop nanocomposites of PCL with layered silicates. PCL/clay nanocomposites were prepared through kind of methods. For example, in situ polymerization, melt intercalation and extrusion, and solution and casting also have used kind of silicate layers and modified clay with different their modifiers.

Keywords
Nanocomposite • Polycaprolactone • Silicate layers

1 Polycaprolactone

Poly(ε-caprolactone) PCL is linear polyester manufactured usually by ring-opening polymerization of ε-caprolactone in the presence of metal alkoxides (aluminum isopropoxide, tin octoate, etc.) [1–3]. Molecular structure of PCL is presented in Fig. 2.1. The PCL chain is flexible and exhibits high elongation at break and low modulus. Its physical properties and commercial availability make it very attractive not only as a substitute material for nondegradable polymers for commodity applications but also as a specific plastic of medicine and agricultural areas [4]. PCL is widely used as a PVC solid plasticizer or for polyurethane applications, as polyols, but it finds also some applications based on its biodegradable character in domains such as biomedicine (e.g., drugs' controlled release) and environment (e.g., soft compostable packaging). It is a semicrystalline polymer with a degree of crystallinity around 50 %. PCL shows a very low Tg (−61 °C) and a low melting point (65 °C), which could be a handicap in some applications. Therefore, PCL is generally blended [5–8] or modified (e.g., copolymerization, cross-linking [9]. Tokiwa and Suzuki [10]

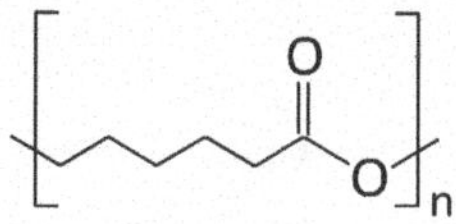

Fig. 2.1 Molecular structure of polycaprolactone

have discussed the hydrolysis of PCL and biodegradation by fungi. They have shown that PCL can be easily enzymatically degraded. According to [7], the biodegradability can be clearly claimed, but the homopolymer hydrolysis rate is very low. The presence of starch can significantly increase the biodegradation rate of PCL [5]. There have been lot of attempts to prepare PCL/OMLS nano-composites with much improved mechanical and material properties than that of neat PCL.

2 Polymer/Silicate Layered Nanocomposites

Recently, the utility of inorganic nanoparticles as additives to enhance the polymer performance has been established [11]. Various nanoreinforcements currently being developed are nanoclay (layered silicates) [12–14], cellulose nanowhiskers [15], ultrafine-layered titanate [16], and carbon nanotubes [17, 18]. Of particular interest are polymer and organically modified layered silicate (OMLS) nanocomposites because of their demonstrated significant enhancement, relative to an unmodified polymer resin, of a large number of physical properties, including barrier, flammability resistance, thermal and environmental stability, solvent uptake, and rate of biodegradability of biodegradable polymers [14]. These improvements are generally attained at lower silicate content compared to that of conventional filler-filled systems. The main reason for these improved properties in polymer/layered silicate nanocomposites is the strong interfacial interactions between matrix and OMLS as opposed to conventional composites [19, 20]. Layered silicates generally have layer thickness in the order of 1 nm and very high aspect ratios (e.g., 10–1,000). A few weight percent of OMLS that are properly dispersed throughout the matrix thus create a much higher surface area for polymer–filler interactions than do conventional composites. On the basis of the strength of the polymer/OMLS interaction, structurally two different types of nanocomposites are thermodynamically achievable (Fig. 2.2): (i) intercalated nanocomposites, where insertion of polymer chains into the silicate structure occurs in a crystallographically regular fashion, regardless of polymer to OMLS ratio, and a repeat distance of few nanometer, and (ii) exfoliated nanocomposites, in which the individual silicate layers are separated in polymer matrix by average distances that totally depend on the OMLS loading. Layered silicates for the preparation of polymer/layered silicate (PLS) nanocomposites belong to the same general family of 2:1 layered or phyllosilicates [21, 22]. Their crystal structure consists of layers made up of two tetrahedrally coordinated silicon atoms fused to an edge-shared octahedral sheet of either aluminum or magnesium hydroxide. The layer thickness is around 1 nm, and the lateral dimensions of these layers may vary from 30 nm to several microns or larger, depending on the particular

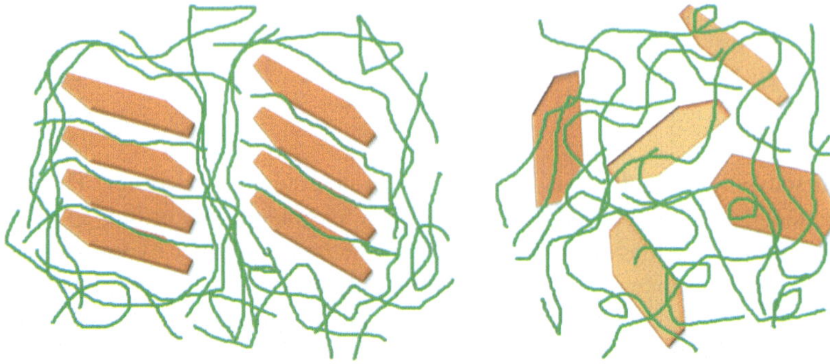

Fig. 2.2 Schematic illustration of two types of structure achievable polymer/layered silicate nanocomposites

layered silicate. Stacking of the layers leads to a regular van der Waals gap between the layers called the interlayer or gallery. Isomorphic substitution within the layers generates negative charges that are counterbalanced by alkali and alkaline earth cations situated inside the galleries. This type of layered silicate is characterized by a moderate surface charge known as the cation exchange capacity (CEC) and generally expressed as mequiv/100 g.

Montmorillonite, hectorite, and saponite are the most commonly used layered silicates. Pristine layered silicates usually contain hydrated Na^+ or K^+ ions [21, 23]. Obviously, in this pristine state, layered silicates are only miscible with hydrophilic polymers. To render layered silicates miscible with biodegradable polymer matrices, one must convert the normally hydrophilic silicate surface to an organophilic one, making the intercalation of many biodegradable polymers possible. Generally, this can be done by ion-exchange reactions with cationic surfactants including primary, secondary, tertiary, and quaternary alkylammonium or alkylphosphonium cations. Alkylammonium or alkylphosphonium cations in the organosilicates lower the surface energy of the inorganic host and improve the wetting characteristics of the polymer matrix and result in a larger interlayer spacing. Additionally, the alkylammonium or alkylphosphonium cations can provide functional groups that can react with the polymer matrix or in some cases initiate the polymerization of monomers to improve the adhesion between the inorganic and the polymer matrix [24, 25].

2.1 Preparation Methods

2.1.1 Intercalation of Polymer and Prepolymer from Solution

This is based on a solvent system in which the polymer or prepolymer is soluble and the silicate layers are swellable. The layered silicate is first swollen in a solvent, such as water, chloroform, or toluene. When the polymer and layered silicate solutions are mixed, the polymer chains intercalate and displace the solvent within the interlayer of the silicate. Upon solvent removal, the intercalated structure remains, resulting in PLS-nanocomposite. The driving force for the polymer intercalation into layered silicate from solution is the entropy gained by desorption of solvent molecules, which compensates for the decreased entropy of the confined, intercalated chains [26]. Using this method, intercalation only occurs for certain polymer/solvent pairs. This method is good for the intercalation of polymers with little or no polarity into layered structures and facilitates production of thin films with polymer-oriented clay intercalated layers.

2.1.2 In Situ Polymerization

In this method, the layered silicate is swollen within the liquid monomer or a monomer solution, so the polymer formation can occur between the intercalated sheets. Polymerization can be initiated either by heat or radiation, by the diffusion of a suitable initiator, or by an organic initiator or catalyst fixed through cation exchange inside the interlayer before the swelling step.

2.1.3 Melt Intercalation

Recently, the melt intercalation technique has become the standard for the preparation of PLS-nanocomposites. During polymer intercalation from solution, a relatively large number of solvent molecules have to be desorbed from the host to accommodate the incoming polymer chains. The desorbed solvent molecules gain one translational degree of freedom, and the resulting entropic gain compensates for the decrease in conformational entropy of the confined polymer chains. Therefore, there are many advantages to direct melt intercalation over solution intercalation. For example, direct melt intercalation is highly specific for the polymer, leading to new hybrids that were previously inaccessible. In addition, the absence of a solvent makes direct melt intercalation an environmentally sound and an economically favorable method for industries from a waste perspective. This process involves annealing a mixture of the polymer and OMLS above the softening point of the polymer, statically or under shear. While annealing, the polymer chains diffuse from the bulk polymer melt into the galleries between the silicate layers.

3 PCL/Silicate Layers Nanocomposites

PCL-based nanocomposite was the first studied nanobiocomposite. In the early 1990s, Giannelis' group from Cornell University started to work on the elaboration of PCL-based nanocomposite by intercalative polymerization [25, 27, 28]. Their work was motivated by previous studies involving polymerization of ε-caprolactam in the presence of layered silicates, which suggested that lactone ROP can be catalyzed by layered silicates. Then, they decided to investigate the intercalation and polymerization of ε-caprolactone within the gallery of layered silicates. The very first PCL-based nanocomposite prepared was based on fluorohectorite,

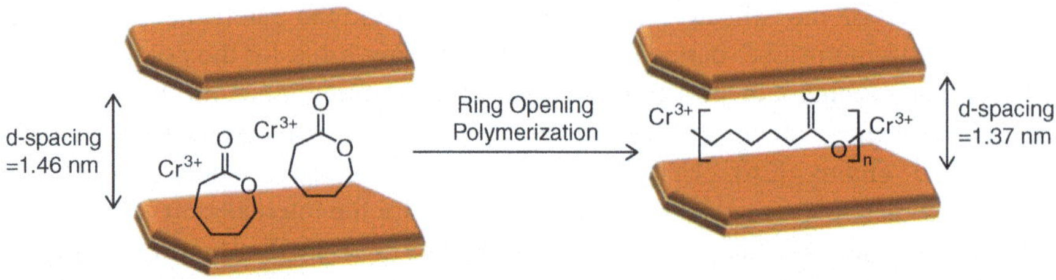

Fig. 2.3 Schematic illustration of reduction of d-spacing of swollen clay with CL monomer after polymerization

a mica-type layered silicate [27]. The ROP of ε-caprolactone (ε-CL) was activated by the surface of the Cr^{3+}-exchanged fluorohectorite. Indeed, the type of interlayer cations (e.g., Cr^{3+}, Cu^{2+}, Co^{2+}, Na^{+}) is important in achieving polymerization since it proceeds through cleavage of the acyl–oxygen bond catalyzed by the interlayer Cr^{3+} ions which present a more acidic character than mono- and divalent cations. The polymer–clay chemical interactions at the interface were proved to be strong and the intercalation of the polymer irreversible. However, authors could also observe the decrease of *d* -spacing from 1.46 to 1.37 nm after polymerization. They attributed this phenomenon to a change in the intercalated molecule organization from the monomer to the polymer (Fig. 2.3).

The observed layer spacing of 1.37 nm correlates as well with the sum of the thickness of the silicate layer (0.96 nm) [21] and the known interchain distance (0.4 nm) in the crystal structure of PCL [27]. Repeated washing with a solvent for PCL did not alter the silicate layer spacing, indicating that the interaction between the intercalated polymer and the silicate surface is strong and that intercalation of the PCL is irreversible.

Similar results were also obtained later by Kiersnowski et al. [29] who prepared the PCL-based composites by in situ polymerization catalyzed by water. Afterwards, Messersmith and Giannelis [25] attempted to prepare PCL-based nanocomposites by in situ polymerization thermally activated and initiated by organic acid. This one constituted the OMMT organomodifier, namely, the protonated form of 12-aminododecanoic acid (NH_3^+ ($C_{11}COOH$)), which was thus present on the clay surface and initiated the ROP by a nucleophilic attack on the ε-caprolactone carbonyl. The resulting PCL was therefore ionically bound to the silicate layers through the protonated amine chain end. XRD results suggested that individual silicate layers were dispersed in the matrix. On the contrary, OMMT layers organomodified with a less polar ammonium (dimethyl dioctadecyl-$N^+(Me)_2(C_{18})_2$ ammonium) [25] or hexadecyltrimethyl-$N^+(Me)_3(C_{16})$-ammonium [30] showed no dispersion in CL or PCL. Subsequently, the interactions occurring at the interface of a PCL/OMMT exfoliated nanocomposite were investigated [31, 32], and the crystallinity, the permeability, and the rheological behavior were examined [25, 28, 31, 33].

Polymerization of lactone monomers can be initiated using a number of different types of catalysts [34–38], including compounds containing labile protons, such as amines, alcohols, and carboxylic acids [34, 35, 37]. Initiation with these molecules has been shown to be the result of nucleophilic attack upon the lactone carbonyl group, resulting in ring opening and formation of a new terminal hydroxyl group [35]. Subsequent propagation then occurs by similar nucleophilic attack by the terminal hydroxyl groups remaining on lactone monomers. By analogy, acid groups ionically bound to the silicate layers at the protonated amine terminus can act as nucleophiles, reacting with CL, which results in addition of one CL unit and production of a terminal hydroxyl group. The reactions occurring between the organic acid group of the OMLS and CL monomer are facilitated and maximized by molecular dispersion of the individual silicate layers in the liquid monomers. Bugatti et al. [39] prepared composites of polycaprolactone with nanohybrids of layered double hydroxide (LDH) with benzoate (Bz), 2,4-dichlorobenzoate (BzDC), and para- and ortho-hydroxybenzoate (p-BzOH, o-BzOH) anions, all having antimicrobial activity. X-ray diffraction analyses of the samples indicated that both the nature and the position of the aromatic ring substituents affect the value of the interlayer distance and the hydrogen bonds of the nanohybrids. X-ray diffraction analysis of all the composites indicated that the LDH samples containing BzDC anion, the less interacting anion with LDH, are mainly exfoliated into the polymeric matrix, whereas those containing p-BzOH anion, the most interacting anion with LDH, remained almost unchanged, giving rise to microcomposites. Intermediate behavior was found for LDH modified with Bz and o-BzOH anions, because they produced partly exfoliated and partly intercalated nanocomposites. All the composites showed an improvement of the elastic modulus, particularly evident at low filler concentration, where they followed the mixing rule. The sample with the best filler dispersion (PCL/BzDC) showed a higher than expected modulus at low filler concentration. An improvement, or at least maintenance, of all the mechanical parameters was obtained for all the samples. The composite isotherms follow the same trend of PCL, although showing a higher sorption in all the activity range, due to the higher hydrophilicity of the inorganic lamellae. At variance, the thermodynamic diffusion parameter, at zero vapor concentration, is significantly lower and decreases on increasing the inorganic concentration for all the composites. A comparison of the molecular release of the antimicrobial benzoate molecules directly dispersed in PCL and in the nanohybrid was shown, indicating the much slower release in the latter case.

3.1 Preparation Methods

3.1.1 Solution Route

PCL-based nanocomposites have also been produced by dissolving the polymer in hot chloroform in the presence of OMLS [40]. SAXS and XRD results revealed that the silicate layers forming the clay could not be dispersed individually in the PCL matrix. In other words, the silicate layers existed in the form of tactoids, consisting of several stacked silicate layers. These are responsible for the formation of special geometrical structures in the blends, which leads to the formation of superstructures in the thickness of the blended film.

Wu et al. have prepared the poly(ε-caprolactone)/Na^+-MMT nanocomposites [41]. Few studies about original clay, such as sodium montmorillonite (Na^+-MMT), were reported. Na^+-MMT has a more polar surface than organically modified clay does, and it almost cannot be exfoliated in polar organic liquids, such as DMF, DMA, and DMSO, because of low Gibbs energy of salvation [42]. In this work, two-step method has been employed to disperse Na^+-MMT in DMF or DMA and applied this method in preparation of PCL/Na^+-MMT nanocomposites. TEM result indicated that the average lateral dimension of the Na^+-MMT particles in PCL composites was around 300 nm and the thickness was less than 20 nm. In addition, this two-step method was also efficient with DMA. DSC results indicated a nucleating effect of Na^+-MMT on PCL. The crystallization behavior of PCL depended on the balance between the nucleating effect and the suppression of the PCL segment mobility. TGA showed an enhanced thermal resistance of PCL by adding Na^+-MMT. Biodegradability study proved that the silicate layers hindered the biodegradability of PCL and acted as a barrier rather than a catalyst. The improvement of biodegradability of PCL induced by organoclays may be caused by the cationic surfactants. Ahmed et al. have studied [43] preparation of poly(ε-caprolactone)/nanoclay composite (PCLNC) films by solvent casting method using a wide range of organoclays: Closite 30B (2.5–10 %). Results indicated the formation of some intercalated nanostructure of PCLNC. Rheological study indicated that the predominating liquid-like properties (viscous modulus, $G'' >$ elastic modulus, G') of neat PCL gradually transformed to solid-like ($G' > G''$) behavior after incorporation of clay in the temperature range of 90–120 °C. A plot of G' vs. G'' provides information on intercalation and microstructure of nanocomposite. Applicability of time–temperature superposition (TTS) principle and van Gurp–Palmen plot (phase angle vs. absolute complex modulus) on rheological data of clay-incorporated PCL were employed and found that the results failed to follow the rules. Incorporation of the nanoclay into PCL matrix increased the crystallization temperature (T_c) and melting temperature (T_m) of neat PCL from 28.7 °C to 32.3 °C and 56.3 °C to 59.2 °C, respectively, due to the nucleating effect, but the glass transition temperature ($T_g = 65$ °C) was remained unaffected. The decrease in crystallinity with increase in clay concentration was confirmed by both XRD and DSC data.

Luduena et al. [44] studied the morphology and mechanical properties of polycaprolactone/clay nanocomposite films prepared by two techniques (casting – exfoliation adsorption; intensive mixing – melt intercalation). Casting, which is a laboratory-scale technique, was selected because it was supposed that the exfoliation of the layered silicate into single layers would be easier since the solvent acts as an exfoliation agent. The other selected technique was chosen because it can be used in the industry. X-ray diffractograms revealed an intercalated–exfoliated mixed structure for both techniques. For casting, the morphology and mechanical properties are influenced by the used solvent and the preparation conditions being the first one the most critical parameter. Otherwise, in the case of intensive mixing, a higher clay dispersion degree was produced by shear forces, and the resultant mechanical properties were superior to those obtained by casting. In both cases, the highest modulus was achieved for 5 wt% of C30B.

Supercritical CO_2 Intercalation PCL

Supercritical CO_2 fluids ($scCO_2$) have attracted much attention because the environmentally friendly, chemically inert, inexpensive, and nonflammable carbon dioxide has potential to be an alternative to substitute for organic solvents to reduce environmental pollutions. The $scCO_2$ used as a medium to prepare clay nanocomposites drew attention recently [45–47]. Polymer/clay nanocomposites have been reportedly prepared by a variety of methods including melt intercalation, in situ polymerization, and solution intercalation. The latter method is limited to certain polymer/solvent pairs, in which the polymer is soluble and the silicate layers are swellable [48, 49]. Monomers are mostly soluble in common solvents, conventional solution intercalation of the monomer in clay is easily obtained, and the polymer-intercalated clay can be easily prepared via in situ polymerization of the intercalated monomer. Via this method, poly(methyl methacrylate) or polystyrene/clay nanocomposites could be prepared in scCO2 in the presence of a fluorinated surfactant-modified [46] or a poly(dimethylsiloxane) surfactant-modified [50] or organomodified [45, 47] clays. Because of very low solubility of polymers in $scCO_2$, due to lack of strong interactions between CO_2 and polymers and low entropy driving force for mixing CO_2 and polymers, the $scCO_2$ intercalation of polymer in clay is qualitatively different from the conventional solution intercalation. Although very few polymers were reportedly soluble in $scCO_2$, some polymers with carbonyl groups [51, 52], ether linkages [53], or C–F linkages [54, 55] have specific interactions with CO_2 that could cause depression of melting temperatures of polymers in CO_2. In a research work, Shieh et al. investigated the effects of high-pressure CO_2 treatments on morphology of PCL [56]. They found that PCL could melt at 8.5 and 30.8 MPa of CO_2 at 35 °C, a much lower temperature than T_m of PCL. These high pressures of CO_2 could assist melting of PCL because of the presence of CO_2–carbonyl group interactions. In the other work, [57] they investigate $scCO_2$ intercalation of this carbonyl groups containing PCL in montmorillonite clays, both unmodified and organomodified. Wide-angle X-ray diffraction patterns find that PCL is slightly intercalated in unmodified montmorillonite clay (Na^+-MMT) but considerably intercalated in organomodified montmorillonite by stearyltrimethylammonium chloride $CH_3(CH_2)_{17}N^+(Cl^-)(CH_3)_3$ (OMMT). The interlayer spacing in OMMT increases considerably from 1.94 nm in OMMT to 3.58 nm in the OMMT/PCL 10/90 sample. PCL8 having molecular weight of 80,000 is harder to intercalate into OMMT than PCL1 having molecular weight of 10,000. Higher $scCO_2$ pressures at a temperature allow larger intercalations of PCL in OMMT to exhibit larger interlayer spacing in OMMT. The interlayer spacing in OMMT, however, is not clearly found to relate with the CO_2 temperature at a given pressure. TGA data show that OMMT enhances the thermal stability of PCL1, with a higher content of OMMT giving a higher amount of PCL1 residue. DSC data find that the PCL1-intercalated OMMT expedites the melt crystallization rate of PCL1 from the melt but suppresses the crystallinity of PCL1. Study of Avrami's rate constants *k* and exponent n finds that the PCL1-intercalated OMMT enhances the isothermal crystallization rate of PCL1 and that the crystal growth dimension is 3 for pure PCL1 but decreases with increasing OMMT content in the blends. Modulus data find that the PCL1-intercalated OMMT is an effective reinforcement for PCL8.

3.1.2 Melt Intercalation Route

Dubois' group (Mons, Belgium) has worked on PCL nanocomposites. They were interested in the in situ ROP of ε-CL and in the melt intercalation route. They demonstrated that the formation of PCL-based nanocomposites depends not only on the ammonium cation and related functionality but also on the elaboration route. Recently, Di et al. [58] reported the preparation of PCL/OMLS nanocomposites in the molten state, using a twin-screw extruder. They used two different types of OMLS for the preparation of nanocomposites and attempted to determine the dependence of OMLS intercalation and/or exfoliation on the processing conditions and types of OMLS and also the thermal and rheological behavior of the prepared nanocomposites. Nanocomposites were prepared using a Haake co-rotating twin-screw extruder, which was operated at 100 °C and 180 °C with a screw speed of 100 rpm, and the residence time was 12 min. XRD patterns clearly revealed that the delamination of silicate layers in the PCL matrix was directly related to the type of OMLS, the OMLS content, and the processing temperature. The strong interaction between the organic surfactants covering the clay layers and the PCL matrix molecules was favored in the exfoliation process. Processing at low temperatures resulted in high stress in comparison with that at high temperatures, and this helped with the fracturing of the OMLS particles and caused a good dispersion of them in the PCL matrix. A higher OMLS content hybrid required more processing time for achieving an exfoliation structure than a lower OMLS content hybrid. Lepoittevin et al. [59] also used same method for the production of PCL/OMLS nanocomposites.

Delamini et al. [60] have studied the effect of the compatibility of organoclays with structurally different semicrystalline polymer matrices. Using the melt-mixing technique, PCL and ethylene vinyl acetate (EVA) matrices were used with various amounts of C20A to develop nanocomposites. The results show that the structure of a polymer matrix plays a significant role towards compatibilization with the silicate layers of the clay. Polycaprolactone nanocomposites have higher tensile strain, tensile stress, and Young's modulus values relative to EVA nanocomposites at identical organoclay loading. Campbell et al. [61] have reported the preparation and characterization of a poorly water-soluble drug (ibuprofen)-loaded PCL layered silicate nanocomposite using hot-melt extrusion. Cloisite 20A and is a natural montmorillonite modified with a dimethyl, dehydrogenated tallow, quaternary ammonium surfactant were used. In all cases, an intercalated and partially exfoliated composite morphology was attained. Ibuprofen molecules were well dispersed and distributed throughout the PCL matrix. Addition of clay platelets can be used to manipulate the mechanical properties of PCL and is dependent on the aspect ratio of the clay particles. The incorporation of both ibuprofen and organoclays alters the crystallization behavior of PCL. PCL crystallization is disrupted as polymer chain dynamics is restricted for those chains that diffused between clay platelets or are tethered to clay particles; thus, an increased fraction of PCL is in the amorphous phase. The constrained mobility of PCL chains also resulted in an increase in Tg by up to 16 °C. Furthermore, the storage modulus for all composites decreased relative to neat PCL, by about two orders of magnitude for the composites with 10 wt% additives. The release of ibuprofen from PCL was retarded when both organoclays was dispersed in the PCL–ibuprofen blend. That the mechanical properties of such polymer matrices can be manipulated by addition of nanoclays may present opportunities for easier subsequent milling or compression of drug-loaded biopolymers.

In the new work, Luduena and coworker have studied on the effect of unmodified and several organomodified montmorillonites on the morphology, mechanical properties, and thermal behavior of polycaprolactone-based nanocomposites prepared by melt intercalation [62]. The study was centered on the analysis of the clay characteristics that have influence on the final properties of PCL/clay nanocomposites. Matrix and nanocomposites with 5 wt% of each clay were prepared in a double-screw extruder using a temperature profile of 60–90–120 °C at a screw rotating speed of 150 rpm; the residence time was 1 min. Then, films were obtained by compression molding (100 °C, 10 min without pressure and 10 min at 50 bar, mold cooling with water). The neat MMT (CNa^+) and different organomodified MMT, C25A, C30B, C10A, C93A, and C20A were used. The Young's modulus of polymer clay nanocomposites depends on both the clay dispersion degree and the clay content inside the matrix. It was shown that organomodified clays improve the effectiveness enhancing the rigidity of the matrix. This result was a consequence of the higher dispersion degree of the clay, but it was not found a clear trend related to the polymer/clay compatibility which was attributed to the degradation of clay organomodifiers during processing. The C20A clay showed the highest thermal stability and good compatibility with PCL which led to the nanocomposite with the highest clay dispersion degree and hence optimal mechanical performance. The theoretical parameters describing the clay dispersion degree that were obtained from the effective micro-mechanical modeling supported these results. The following studies will be performed only for the PCL and the nanocomposites with C20A, C30B (which showed the best performance), and CNa^+ (used as reference). As a general conclusion, when polymer/organomodified clay nanocomposites are prepared by techniques such as extrusion, in which temperature and shear forces are involved, the degradation of the clay organomodifiers can take place. In this case, improving polymer–clay compatibility may not be the main factor to achieve the best mechanical performance.

In the other work, Fukushima et al. reported PLA and PCL nanocomposites prepared by adding two organically modified montmorillonites (C30B, and NANOFIL 804), and one unmodified sepiolite (PANGEL S9) was obtained by melt blending [63]. The highest thermal–mechanical enhancements were obtained for PLA nanocomposites compared to PCL ones probably due to higher polymer/filler interactions. Considerable thermomechanical improvements of the PLA and

PCL matrices were found by the incorporation of C30B and SEPS9 because of the good dispersion and high polymer/filler interaction. In the case of C30B, the above features can be ascribed to the high organic modifier–OH groups' availability, whereas in the case of SEPS9, they could depend on the high surface hydroxyl group concentration promoting polymer/fillers interactions. They believe that especially sepiolites can be regarded as potentially interesting materials for the enhancement of biodegradable polymers' properties due to their easy dispersibility without the need of organic modifiers and compatibilizers and due to the considerable increases in the thermal and thermomechanical properties obtained for PLA and PCL.

Masterbatch Process

In order to improve the clay dispersion via melt blending techniques, several authors [30, 64–66] report on the use of masterbatches (polymer/clay blends with a high weight content of clay, prepared by melt mixing) that are redispersed (diluted) into polymer matrices [64, 66], but this technique generally leads to semi-intercalated/semi-exfoliated morphology. An interesting process for masterbatch production has been previously described by Dubois et al. [67], which leads to a more efficient clay delamination after melt redispersion in a host matrix. First, a highly filled polymer/clay nanocomposite is synthesized by in situ intercalative polymerization of the monomer in the presence of a high amount of clay (>10 wt%). The polymer can even be anchored onto the clay nanosheet surface when initiating species are borne by the ammonium organomodifier fixed at the clay sheet surface. The final nanocomposite is obtained by melt blending a suitable amount of the resulting masterbatch into a polymer matrix in order to obtain a nanocomposite containing a few percent of well-dispersed clay. An exfoliated morphology is generally obtained in the final nanocomposite, as long as the matrix is compatible with the intercalated compatibilizer. Dubois et al. were the first to describe the synthesis of such masterbatches, and they mainly focused their study on the poly(ε-caprolactone) (PCL)/clay system. They conducted their synthesis in bulk, with a tin(IV) initiator, and a low monomer conversion was used in order to reach a high inorganic content.

This masterbatch process [67], or equivalent [68], yielded intercalated–exfoliated structures that are rather difficult to reach by direct melt blending. This process also turned out to be a good way to compatibilize and thus to reinforce other thermoplastics like conventional polymers – SAN [30, 64], PVC [67], PC [69], PP, PE, PS, and ABS [70], or biopolymers – PBS and PBAT [71]. Conversely, PCL was blended by a reactive process with thermoplastic–clay systems [72] to improve the properties of the final material. A remarkable study on the nanocomposite preparation of oligo-PCL/OMMT by simple mechanical mixing was reported by Maiti [73]. Different types of clays having different aspect ratios (hectorite, mica, smectite) organomodified with various phosphonium cations were selected to investigate their influence on miscibility with oligo-PCL (o-PCL). The alkyl phosphonium cations were *n*-octyltri-*n*-butylphosphonium ($P^+(But)_3(C_8)$), *n*-dodecyltri-*n*-butylphosphonium (($P^+(But)_3(C_{12})$), *n*-hexadecyltri-*n*-butylphosphonium

($P^+(But)_3(C_{16})$), and methyltriphenylphosphonium ($P^+(Me)(ph)_3$). Immiscible, intercalated, and exfoliated nanostructures were observed in o-PCL nanocomposites, depending on the nature of the organic modifier as well as the aspect ratio. According to Maiti, when o-PCL is immiscible with a certain organic modifier, it cannot intercalate into the silicate gallery, while for a short-chain miscible modifier, o-PCL intercalates and, in the case of a long-chain modifier, the modifier orients itself away from the silicate surface and is solubilized into the o-PCL phase, resulting in the collapse of the silicate gallery. Considering the effect of the aspect ratio with a given organomodifier, when the aspect ratio is low, combined with a high CEC, the organic modifier is favored to diffuse out the gallery and to interact with o-PCL leading to exfoliated structure. For higher aspect ratio, i.e., for larger lateral dimensions of the silicate layers, the organic modifier hardly accesses outside the gallery, and thus, the o-PCL must intercalate. Other works attempted to better understand the mechanism of intercalation/exfoliation process either by melt intercalation or in situ ε-CL polymerization by molecular dynamics simulations [74–77].

In the other work, exfoliated nanocomposites were prepared by dispersion of poly (ε-caprolactone)-grafted montmorillonite nanohybrids used as masterbatches in chlorinated polyethylene (CPE) [78]. The PCL-grafted clay nanohybrids with high inorganic content were synthesized by in situ intercalative polymerization of ε-caprolactone between silicate layers organomodified by alkylammonium cations bearing two hydroxyl functions (Closite 30B). The polymerization was initiated by tin alcoholate species derived from the exchange reaction of tin(II) bis (2-ethylhexanoate) with the hydroxyl groups borne by the ammonium cations that organomodified the clay. These highly filled PCL nanocomposites (25 wt% in inorganics) were dispersed as masterbatches in commercial chlorinated polyethylene by melt blending. CPE-based nanocomposites containing 3–5 wt% of inorganics have been prepared. The formation of exfoliated nanocomposites was assessed both by wide-angle X-ray diffraction and transmission electron microscopy. The Young's modulus of CPE is increased by a decade when a PCL-grafted clay masterbatch is exfoliated to reach 5 wt% of clay in the resulting nanocomposite.

In other work, pre-exfoliated nanoclays were prepared through a masterbatch process using supercritical carbon dioxide as solvent and poly(ε-caprolactone) as organic matrix [79]. In situ polymerization of ε-caprolactone in the presence of large amount of clay was conducted to obtain these easily dispersible nanoclays, collected as a dry and fine powder after reaction. Dispersion of these pre-exfoliated nanoclays in chlorinated polyethylene was also investigated. All the results confirm the specific advantages of supercritical CO_2 towards conventional solvents for filler modification. In this study, they have in situ synthesized PCL/clay masterbatches in supercritical carbon dioxide. This unique medium allows to reach a very high clay loading in the masterbatch. Also, the product obtained after depressurization is an easily recoverable fine powder. Another advantage of using supercritical CO_2 is its capacity to extract the residual monomer during depressurization, leading directly to a ready-to-use dry powder. Three types of clays have been tested (MMT-Na^+, Closite 20A, 30B). They all lead to intercalated masterbatches with a clay loading ranging from 32.5 to 66 wt% and an upper interlayer distance of around 3.2–3.6 nm. The PCL crystallinity has been studied by DSC, and it appears that the crystallization behavior is different when the polymer is free or intercalated/partially grafted between high amounts of clay sheets. This phenomenon has been explained by the polymer confinement due to the high clay content, as well as the partial polymer grafting onto the clay surface, which prevents the polymer from efficiently crystallizing.

In the new work [80], three commercial montmorillonites, Nanofil5, Nanofil2, and Cloisite 30B, were used to prepare organomodified montmorillonite–poly(ε-caprolactone) composites by melt intercalation. Montmorillonite nanocomposites were prepared using a specific extrusion profile from a 30 wt% masterbatch of organomodified clay, which was then diluted at 1 %, 3 %, and 5 %. Clay dispersion was investigated by XRD and TEM. Results showed exfoliated/intercalated morphologies and rheological tests and highlighted a better dispersion degree for Nanofil2. Owing to the selection of the screw profile, nanocomposite morphology was achieved in this study, regardless of the polar (Cloisite 30B) or nonpolar nature of the alkyl chains of the clay modifier. The incorporation of organomodified clays allowed the thermal stability to be improved in comparison with the virgin PCL. However, due to PCL hydrolysis caused by the presence of hydroxyl groups in the modifier, a lower thermal stability was noticed for Cloisite 30B composition in comparison with that of Nanofil2, in which modifier contains only nonpolar groups and which seems to create a better barrier effect and promotes catalytic activity due to its acidic surface after its own thermal degradation. Young's modulus was also significantly increased by the incorporation of the organomodified clays. The marked increase of Young's modulus is ascribed to the high degree of dispersion and the relatively low value of residence time using twin-screw extrusion.

In the new work [133], poly(ε-caprolactone) has been grafted on organoclay surface by ring-opening polymerization of ε-caprolactone (ε-CL) in the presence of Sn $(Oct)_2$ as a masterbatch. The polymerization was carried out at 100 °C for 48 h. The surfactant dodecyl dimethyl-2-hydroxyethylammonium bromide (DDHAB) was prepared from the reaction of 1-bromododecane and diethylethanolamine. Layered silicate/poly(PHB/PCL-PEG-PCL) urethanes nanocomposites have been prepared using poly(ε-caprolactone)-grafted organoclay (OMM-PCL) nanohybrids using dispersion technique to promote clay delamination. The DSC study showed that the OMM-PCL organoclay enhanced the crystallization rate of the crystallizable PHB and PCL-PEG-PCL segments. The results also showed that the Tg of PCL-PEG-PCL soft segments in CPNs increased as the OMM-PCL content increased.

3.1.3 In Situ Polymerization Route

Contrary to Messersmith and Giannelis [25, 27], PCL-based nanocomposites were prepared by in situ ROP according to a "coordination–insertion" mechanism [81–87], as for PLA [88, 89]. This reaction consists in swelling the OMMT organomodified by alkylammonium bearing hydroxyl groups (MMT-$N^+(Me)_2(EtOH)(C_{16})$ or MMT-$N^+(Me)(EtOH)_2$(tallow)) and then adding an initiator/activator such as tin(II) octoate ($Sn(Oct)_2$), dibutyltin- (IV) dimethoxide ($Bu_2Sn(OMe)_2$), or triethyl aluminum ($AlEt_3$). The ammonium is thus activated and can yield surface-grafted PCL chains. Every hydroxyl function generates a PCL chain. Consequently, the higher the hydroxyl groups content, the lower the PCL average molar masses. It is worth noting that, in the presence of tin(IV) catalysts, since they are more efficient towards ε-CL ROP, the preparation took place in milder conditions compared to $Sn(Oct)_2$ [85]. Moreover, in all cases, the nanocomposites exhibited a continuous decrease of molar masses with clay concentration. This can be explained by OH functions, which can act both as co-initiator and chain transfer agent. This in situ polymerization process led to well-exfoliated PCL-based nanocomposites with 3 wt% of clay, while with higher content (10 wt%) partially exfoliated/partially intercalated structures were observed. Pantoustier et al. [85] used the in situ intercalative polymerization method for the preparation of PCL-based nanocomposites. They compared the properties of nanocomposites prepared using both pristine MMT and x-aminododecanoic acid-modified MMT. For nanocomposite synthesis, the desired amount of pristine MMT was first dried under vacuum at 70 °C for 3 h. A given amount of CL was then added to a polymerization tube under nitrogen, and the reaction medium was stirred at room temperature for 1 h. A solution of initiator ($Sn(Oct)_2$) or $Bu_2Sn(Ome)_2$) in dry toluene was added to the mixture in order to reach a [monomer]/[Sn] molar ratio equal to 300. The polymerization was then allowed to proceed for 24 h at room temperature.

The polymerization of CL with pristine MMT gives PCL with a molar mass of 4,800 g mol^{-1} and a narrow distribution. For comparison, the authors also conducted the same experiment without MMT but found that there is no CL polymerization. These results demonstrate the ability of MMT to catalyze and to control CL polymerization, at least in terms of a molecular weight distribution that remains remarkably narrow. For the polymerization mechanism, the authors assume that the CL is activated through interaction with the acidic site on the clay surface. The polymerization is likely to proceed via an activated monomer mechanism using the cooperative functions of the Lewis acidic aluminum and Brønsted acidic silanol functionalities on the initiator walls. On the other hand, the polymerization of CL with the protonated x-aminododecanoic acid-modified MMT yields a molar mass of 7,800 g mol^{-1} with a monomer conversion of 92 % and again a narrow molecular weight distribution. In another publication [82], the same group prepared PCL/MMT nanocomposites using dibutyltin dimethoxide as an initiator/catalyst in an in situ ring-opening polymerization of CL.

In other new work, Tasdelen has reported the preparation of poly(ε-caprolactone)/clay nanocomposites by copper(I)-catalyzed azide/alkyne cycloaddition (CuAAC) "click" reaction [90]. In this method, ring-opening polymerization of ε-caprolactone using propargyl alcohol as the initiator has been performed to produce alkyne-functionalized PCL, and the obtained polymers were subsequently attached to azide-modified clay layers by a CuAAC "click" reaction. The advantage of this technique is that many kinds of polymer chains with quantitative efficiency are easily attached on the surface of silicate layers [91–93]. As the first step of this strategy, alkyne-functionalized PCL was synthesized by ROP of ε-caprolactone with tin(II) 2-ethylhexanoate as a catalyst and propargyl alcohol as an initiator. Subsequently, azido-functional montmorillonites were prepared by the conversion of hydroxyl groups of Cloisite 30B clay into azides. Finally, the alkyne-PCL is attached onto the surface and into the interlayer of the organoclay, using the CuAAC reaction between azides and alkynes. The random dispersion of clay layers in the polymer matrix was confirmed by XRD and TEM measurements. TGA traces showed that the nanocomposites have higher thermal stabilities relative to that of the pristine PCL.

Gorrasi et al. reported the preparation of poly(ε-caprolactone) chains grafted onto montmorillonite modified by a mixture of nonfunctional ammonium salts and ammonium-bearing hydroxyl groups [94]. The clay content was fixed to 3 wt%, whereas the hydroxyl functionality was 25 %, 50 %, 75 %, and 100 %, obtaining an intercalated or exfoliated system.

Na^+-MMT was intercalated by given mixtures of monohydroxyl-functionalized ammonium cation, $(CH_3)_2(C_{16}H_{33})N^+(CH_2CH_2OH)$, with nonfunctionalized ammonium cation, $(CH_3)_3N^+(C_{16}H_{33})$. This co-intercalation with various monohydroxyl-functionalized ammonium relative contents equal to 25 %, 50 %, 75 %, and 100 % was performed in water at 85 °C by an ionic exchange reaction of the interlayer sodium cations with the corresponding ammonium iodide salts. The transport properties of water and dichloromethane vapors and the mechanical properties were investigated. The mechanical and dynamic mechanical analyses showed improvement of the nanocomposite elastic modulus in a wide temperature range. Interestingly, for the higher hydroxyl contents (50 %, 75 %, and 100 %), the decrease of modulus at higher temperature, due to the PCL crystalline melting, did not lead to the loss of mechanical consistence of the samples. Consequently, they revealed a measurable modulus up to 120 °C, a much higher temperature with respect to pure PCL. Water sorption was investigated in the entire activity range, and a lower sorption was observed on increasing the hydroxyl content, up to the sample with 100 % hydroxyl content, which turned to be completely impermeable, even in liquid water. The sample with 75 % hydroxyl content showed a threshold activity ($a = 0.4$) below which was impermeable to water vapor. Also, the diffusion parameters decreased when the hydroxyl content increased, up to the 100 % sample, which showed zero diffusion. The diffusion parameters of an organic vapor, dichloromethane, also exhibited a decreasing value on increasing the hydroxyl content in the nanocomposites.

3.2 Characterization and Properties

3.2.1 Crystallinity Behavior

Both the crystallinity and the crystallite size decreased because of the dispersed silicate layers that represent physical barriers and hinder PCL crystal growth. Homminga et al. [95] have reported that combination of surfactant-modified montmorillonite (Closite 15A) silicate layers, poly(ε-caprolactone), and the adopted melt processing procedure results in intercalated nanocomposites in which the silicate layers act as nucleating agents for the crystallization of the PCL matrix and by which in turn the overall crystallization rate increases. At a sufficiently high MMT concentration and degree of supercooling, the polymer-swollen silicate layer stacks disturb crystal growth, resulting in a decrease in the overall crystallization rate. Simultaneous, time-resolved, synchrotron, small- and wide-angle X-ray scattering experiments reveal that when the retarding effect is absent at a sufficiently high temperature, the final semicrystalline structures of pure PCL and its nanocomposites are identical. The poorer nucleation in the case of pure PCL, however, results in a timewise smearing of primary and secondary crystallization, whereas in the nanocomposites, these events are well separated due to a nucleation-induced, efficient, and rapid primary crystallization. Secondary crystallization involves the insertion of new lamellar crystals in between the already existing ones.

3.2.2 Permeability Behavior

The dispersion of high aspect ratio platelets also reduced the water permeability, nearly by an order of magnitude at 4.8 vol.% silicate [25]. Tortora et al. [33], who examined the water and dichloromethane permeability, assumed that the diffusion path of the polar water molecules is slowed down compared to dichloromethane vapor, not only because of the physical barrier of the clay layers, but also because of the hydrophilic character of the platelets. Gorrasi et al. have reported [96] the preparation of nanocomposites of polycaprolactone with MMT-Na^+ by melt blending; exfoliated nanocomposites were obtained by in situ polymerization of ε-caprolactone with organo-montmorillonite; intercalated nanocomposites were obtained either by melt blending or by in situ intercalative polymerization. Transport properties of water vapor and dichloromethane were measured. The sorption curves of water vapor in all the composite samples follow the dual-sorption behavior. Montmorillonite presents specific sites on which the water molecules are absorbed. The amount of absorbed solvent derived from the linear part of the curve increases on increasing the MMT content, particularly for the microcomposites obtained from the unmodified MMT-Na^+. The diffusion parameters depend on the amount of vapor sorbed; therefore, the diffusion parameter D_0 was derived by extrapolation to zero vapor concentration and compared to the value of the pure PCL. The microcomposites as well as the intercalated nanocomposites have diffusion parameters very near to PCL, while the exfoliated nanocomposites show much lower values, even at low montmorillonite content. This is an indication that the water molecules on specific sites are not immobilized but can jump from one site to another. Only in the case of the exfoliated samples the inorganic platelets, dispersed in a not ordered distribution, can constitute a barrier to the path of the hydrophilic molecules. The sorption curves of dichloromethane are similar to the pure PCL, showing that no specific sites of MMT are occupied by dichloromethane. In this case, the value of sorbed solvent at low activity is mainly dominated by the amorphous fraction present in PCL. At high vapor activity, all curves show an exponential increase, due to plasticization of the polymer. The diffusion parameters of the microcomposites are very close to PCL, while the exfoliated nanocomposites also in this case show much lower values. For the organic solvent, also the intercalated samples show lower diffusion parameters confirming that it is not the content of clay alone but the type of dispersion of the inorganic component in the polymer phase that is important for improving the barrier properties of the samples. The structural characterization and transport properties of blends of a commercial high molecular weight poly(ε-caprolactone) with different amounts of a montmorillonite–poly (ε-caprolactone) nanocomposite containing 30 wt% clay were studied [97]. Modified MMT by 12-aminolauric acid was used. Two different vapors were used for the sorption and diffusion analysis – water as a hydrophilic permeant and dichloromethane as an organic permeant – in the range of vapor activity between 0.2 and 0.8. The blends showed improved mechanical properties in terms of flexibility and drawability as compared with the starting nanocomposites. The permeability (P), calculated as the product of the sorption (S) and the zero-concentration diffusion coefficient (D_0), showed a strong dependence on the clay content in the blends. It greatly decreased on increasing the montmorillonite content for both vapors. This behavior was largely dominated by the diffusion parameters.

3.2.3 Viscoelastic Behavior

Eventually, the linear viscoelastic behavior of the nanocomposites with various OMMT contents was examined. A "pseudo solid-like" behavior was clearly seen at silicate loading greater than 3 wt%, suggesting that domains were formed wherein some long-range order structure was preserved and the silicate layers were oriented in some direction. Furthermore, the nonterminal effect was more pronounced with increasing clay content. These long-range order and domain structures were hence likely to become better defined when the mean distance between the layers becomes less than the lateral dimensions of the silicate layers and thus forcing some preferential orientation between the layers.

3.2.4 Morphological Studies

Ogata et al. applied the exfoliation–adsorption method for the production of poly (ε-caprolactone) (PCL) biodegradable nanocomposites [40] using montmorillonite modified with distearyldimethylammonium cations. The composites were prepared by dissolving PCL in hot chloroform in presence of a given amount of the modified clay and then vaporizing the solvent to obtain homogeneous films. However, under those conditions, it was found that no intercalation took place in the presence of whatever polyester. It is worth to point out that the organomodified clay rather formed a remarkable geometric structure in the filled polymers where tactoids consisting of several silicate monolayers form a superstructure in the thickness direction of the film. Such structural features have been found on one hand to substantially increase the Young's modulus of the PLA-based composites (which is almost doubled with 5 wt% of organomodified clay) and on the other hand to enhance both storage and loss moduli determined by dynamic mechanical analysis (DMA) carried out on the organoclay-filled PCL.

Further morphological observations were carried out by scanning probe microscopy (SPM), while surface analysis was examined by X-ray photoelectron spectroscopy (XPS) and Fourier transform infrared spectroscopy in the reflection–absorption mode (FT-IRAS) [87]. Taking into account the structure, the thermal stability increased and the water permeability decreased since the well-dispersed fillers with high aspect ratio acted as barriers to oxygen and volatile degradation products [85, 96]. In contrast, nanocomposites filled with non-hydroxyl functional clays exhibited only intercalated structures [81, 82, 86]. Since ε-CL polymerization is initiated by OH groups, polymer chain lengths can be predetermined and controlled by the clay loading. Thus, the clay content is limited to a certain range of concentrations to prevent from obtaining too short PCL chain lengths. Nevertheless, this can be modulated by tuning the number of OH groups, e.g., by modifying the clay surface by a mixture of nonfunctional alkylammonium and monohydroxylated ammonium cations [83, 84, 87]. Thus, using this interesting in situ intercalative process, the inorganic content, the quantitative surface grafting, the number of polyester chains per clay surface, as well as the polymer chain length and molecular weight distribution are well controlled [83]. Viville et al. [87] also studied the morphology of PCL-grafted chains on the silicate layer surface depending on the OH content. They showed that the grafting density drastically increased as the proportion of OH-substituted alkylammonium cations used to organomodify the clay increased. Since separate polymer islands were formed in the low OH systems, they assumed that a phase separation process occurred between the ammonium ions induced by the polymerization reaction. Homogeneous coverage and subsequent thickening only take place from 50 % OH content. When this situation was achieved, adjacent platelets become fully independent of each other, which greatly favored exfoliation. The "coordination–insertion" mechanism, i.e., in situ intercalation catalyzed by initiators, was compared to the thermally activated in situ intercalation with various OMMT [86]. Messersmith and Giannelis' results [25] stating that large catalytic surface of montmorillonite can contribute to polymerization of ε-CL were confirmed. Exchanged cations bearing protic functions like NH_3^+, OH, and COOH significantly favored the polymerization and led to similar structures to those obtained by the "coordination–insertion" mechanism. Nevertheless, the PCL molecular weights remained low and the polydispersity index at high conversion reached values higher than 2, confirming that the in situ intercalation in the presence of OH groups and initiators provides better polymerization control. Eventually, the melt intercalation route led to intercalated or intercalated–exfoliated structures when PCL was associated with OMMT bearing quaternized octadecylamine (MMT-$NH_3^+(C_{18})$), di(hydrogenated tallow) dimethylammonium (MMT-$N^+(Me)_2(tallow)_2$), dimethyl 2-ethylhexyl(hydrogenated tallow) (C25A), or methyl bis(2-hydroxyethyl) (hydrogenated tallow) (C30B) [59, 85, 98]. On the contrary, MMT-Na and MMT organomodified with ammonium-bearing 12-aminododecanoic acid (MMT-$NH_3^+(C_{11}COOH)$) formed microcomposites since no change of interlayer gap was observed whereas the in situ intercalation showed exfoliation in the case of MMT-$NH_3^+(C_{11}COOH)$ [25, 85, 86]. Therefore, contrary to the in situ intercalative process, complete exfoliation was not reached by the melt intercalation route, whatever the OMMT considered.

Ko et al. [99] have prepared the PCL/silicate layers with Closite 30A as organoclay, and in this work, the viscosity effect of matrix polymer on melt exfoliation behavior of an organoclay in PCL was investigated. The viscosity of matrix polymer was controlled by changing the molecular weight of poly(ε-caprolactone), the processing temperature, and the rotor speed of a mini-molder. Applied shear stress facilitates the diffusion of polymer chains into the gallery of silicate layers by breaking silicate agglomerates down into smaller primary particles. When the viscosity of PCL is lower, silicate agglomerates are not perfectly broken into smaller primary particles. At higher viscosity, all of silicate agglomerates are broken down into primary particles and finally into smaller nanoscale building blocks. It was also found that the degree of exfoliation of silicate layers is dependent upon not only the viscosity of matrix but also the thermodynamic variables.

Miltner et al. [100] have studied the benefits and limitations of several methods available for evaluating nanofiller dispersion, as applied to poly(ε-caprolactone) nanocomposites. Bentone®108 is a natural hectorite exchanged with dimethyl bis (hydrogenated tallow alkyl) ammonium, Nanofil®15 is a natural montmorillonite exchanged with dimethyl dioctadecyl ammonium, and Nanofil®SE3010 is based on a similar compound, but its exact composition remains proprietary information of Süd-Chemie. Cloisite 10A and Cloisite 30B were used.

The nanocomposites of this work, with filler contents up to 10 wt%, were prepared by melt mixing at 130 °C using a batch-operated lab-scale twin-screw DSM Xplore Micro-Compounder (15 cm^3, N2 purge, screw rotation speed of 170 rpm). Unless otherwise specified, the residence time within the extruder was 30 min. All compositions are expressed in terms of the inorganic filler content, as determined from thermogravimetric analysis under nitrogen. Subsequent to melt mixing, all nanocomposites were compression molded at 140 °C under 100 bar pressure. Analyses resulted that dynamic rheometry effectively indicates the formation of a percolating silicate network, but although the extent of network formation must depend on the degree of silicate exfoliation, its occurrence can definitely not be considered an unambiguous fingerprint of an exfoliated state. The secant modulus is without a doubt fairly sensitive to the extent of silicate dispersion, but as it depends on the overall structure of the nanocomposite, it may be influenced by aspects such as an altered degree of crystallinity and processing-induced filler orientation; all three can contribute to a qualitative assessment of the filler dispersion and the main factors affecting it: the intrinsic affinity between the matrix polymer and the nano-sized filler particles as well as the choice of nanocomposite processing conditions. It was demonstrated that PCL nanocomposites containing layered silicates with a higher intrinsic affinity for the matrix polymer show a more pronounced solid-like behavior at low frequency in dynamic rheometry experiments as well as a higher excess heat capacity during quasi-isothermal crystallization of the polymer matrix. Accordingly, those nanocomposites displaying the highest dispersion levels were also found to exhibit superior improvement in mechanical properties such as stiffness. From a direct comparison between the various methods, the innovative C_p^{excess} approach was evidenced to offer a valuable alternative to the commonly employed characterization tools, providing accurate data with regard to nanofiller dispersion in addition to a more fundamental insight into interphase formation and into the way the matrix polymer is affected in the vicinity of dispersed nanofiller particles. The approach was also shown to retain good sensitivity at low nanofiller loadings, where mechanical and rheological approaches were found unsuitable to reliably discriminate between slight changes in the silicate dispersion state.

Homminga et al. demonstrated the influence of melt processing on the final polymer/layered silicate nanocomposite morphology [101]. In particular, the role of shear forces on the transformation of the original large clay agglomerates is of interest. Several polymer nanocomposites were prepared by melt extrusion, involving polycaprolactone, poly(ethylene oxide), polyamide-12 or polyamide-6 as the matrix polymer, and C30B and C15A as organoclays. The nanocomposite morphology was characterized by X-ray diffraction and transmission electron microscopy and the clay tactoid morphology with polarized optical microscopy and scanning electron microscopy. The development of the tactoid and nanocomposite morphology during melt mixing under shear was studied by time-resolved optical microscopy in conjunction with a rheometer and synchrotron X-ray scattering together with a Couette-type flow cell. The shear forces in the melt preparation of polymer/layered mineral nanocomposites facilitate the breakup of large-sized agglomerates, whereas the extent of further exfoliation of the mineral layers is determined by the compatibility between the polymer matrix and the mineral layers rather than by shear forces. Results showed PCL nanocomposites were found to have an intercalated/partially exfoliated morphology. In the study of the breakup of μm-sized clay tactoids during melt mixing, the following observations were done. Without applied shear forces, a gradual disappearance of the clay agglomerates was observed, while with shear forces applied, a fast breakup of the clay agglomerate morphology was noticed.

Chung et al. prepared poly(ε-caprolactone) nanocomposites, PCL/C25A and PCL/C30B, with organoclays having nonpolar and polar organic modifiers, respectively, by melt-mixing method and additional heat treatment [102]. WXRD analysis revealed that both nanocomposites were exfoliated, irrespective of the OMMT polarity. However, WXRD failed to show the degree of exfoliation of the nanocomposites, because the d_{001} peaks disappeared. Thus, dynamic mechanical analysis (DMA) was carried out to compare the degree of exfoliation of the PCL nanocomposites. From dynamic mechanical analysis (DMA), PCL/C30B showed higher elasticity, storage moduli, viscosity, and activation energy than PCL/C25A, indicating that PCL/C30B had a more exfoliated structure than PCL/C25A. This is due to the polar interaction in PCL/C30B, as verified by the plots of a_T versus temperature. Thus, it was confirmed that DMA provides an alternative approach to evaluating the degree of exfoliation of nanocomposites.

3.2.5 Mechanical Properties

However, the tensile and thermal properties were improved. For instance, the modulus increased from 210 MPa for unfilled PCL to 280 MPa or 400 MPa with 3 wt% of MMT-$NH_3^+(C_{18})$, MMT-$N^+(Me)_2(tallow)_2$, or C25A and 10 wt% of C30B, respectively, attesting for an almost twofold increase of the PCL rigidity in the latter case [59, 98]. Chen and Evans [103] demonstrated on similar systems that the elastic modulus trends with clay volume fraction may be interpreted using well-established theory for conventional composites, namely, the Hashin–Shtrikman bounds. At OMMT content higher than 5 wt%, the elongation at break dropped off due to clay aggregation [59, 98]. Dynamic mechanical measurements also revealed that with 1 wt% clay, nanocomposite materials exhibited a pseudo solid-like behavior [82].

3.2.6 Thermal Behavior

The nanocomposites showed an improved thermal stability, which is consistent with an effective barrier against permeation of molecular oxygen and evolved combustion gas formed by the silicate sheets. The weight loss due to the formation of volatile degradation products was monitored as a function of temperature. Flame retardancy was remarkable and related to the deposition of an insulating and incombustible char whenever the PCL nanocomposites are exposed to the flame. The detailed study of PCL melt-intercalated nanocomposites with natural Na^+-MMT- and HTA-based quaternary ammonium cations was conducted by the same author [59]. The nanocomposites were found to be stable and burned without droplets during visual burning examination. However, Kwak and Oh [104] demonstrated that PCL chains can diffuse further into the silicate gallery due to additionally subjecting the samples to heat during the analyses, and finally, extended exfoliation is achieved. The 50 % weight loss temperature is shifted by 60 °C towards higher temperature on the addition of 1 wt% of clay, whereas the temperature shift is only 30 °C at 10 wt%. Thus, PCL nanocomposites combine high stiffness, good ductility, and improved thermal stability at low clay content (<5 wt%). Only Di et al. [58] reached exfoliated state in the case of PCL/C30B systems prepared by direct melt intercalation with 2–5 wt% of clay. Obviously, they reported great enhancements of mechanical and thermal properties as well as a pseudo solid-like rheological behavior caused by the strong interactions between the organoclay layers and PCL and by the good dispersion of exfoliated organoclay platelets. Moreover, the isothermal crystallization behavior [105] revealed that the well-dispersed organoclay platelets act as nucleating agents but also affected the crystals' quality by the restricted chains mobility.

3.2.7 Biodegradability

Tetto et al. [106] first reported results on the biodegradability of nanocomposites based on PCL, reporting that the PCL/OMLS nanocomposites showed improved biodegradability compared to pure PCL. The improved biodegradability of PCL after nanocomposite formation may be due to a catalytic role of the OMLS in the biodegradation mechanism, but this is not clear. The biodegradability of PCL-based nanocomposites was studied under composting conditions after preparation by in situ polymerization and extrusion. The composites filled with 10 % clay showed highest modulus, but a decrease in elongation was observed. Biodegradation studies were carried out under soil and marine environments by measuring the evolved carbon dioxide and weight loss, respectively. The data at the incubation of 22 °C have a minimal weight loss for both pure PCL and PCL/clay (50/50 %) composites in the absence of any nutrients of exposure. PCL/clay nanocomposites were obtained by melt blending of PCL [70] with OMS and from in situ ring-opening polymerization of caprolactone with OMS. In the TGA, it was observed that the clays alone can undergo degradation in two steps, a Hofmann elimination, giving a trialkylammonium cation, followed by the loss of the amine leaving only a proton as the counterion. The detailed investigation on the biodegradation of PCL nanocomposites is required to conclude the role of modifier, clay dispersion on the mechanism of bioconsumption.

Among natural fillers, sepiolites, which are layered hydrated magnesium silicates characterized by a needlelike morphology and very high surface area (BET 374 m^2/g), were proven to play significant reinforcement actions in some polymers, as well as adhesion with polymer matrices like polypropylene [107], due to the large concentration of surface silanols that are easily available for coupling reactions. In the new research work [108], PLA and PCL nanocomposites prepared by adding 5 wt% of a sepiolite (SEPS9) were degraded in compost, leading to effective degradation for all samples. PLA and PLA/SEPS9 seem to be mainly degraded by a bulk mechanism, showing a significant level of polymer degradation; however, the presence of SEPS9 particles partially delays the degradation probably due to a preventing effect of these particles on polymer chain mobility and/or PLA/enzymes miscibility. PCL and PCL/SEPS9 showed a preferential surface mechanism of degradation, and in contrast to PLA, sepiolite does not present a considerable barrier effect on the degradation of PCL.

3.2.8 NMR Analyses

Very recently, solid-state NMR has emerged as a tool to characterize clay/polymer nanocomposites in complement to data from classical methods (XRD, TEM) since it is a powerful technique for probing the molecular structure, conformation, and dynamics of species at interfaces. Therefore, solid-state NMR was used in PCL-based nanocomposites to investigate how the surfactant conformation and mobility are changed by the polymer adsorption and how the polymer motion is perturbed after intercalation in the nanocomposites [74, 109, 110]. Finally, Calberg et al. [74] validated this characterization method to determine the structure of PCL-based nanocomposites since they demonstrated that there was a correlation between variations in the proton relaxation times $T_1(H)$ and the quality of clay dispersion. ^1H-NMR spectra of CL, PCL, and PCLC2 (a nanocomposite containing 2 wt% of OMLS) have shown the corresponding chemical shifts clearly demonstrating conversion from CL to PCL. Complete conversion of CL to PCL was assumed because residual CL was not detected in the NMR spectra of any of the composites. The peak broadening effect seen in XRD pattern (not reported here) of

PCLC2 is believed to be due to the strong attachment of PCL chains to the silicate layers, resulting in partial solid-like behavior [25].

4 Nanocomposites Based on PCL/Polymer/Clay

Namazi et al. prepared the starch-g-polycaprolactone (starch-g-PCL) nanocomposites with graft polymerization through in situ ring-opening polymerization of ε-caprolactone in the presence of starch and $Sn(Oct)_2$ (Tin(II) 2-ethyl hexanoate) as an initiator/catalyst [111–113]. A surface-modified montmorillonite by dimethyl (hydrogenated tallow alkyl) ammonium cation was used (Closite15A). Results of XRD analyses showed the best diffusion and intercalation of copolymer into galleries of organoclay that occurs through solution intercalation in comparison with in situ method. Because organoclay is compatible with starch-g-PCL copolymer, it is not compatible with pure starch in in situ method. TGA investigations showed that with grafting, caprolactone on starch thermal stability of both grafted starch and also their nanocomposites improved in comparison with thermal stability of pure starch.

In other works reported by Namazi and Mosadegh [114–116], two types of clay, MMT and C15A, have been used for the preparation of the starch-g-PCL nanocomposites, and investigations showed different and special properties. In the first method, starch/MMT nanocomposite with solution route prepared then ε-caprolactone monomer and initiator were added to this sample. After ROP polymerization, starch/MMT/PCL nanocomposite was obtained. Starch/PCL nanocomposite was prepared with hydrophobic C15A. This composite was produced with monomer swelling in Closite 15A and then polymerization in the presence of starch and initiator. These silicate layers show different behavior when clays introduce into starch biopolymer. The XRD results showed that the biopolymer starch with hydrophilic property is compatible with montmorillonite. The basal peak in MMT ($2\theta = 7.13$, d-spacing = 12.38 A°) has broadened and shifted to low angle in starch/MMT samples, which is an indication of intercalation/exfoliation state. SEM images showed the change in morphology of granules of starch in composite samples. Granules were destroyed due to gelatinization affected by water, temperature, and polymerization reaction on surface of starch. By adding mineral clay through biopolymer matrix, thermal behavior of starch has changed. From DSC thermograms of the nanocomposite, it was possible to see that the melting of crystalline (Tm) increased except starch/MMT sample. Decrease in Tm with addition of MMT, around 125 °C, was attributed to the reduced crystalline size and presence of crystal imperfections due to compatibility of the MMT with starch which suppressed the crystallization where the silicate platelet prevented the amylase chains to reorganize. The water uptake behavior of biodegradable layered silicate/starch–polycaprolactone blend nanocomposites was evaluated by Perez et al. [117]. Three different commercial layered silicates (Cloisite Na+, Cloisite 30B, and Cloisite 10A) were used as reinforcement nanofillers. Nanocomposites were prepared by melt intercalation using an intensive mixer. Filler content was in the range from 0 to 7.5 wt%. Tests were carried out in two different environments: 60 % and 90 % relative humidity using glycerol solutions. The intercalated structure (determined by wide-angle X-ray diffraction) showed a decrease in water absorption as a function of clay content probably due to the decrease of the mean free path of water molecules. The diffusion coefficient decreased with clay incorporation, but a further increase in the clay content did not show an important effect on this parameter. Elongation at break increased with exposure showing matrix plasticization. Mechanical properties of the nanocomposites deteriorated after exposure whereas they remained almost constant in the case of the neat matrix.

Vertuccio et al. reported that sodium montmorillonite was incorporated into a poly(ε-caprolactone)–starch blend by means of a ball mill [118]. Samples in powder form were milled at ambient temperature in a ball mill. Milling was conducted for three different times (4, 7, 10 h) at the rotation speed of 580 rpm. Scanning electron microscopy and X-ray characterization show that the milling process can improve the compatibilization between the PCL and the starch phases while promoting the dispersion of clay minerals at nanometric level. The milling time strongly influences the mechanical and barrier properties. In particular, the best results in terms of elastic modulus and permeability coefficient were achieved with a complete delamination of the pristine clay structure. In summary, the milling process not only has demonstrated to be a promising compatibilization method for immiscible PCL–starch blends, but it can be also used to improve the dispersion of nanoparticles into the polymer blends. The addition of 3%wt. of clay produces an increase of the elastic modulus of the PCL–starch blend. Moreover, going from the microcomposite to the partially exfoliated sample up to the exfoliated one, the modulus increases of about 50 %, 74 %, and 93 %, respectively. Such a result indicates that the complete delamination of the clay structure is a crucial parameter for improving the mechanical properties.

Hoidy et al. [119] have reported that poly(L-lactic acid) (PLLA)/(PCL) and two types of organoclay (OMMT) including a fatty amide and octadecylamine montmorillonite (FA-MMT and ODA-MMT) were employed to produce polymer nanocomposites by melt blending and solution casting. It is found that the highest tensile strength was observed at 80/20 (PLLA/PCL) blend. In addition to improvement of mechanical properties, the addition of OMMT to the PLLA/PCL blends significantly enhanced the thermal stability of PLLA/PCL blend when adding ODA-MMT. However, the thermal stability of PLLA/PCL blend decreased after adding FA-MMT. This may be due to the weak interaction between FA-MMT and the organic medium. The best ratios to obtain the highest basal spacing were observed when the OMMT content was 2 % and 3 % for ODA-MMT and FA-MMT, respectively; further amount of organoclay could cause brittleness of nanocomposites.

In the other work, Wu et al. [120] have studied on crystallization and thermoelectric properties of PCL/poly(vinyl butyral) (PVB)/montmorillonite nanocomposites containing carbon black (CB) as a function of a small amount of amorphous PVB content and a wide range of molecular weight of PVB. X-ray diffraction data of PCL/PVB/MMT nanocomposites indicates that most of the swellable silicate layers are exfoliated and randomly dispersed into PCL/PVB system. The band spacings of PCL spherulites in PCL/PVB/MMT nanocomposites decrease with increasing PVB content, and this indicates that increasing the PVB content greatly shortens the period of lamellar twisting. The presence of 1 wt% MMT and higher molecular weight of PVB also shorten the period of PCL lamellar twisting. Nucleation and crystallization parameters, such as growth rate G and Avrami exponent n, can be determined by using POM and DSC isothermally crystallized at 41 °C.

Yoshioka and coworker [121] have reported the preparation of cellulose acetate/ layered silicate grafted poly(ε-caprolactone) [(CA/layered silicate)-g-PCL] nanocomposites by in situ polymerization of ε-caprolactone in the presence of cellulose acetate (CA) and organically modified layered silicate (OMLS). When SPN (synthetic hectorite with polyhydroxypropyl ($n = 25$), diethyl, methylammonium cation as modifier) having one hydroxyl group in its modifier was used, the silicate layers could not be dispersed thoroughly and however existed as aggregates consisting of several silicate layers. Among them, the crystal growth of PCL developed by transcrystallization, where the crystal growth was restricted in the confined space. When Cloisite 30B having two hydroxyl groups within the modifier was used, the silicate layers forming the clay were dispersed completely in the composite, and random orientation of the OMLS was observed.

The above poly(ε-caprolactone)-based nanocomposite synthesis has been applied by Chen et al. [122] to produce novel segmented polyurethane/clay nanocomposites articulated on diphenylmethane diisocyanate, butanediol, and preformed polycaprolactone diol. Even if the mechanism proposed for the chemical link between the nanofiller surface and the polymer does not appear appropriate (ammonium salts are not known to induce ε-caprolactone ring-opening polymerization), they succeeded in producing a material where the nanofiller acts as a multifunctional chain extender inducing the formation of star-shaped segmented poly(urethane).

5 PCL Nanocomposites Based on Other Nanofillers

Calandrelli et al. [123] have reported the preparation and characterization of new nanocomposites based on polycaprolactone reinforced with nanostructured silica, projected for biomedical applications in orthopedic, maxillofacial, and dental surgery fields. The chemical modification of the PCL, necessary to realize a stable interface with the nanosilica, has been carried out directly in the processing equipment and has allowed increasing the mechanical characteristics of the polyester without worsening its biocompatibility characteristics. Young's modulus was increased to a level by far superior to the neat PCL. The improvement of mechanical properties of the nanocomposites compared with the pure PCL can allow their potential use as scaffolds or membrane barriers.

Huang et al. [124] have reported the preparation of poly-2-hydroxyethylmethacrylate (PHEMA)/polycaprolactone nanocomposite based on nano-sized hydroxyapatite (nano-HA) as reinforcing filler. Prepared nano-HA particles had a rodlike morphology, 20–30 nm in width and 50–80 nm in length.

Porous nanocomposite scaffolds were then produced using a porogen leaching method. An in vitro study found that the nanocomposites were bioactive as indicated by the formation of a bone-like apatite layer after immersion in simulated body fluid. Furthermore, the nanocomposites were able to support the growth and proliferation of primary human osteoblast (HOB) cells. HOB cells developed a well-organized actin cytoskeletal protein on the nanocomposite surface. The results demonstrate the potential of the nanocomposite scaffolds for tissue engineering applications for bone repair.

Johari et al. [125] have shown that biodegradation and biocompatibility of novel poly(ε-caprolactone)/nano-fluoridated hydroxyapatite (PCL–FHA) scaffolds. The FHA nanopowders were prepared via mechanical alloying method and had a chemical composition of Ca_{10} $(PO_4)_6OH_{2-x}F_x$ (where x values were selected equal to 0.5 and 2.0). In order to fabricate PCL–FHA scaffolds, 10, 20, 30, and 40 wt% of the FHA were added to the PCL. The PCL–FHA scaffolds were produced by the solvent casting/particulate leaching using sodium chloride particles (with diameters of 300–500 μm) as the porogen. In vitro degradation of PCL–FHA scaffolds was studied by incubating the samples in phosphate-buffered saline at 37 °C and pH 7.4 for 30 days. Moreover, biocompatibility was evaluated by MTT assay after seeding and culture of osteoblast-like cells on the scaffolds. Results showed that the osteoblast-like cells attached to and proliferated on PCL–FHA and increasing the porosity of the scaffolds increased the cell viability. Also, degradation rate of scaffolds were increased with increasing the fluorine content in scaffolds composition.

Deng at al. have reported [126] the study to improve the tensile strength and elastic modulus of nano-apatite/poly(ε-caprolactone) composites by silane modification of the nano-apatite fillers. Three silane coupling agents (KH560, KH570, and KH792) were used to modify the surfaces of nano-apatite particles, and composites of silanized apatite and PCL were prepared by a technique incorporating solvent dispersion, melt blending, and hot pressing. The results showed that the silane coupling agents successfully modified the surfaces of nano-apatite fillers, and the crystallization temperatures of the silanized apatite/PCL composites were higher than that of the non-silanized control material, although the melting temperature of the composites remained almost unaffected by silanization. The ultimate tensile strength and elastic modulus of the silanized composites reached 22.60 MPa and 1.76 GPa, as a result of the improved interfacial bonding and uniform dispersion of nano-apatite fillers.

A novel bone scaffold was fabricated by electrospinning of poly (ε-caprolactone) filled with nano-HA [127]. That the coupling agent γ-glycidoxypropyltrimethoxysilane (A-187) was employed to treat HA particles previously was essential to disperse HA well in PCL matrix. The good dispersion of HA in PCL could enhance tensile strength and modulus of the scaffold.

In other work, the novel 3D scaffolds made of nano nonstoichiometric apatite and poly(ε-caprolactone) bioactive composites were fabricated using a prototyping controlled process [128]. The pore size, morphology, and interconnectivity were well controlled by the PCP technology.

The prepared ns-AP/PCL composite scaffolds had a high porosity of 76 %, with open and interconnected pores size ranging from 400 to 500 μm.

Makarov and coworker have prepared the calcium phosphate–PCL nanocomposite powders with high ceramic volume fractions (80–95 %) by a nonaqueous chemical reaction in the presence of the dissolved polymer [129]. Depending on the reagents' combination used (calcium acetate, phosphoric acid, and sodium methoxide), different Ca phosphates (dicalcium phosphate (DCP)–PCL or Ca-deficient HA (CDHA), both considered bioresorbable) were obtained. The process leading to the formation of DCP allowed full incorporation of the dissolved polymer in the composite whereas roughly 50 % of the total polymer amount was incorporated in the CDHA–PCL material. Despite the very low PCL fractions, a uniform distribution of the polymer phase has been achieved. The CDHA–PCL composite powders obtained were high pressure consolidated into dense materials at room temperature at 2.5 GPa. The advantage of the process is the possibility of incorporating biomolecules (drugs, growth factors) already during fabrication. CDHA-based composites nominally containing 11 and 24 vol.% PCL exhibited high strengths and supported the attachment and proliferation of endothelial and osteoblastic cell lines. These properties make the synthesized nanocomposites attractive materials for bioresorbable scaffold fabrication.

Composite materials based on poly(ε-caprolactone) and carbon nanofibers (CNFs) were processed by solvent casting and electrospinning [130]. Neat and composite PCL films were obtained by solvent casting. In the first case, PCL granules were added to $CHCl_3$, the mixture was magnetically stirred at room temperature (RT) after it was completely dissolved. Thereafter, the polymeric solution was cast on a Teflon support and air-dried for 48 h at RT and for further 48 h in vacuum. Nanohybrid PCL/CNFs mats of different composition (1, 3, and 7 wt%) were produced by means of electrospinning technique following a two-step procedure. The selected amount of carbon nanostructure was dispersed in a mixture of THF:DMF (1:1) and ultrasonicated for 1 h. Then, PCL was added (14 % w/v) and the resulting suspension magnetically stirred at room temperature until polymer dissolution was completed. The mixture was poured into a glass syringe fitted with a metallic blunt tip needle (22G), connected to the positive output of a high-voltage power supply, set at 12 kV. A controlled flow was achieved by means of a syringe pump, running at 0.4 mL/h. Electrospun mats were collected in air at RT onto an aluminum target (Ø/10 cm) fixed at 15 cm from the tip of the needle; then samples were then vacuum dried for 48 h. The composite films showed a good dispersion in the PCL matrix, while electrospun samples consisted of homogeneous and uniform fibers up to 3 wt% CNFs with average fiber diameter ranging between 0.5 and 1 μm. Composite films and mats revealed an increased crystallization temperature with respect to the neat PCL matrix. A stiffness increase was achieved in PCL films depending on the CNF content, while mechanical properties of mats were only slightly affected by CNF introduction.

In other new work [131], the nanocomposites of poly(ε-caprolactone) and tungsten trioxide (WO_3) were prepared by solvent casting using 5 % and 10 % of WO_3 nanoparticles. Photodegradation of PCL/WO_3 nanocomposites was studied exposing the samples to a xenon lamp, which simulates the UV spectrum of the sun. The results obtained showed that due to the incorporation of WO_3 nanoparticles, the nanocomposites exhibit higher thermal stability together with higher photodegradation efficiency.

Chen and Liu reported preparation of nanocomposites comprising of poly(ε-caprolactone) and attapulgite (AT) by solution mixing [132]. In this work, SEM observation showed that AT nano-rods were well dispersed in the PCL matrix in a nanometer scale at lower AT content and a severe aggregation would happen at 7 wt% AT content. The rheological analysis found that up to 1 wt%, AT nano-rods did not have large effect on the complex viscosity of PCL. However, the nanocomposite with 7 wt% AT displayed notable shear-thickening behavior at low frequency and had much higher complex viscosity than other samples. These results have been attributed to the aggregation of AT. The storage modulus of the nanocomposites showed a monotonic increase with AT content in low-frequency region due to the interaction between the PCL matrix and AT, which was consistent with the slope changes found in the Cole–Cole plot. The apparent activation energies of flow were determined by considering the Arrhenius relationship. It has been found that the activation energy is nearly AT content independent. AT also can be used as an effective nucleating agent and has effects on the growth of crystallites in the crystallization process of PCL matrix.

References

1. Chiellini E, Solaro R (1996) Adv Mater 8:305
2. Albertsson AC, Varma IK (2002) Adv Polym Sci 157:1
3. Okada M (2002) Prog Polym Sci 27:87
4. Nakayama A, Kawasaki N, Maeda Y, Arvanitoyannis I, Ariba S, Yamamoto N (1997) Appl Polym Sci 66:741
5. Bastioli C, Cerutti A, Guanella I, Romano GC, Tosin M (1995) Environ Polym Degrad 3:81
6. Bastioli C (1998) Polym Degrad Stabil 59:263
7. Bastioli C (1998) Macromol Symp 135:193
8. Averous L, Moro L, Dole P, Fringant C (2000) Polymer 41:4157
9. Koenig MF, Huang SJ (1994) Polym Degrad Stabil 45:139
10. Tokiwa Y, Suzuki T (1977) Nature 270:76
11. Sinha Ray S, Bousmina M (2005) Prog Mater Sci 50:962
12. Giannelis EP (1996) Adv Mater 8:29
13. Biswas M, Sinha Ray S (2001) Adv Polym Sci 155:167
14. Sinha Ray S, Okamoto M (2003) Prog Polym Sci 28:1539
15. Mohanty AK, Drzal LT, Misra M (2003) Polym Mater Sci Eng 88:60
16. Hiroi R, Sinha Ray S, Okamoto M, Shiroi T (2004) Macromol Rapid Commun 25:1359
17. Mitchell CA, Bahr JL, Arepalli S, Tour JM, Krishnamoorti R (2002) Macromol 35:8825
18. Andrews R, Wisenberger MC (2004) Curr Opin Solid State Mater Sci 8:31
19. Chen JS, Poliks MD, Ober CK, Zhang Y, Wiesner U, Giannelis EP (2002) Polymer 43:4895
20. Namazi H, Ahmadi H (2011) J Power Sources 196:2573
21. Grim RE (1953) Clay mineralogy. McGraw-Hill, New York
22. Brindly SW, Brown G (1980) Mineralogical Society, London
23. Krishnamoorti R, Vaia RA, Giannelis EP (1996) Chem Mater 8:1728

24. Blumstein A (1965) J Polym Sci A 3:2665
25. Messersmith PB, Giannelis EP (1995) J Polym Sci Polym Chem 33:1047
26. Vaia RA, Giannelis EP (1997) Macromolecules 30:7990
27. Messersmith PB, Giannelis EP (1993) Chem Mater 5:1064
28. Krishnamoorti R, Giannelis EP (1997) Macromolecules 30:4097
29. Kiersnowski A, Dabrowski P, Budde H, Kressler J, Piglowski J (2004) Eur Polym J 40:2591
30. Kiersnowski A, Piglowski J (2004) Eur Polym J 40:1199
31. Pucciariello R, Villani V, Belviso S, Gorrasi G, Tortora M, Vittoria V (2004) J Polym Sci Polym Phys 42:1321
32. Pucciariello R, Villani V, Langerame F, Gorrasi G, Vittoria V (2004) J Polym Sci Polym Phys 42:3907
33. Tortora M, Vittoria V, Galli G, Ritrovati S, Chiellini E (2002) Macromol Mater Eng 287:243
34. Hall HK, Schneider HK (1958) J Amer Chem Soc 80:6409
35. Wilson DR, Beaman RG (1970) J Polym Sci Part A 8:2161
36. Johns DB, Lenz RW, Luecke A (1984) In: Ivin KJ, Saegusa T (eds) vol 1. Elsevier Applied Science, NewYork, p 461
37. Cerrai P, Tricoli M, Andruzzi F, Paci M (1989) Polymer 30:338
38. Knani D, Gutman AL, Kohn DH (1993) J Polym Sci Part A Polym Chem 31:1221
39. Bugatti V, Costantino U, Gorrasi G, Nocchetti M, Tammaro L, Vittoria V (2010) Euro Polym J 46:418
40. Jimenez G, Ogata N, Kawai H, Ogihara T (1997) J Appl Polym Sci 64:2211
41. Wu T, Xie T, Yang G (2009) Appl Clay Sci 45:105
42. Olejnik S, Posner AM, Quirk JP (1974) Clay Clay Miner 22:361
43. Ahmed A, Auras R, Kijchavengkul T, Varshney SK (2012) J Food Eng 111:580
44. Luduena LN, Alvarez VA, Vazquez A (2007) Mater Sci Eng A 460–461:121
45. Zerda AS, Caskey TC, Lesser AJ (2003) Macromol 36:1603
46. Zhao Q, Samulski ET (2005) Macromol 38:7967
47. Li J, Xu Q, Peng Q, Pang M, He S, Zhu C (2006) J Appl Polym Sci 100:671
48. Greenland DJ (1963) J Colloid Sci 18:647
49. Zhao Q, Samulski ET (2003) Macromol 36:6967
50. Zhao Q, Samulski ET (2006) Polymer 47:663
51. Briscoe BJ, Kelly CT (1995) Polymer 36:3099
52. Kazarian SG, Vincent MF, Bright FV, Liotta CL, Eckert CA (1996) J Am Chem Soc 118:1729
53. Su B, Lv X, Yang Y, Ren Q (2006) J Chem Eng Data 51:542
54. Shah VM, Hardy BJ, Stern SA (1993) J Polym Sci B Polym Phys 31:313
55. Dinoia TP, Conway SE, Lim JS, McHugh MA (2000) J Polym Sci B Polym Phys 38:2832
56. Shieh YT, Yang HS (2005) J Supercrit Fluids 33:183
57. Shieh YT, Lai JG, Tang WL, Yang CH, Wang TL (2009) J Supercrit Fluids 49:385
58. Di Y, Iannace S, Di Maio E, Nicolais L (2003) J Polym Sci Polym Phys 41:670
59. Lepoittevin B, Devalckenaere M, Pantoustier N, Alexandre M, Kubies D, Calberg C (2002) Polymer 43:4017
60. Dlamini DS, Mishra SB, Mishra AK, Mamba BB (2011) J Inorg Organomet Polym 21:229
61. Campbell KT, Craig DQM, McNally T (2010) J Mater Sci Mater Med 21:2307
62. Luduena LN, Kenny JM, Vazquez A, Alvarez VA (2011) Mater Sci Eng A 529:215
63. Fukushima K, Tabuani D, Camino G (2009) Mater Sci Eng C 29:1433
64. Kim SW, Jo WH, Lee MS, Ko MB, Jho JY (2001) Polymer 42:9837
65. Pucciariello R, Villani V, Gorrasi G, Vittoria V (2005) J Macromol Sci Phys 44:79
66. Ren J, Huang Y, Liu Y, Tang X (2005) Polym Test 24:316
67. Lepoittevin B, Pantoustier N, Devalckenaere M, Alexandre M, Calberg C, Jerome R (2003) Polymer 44:2033
68. Shibata M, Teramoto N, Someya Y, Tsukao R (2007) J Appl Polym Sci 104:3112
69. Gonzalez I, Eguiazabal JI, Nazabal J (2006) Polym Eng Sci 46:864

70. Zheng X, Wilkie CA (2003) Polym Degrad Stab 82:441
71. Pollet E, Delcourt C, Alexandre M, Dubois P (2006) Eur Polym J 42:1330
72. Kalambur SB, Rizvi SS (2004) Polym Int 53:1413
73. Maiti P (2003) Langmuir 19:5502
74. Calberg C, Jerome R, Grandjean J (2004) Langmuir 20:2039
75. Gardebien F, Gaudel-Siri A, Bredas J-L, Lazzaroni R (2004) J Phys Chem B 108:10678
76. Gardebien F, Bredas J-L, Lazzaroni R (2005) J Phys Chem B 109:12287
77. Namazi H, Fathi F, Dadkhah A (2011) Sci Iran C 18:439
78. Benali S, Peeterbroeck S, Brocorens P, Monteverde F (2008) Eur Polym J 44:1673
79. Urbanczyk L, Calberg C, Stassin F, Alexandre M, Jerome R, Jerome C, Detrembleur C (2008) Polymer 49:3979
80. Labidi S, Azema N, Perrin D, Cuesta JML (2010) Polym Degrad Stab 95:382
81. Kubies D, Pantoustier N, Dubois P, Rulmont A, Jerome R (2002) Macromol 35:3318
82. Lepoittevin B, Pantoustier N, Devalckenaere M, Alexandre M, Kubies D, Calberg C (2002) Macromol 35:8385
83. Lepoittevin B, Pantoustier N, Alexandre M, Calberg C, Jerome R, Dubois P (2002) Macromol Symp 183:95
84. Lepoittevin B, Pantoustier N, Alexandre M, Calberg C, Jerome R, Dubois P (2002) J Mater Chem 12:3528
85. Pantoustier N, Lepoittevin B, Alexandre M, Kubies D, Calberg C, Jerome R (2002) Polym Eng Sci 42:1928
86. Pantoustier N, Alexandre M, Degée P, Kubies D, Jerome R, Henrist C (2003) Compos Interfaces 10:423
87. Vivlle P, Lazzaroni R, Pollet E, Alexandre M, Dubois P, Borcia G (2003) Langmuir 19:9425
88. Paul MA, Alexandre M, Degée P, Calberg C, Jerome R, Dubois P (2003) Macromol Rapid Commun 24:561
89. Paul MA, Delcourt C, Alexandre M, Degée P, Monteverde F, Rulmont A (2005) Macromol Chem Phys 206:484
90. Tasdelen MA (2011) Eur Polym J 47:937
91. Chen JC, Xiang JM, Cai ZW, Yong H, Wang HD, Zhang LH (2010) J Macromol Sci Pure Appl Chem 47:655
92. Chen JC, Wang HD, Luo WQ, Xiang JM, Zhang LH, Sun BB (2010) Colloid Polym Sci 288:173
93. Ye YS, Yen YC, Cheng CC, Syu YJ, Huang YJ, Chang FC (2010) Polymer 51:430
94. Gorrasi G, Tortora M, Vittoria V, Pollet E, Alexandre M, Dubois P (2004) J Polym Sci Part B Polym Phys 42:1466
95. Homminga D, Goderis B, Dolbnya I, Groeninckx G (2006) Polymer 47:1620
96. Gorrasi G, Tortora M, Vittoria V, Pollet E, Lepoittevin B, Alexandre M (2003) Polymer 44:2271
97. Gorrasi G, Tortora M, Vittoria V, Galli G, Chiellini EJ (2002) Polym Sci Part B Polym Phys 40:1118
98. Pantoustier N, Alexandre M, Degée P, Calberg C, Jerome R, Henrist C (2001) e-Polym
99. Ko MB, Jho JY, Jo WH, Lee MS (2002) Fibers Polym 3:103
100. Miltner HE, Watzeels N, Block C, Gotzen NA, Assche GV, Borghs K, Durme KV, Mele BV, Bogdanov B, Rahier H (2010) Eur Polym J 46:984
101. Homminga D, Goderis B, Hoffman S, Reynaers H, Groeninckx G (2005) Polymer 46:9941
102. Chung JW, Oh KS, Kwak SY (2007) Macromol Mater Eng 292:627
103. Chen B, Evans JRG (2006) Macromol 39:747
104. Kwak SY, Oh KS (2003) Macromol Mater Eng 288:503
105. Di Maio E, Iannace S, Sorrentino L, Nicolais L (2004) Polymer 45:8893
106. Tetto JA, Steeves DM, Welsh EA, Powell BE (1999) ANTEC'99. 1628
107. Tartaglione G, Tabuani D, Camino G, Moisio M (2008) Compos Sci Technol 68:451
108. Fukushima K, Tabuani D, Abbate C, Arena M, Ferreri L (2010) Polym Degrad Stab 95:2049

109. Hrobarikova J, Robert J-L, Calberg C, Jerome R, Grandjean J (2004) Langmuir 20:9828
110. Urbanczyk L, Hrobarikova J, Calberg C, Jerome R, Grandjean J (2006) Langmuir 22:4818
111. Namazi H, Dadkhah AJ (2008) App Polym Sci 110:2405
112. Namazi H, Mosadegh M, Dadkhah A (2009) Carbohyd Polym 75:665
113. Namazi H, Dadkhah A (2010) Carbohydr Polym 79:731
114. Namazi H, Mosadegh M (2011) J Polym Environ 19:980
115. Namazi H, Mosadegh M (2011) Nova Science, New York
116. Namazi H, Fathi F, Heydari A (2012) InTech Publisher
117. Perez CJ, Alvarez VA, Mondragon I, Vazquez A (2008) Polym Int 57:247
118. Vertuccio L, Gorrasi G, Sorrentino A, Vittoria V (2009) Carbohyd Polym 75:172
119. Hoidy WH, Al-Mulla EAJ, Al-Janabi KW (2010) J Polym Environ 18:608
120. Wu TM, Cheng JC, Yan MC (2003) Polymer 44:2553
121. Yoshioka M, Takabe K, Sugiyama J, Nishio Y (2006) J Wood Sci 52:121
122. Chen TK, Tien YI, Wei KH (1999) J Polym Sci Part A Polym Chem 37:2225
123. Calandrelli L, Annunziata M, Ragione FD, Laurienzo P, Malinconico M, Oliva A (2010) J Mater Sci Mater Med 21:2923
124. Huang J, Lin YW, Fu XW, Best SM, Brooks RA, Rushton N, Bonfield W (2007) J Mater Sci Mater Med 18:2151
125. Johari M, Fathi MH, Golozar MA, Erfani E, Samadikuchaksaraei A (2012) J Mater Sci Mater Med 23:763
126. Deng C, Weng J, Duan K, Yao N, Yang XB, Zhou SB, Lu X, Qu SX, Wan JX, Feng B, Li XH (2010) J Mater Sci Mater Med 21:3059
127. Li L, Li G, Jiang G, Liu X, Luo L, Nan K (2012) J Mater Sci Mater Med 23:547
128. Ye L, Zeng X, Li H, Ai Y (2010) J Mater Sci Mater Med 21:753
129. Makarov C, Gotman I, Jiang X, Fuchs S, Kirkpatrick CJ, Gutmanas EY (2010) J Mater Sci Mater Med 21:1771
130. Armentano I, Del Gaudio C, Bianco A, Dottori M, Nanni F, Fortunati E, Kenny JM (2009) J Mater Sci 44:4789
131. Machado AV, Botelho G, Silva MM, Neves IC, Fonseca AM (2011) J Polym Res 18:1743
132. Liu Q, Chen D (2008) Euro Polym J 44:2046
133. Naguib HF, Abdel Aziz, MS, Sherif SM, Saad GR (2012) Appl Clay Sci 57:55

Polylactic Acid (PLA) Layered Silicate Nanocomposites

3

Mahsa A. Tehrani, Abozar Akbari, and Mainak Majumder

Contents

Abstract

The depletion of fossil fuels and increased environmental awareness among peoples over the disposal of short-life plastics attracted the use of biodegradable and bio-based plastics for different application. Over the past decade, scientist has been showing sustainable research interest on biodegradable polymers as one of the solution to alleviate solid waste disposal problem and less dependency on petroleum-based plastics. PLA has attracted the attention of polymer scientist

M.A. Tehrani
School of Industrial Technology, Universiti Sains Malaysia, Penang, Malaysia
e-mail: tehrani.mahsa@gmail.com

A. Akbari (✉)
Enhanced Polymer Research Group (EnPRO), Department of Polymer Engineering, Faculty of Chemical Engineering, Universiti Teknologi Malaysia, Johor Bahru, Johor, Malaysia
e-mail: Akbari.Abouzar@gmail.com

M. Majumder
Department of Mechanical and Aerospace Engineering, Monash University, Clayton, VIC, Australia
e-mail: Mainak.Majumder@monash.edu

J.K. Pandey et al. (eds.), *Handbook of Polymernanocomposites. Processing, Performance and Application – Volume A: Layered Silicates*, DOI 10.1007/978-3-642-38649-7_5,

recently as a potential biopolymer to substitute the conventional petroleum-based plastics. The chapter aims to highlight on the recent developments in preparation and characterization of PLA/layered silicate nanocomposites.

Keywords
Biodegradation • Biopolymer • Clay • Layered silicate • Nanocomposite • Polylactic acid

1 Introduction

In recent years, the uses of the renewable resources as chemical feedstocks for the synthetic polymeric materials have attracted considerable attention. The reason for such activity is due to the finite nature of traditional petrochemical-derived composites, in addition to economic and environmental concerns. The use of fossil-based raw materials for the synthesis of industrial and domestic chemicals is limited due to high-energy processing required. Fossil-based by-products are also responsible for the emission of carbon dioxide (CO_2), while biological break of the plastics releases carbon dioxide and methane and heat-trapping greenhouse gases, which lead to a rise in global temperature. Similarly petrochemical-based materials are often nonbiodegradable, thus significantly damaged our environments since they do not degrade in a natural environment and their disposal poses a serious problem [1, 2].

Plastic materials obtained from renewal source are a new generation of materials capable of reducing significantly the environmental impact in terms of satisfying certain technical requirement and being fully biodegradable. Therefore, these materials offer a promising alternative to the traditional polymeric materials when recycling is not cost-effective or technically impassible [3]. The environmental concerns and sustainability issues associated with petrochemical-based polymers have driven considerable engineering and scientific efforts devoted to the discovery, development, and modifications of biodegradable and renewably derived polymers over the past several decades [4, 5].

2 Biopolymer

The preservation and management of our diverse resource are fundamental political tasks to foster sustainable development in the twenty-first century. Sustainable economic growth requires safe and sustainable resources for industrial production, a long-term and confident investment and financed system, ecological safety, and sustainable life and work perspectives for the public. Fossil resources are not regarded as sustainable; however, their availability is more than questionable in the long term. Because of the increasing price of fossil resources, the feasibility of their utilization is declining [6].

It is essential to establish solution which reduces the rapid consumption of fossil resources such as petroleum, natural gas, coal, and minerals, which are not renewable. Biopolymers are expected to be an alternative for conventional plastics due to limited resources and soaring petroleum prices which will limit use of plastic based on petroleum in the near future. Biopolymers are renewable, because they are made from plant materials and growing enormously [7]. These plant materials come from agricultural waste of food crops. Biopolymers have recently attracted increasing attention, predominantly due to environmental concerns and the depletion of petroleum resources.

Commercial polymer products derived from renewable resources are mainly carbohydrates, namely, cellulose, which include virgin and regenerated cellulose; starch including the directly starch-based products where starch is used more or less unmodified and more recently developed fermented starch polymers, where the starch first has been fermented; and then polymerized and sugar-based materials. There are also a range of biodegradable polymers derived from fossil resources, such as polycaprolactone (PCL) and certain co-polyesters. There are also commercially biodegradable polymers, and these are often blendes of fossil derived and renewable resources. Furthermore, there are the still very small groups of protein and fatty acid-based polymers.

The most common biodegradable syntactic polymers are aliphatic polyester such as poly(lactic acid) (PLA), poly(glycolic acid) (PGA), polycaprolactone (PCL), and polyhydroxybutyrate (PHB). Among these biopolymers, PLA as biodegradable polymer is of increasing commercial interest due to its exclusive characteristics that included its desired mechanical strength, thermal plasticity, and biocompatibility [7–9].

3 Poly(Lactic Acid) (PLA)

Poly(lactic acid) is biocompatible with nontoxic degradation products (at low concentrations), making it a natural choice for many biomedical applications. PLA is a linear, aliphatic thermoplastic polyester which can be either semicrystalline or amorphous depending on the stereopurity of the polymer backbone [10]. PLA presents the major advantage to enter in the natural cycle implying its return to the biomass. The life cycle of PLA is shown in Fig. 3.1. It is known that agricultural raw materials such as sugarcane or corn can be used as basic materials in PLA production. However, waste biomass such as whey and cellulose waste can also be utilized. It can be produced by direct polycondensation of lactic acid or by using the combination of fermentation and polymerization from 100 % renewable resources of corn and sugar beets [8, 10]. Preparation of PLA can be conducted by (1) ring-opening polymerization (ROP) of the dehydrated ring-formed dimer or dilactide, (2) polycondensation and manipulation of the equilibrium between lactic acid and polylactide by removal of reaction water using drying agents, or (3) polycondensation and linking of lactic acid monomers. Due to the existence of a chiral carbon in lactic acid, the repeating unit of PLA can have two different configurations (D-(dextro) or L-(levo)) and the relative amount and distribution of these stereoisomers influences various properties of the resulting PLA. In general, PLA built with L-stereoisomer monomer

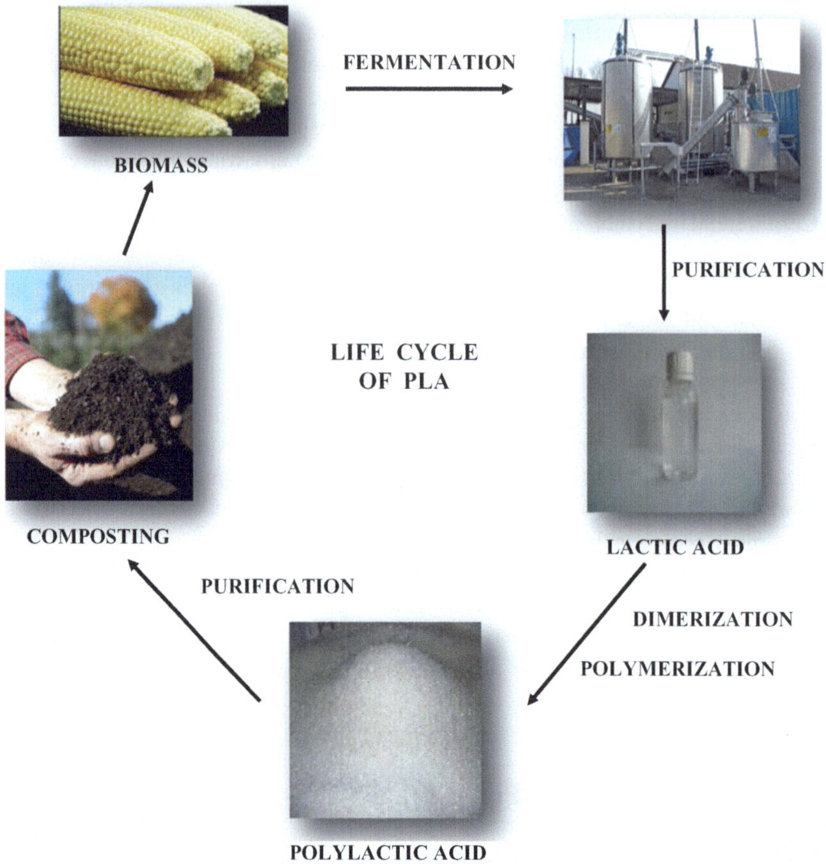

Fig. 3.1 Life cycle of PLA

is referred to as poly(L-lactic acid) (PLLA), whereas PLA which contains both D- and L-stereoisomer monomers is referred to as (PDLLA) [11].

In past decade PLA has been used for many applications including grocery and compositing bags, automatable panels, textiles, and bio-absorbable medical materials. Dannon and McDonald's (Germany) pioneered use of PLA as a packaging material in yogurt cups and cutlery [12]. NatureWorks has used PLA for range of packaging applications such as high-value films, rigid thermoformed containers, and coated papers [13]. Ingeo, a PLA-based fiber, has been designed for apparel, furnishing, and nonwoven applications [14]. BASF's Ecovio, which is a derivative of petrochemical-based biodegradable Ecoflex and contains 45 wt% PLA, has been used to make carrier bags, compostable can liners, mulch film, and food wrapping. Commercially available PLA films and packages have been found to provide mechanical properties better than polystyrene (PS) and comparable to PET [12].

Table 3.1 Mechanical properties of PLA (Material datasheet by Biomer for L9000)

Properties	Values
Young's modulus (MPa)	3,600
Tensile strength (MPa)	70
Elongation at break (%)	2.4
Flexural strength (N/mm^2)	98
Impact strength (kJ/m^2)	16.5
Notched impact strength (kJ/m^2)	3.3
MFI (g/10 min)	3–6
Density (g/cm^3)	1.25
Moisture absorption (%)	0.3

The typical values of mechanical properties of such a PLA polymer are displayed in Table 3.1. Despite its origin form the nature, PLA's good stiffness and strength as well as good processability has enabled it to contest with other existing chemically based commodity plastics. Previous study on the mechanical properties of neat PLA by Jacobsen et al. also confirmed that PLA has great potential to be a substitute polymer for petroleum-based plastic [15]. The respective values of mechanical properties of PLA with comparison of other petroleum-based plastic, e.g., polypropylene (PP) [16], polystyrene (PS) [17], high-density polyethylene (HDPE) [18], and polyamide (PA6) [19], are shown in Fig. 3.2.

Although some properties of PLA such as stiffness and strength caused it to be considered as a suitable substitute for polymer-based petroleum, some properties such as high cost and low elongation at break and brittleness had restrict its use in a wide range of application. Therefore, modification of these biopolymers through innovative technology is a formidable task for materials scientists. Copolymerization, blending, and filling techniques have been used to overcome the PLA drawbacks, which among these methods, the incorporation of a small amount (1–10 wt%) of organically modified nanoscale particles into PLA to produce nanocomposites has attained the highest interest due to its lower cost in comparison with other techniques [20].

4 PLA Nanocomposites

In recent decade, the relatively poor performance/functionality of the current commercially available biodegradable polymers has brought into concern. PLA nanocomposites hold the future in biopolymer nanocomposites as there is an urge for the development of "green" technology from sustainable and renewable resources. Nano-sized filler-based technology has enabled the application of both organic and inorganic fillers with at least one of its dimension in the scale of nanometer, into PLA leading to the formation of PLA nanocomposites. Recently, various inorganic nanoparticles such as layered silicates, cellulose whiskers, and carbon nanotubes have been selected as additives to improve the performance of

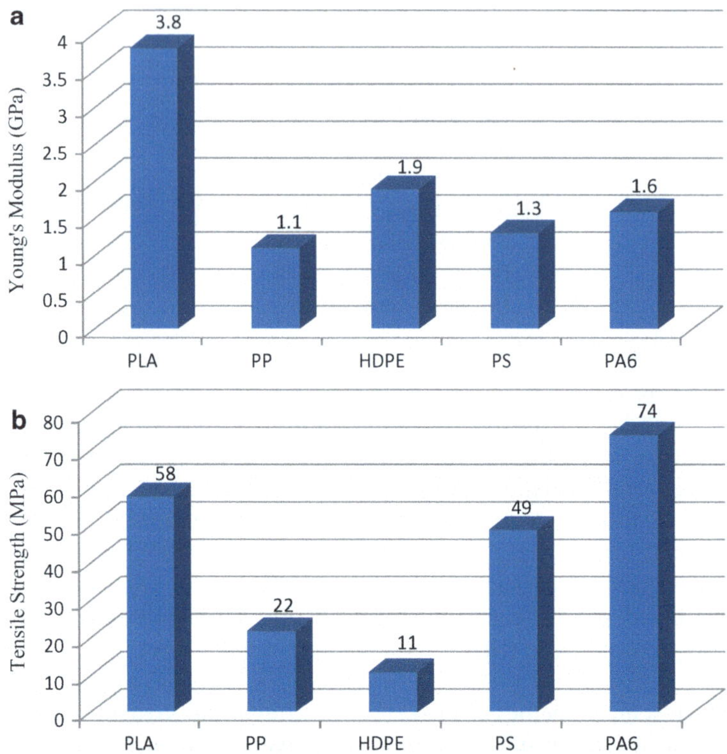

Fig. 3.2 Mechanical properties of PLA and other commodity plastics: (**a**) Young's modulus (**b**) Tensile strength

polymers. Nanocomposites consisting of organically modified layered silicate are of particular interest since they regularly show considerably enhanced mechanical, thermal, barrier, and flame retardance properties comparing to those of original polymer [21–23].

5 PLA/Layered Silicate Nanocomposite

The incorporation of layered silicates in polymer matrixes was first investigated more than 40 years ago. Layered silicates are clay minerals which, in its pristine form, are hydrophilic in nature, and this property makes them very difficult to be well dispersed in PLA. Clays contain layered silica sheets in which negative charges exist on the surface of silicate layers. These negative charges are balanced by the charge of cations (typically sodium cations) within the gallery space. Thus, the gallery is quite hydrophilic (Fig. 3.3) [24]. The most common strategy to overcome this difficulty is to replace the interlayer MMT cations with quaternized

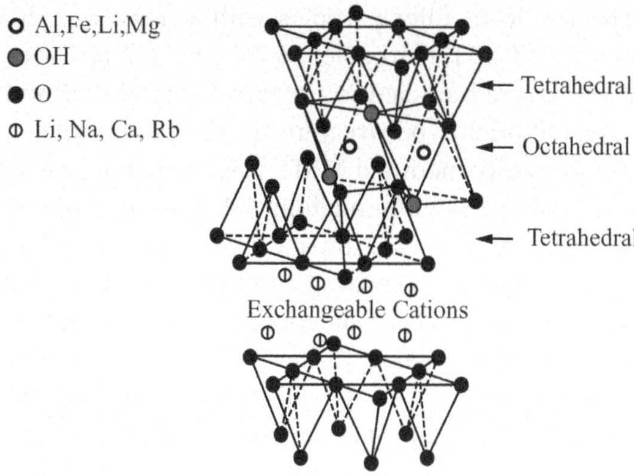

Fig. 3.3 Clay structure [24]

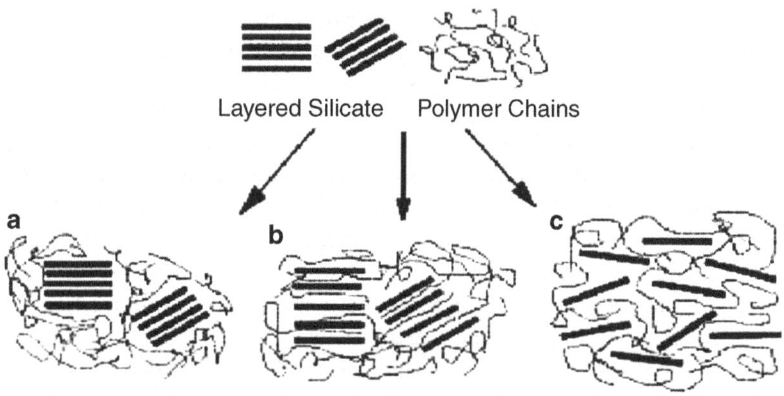

Fig. 3.4 (**a**) Microcomposite structure (*separated phase*); (**b**) Intercalated nanocomposite structure. (**c**) Exfoliated nanocomposite structure [24]

ammonium cations, preferably with long alkyl chains. The presence of hydrophobic chain within the interlayer of MMT enables PLA chains to diffuse resulting in separation of the layers [24].

In the last two decades, layered silicates were more interested as an additive in polymeric materials due to its nanoscale size and intercalation/exfoliation properties. They are mostly prepared by three different methods: in situ polymerization [24], solution intercalation [25, 26], and melt compounding [27, 28] which the last one is the simplest and most frequent technique for industrial applications. In nanocomposites, if the polymer chains penetrate the space between clay layers, but do not separate them thoroughly as the registry between them is maintained, the nanocomposite is described as intercalated, while if the clay layers are thoroughly detached from each other by the polymer chains while the original registry is lost, the nanocomposite is called exfoliated (Fig. 3.4) [24].

In the meanwhile, special attention has been paid to nanocomposites that consist of polymers and montmorillonite (MMT)-layered silicates regarding to the high aspect ratio of filler particles with at least one dimension in the nanometer scale which leads to remarkable improvement in polymer properties. The frequently used MMT-layered silicate is organically modified and consists of sheets arranged in a layered mica-type structure. In situ polymerization of D,L-lactide in the presence of organically modified MMT was carried out by using supercritical carbon dioxide as a polymerization medium. A considerable improvement in toughness and stiffness as well as impact resistance was fairly obvious for the PLA/clay nanocomposites compared with the unfilled matrix [29]. In another approach, PLA/layered silicate nanocomposites has been prepared via simple melt extrusion of PLA and organically modified MMT which the resultant nanocomposite exhibited improved biodegradability, storage modulus, flexural properties, gas permeability, and heat distortion temperature compared to that of virgin PLA [30].

Chang et al. [31] produced PLA/modified MMT nanocomposite by solution intercalation methods. They found an increase in tensile modulus but a decrease in tensile strength by addition of 4(%w/w) MMT into PLA matrix. The similar PLA-based nanocomposite by 2–3(%w/w) filler content showed an increase in storage modulus up to 40 % [32].

Jiang et al. applied melt-compounding method using a twin-screw extruder and injection molding to prepare test specimens form PLA and MMT. They observed a good dispersion of MMT nanofiller at the concentration of 5(wt%) and MMT intercalated by PLA through the nanocomposite [33]. They also revealed that there are more agglomerates of MMT in higher concentrations of MMT, which caused premature fracture in mechanical testing. At low particle concentration, an increase in strain at break and tensile strength of the composite was observed for up to 2.5(wt%) and 5(wt%) MMT content, respectively.

There is another investigation which was done in order to determine the influence of an organically modified montmorillonite on the crystallization and on the gas barrier properties of PLA [34]. It was shown that by small amount loading of organo-montmorillonite (OMMT) into the PLA, a reduction in cold crystallization temperature for heating thermal treatments and also the crystallization half-time for isothermal ones took place. Furthermore, the decrease of permeability for the annealed nanocomposite film containing 4 wt% of OMMT was indicated regarding to the amorphous reference PLA film rise from the respective contribution of the fillers and of the polymer crystalline phase.

In a sequence research papers by Ray et al., preparation of a novel nanocomposite by melt extrusion of PLA and MMT modified with octadecylammonium cation (C18-MMT) was studied [35]. The intercalated, stacked, and randomly distributed clay through the matrix was exhibited after WAXD analyses and TEM observations. The intercalated PLA/C18-MMT nanocomposites demonstrated considerable improvement of mechanical properties in both solid and melt phases as compared to that of neat PLA. Additionally, an increase in biodegradation rate of neat PLA was apparent after the nanocomposite preparation due to the barrier properties improvement of the aliphatic polyester after incorporation of the clay.

A series of studies by Ray et al. [29, 36, 37] on preparation of PLA/clay nanocomposites via melt extrusion demonstrated a considerable improvement in flexural properties, biodegradability, and permeability of pure PLA. It is proved that the parallel property improvements in those nanocomposites originate from interfacial interaction between the organically modified layered silicate and polymer matrix.

Incorporation of organically modified clays into the polymeric matrix can intensify thermal degradation of PLA resulting in a loss of molecular weight [10]. The reduction in molecular weight of polyesters is mostly caused by the random hydrolysis of the ester linkage. There are two types of transesterification which often take place, intramolecular and intermolecular. The former, which is named "backbiting" too, leads to the formation of cyclic polylactide oligomers and polymer degradation, while the intermolecular transesterifications influence the sequence of different polymeric segments.

Najafi et al. introduced various chain extenders such as tris(nonylphenyl) phosphite (TNPP), polycarbodiimide (PCDI), and Joncryl to control the rate of thermal degradation of PLA [10]. They revealed that thermal degradation increases after clay addition by consumption of small molecules like lactic acid and moisture. The catalytic effect of montmorillonites on the biodegradation or hydrolytic degradation of various aliphatic polyesters because of high hydrophilicity of these nanoparticles has also previously reported [38–40].

Zhou et al. [41] studied on hydrolytic degradation of amorphous and semicrystalline PLA-based materials containing unmodified and organically modified montmorillonites. They revealed that hydrophilicity and the effective pH of the organically modified nanofillers in temperature of 50–70 °C extensively increased the rate of degradation for nanocomposites. The similar increase in the rate of PLA biodegradation has been observed by Lee et al. [42] and Paul et al. [38] by loading nanoclays due to easier permeability of water into the matrix and stimulating the hydrolytic degradation. However, other researchers have proved that nanoclays may hinder the degradation of aliphatic polyesters during either hydrolytic degradation or biodegradation contributed to the enhanced barrier properties of the layered silicate nanocomposites [42–46]. Similarly, in another study, the degradation rate for pristine nanoclay-based microcomposites was inferior to that of the virgin polymers due to the lower catalytic effect of carboxyl groups after neutralization with the hydrophilic alkaline fillers [11].

In a study by Fukushima et al. [47], the effect of sepiolite on the biodegradation of nanocomposites was investigated. Sepiolite is a type of nanoclay which is made by a complex of magnesium silicate. Sepiolite incorporation clearly presented high dispersion level of small bundles and single dispersed needle into the PLA and PCL matrices. It was dedicated that the unmodified sepiolite caused a significant level of degradation in compost of PLA and PCL at 58 °C by a special mechanism of bulk degradation. However, the presence of sepiolite particles seemed to delay the PLA matrix degradation to some extent.

According to the attempts of Fukushima et al. [47, 48], the biodegradation effect of two modified montmorillonites in compost of PLA and PCL revealed that the addition of montmorillonites catalyzed the biodegradation of PLA, while in the case of PCL, it was delayed. This is disagreement resulted from the different degradation mechanism of the polymers and also the dispersion level of the silicate platelets in the polymer matrix. The above groups also reported that the addition of modified montmorillonite and unmodified sepiolite delayed the degradation of PLA at 37 °C, particularly for montmorillonite due to its inducing PLA crystallization effect. The water isolation between the filler pores might be responsible for the late polymer degradation owing to a reduced amount of water which is accessible for hydrolysis in the polymer matrix. This effect is less important in the case of using fluorohectorite filler due to its lower hydrophilic properties.

In a recent novel approach, two structurally different additives, sepiolite and organically modified montmorillonite, were incorporated into PLA by combination of electrospinning and extrusion/injection molding process in order to improve the fire retardancy behavior of PLA [21]. It was shown that the severe reduction in molecular weight of hybrid composite had affected negatively on the thermal stability and the fire performance.

The highest thermal stability for silicate-filled PLA (sodium and organo-modified Cloisite) was obtained with the nanocomposite containing Na^+/MMT at its maximum concentration (6 wt%) [49].

The addition of kaolinite (type of layered silicate clay) and MMT nanoclays in Fukushima et al.'s work also increased the PLA degradation rate, especially for kaolinite, while MMT demonstrated a catalytic effect on degradation of composite only at the end of degradation process. Initially, MMT tend to delay the degradation of PLA possibly due to its higher dispersion through the polymer matrix comparing to kaolinite and also causing a barrier effect towards water hydrolysis and microbial attack through chain-end hydroxyl groups [50].

Rectorite is a frequently interstratified clay mineral of dioctahedral smectite-like layer (expansible) and dioctahedral mica-like layer (non-expansible) in an equal fraction. Comparing to MMT, it has more aspect ratio and longer interlayer distance, and its special structure is promising to produce intercalated/exfoliated nanocomposites with improved properties. In an attempt by Li et al., organically modified rectorite (OREC) has been used to improve the toughness of PLA via melt extrusion process [22]. They revealed that OREC addition had a remarkable effect on the increase of elongation at break as well as the overall crystallization rate of composite.

Cloisite also is another type of OMMT which have two hydroxyl groups. Najafi et al. [51] introduced Cloisite into PLA by melt mixing, using a twin-screw extruder at the presence of Joncryl chain extender [51]. Depending on the processing conditions on clay dispersion, the resultant nanocomposites exhibited considerably reduced permeability and increased molecular weight in which the latter leads to enhanced modulus, toughness, and drawability of nanocomposite.

In a recent study by Sermsantiwanit and Phattanarudee [49], two types of silicate, sodium and organo-modified Cloisite, were applied into PLA via emulsification–diffusion method. According to the SEM micrographs, creation of a spherical shape with different sizes, depending on silicate type and concentration,

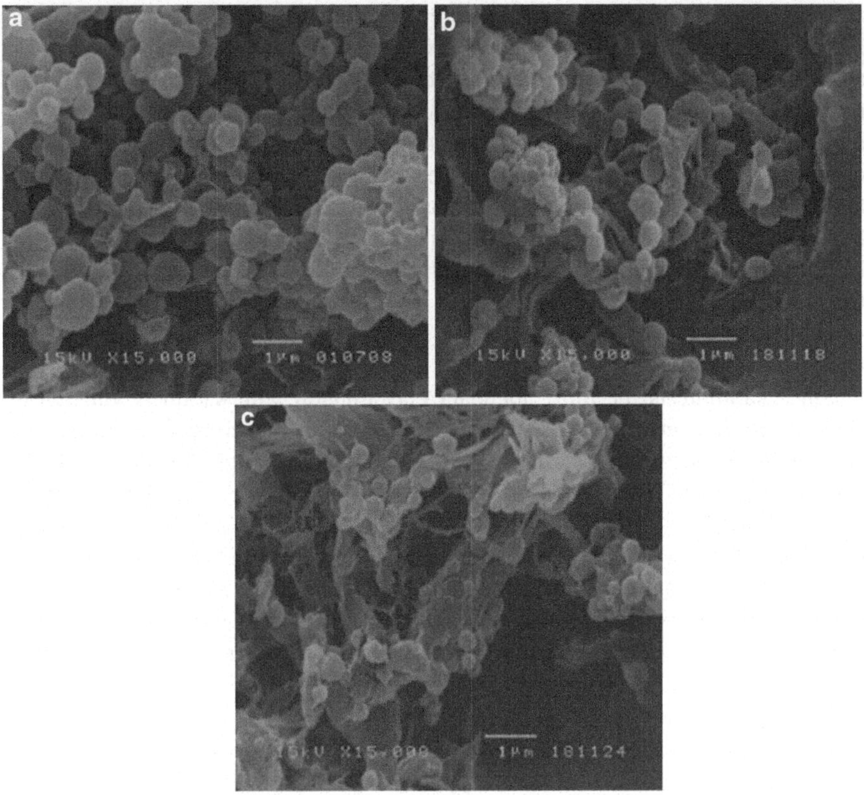

Fig. 3.5 Particle morphology of the latex particles (**a**) unfilled PLA, (**b**) PLA with Cloisite 30B (at 1 wt%) and (**c**) PLA with Cloisite Na^+ (at 1 wt%) [49]

was obvious (Fig. 3.5). In addition, they resulted by XRD and TEM tests that the silicate layers were extensively exfoliated and partially intercalated through the PLA matrix. It could be concluded that the high organic modifier –OH groups' availability of Cloisite through PLA matrix revealed high thermal–mechanical enhancements for PLA nanocomposites.

In situ coordinative polymerization of L,L-lactide in bulk was carried out to produce exfoliated poly(L-lactide)/OMMT nanocomposite by Paul et al. [25]. The effect of chain grafting on the nanocomposite thermal properties and morphology was investigated by comparing two different organo-modified clays (Cloisite 25A and Cloisite 30B), bearing or not bearing hydroxyl functions. The difference in nanocomposite structure, or in more details exfoliation of clay platelets, can be in charge for the improved thermal stability of the nanocomposite compare to both original PLA and even intercalated nanocomposites. Cloisite 30B was more effective in the above area since it could increase the possibility of polymer chain grafting onto the clay surface.

In another research, Chaloupi [52] used melt-compounding method to produce PLA-nanoclay, without and with chain extender nanocomposites. They have seen a rapid thermal degradation of PLA in the presence of organo-modified clay from rheological point of view. In addition, thermogravimetric analysis showed a reduction in thermal stability of the matrix by incorporation of clay, while the chain extender addition enhanced the onset temperature of thermal degradation due to the formation of less chain tails per mass. It has also exhibited an improved and homogenous dispersion of Cloisite by using Joncryl chain extender during master batch approach as well as lowest permeability and the highest mechanical properties of the resultant nanocomposites. On the other hand, nanocomposites which prepared by direct mixing method presented the worst morphology among the other samples.

Since kaolinite is the most abundant clay mineral with kaolin production of about 20 million tons per year, its utilization as filler for nanocomposites production attained much attention during these years. Kaolinite is a 1:1 phyllosilicate with the chemical composition of $Al_2Si_2O_5(OH)_4$ which the layers linked together by hydrogen bonds in tetrahedral and octahedral sheets. The addition of kaolinite nanotubes (1 wt%) into the PLA considerably improved mechanical properties of nanocomposite, e.g., tensile strength and Young's modulus. Moreover, the poorly ordered kaolinite was preferable than well-ordered one in terms of mechanical properties improvement. The reason behind this improvement is excellent dispersion of the mineral fillers into the polymer matrix as well as interaction of mineral hydroxyl groups with the polymer functional groups [53].

In related research by Matusik et al. [53], kaolinite nanotubes and platy kaolinites of different structural orders were applied as fillers into the PLA matrix. As a consequence, mechanical properties of composites, tensile strength, and Young's modulus increased significantly by incorporation of the nanofiller, and this trend was higher in the case of the nanotubular kaolinite due to the excellent dispersion characteristics. Moreover, interaction of mineral hydroxyl groups with polymer functional groups could be also responsible for mechanical properties improvement.

It has been reported in previous researches that sepiolites can be considered as potentially attractive materials for the enhancement in properties of biodegradable polymers since they are easily dispersible without aid of any organic modifiers and compatibilizers [28]. They also cause a remarkable increase in thermal and thermomechanical properties of PLA-based composites. For instance, PLA/sepiolite nanocomposite prepared from melt blending process in Fukushima et al.'s study revealed remarkable thermomechanical improvements due to the good dispersion and high polymer/filler interaction which is fairly dependent on the high surface hydroxyl group concentration.

Furthermore, unmodified and organic modified sepiolite well dispersed through the PLA matrix by solution casting method [54]. Sepiolite nanofibers exhibited a randomly orientation with cross-link among them. The thermal studies' results proved higher improvement in thermal stability of nanocomposites by addition of unmodified sepiolite than the organic modified sepiolite. The reason behind this is good dispersion and strong interfacial interaction between silanol groups (Si–OH) on sepiolite and the ester groups of PLA while the modified sepiolite introduced inferior thermal stability for nanocomposites due to hindering effect of modifiers for interaction between PLA and sepiolite.

6 Future Direction

In spite of many investigations by scientists, the current PLA available on the market are faced to a considerable challenge over conventional plastics. Although PLA and PLA composites have good mechanical properties and biodegradability compared with conventional plastics, in current market, high price of PLA could not encourage manufacturers to replace conventional plastics with PLA. The high price of PLA is related to the deriving process of lactic acid from corn or sugar beet resources and also the polymerization process to produce PLA. It can be concluded that one of the main future challenge is finding a new cheap method to product PLA. On the other side, the limited toughness of PLA is another challenge that researchers should develop a new biodegradable impact modifier which has a minimal defect on biodegradable properties of PLA. The incorporation of nanoclay goes some method to redressing cost equation along with improved performance of PLA however, will have to consider several performances such as expected degradability, flame retardancy, resistance to moisture absorption, and embrittlement are the major setbacks to be tackled.

References

1. Nick T, Mark J (2004) Low environmental impact polymer, 1st edn. Rapra Technology, Shrewsbury
2. Koh HC, Park JS, Jeong MA, Hwang HY, Hong YT, Ha SY, Nam SY (2008) Preparation and gas permeation properties of biodegradable polymer/layered silicate nanocomposite membranes. Desalination 233:201–209
3. Lluts C, Jose LF, Marta PV, Jose ML, Gimenez E (2006) Optimization of biodegradable nanocomposites based on a PLA/PCL blends for food packaging applications. Macromol Symp Belgium 233:191–197
4. Eling B, Gogolewski S, Pennings AJ (1982) Biodegradable materials of poly(L-lactic acid): 1. Melt-spun and solution spun fibers. Polymer 23:1587–1593
5. Schmack G, Tändler B, Vogel R, Beyreuther R, Jacobsen S, Fritz H-G (1999) Biodegradable fibers of poly (L-lactide) produced by high-speed melt spinning and spin drawing. J Appl Polym Sci 73:2785–2797
6. Birgit K, Patrick RG, Michael KW (2006) Biorefineries – industrial processes and products: status quo and future directions. Wiley-VCH, Weinheim
7. Mueller RJ (2006) Biological degradation of synthetic polyesters – enzymes as potential catalysts for polyester recycling. Process Biochem 41(10):2124–2130
8. Anderson KS, Schreck KM, Hillmyer MA (2008) Toughening polylactide. Polym Rev 48:85–108
9. Fukushima K, Tabuani D, Dottori M, Armentano I, Kenny JM, Camino G (2011) Effect of temperature and nanoparticle type on hydrolytic degradation of poly(lactic acid) nanocomposites. Polym Degrad Stab 96:2120–2129

10. Pluta M, Jeszka JK, Boiteux G (2007) Polylactide/montmorillonite nanocomposites: structure, dielectric, viscoelastic and thermal properties. Eur Polym J 43:2819–2835
11. Najafi N, Heuzey MC, Carreau PJ, Paula MW (2012) Control of thermal degradation of polylactide (PLA)-clay nanocomposites using chain extenders. Polym Degrad Stab 97:554–565
12. Auras R, Harte B, Selke S (2004) An overview of polylactides as packaging materials. Macromol Biosci 4:835–864
13. Natureworks LLC Website (2007) Technology focus report: toughened PLA. http://www.natureworkspla.com. Accessed June 2011
14. Antonio N, Akihiro F, Toshiaki S, Yoshiaki H, Hiroyuki Y (2009) Production of microfibrillated cellulose (MFC)-reinforced polylactic acid (PLA) nanocomposites from sheets obtained by a papermaking-like process. Compos Sci Technol 69:1293–1297
15. Jacobsen S, Fritz HG (1999) Plasticizing polylactide – the effect of different plasticizers on the mechanical properties. Polym Eng Sci 39:1303–1310
16. Lim JW, Hassan A, Rahmat AR, Wahit MU (2006) Morphology, thermal and mechanical behavior of polypropylene nanocomposites toughened with poly (ethylene-co-octene). Polym Int 55:204–215
17. Hasegawa N, Okamoto H, Kawasumi M, Usuki A (1999) Preparation and mechanical properties of polystyrene-clay hybrids. J Appl Polym Sci 74:3359–3364
18. Gupta K, Rana SK, Deopura B (1992) Mechanical properties and morphology of high-density polyethylene/linear low-density polyethylene blend. J Appl Polym Sci 46:99–108
19. Wahit MU (2006) Rubber toughened polyamide 6/polypropylene nanocomposites: mechanical, thermal and morphological properties. PhD thesis, University Technology Malaysia, Skudai
20. Alexandre M, Dubois P (2000) Polymer-layered silicate nanocomposites: preparation, properties and uses of a new class of materials. Mater Sci Eng 28:1–63
21. Wu D, Wu L, Zhang M (2006) Rheology and thermal stability of polylactide/clay nanocomposites. Polym Degrad Stab 91:3149–3155
22. González A, Dasari A, Herrero B, Plancher E, Santarén J, Esteban A, Lim SH (2012) Fire retardancy behavior of PLA based nanocomposites. Polym Degrad Stab 97:248–256
23. Li B, Dong FX, Wang XL, Yang J, Wang DY, Wang YZ (2009) Organically modified rectorite toughened poly(lactic acid): nanostructures, crystallization and mechanical properties. Eur Polym J 45:2996–3003
24. Akbari A, Talebanfard S, Hassan H (2010) The effect of the structure of clay and clay modifier on polystyrene-clay nanocomposite morphology: a review. Polym-Plast Technol Eng 49:1433–1444
25. Paul MA, Alexandre M, Degée P, Calberg C, Jérôme R, Dubois P (2003) Exfoliated polylactide/clay nanocomposites by in-situ coordination-insertion polymerization. Macromol Rapid Commun 24:561–566
26. Chiang MF, Wu TM (2010) Synthesis and characterization of biodegradable poly(L-lactide)/layered double hydroxide nanocomposites. Comp Sci Tech 70:110–115
27. McLauchlin AR, Thomas NL (2009) Preparation and thermal characterisation of poly(lactic acid) nanocomposites prepared from organoclays based on an amphoteric surfactant. Polym Degrad Stab 94:868–872
28. Fukushima K, Tabuani D, Camino G (2009) Nanocomposites of PLA and PCL based on montmorillonite and sepiolite. Mater Sci Eng C 29:1433–1441
29. Ray SS, Maiti P, Okamoto M, Yamada K, Ueda K (2002) New polylactide/layered silicate nanocomposites. 1. Preparation, characterization, and properties. Macromolecules 35:3104–3110
30. Urbanczyk U, Ngoundjo F, Alexandre M, Jérôme C, Detrembleur C, Calberg C (2009) Synthesis of polylactide/clay nanocomposites by in situ intercalative polymerization in supercritical carbon dioxide. Eur Polym J 45:643–648
31. Ray SS, Yamada K, Okamoto M, Ueda K (2003) New polylactide-layered silicate nanocomposites. 2. Concurrent improvements of material properties, biodegradability and melt rheology. Polymer 44:857–866

32. Chang JH, Yeong U, Cho D, Giannelis EP (2003) Poly(lactic acid) nanocomposites: comparison of their properties with montmorillonite and synthetic mica. Polymer 44:3715–3720
33. Maiti P, Yamada K, Okamoto M, Ueda K, Okamoto K (2002) New polylactide/layered silicate nanocomposites: role of organoclays. Chem Mater 14:4654–4661
34. Jiang L, Zhang J, Wolcott MM (2007) Comparison of polylactide/nano-sized calcium carbonate and polylactide/montmorillonite composites: reinforcing effects and toughening mechanisms. Polymer 48:7632–7644
35. Ray SS, Yamada K, Okamoto M, Ueda K (2002) Polylactide-layered silicate nanocomposite: a novel biodegradable material. Nano Lett 2:1093–1096
36. Ray SS, Yamada K, Okamoto M, Ogami A, Ueda K (2003) New polylactide/layered silicate nanocomposites. 3. High-performance biodegradable materials. Chem Mater 15:1456–1465
37. Ray SS, Yamada K, Okamoto M, Fujimoto Y, Ogami A, Ueda K (2003) New polylactide/layered silicate nanocomposites. 5. Designing of materials with desired properties. Polymer 44:6633–6646
38. Paul MA, Delcourt C, Alexandre M, Degée P, Monteverde F, Dubois P (2005) Polylactide/montmorillonite nanocomposites: study of the hydrolytic degradation. Polym Degrad Stabil 87:535–542
39. Ray S, Yamada K, Okamoto M, Ueda K (2003) Control of biodegradability of polylactide via nanocomposite technology. Macromol Mater Eng 288:203–208
40. Ratto JA, Steeves DM, Welsh EA, Powell BE (1999) A study of polymer/clay nanocomposites for biodegradable applications. Proc SPE Ann Tech Conf 45:1628–1632
41. Zhou Q, Xanthos M (2008) Nanoclay and crystallinity effects on the hydrolytic degradation of polylactides. Polym Degrad Stabil 93:1450–1459
42. Lee SR, Park HM, Lim H, Kang T, Li X, Cho WJ, Ha CS (2002) Microstructure, tensile properties, and biodegradability of aliphatic polyester/clay nanocomposites. Polymer 43:2495–2500
43. Someya Y, Kondo N, Shibata M (2007) Biodegradation of poly(butylene adipate-co-butylene terephthalate)/layered-silicate nanocomposites. J Appl Polym Sci 106:730–736
44. Wu T, Wu C (2006) Biodegradable poly (lactic acid)/chitosan-modified montmorillonite nanocomposites: preparation and characterization. Polym Degrad Stab 91:2198–2204
45. Wu K, Wu C, Chang J (2007) Biodegradability and mechanical properties of polycaprolactone composites encapsulating phosphate-solubilizing bacterium *Bacillus*. Process Biochem 42:669–675
46. Fukushima K, Abbate C, Tabuani D, Gennari M, Rizzarelli P, Camino G (2010) Biodegradation trend of poly(ε-caprolactone) and nanocomposites. Mater Sci Eng C 30:566–574
47. Fukushima K, Tabuani D, Abbate C, Arena M, Ferreri L (2010) Effect of sepiolite on the biodegradation of poly(lactic acid) and polycaprolactone. Polym Degrad Stab 95:2049–2056
48. Fukushima K, Abbate C, Tabuani D, Gennari M, Camino G (2009) Biodegradation of poly (lactic acid) and its nanocomposites. Polym Degrad Stab 94:1646–1655
49. Sermsantiwanit K, Phattanarudee S (2012) Preparation of bio-based nanocomposite emulsions: effect of clay type. Prog Org Coat 74:660–666
50. Fukushima K, Giménez E, Cabedo L, Lagarón JM, Feijoo JL (2012) Biotic degradation of poly(DL-lactide) based nanocomposites. doi: 10.1016/j.polymdegradstab.2012.05.029
51. Najafi N, Heuzey MC, Carreau PJ (2012) Polylactide (PLA)-clay nanocomposites prepared by melt compounding in the presence of a chain extender. Compos Sci Technol 72:608–615
52. Chaloupi NN (2011) Development of polylactide-clay nanocomposites for food packaging applications. PhD thesis, Université de Montréal
53. Matusik J, Stodolak E, Bahranowski K (2011) Synthesis of polylactide/clay composites using structurally different kaolinites and kaolinite nanotubes. Appl Clay Sci 51:102–109
54. Liu M, Pu M, Ma M (2012) Preparation, structure and thermal properties of polylactide/sepiolite nanocomposites with and without organic modifiers. Compos Sci Technol. http://dx.doi.org/10.1016/j.compscitech.2012.05.017

Recent Progress in the Development of Starch-Layered Silicate Nanocomposites

4

Yi-Lin Chung and Hsi-Mei Lai

Contents

Abstract

Biobased composite materials have been of great interests in recent years because of the need for environmentally friendly alternatives to traditional petroleum-based plastic materials. Among potential renewable materials in eco-friendly markets, starches are considered relatively abundant and inexpensive biopolymers with good biodegradability and biocompatibility. In the presence of plasticizers, starches can be engineered using conventional thermoplastic techniques, e.g., extrusion and injection molding. However, starch-based materials have some limitations due to their inadequate mechanical properties and moisture sensitivity. The blending of nano-sized layered silicates into a starch matrix provides a range of improved properties (increased stiffness, barrier, and thermal stability). In the recent years, new modification reactions and engineering methods have been developed to increase the interfacial interaction between starch polymers and layered silicates. Using novel additives such as plasticizers and compatibilizers could facilitate the dispersion of layered silicates in the polymeric matrix and lead to

Y.-L. Chung • H.-M. Lai (✉)
Department of Agricultural Chemistry, National Taiwan University, Taipei, Taiwan
e-mail: hmlai@ntu.edu.tw

J.K. Pandey et al. (eds.), *Handbook of Polymernanocomposites. Processing, Performance and Application – Volume A: Layered Silicates*, DOI 10.1007/978-3-642-38649-7_14,

starch-layered silicates with improved performance. This article covers important topics in the field of starch-layered silicate materials and also reports the most recent progress in the technologies and performance of starch-layered silicate nanocomposites.

Keywords
Biocomposites • Biomaterials • Bioplastics • Clay • Green

1 Introduction

The depletion of fossil fuel resources and the slow degradation rate of petroleum-based plastic materials have driven a strong global interest in renewable polymers and chemicals derived from natural resources. Among potential renewable biopolymers in eco-friendly markets (Fig. 4.1), starch and polylactide (PLA) are the dominant biopolymers in current global market of biobased plastics [50, 57]. In the future, starch, PLA (produced by polymerization of lactic acid from sugar fermentation), biobased polyethylene (PE) (produced from bioethanol through chemical dehydration), polyhydroxyalkanoates (PHAs) (a microbial polymer), and biobased epoxy resin (epichlorohydrin, produced from biodiesel glycerol) are expected to be the major types of biobased plastics [57]. The current markets, manufacturing, properties, and applications of different biopolymers are summarized by Endres and Siebert-Raths [23]. Although much research has been done on starch-based materials, their applications still have limitations due to the mechanical properties and moisture sensitivity of starches. Recently, many researchers have reported that the blending of nano-sized layered silicates into starch matrix could provide an economical and environmental-friendly solution to improve the stiffness, barrier, and thermal stability of starch-based materials.

Layered silicates are relatively cheap and effective reinforcing nanofillers in comparison to other nanoparticles, e.g., silica nanoparticles and carbon nanotubes. Polymer-layered silicate nanocomposites (PLSNs) have attracted great attentions since the pioneering work of the Toyota research group on the pronounced improvements of the mechanical and thermal properties of polymers by blending a low amount of layered silicates. In the early 1990s, they discovered that direct intercalation of the polymer melt into the layered silicates without using solvents was possible [30, 43, 62, 63] that making the process more environmental friendly. Lots of articles, patents, and commercial products have been published and launched over the past 20 years in various fields and applications [52, 55, 61]. The important value-added properties of PLSNs are the enhanced mechanical properties, gas barrier performance, fire/heat resistance, weight reduction, transparency, and cost-effectiveness in comparison to conventional polymer composites. The dispersion of nanofillers in a polymer matrix plays an important role in the enhancement of load transfer across the nanofiller/polymer interface [16]. The restriction of polymer chains in the interfacial region can also change the molecular mobility and crystal structure of the polymer matrix, which

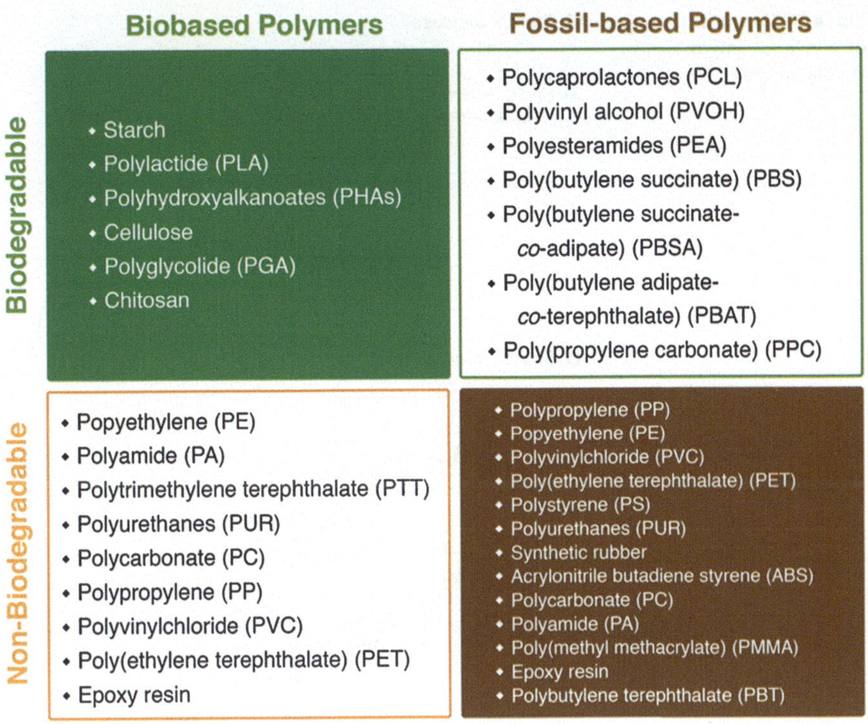

Fig. 4.1 Main biobased and fossil-based polymers in the current and emerging bioplastic markets

can in turn strengthen the bulk properties of the materials. The strong aggregation tendency of nanofillers not only prevents the homogeneous dispersion of layered silicates but also destroys the enhancement of properties gained by the nanoeffect. Thus, the development of a suitable modified layered silicate/polymer structure, an efficient engineering procedure, and the addition of miscible compatibilizers and plasticizers to achieve a high percentage of nanodispersion of layered silicates are crucial for preparing high-performance PLSNs materials.

This article provides an overview of the current progress on the technology and performance of starch-layered silicate nanocomposites. We address the novel engineering strategies and modification methods required to successfully exfoliate or intercalate layered silicates into a starch matrix and further produce starch-layered silicate nanocomposites with improved properties.

2 Starch Fundamentals

Starch is a dominant carbohydrate reserve material of higher plants that is synthesized in a granular form in special organelles, including amyloplasts and chloroplasts. It consists of two types of α-D-glucose biopolymers: amylose

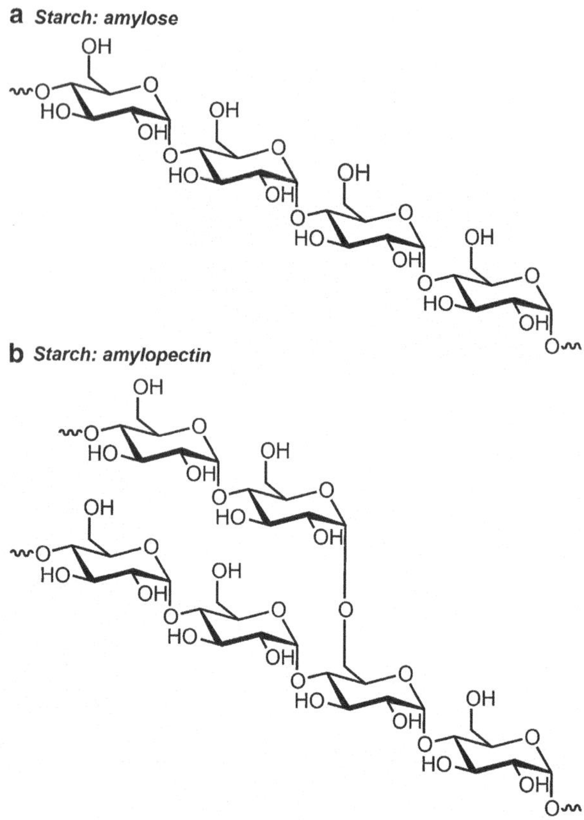

Fig. 4.2 Chemical structures of starch: (**a**) amylose and (**b**) amylopectin

and amylopectin (Fig. 4.2). Amylose is an essentially linear polymer composed of anhydroglucose units connected through α-(1–4)-linkages with a few α-(1–6)-linkages, and its molecular weight is approximately 10^5 to 10^6 Da. Amylopectin is a highly branched macromolecule with α-(1–4)-linked D-glucose backbones and about 5 % of α-(1–6)-linked branching points. The molecular weight of amylopectin is around 10^6 to 10^8 Da, and it has an average branched chain length degree of polymerization (DP) of 17–26 [48]. Most starch granules are semicrystalline with 15–45 % crystallinity [72]. Inside the starch granules, amylose and the branching points of amylopectin are deposited in the amorphous regions, while the short-chain segments of amylopectin form the crystalline regions.

Normal starch contains about 20–30 % amylose and 80–70 % amylopectin. There are also starch types with different ratios of amylose to amylopectin. The amylose content of starch can vary from <1 % to 80 % depending on the plant sources. Waxy starch (e.g., waxy maize and sticky rice) contains little or no amylose, whereas genetically modified high-amylose starch consists of >50 % amylose. Starches with different amylose contents have different thermal, rheological, and processing properties [68]. High-amylose starch has a high tendency to retrograde and produce films with better mechanical and gas barrier properties

[29, 34, 44, 53, 59, 67]. However, the high price [23], high gelatinization temperature [4], and fast retrogradation [70] are challenges for the applications of high-amylose starches. Mondragon et al. [38] investigated the influence of amylose and amylopectin on the morphology and properties of plasticized starch films reinforced with montmorillonite (MMT). The incremental improvement of Young's modulus followed the order of waxy > normal > high-amylose starch. For high-amylose starch, the higher degree of retrogradation led to the stronger intermolecular interaction that was expected to limit the exfoliation process of layered silicates in a starch matrix. The short-branched chains of waxy starch were hypothesized to favor the interactions between starch and layered silicates and thus resulted in an obvious improvement of Young's modulus.

Starch gelatinization is the disruption of molecular order within a starch granule, which manifests as irreversible changes in properties such as granular swelling, native crystalline melting, loss of birefringence, and starch solubilization [2]. Dissolution of starch molecules can be obtained either by a solvent-casting process or by extrusion, which is the most widely used engineering method for the production of starch-based materials at low moisture content. The extrusion process involves high shear and high pressure to tear apart the starch granules, allowing plasticizers penetrate into the interior starch molecules as well as inducing starch gelatinization. The shear stress can also result in the degradation and decomposition of amylopectin and amylose [31]. Starch granules undergo a complicated phase transition during thermal-mechanical processing. The phenomena observed due to this transition include plasticizer diffusion, granular fragmentation and disruption, starch gelatinization, crystalline melting, and starch degradation [3]. Most starch extrusion processes require the addition of nonvolatile plasticizers to broaden the processing window since the melting temperature of dry starch (T_m=220–240 °C) [54] is very close to its degradation temperature (220 °C) [58]. Glycerol and water are commonly used plasticizers for starch-based materials. Other plasticizers such as sorbitol [10, 32], urea [6], urea/ethanolamine [27, 71], urea/formamide [41], formamide/ethanolamine [26], triacetin [40], and hydroxyalkylformamide [17, 18] have been reported in recent studies of starch-layered silicate nanocomposites. The innovative aspect here is that the plasticizers are not only used to lower the melting temperature of starch polymer but also function as compatibilizers for increasing the interaction between layered silicates and starch.

3 Nanostructure and Surface Chemistry of Modified Layered Silicates

The layers of 2:1-type layered silicates are made up of two tetrahedral silica sheets and a central octahedral alumina sheet. The sandwich triple layer is around 1 nm in thickness and from tens of nanometers to several microns in the lateral dimensions. Because of isomorphic substitution of aluminosilicate elements with other metal ions (e.g., Al^{3+} replaced by Mg^{2+} or Si^{4+} replaced by Al^{3+}), the net charge of layered sheets is negative and is counter balanced by the cations (e.g., Na^+, Ca^{2+}, K^+)

absorbed in the interlayer space. These cations can be exchanged with other cations by ion-exchange reaction. Cation-exchange capacity (CEC) is an indicator of the total of the exchangeable cations in the layered silicates. Montmorillonite (sodium MMT) is the most commonly used 2:1-type layered silicates for the starch-based nanocomposites because of its high CEC and expansion/swelling properties. Other layered silicates such as hectorite (lithium MMT) [5, 65], fluorohectorite [22], calcium bentonite (calcium MMT) [35], kaolinite [37, 39, 65], sepiolite [11], and layered double hydroxide (LDH) [15, 65] have been investigated in the starch-layered silicate composites. Wilhelm et al. [65] have shown that hectorite could form intercalated and exfoliated nanocomposites with starch, whereas kaolinite, brucite, and LDH formed conventional composites structure in this system. Based on the change in glass transition temperature, hectorite is more miscible with the thermoplastic starch matrix than other layered silicates. Hectorite and kaolinite displayed a similar enhancement of storage modulus at room temperature although the structure of hectorite and kaolinite in the starch matrix was quite different. MMT generally provided a better improvement in the mechanical properties compared with hectorite [5] and fluorohectorite [22].

Strong interaction between layered silicates and polymers is required to produce high-performance PLSNs because the poor interfacial interaction usually leads to the phase separation and a weak load transfer between polymeric matrix and fillers. Introducing an organic moiety into the interlayer by ion-exchange reaction is a common approach to modify surface affinity of layered silicates. Most of the commercial organoclays are modified MMT with cationic surfactants such as tertiary or quaternary alkylammonium cations. The hydrocarbon chains and benzyl groups of the intercalated surfactants improve the compatibility of layered silicates with hydrophobic polymeric matrix. Since starch is a hydrophilic polymer, hydrophobic organoclays lead to the formation of microscale aggregation of layered silicates in the native starch-based composites [5, 8, 46, 47]. However, organoclays showed a better dispersion and greater reinforcing effect in the acetylated starch matrix in comparison to unmodified MMT [49, 69]. Gao et al. [24] reported that the layered silicate with medium hydrophilic properties was more suitable for the preparation of hydroxypropyl distarch phosphate-based nanocomposites. The results addressed the importance of matching the hydrophobicity of layered silicates with the polymeric matrix for successful formation of exfoliated starch-based nanocomposites.

Chivrac et al. [12] developed a new strategy by using cationic starch as a layered silicate modifier for better matching the polarity of the starch matrix and thus to promote the exfoliation process of layered silicates. Morphology analysis (x-ray diffraction (XRD) and transmission electron microscopy (TEM)) and tensile tests confirmed the nanoscale dispersion and the enhanced mechanical properties. The modified surface polarity established by the cationic starch increased the wetting interactions between natural layered silicates and native starch matrix. Chitosan, a natural cationic polysaccharide found in exoskeletons of crustaceans and insects and the cell wall of fungi and microorganisms, can be intercalated into MMT by ion-exchange reaction. The interlayer spacing of MMT expanded from

1.20 nm to 1.45 nm (monolayer) or 2.09 nm (bilayer), depending on the adsorption number of chitosan chains [19]. A family of intercalated LDH with polysaccharides, including alginate, pectin, and ι-carrageenan, has been synthesized by coprecipitation [20]. Alginate and pectin were intercalated as a monolayer of the biopolymer, whereas ι-carrageenan was intercalated as a bilayer or a double helix in LDH. These interesting examples of modified layered silicates with natural biopolymers show promise for in applications requiring nontoxic biodegradable organoclays and for generally improving the compatibility and performance of starch-based composites and blends.

4 Plasticizers and Compatibilizers for Starch-Layered Silicate Nanocomposites

Starch and plasticizers both have the tendency to migrate into interlayer galleries as they both encounter polar-polar attractive forces within layered silicates. It is well known that glycerol is favored to form the intercalated structure with layered silicates because of its small molecular size and hydrophilic nature [5, 7, 10, 33, 45, 66]. For glycerol content higher than 10 %, the formation of an intercalated structure with *d*-spacing increased from 1.2 nm to 1.8 nm is mainly attributed to the glycerol intercalation [9]. When the glycerol content is below 5 % or water is the only plasticizer, the exfoliation extent in the starch matrix increased [21, 22, 60]. Pandey and Singh [45] investigated the sequence of the addition of starch, glycerol, and layered silicates. It was concluded that the better nanoclay dispersion was obtained if glycerol was added after mixing clay in the starch matrix. A possible explanation for these results is that the samples containing a higher glycerol concentration had an increase in starch-glycerol interactions, which competed with interactions between starch, glycerol, and layered silicate surface and could disturb the exfoliation process of layered silicates [7].

Chivrac et al. [10] investigated the effects of plasticizer type (glycerol and sorbitol) on the exfoliation of layered silicates into starch-based nanocomposites. Both sorbitol and glycerol led to the formation of intercalated structure in the starch-MMT nanocomposites. A completely exfoliated morphology was obtained with cationic starch-modified MMT as the nanofillers and glycerol as the plasticizer. For the starch nanocomposites with sorbitol as the plasticizer, the intercalated-exfoliated structures with small tactoids of layered silicates were observed with cationic starch-modified MMT as the nanofiller. The interaction between sorbitol and MMT/cationic starch-modified MMT may not favor the exfoliation process, which leads to the decrease of the exfoliation extent in the starch matrix.

To overcome the limitations of polyol plasticizers, the amine compounds such as urea [6], urea/ethanolamine [27, 71], urea/formamide [41], and formamide/ethanolamine [26] were used for preparing plasticized thermoplastic starch. The NH_2 groups in the plasticizers can form strong interactions with both of the plasticized starch and the primary or quaternary ammonium molecules in the organoclay [6]. The strong interfacial interaction built by NH_2 groups promoted the exfoliation process of organoclay, which led to the starch-layered silicate nanocomposites with significantly enhanced mechanical properties, thermal stability, and water resistance (Table 4.1) [26, 27, 41]. Besides, the incorporation of citric acid-modified MMT into the urea/formamide [28] or glycerol [36, 42] plasticized starch matrix provided the starch-MMT nanocomposites with the improved tensile modulus and strength. Mbey et al. [37] reported that the dimethyl sulfoxide (DMSO) intercalated kaolinite also had a better interaction and dispersion within the starch matrix than the unmodified kaolinite. Considering the toxicity of plasticizers and compatibilizers, Dai et al. [17, 18] focused their attention on the use of two physiologically harmless compounds, N-(2-hydroxyethyl)formamide (HF) and N,N-bis(2-hydroxyethyl)formamide (BHF), in the starch-MMT systems. Fourier transform infrared spectroscopy (FTIR) and XRD showed that HF and BHF acted as a plasticizer for thermoplastic starch as well as a swelling agent for MMT. The starch-MMT nanocomposites with HF/BHF additives displayed an improved tensile strength and water resistance in comparison to the neat thermoplastic starch.

Nejad et al. [40] investigated the effects of the addition of MMT and commercial available organoclay on the mechanical properties of the triacetin plasticized starch mixed esters. It is known that esterification of hydroxyl groups of starch increases the hydrophobicity of starch and improves the water resistance of starch-based materials. But when changing the starch esters from acetate to laurate, a reduction in the Young's modulus was observed [25]. The addition of commercial organoclay in the triacetin plasticized starch mixed esters formed a homogenous distribution of clay tactoids with 4–5 layered sheets, which led to improvement of the composite modulus and strength (Table 4.1). This study provides a new strategy to improve/fine-tune the hydrophobicity and mechanical properties of starch-based materials by using a suitable combination of starch esters, organoclay, and plasticizers.

5 Engineering Methods and Performance of Starch-Layered Silicate Nanocomposites

Solvent casting and melt processing are the main engineering methods for preparing starch-layered silicate nanocomposites. Solvent casting is a quick and convenient preparation method for laboratory experiments. Melt processing, such as conventional extrusion, is a viable large-scale engineering method, especially suited for industrial application, because of the absence of solvent, high throughput rate, and high efficiency. The melt processing of starch-layered silicate nanocomposites generally includes two stages. First, the starch, plasticizers, and layered silicates are mixed mechanically with a high-speed mixer and then stored for several hours to obtain a homogeneous plasticized starch matrix and swollen layered silicates. At the second stage, the mixture is introduced into a melt-mixing device such as an extruder or mixer. The thermal-mechanical treatment leads to starch gelatinization, diffusion of starch chains into the interlayer gallery, layered silicates delamination, and starch degradation. Adding additional components and procedural changes during the mixing and extrusion steps can help to further increase the dispersion and interaction

Table 4.1 Effect of different types of plasticizers on performance and morphology of starch-layered silicate nanocomposites

Plasticizers	Layered silicates	Starch	Morphology	Mechanical properties (times)[a] Modulus	Strength	Elongation	Moisture resistance	Thermal stability	References
Water	MMT	HAMS[g]	Intercalated-exfoliated	1.25~3.3	1~3	0.6~0.9	–	–	Dean et al. [22]
Glycerol	MMT	Corn starch	Intercalated-exfoliated	0.84~2.98	1.33~1.87	0.93~1.2	–	Improved	Aouada et al. [1]
Sorbitol	MMT/cationic starch-MMT	Wheat starch	Intercalated	1.4	–	0.47	–	–	Chivrac et al. [10]
	MMT	HAMS	Intercalated-exfoliated	–	–	–	–	–	Liu et al. [32]
Triacetin	MMT	Starch esters	Aggregated-intercalated	1.09~1.82	0.8~1.38	0.5~4.7	–	–	Nejad et al. [40]
	Dellite 43B[d]	Starch esters	Intercalated-exfoliated	1.38~2.67	1.08~1.72	0.47~0.79	–	–	Nejad et al. [40]
	Dellite 67G[e]	Starch esters	Intercalated-exfoliated	1.4	1.26	0.76	–	–	Nejad et al. [40]
Urea	Ammonium-MMT	Potato starch	Exfoliated	–	–	–	–	–	Chen et al. [6]
Urea/ethanolamine	Ethanolamine-MMT	Corn starch	Exfoliated	2.04~4.9	1.6~3.94	0.32~1.24	Improved (water absorption)	–	Huang & Yu, [27]
	MMT	Potato starch	Intercalated-exfoliated	–	–	–	Improved (WVP[h])	No Difference	Zeppa et al. [71]
	Cloisite 30B[f]	Potato starch	Aggregated-intercalated	–	–	–	Improved (WVP)	No Difference	Zeppa et al. [71]

(*continued*)

Table 4.1 (continued)

Plasticizers	Layered silicates	Starch	Morphology	Mechanical properties (times)[a]			Moisture resistance	Thermal stability	References
				Modulus	Strength	Elongation			
Urea/formamide	Urea-MMT	Corn starch	Intercalated-exfoliated	1.6~8.4	1.67~6	0.45~0.91	Improved (water absorption; WVP)	Improved	Ning et al. [41]
	Citric acid-MMT	Corn starch	Exfoliated	–	3~5.53	0.92~1.23	–	–	Huang et al. [28]
Formamide/ethanolamine	Ethanolamine-MMT	Corn starch	Intercalated	3.07	1.34	0.89	Improved (water absorption)	Improved	Huang et al. [26]
HF[b]	MMT	Corn starch	Intercalation-exfoliation	–	1.24~1.76	1.25~1.72	Improved (water absorption)	Improved	Dai et al. [17]
BHF[c]	MMT	Corn starch	Intercalation-exfoliation	–	1.17~1.25	0.91~0.96	Improved (water absorption)	–	Dai et al. [18]

[a]Enhancement of mechanical properties compared with neat plasticized thermoplastic starch materials
[b]HF: N-(2-hydroxyethyl)formamide
[c]BHF: N, N-bis(2-hydroxyethyl)formamide
[d]Dellite 43B: MMT modified with dimethyl-benzyl-hydrogenated tallow ammonium
[e]Dellite 67G: MMT modified with dimethyl-dihydrogenated tallow ammonium
[f]Cloisite 30B: MMT modified with methyl tallow bis-2-hydroxylethyl quaternary ammonium
[g]HAMS: high-amylose starch
[h]WVP: water vapor permeability

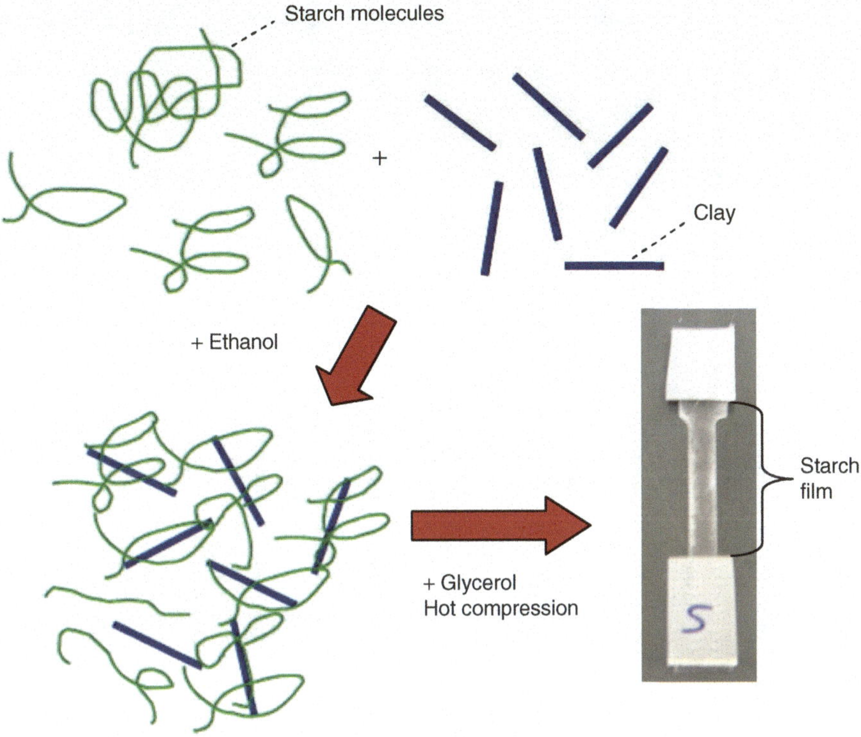

Fig. 4.3 Schematic representation of the preparation procedure for well-dispersed starch-layered silicate nanocomposites by blending a diluted clay dispersion to a starch suspension followed by coprecipitated in ethanol (Source: [14])

among different components [28, 56]. However, since sheer force and thermal treatment can degrade the polymer chains [9], it is necessary to balance the processing parameters to increase the layered silicate exfoliation, minimize the starch chain degradation, and maximize procedural efficiency.

Chung et al. [14] developed a new processing strategy by adding dilute MMT and Laponite dispersions to a starch suspension followed by coprecipitation in ethanol to prepare well-dispersed starch-layered silicate nanocomposites. A schematic of the synthesis is shown in Fig. 4.3. A low concentration suspension of layered silicates suspension (0.2–1.4 wt%) was prepared to promote clay dispersion. A low concentration starch solution was prepared to decrease the viscosity of film-forming solution but increase the miscibility of layered silicates and starch. During precipitation, the well-dispersed clay layers were trapped among starch molecules. The glycerol was added after the precipitation to prevent the intercalation of glycerol and promote starch-clay interactions. These starch-clay nanocomposites showed the pronounced improvements in their modulus and strength without evidence of decreased elongation at break (Table 4.2). Aouada et al. [1] developed a simple method based on the combination of the

Table 4.2 Effect of engineering methods on mechanical properties and morphology of starch-layered silicate nanocomposites

References	Solvents	Specific equipments	Polymers	Layered silicates	Plasticizers	Morphology	Mechanical properties (times)[a]		
							Modulus	Strength	Elongation
Chung et al. [14]	Water/ ethanol	–	Corn starch	MMT	Glycerol	Intercalated-exfoliated	1.65	1.31	0.94
Aouada et al. [1]	Water	Batch mixer	Corn starch	MMT	Glycerol	Intercalated-exfoliated	0.84~2.98	1.33~1.87	0.93~1.2
Chivrac et al. [13] (SICO)[b]	Water	Batch mixer	Corn starch	Cationic starch-MMT	Glycerol	Exfoliated	1.61	–	1.06
Chivrac et al. [13] (EXAD)[c]	Water	–	Corn starch	Cationic starch-MMT	Glycerol	Intercalated-exfoliated	1.68	–	1.02
Vertuccio et al. [64]	No	High-energy ball mill	HAMS[d]; PCL	MMT	No	Exfoliated	1.7	–	0.54
Raquez et al. [51]	No	Twin-screw extruder	Corn starch; PBAT	MMT	Glycerol	Intercalated	1.66	1.61	0.95
Raquez et al. [51]	No	Twin-screw extruder	Corn starch; PBAT	Cloisite 30B[e]	Glycerol	Intercalated-exfoliated	1.71~1.92	1.98~2.27	1.09~1.14
Namazi & Mosadegh, [39]	Water	–	Potato starch; PCL	MMT	–	Intercalated-exfoliated	–	–	–

[a]Enhancement of mechanical properties compared with neat plasticized thermoplastic starch materials
[b]SICO: shear induced clay organo-modification
[c]EXAD: exfoliation/adsorption process
[d]HAMS: high-amylose starch
[e]Cloisite 30B: MMT modified with methyl tallow bis-2-hydroxylethyl quaternary ammonium

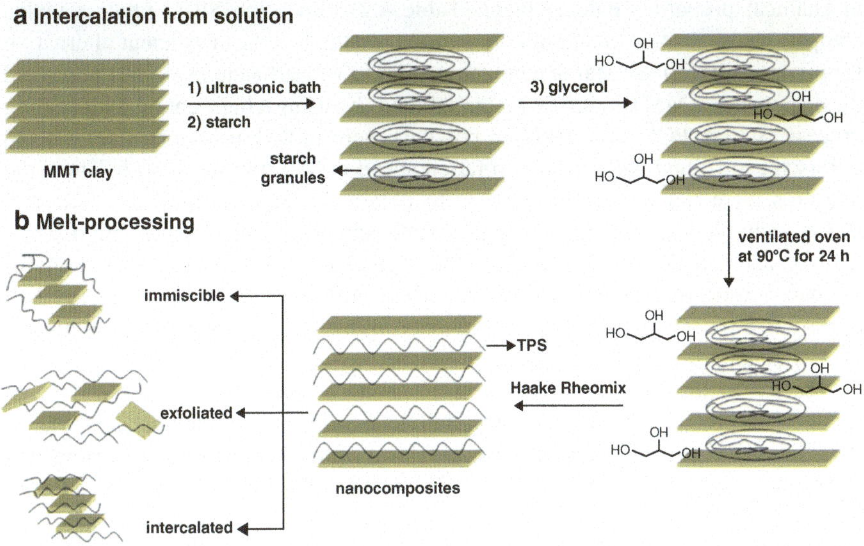

Fig. 4.4 Schematic representation of the processing method based on the combination of (**a**) the intercalation from solution and (**b**) the melt-processing preparation method (Source: [1])

solution and the melt-processing preparation method to prepare the highly exfoliated and compatible starch-MMT nanocomposites. Figure 4.4 illustrates this method. In the first step, the MMT dispersion was prepared in an ultrasonic bath, and then the starch was dispersed into the MMT followed by adding the glycerol. The mixture was heated at 90 °C for 24 h, allowing the vaporization of water and the diffusion of glycerol into starch granules. In the second step, the mixture was processed in a batch mixer at 120 °C for 20 min, and the homogeneous molten starch matrixes with various degrees of MMT dispersions were obtained. XRD results confirmed that the structure of the layered silicates platelets of starch nanocomposites was completely exfoliated with both 1 % and 2 % MMT. For samples, with 3 % and 5 % MMT, the intercalated and partially exfoliated structures were observed. The starch-MMT nanocomposites prepared by this combination method displayed significant enhancements in the mechanical and thermal properties with a low amount of native MMT addition (Table 4.2).

"Shear induced clay organo-modification" (SICO) process is a fast and simple procedure developed by Chivrac et al. [13] to prepare cationic starch-modified MMT. This process is based on in situ organo-modification induced by a batch mixer with a rotor speed from 50 rpm to 100 rpm at 70 °C for 40 min, allowing a fast preparation of pre-exfoliated cationic starch-modified MMT. The cationic starch-modified MMT paste prepared from the SICO process and the glycerol plasticized starch were directly mixed in the internal batch mixer and then hot pressed at 110 °C for 15 min to obtain the starch nanocomposites. The organoclay prepared from either SICO or a conventional exfoliation/adsorption (EXAD) process could be exfoliated into the plasticized starch matrix and led to the same mechanical properties enhancement (Table 4.2). Since the SICO process yields comparable properties to that of conventional methods, it is an efficient alternative for preparing exfoliated starch nanocomposites [13]. Vertuccio et al. [64] improved the dispersion of MMT in the PCL/starch blends by using a high-energy ball milling process. Starch, PCL, and MMT in powders were milled at room temperature in a Retsch centrifugal ball mill at a rotation speed of 580 rpm for 4–10 h. Then, the mixtures were hot pressed at 120 °C to obtain the layered silicates reinforced PCL/starch blends. The morphology (XRD and scanning electron microscopy (SEM)) and tensile test showed that the homogeneity and Young's modulus of the nanocomposites increased with increasing milling time. The high-energy ball milling is a promising solvent-free engineering method for improving the dispersion of nanoparticles into the polymer blends.

Reactive extrusion has also been used to prepare biodegradable layered silicate-reinforced starch-PBAT (poly(butylene adipate-co-terephthalate)) blends [51]. In situ, chemically modified thermoplastic starch (MTPS) was prepared first with maleic anhydride (MA) and MMT/Cloisite 30B. Then, the layered silicate-reinforced MTPS was reactively melt blended with PBAT through transesterification reactions promoted by MA-derived acidic moieties grafted onto the starch backbone. The PBAT-*g*-MTPS blends with either MMT or Cloisite 30B showed significant improvements in mechanical properties, water vapor permeability, and oxygen barrier (Table 4.2). The explanation for the more significant increased tensile properties of the Cloisite 30B-reinforced starch nanocomposites is that the hydroxyl groups of Cloisite 30B had stronger hydrogen interactions with MA-derived acid moieties on the MTPS and also reacts with MTPS and PBAT through esterification reactions. This esterification reaction grafted some PBAT-*g*-MTPS polymers onto the surface of the organo-modified layered silicates, allowing a stronger interfacial interaction between layered silicates and starch matrix.

Namazi and Mosadegh [39] prepared layered silicate-reinforced starch-PCL nanocomposites through bulk (Fig. 4.5a) and in situ (Fig. 4.5b) polymerization methods. XRD showed that the processing procedure for grafting PCL onto starch through bulk polymerization produced starch-PCL nanocomposites with intercalated-exfoliated structure. The mechanical properties and water resistance of the nanocomposites prepared by the grafting polymerization are of interest in further investigations.

6 Conclusions

Recent studies have shown that the incorporation of a low amount of well-dispersed layered silicates into a starch matrix significantly enhances the mechanical properties, water resistance, and thermal stability of starch-based materials. Since starch and layered silicates are both hydrophilic, well-dispersed starch-layered silicate nanocomposites can be obtained by processing native starch and MMT with industrially viable and efficient engineering procedures. A suitable plasticizer used in the optimal concentration can create a favorable environment for strong

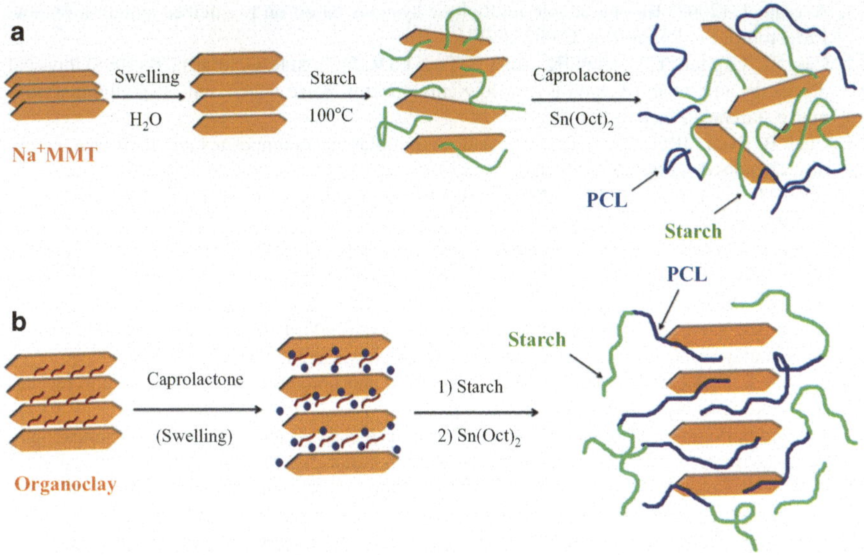

Fig. 4.5 Preparation routes for starch-*g*-PCL nanocomposites through (**a**) bulk and (**b**) in situ polymerization methods (Source: [39])

interactions between layered silicates and starch, leading to the formation of exfoliated structure. The further development of simple and fast engineering procedures to efficiently exfoliate layered silicates and disperse clay platelets into starch-based materials will be an important factor in realizing the commercial viability of layered silicate-reinforced starch nanocomposites.

7 Future Prospective

Most of the studies have focused on the performance of starch-layered silicate nanocomposites but seldom addressed the potential impact of the ecotoxicity of plasticizers and the modification process of layered silicates on the environment and human health. One of the important future research directions in this field is the safety evaluation and life cycle analysis of the high-performance starch-layered silicate nanocomposites to elaborate safe green starch-layered silicate nanocomposites in a wide range of applications.

References

1. Aouada FA, Mattoso LHC, Longo E (2011) New strategies in the preparation of exfoliated thermoplastic starch-montmorillonite nanocomposites. Ind Crop Prod 34:1502–1508
2. Atwell WA, Hood LF, Lineback DR, Varriano-Marston E, Zobel HF (1988) The terminology and methodology associated with basic starch phenomena. Cereal Foods World 33:306–311

3. Averous L (2004) Biodegradable multiphase systems based on plasticized starch: a review. J Macromol Sci Polym Rev C44:231–274
4. Case SE, Capitani T, Whaley JK, Ahi YC, Trzasko R, Jeffcoat R, Goldfarb HB (1998) Physical properties and gelatin behavior of a low-amylopectin maize starch and other high amylose maize starches. J Cereal Sci 27:301–314
5. Chen B, Evans JRG (2005) Thermoplastic starch-clay nanocomposites and their characteristics. Carbohydr Polym 63:455–463
6. Chen M, Chen B, Evans JRG (2005) Novel thermoplastic starch-clay nanocomposite foams. Nanotechnology 16:2334–2337
7. Chiou BS, Wood D, Yee E, Imam SH, Glenn GM, Orts WJ (2007) Extruded starch-nanoclay nanocomposites: effects of glycerol and nanoclay concentration. Polym Eng Sci 47:1898–1904
8. Chiou BS, Yee E, Wood D, Shey J, Glenn G, Orts W (2006) Effects of processing conditions on nanoclay dispersion in starch-clay nanocomposites. Cereal Chem 83:300–305
9. Chivrac F, Pollet E, Averous L (2009) Progress in nano-biocomposites based on polysaccharides and nanoclays. Mater Sci Eng Rep 67:1–17
10. Chivrac F, Pollet E, Averous L (2010) Shear induced clay organo-modification: application to plasticized starch nano-composites. Polym Adv Technol 21:578–583
11. Chivrac F, Pollet E, Dole P, Averous L (2010) Starch-based nano-biocomposites: plasticizer impact on the montmorillonite exfoliation process. Carbohydr Polym 79:941–947
12. Chivrac F, Pollet E, Schmutz M, Averous L (2008) New approach to elaborate exfoliated starch-based nanobiocomposites. Biomacromolecules 9:896–900
13. Chivrac F, Pollet E, Schmutz M, Averous L (2010) Starch nano-biocomposites based on needle-like sepiolite clays. Carbohydr Polym 80:145–153
14. Chung YL, Ansari S, Estevez L, Hayrapetyan S, Giannelis EP, Lai HM (2010) Preparation and properties of biodegradable starch-clay nanocomposites. Carbohydr Polym 79:391–396
15. Chung YL, Lai HM (2010) Preparation and properties of biodegradable starch-layered double hydroxide nanocomposites. Carbohydr Polym 80:525–532
16. Crosby AJ, Lee JY (2007) Polymer nanocomposites: the "nano" effect on the mechanical properties. Polym Rev 47:217–229
17. Dai H, Chang PR, Geng F, Yu J, Ma X (2009) Preparation and properties of thermoplastic starch/montmorillonite nanocomposite using N-(2-hydroxyethyl)formamide as a new additive. J Polym Environ 17:225–232
18. Dai H, Sheng X, An L, Liu N, Yu J, Ma X (2012) Preparation and properties of thermoplastic starch/montmorillonite nanocomposites using N, N-bis(2-hydroxyethyl)formamide as a new additive. Polym Compos 33:225–231
19. Darder M, Colilla M, Ruiz-Hitzky E (2003) Biopolymer-clay nanocomposites based on chitosan intercalated in montmorillonite. Chem Mater 15:3774–3780
20. Darder M, Lopez-Blanco M, Aranda P, Leroux F, Ruiz-Hitzky E (2005) Bio-nanocomposites based on layered double hydroxides. Chem Mater 17:1969–1977
21. Dean KM, Do MD, Petinakis E, Yu L (2008) Key interaction in biodegradable thermoplastic starch/poly(vinyl alcohol)montmorillonite micro- and nanocomposites. Compos Sci Technol 68:1453–1462
22. Dean K, Yu L, Wu DY (2007) Preparation and characterization of melt-extruded thermoplastic starch/clay nanocomposites. Compos Sci Technol 67:413–421
23. Endres HJ, Siebert-Raths A (2011) Engineering biopolymers. Markets, manufacturing, properties and applications. Hanser Publishers, Munich
24. Gao W, Dong H, Hou H, Zhang H (2012) Effects of clays with various hydrophilicities on properties of starch-clay nanocomposites by film blowing. Carbohydr Polym 88:321–328
25. Gros AT, Feuge RO (1962) Properties of the fatty acid esters of amylose. J Am Oil Chem Soc 39:19–24

26. Huang MF, Yu JG, Ma XF, Jin P (2005) High performance biodegradable thermoplastic starch-EMMT nanoplastics. Polymer 46:3157–3162
27. Huang M, Yu J (2006) Structure and properties of thermoplastic corn starch/montmorillonite biodegradable composites. J Appl Polym Sci 99:170–176
28. Huang M, Yu J, Ma X (2006) High mechanical performance MMT-urea and formamide-plasticized thermoplastic cornstarch biodegradable nanocomposites. Carbohydr Polym 63:393–399
29. Kester JJ, Fennema OK (1986) Edible films and coatings: a review. Food Technol 40:47–59
30. Kojima Y, Usuki A, Kawasumi M, Okada A, Fukushima Y, Kurauchi T, Kamigaito O (1993) Mechanical properties of nylon 6-clay hybrid. J Mater Res 8:1185–1189
31. Lai LS, Kokini JL (1991) Physicochemical changes and rheological properties of starch during extrusion (a review). Biotechnol Progr 7:251–266
32. Liu H, Chaudhary D, Yusa S, Tade M (2011) Glycerol/starch/Na^+-montmorillonite nanocomposites: a XRD, FTIR, DSC and ^1H NMR study. Carbohydr Polym 83:1591–1597
33. Liu H, Chaudhary D, Yusa SI, Tade MO (2011) Preparation and characterization of sorbitol modified nanoclay with high amylose bionanocomposites. Carbohydr Polym 85:97–104
34. Lourdin D, Della VG, Colonna P (1995) Influence of amylose content on starch film and foams. Carbohydr Polym 27:261–270
35. Magalhaes NF, Andrade CT (2010) Calcium bentonite as reinforcing nanofiller for thermoplastic starch. J Brazil Chem Soc 21:202–208
36. Majdzadeh-Ardakani K, Navarchian AH, Sadeghi F (2010) Optimization of mechanical properties of thermoplastic starch/clay nanocomposites. Carbohydr Polym 79:547–554
37. Mbey JA, Hoppe S, Thomas F (2012) Cassava starch-kaolinite composite film. Effect of clay content and clay modification on film properties. Carbohydr Polym 88:213–222
38. Mondragon M, Mancilla JE, Rodriguez-Gonzalez FJ (2008) Nanocomposites from plasticized high-amylopectin, normal and high-amylose maize starches. Polym Eng Sci 48:1261–1267
39. Namazi H, Mosadegh M (2011) Preparation and properties of starch/nanosilicate layer/polycaprolactone composites. J Polym Environ 19:980–987
40. Nejad MH, Ganster J, Bohn A, Volkert B, Lehmann A (2011) Nanocomposites of starch mixed esters and MMT: improved strength, stiffness and toughness for starch propionate acetate laurate. Carbohydr Polym 84:90–95
41. Ning W, Xingxiang Z, Na H, Haihui L (2010) A facile method for preparation of thermoplastic starch/urea modified montmorillonite nanocomposites. J Compos Mater 44:27–39
42. Ning W, Xingxiang Z, Na H, Shihe B (2009) Effect of citric acid and processing on the performance of thermoplastic starch/montmorillonite nanocomposites. Carbohydr Polym 76:68–73
43. Okada A, Kawasumi M, Kurauchi T, Kamigaito O (1987) Synthesis and characterization of a nylon 6-clay hybrid. Polym Prepr 28:447–448
44. Palviainen P, Heinarnaki J, Myllarinen P, Lahtinen R, Yliruusi J, Forssel P (2001) Corn starches as film formers in aqueous-based film coating. Pharm Dev Technol 6:351–359
45. Pandey JK, Singh RP (2005) Green nanocomposites from renewable resources: effect of plasticizer on the structure and material properties of clay-filled starch. Starch/Staerke 57:8–15
46. Park HM, Lee WK, Park CY, Cho WJ, Ha CS (2003) Environmentally friendly polymer hybrids. Part I Mechanical, thermal, and barrier properties of thermoplastic starch/clay nanocomposites. J Mater Sci 38:909–915
47. Park HM, Li X, Jin CZ, Park CY, Cho WJ, Ha CS (2002) Preparation and properties of biodegradable thermoplastic starch/clay hybrids. Macromol Mater Eng 287:553–558
48. Perez S, Bertoft E (2010) The molecular structures of starch components and their contribution to the architecture of starch granules: a comprehensive review. Starch/Staerke 62:389–420
49. Qiao X, Jiang W, Sun K (2005) Reinforced thermoplastic acetylated starch with layered silicates. Starch/Staerke 57:581–586

50. Queiroz AUB, Collares-Queiroz F (2009) Innovation and industrial trends in bioplastics. J Macromol Sci Polym Rev 49:65–78
51. Raquez JM, Nabar Y, Narayan R, Dubois P (2011) Preparation and characterization of maleated thermoplastic starch-based nanocomposites. J Apply Polym Sci 122:639–647
52. Ray SS, Okamoto M (2003) Polymer/layered silicate nanocomposites: a review from preparation to processing. Prog Polym Sci 28:1539–1641
53. Rindlav-Westling A, Stading M, Hermansson AM, Gatenholm P (1998) Structure, mechanical and barrier properties of amylose and amylopectin films. Carbohydr Polym 36:217–224
54. Russell PL (1988) Gelatinisation of starches of different amylose/amylopectin content. A study of differential scanning calorimetry. J Cereal Sci 6:133–145
55. Sanchez C, Belleville P, Popall M, Nicole L (2011) Applications of advanced hybrid organic–inorganic nanomaterials: from laboratory to market. Chem Soc Rev 40:696–753
56. Sharif A, Aalaie J, Shariatpanahi H, Hosseinkhanli H, Khoshniyat A (2011) Study on the structure and properties of nanocomposites based on high-density polyethylene/starch blends. J Polym Res 18:1955–1969
57. Shen L, Worrell E, Patel M (2009) Present and future development in plastics from biomass. Biofuel Bioprod Biorefin 4:25–40
58. Shogren RL (1992) Effect of moisture content on the melting and subsequent physical aging of corn starch. Carbohydr Polym 19:83–90
59. Soest JJG, Borger DB (1997) Structure and properties of compression-molded thermoplastic starch materials from normal and high-amylose maize starches. J Apply Poly Sci 64:631–644
60. Tang X, Alavi S, Herald T (2008) Barrier and mechanical properties of starch-clay nanocomposite films. Cereal Chem 85:433–439
61. Vaia RA, Giannelis EP (2001) Polymer nanocomposites: status and opportunities. MRS Bulletin 26:394–401
62. Vaia RA, Ishii H, Giannelis EP (1993) Synthesis and properties of two-dimensional nanostructures by direct intercalation of polymer melts in layered silicates. Chem Mater 5:1694–1696
63. Vaia RA, Vasudevan S, Krawiec W, Scanlon LG, Giannelis EP (1995) New polymer electrolyte nanocomposites: melt intercalation of poly(ethylene oxide) in mica-type silicates. Adv Mater 7:154–156
64. Vertuccio L, Gorrasi G, Sorrentino A, Vittoria V (2009) Nano clay reinforced PCL/starch blends obtained by high energy ball milling. Carbohydr Polym 75:172–179
65. Wilhelm HM, Sierakowski MR, Souza GP, Wypych F (2003) The influence of layered compounds on the properties of starch/layered compound composites. Polym Int 52:1035–1044
66. Wilhelm HM, Sierakowski MR, Souza GP, Wypych F (2003) Starch films reinforced with mineral clay. Carbohydr Polym 52:101–110
67. Wolff IA, Cluskey DJE, Gundrum LJ, Rist CE (1951) Preparation of films from amylose. Ind Eng Chem 43:915–919
68. Xie F, Halley PJ, Averous L (2012) Rheology to understand and optimize processability, structures and properties of starch polymeric materials. Prog Polym Sci 37:595–623
69. Xu Y, Zhou J, Hanna MA (2005) Melt-intercalated starch acetate nanocomposite foams as affected by type of organoclay. Cereal Chem 82:105–110
70. Young AH (1984) Fraction of starch. In: Whistler R, BeMiller JN, Paschall EF (eds) Starch chemistry and technology, 2nd edn. Academic, New York, p 264
71. Zeppa C, Gouanve F, Espuche E (2009) Effect of a plasticizer on the structure of biodegradable starch/clay nanocomposites: thermal, water-sorption, and oxygen-barrier properties. J Appl Polym Sci 112:2044–2056
72. Zobel HF (1988) Molecules to granules: a comprehensive review. Starch/Staerke 40:1–7

Structure-Property Correlations of Poly(ethylene oxide) Nanohybrids with Layered Silicates and Silica Nanoparticles

5

Engin Burgaz

Contents

Abstract

Structure and properties of solution intercalated poly(ethylene oxide) (PEO) nanohybrids with silica and clay nanoparticles in comparison with those of PEO/clay nanocomposites are reviewed. Herein, the structure, preparation, morphological behavior, and performance characteristics (mechanical and thermal properties) of PEO nanohybrids are discussed. One distinct approach is presented here: The simultaneous use of hydrophilic fumed silica, unmodified clay and an adsorbing polymer in solution leads to phase separation based on the bridging of particles by polymer chains. PEO/clay/silica hybrids display good thermal stability and show significant property improvements (much lower crystallinity, higher modulus and enhanced thermomechanical properties) compared to PEO/clay hybrids and pure PEO. The reinforcement effect and reduction of crystallinity as a function of fumed silica loading strongly depend on the morphological behavior of nanohybrids.

E. Burgaz
Department of Materials Science and Engineering, Ondokuz Mayis University, Atakum Samsun, Turkey
e-mail: eburgaz@omu.edu.tr

J.K. Pandey et al. (eds.), *Handbook of Polymernanocomposites. Processing, Performance and Application – Volume A: Layered Silicates*, DOI 10.1007/978-3-642-38649-7_8,

1 Introduction

Nanocomposites have been defined as new class of composite materials which have morphological dimensions on the order of several nanometers and have unusual combination of material properties such as higher strength and stiffness and lighter weight compared to conventional composites [1, 2]. During the past 20 years, researchers from both academia and industry have worked on the field of polymer/clay nanocomposites in a great deal because the knowledge gathered from these R&D efforts led to the development of new hybrid materials in which the properties are greatly improved in comparison with the pure polymer and traditional polymer composites. The enhanced properties for polymer/clay nanocomposites can be defined as increased mechanical properties such as higher strength and modulus [3–5], increased thermal stability and decreased thermal degradation [6], improved permeability and barrier properties [7] and decreased flammability [8]. Polymer/clay nanocomposites can be synthesized by three common methods: (1) Mixing the layered silicate with the polymer and heating the mixture above the glass transition temperature of the polymer (the melt blending) [9, 10]. (2) The intercalation of a suitable monomer into layered silicates followed by polymerization (*in-situ* polymerization) [11, 12]. (3) Polymer intercalation from solution (the solution blending) [13]. In the field of polymer/clay nanocomposites, researchers have incorporated layered silicates into many different types of polymers in order to synthesize polymer/clay nanocomposites with novel and improved properties. PEO which is a highly efficient polymeric material in so many different applications such as denture adhesives, packaging films, thickening of water-based paints, friction reduction, purification of biological materials, and pharmaceutical drugs has been used as a very successful matrix material in the fabrication process of polymer/clay nanocomposites. PEO can be also used as a potential biomaterial due to its attractive properties such as hydrophilicity, biocompatibility, and versatility. Besides these important applications, PEO has been used in a great extent as a solid polymer electrolyte material in rechargeable solid-state lithium-ion polymer batteries [14, 15].

In the field of polymer electrolytes, PEO has been known as the most widely studied solid electrolyte material since the chemical and physical structure of PEO is very suitable for fast ion diffusion [16, 17], and also different types of salts can be dissolved in the PEO matrix [18]. Recently, the microscopic picture for the ion transport mechanism in PEO was explained by neutron scattering and dielectric spectroscopy experiments in combination with molecular dynamics simulations such that PEO chains form ethylene oxygen cages over several monomer units and the lithium ions "jump" from one cage to another [19]. PEO has a considerable amount of crystalline structure, and this property leads to low conductivity values at room temperature which is not very beneficial in terms of rechargeable solid-state lithium-ion polymer battery applications. Researchers have used many different methods in order to decrease the amount of PEO crystalline phase, but at the same time, the mechanical properties such as flexibility and mechanical stability in a wide temperature range were not deteriorated [17, 20–22]. One of the most effective methods has been the addition of clay or silica nanoparticles to the PEO electrolye material for reducing crystallinity [23–26]. It was also shown that nanotubes can be packaged within insulating clay layers to form effective 3D nanofillers and these hybrid nanofillers increase the lithium ion conductivity of PEO electrolyte by almost two orders of magnitude [27].

In the early 1990s, Aranda and Ruiz-Hitzky showed that the intercalation of PEO molecules within the silicate layers reduces polymer crystallization, increases ionic conductivity, and thus leads to nanocomposite materials that are potentially interesting as ionic conductors and useful for membrane and solid electrolyte applications [14, 15]. The solution intercalation is a good method to intercalate polymers with little or no polarity into layered structures, and produces thin polymer films with oriented clay layers [28, 29]. X-ray diffraction results confirmed that high molecular mass fractions of poly(ethylene glycol)s intercalate preferentially into layered slicates during solution casting [29, 30]. PEO/clay nanocomposites were prepared and characterized via melt extrusion and the results showed that the mechanical properties of the nanocomposites based on the high molar mass PEO matrix were superior when compared with the lower molar mass PEO [30]. Several research groups have studied the conformational properties of PEO within layered silicates [28, 31–34]. It was shown that PEO polymer chains preserves their helical structure [32], and other researchers concluded that the PEO conformation in layered silicates is a distorted helical structure as a result of polymer intercalation [28, 31]. Rohr and coworkers studied the conformation of poly(ethylene oxide) within the clay layers by two dimensional double-quantum NMR and found out that the conformation of the "–OC–CO–" bonds of PEO is 90 ± 5 % gauche, inducing constraints on the chain conformation in the silicate interlayer [33].

Several research groups have investigated the crystallization behavior of PEO/clay nanocomposites [24, 31, 35–38]. The crystalline structure and morphology of PEO/clay nanocomposites were described in a model such that an interphase layer exists between the layered silicates and PEO lamellar crystals in which the secondary crystallization occurs [36]. Amorphous "crown ether" type of backbone polymer conformations in the vicinity of the silicate layers were promoted by the coordination of PEO molecules with layered silicates. Thus, layered silicates retard the crystalline growth front and this phenomenon leads to crystalline structures with smaller nonspherical shapes and jagged edges [37]. It was previously shown that at lower clay concentrations, the increased amount of nucleating sites enhances the crystallization kinetics, while at higher clay concentrations, the crystallization slows down and the silicate layers act as non-crystallizable barriers in the crystallization of the PEO matrix [39]. The melting point and the degree of crystallinity initially increase with the incorporation of lower amount of clay into PEO since the silicate layers act as nucleation centers and, then they decrease with further increase in the clay concentration [31, 40]. However, the structure of layered silicates during or after the crystallization of PEO was not completely understood [37, 38].

Several research groups have used hydrophilic, hydrophobic and mesoporous nanosilica in order to understand the effect of filler on the morphology, ionic conductivity and mechanical properties of PEO/silica nanocomposites [25, 41–43]. In addition to these works, other researchers have studied the molecular interactions of surface modified nanosilica with low molecular weight PEO oligomers [44] and nanosilica that was prepared via sol-gel method [45] with PEO polymer chains in order to reveal the mechanism for the reduction in chain motion and crystallization. PEO interacts more strongly with hydrophilic fumed silica as compared to hydrophobic silica due to hydrogen bonding interactions of PEO ether oxygens with acidic silanols. These hydrogen bonding interactions clearly prevents chain mobility and crystallization [44], and also increases thermal stability and mechanical properties of PEO/silica nanocomposites [26, 44, 46]. Strong dynamic hydrogen bonding forces and stable low viscosity solutions can be formed between the hydrophilic silica and various solvents such as ethylene glycol and its oligomers and short-chain alcohols [47]. The structure of this stable solution can be preserved after solution casting and this might lead to the uniform distribution of silica particles without any aggregation in PEO nanocomposites which in fact has a positive effect in the reduction of crystallinity and increase of thermal stability and mechanical properties of PEO nanocomposites. Unlike surface modified silica particles, bare hydrophilic nanosilica particles form a physical polymer-particle network structure due to strong PEO-nanoparticle interactions which leads to modulus increase in PEO/silica nanocomposites [48, 49].

Nanolayered silicates and hydrophilic nanosilica particles have different affinities toward PEO chains due to their different surface properties. Due to this surface chemistry difference, PEO is strongly adsorbed on the surface of hydrophilic nanosilica, while it is weakly adsorbed on the surface of nanoclay. Polymer bridging interactions between particles can be formed if an adsorbing polymer is present in a colloidal suspension. The particle volume fraction and surface chemistry and molecular weight of the polymer have a great influence on the strength and the magnitude of polymer bridging interactions [48–51]. The rheology and microstructure of PEO/silica nanocomposites were studied by viscosity and X-ray scattering methods in order to understand the effect of filler type, size and concentration on the strength and mechanism of polymer bridging interactions [48–50]. The roles of particle-to-monomer diameter ratio, polymer molecular weight, strength of monomer-particle attractions on polymer bridging interactions and polymer-particle network formation have been established in terms of theory [51–53]. Computer simulations were used to understand the molecular basis of the rheology changes in polymer melts when loaded with filler particles, specifically when the polymer is an adsorbing polymer [54, 55]. Bridging of individual particles or particle clusters via polymer chains can be formed *via* these specific polymer-particle interactions and, in some cases, if the volume fraction of nanoparticles is high enough, these polymer-particle cluster system might lead to the formation of a physical particle-polymer network [56].

This chapter reviews the structure and properties of PEO hybrids with sodium montmorillonite (Na^{+}MMT) and high surface area hydrophilic fumed silica, and focuses on the comparison of properties of PEO/clay/silica nanohybrids with those of PEO/clay nanocomposites. The process of polymer-particle bridging interactions was exploited to fabricate new nanohybrid materials with improved thermal stability and mechanical properties, and reduced crystallization behavior [56]. The goal of this chapter is not to present a complete literature review in the topic. Interested readers are referred to a number of recent reports that provide a comprehensive record of the substantial work published in this field [30–61]. However, careful consideration of the work reported can identify promising approaches and further advance the design of a new generation of polymer nanocomposites with improved performance.

2 Structural Characteristics of PEO Nanohybrids

2.1 Dispersion Medium and Casting Process

Solution intercalation, steady-state annealing and melt intercalation are the commonly used methods in the preparation of PEO/clay nanocomposites. In the solution casting method, the degree of polarity of solvents is the main factor in the intercalation of polymers into layered silicates [32]. The driving force for the polymer intercalation into layered silicates in solution comes from the entropy increase by the removal of solvent molecules, which compensates for the entropy decrease of the intercalated polymer chains [31]. The crystallization behavior of PEO and the state of dispersion of silicate layers or fumed silica particles critically depend on the processing protocol followed for the preparation of nanohybrids. By optimizing the dispersion medium and the overall processing conditions, homogenous, mechanically and dimensionally stable free-standing films can be easily synthesized to a very good extent. The chemical nature of solvent is very important to facilitate the insertion of PEO molecules between the silicate layers, thus the polarity of dispersion medium is the determining factor for intercalation process. The high polarity of water causes swelling of Na^{+}MMT and leads to the cracking of films. Methanol is not suitable as a solvent for high molecular weight PEO, whereas methanol/acetonitrile or water/methanol solvent mixtures appear to be as useful for PEO intercalations [32, 56]. Previously, PEO/clay nanocomposites were prepared by a solution intercalation method using chloroform as a solvent [31, 35]. On the other hand, acetonitrile was used as a dispersion medium for the preparation of PEO/PEO-grafted fumed silica composite electrolytes [25]. For the well dispersion and solution casting of PEO/clay/silica nanohybrids, the affinity of PEO towards both silica and clay nanoparticles, polarity differences between nanoparticles and solvent or solvent mixtures and the surface properties of nanoparticles should be taken into consideration. In contrast, homogeneous and well-dispersed PEO/clay/silica nanohybrid films can be prepared by casting at room temperature using a deionized water/methanol (3:1) solvent mixture [56]. Dimensionally stable free-standing films can be formed if the resulting PEO nanohybrid materials show good stability toward treatment with different solvents at room temperature for more than 1 day period of time. Previous experiments showed that intercalation rates in Na^+-montmorillonite vary inversely with the average molecular weight of PEO. Thus, the maximum amount of intercalated PEO of high molecular weight was found for PEO which has a molecular weight of 10^5 gr/mol [32].

2.2 Morphological Behavior

Polymer bridging interactions between particles (either nanoclay or nanosilica, in our case) can be formed if an adsorbing polymer is present in a colloidal suspension. Bridging of individual particles or particle clusters via polymer chains can be formed by these specific polymer-particle interactions and, in some cases, if the volume fraction of nanoparticles is high enough, this polymer-particle cluster system might lead to the formation of a physical particle-polymer network [56]. Figure 5.1 shows the schematic showing the polymer bridging interactions (a) between aggregates of spherical nanoparticles (b) between intercalated aggregates of layered silicates.

X-ray diffraction (XRD) studies of PEO/clay nanocomposites revealed that polymer chains are intercalated inside clay galleries, and schematic structural models have been proposed about the intercalation process [14, 15, 32]. Ruiz-Hitzky and coworkers proposed a structural model for the intercalated PEO-cation complex such that the axes of the helical polymer chains are positioned parallel to the plane of silicate layers [14, 15]. This structural arrangement of the intercalated PEO that is in interaction with the interlayer cations was further confirmed by IR, ^{13}C and ^{23}Na NMR spectroscopic data [32]. The clay gallery size of PEO/clay/silica and PEO/clay nanohybrids is compared in Fig. 5.2. XRD results of PEO/ clay nanocomposites showed that the interlayer distance of clay in the nanocomposites increases systematically compared to that of pure Na^+MMT [14, 15, 32]. In Fig. 5.2, the difference between the d-spacing of nanoclay (d_{001} = 17.45 Å) in PEO nanohybrids and the thickness of layered silicate (~9.5 Å) in pure clay was calculated about 8 Å [32, 56]. It was previously shown that an interlayer distance of 8 Å is compatible with the intercalation of two monolayers of PEO in a planar zigzag conformation or one PEO monolayer in a helicoidal conformation [14, 32, 62]. The intercalation of PEO into transition metal oxides or inorganic clay layers can facilitate the mobility of lithium ions, and consequently these systems can be used efficiently as solid electrolytes for high-energy batteries [14, 15, 32]. In Fig. 5.2, increasing the amount of fumed silica in PEO/clay/silica nanohybrids does not really change the intercalation ability of polymer molecules compared to the case in which the clay amount is systematically increased in PEO/clay nanocomposites. For instance, the sample containing 20 wt.% clay has a smaller clay gallery size compared to that of the nanohybrid containing 10 wt.% clay and 10 wt.% silica (Sample 10–10). Thus, the concentration increase of silica rather than nanoclay provides better intercalation for polymer chains within the silicate layers. Intercalation of polymer chains inside clay galleries is provided by possible bonding mechanisms such as the H-bonding between

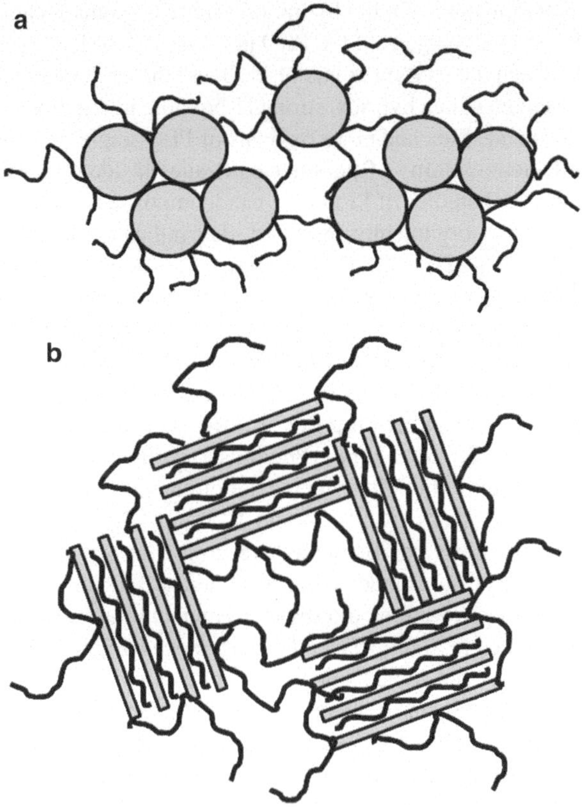

Fig. 5.1 Schematic showing the polymer bridging interactions (**a**) between aggregates of spherical nanoparticles (**b**) between intercalated aggregates of clay nanoparticles. As the volume faction of nanoparticles increases, particle-particle aggregations as well as polymer-particle interactions can be generated in the presence of an adsorbing polymer (i.e., a polymer that is absorbed by the particles) (Reproduced with permission from Ref. [56]. Copyright 2011 Elsevier Ltd)

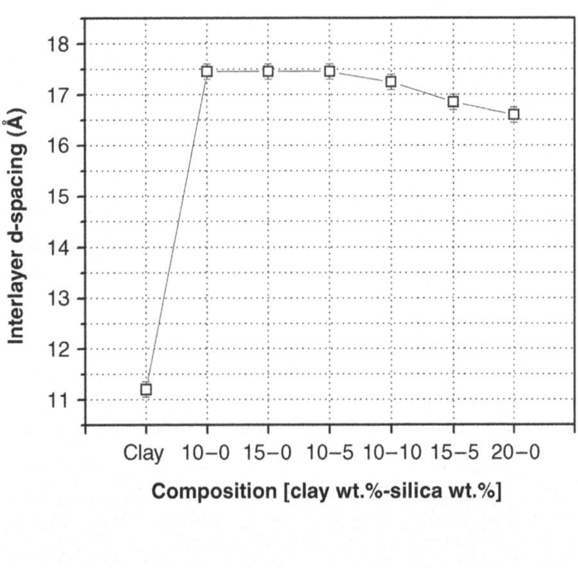

Fig. 5.2 Interlayer spacing of clay as a function of sample composition for pure clay (Na^{+}MMT), and PEO nanohybrids (Reproduced with permission from Ref. [56]. Copyright 2011 Elsevier Ltd)

water molecules in the hydration sphere of exchangeable cations on montmorillonite and the oxygens of the PEO polymer [63–65], and direct ion–dipole interactions between the oxygen atoms of PEO and the exchangeable cations [66]. However, in PEO/clay/silica hybrids, strong H-bonding interactions between the surface silanols of fumed silica and ether oxygens of PEO might provide extra improvement in the interaction ability of polymer molecules inside clay galleries. Ratna et al. showed that XRD studies of PEO/clay nanocomposite samples containing 5 % and 10 % by weight of organically modified clay indicates that the d-spacing increases up to 9.5 nm as a result of sonication [31]. Application of ultrasonication provides a swelling mechanism for layered silicates so that polymer chains are preferentially intercalated inside clay galleries, and also it gives more energy to layered silicates in order to well disperse and separate clay tactoids in solution [67].

Vibrational studies of pure PEO and PEO-salt complexes in the 1,500–500-cm^{-1} region allowed the proposal of different polymer chain conformations inside clay galleries [28, 31, 32, 68]. Previous experimental results [15, 32] showed that PEO should fulfill the following two requirements to have a helical conformation: (1) the presence of two bands around 945 and 850 cm^{-1} that are responsible for CH_2 rocking vibrations of methylene groups in the gauche conformation and (2) the absence of the characteristic band near 1,320 cm^{-1} which is the signature for CH_2 stretching vibrations of ethylene groups in the trans conformation. In Fig. 5.3, the attenuated total reflectance Fourier transform infrared (ATR FT-IR) spectra of pure PEO, clay and fumed silica and PEO/clay/silica and PEO/clay hybrids in the region of 1,350–800 cm^{-1} are shown. The peaks related to the symmetric and asymmetric stretching modes of C–O–C groups are present around 1,150, 1,100 and 1,060 cm^{-1} [68]. The characteristic peaks due to the Si–O stretching (out-of-plane) and Si–O stretching (in-plane) vibrations of clay that are located at 1,115 and 991 cm^{-1}, respectively [69]. The peak responsible for the Si–O–Si bulk mode of fumed silica is centered at 1,070 cm^{-1} [70]. In Fig. 5.2, the major peak related to C–O–C groups of PEO broadens and shifts to different wavenumbers due to the coordination of sodium cations of layered silicates with the ether oxygens of PEO [68], and also H-bonding interactions between the ether oxygen of PEO molecules and the surface silanols of fumed silica. Also, previous FT-IR results of PEO/clay nanocomposites confirmed the presence of ion dipole interactions between the interlayer cation of clay and the oxygen atoms of oxyethylene units [28, 31]. Ratna et al. confirmed the peaks related to CH_2 rocking vibrations of methylene groups in the gauche conformation required for the helical conformation, and the absence of the characteristic PEO peak that is responsible for CH_2 vibrations of the ethylene group in the trans conformation [31]. The ATR FT-IR spectroscopy results in Fig. 5.2 suggest that PEO chains of both pure PEO and PEO nanohybrids have a helical conformation. However, previous work on PEO/clay nanocomposites showed that the helical structure of PEO in the composite is distorted/perturbed as a result of intercalation [31].

Previously, small angle X-ray scattering (SAXS) studies about PEO/clay nanocomposites tried to figure out the interaction mechanism of PEO crystallites with clay layers, and consequently the effects of morphology on the crystallization

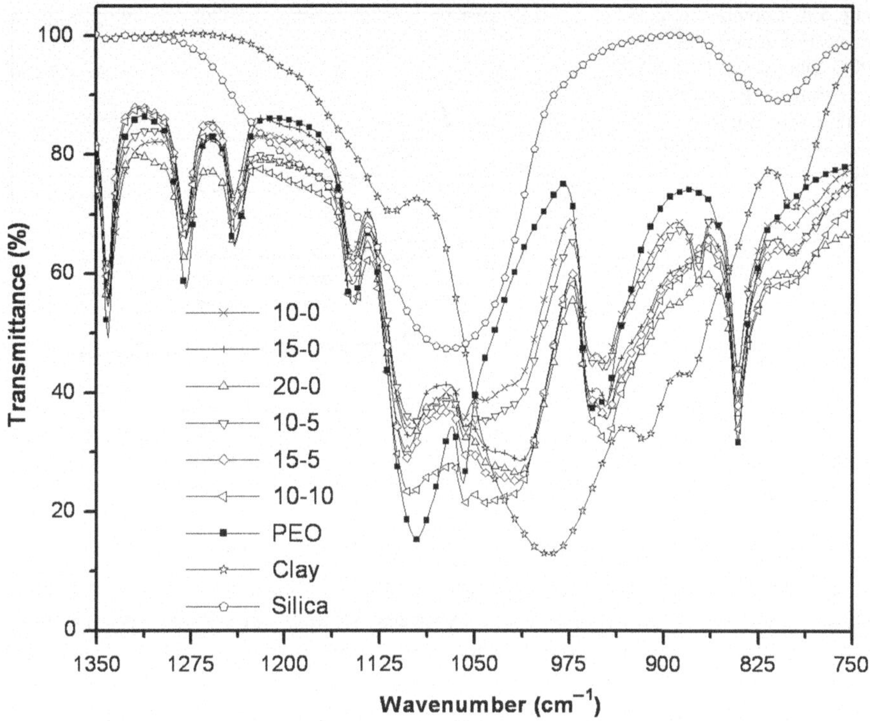

Fig. 5.3 ATR FT-IR spectroscopy of pure PEO, pure clay (Na^+MMT), pure fumed silica and PEO nanohybrids in the region of 1,350–750 cm^{-1} (Reproduced with permission from Ref. [56]. Copyright 2011 Elsevier Ltd)

process [35, 36, 62]. It was previously shown from SAXS experiments that clay minerals would exist in the form of a tactoid in the PEO-clay blend, and these tactoids are stacked in the thickness direction of the polymer film where the PEO lamellae is inserted between the parallel tactoids [35]. Previously, a schematic diagram about an interphase PEO layer which exists between silicate layers and PEO lamellar crystals was proposed, and it was speculated that this interphase layer has a finite size and the secondary PEO crystallization may take place in this interphase [36]. Lorentz-corrected SAXS profiles of PEO/clay and PEO/clay/silica hybrids are shown in Fig. 5.4. The second and third order diffraction peaks are the higher order peaks which confirm the lamellar morphology of the PEO crystalline structure and also show that the crystalline lamella is highly ordered in the PEO material. SAXS results on PEO/clay nanocomposites showed that free montmorillonite layers coexist with open aggregates and tactoids with different features depending on the filler proportion and preparation method [62]. In PEO/clay/silica nanohybrids of Fig. 5.4, as the fumed silica is added and the total concentration of nanoparticles exceeds 15 wt.%, the second order diffraction peaks in PEO nanohybrids gradually disappears, and the presence of fumed silica in addition to

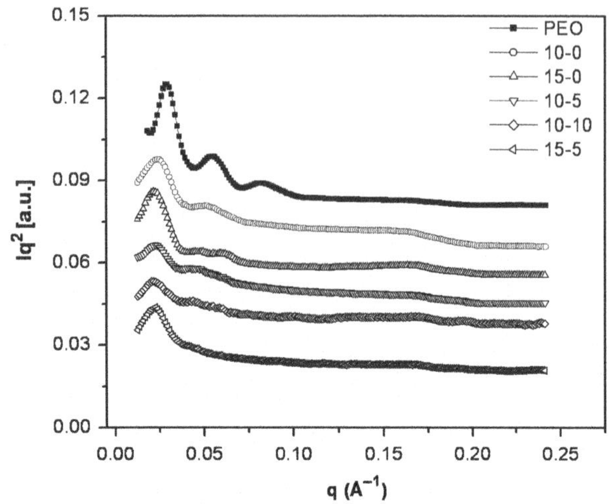

Fig. 5.4 Lorentz corrected intensities of pure PEO and PEO nanohybrids from SAXS experiments (Reproduced with permission from Ref. [56]. Copyright 2011 Elsevier Ltd)

silicate layers clearly disrupts the packing of PEO crystalline lamella. It was also shown that the presence of silica particles in the silica-poly(oxyethylene)-montmorillonite hybrid material hinders the uniformity of parallel staking of silicate layers [62]. In Fig. 5.4, the first order diffraction peak shifts to smaller q values as a function of clay and fumed silica addition, verifying the increase of long period of the crystalline lamella [56]. In PEO/poly(vinyl acetate) blends, it was previously shown that the long period slightly increases with increasing poly(vinyl acetate) content, and this result was associated with the increase of crystalline domain or the interlamellar amorphous layer thickness [71].

The morphology (degree of intercalation and the aspect ratio of clay particles) in polymer nanocomposites can be closely related to the increase of modulus and enhancement of mechanical properties [72]. Figure. 5.5 shows scanning electron microscopy (SEM) images of a PEO/clay/silica hybrid containing 10 wt.% clay and 5 wt.% silica before any melt crystallization. As clearly shown in Fig. 5.5a, the segregation of clay and silica aggregates in the interspherulitic region (amorphous region), and the uniform distribution of nanosilica aggregates (1–2 μm) inside spherulites are observed. Moreover, the overall density of nanoparticle aggregates in the amorphous region is much higher compared to that inside spherulites. Previous TEM studies on intercalated polypropylene nanocomposites reported that higher density of clay particles in the interspherulitic region compared to the density inside spherulites could lead to higher modulus values [73]. Since modulus is determined by the stiffness of intercrystalline regions, the segregation of silica clusters in the amorphous region which act as physical cross-link sites via strong H-bonds could increase the modulus more than that of the system containing only 15 wt.% clay. In agreement with previous experimental results [73], the excess silica and clay nanoparticles in the interspherulitic region might be due to the result of segregation of the clay and

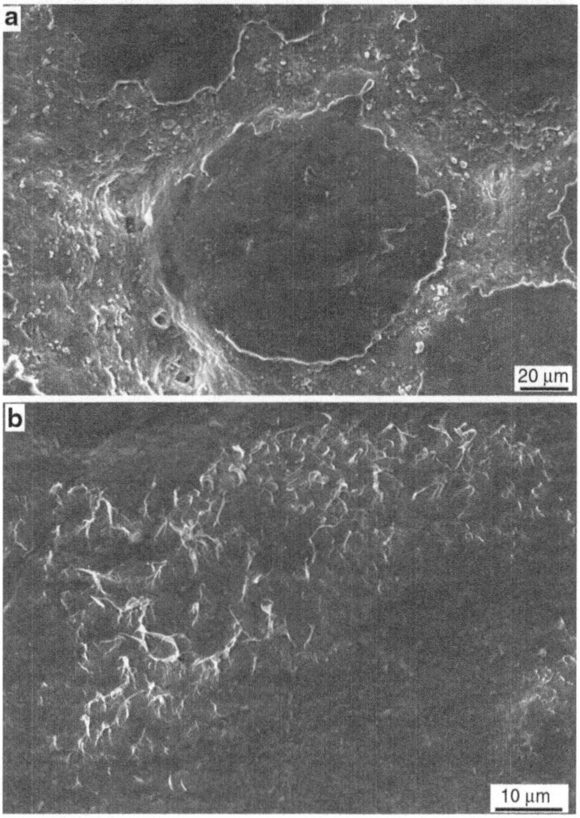

Fig. 5.5 (**a**) SEM of nanohybrid with 10 wt.% clay and 5 wt.% silica showing the presence of particle aggregates both inside and outside of a spherulite, (**b**) SEM of nanohybrid with 10 wt.% clay and 5 wt.% silica showing the presence of edge-to-face aggregation of intercalated clay nanoparticles which forms a polymer-clay network structure (Reproduced with permission from Ref. [56]. Copyright 2011 Elsevier Ltd)

silica nanoparticles during the PEO crystalline growth. A clay-polymer network structure via the edge-to-face aggregation of intercalated clay particles is shown in Fig. 5.5b. This type of polymer-clay network structure was previously observed in Nafion-clay studies [74–76].

3 Performance Characteristics of PEO Nanohybrids

3.1 Mechanical Properties

Previous dynamic mechanical analysis (DMA) on PEO/clay nanocomposites [31, 35] revealed that the storage modulus, E' of nanohybrids is considerably larger compared to that of pure PEO at all temperatures. Ogata et al. showed that E' at a given temperature above T_g increases with the increasing clay content. The DMA of PEO/clay and PEO/clay/silica nanohybrids is shown in Fig. 5.6. The storage modulus results of Fig. 5.6a agree well with those of previous PEO/clay nanocomposites. These results imply that layered silicates restrict the segmental

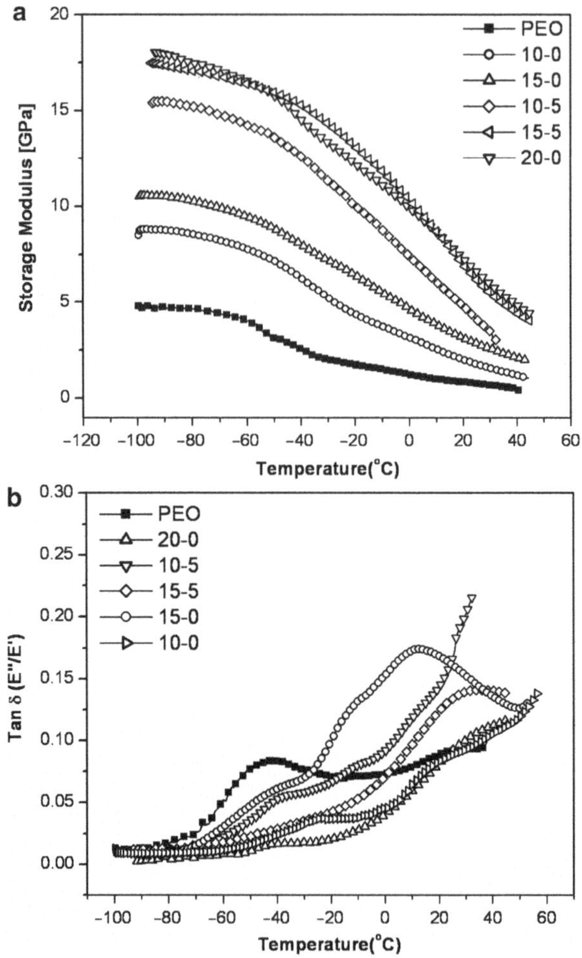

Fig. 5.6 (a) DMA of pure PEO and PEO nanohybrids. Storage modulus as a function of temperature, (b) DMA of pure PEO and PEO nanohybrids. Tan δ as a function of temperature (Reproduced with permission from Ref. [56]. Copyright 2011 Elsevier Ltd)

motions of the PEO amorphous chains as a function of clay concentration in the hybrids. In the glass transition region of PEO (Fig. 5.6b), *tan δ* peaks of the nanocomposites are slightly shifted to higher temperatures due to different degrees of PEO-clay and PEO-fumed silica interactions which restrict the mobility of polymer chains. It was previously shown that a slight shift in the *tan δ* peak towards higher temperatures can be attributed to the confinement of polymer chains as a result of intercalation inside clay galleries [31]. In Fig. 5.6b, the PEO nanohybrid containing 15 wt.% clay has a second relaxation mode that occurs at ~10 °C. This second relaxation might arise due to the coordination of Na^+ ions of clay with PEO molecules. In previous DMA data of PEO/clay nanocomposites [35], the same type of second relaxation is clearly visible for the PEO/clay sample containing 15 wt.% clay, but at a much higher temperature that is around 30 °C.

Similarly, in Nafion/clay studies [74–76], a partial ion exchange reaction between Nafion and Na+ MMT leading to the formation of the Na+ -form of Nafion was shown to be the main reason for the profound displacement and suppression of the α relaxation *tan* δ peak. The PEO/clay/silica nanohybrid containing 10 wt.% clay and 5 wt.% silica (Sample 10–5) has substantially higher storage modulus within the entire temperature range compared to the 15 wt.% clay system (Sample 15–0). This result clearly shows that the presence of fumed silica introduces extra reinforcement into the PEO/clay/silica system. It was previously shown that strong polymer-particle interactions such as H-bonding, etc. result in a physical polymer-particle network, accompanied with the strongest reinforcement in bare silica filled nanocomposites [48, 49]. The adsorption mechanism between PEO chains and silica particles is through H-bonding of ether oxygens of PEO with the surface silanols of fumed silica [77]. Polymer-particle physical network can be formed due to strong polymer-particle interactions and small particle sizes at about 2 vol.% silica loadings. For this low filler networking threshold, a mechanism was proposed such that silica nanoparticles are surrounded by polymer shells, first and then they are strongly bridged by PEO molecules leading to a temporary polymer-particle network [49].

In comparison with other thermoplastic polymers, there are fewer studies [22, 29] on the mechanical properties of PEO nanocomposites since they are not used for high-strength applications. It was previously shown that tensile strength initially increases then reduces, however elastic modulus of PEO films continuously increases as a function of clay loading [24, 31]. It was also previously shown that the elongation at break continuosly decreases and PEO films become more brittle with clay loading [31]. Tensile strength and elastic modulus values of PEO/clay/silica hybrids that are determined from stress–strain plots are presented in Fig. 5.7. The addition of 5 wt.% fumed silica to PEO/clay nanocomposites containing either 10 wt.% or 15 wt.% clay clearly increases the elastic modulus more than 5 wt.% clay addition to the same PEO/clay nanocomposites. Thus, this makes PEO/clay/silica nanocomposites much stiffer compared to PEO/clay nanocomposites. Previous tensile tests on PEO/clay nanocomposites showed that tensile strength gains a maximum value of 13.7 MPa at around 12.5 wt% of clay, and its value decreases sharply up to 20 wt% of clay loading, and above 20 wt% of clay loading, the nanocomposite sample show almost the same strength as that of the unmodified PEO [31]. These results in accordance with the results of Fig. 5.7 indicate that incorporation of layered silicates up to an optimum level resulted in a considerable reinforcing effect. In agreement with previous tensile studies of PEO/clay nanocomposites [24, 31], it is evident from Fig. 5.7 that both tensile strength and elastic modulus increases as the clay amount increases in the hybrids. However, as fumed silica is added to PEO/clay hybrids, both tensile strength and elastic modulus first decrease and reaches to a plateau region. Consistent with DMA measurements, the tensile tests at room temperature show much higher values of elastic modulus for PEO/clay/silica nanohybrids, while the elongation at break systematically decreases with clay and fumed silica loadings, indicating a more brittle response of the hybrids [56].

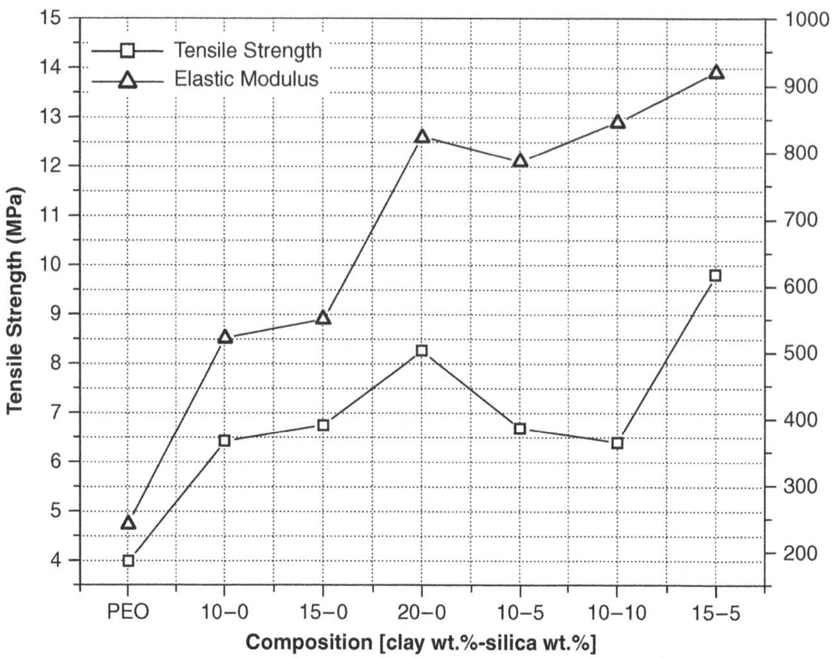

Fig. 5.7 The effect of clay and fumed silica loading on the tensile stress and elastic modulus values of PEO nanohybrids (Reproduced with permission from Ref. [56]. Copyright 2011 Elsevier Ltd)

3.2 Thermal Properties

Previous thermal studies on PEO/clay nanocomposites reported that a decrease in PEO crystallinity in case of nanocomposite, was confirmed by a decrease in the heat of melting and spherulite size [30, 31, 37]. Crystallization of PEO is found to be inhibited with the addition of clay, exhibiting a decrease of spherulite growth rate and crystallization temperature. However, the overall crystallization rate of PEO increases with clay loading as a result of extra nucleation sites in the bulk PEO matrix [35, 37]. Previous differential scanning calorimetry (DSC) studies on PEO/clay nanocomposites revealed that the melting point and heat of crystallisation initially increases at lower concentration up to 2.5 % of clay loading and decreases with further increase in clay concentration [31]. On the other hand, for PEO/silica nanocomposites, the inhibition of crystallization and the tendency for chain reorganization after melting were least for the hydrophobic silica and increased with increasing specific surface area for the hydrophilic silica [44]. Thus, hydrophilic silica is a much more effective nanoparticle in the inhibition of crystallization compared to hydrophobic silica. DSC studies also showed that the interaction of surface silanols of hydrophilic silica decreased both the melt entropy and crystal size and perfection [44]. PEO hybrids with different clay and fumed silica

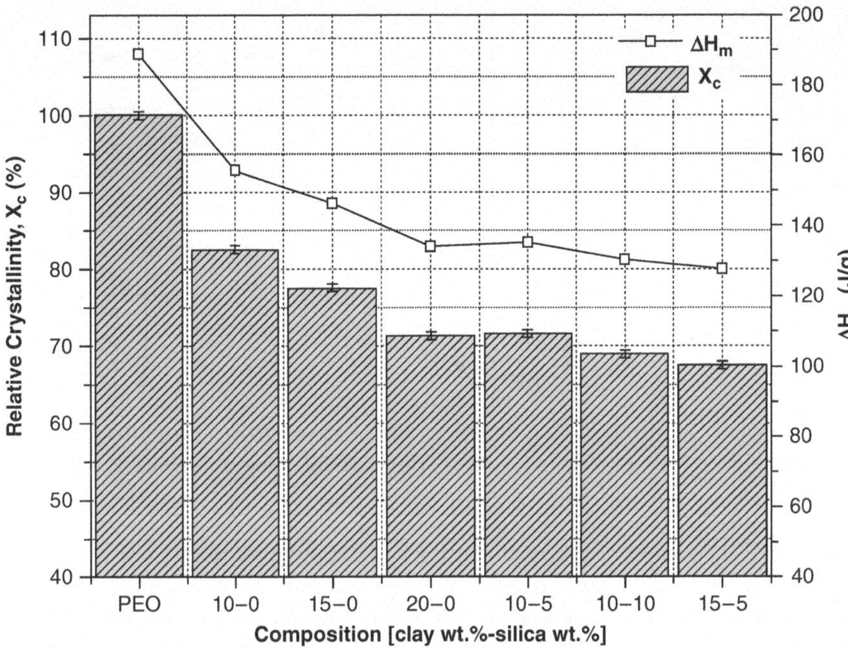

Fig. 5.8 X_c and ΔH_m of pure PEO and PEO nanohybrids as a function of sample composition (Reproduced with permission from Ref. [56]. Copyright 2011 Elsevier Ltd)

concentrations were subjected to (DSC) analysis, and the relative crystallinity (X_c) and enthalpy of melting (ΔH_m) are compared in Fig. 5.8. The addition of 5 wt.% fumed silica to PEO/clay system containing either 10 or 15 wt.% clay reduces ΔH_m and X_c more than the addition of 5 wt.% clay to the same PEO/clay systems. The decrease of X_c with the addition of clay and fumed silica can be explained by two possible reasons: The clay acts as a blockade for the crystalline growth front and thus decreases the rate of crystallization [31, 37]. PEO interacts more strongly with hydrophilic fumed silica as compared to hydrophobic silica because of the hydrogen bonding of ether oxygens of PEO with acidic silanols, preventing chain mobility and crystallization [44]. The strong H-bonding interactions between fumed silica and PEO chains could form a particle-polymer network structure thus effectively reduces the number of PEO chains that form the crystallization regions in PEO/clay/silica hybrids. For the crystallization behaviour of PEO/clay nanocomposites, Chen et al. [40] reported that crystallinity initially increases with incorporation of clay and decreases with further addition of clay due to geometric obstacles created by randomly oriented layered silicates. On the other hand, Loyens et al. [23] reported reduction of crystallinity even at clay concentration less than 1 wt%. For the PEO/clay/silica system in Fig. 5.8, relative crystallinity and the enthalpy of melting reduce systematically up to 20 wt.% clay loadings to pure PEO, then their values reach to a plateau region as a function of fumed silica addition to PEO/clay nanocomposites.

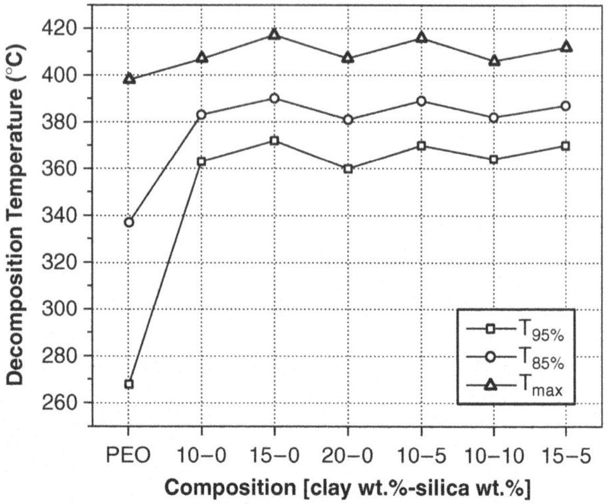

Fig. 5.9 Decomposition temperatures of pure PEO and PEO nanohybrids (Reproduced with permission from Ref. [56]. Copyright 2011 Elsevier Ltd)

The well-dispersion of clay nanoparticles in the polymer matrix is expected to improve mechanical strength, thermal stability and molecular barrier of the polymer which are desirable properties for solid polymer electrolyte applications [30–32]. Loyens et al. compared the thermal stability and decomposition temperatures of PEO nanocomposites with unmodified clay (Cloisite Na^+) and organically modified clay, and found out that increasing the Cloisite Na^+ content induces a decrease of the decomposition temperature. In contrast, organically modified clay nanocomposites show an overall trend of increasing decomposition temperatures with silicate concentration [30]. The positive influence of clay addition on the thermal stability and decomposition temperature is often assigned to the increased pathway for volatile decomposition products [78].

The incorporation of hydrophilic fumed silica into PEO is also beneficial in improving the thermal stability and mechanical properties due to H-bonding interactions between ether oxygens of PEO with surface silanols of fumed silica [26, 41, 46]. Previous XRD, ESR (Electron Spin Resonance) and TGA (Termogravimetric Analysis) measurements of PEO/layered fluoromica clay showed that interactions between PEO segments are negligible and cooperative segmental motions are reduced, and the H-Bonding is hindered in the clay galleries [79]. Enhancement of thermal stability of PEO/clay nanocomposites with respect to the pure polymer has been reported in the literature [56, 57, 80] and this delayed thermal degradation may be caused from favorable interactions and stabilization of PEO chains with layered silicates. TGA of PEO/clay/silica hybrids is shown in Fig. 5.9. The values of decomposition temperatures increase with the addition of 5 wt.% fumed silica to a PEO/clay nanocomposite containing 10 wt.% clay. Similar to the findings of Loyens et al. [30], further addition of 5 wt.% fumed silica to the sample containing 10 wt.% clay decreases the decomposition temperatures. TGA experiments of PEO/Cloisite Na^+ and PEO/laponite nanocomposites showed that Cloisite Na^+ seems to provide a better barrier at silicate concentrations less than 2.5 wt.% compared to laponite since the lower aspect ratio of laponite is less effective in providing a good barrier for the diffusion of decomposed molecules [30, 81]. In PEO/clay nanocomposites, at lower clay loadings, the organically modified clays display the strongest increase of the decomposition temperature since the presence of the organic modifier between the silicate layers may provide an additional hindrance for the diffusion of volatile decomposition products [81]. In PEO/clay/silica hybrids, the system containing 15 wt.% clay and 5 wt.% silica has a higher thermal stability and lower crystallinity compared to the system containing 20 wt.% clay due to H-Bonding interactions between fumed silica nanoparticles and PEO molecules. The improvement of thermal stability and the decrease in relative crystallinity of PEO/clay/silica hybrids compared to PEO/clay nanocomposites show that the addition of 5 wt. % fumed silica to a PEO/clay nanocomposite with 10 or 15 wt. % clay content is more beneficial than the addition of 5 wt.% clay to the same PEO/clay nanocomposites.

4 Conclusions

Clearly, PEO/clay/silica nanohybrids show significant property improvements compared to PEO/clay nanocomposites and pure PEO. The system containing 10 wt.% clay and 5 wt.% silica has substantially higher modulus and much lower crystallinity compared to the 15 wt.% clay system. The reduction of crystallinity in PEO/clay/silica hybrids is due to the fact that the presence of fumed silica in addition to silicate layers clearly disrupts the stacking of PEO crystalline lamella, and strong H-bonding interactions between fumed silica and PEO chains reduce the number of crystallizing PEO chains. Higher density of silica aggregates in the amorphous region compared to that inside spherulites, and the formation of physical cross-link sites by fumed silica via strong H-bonds were shown to be the main reasons for the reinforcement effect in PEO/clay/silica hybrids. Lastly, PEO/clay/silica hybrids display good thermal stability and are much stiffer compared to pure PEO and PEO/clay nanocomposites. Since, all of these properties are very important in the performance of solid polymer electrolytes for rechargeable polymer lithium battery applications, these results can be effectively used in the synthesis of PEO electrolytes with improved properties.

For future studies, simultaneous incorporation of clay and silica nanoparticles surface-modified with different functional groups or other types of nanoparticles with different chemical nature in a PEO matrix can be performed to obtain enhanced physical and mechanical properties. It will be also interesting to systematically prepare and characterize polymer electrolytes from these different binary nanoparticle combinations in order to see the outcome on the performance of solid polymer electrolytes. Furthermore, the simultaneous use of silica and clay or different types of nanoparticles in a polymer matrix which is chemically or physically compatible with the surface of nanoparticles might also improve the physical and mechanical properties of polymer nanocomposites. Thus, the concept of using binary mixtures of nanoparticles in a polymer matrix can be systematically extended to other types of polymers, and consequently to different useful applications.

Acknowledgments This work was supported by Ondokuz Mayis University under Grant No. PYO.MUH.1901.10.007.

References

1. Gleiter H (1992) Nanostructured materials. Adv Mater 4:474
2. Ziolo RF, Giannelis EP, Weinstein BA, O'Horo MP, Ganguly BN, Mehrotra V, Russell MW, Huffman DR (1992) Matrix-mediated synthesis of nanocrystalline ggr-Fe_2O_3: A new optically transparent magnetic material. Science 257:219
3. LeBaron PC, Wang Z, Pinnavaia TJ (1999) Polymer-layered silicate nanocomposites: an overview. Appl Clay Sci 15:11
4. Okada A, Kawasumi M, Usuki A, Kojima Y, Kurauchi T, Kamigaito O (1990) Nylon-6 clay hybrid. Mater Res Soc Symp Proc 171:45
5. Giannelis EP, Krishnamoorti R, Manias E (1999) Polymer-silicate nanocomposites: model systems for confined polymers and polymer brushes. Adv Polym Sci 138:107
6. Gilman JW (1999) Flammability and thermal stability studies of polymer layered-silicate (clay)nanocomposites. Appl Clay Sci 15:31
7. Messersmith PB, Giannelis EP (1995) Synthesis and barrier properties of poly (ε-caprolactone)-layered silicate nanocomposites. J Polym Sci Part A Polym Chem 33:1047
8. Gilman JW, Jackson CL, Morgan AB, Harris R, Manias E, Giannelis EP, Wuthenow M, Hilton D, Phillips H (2000) Flammability properties of polymer-layered-silicate nanocomposites. Polypropylene and polystyrene nanocomposites. Chem Mater 12:1866
9. Giannelis EP (1996) Polymer layered silicate nanocomposites. Adv Mater 8:29
10. Fisher H, Gielgens L, Koster T (1998) Nanocomposites from polymers and layered minerals, TNO-TPD Report
11. Usuki A, Kojima Y, Kawasumi M, Okada A, Fukushima Y, Kurauchi T, Kamigaito O (1993) Synthesis of nylon 6-clay hybrid. J Mater Res 8:1179
12. Lan T, Pinnavaia TJ (1994) Clay-reinforced epoxy nanocomposites. Chem Mater 6:2216
13. Chang JH, An YU, Cho DH, Giannelis EP (2003) Poly (butylene terephthalate)/organoclay nanocomposites prepared by in situ interlayer polymerization and its fiber (II). Polymer 44:3715
14. Ruiz-Hitzky E, Aranda P (1990) Polymer-salt intercalation complexes in layer silicates. Adv Mater 2:545
15. Ruiz-Hitzky E, Aranda P, Casal B, Galvan JC (1995) Nanocomposite materials with controlled ion mobility. Adv Mater 7:180
16. Abraham KM, Jiang Z, Carroll B (1997) Highly conductive PEO-like polymer electrolytes. Chem Mater 9:1978
17. Labreche C, Levesque I, Prud'homme J (1996) An appraisal of tetraethylsulfamide as plasticizer for poly (ethylene oxide)-LiN (CF_3SO_2) 2 rubbery electrolytes. Macromolecules 29:7795
18. Lightfoot P, Mehta MA, Bruce PG (1993) Crystal structure of the polymer electrolyte poly (ethylene oxide) 3: $LiCF_3SO_3$. Science 262:883
19. Do C, Lunkenheimer P, Diddens D, Gotz M, Weiss M, Loidl A, Sun XG, Allgaier J, Ohl M (2013) Li^+ transport in poly(ethylene oxide) based electrolytes: neutron scattering, dielectric spectroscopy, and molecular dynamics simulations. Phys Rev Lett 111:018301
20. Munshi MZA, Owens BB (1988) Performance of polymer electrolyte cells at +25 to +100 °C. Solid State Ionics 26:41
21. Huq R, Farrington GC, Koksbang R, Tonder PE (1992) Influence of plasticizers on the electrochemical and chemical stability of a Li^+ polymer electrolyte. Solid State Ionics 57:277
22. Croce F, Appetecchi GB, Persi L, Scrosati B (1998) Nanocomposite polymer electrolytes for lithium batteries. Nature 394:456

23. Loyens W, Jannasch P, Maurer FHJ (2005) Melt-compounded salt-containing poly (ethylene oxide)/clay nanocomposites for polymer electrolyte membranes. Polymer 46:7334
24. Abraham TN, Siengchin S, Ratna D, Karger-Kocsis J (2010) Effect of modified layered silicates on the confined crystalline morphology and thermomechanical properties of poly (ethylene oxide) nanocomposites. J Appl Polym Sci 118:1297
25. Zhang H, Maitra P, Wunder SL (2008) Preparation and characterization of composite electrolytes based on PEO (375)-grafted fumed silica. Solid State Ionics 178:1975
26. Xie J, Duan RG, Han Y, Kerr JB (2004) Morphological, rheological and electrochemical studies of poly (ethylene oxide) electrolytes containing fumed silica nanoparticles. Solid State Ionics 175:755
27. Tang CY, Hackenberg K, Fu Q, Ajayan PM, Ardebili H (2012) High ion conducting polymer nanocomposite electrolytes using hybrid nanofillers. Nano Letters 12:1152
28. Shen Z, Simon GP, Cheng YB (2002) Comparison of solution intercalation and melt intercalation of polymer–clay nanocomposites. Polymer 43:4251
29. Chen B, Evans JRG (2004) Preferential intercalation in polymer-clay nanocomposites. J Phys Chem B 108:14986
30. Loyens W, Jannasch P, Maurer FHJ (2005) Melt-compounded salt-containing poly (ethylene oxide)/clay nanocomposites for polymer electrolyte membranes. Polymer 46:903
31. Ratna D, Divekar S, Samui AB, Chakraborty BC, Banthia AK (2006) Poly (ethylene oxide)/ clay nanocomposite: thermomechanical properties and morphology. Polymer 47:4068
32. Aranda P, Ruiz-Hitzky E (1992) Poly (ethylene oxide)-silicate intercalation materials. Chem Mater 4:1395
33. Harris DJ, Bonagamba TJ, Schmidt-Rohr K (1999) Conformation of poly (ethylene oxide) intercalated in clay and MoS_2 studied by two-dimensional double-quantum NMR. Macromolecules 32:6718
34. Kwiatkowski J, Whittaker AK (2001) Molecular motion in nanocomposites of poly (ethylene oxide) and montmorillonite. J Polym Sci Part B Polym Phys 39:1678
35. Ogata N, Kawakage S, Ogihara T (1997) Structure and thermal/mechanical properties of poly (ethylene oxide)-clay mineral blends. Polymer 38:5115
36. Sun L, Ertel EA, Zhu L, Hsiao BS, Avila-Orta CA, Sics I (2005) Reversible de-intercalation and intercalation induced by polymer crystallization and melting in a poly (ethylene oxide)/ organoclay nanocomposite. Langmuir 21:5672
37. Strawhecker K, Manias E (2003) Crystallization behavior of poly (ethylene oxide) in the presence of Na+ montmorillonite fillers. Chem Mater 15:844
38. Harrats C, Groeninckx G (2008) Features, questions and future challenges in layered silicates clay nanocomposites with semicrystalline polymer matrices. Macromol Rapid Commun 29:14
39. Homminga D, Goderis B, Dolbnya I, Reynaers H, Groeninckx G (2005) Crystallization behavior of polymer/montmorillonite nanocomposites. Part I. Intercalated poly (ethylene oxide)/montmorillonite nanocomposites. Polymer 46:11359
40. Chen HW, Chang FC (2001) The novel polymer electrolyte nanocomposite composed of poly (ethylene oxide), lithium triflate and mineral clay. Polymer 42:9763
41. Maitra P, Ding J, Huang H, Wunder SL (2003) Poly (ethylene oxide) silananted nanosize fumed silica: DSC and TGA characterization of the surface. Langmuir 19:8994
42. Reddy CVS, Wu GP, Zhao CX, Jin W, Zhu QY, Chen W, Mho S (2007) Mesoporous silica (MCM-41) effect on (PEO + LiAsF6) solid polymer electrolyte. Curr Appl Phys 7:655
43. Tominaga Y, Asai S, Sumita M, Panero S, Scrosati B (2005) A novel composite polymer electrolyte: Effect of mesoporous SiO_2 on ionic conduction in poly(ethylene oxide)–$LiCF_3SO_3$ complex. J Power Sour 146:402
44. Ding J, Maitra P, Wunder SL (2003) Characterization of the interaction of poly (ethylene oxide) with nanosize fumed silica: Surface effects on crystallization. J Polym Sci Part B Polym Phys 41:1978

45. Kweon JO, Noh ST (2001) Thermal, thermomechanical, and electrochemical characterization of the organic–inorganic hybrids poly(ethylene oxide) (PEO)–silica and PEO–silica–$LiClO_4$. J Appl Polym Sci 81:2471
46. Liu Y, Lee JY, Hong L (2002) Functionalized SiO_2 in poly(ethylene oxide)-based polymer electrolytes. J Power Sour 109:507
47. Raghavan SR, Walls HJ, Khan SA (2000) Rheology of silica dispersions in organic liquids: new evidence for solvation forces dictated by hydrogen bonding. Langmuir 16:7920
48. Aranguren MI, Mora E, DeGroot JV, Macosko CW (1992) Effect of reinforcing fillers on the rheology of polymer melts. J Rheol 36:1165
49. Zhang Q, Archer LA (2002) Poly (ethylene oxide)/silica nanocomposites: structure and rheology. Langmuir 18:10435
50. Anderson BJ, Zukoski CF (2009) Rheology and microstructure of entangled polymer nanocomposite melts. Macromolecules 42:8370
51. Hooper JB, Schweizer KS (2006) Theory of phase separation in polymer nanocomposites. Macromolecules 39:5133
52. Khalatur PG, Zherenkova LV, Khokhlov AR (1997) Aggregation of colloidal particles induced by polymer chains: The RISM integral equation theory. Physica A 247:205
53. Hooper JB, Schweizer KS (2005) Contact aggregation, bridging, and steric stabilization in dense polymer-particle mixtures. Macromolecules 38:8858
54. Allegra G, Raos G, Vacatello M (2008) Theories and simulations of polymer-based nanocomposites: from chain statistics to reinforcement. Prog Polym Sci 33:683
55. Salaniwal S, Kumar SK, Douglas JF (2002) Amorphous solidification in polymer-platelet nanocomposites. Phys Rev Lett 89:258301
56. Burgaz E (2011) Poly (ethylene oxide)/clay/silica nanocomposites: morphology and thermomechanical properties. Polymer 52:5118
57. Wu JH, Lerner MM (1993) Structural, thermal, and electrical characterization of layered nanocomposites derived from Na-montmorillonite and polyethers. Chem Mater 5:835
58. Schmidt G, Malwitz MM (2003) Properties of polymer–nanoparticle composites. Curr Opin Colloid Interface Sci 8:103
59. Zeng QH, Yu AB, Lu GQ, Paul DR (2005) Clay-based polymer nanocomposites: research and commercial development. J Nanosci Nanotech 5:1574
60. Ruiz-Hitzky ER, Darder M, Aranda P (2005) Functional biopolymer nanocomposites based on layered solids. J Mater Chem 15:3650
61. Lim SK, Kim JW, Chin I, Kwon YK, Choi HJ (2002) Preparation and interaction characteristics of organically modified montmorillonite nanocomposite with miscible polymer blend of poly(ethylene oxide) and poly(methyl methacrylate). Chem Mater 14:1989
62. Hikosaka MY, Pulcinelli SH, Santilli CV, Dahmouche K, Craievich AF (2006) Montmorillonite (MMT) effect on the structure of poly (oxyethylene)(PEO)–MMT nanocomposites and silica–PEO–MMT hybrid materials. J Non-Crystalline Solids 352:3705
63. Theng BKG (1982) Clay polymer interactions - summary and perspectives. Clays Clay Miner 30:1
64. Bujdak J, Hackett E, Giannelis EP (2000) Effect of layer charge on the intercalation of poly (ethylene oxide) in layered silicates: implications on nanocomposite polymer electrolytes. Chem Mater 12:2168
65. Parfitt RL, Greenland DJ (1970) The adsorption of poly (ethylene glycols) on clay mineral. Clay Miner 8:305
66. Aranda P, Ruiz-Hitzky E (1999) Poly(ethylene oxide)/$NH4^+$-smectite nanocomposites. Appl Clay Sci 15:119
67. Chen C, Tolle TB (2004) Fully exfoliated layered silicate epoxy nanocomposites. J Polym Sci Polym Phys 42:3891
68. Reddy MJ, Chu PP, Rao UVS (2006) Study of multiple interactions in mesoporous composite PEO electrolytes. J Power Sources 158:614

69. Patel HA, Somani RS, Bajaj HC, Jasra RV (2007) Preparation and characterization of phosphonium montmorillonite with enhanced thermal stability. Appl Clay Sci 35:194
70. Snyder MQ, McCool BA, DiCarlo J, Tripp CP, DeSisto WJ (2006) An infrared study of the surface chemistry of titanium nitride atomic layer deposition on silica from $TiCl_4$ and NH_3. Thin Solid Films 514:97
71. Fragiadakis D, Runt J (2010) Microstructure and dynamics of semicrystalline poly (ethylene oxide)– poly (vinyl acetate) blends. Macromolecules 43:1028
72. Nam PH, Maiti P, Okamoto M, Kotaka T, Hasegawa N, Usuki A (2001) A hierarchical structure and properties of intercalated polypropylene/clay nanocomposites. Polymer 42:9633
73. Maiti P, Nam PH, Okamoto M, Hasegawa N, Usuki A (2002) Influence of crystallization on intercalation, morphology, and mechanical properties of polypropylene/clay nanocomposites. Macromolecules 35:2042
74. Alonso RH, Estevez L, Lian H, Kelarakis A, Giannelis EP (2009) Nafion–clay nanocomposite membranes: morphology and properties. Polymer 50:2402
75. Burgaz E, Lian H, Alonso RH, Estevez L, Kelarakis A, Giannelis EP (2009) Nafion–clay hybrids with a network structure. Polymer 50:2384
76. Kelarakis A, Alonso RH, Lian H, Burgaz E, Estevez L, Giannelis EP (2010) In: Wang Q, Zhu L (eds) ACS symposium series, vol 1034, Functional polymer nanocomposites for energy storage and conversion. American Chemical Society, Washington, DC, p 171
77. Rubio J, Kitchener JA (1976) The mechanism of adsorption of poly (ethylene oxide) flocculant on silica. J Colloid Interface Sci 57:132
78. Alexandre M, Dubois P (2000) Polymer-layered silicate nanocomposites: preparation, properties and uses of a new class of materials. Mater Sci Eng 28:1
79. Miwa Y, Drews AR, Schlick S (2009) Reduced cooperative segmental motion and hindered hydrogen bonding of poly(ethylene oxide) intercalated in layered silicates: deductions revealed by simulating ESR spectra of spin-labeled PEO and by XRD, DSC, and TGA measurements. Macromolecules 42:8907
80. Sehaqui H, Kochumalayil J, Liu A, Zimmermann T, Berglund LA (2013) Multifunctional nanoclay hybrids of high toughness, thermal, and barrier performances. ACS Appl Mater Inter 5:7613
81. Loyens W, Jannasch P, Maurer FHJ (2005) Poly(ethylene oxide)/laponite nanocomposites via melt-compounding: effect of clay modification and matrix molar mass. Polymer 46:915

Recent Developments in Cellulose and Cellulose Derivatives/Clay Nanocomposites

6

Maria do Carmo Gonçalves and Marcia Maria Favaro Ferrarezi

Contents

Abstract

This chapter describes advances in the preparation, characterization, and application of polymer nanocomposites made up of cellulose or cellulose derivatives as matrixes and layered silicates, especially montmorillonite, as nano-sized fillers. Cellulose-based nanocomposites have been shown to be interesting polymer systems, which could find innumerous applications, such as films and membranes, textile fibers, and tissue scaffolds. This review discusses the different strategies of preparing these eco-friendly nanocomposites, including intercalation methods and filler surface modification, as well as the most recent available experimental results. Changes of cellulose and cellulose derivative properties, such as mechanical reinforcement, barrier effect, fire retardation, water absorption, and thermal stabilization, are described. This chapter concludes with a brief discussion of what still should be overcome for cellulose

M.C. Gonçalves (✉) • M.M.F. Ferrarezi
Institute of Chemistry, University of Campinas (UNICAMP), Campinas, SP, Brazil
e-mail: maria@iqm.unicamp.br

J.K. Pandey et al. (eds.), *Handbook of Polymernanocomposites. Processing, Performance and Application – Volume A: Layered Silicates*, DOI 10.1007/978-3-642-38649-7_10,

and cellulose derivative nanocomposites to be applied as nontoxic and renewable-based materials so as to expand their applications and become interesting options to replace conventional fossil resource-based polymers.

1 Introduction

The development of eco-friendly materials has been in focus in the last decade in both scientific and commercial fields, due to the awareness about the probability of increasing price and dwindling supply of petroleum-based products and the environment impact of the polymers daily used. There is a great expectation on polymeric materials that are renewable based, biodegradable, and recyclable or some combination thereof. Following this tendency, biopolymers have received considerable attention as an interesting alternative to replace conventional petroleum-based plastics. Several biopolymer definitions have been documented, and it is well accepted that biopolymers are materials that are obtained from natural sources and they often have the capacity to decompose into water, inorganic compounds, biomass, and carbon dioxide and/or methane, depending on the degradation conditions and medium, because of the action of either microorganisms or enzymes. Biopolymers can be extracted from biomass (proteins and polysaccharides), produced from renewable resources (biodegradable polyester), and also obtained by microbial production (polyhydroxyalkanoates) [1, 2]. Among the innumerous available biopolymers, polysaccharides, e.g., cellulose, starch, and chitin, are worth particular attention, as this polymer group includes the most abundant macromolecules in the biosphere, highlighting cellulose.

Cellulose is a linear macromolecule with a uniform distribution of hydroxyl groups, which are responsible for the formation of highly packed structures with high tensile strength. These numerous hydroxyl functions presented in cellulose chains makes this polymer non-fusible and poorly soluble in common solvents [3].

Chemical modifications of cellulose chains are usually applied to produce thermoplastic cellulose, by the replacement of the hydroxyl groups, for instance, by acetate or methyl functions yielding cellulose derivatives. Therefore, cellulose derivatives are often considered to be semisynthetic polymers. In fact, cellulose nitrate was the first reported semisynthetic polymer that led to the discovery of plastics and after that to the invention of synthetic fibers.

Although cellulose and majority of cellulose derivative materials present excellent optical clarity and stiffness, other properties, such as moisture sensitivity, high brittleness, and low dimensional stability under high temperature and humidity, should be enhanced. The ordinary approach to prepare polymer materials with outstanding properties is the preparation of multiphase materials, e.g., blends and composites.

The incorporation of inorganic particles into polymer matrixes has long been studied [4, 5] and the results show improved mechanical, thermal, and barrier polymer properties due to the presence of these dispersed particles. Despite this, since the studies developed by Toyota research group [6] and Vaia et al. [7], the investigation of nanoscale inorganic particles in polymer matrixes has increased, as these authors demonstrated that the very low loading of nanoparticles can give remarkable property enhancement and also that thermoplastic polymer nanocomposites can be prepared in the melt state.

Over the last 20 years, intensive research has been focused on nanocomposites based on layered silicates. The property gains attributed to the polymer-layered silicate nanocomposites are dependent on the filler nanometer size and on the high surface area of the dispersed clay associated to strong interfacial interaction between clay and polymer matrix. For example, in the case of nanocomposites made up of MMT, exfoliated MMT can reach surface areas in the range of 600 and 800 m^2/g, which are significant surface areas and which can be responsible for effective tension transfer between the composite phases during material solicitations. Indeed, improvements in properties, such as tensile strength and elastic modulus, heat distortion temperature, and gas barrier, can be reached by adding low clay fractions to polymers, without negatively affecting the optical homogeneity. Nanocomposites are also usually lighter than conventional composites due to weight filler fraction usually being less than 10 wt% [8–11]. In most recent researches, the polymer matrix has been made up of synthetic polymers, such as polyamides, polyethylene, poly(methyl methacrylate), polystyrene, or rubbers, among others [10–12].

Another important aspect to be considered is that the clays are hydrophilic in their pristine form, which permits electrostatic interactions with ionic polymers but hinders effective dispersion throughout hydrophobic polymer matrices. Substituting the cations present in the clay interlayer gallery by quaternized ammonium or phosphonium cations, preferably with long alkyl chains [8], is the most applied method to overcome this difficulty.

Cellulose and cellulose derivative nanocomposites have been obtained with innumerous nanoparticle types, such as inorganic (layered silicates [8] and hydroxyapatite [13]), metallic (silver [14]), carbon-type (carbon nanotube [15] and graphene [16]), and organic (cellulose nanocrystal [17]) nanoparticles. Among these available nanoparticles, the majority of the studies has focused on the use of layered silicates, such as montmorillonite, hectorite, and saponite, due to their availability, versatility, thermal and chemical stability, and eco-friendly characteristics.

The following topics of the chapter present advances in the preparation, characterization, and application of polymer nanocomposites made up of cellulose or cellulose derivatives and layered silicates.

2 Cellulose and Cellulose Derivative Structures

2.1 Cellulose

Cellulose was first investigated by Braconnot in 1819 [18] and by Anselme Payen in 1838 [19]. Its structure was found predominantly in plants and sea animals, but some algae, fungi, and bacteria species are able to produce

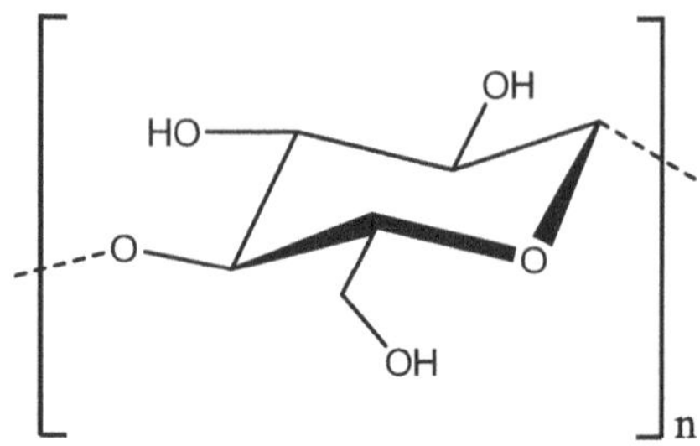

Fig. 6.1 Cellulose molecular structure

cellulose [20]. World production of natural fibers over the period of 2003 and 2005 was approximately 32 million tons [21], showing an abundant source of raw material. In 2009, the biggest producers of cellulose around the world were United States, Canada, China, and Brazil [22]. The main industrial applications of cellulose are as paper and textile fibers. Also, cellulose can be applied as water-soluble binders and adhesives, food thickeners and stabilizers, building insulation, and as biofuel source [23].

Generally, cellulose is presented in plants in the form of microfibrils, which are bundles of long cellulose chains stabilized by both inter- and intramolecular hydrogen bonds. These chain arrangements present a diameter between 3 and 30 nm and represent a packing of 30–100 cellulose chains. These microfibrils are the lowest cellulose chain agglomerates and form cellulose crystals linked by amorphous domains. These microfibrils, in plants, are immersed in a matrix made up of hemicellulose and lignin [3, 17, 24].

Formally, cellulose is a polymer formed by aldehyde sugars units, namely, D-anhydroglucopyranose ($C_6H_{11}O_5$), linked by β (1→4) linkages and assembled in groups of two, called cellobiose. Figure 6.1 represents the molecular structure of cellulose. The cellulose weight-average molecular weight is often reported in the range of 162,000–2,400,000 g mol^{-1}. The adjacent rings can form hydrogen bonds, which hinder the chain-free rotation and make cellulose insoluble in ordinary solvents. Hydronium ions from strong acids can penetrate the cellulose amorphous regions, promoting the hydrolytic cleavage of glycosidic links [3, 17, 25, 26]. Although few, the solvents used are metal complexes such as cupriethylenediamine hydroxide, aqueous salt complexes such as saturated zinc chloride and calcium thiocyanate, and nonaqueous salt complexes such as lithium chloride (LiCl)/dimethylacetamide (DMAc). However, the dissolution of cellulose often causes some degree of chain scission. A reduction in the molecular weight due to chain scission is not however likely to cause a significant reduction in the final properties due to the high cellulose molecular weight [27].

While the intramolecular interactions are responsible for cellulose stiffness, intermolecular hydrogen bonds and van der Waals interactions interfere in the chain arrangements. Cellulose stability is then obtained as a result of the formation of dense hydrogen bond networks [28].

There are different cellulose polymorph structures reported in literature, e.g., *Types I*, *II*, *II*, and *IV*, which can be exchanged during some chemical or thermal treatments. The native cellulose (natural form) is Type *I*, described by a triclinic (cellulose I_α) cell unit and a monoclinic (cellulose I_β) cell unit, both presenting same conformation of the polysaccharide chains, but distinct hydrogen link pattern. Both crystalline forms co-exist in cellulose *I*, where their ratio is dependent on the source. Also, cellulose I_α can be transformed into cellulose I_β by annealing. Cellulose *II* is obtained by mercerization or regeneration of cellulose *I*. Literature reports that Type *II* is more stable than Type *I*. Cellulose *III* is obtained from cellulose *I* or *II* by ammonia or ethylene diamine treatments and cellulose *IV* can be obtained from cellulose *III* in a glycerol medium at high temperatures [29].

Cellulose and cellulose derivatives also form liquid crystals in a variety of different supermolecular structures, such as thermotropic and lyotropic liquid crystalline phases.

2.2 Cellulose Derivatives

With the aim of obtaining materials with different properties, cellulose is converted into cellulose derivatives. Thus the industrial applications are greatly expanded from the polysaccharide itself. Take, for example, when cellulose is converted to esters, it produces polymers that are able to be processed into several forms, such as three-dimensional objects, films, and fibers. These polymers are applied in diverse areas, such as laminates, optical films, textile fibers, and coatings for pharmaceuticals and foods [30]. In terms of membrane and encapsulation applications, cellulose derivatives can provide sophisticated network structures through aggregation-induced phase separation. The most applied and studied cellulose derivatives are esters and ethers. The most common structures are shown in Fig. 6.2.

Cellulose acetate (CA), cellulose acetate propionate (CAP), and cellulose acetate butyrate (CAB) are examples of polymers produced by the esterification of cellulose materials such as cotton, recycled paper, wood cellulose, and sugarcane. Cellulose acetate is the most studied and applied cellulose ester derivative [27]. In 2008, CA global production was over 800,000 metric tons [31]. It is obtained through the reaction of cellulose chains and acetic anhydride, in which hydroxyl groups are replaced by acetate groups. Cellulose ester derivatives can be prepared with different substitution degrees (SD), which is related to the average number of acetyl groups per glucose units, and can range between 0 and 3. Also, the properties of cellulose ester derivatives depend on the degrees of polymerization (PD) and substitution. For example, low SD levels increase CA solubility in polar solvents, while high SD levels decrease its vapor permeability [30]. In relation to solution behavior, CA molecules are practically never completely molecularly dispersed in solution but rather exist as complex molecular associates, which depend on the strength and the amount of intra- and intermolecular interactions [32]. These hydrogen bonds bridge neighboring cellulose structural unit segments, thus making CA molecules particularly rigid [32–34].

On the other hand, cellulose ethers are produced by an alkylation reaction. For example, sodium carboxymethyl cellulose (CMC), the most used cellulose ether

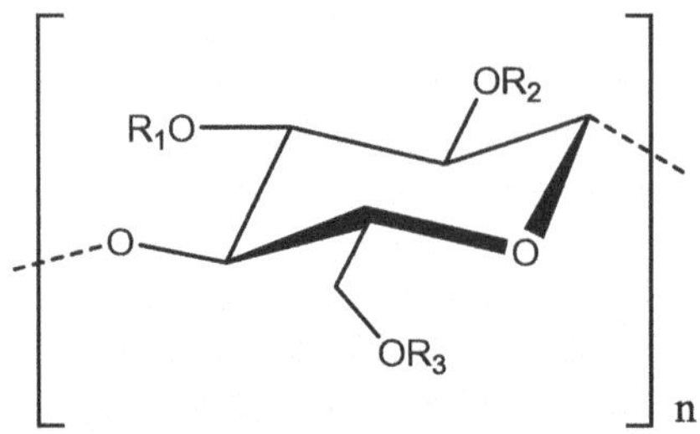

CELULLOSE ESTERS	CELULLOSE ETHERS
Cellulose acetate	Carboxymethyl cellulose
$R_1, R_2, R_3 = COCH_3$ or H	$R_1, R_2, R_3 = CH_2COOH$ or H
Cellulose acetate butyrate	Hydroxyethyl cellulose
$R_1, R_2, R_3 = COCH_3$ or $CO(CH_2)_2CH_3$ or H	$R_1, R_2, R_3 = CH_2CH_2OH$ or H
Cellulose acetate propionate	Methyl hydroxypropyl cellulose
$R_1, R_2, R_3 = COCH_3$ or $COCH_2\,CH_3$ or H	$R_1, R_2, R_3 = CH_2CH(OH)CH_3$ or CH_3 or H

Fig. 6.2 Cellulose derivative molecular structures

derivative, is obtained through the reaction between chloroacetic acid and cellulose. Other important cellulose ethers are ethylhydroxyethyl cellulose, hydroxypropyl cellulose, and methylhydroxypropyl cellulose. Cellulose ether properties also depend on PD and SD.

Substantial activities around novel cellulose derivatives can be noted despite the long existence of cellulose derivatives. These include performance improvements and innovative approaches to expand their unique characteristics for commercial applications [30].

3 Layered Silicate Nanocomposites

Hybrid materials are prepared with the purpose of combining different components in such a way that properties of the resulting material are enhanced in terms of each component alone.

Although studies of polymer/clay nanocomposites started decades ago, the preparation of cellulose and cellulose derivative nanocomposites is relatively new. Nevertheless, the interest in these recent polymer nanocomposites is also driven by the possibility of excellent physical property enhancements at low filler levels.

3.1 Cellulose/Layered Silicate Nanocomposites

Fewer studies of cellulose/layered silicate nanocomposites are found in literature [35–46] than those reported for other polymeric matrices. The most relevant investigations explore the use of nanoparticulated clay as a low-cost flame-retardant filler for the production of regenerated cellulose fibers and the combination of clay and cellulose (cellulose fibers and cellulose micro- and nanocrystals) as reinforcement fillers for polymeric matrices. In addition, cellulose is a lightweight material; thus the use of lightweight fillers is an appropriate strategy when developing cellulose nanocomposites.

White [35] produced cotton/clay nanocomposites using three different intercalation and exfoliation methods, differing in the solvent types and clay pretreatment procedures. The authors emphasized that, as cotton is a highly crystalline biomaterial and its chains do not melt before the decomposition onset, the experimental procedures followed two sequential steps: clay exfoliation in a solvent in which the polymer is soluble and then addition of the polymer. In Method 1, pristine montmorillonite (MMT) was dispersed in a 4-methylmorpholine-*N*-oxide (MMNO) aqueous solution; afterwards cotton was added to the solution. In Method 2, MMT pretreated with dodecylamine ammonium salt was dispersed in the same solvent, and then, cotton was added. In Method 3, the pretreated MMT was dispersed in a LiCl/DMAc solution, previous to the cotton addition. The results show that only Method 2 is able to produce exfoliated cotton/organophilic MMT nanocomposites, as the pretreatment favors the silicate sheet separation and consequently the polymer chain penetration. The nanocomposite degradation temperatures increase by 45 °C for nanocomposites containing montmorillonite in amounts as small as 1 wt%. This is an interesting property since flammability is one of the major hindrances of cellulosic fibers. After, Delhom et al. [36] studied the flame-retardant property of cotton/clay nanocomposites in a sequence group study. The cellulose nanocomposites were prepared by solution, using the organophilic MMT pretreated with dodecylamine ammonium salt and MMNO as a solvent (Method 2 from the White [35] study). The authors suggested that MMT addition can alter the cellulose crystalline structure. Also, the increase in the char yield is an indication of flame resistance, as char formation can prevent material from combustion. The nanocomposites show significant improvements in thermal properties, higher elastic modulus, and an increase around 80 % in the ultimate stress in relation to the pure cotton fiber.

Cerruti et al. [37] also prepared cellulose nanocomposites applying precipitation from MMNO/water solutions in the presence of trimethylstearylammonium salt modified MMT. The authors focused on the effect of the mixing time on the clay exfoliation and dispersion degree on the matrix. The nanocomposites present higher degradation temperatures than pure cellulose, attributed to the organophilic MMT ability in terms of hindered transfer of heat, oxygen, and degraded volatiles. Chemiluminescence analysis shows that MMT suspensions are able to catalyze cellulose degradation leading to chain scission and formation of carbonyl groups on the polymer backbone when the materials are submitted for a long period of time in

MMNO/water solutions. On the other hand, a prolonged mixing time between cellulose chains and clay can favor a more effective polymer diffusion into silicate layers. The results above show that cellulose, at temperatures lower than 100 °C, is less stable, whereas at higher temperatures, the hindering effect of oxygen transfer dominates, thus increasing cellulose thermo-oxidative stability.

Jang et al. [38] prepared cellulose/mica hybrid films by solution dispersion of the components in MMNO followed by solution intercalation. The authors studied the effect of two different organo-micas on polymer properties, one treated with dodecyltriphenylphosphonium ($C_{12}PPh\text{-}Cl^-$) and another with hexadecylammonium chloride ($C_{16}\text{-}Cl^-$). The authors show that each organo-mica acts more effectively on a specific property, i.e., while cellulose modulus and tensile strength are improved by the addition of C_{12}PPh-mica, barrier properties are higher in the presence of C_{16}-mica.

Recently, Mahmoudian et al. [39] prepared regenerated cellulose (RC)/organophilic montmorillonite (OMMT) nanocomposites by solution casting using the ionic liquid, 1-butyl-3-methylimidazolium chloride, as a solvent. MMT was organically modified with octadecylamine. The addition of 6 wt% OMMT improves the thermal stability and the char yield; increases the tensile strength and the elastic modulus in 12 % and 40 %, respectively; and enhances the gas-barrier properties and water absorption resistance of regenerated cellulose. The authors point out that the development of RC/OMMT nanocomposites is expected to play an important role in packaging and membrane technologies.

On the other hand, the preparation of cellulose/clay biocomposites to be used as scaffolds for tissue regeneration has been studied by Haroun et al. [40]. The biocomposites were prepared by blending gelatin with cellulose in the presence of MMT in aqueous suspensions and by cross-linking the biocomposites with glutaraldehyde or *N,N*-methylene-bisacrylamide. The use of pristine MMT changes the scaffold structural stiffness and decreases its water absorption behavior. The MMT also causes a barrier effect, inhibiting the interaction between gelatin-cellulose macromolecules and water molecules. Also, MMT improves the cytocompatibility between the scaffolds and osteoblasts.

The combination of layered silicates and cellulose fibers forms a new generation of materials to be used as reinforcement fillers in polymeric materials. Studies in literature show that some synergetic effects between clays and biofillers on the composite properties can be obtained [41]. Mariano et al. [42] prepared nanocomposites by the technique of co-coagulating natural rubber (NR) latex, cellulose II (regenerated cellulose), and pristine Na-MMT aqueous suspension. The combination of cellulose and clay causes an increase of the storage modulus of the composite, reflecting a strong interaction between clay sheets, cellulose, and rubber chains, and consequently improved mechanical properties are achieved. The authors associated these property gains to MMT presence and to the nanocrystalline character of cellulose II. Zine et al. [43] also studied rubber/cellulose II/pristine Na-MMT nanocomposites by the same co-coagulation methodology. In this case, styrene butadiene rubber (SBR) was used as the matrix and cellulose II was incorporated at 15 phr, with clay varying from 0 to 7 phr. The nanocomposites with 5 phr of clay show improved tensile properties when compared to vulcanizated SBR gum.

In the same research line of the combined use of cellulose fibers and layered silicates as polymer fillers, Lindström et al. [44] studied nanoclay plating of cellulose fiber surface in order to use these combined fibers in polymer composites. The plating method consists of the deposition of montmorillonite platelets on the cellulose fiber surface to promote the reduction of the hydrogen bonding interactions between the fibers, which are a basic problem when preparing cellulose-reinforced composites. Firstly the cellulose fiber surface is recharged using a cationic polyelectrolyte and then it adsorbs a negatively charged nanoclay. The results show that the clay platelets are able to decrease the interaction strength between the cellulose fibers. In a sequential study, Thunwall et al. [45] used undelaminated sodium bentonite or delaminated montmorillonite (obtained from the bentonite) for the treatment of the cellulose fibers in order to improve the dispersion of the fibers in polypropylene or poly(lactic acid) matrices. The results show that mineral treated fibers are well dispersed throughout the PP matrix; however, no significant effect on the composite tensile strength and elastic modulus is observed. The combination of the fiber treatment with mineral platelets and a debonding agent (tertiary amine) seems to be a possible route to prepare composites from hydrophobic matrices; however, further investigation is advisable.

Pandey et al. [41] modified layered silicate with micro- and nanoparticles of cellulose, investigating possible synergistic effects between these materials. Pristine montmorillonite was previously intercalated with cellulose crystals, obtained from wood flour hydrolysis, in an aqueous solution. The cellulose/modified clay was used as a filler to prepare PP composites. The results show that the tensile strength is higher for the composites made up of PP/cellulose/modified clay (5 wt% filler concentration) than the PP/pristine clay or PP/cellulose particle composites. On the other hand, filler agglomeration throughout the matrix is significantly increased when high concentrations of cellulose/modified clay are added.

Ho et al. [46] prepared composites of nanofibrillated cellulose with cationic functional surface groups (trimethylammonium-modified nanofibrillated cellulose) and various anionic layered silicates (montmorillonite, vermiculite, and mica). The interactions between modified cellulose and layered silicates were expected to be effective due to the electrostatic interaction of the cationic cellulose fibrils and the anionic silicate layers. The composites were prepared by high-shear homogenization, filtration, and hot-press processes. The results show that montmorillonite better interacts with cationic fibrils compared with vermiculite and mica platelets, probably due to its smaller crystallite size and platelet dimensions, as well as rougher platelet surface. In fact, while MMT layers cover the cellulose nanofibrils and form a denser structure, vermiculite and mica layers are only attached on the modified nanofibrillated cellulose network and exhibit flat morphologies. In this study, the cationic groups at the cellulose fiber surface are essential for compatibility of cellulose and negatively charged silicate layers of clays and micas.

3.2 Cellulose Derivative/Layered Silicate Nanocomposites

Cellulose derivative composites have been combined with plant-based fillers, due to the structural similarity between the cellulose derivative chains and natural fillers (such as flax, sisal, jute, and curaua). These materials have been used in different applications, from construction to the automotive industry [47]. This similarity is responsible for the good adhesion between the components [28, 30, 48–51]. Also, cellulose esters/natural fiber composites are renewable and in some cases become biodegradable materials [30]. When these composites are prepared by conventional extrusion process, the processing temperature, which is usually near to the degradation temperature of the lignocellulosic fibers, is the main drawback [51]. In this case, the substitution of cellulose fibers by layered silicates permits temperature use range to expand during the conventional industrial processes. As mentioned before, the main motivation of adding layered silicates to a polymer matrix is to enhance the polymer stiffness, strength, toughness, and thermal stability as well as reduced gas permeability and coefficient of thermal expansion.

In the last 10 years, several studies of cellulose derivatives/clay nanocomposites are found in literature [52–69]. Mohanty et al. [52–55] were one of the first groups to develop these materials. Initially [52], the authors prepared cellulose acetate (CA)/organoclay nanocomposites by melt compounding, using triethyl citrate (TEC) as an environmentally friendly plasticizer to optimize the processing conditions. The MMT was modified with methyl tallow bis(2-hydroxyethyl) quaternary ammonium. The results show that the nanocomposites exhibit lower vapor permeability properties than the pure polymer. In a second study [53], maleic anhydride grafted cellulose acetate butyrate (CAB-*g*-MA) was used as a compatibilizer for CA/TEC/MMT nanocomposite. The CAB-*g*-MA presence improves the organoclay dispersion level and consequently, the samples exhibit higher mechanical properties than the non-compatibilized samples. The addition of 5 wt% of the compatibilizer increases the nanocomposite tensile strength and elastic modulus by 20 % and 68 %, respectively, and the flexural strength and modulus by 20 % and 25 %, respectively. The gains are associated to a nanostructural arrangement of the inorganic network in the polymeric matrix. According to the above results, the best mechanical property enhancements are obtained with the exfoliated morphology, which is attained at 5 wt% CAB-*g*-MA. The mechanical properties decrease at higher compatibilizer content. In sequential studies [54, 55], the processing method and conditions as well as the clay amount to prepare CA nanocomposites were optimized. The results show that higher screw rotation during extrusion combined with a sequential injection molding process promotes additional shear rates, which results in a more homogeneous mixing and favors the clay exfoliation. The CA nanocomposite tensile strength and elastic modulus increase around 38 % and 33 %, respectively, after incorporating 5 wt% organoclay. The CA heat deflection temperature also increases with the addition of clay as reinforcing agent. Also, the CAB-*g*-MA compatibilized nanocomposites reach higher clay exfoliation level with a shorter extrusion residence time than in the system without a compatibilizer.

Yoshioka et al. [56] studied CA/MMT/poly(ε-caprolactone) nanocomposites. The materials were prepared by in situ polymerization of poly(ε-caprolactone) in the presence of cellulose acetate and organically modified layered silicates. The procedure involved the previous clay dispersion in the ε-caprolactone monomer, then CA addition followed by the monomer polymerization at high temperature. The nanocomposites were prepared by hot-press molding. Three-layered silicate types were used in this study: natural montmorillonite, synthetic hectorite, and synthetic fluoromica modified with several kinds of ammonium salts. The changes in the clay gallery distances depend on the number of hydroxyl groups in each organomodifier; almost complete exfoliation is achieved when organomodifiers with two hydroxyl groups are used. On the other hand, when comparing the nanocomposites with highly exfoliated structures and the ones which present intercalated structures with several stacked silicate layers, the authors suggested that the latter presents higher functionality, probably due to the clay distribution arrangement. Similar conclusions were reached by Chujo [57]. Chujo's results [57] indicate that polyamide 6/organophilic MMT nanocomposites present a maximum Izod impact strength when intercalated MMT structures are developed, whereas their tensile strength and elastic modulus are enhanced as MMT exfoliation is reached.

Petersson et al. [58] evaluated the behavior of two fillers separately, organophilic smectite modified with dimethyl benzyl hydrogenated tallow and cellulose nanocrystals (CN), as cellulose acetate butyrate (CAB) reinforcements. The fillers were previously dispersed in ethanol solution (low polarity medium) and then, the nanocomposites were prepared by solvent casting. Nanocomposites made up of MMT are less transparent than the ones made up of cellulose nanocrystals, what is explained by the presence of intercalated structure and stacks. Both nanofillers (5 wt%) increase the storage modulus of the CAB matrix over a wide temperature range. At 80 °C, both nanocomposites present an increase of 60 % in the storage modulus. However, CAB/CN nanocomposites present a lower glass transition temperature ($T_g = 152$ °C) than pure CAB ($T_g = 158$ °C), while for CAB/MMT nanocomposite there is no change of this transition value.

Romero et al. [59, 60] prepared CA nanocomposites by solution and melt intercalation processes. A good control of the nanocomposite structure and morphology homogeneity can be achieved by the solution method, whereas the melt technique is particularly attractive due to its suitability with conventional industrial processes. The solution intercalation method was carried out in acetone, acetic acid, acetone/water, and acetic acid/water solutions [59] and the melt intercalation process was carried out in a twin-screw extruder using triethyl citrate as a plasticizer [60]. In these studies, three layered silicates were used to prepare CA nanocomposites: pristine montmorillonite (MMT), organic-modified montmorillonite with methyl tallow bis-2-hydroxyethyl quaternary ammonium salt (OMMT), and a long chain quaternary alkyl ammonium salt chemically modified montmorillonite (v-MMT). The results show that the clay delamination in solution is obtained when the solvent presented favorable interactions with the clay, where the polar interactions and hydrogen bonds are responsible for the stability of the clay suspensions in the solvents. Also, water addition in the acetone and acetic acid solutions promotes further clay swelling and delamination. Among the nanocomposites obtained by solution, the one prepared in acetic acid/water solution shows the highest clay delamination. In this case, the high polarity of the solvents and the formation of hydrogen bonds with the layered silicate platelets are responsible for the significant clay gallery expansion, showing that clay exfoliation can be achieved in polymer nanocomposites by solution intercalation method even when pristine MMT is used. Figure 6.3 shows some CA nanocomposite morphologies obtained by transmission electron microscopy. CA/MMT prepared by solution process presents long clay particles (Fig. 6.3a), whereas CA/TEC/MMT prepared by extrusion presents short and well-dispersed clay tactoids (Fig. 6.3c). The morphology of the nanocomposite prepared with OMMT by solution method shows swollen tactoids and dispersed clay layers (Fig. 6.3b). On the other hand, CA/TEC/OMMT nanocomposite prepared by melt intercalation shows the formation of a three-dimensional network containing clay tactoids, small intercalated stacks, and exfoliated clay layers (Fig. 6.3d). Also, TEM micrographs show higher content of individual clay platelets for the CA/TEC/OMMT (Fig. 6.3d) compared to CA/TEC/v-MMT (Fig. 6.3e), probably due to a higher swelling effect of TEC molecules in the OMMT gallery than in the one of v-MMT. Dynamic-mechanical analyses of the CA nanocomposites were also carried out. The CA glass transition temperatures are 22 °C and 4 °C higher for the CA/MMT and CA/OMMT, respectively, than pure CA, when the nanocomposites are prepared in solution. Also, increases in the storage modulus of these nanocomposites were observed. The other nanocomposites present no differences on these parameters. The increase of the glass transition temperature is attributed to the motion restriction of the polymer chains, therefore, indicating favorable interactions between CA and the silicate layers that are promoted in solution. The permeability of water vapor of the nanocomposites prepared by melt intercalation was also studied. The results show that CA/TEC/OMMT presents the highest reduction (30 %) of this property, probably due to the highest aspect ratio of the OMMT, when compared to the other studied MMT.

Wang and Wang [61] prepared superabsorbent nanocomposites made up of sodium carboxymethyl cellulose (CMC) and attapulgite by radical solution polymerization. The nanocomposites were prepared by radical solution polymerization of sodium carboxymethyl cellulose, acrylic acid (NaA), and attapulgite (APT) using ammonium persulfate as an initiator and *N*,*N*-methylenebisacrylamide as a cross-linker. The results show that acrylic acid is successfully grafted on CMC backbone and that attapulgite nanofibers are uniformly dispersed in the CMC-g-PNaA matrix. The nanocomposite thermal stability and swelling capability are enhanced with the incorporation of 5, 10, or 20 wt% APT. The nanocomposites deflate in an acidic medium and swell in a basic medium, exhibiting excellent pH-sensitive characteristics.

Tunç and Duman [62] prepared methyl cellulose (MC)/pristine MMT nanocomposite films using different procedures, including different mixing equipments (such as homogenizator, magnetic stirrer, and ultrasonic treatment), speed,

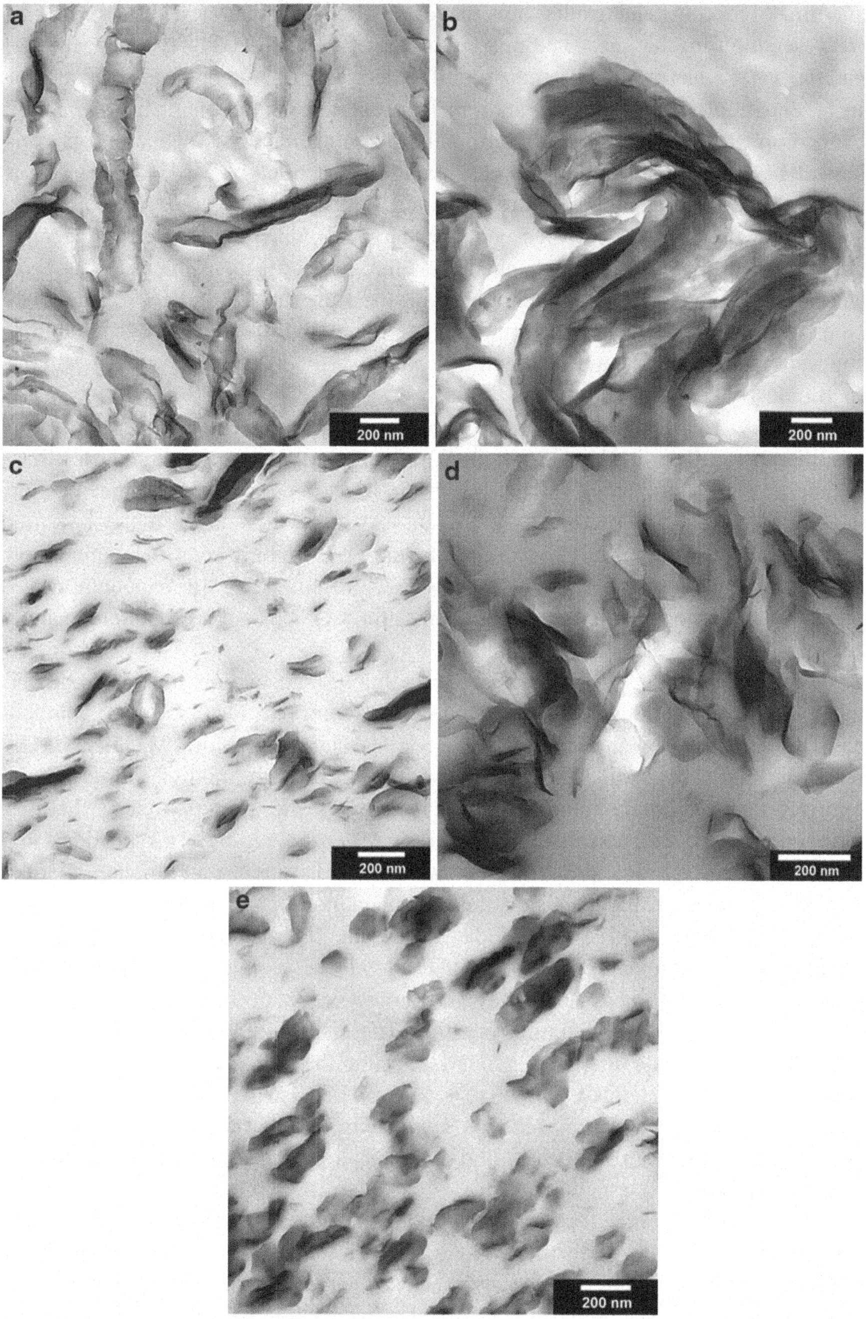

Fig. 6.3 TEM micrographs of CA nanocomposites: (**a**) CA/MMT (acetic acid/water solution), (**b**) CA/OMMT (acetic acid/water solution), (**c**) CA/TEC/MMT (extrusion), (**d**) CA/TEC/OMMT (extrusion) and (**e**) CA/TEC/v-MMT (extrusion)

and time of mixing. The results show that as the MMT content increases (from 10 to 60 wt%), the film thickness and opacity increase by 40 % and 83 %, respectively, and the water adsorption and water solubility decrease 40 % and 43 %, respectively. In a sequential study [63], the authors incorporated carvacrol in the MC/MMT nanocomposites to produce antimicrobial packaging materials. The authors conclude that the antimicrobial agent release rate is controlled by MMT concentration and by the film storage temperature. The increase in the MMT concentration (from 10 to 60 wt%) causes a decrease in the carvacrol release under some specific conditions, probably due to the change of the carvacrol diffusion pathway, i.e., MMT platelets are oriented in parallel with the film surface and carvacrol molecules have to circumvent these impermeable platelets to be released. The authors suggest that these films have a potential to be active food packaging materials in the food industry.

Rodriguez et al. [64] developed cellulose acetate nanocomposite films using modified nanofillers, such as montmorillonite/alkylammonium and montmorillonite/chitosan, which were synthesized using a cationic exchange process. The nanocomposites were prepared by solvent casting which favored the polymer intercalation inside the montmorillonite layers. The highest level of intercalation is obtained with organoclays modified with chitosan. Also, the nanocomposites present a reduction in the oxygen transmission rate, but not to the water vapor, compared to pure cellulose acetate films.

de Lima et al. [65] prepared cellulose acetate/organophilic MMT nanocomposites by melt extrusion, using two different plasticizers: di-octyl phthalate (DOP) and triethyl citrate (TEC). The commercial organophilic MMT (OMMT) used in this study was reported to be modified with methyl tallow bis-2-hydroxyethyl quaternary ammonium. The nanocomposites present mixed structures with intercalated and exfoliated structures. Also, the plasticizer type does not influence the mechanical properties, showing that TEC can be an efficient alternative CA plasticizer and can substitute DOP, a non-eco-friendly plasticizer, as initially indicated by Misra et al. [52]. Subsequently [66], the authors prepared CA/poly(epichlorohydrin) (PePi)/OMMT blend nanocomposites also by melt extrusion. The combined use of clay and rubber was adopted to obtain a balance between toughness and strength properties. The results show that polymer chains are able to penetrate the clay gallery, causing the exfoliation of the OMMT platelets. OMMT platelets are found in CA phase, but also at the CA/PEPi interface, surrounding the PEPi phase, as can be seen in Fig. 6.4. It was concluded however that the concomitant use of rubber and clay does not produce a desirable stiffness/toughness balance.

Flexible transparent nanocomposite films with heat-resistant and high gas-barrier properties were developed by the group of Mizukami [67–69]. Nanocomposite films made up of water-soluble polymers, such as carboxymethyl cellulose, sodium polyacrylate, and synthetic Na^+- saponite, were prepared by solvent-casting technique. The polymers with sodium anions act as binders, connecting the clay platelets through ionic interaction; consequently uniform structures comprising dense arrangements of clay platelets and polymer chains between these platelets

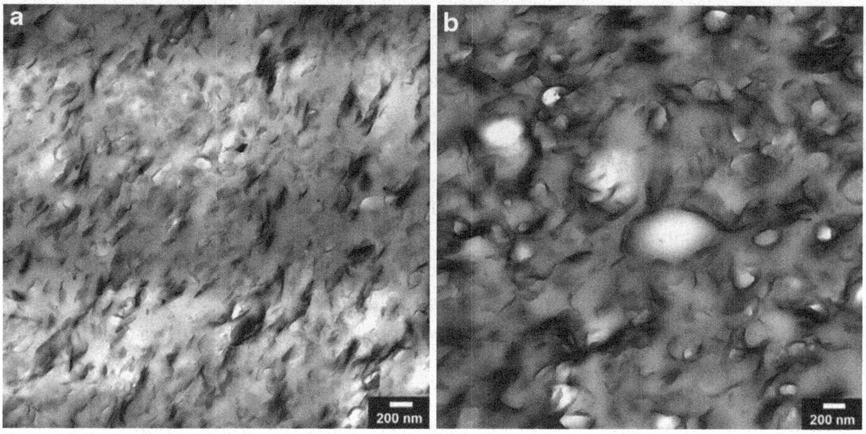

Fig. 6.4 TEM micrographs of the nanocomposites: (**a**) CA-TEC/OMMT and (**b**) CA-TEC/PEPi/OMMT

are formed. As a result, the polymer thermal stability of the nanocomposites increases. It was also reported that a uniform disperse phase structure throughout the film is achieved with the addition of 20–80 wt% clay. Indeed, films with 80 wt% clay exhibit excellent optical and physical properties and the authors suggest that these materials have potential to be used as optoelectronic displays.

4 Final Remarks

As well known, the properties of nanostructured composites are extremely structure-size dependent. A fundamental understanding of the formation, structure, and quality of the interface as well as the properties and their relations, on various length scales, are necessary to analyze the exceptional properties observed at a nanoscale and to use them at a macroscale.

Literature has shown scientists that it is possible to obtain intercalated and exfoliated nanocomposites from cellulose derivatives/clay by solution and melt intercalation methods. However, obtaining materials from cellulose and clay combinations is little attempted, and from these few attempts, real nanocomposites are still not evident.

Experimental results indicate that the outcome of polymer intercalation depends critically on silicate functionalization and component interactions. This is necessary to eliminate the chemical incompatibility between the components and fundamental to obtain significant properties gains.

In terms of water-insoluble cellulose derivatives/clay nanocomposites prepared by solution intercalation technique, the appropriate choice of solvent promotes the solution intercalation. However, the use of organic solvents and some ionic liquids can cause the process to become both highly costly and environmentally unfriendly.

In relation to the melt intercalation technique, thermal and mechanical effects and high residence time can enhance cellulose derivative degradation. A carbohydrate-based matrix often needs the plasticizer incorporation to melt and also to control its thermal degradation. Nevertheless, results from literature state that when excessive quantities of plasticizer are used, there is a negative effect due to strong interactions between these and the layered silicates, which disturb the intercalation and/or exfoliation process. Therefore, to minimize polymer chain degradation and to promote intercalated and/or exfoliated clay structures, the development of a process parameter balance and use of adequate compatibilizers and/or plasticizers are required. One way to overcome compatibilizer limitations is to develop, as per literature, carbohydrate compatibilizers, which can promote an appropriate modification of the clay/matrix interface [56].

Based on literature, the authors would like to highlight three important aspects in terms of future development of cellulose/clay-based materials: Firstly that well-conducted cellulose source harvesting and processing quantity management is necessary for reliable and reproducible results due to the fact that cellulose generally comes from vegetable fibers which usually have different compositions and molecular weights. Secondly, there is a need to further promote studies into swelling and degradation in long-term applications. Last but not least, to encourage more extensive research into cellulose/clay as well as cellulose derivatives/clay nanocomposites, albeit at times demanding that changes in processes, solvents, and additives be made, which will lead towards greener and more eco-friendly materials.

References

1. Chivrac F, Pollet E, Avérous L (2009) Progress in nano-biocomposites based on polysaccharides and nanoclays. Mater Sci Eng R 67:1–17
2. Mohanty AK, Misra M, Drzal LT (2002) Sustainable bio-composites from renewable resources: opportunities and challenges in the green materials world. J Polym Environ 10:19–26
3. Hon DN-S, Shiraishi N (1990) Wood and cellulose chemistry. Marcel Dekker, New York
4. Blumstein A (1965) Polymerization of adsorbed monolayers: II. Thermal degradation of the inserted polymers. J Polym Sci A 3:2665–2673
5. Theng BKG (1979) Formation and properties of clay–polymer complexes. Elsevier, Amsterdam
6. Okada A, Kawasumi M, Usuki A, Kojima Y, Kurauchi T, Kamigaito O (1990) Synthesis and properties of nylon-6/clay hybrids. In: Schaefer DW, Mark JE (eds) Polymer based molecular composites. MRS symposium proceedings, vol 171. Materials Research Society, Pittsburgh, pp 45–50
7. Vaia RA, Ishii H, Giannelis EP (1993) Synthesis and properties of two-dimensional nanostructures by direct intercalation of polymer melts in layered silicates. Chem Mater 5:1694–1696
8. Ray SS, Boumina M (2005) Biodegradable polymers and their layered silicate nanocomposites: in greening the 21st century materials world. Prog Mater Sci 50:962–1079
9. Pavlidou S, Papaspyrides CD (2008) A review on polymer–layered silicate nanocomposites. Prog Polym Sci 33:1119–1198

10. Alexandre M, Dubois P (2000) Polymer-layered silicate nanocomposites: preparation, properties and uses of a new class of materials. Mater Sci Eng 28:1–63
11. Ray SS, Okamoto M (2003) Polymer/layered silicate nanocomposites: a review from preparation to processing. Prog Polym Sci 28:1539–1641
12. Tjong SC (2006) Structural and mechanical properties of polymer nanocomposites. Mater Sci Eng R 53:73–197
13. He M, Chang C, Peng N, Zhang L (2012) Structure and properties of hydroxyapatite/cellulose nanocomposite films. Carbohydr Polym 87:2512–2518
14. Pinto RJB, Marques PAA, Neto CP, Trindade T, Daina S, Sadocco P (2009) Antibacterial activity of nanocomposites of silver and bacterial or vegetable cellulosic fibers. Acta Biomater 5:2279–2289
15. Kim D-H, Park S-Y, Kim J, Park M (2010) Preparation and properties of the single-walled carbon nanotube/cellulose nanocomposites using *n*-methylmorpholine-*n*-oxide monohydrate. J Appl Polym Sci 117:3588–3594
16. Zhang X, Liu X, Zheng W, Zhu J (2012) Regenerated cellulose/graphene nanocomposite films prepared in DMAC/LiCl solution. Carbohydr Polym 88:26–30
17. Habibi Y, Lucia L, Rojas O (2010) Cellulose nanocrystals: chemistry, self-assembly, and applications. Chem Rev 110:3479–3500
18. Braconnot H (1819) Verwandlungen des holzstoffs mittelst schwefelsäure in gummi, zucker und eine eigne säure, und mittelst kali in ulmin. Ann Phys 63:347–371
19. Payen A, Hebd C (1838) Mémoire sur la composition du tissu propre des plantes et du ligneux. Seances Acad Sci 7:1052–1056
20. Nakashima K, Nishino A, Horikawa Y (2011) The crystalline phase of cellulose changes under developmental control in a marine chordate. Cell Mol Life Sci 68:1623–1631
21. Siqueira G, Bras J, Dufresne A (2010) Cellulosic Bionanocomposites: a review of preparation, properties and applications. Polymers 2:728–765
22. http://www.bracelpa.org.br/bra/estatisticas/pdf/booklet/booklet.pdf
23. Lejeune A, Deprez T (eds) (2010) Cellulose: structure and properties, derivatives and industrial uses. Nova, Hauppauge
24. Brown RM Jr, Saxena IM, Kudlicka K (1996) Cellulose biosynthesis in higher plants. Trends Plant Sci 1:149–156
25. Samir MASA, Alloin F, Dufresne A (2005) Review of recent research into cellulosic whiskers, their properties and their application in nanocomposite field. Biomacromolecules 6:612–626
26. Moon R, Martine A, Nairn J, Simonsen J, Youngblood J (2011) Cellulose nanomaterials review: structure, properties and nanocomposites. Chem Soc Rev 40:3941–3994
27. Lewin M (ed) (2006) Handbook of fiber chemistry. CRC Press, New York
28. Huber T, Müssig J, Curnow O, Pang S, Bickerton S, Staiger MP (2012) A critical review of all-cellulose composites. J Mater Sci 47:1171–1186
29. O'Sullivan AC (1997) Cellulose: the structure slowly unravels. Cellulose 4:173–207
30. Edgar KJ, Buchanan CM, Debenham JS, Rundquist PA, Seiler BD, Shelton MC, Tindall D (2001) Advances in cellulose ester performance and application. Prog Polym Sci 26:1605–1688
31. Puls J, Wilson SA, Hölter D (2011) Degradation of cellulose acetate-based materials: a review. J Polym Environ 19:152–165
32. Kawanishi H, Tsunashima Y, Okada S, Horii F (1998) Change in chain stiffness in viscometric and ultracentrifugal fields: cellulose diacetate in *N,N*-dimethylacetamide dilute solution. J Chem Phys 108:6014–6025
33. Schulz L, Seger B, Burchard W (2000) Structures of cellulose in solution. Macromol Chem Phys 201:2008–2022
34. Kawanishi H, Tsunashima Y, Okada S, Horii F (2000) Dynamic light scattering study of structural changes of cellulose diacetate in solution under coquette flow. Macromolecules 33:2092–2097
35. White LA (2004) Preparation and thermal analysis of cotton-clay nanocomposites. J Appl Polym Sci 92:2125–2131

36. Delhom CD, White-Ghoorahoo LA, Pang SS (2010) Development and characterization of cellulose/clay nanocomposites. Comp Part B 41:475–481
37. Cerruti P, Ambrogi V, Postiglione A, Rychlý J, Matisová-Rychlá L, Carfagna C (2008) Morphological and thermal properties of cellulose-montmorillonite nanocomposites. Biomacromolecules 9:3004–3013
38. Jang S-W, Kim J-C, Chang J-H (2009) Preparation and characterization of cellulose nanocomposite films with two different organo-micas. Cellulose 16:445–454
39. Mahmoudian S, Wahit MU, Ismail AF, Yussuf AA (2012) Preparation of regenerated cellulose/montmorillonite nanocomposite films via ionic liquids. Carbohydr Polym 88:1251–1257
40. Haroun AA, Gamal-Eldeen A, Harding DRK (2009) Preparation, characterization and in vitro biological study of biomimetic three-dimensional gelatin–montmorillonite/cellulose scaffold for tissue engineering. J Mater Sci Mater Med 20:2527–2540
41. Pandey JK, Lee S, Kim H-J, Takagi H, Lee CS, Ahn SH (2012) Preparation and properties of cellulose-based nanocomposites of clay and polypropylene. J Appl Polym Sci 125: E651–E660
42. Mariano RM, Picciani PHS, Nunes RCR, Visconte LLY (2011) Preparation, structure, and properties of montmorillonite/cellulose II/natural rubber nanocomposites. J Appl Polym Sci 120:458–465
43. Zine CLG, Conceição AJ, Visconte LLY, Ito EN, Nunes RCR (2011) Styrene–butadiene rubber/cellulose II/clay nanocomposites prepared by co-coagulation-mechanical properties. J Appl Polym Sci 120:1468–1474
44. Lindström T, Banke K, Larsson T, Glad-Nordmark G, Boldizar A (2008) Nanoclay plating of cellulosic fiber surfaces. J Appl Polym Sci 108:887–891
45. Thunwall M, Boldizar A, Rigdahl M, Banke K, Lindström T, Tufvesson H, Högman S (2008) Processing and properties of mineral-interfaced cellulose fibre composites. J Appl Polym Sci 107:918–929
46. Ho TTT, Ko YS, Zimmermann T, Geiger T, Caseri W (2012) Processing and characterization of nanofibrillated cellulose/layered silicate systems. J Mater Sci 47:4370–4382
47. John MJ, Thomas S (2008) Biofibres and biocomposites. Carbohydr Polym 71:343–364
48. Bledzki AK, Reihmane S, Gassan J (1996) Properties and modification methods for vegetable fibers for natural fiber composites. J Appl Polym Sci 59:1329–1336
49. Mohanty AK, Wibowo A, Misra M, Drzal LT (2004) Effect of process engineering on the performance of natural fiber reinforced cellulose acetate biocomposites. Compos Part A 35:363–370
50. Alvarez VA, Kenny JM, Vazquez A (2004) Creep behavior of biocomposites based on sisal fiber reinforced cellulose derivatives/starch blends. Polym Compos 25:280–288
51. Gutierrez MC, De Paoli M-A, Felisberti MI (2012) Biocomposites based on cellulose acetate and short curauá fibers: effect of plasticizers and chemical treatments of the fibers. Compos Part A 43:1338–1346
52. Park H-M, Misra M, Drzal LT, Mohanty AK (2004) "Green" nanocomposites from cellulose acetate bioplastic and clay: effect of eco-friendly triethyl citrate plasticizer. Biomacromolecules 5:2281–2288
53. Park H-M, Liang X, Mohanty AK, Misra M, Drzal LT (2004) Effect of compatibilizer on nanostructure of the biodegradable cellulose acetate/organoclay nanocomposites. Macromolecules 37:9076–9082
54. Wibowo AC, Misra M, Park H-M, Drzal LT, Schalek R, Mohanty AK (2006) Biodegradable nanocomposites from cellulose acetate: mechanical, morphological, and thermal properties. Compos Part A 37:1428–1433
55. Park H-M, Mohanty AK, Drzal LT, Lee E, Mielewski DF, Misra M (2006) Effect of sequential mixing and compounding conditions on cellulose acetate/layered silicate nanocomposites. J Polym Environ 14:27–35
56. Yoshioka M, Takabe K, Sugiyama J, Nishio Y (2006) Newly developed nanocomposites from cellulose acetate/layered silicate/poly(e-caprolactone): synthesis and morphological characterization. J Wood Sci 52:121–127

57. Chujo K (2002) On the relation between the degree of delamination and the mechanical properties of polymer/clay nanocomposites. Seikei-Kakou 14:686–687
58. Petersson L, Mathew AP, Oksman K (2009) Dispersion and properties of cellulose nanowhiskers and layered silicates in cellulose acetate butyrate nanocomposites. J Appl Polym Sci 112:2001–2009
59. Romero RB, Leite CAP, Gonçalves MC (2009) The effect of the solvent on the morphology of cellulose acetate/montmorillonite nanocomposites. Polymer 50:161–170
60. Romero RB, Kaneko MLQA, Gonçalves MC (2008) Mechanical and morphological properties of cellulose acetate/organoclay nanocomposites. In: Proceedings of the 24th polymer processing society, Salerno, Italy, June 15–19
61. Wang W, Wang A (2010) Nanocomposite of carboxymethyl cellulose and attapulgite as a novel pH-sensitive superabsorbent: Synthesis, characterization and properties. Carbohydr Polym 82:83–91
62. Tunç S, Duman O (2010) Preparation and characterization of biodegradable methyl cellulose/montmorillonite nanocomposite films. Appl Clay Sci 48:414–424
63. Tunç S, Duman O (2011) Preparation of active antimicrobial methyl cellulose/carvacrol/montmorillonite nanocomposite films and investigation of carvacrol release. LWT Food Sci Technol 44:465–472
64. Rodriguez FJ, Galotto MJ, Guarda A, Bruna JE (2012) Modification of cellulose acetate films using nanofillers based on organoclays. J Food Eng 110:262–268
65. de Lima JA, Pinotti CA, Felisberti MI, Gonçalves MC (2012) Morphology and mechanical properties of nanocomposites of cellulose acetate and organic montmorillonite prepared with different plasticizers. J Appl Polym Sci 124:4628–4635
66. de Lima JA, Pinotti CA, Felisberti MI, Gonçalves MC (2012) Blends and clay nanocomposites of cellulose acetate and poly(epichlorohydrin). Compos Part B 43:2375–2381
67. Ebina T, Mizukami F (2007) Flexible transparent clay films with heat-resistant and high gas-barrier properties. Adv Mater 19:2450–2453
68. Tetsuka H, Ebina T, Nanjo H, Mizukami F (2007) Highly transparent flexible clay films modified with organic polymer: structural characterization and intercalation properties. J Mater Chem 17:3545–3550
69. Tetsuka H, Ebina T, Tsunoda T, Nanjo H, Mizukami F (2007) Flexible organic electroluminescent devices based on transparent clay films. Nanotechnology 18:355701/1–4

Thermal and Rheological Properties of Poly(ethylene-co-vinyl acetate) (EVA) Nanoclay

7

Vinicius Pistor and Ademir José Zattera

Contents

Abstract
Poly(ethylene-co-vinyl acetate) (EVA) is a copolymer of utmost importance in nanocomposite applications. Clay dispersed in an EVA matrix was tested using different types of clays with hydrated silicate layers. Several observations have led to the unexpected physical point of view of (clay/polymer) interaction. Due to the polar nature of the acetate group, a strong interaction would be expected when using a polar group-bearing surfactant such as organophilic modified montmorillonite (OMMT – 30B). Through research using techniques that explore the rheology in the molten state, it was evident that a stronger interaction is possible for nonpolar group-modified organophilic clays. This strong interaction is the result of a complex environment of physical entanglements that show a pseudo-solid behavior among the nanoclay dispersed in the EVA matrix. In this chapter, we will emphasize the influence of processing and type of organic modifier in natural organophobic clay (Cloisite® Na^+) dispersed in an

V. Pistor (✉) • A.J. Zattera
Laboratory of Polymers (LPOL), Center for Exact Sciences and Technology (CCET), Caxias do Sul University (UCS), Caxias do Sul, RS, Brazil
e-mail: pistorv@yahoo.com.br; ajzattera@terra.com.br

J.K. Pandey et al. (eds.), *Handbook of Polymernanocomposites. Processing, Performance and Application – Volume A: Layered Silicates*, DOI 10.1007/978-3-642-38649-7_17,

EVA matrix. The characterization techniques will be discussed by the study of melt state rheology and thermal behavior of these materials. The relaxation phenomena and creep compliance behavior as well as a correlation between the processing and final properties are detailed for EVA and OMMT nanocomposites.

Keywords
EVA • Melting behavior • Nanoclay • Physcical interactions

Abbreviations

EVA	Poly(ethylene-co-vinyl acetate)
NLREG	Nonlinear regularization
OMMT	Organophilic modified montmorillonite
PVAc	Poly(vinyl acetate)
TEM	Transmission electron microscopy
VAc	Vinyl acetate
XRD	X-ray diffraction

1 Introduction

The ethylene-co-vinyl acetate (EVA) copolymer belongs to a class of statistical copolymers of intermediary properties between its two constituent homopolymers, polyethylene and poly(vinyl acetate) (PVAc). Such intermediary properties derive from its complex morphology, which comprises a crystalline phase (with ethylene units), an interface region (with ethylene moieties and VAc moieties), and a complex amorphous phase (with ethylene moieties and non-crystallized VAc units) [1]. Based on the growing rise in nanocomposites research, various studies involving EVA and clays were conducted resulting in different approaches described in the literature so as to form a rich database [2–12]. The most important issue related to the EVA-clay nanocomposites is perhaps the relationship between the VAc content and the kind of clay and the effects of the variation of each of them. Since the VAc content is responsible for the EVA crystallinity, the stiffness of the EVA copolymer depends on the size of the ethylene crystallizable phase.

Zhang et al. [4] have evaluated the effect of different kinds of clay in EVA matrices containing 28, 40, 50, and 80 wt% VAc. The authors observed that organic modifier-containing clays favored the passage of EVA chains between the clay platelets caused by the increased basal spacing. They also observed that the EVA polarity is an important factor in the nanocomposites morphology and properties, with the 28 wt% EVA exhibiting more evident improvements in the composites thermal and mechanical properties. The VAc concentration can determine the extent of EVA hydrophilicity or hydrophobicity, and in this case the type of organic modifier used in the clay should be considered. Nonpolar surfactant-modified clays are more adequate for low VAc contents, while polar modifier-containing clays such as the 30B clay can lead to a larger interaction with higher VAc contents [9]. A further relevant factor is the way of dispersion that each kind of clay will present when added to the polymer. The most commonly observed types of morphology are tactoid shaped where the most probable ones are equivalent to the formation of microcomposites. Clay intercalation with the polymer matrix and clay exfoliation promote the detachment of the clay and thus increases its dispersion [10]. Clay polar surfactant modifiers can render exfoliation more efficient [2].

For EVA-clay nanocomposites processed in a twin-screw extruder, it has been observed that the rise in clay concentration promotes the increase in the storage (G') and loss (G'') moduli while increasing the nanocomposites viscosity (η) in the melt state. This rise is attributed to the increased interlamellar distance of the clay layers [6] and the higher interaction and ease of exfoliation in the case of hydrophilic modifiers [11]. Besides, depending on the surfactant type, different dispersion forms can occur and through dispersion, the balance existing between physical interactions in nanocomposites will take into account other factors besides polarity [12].

This chapter will investigate the flow behavior and the physical interactions in the melt state for nanocomposites containing the montmorillonite clays Cloisite® Na^+, Cloisite® 15A, and Cloisite® 30B. The study will be presented in two steps: results of different concentrations of the 30B clay dispersed in the EVA matrix and results comparing the influence of the Na^+, 15A, and 30B clays on the concentration with the best results found for the addition of the 30B clay. Oscillatory rheometry analyses were conducted as well as flow and power law curves.

Using the nonlinear regularization program (NLREG) [13], data such as relaxation $H(\tau)$ and retardation $L(\tau)$ spectra, relaxation modulus $G(t)$, and creep compliance $J(t)$ were obtained by evaluating the viscoelastic properties. The goal of the comparison between these studies is to present a solid basis for understanding which clay better interacts with EVA and how the incorporation of these nanoparticles influences the melt flow thermal stability.

2 Preparation Methods of EVA Nanocomposites

The materials used in this study were poly(ethylene-co-vinyl acetate) (EVA) (UE-2824/32) containing 28 wt% vinyl acetate (VAc), supplied by Polietilenos União S/A (SP/BR); the natural organophobic clay (Cloisite® Na^+); and organophilic modified montmorillonite (OMMT) clay with quaternary ammonium salt-containing chain segments with ~65 % C18, ~30 % C16, and ~5 % C14 (Cloisite® 15A; dimethyl, dehydrogenated tallow, quaternary ammonium) and OMMT clay with two terminal hydroxyls (Cloisite® 30B; methyl, tallow, bis-2-hydroxyethyl, quaternary ammonium), supplied by Southern Clay Products.

Figure 7.1 is an illustration of the materials utilized for obtaining the nanocomposites.

EVA nanocomposites of varied montmorillonite clay concentration (0, 2, 5, and 10 parts per hundred – phr of Cloisite® 30B) and EVA nanocomposites containing 5 phr of Cloisite Na+, 15A and 30B were prepared to compare the different

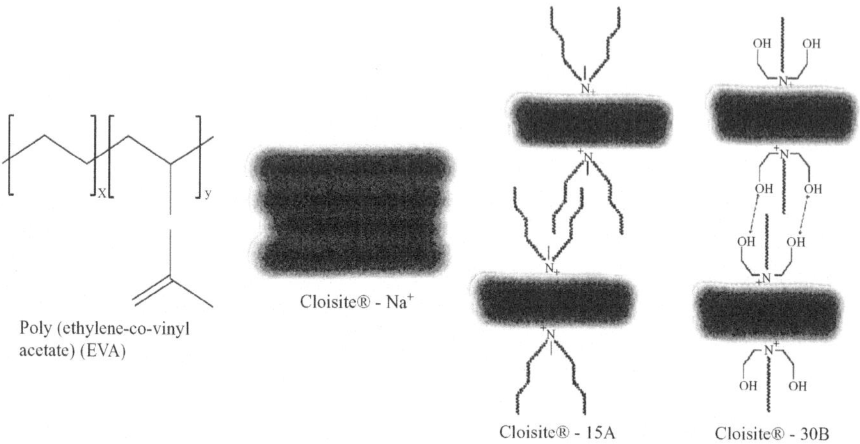

Fig. 7.1 Chemical structure of EVA and schematics of the three different types of clay used in this work

salt-containing chain segments. After drying the clays in an oven with air circulation at a temperature of 60 °C for 8 h, the nanocomposites were processed in a molten mixture process using a corotating twin-screw extruder (MH-COR-20-32-LAB, MH Equipment; D = 20 mm, L/D = 32) with eight heating zones and temperature profiles (100, 120, 140, 160, 130*, 160, 160, and 160 °C – barrel to matrix, respectively). The speed processing of 400 rpm was used to process the nanocomposites. In zone 5* degassing with the aid of a vacuum was used for the elimination of volatiles.

3 Characterization Methods of EVA Nanocomposites

X-ray diffractograms were collected on a sample holder mounted in a Shimadzu diffractometer (XRD-6000) with monochromatic Cu Kα radiation ($\lambda = 0{,}15418$ nm), the generator working at 40 kV and 30 mA. Intensities were measured in the range $1 < 2\theta < 12°$, typically with step scans 0.05° and 2 s/step (1.5° min^{-1}).

The samples were prepared using cryo-ultramicrotomy. Samples were mounted on cryo-pins and frozen in liquid nitrogen. Analyses were performed in a JEOL model JEM 1200 EX II transmission electronic microscope with energy filter and magnification, at 80 kV voltage.

The samples were analyzed in an Anton Paar oscillatory rheometer (Physica MCR 101), with parallel plates of 25 mm diameter and 1 mm gap between the plates, test temperature of 160 °C, frequency range of 0.1–100 Hz, shear stress of 200 Pa, and nitrogen flow of 1 m^3/h. From the rheological analysis, the relaxation ($H(\tau)$) spectra were collected using a nonlinear regularization (NLREG) program [11–14].

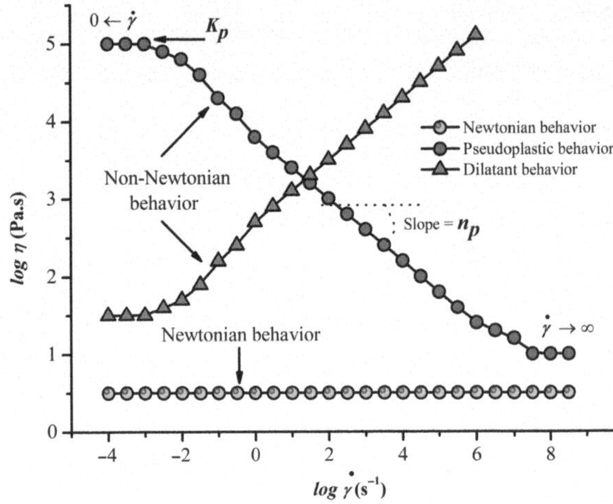

Fig. 7.2 Curves for viscosity as a function of the shear rate for the three basic flow components

Considering the shearing flow for Cartesian coordinates x and y, in which the ratio between the shear stress τ_{xy} and the shear rate ($\dot{\gamma}_{xy}$) is constant, this is the behavior of a Newtonian fluid. However, molten polymers will show a Newtonian fluid behavior either when the shear rate is very low or tends to the infinite, i.e., $\dot{\gamma} \rightarrow 0$ or $\dot{\gamma} \rightarrow \infty$. For polymers, the intermediate values between 0 and ∞ are not constant; that is why the ratio between τ_{xy} and $\dot{\gamma}_{xy}$ describes the non-Newtonian behavior $(\eta(\dot{\gamma}))$. Such representation can be observed in Fig. 7.2. Basically, if the viscosity is kept constant with the increase in γ, the observed behavior is that of a Newtonian fluid; when η increases, this means a dilatant behavior, and if η diminishes, the polymer will exhibit a pseudoplastic behavior. In both cases the viscosity can be represented by the power law in Eq. 7.1:

$$\tau(\dot{\gamma}) = K_p \dot{\gamma} n_p \tag{7.1}$$

where, in the region of the Newtonian plateau, as illustrated in Fig. 7.2, k_p is known as consistency and n_p is the power law index. The values of n_p are considered as a measurement of the pseudoplasticity of polymers. In other words, for $n_p = 1$ there is no variation in viscosity and the material will show the behavior of a Newtonian fluid. On the other hand, for $n_p < 1$ or $n_p > 1$, a pseudoplastic and dilatant behavior will be observed, respectively.

These figures can be obtained from the slope of the $\log \eta$ versus $\log \dot{\gamma}$ curve. It should be emphasized, however, that the adjustment is valid solely for the region after the Newtonian plateau. Power law indices can be obtained through oscillatory rheometry analysis for low shear rates by means of the Cox-Merz rule, considering a plate-plate geometry as illustrated in Fig. 7.3 [15].

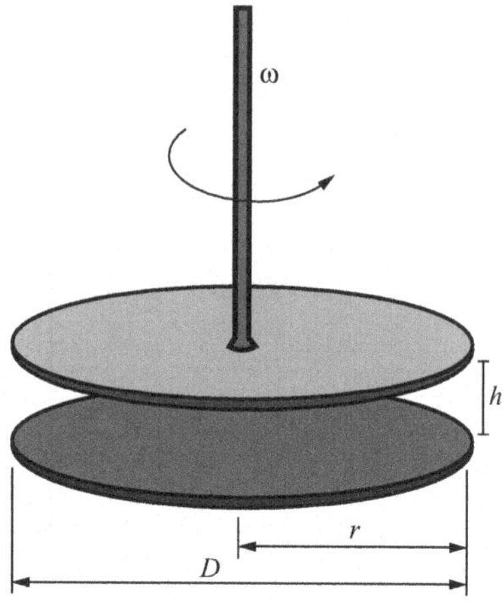

Fig. 7.3 Schematics of a plate-plate geometry utilized for performing oscillatory rheometry analyses

By assuming stationary state conditions, the complex viscosity (η*) can be obtained by the angular frequency (ω) and can be considered as equivalent to the viscosity (η) obtained through the shear rate, $\dot{\gamma} = \omega$.

Thus,

$$\eta(\dot{\gamma}) = |\eta^*(\omega)|_{\dot{\gamma}=\omega} \text{ where, } |\eta^*(\omega)| = \left[\eta'^2(\omega) + \eta''^2(\omega)\right]^{1/2} \tag{7.2}$$

where η' and η'' are the actual and imaginary viscosity obtained through the oscillatory experiment.

By utilizing the constants of the plate-plate geometry as illustrated in Fig. 7.2, the shear stress in the rheometer can be defined by Eq. 7.3:

$$\tau = \frac{3 \cdot M}{2 \cdot \pi \cdot r^3} \tag{7.3}$$

where M is the torque and r is the geometry radius. The shear rate ($\dot{\gamma}$) of the plate-plate rheometer depends on the external radius of the turning plate and can be determined by Eq. 7.4:

$$\ln\dot{\gamma} = r \cdot \frac{\omega}{h} \tag{7.4}$$

where h is the distance between plates and ω is the angular frequency.

In an experiment at low deformation amplitudes, the modulus of relaxation can be obtained at all deformation levels [16]. However, several physical functions, such as the relaxation ($H(\tau)$) and retardation spectra ($L(\tau)$), cannot be calculated directly by the experiments. Notwithstanding, these functions can be obtained based on a experimentally measurable quantity by the resolution of an inverse problem [11, 13]. Such viscoelastic functions can be obtained through the storage (G') and loss (G'') moduli by Eqs. 7.5 and 7.6:

$$G'(\omega) = G_\theta + \int_{-\infty}^{\infty} H(\tau)\frac{\omega^2\tau^2}{1+\omega^2\tau^2} d \ln \tau \quad (7.5)$$

$$G''(\omega) = \int_{-\infty}^{\infty} H(\tau)\frac{\omega^2\tau^2}{1+\omega^2\tau^2} d \ln \tau \quad (7.6)$$

where G_θ is the modulus of equilibrium, $H(\tau)$ the function of the relaxation spectrum, and τ is the relaxation time or retardation time.

The relaxation spectrum results from the sum of Maxwell infinite elements which describe the mechanical behavior of the elastic portion of the polymer, as well as its ability to store energy [11, 16]. The estimation of the retardation spectrum is performed through the interrelations of the spectra described by Ferry and results from the sum of the infinite elements of Voigt which are in charge of the entire viscous portion of the polymer [16] (Eqs. 7.7 and 7.8):

$$H(\tau) = \frac{L(\tau)}{\left[J_g + \int_{-\infty}^{\infty} \frac{L(\tau)}{(1-\tau)/\tau} d \ln \tau - \frac{\tau}{\eta_0}\right]^2 + \pi^2 L(\tau)^2} \quad (7.7)$$

$$L(\tau) = \frac{H(\tau)}{\left[G_\theta - \int_{-\infty}^{\infty} \frac{H(\tau)}{\tau/(\tau-1)} d \ln \tau\right]^2 + \pi^2 H(\tau)^2} \quad (7.8)$$

Thus, the creep compliance function for viscoelastic liquids is approximately given by

$$J(t) = J_g + \int_{-\infty}^{\infty} l(\tau)\left(1 - e^{\,t/\tau}\right) d \ln \tau + \frac{t}{\eta^0} \quad (7.9)$$

Because the creep compliance ($J(t)$) function decays slowly over time, using an interface similar to the stress relaxation modulus ($G(t)$) requires that $J(t)$ relates to a decreasing function. Thus, $G(t)$ can be obtained from the following relationship:

$$G(t) = G_e + \int_{-\infty}^{\infty} \text{In}(\tau) e^{\,t/\tau} d \ln \tau \quad (7.10)$$

4 Properties of EVA Nanocomposites

In order to understand the effects of the clays dispersed in an EVA matrix, the results were segregated into two parts. In the first part, 2, 5, and 10 parts per hundred (phr) of the organophilic modified montmorillonite were incorporated into the EVA matrix. In the second part, a comparative between nanocomposites containing 5 phr of the Na^+, 15A, and 30B clays is illustrated. Figure 7.4 depicts images of transmission electronic microscopy (TEM) for the nanocomposites containing different amounts (in phr) of the 30B clay. The incorporation of 2 phr led to a predominantly tactoid morphology, possibly because this amount is insufficient for dispersion and exfoliation. For the samples containing 5 and 10 phr of OMMT, partial intercalation and exfoliation were the predominant characteristics observed, and these samples also contained a small tactoid fraction.

One of the main factors required for exfoliation to occur in EVA-30B clay nanocomposites is the strong interaction of the vinyl acetate group VAc with the hydroxyl groups of the clay surfactant [17].

Figure 7.5 illustrates the TEM analysis for the nanocomposites containing Na^+, 15A, and 30B clays. The Na^+ clay-containing nanocomposite shows features of microcomposite agglomerates [18]. In relation to the 15A nanocomposite, intercalated structures (that are visually observed from the dispersed lower platelets) in the EVA matrix were ascertained. With the incorporation of the 30B nanofiller, the predominant morphology observed was the exfoliated one, corroborating with the viscoelastic properties observed in this study.

Since the platelet separation produces more percolated clay in the matrix, one of the factors related to the strong interaction of the 30B clay is the presence of terminal OHs in the surfactants used to modify the clay surface. Polar surface agent-modified clays make the exfoliation more efficient [2]. Also, for the nanocomposites containing 15A and 30B nanofiller (intercalated and exfoliated morphologies, respectively), the presence of further morphological characteristics can be observed. This is due to some factors like the polar surfactants (more compatible with hydrophilic polymers) [2, 9] and the polarity [4] that contributes to the formation of a determined morphology.

4.1 Melt State Behavior of EVA/Nanoclays

Figure 7.6a, b shows the figures for storage (G') and loss (G'') moduli determined for the present studies. Figure 7.6a corresponds to the addition of different parts of montmorillonite clay 30B to the EVA matrix, while Fig. 6b represents the results for a comparative associated with the influence of the type of surface agent utilized for the modification of the sodium clay (Na^+).

In the molten state, polymers are free of molecular ordering and of any structure or interaction typical of crystalline solids. In this fluidity state, values of G'' should be higher than those of G' since the high molecular vibration promotes dissipation of energy [11, 12].

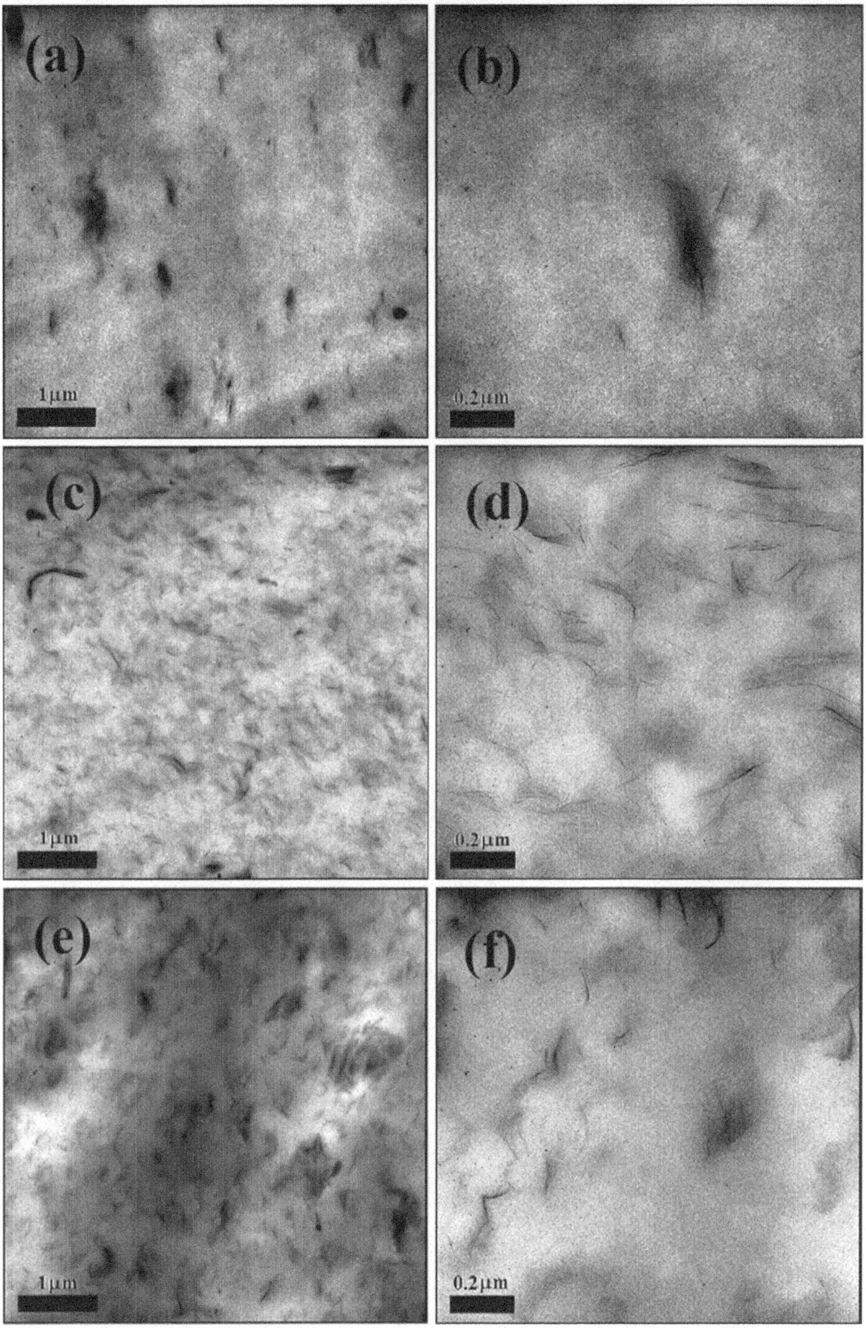

Fig. 7.4 TEM images of nanocomposites where (**a**, **b**) EVA 2 phr OMMT, (**c**, **d**) EVA 5 phr OMMT, and (**e**, **f**) EVA 10 phr OMMT

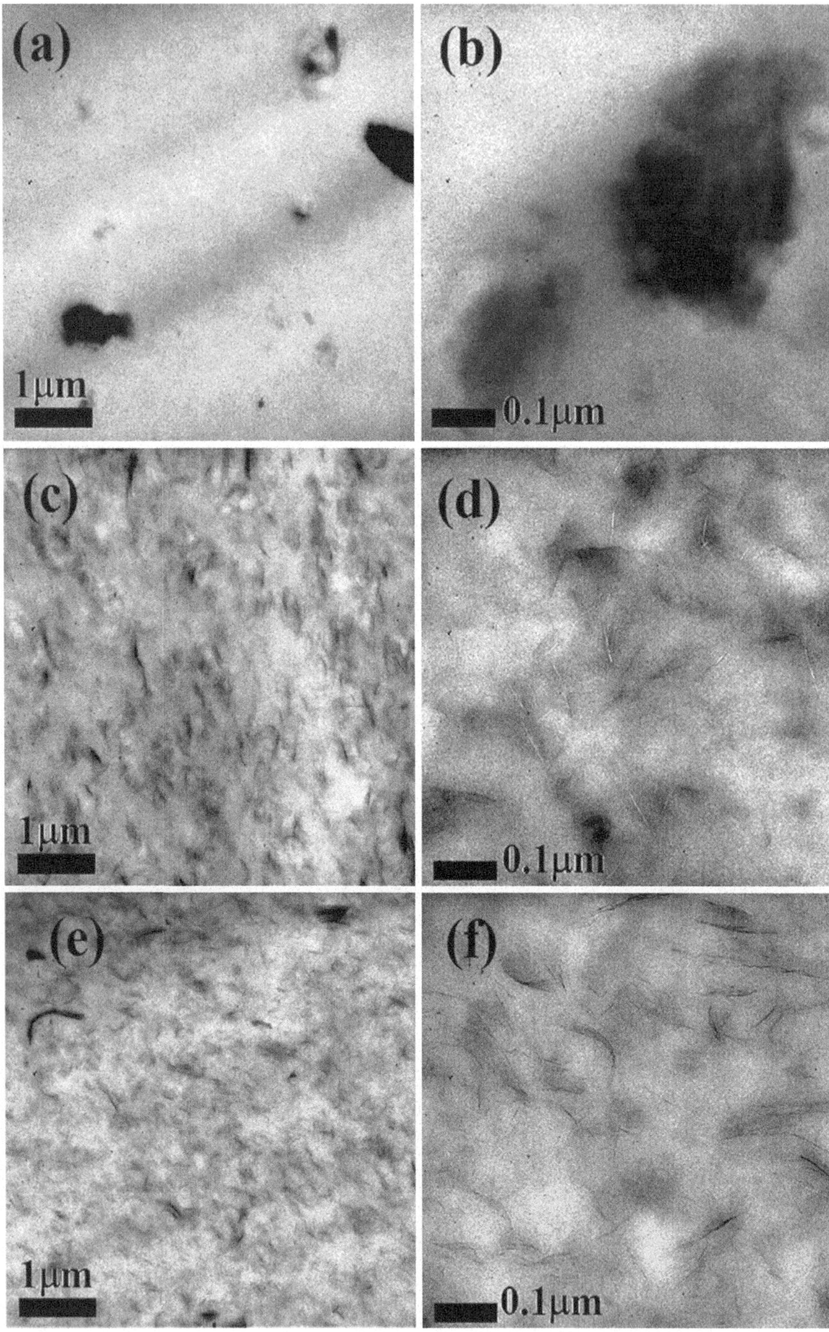

Fig. 7.5 TEM images of the nanocomposites where (**a**, **b**) EVA/Na^+ 5 phr, (**c**, **d**) EVA/15A 5 phr, and (**e**, **f**) EVA/30B 5 phr

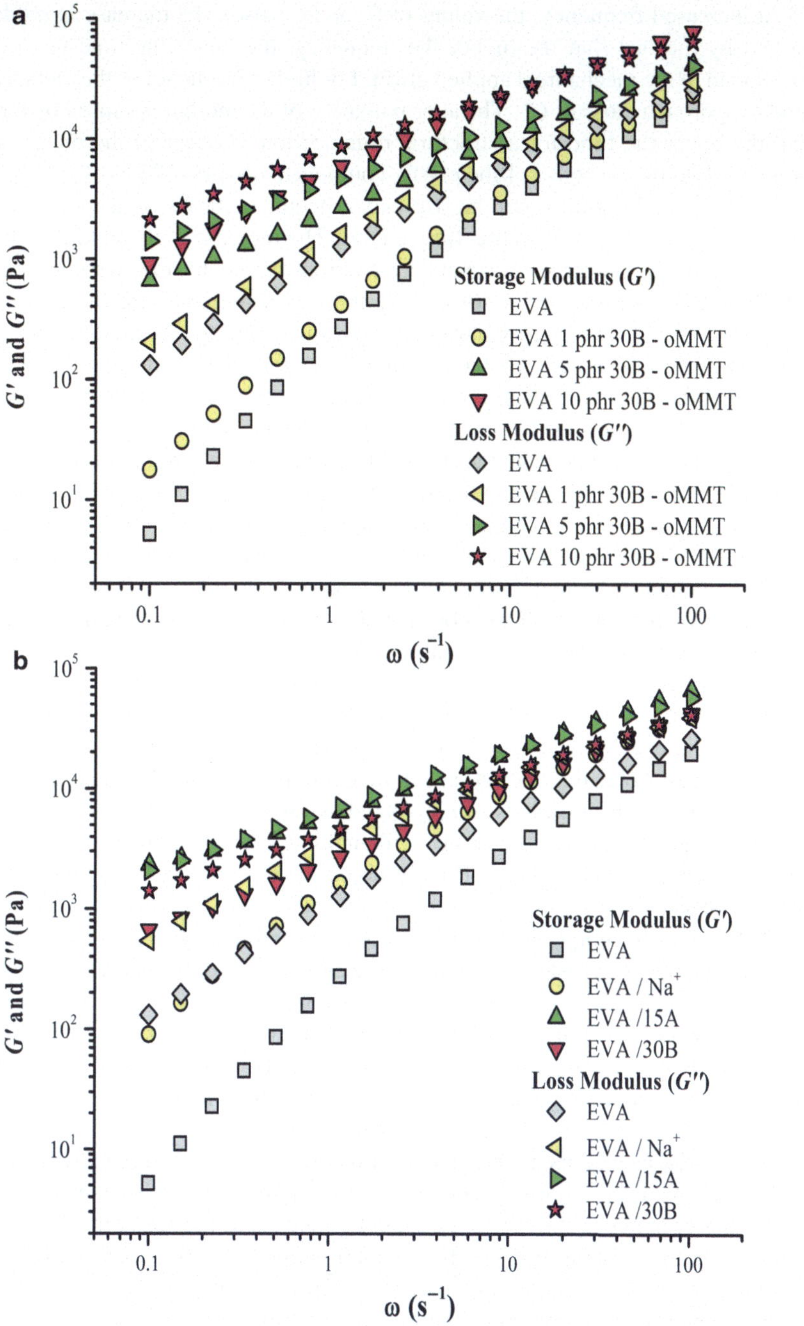

Fig. 7.6 Curves of storage (G') and loss (G'') moduli where (**a**) presents the variation in the concentration of the 30B clay and (**b**) comparative between the Na^+, 15A, and 30B clays

At increased frequency, the values of G' and G'' rise. The increase in moduli is caused by the fact that the higher the frequency, the lower the time for a chain to respond to the mechanical applied effort. For higher frequencies, the values of G' tend to approach those of G''. The approximation of the moduli is caused by the fact that the lower the time for a structural organization to occur in the material, the larger will be the response of the elastic component, that is, G'.

In Fig. 7.6a, the addition of the 30B clay reflected in the rise and approximation of the values of G' and G'' at low frequencies. The increase in the moduli values is associated with the effect of spatial restriction caused by the presence of clay which in this case can be confirmed by the fair dispersion and the exfoliation shown in Fig. 7.4. The approximation of the values of G' and G'' also suggests that there is more conservation of molecular vibration energy throughout the frequency range studied. Energy conservation in this case can be the result of a network of interactions between the clay nanoplates and the EVA matrix.

In Fig. 7.6b, it is observed that the addition of the three types of clay (Na^+, 15A, and 30B) increased and approximated the values of G' and G'' at low frequencies as shown in the samples containing different amounts of the 30B clay. The increase in modulus for the different types of clay follows a logical trend of interaction, that is, the Na^+-containing sample exhibited the lowest modulus increment caused by it having clay clusters which do not spatially hinder the EVA chains' molecular movement ability. The incorporation of the 30B clay resulted in values higher than those found for Na^+ and lower than those found for the addition of the 15A clay. The higher values of modulus found for the 15A clay suggest stronger molecular restriction of the polymer chain in the molten state. Such stronger molecular restriction can be consequent to the dispersion accompanied by higher concentrations of clay platelets intercalated with the EVA chains, while for the 30B clay, the more evident kind of dispersion was exfoliation. Intercalation can lead to more difficult molecular rearrangements than the exfoliated platelets.

Differences in morphology, when associated with the type of surfactant used, can be related to the existence of a disentanglement point of the EVA chains and the chain segments of the 15A nanoclay. Since the 30B nanoclay bears a small number of carbon segments at the surface of the platelets, the increase in the G' modulus as the frequency increases follows a linear trend. However, the higher stability observed with the increase in moduli for both modified nanoclays may be due to the physical interaction between the chain segments of the 15A clay and the terminal OHs, as well as an interaction of the acetate groups of the 30B nanofiller and the EVA matrix. These interactions corroborate the results obtained from the DMA analysis, hence the increase in the T_g values and the reduction in the tan δ height peak in the literature [12]. The approximation in the G' and G'' values at low frequencies occurred in the same way as in Fig. 7.6a, suggesting also in this case higher capacity of vibrational energy conservation. This phenomenon can be observed through the ratio between the loss and storage moduli in Fig. 7.7a, b, which is called the loss factor or loss tangent (tan δ) (Eq. 7.11) [16].

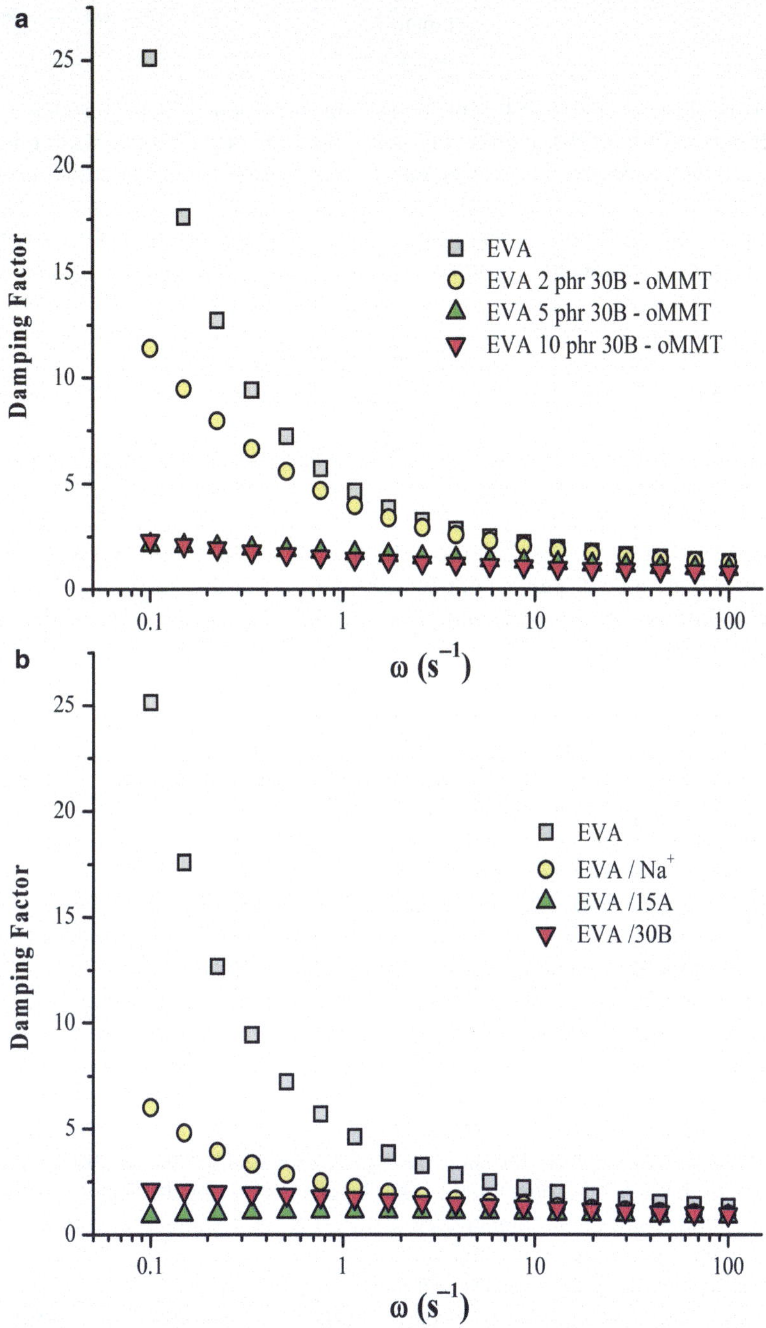

Fig. 7.7 Loss factor curves for the samples studied where (**a**) shows the variation in the concentration of the 30B clay and (**b**) comparative between the Na^+, 15A, and 30B clays

$$\tan\delta = {}^{G''}\!/_{G'} \tag{7.11}$$

For the dimensional stability of the system under study, it is considered that the closer the ratio of the moduli is to unity, the lower the energy loss in relation to its storage capacity, i.e., lower energy dissipation in the system impaired handling of the polymer chains [19]. Hyun et al. [11, 20] carried out rheological measurements on the nanocomposites and reported that the particle dispersion level can be considered as nonassociated, weakly associated, and strongly associated for ($\tan\delta > 3$) ($1 < \tan\delta < 3$) and ($\tan\delta < 1$), respectively. In Fig. 7.7a, it was observed that at frequencies of up to 1 s^{-1}, only the samples containing 5 and 10 phr of OMMT showed strongly associated particle dispersion [20]. In Fig. 7.7b, the same was observed for the 15A clay-containing sample.

Obtaining flow curves observed in Fig. 7.8a, b makes it possible to better understand the phenomena arising from processing, since the viscosity constitutes a function of the shear rate ($\dot{\gamma}$). The increase in viscosity with respect to increasing levels of OMMT (Fig. 7.8a) and the addition of 15A clay (Fig. 7.8b) may also be associated with good dispersion of the clays in the matrix, by the strong interaction of the acetate groups with OH end groups in the 30B clay [4, 17] and the chain segments of the 15A clay [12].

Chaudhary et al. [9], studying the incorporation of 15A and 30B organophilic clays in an EVA matrix, reported that the modified clay 15A is the best alternative when a hydrophobic polymer (like EVA) is used. The long aliphatic chains of the acetate vinyl content of the EVA matrix lead to a better attraction between the galleries of the clays and the polymeric chains. In addition, the modified 30B nanoclay bears bis-(2-hydroxyethyl) tallow in its structure, leading to better interaction with hydrophilic polymers.

Considering the 30B clay and the results illustrated in Fig. 7.8, the VAc content in EVA is a crucial factor in terms of the structure of EVA-clay nanocomposites since a rise in VAc content increased both the polarity and flexibility of EVA chains by promoting the passage of chains between the galleries of the clay, leading to the formation of intercalated structures [4].

Since in this work the VAc content was held fixed, the results depend uniquely from the amount of incorporated clay, and therefore, the rise in modulus and viscosity for the different concentrations of the 30B clay result from the good dispersion and hydrophilic interactions between the OH group-containing surfactant and the EVA VAc group. On the other hand, considering the low VAc content, the best interaction or the higher increment observed for the 15A clay could be linked to stronger physical entanglement effects of the ethylene moieties of EVA with the surfactant chain moieties in the clay.

In both cases, the increase in viscosity in the presence of nanoclays dispersed in a twin-screw extruder may reflect the non-Newtonian behavior becoming more pronounced, probably due to fair dispersion and deeper strain-hardening effect in the isothermal elongational flow as compared to neat EVA [6].

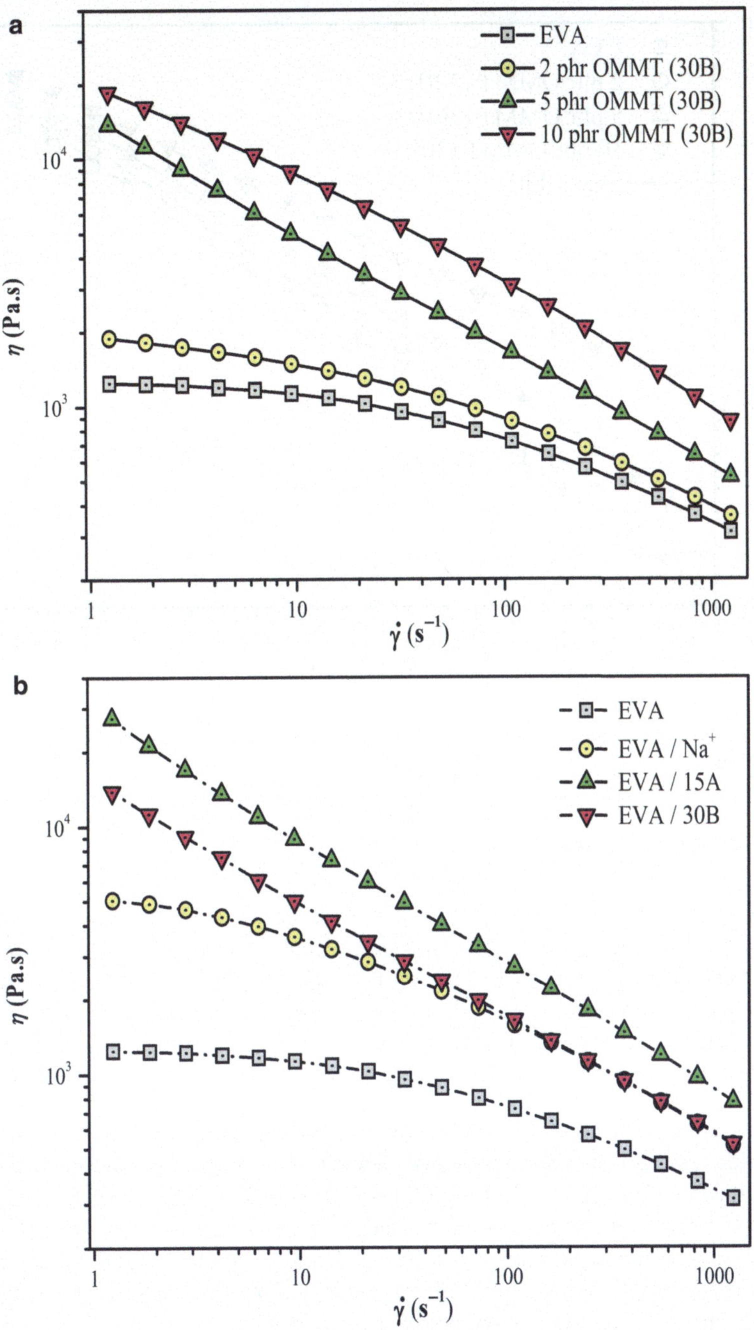

Fig. 7.8 Kinematic viscosity curves ($\eta(\dot{\gamma})$) obtained by the Cox-Merz rule where (**a**) shows the variation of the concentration of the 30B clay and (**b**) comparative between the Na^+, 15A, and 30B clays

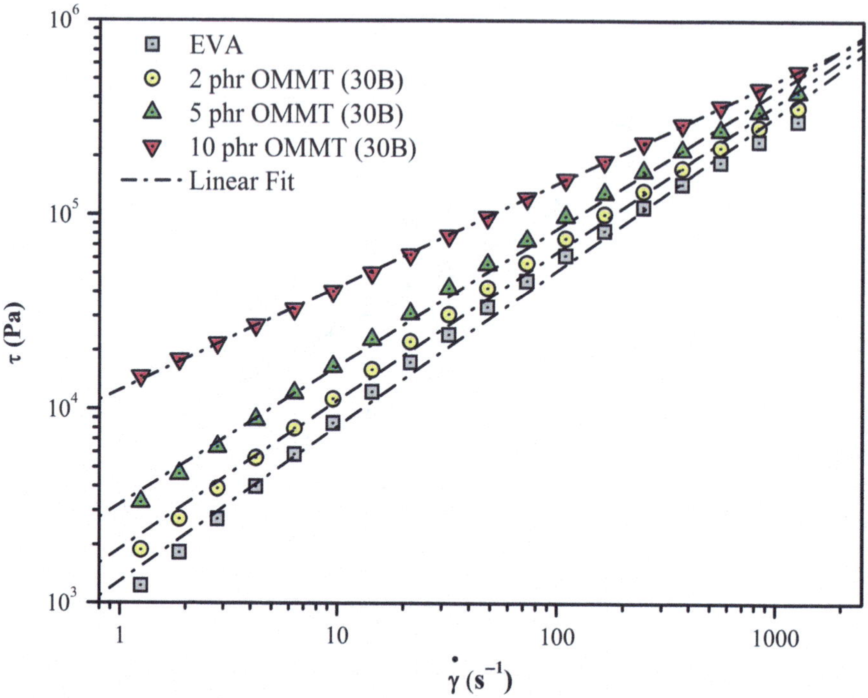

Fig. 7.9 Linear adjustment drawn for obtaining the parameters of the power law for EVA-30B clay samples

Table 7.1 Parameters obtained by the power law

Case 1				Case 2			
Sample	k_p	n_p	r	Sample	k_p	n_p	r
EVA	5805.32	0.67	0.9966	**EVA**	5805.32	0.67	0.9966
30B – 2 phr	8656.27	0.62	0.9984	**Na$^+$**	5759.06	0.66	0.9963
30B – 5 phr	19507.22	0.56	0.9985	**30B**	19507.22	0.56	0.9985
30B – 10 phr	21022.13	0.53	0.9971	**15A**	24552.03	0.49	0.9998

The nanocomposites flow behavior was investigated with the aid of the Cox-Merz rule for obtaining the power law parameters through the linear adjustment illustrated in Fig. 7.9. Through linearization of Eq. 7.1, by isolating the ln τ versus ln $\dot{\gamma}$ terms, the figures for power indices (n_p) and consistency index (k_p) can be obtained by the angular and linear coefficients, respectively. The linear adjustment drawn for the samples from different clays follows the same profile as that of Fig. 7.9, and thus, its illustration can be dispensed with.

Adjustments were drawn at confidence intervals of 95 % and with correlation coefficient (r) close to unity, i.e., $r \approx 1$. Table 7.1 lists the data obtained by the power law. The consistency index (k_p) is the value of viscosity where the shear rate is equal to unity, while the power index (n_p) is a measure of polymer pseudoplasticity, being therefore a consequence of both the disentanglement effect and molecular orientation besides the generation of temperature caused by shear [21]. The closer the value of n_p is to zero, the more pseudoplastic the material is, any slight variation in shear implying in viscosity variations, which impair processing.

The data obtained confirm the k_p rise and the n_p reduction with the increase in the 30B clay content. The k_p figures stand for the stationary viscosity, and the rise associated with the clay content means that there is more movement restriction resulting from nanoconfinement effects and spatial hindrance between the exfoliated clay and the polymer chains.

The progressive k_p rise reflects in increase in the pseudoplastic effect, i.e., n_p reduction as observed in the nanocomposites. The pseudoplastic effect reflects in the occurrence of physical entanglements which are a function of the molecular weight and ramifications present in the material. Further to clay incorporation, the rise in the pseudoplastic effect can be dependent on the nanoconfinement effect which occurred from the interaction of clay with the EVA chains. In this case the way by which the clay will be dispersed and the nanoparticle size will affect directly the level of interaction with the polymer. The more exfoliated clay can lead to better interaction due to factors such as increased surface contact area, percolation of the clay dispersed in the matrix, and a larger volume of nanoparticles dispersed per unit area [10]. These relationships are corroborated by Fig. 7.5, where by increasing the OMMT content, larger distribution of the dispersed clay per unit area can be observed along with deeper exfoliation of the clay also generating a larger contact area.

As for the Na^+ clay addition, no significant variation in the k_p and n_p parameters could be observed since dispersion is not good and clay is located as agglomerates. These agglomerates do not interfere with the EVA flow conditions. For the 15A clay-containing samples, k_p and n_p values were higher than those observed for the 30B clay-containing nanocomposites.

It is also worth noting that consequent to the incorporation of 5 phr of 15A clay, the increment associated with k_p and n_p outweighs the data obtained for the nanocomposite containing 10 phr of the 30B clay. Such difference related to the concentration of 30B and 15A clays suggests that the physical interaction effects in EVA-clay nanocomposites are more pronounced when utilizing a hydrophobic surfactant-bearing clay. This relationship is based on the study of the 28 wt% VAc EVA copolymer.

4.2 Physical Interactions in the EVA/Nanoclay Nanocomposites

According to Matsuoka [22], in the molten state, there are three typical relaxation stages, illustrated in Fig. 7.10. The unlinking stage is associated with chain conformation (1st stage), chain slippage (2nd stage) that is due to the higher number of degrees of freedom and the free volume, and correlated blob slippage (3rd stage).

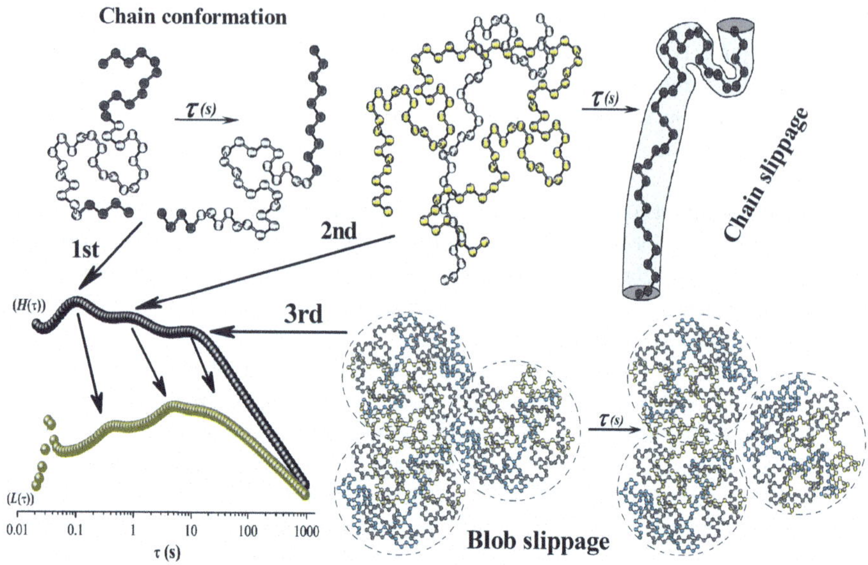

Fig. 7.10 Schematics of the relaxation and retardation steps in the molten state

For each relaxation ($H(\tau)$) stage, there is a closely related retardation ($L(\tau)$) stage. The retardation phenomenon associated with each relaxation step is always out of phase for longer times. The reason is that relaxation is related to the elastic portion while retardation is associated with the viscous portion which in terms of mechanical response takes longer times to return to the equilibrium state. The factors influencing each relaxation stage in the molten state are:

- The first stage depends primarily on the chemical nature of atoms and on the molecular weight. Short chains exhibit higher degrees of freedom and can undergo relaxation more easily. Chemical bonds, such as nonpolar and saturated, are more easily rearranged due to weaker linking forces if compared to polar links, such as those of amide or ester groups besides the presence of unsaturations in the main chain and aromatic rings.
- The second relaxation stage will depend on chain size, branching, and intermolecular interaction forces. Flow can be impaired by the physical entanglement between chains or by the intermolecular forces such as hydrogen bonds which can hinder the mobility of the moieties.
- The third relaxation stage will depend on the level of molecular entanglement, i.e., the higher the entanglement and the more intertwined the chains are, the higher the resistance to flow. Physical entanglements can therefore hinder the second relaxation stage creating the displacement of entanglements or balls which are not able to disrupt in view of their topology.

Taking into consideration the existence of different relaxation phenomena, accordingly the time required for the occurrence of each phenomenon will be different, and since there is a distribution of molecular weight intrinsic to the material, there will be also the distribution of relaxation times. Such distribution is illustrated by the $H(\tau)$ and $L(\tau)$ spectra in Fig. 7.11a, b.

As regards EVA, two distinctive relaxation peaks can be seen at 0.1 and 1 s. The first peak is associated with the first relaxation stage, i.e., the molecular conformation. The second peak refers to the slipping of chains across the direction of the force applied in the oscillatory analysis. Following the addition of OMMT (Fig. 7.11a), peak broadening was observed in the first two stages of relaxation ($\tau \approx 0.1$ s and $\tau \approx 1$ s) and the appearance of the third relaxation phenomenon at $\tau \approx 10$ s resulting from the increase in OMMT content.

The longer relaxation times and the large distribution are related to the elastic component of the system [11, 12, 16], i.e., following OMMT addition, it appears that the system becomes structurally stronger, thereby increasing its ability to store energy and enabling the cooperative effects of more prolonged relaxation and retardation over time.

The appearance of the third relaxation stage suggests that there is a network of interactions which impairs flow and causes the movement of chain entanglements which are restricted by the spatial hindrance of the clay platelets.

In Fig. 7.11b, with incorporation of the Na^+ clay, a distinct behavior in relation to the relaxation phenomenon was not observed, suggesting that the chains have the same degrees of freedom for the relaxation phenomenon when compared with the pure EVA matrix. However, when the 15A clay was incorporated, the appearance of a third relaxation phenomenon (related to correlated blob slippage) was also noted. For the 30B nanocomposite, the third relaxation peak is found at $\tau \approx 10$ s and for the nanocomposite containing 15A, at $\tau \approx 30$ s. The difference observed in the third relaxation time for the 15A and 30B clay-containing nanocomposites could be associated with the different ways of dispersion observed in the TEM images (Fig. 7.5). In the case of an intercalated morphology, it is possible that the slippage of the chains along the flow be impeded, since the platelets inhibit the chain movement. This phenomenon explains the higher shift of the third relaxation phenomenon observed in the nanocomposites containing the 15A clay, suggesting a more intercalated system in relation to the 30B nanocomposite.

In order to better understand the relationship between dimensional stability and creep in the systems studied, Figure 7.12a and b illustrates the curves for the stress relaxation modulus ($G(t)$) and creep compliance ($J(t)$). Higher OMMT contents (Fig. 7.12a) and addition of different clays (Fig. 7.12b) promoted a significant increase in $G(t)$, corroborating the relaxation times shown. This rise in $G(t)$ occurs as a result of increased structural rigidity caused by the presence of the clays forming a physical network between the platelets which can act as an energy dissipation mechanism [11]. These features in the case of the EVA/30B nanocomposites of increasing OMMT levels may also be associated with the strong interaction between the VAc and clay terminal hydroxyl groups [3, 5]. This further contributes to the formation of a physical network which is more rigid due to the exfoliated and intercalated clay morphology. This effect is still more pronounced for the addition of the 15A clay since the intercalated morphology inhibits free flow by chain interpenetration between the clay platelets.

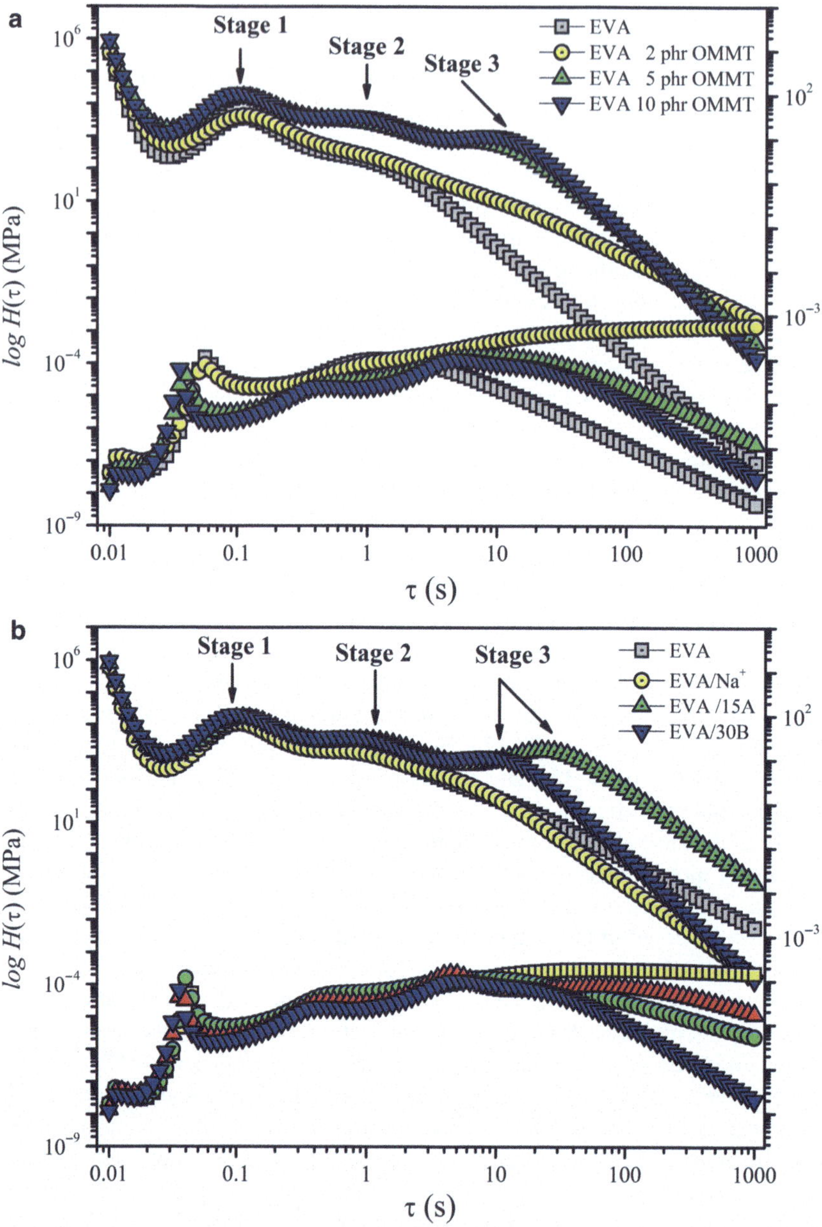

Fig. 7.11 Relaxation and retardation spectra where (**a**) shows the variation in the concentration of the 30B clay and (**b**) comparative between the Na^+, 15A, and 30B clays

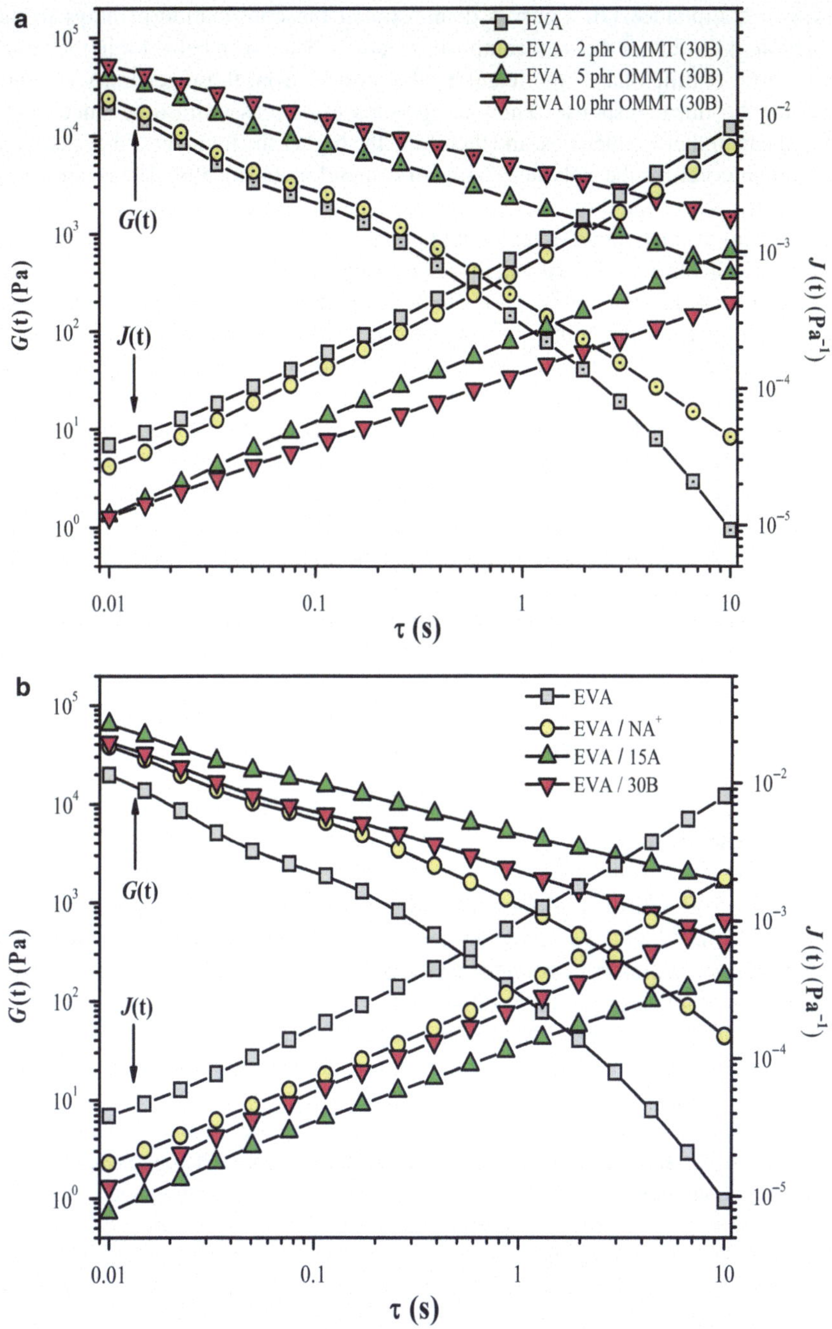

Fig. 7.12 Relaxation modulus and creep compliance curves where (**a**) shows the variation in the concentration of the 30B clay and (**b**) comparative between the Na^+, 15A, and 30B clays

Creep compliance ($J(t)$) curves demonstrated the contribution of clays to the reduction of fluidity in the nanocomposites studied. The creep behavior in the linear viscoelastic regimen has some features which differ from those generally encountered for the linear response. Thus, creep tends to increase with time since lower times mean higher frequencies, and therefore, the higher the frequency, the smaller is the time interval for the polymer chains to respond mechanically. The relationship between the relaxation properties and the creep properties will depend entirely of the summing up of cooperative relaxation phenomena. The creep behavior can be also described as inversely proportional to the viscosity since it depends on the retardation or the deformation which promotes chain flow. In other words the higher the physical entanglement effect, the higher the viscosity and the elastic response of the material, while in the absence of physical entanglements, chain flow and creep will be higher.

In Fig. 7.12a, the addition of the 30B clay reduced $J(t)$, while in Fig. 7.12b, the different types of clay also exhibited reduced $J(t)$ chiefly for the 15A sample. Reduced $J(t)$ corroborates the further results of this work and suggests that the fair interaction of the clays incorporated into EVA chiefly in terms of the rise in viscosity occurs as a result of what can be described as a pseudo-solid behavior. On its turn, the pseudo-solid behavior is dependent on the type of dispersion and the homogeneity of the clay distributed along the matrix [2, 11].

This could explain the higher resistance to flow for the 15A clay-containing nanocomposite as compared with the 30B clay. In the 15A clay, intercalation, which more deeply inhibited the molecular movement, is predominant, while in the 30B clay, exfoliation makes that the stronger flow inhibition is caused by spatial hindrance and by interactions between OH and VAc groups.

5 Conclusions

The study of the influence of the 30B clay concentration and the addition of Na^+, 15A, and 30B clays to a poly(ethylene-co-vinyl acetate) (EVA) matrix was performed by oscillatory rheometry analysis in order to evaluate flow properties and physical interactions in the molten state.

TEM analyses showed that the NA^+ sample formed a microcomposite, the sample containing nanoclay 15A a more intercalated system, and the 30B clay-containing nanocomposite a more exfoliated system.

Oscillatory rheometry data confirm that by increasing the content of the 30B clay, there was a rise in storage modulus, loss modulus, and viscosity. The same was observed for samples containing Na^+, 15A, and 30B clay, respectively, incorporated into EVA.

The power law parameters demonstrated that clay addition favors the rise in stationary viscosity and thus increases the nanocomposites pseudoplasticity. Figures for k_p and n_p demonstrated that the addition of 5 phr of 15A clay promoted a higher increment than even the addition of 10 phr of the 30B clay. The rise in viscosity and the higher flow restriction for the EVA/15A nanocomposite are attributed to the different ways of interaction between EVA chains and the nonpolar surface agent utilized for clay modification.

Data collected from the viscoelastic functions *H*(τ), *L*(τ), *G*(*t*), and *J*(*t*) confirmed this reasoning and demonstrated that the 30B clay exhibits strong interaction with the EVA matrix resulting from the physical interactions of the clay surface agent OH group and the EVA VAc groups. However, the physical interactions in the EVA/15A clay sample were stronger than those of the EVA/30B clay sample since the physical entanglement effects between the 15A clay modifier are favored by the intercalated form and by the higher ease of physical entanglements with EVA.

6 Future Developments in the Area EVA/Nanoclay Nanocomposites

The future development of EVA/nanoclay nanocomposites is mainly in the footwear industry. The EVA has high impact strength at low temperatures, can be transparent and flexible, has good elasticity, and is compatible with other thermoplastics. These characteristics are supported in manufacture of footwear. The presence of the clay contributes even more with the mechanic strength, is less abrasive, and improves the flow stability of the polymer. Moreover, it is possible to study the preparation of nanocomposites with other types of clay. These researches can improve the dispersion techniques and the polymer/clay interaction.

References

1. Yamaki SB, Prado EA, Atvars TDZ (2002) Phase transitions and relaxation processes in ethylene-vinyl acetate copolymers probed by fluorescence spectroscopy. Eur Polym J 38:1811–1826. doi:10.1016/S0014-3057(02)00067-8
2. Marini J, Branciforti MC, Lotti C (2009) Effect of matrix viscosity on the extent of exfoliation in EVA/organoclay nanocomposites. Polym Adv Technol 21:408–417. doi:10.1002/pat.1444
3. Riva A, Zanetti M, Braglia M, Camino G, Falqui L (2002) Thermal degradation and rheological behaviour of EVA/montmorillonite nanocomposites. Polym Degrad Stab 77:299–304. doi:10.1016/S0141-3910(02)00065-4
4. Zhang W, Chen D, Zhao Q, Fang Y (2003) Effects of different kinds of clay and different vinyl acetate content on the morphology and properties of EVA/clay nanocomposites. Polymer 44:7953–7961. doi:10.1016/j.polymer.2003.10.046
5. Costache MC, Jiang DD, Wilkie CA (2005) Thermal degradation of ethylene-vinyl acetate copolymer nanocomposites. Polymer 46:6947–6958. doi:10.1016/j.polymer.2005.05.084
6. La Mantia FP, Dintcheva NT (2006) Eva copolymer-based nanocomposites: Rheological behavior under shear and isothermal and non-isothermal elongational flow. Polym Test 25:701–708. doi:10.1016/j.polymertesting.2006.03.003
7. Mishra SB, Luyt AS (2009) Effect of organic peroxides on the morphology and properties of EVA/Cloisite 15A nanocomposites. J Appl Polym Sci 112:218–225. doi:10.1002/app.29400
8. Sureshkumar MS, Filippi S, Polacco G, Kazatchkov I, Stastna J, Zanzotto L (2010) Internal structure and linear viscoelastic properties of EVA/asphalt nanocomposites. Eur Polym J 46:621–633. doi:10.1016/j.eurpolymj.2009.12.024
9. Chaudhary DS, Prasad R, Gupta RK, Bhattacharya SN (2005) Morphological influence on mechanical characterization of ethylene-vinyl acetate copolymer–clay nanocomposites. Polym Eng Sci 45:889–897. doi:10.1002/pen.20349

10. Pavlidou S, Papaspyrides CD (2008) A review on polymer–layered silicate nanocomposites. Prog Polym Sci 33:1119–1198. doi:10.1016/j.progpolymsci.2008.07.008
11. Pistor V, Lizot A, Fiorio R, Zattera AJ (2010) Influence of physical interaction between organoclay and poly(ethylene-co-vinyl acetate) matrix and effect of clay content on rheological melt state. Polymer 51:5165–5171. doi:10.1016/j.polymer.2010.08.045
12. Pistor V, Ornaghi HL Jr, Ferreira CA, Zattera AJ (2011) Performance of poly(ehtylene-co-vinyl acetate) nanocomposites using distinct clays. J App Polym Sci 125:E462–E470. doi:10.1002/app.34804
13. Roths T, Marth M, Weese J, Honerkamp J (2001) A generalized regularization method for nonlinear ill-posed problems enhanced for nonlinear regularization terms. Comput Phys Commun 139:279–296. doi:10.1016/S0010-4655(01)00217-X
14. Weese J (1993) A regularization method for nonlinear ill-posed problems. Comput Phys Commun 77:429–440. doi:10.1016/0010-4655(93)90187-H
15. Riande E, Díaz-Calleja R, Prolongo MG, Masegosa RM, Salom C (2000) Polymer viscoelasticity: stress and strain in practice. Marcel Dekker, New York
16. Ferry JD (1980) Viscoelastic properties of polymers. Wiley, New York
17. Lee HM, Park BJ, Jin Choi H, Gupta RK, Bhattachary SNJ (2007) Preparation and rheological characteristics of ethylene-vinyl acetate copolymer/organoclay nanocomposites. Macromol Sci, Part B: Phys 46:261–273. doi:10.1080/00222340601066956
18. Rodríguez-Vázquez M, Liauw CM, Allen NS, Edge M, Fontan E (2006) Degradation and stabilisation of poly(ethylene-stat-vinyl acetate): 1. Spectroscopic and rheological examination of thermal and thermo-oxidative degradation mechanisms. Polym Degrad Stab 91:154–164. doi:10.1016/j.polymdegradstab.2005.04.034
19. Hong-mei Y, Qiang Z, Miao D (2006) Rheological behavior of the melts for polyethylene-based montmorillonite nanocomposites. Chem Res Chin Univ 22:651–657. doi:10.1016/S1005-9040(06)60182-7
20. Hyun YH, Lim ST, Choi HJ, Jhon MS (2001) Rheology of poly(ethylene oxide)/organoclay nanocomposites. Macromolecules 34:8084–8093. doi:10.1021/ma002191w
21. Sperling LH (2006) Introduction to physical polymer science. Wiley, New York
22. Matsuoka S (1992) Relaxation phenomena in polymers. Hanser, Munich

Polypropylene Clay Nanocomposites

8

Kummetha Raghunatha Reddy

Contents

Keywords
Flameretardants • Isotactic polypropylene • Layared silicate nanocomposites • Maleic anhydride • Metallocene

K.R. Reddy
Department of Future Industry-Oriented Basic Science and Materials, Toyota Technological Institute, Tempaku, Nagoya, Japan

School of Science and Technology, Kwansei Gakuin University, Sanda, Hyogo, Japan
e-mail: kr_raghunath@yahoo.com

J.K. Pandey et al. (eds.), *Handbook of Polymernanocomposites. Processing, Performance and Application – Volume A: Layered Silicates*, DOI 10.1007/978-3-642-38649-7_2,

1 Introduction

Recently, research on polymer/layered silicate nanocomposites has been received great interest because they often show superior property improvements with respect to pristine polymers or conventional composites [1–11]. The extent of enhancement in properties depend on the interfacial interaction between the polymer matrix and the layered silicate layers. There are many types of silicate layers which have been used for the filling in polymers, such as montmorillonite (MMT), hectorite, saponite, laponite, halloysite, and kanemite, etc [9]. Unlike traditional filled polymer systems, nanocomposites require relatively low filler loadings to achieve significant improvements in the properties. These materials are called nanocomposites because of the finely dispersed nanometer thickness clay sheets in the polymer matrix. In addition to their properties, the discovery of simple processing methods such as melt compounding provided a potentially commercially viable method to produce nanocomposite materials on a large scale [12].

Isotactic polypropylene (PP) is one of the most important commodity polymers and has been extensively used in several applications, including as interior and exterior decoration materials of automobiles, home electric appliances, etc. PP has a simple chemical structure ($-[CH_2\text{-}CH(CH_3)]_n-$) and is one of the most basic and important polymers from both the industrial and scientific points of view [13]. From the commercial perspective, PP is one of the most important members of the polyolefin family due to its outstanding combination of low cost, low weight, moderate heat distortion temperature of about 100 °C, along with extraordinary usefulness in terms of properties applications and *recycling* [14–30]. To improve the range of its applicability, it is essential to improve its toughness and thermal resistance so that it meets the demands of engineering products. Therefore, it is proposed that the uniform dispersion of layered silicates in the polymer matrix must enhance its mechanical performance [19–30].

This chapter briefly describes the advancement in this area, along with possible future developments. The chapter mainly focuses on the preparation methods of polypropylene clay nanocomposites (PPCNs) and their significant aspects in terms of properties, structure, and morphology behaviors will be discussed.

2 Polypropylene/Layered Silicate Nanocomposites

In the layered silicate nanocomposites, the most commonly used clay is MMT (a smectite type), which contains 1-nm-thick layers, where alumina is located in the center as an octahedral sheet sandwiched between two silica tetrahedral sheets. The clays have negative charges generated on the surface by isomorphic substitution, balanced by cations, which is known as the cation-exchange capacity (CEC). Layered silicate clays are good candidates for the preparation of inorganic–organic hybrid nanocomposites because they are abundant in nature, inexpensive chemically, and thermally stable. The parallel-organized sheets exist in the form of stacks with regular van der Waal's gaps, called the interlayer space of the gallery where

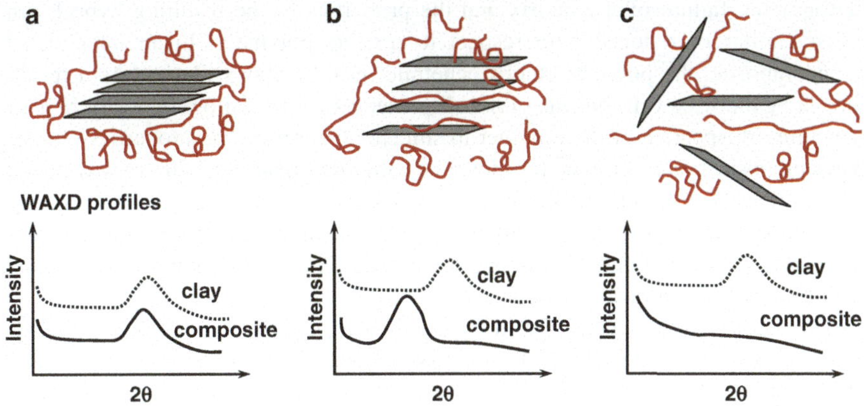

Fig. 8.1 Schematic representations of different states of clay dispersion and their corresponding WAXD profiles. (**a**) Phase-separated microcomposite. (**b**) Intercalated nanocomposite. (**c**) Exfoliated nanocomposite

the cations are located. The replacement of inorganic exchangeable cations with organic *onium* ions on the gallery surfaces decrease the attractive force between the clay galleries and enhance the compatibility between the clay and the polymer matrix [9, 31, 32]. Another method to improve the compatibility between the clay and the polymer matrix is that the polymer matrix can be functionalized either by grafting or by the incorporation of functionalized monomers in the main chains during the synthesis of the polymer. In particular, for non-polar polymers like PP, in addition to those primary modifications, many special efforts have been made, such as etching with acids, plasma modification, irradiation, etc. The use of a functionalized polymer as a master batch is another common method during the melt intercalation method. Maleic anhydride (MA) grafted on polyethylene (PE) or PP (MA-g-PP/PE) has been extensively used as a compatibilizer during nanocomposites preparation in the presence of peroxide as the initiator. This grafted material enhances the dispersion of clay in the polymer matrix and improves the interfacial adhesion between the filler and the polymer matrix, which ultimately helps in enhancing the material properties.

The ultimate aim of nanocomposites preparation is to attain the best properties by achieving a homogeneous and uniform distribution of silicate layers in the polymer matrix. As mentioned earlier, the layered silicates have a layer thickness on the order of 1 nm and very high aspect ratio (e.g., 10–1,000 nm), therefore, creating a large surface area between polymers and clay platelets, and enhancing the *polymer–filler* interaction more than in conventional microcomposites. Based on the extent of the layered silicate distribution in the polymer matrix, i.e., depending on the strength of the interfacial interaction between the polymer matrix and the layered silicate, three different types of structures are expected: phase-separated nanocomposites (microcomposites), intercalated nanocomposites, and exfoliated nanocomposites (Fig. 8.1).

In the phase-separated nanocomposites, the polymer chains are confined in the clay and remain as tactoids. These tactoids act as stress concentrators, causing cartographic failure of the matrix and the properties of the resulting hybrids are either similar or reduced with respect to pristine polymers. In an intercalated nanocomposite, the polymer chains penetrate into the clay galleries. As a result, the clay galleries swell, but the clay still preserves its crystallographically regular structure, irrespective of the polymer to silicate layer ratio. The properties of such nanocomposites are similar to those of ceramic materials. In an exfoliated nanocomposite, the individual silicate layers separated in the polymer matrix by an average distance. In general, polymer clay nanocomposites often exhibit a mixed morphology, in which the three above-mentioned types of dispersion states coexist within the same composite material.

The degree of clay dispersion in the polymer matrix is established using wide-angle X-ray diffraction (WAXD) analysis and transmission electron microscopy (TEM) image observations. WAXD is the most commonly used technique because of its ease of use and possibility to probe the interlayer spacing of the clay in pristine clays, as well as in the composite materials. By measuring the diffraction angle (2θ), one can calculate the average interlayer spacing (d) of a clay using Bragg's equation $\lambda = 2d \sin \theta$, where λ is the wavelength of incident X-ray radiation. By knowing the position and shape of the clay's basal peaks, the extent of clay dispersion in the nanocomposites can be quantified. For example, in the case of exfoliated nanocomposites, the clay layers are homogeneously dispersed in the polymer matrix, which results in the absence of any coherent X-ray diffraction from the silicate layers. On the other hand, in the intercalated nanocomposites, the clay galleries are expanded due to the intercalation of polymer chains into the clay galleries, which result in the appearance of a new reflection or shift of the diffraction peak to the lower 2θ side in the WAXD profiles. In this way, WAXD is, undoubtedly, a convenient and precise method to determine the interlayer spacing of the silicate layers in polymer silicate nanocomposites. More often, TEM is used as a complement to WAXD to directly observe the level of clay dispersion in the intercalated nanocomposite, although how small a specimen area under examination can explain the dispersion states of silicates inside the matrix remains a significant area of research. More recently, small-angle X-ray scattering and Fourier transform infrared spectroscopy (FTIR) measurements have been used to reveal the extent of clay galleries dispersion in the polymer matrix [33, 34].

3 Preparation of Polypropylene/Layered Silicate Nanocomposites

Many efforts have been made to develop PP nanocomposites with organic silicates by utilizing various appropriate preparative methodologies. The preparation methods used for PPCNs are broadly classified into three main categories:

1. In situ polymerization
2. Melt compounding
3. Solution blending

In addition to these common processing methods, there have been several published reports on some special preparative methods. The most common methods used in the preparation of PP nanocomposites have been extensively reviewed. In this chapter, the discussion is limited to the various popular approaches that have been used for the preparation of PP nanocomposites.

3.1 In situ Polymerization

In this method, generally, the organic modified layer silicate (OMLS) is swollen in the monomer or a monomer solution, so that the polymer formation occurs in between the clay galleries. Polymerization can be initiated by either heating or radiation using a suitable initiator, or catalyst is fixed via the cation exchange process inside the inter layer before swelling the clay galleries with monomer. It is believed that an in situ polymerization is an effective method to obtain exfoliated PPCNs compared with melt and solution blending methods. Many efforts have been made to obtain PP nanocomposites via in situ polymerization [20, 35–44]. The in situ polymerization methods used for the preparation of PE nanocomposites can be extended to prepare PP nanocomposites. Mainly, two approaches have been reported for obtaining polyolefin nanocomposites by using the in situ polymerization method. The first route is the direct synthesis of the PP nanocomposites in the presence of the nanosheets/particles, where the clays are placed in contact with the catalytic system (metallocene catalyst/MAO) [35]. In the second route, the catalytic system is fixed on the surface of the clay prior to the initiation of the polymerization reaction [36–38]. Sun and Garces [39] proposed an original method for the preparation of high-performance PP nanocomposites, where no external activators, such as methylaluminoxane or perfluoroarylborates, are needed to initiate the olefin polymerization and high pressures or high temperatures during processing are also not required. The methodology can be readily extended to make other polyolefins, like PE and poly(1-butene) clay nanocomposites. Tudor and co-workers [40] demonstrated the in situ polymerization method. They showed that a soluble metallocene catalyst is intercalated into the layered silicates, followed by polymerization reactions. First, a synthetic hectorite was treated with methylaluminoxane (MAO) (a common cocatalyst used in metallocene catalyst-based polymerizations) in order to remove all acidic protons (acidic protons are also known to be poisonous to the catalyst), which enables the interlayer spacing to receive metallocene catalysts. By the addition of metallocene catalyst ($[Zr(\eta\text{-}C_5H_5)Me(THF)]^+$), a cation exchange reaction occurs between Na^+ in MAO-treated hectorite and the metallocene catalyst; as a result, an increase in the interlayer spacing of silicate clay is possible. In another attempt, by using synthetic fluorinated mica-type clay with a negligible proton in the clay galleries, the catalyst was intercalated directly with clay galleries without the need for MAO treatment. Yang et al. [20] demonstrated the effect of in situ polymer chain functionalization on the stability of the nanocomposite structure by copolymerizing a propylene monomer with 5-hexenyl-9-borabicyclo[3.3.1]nonane (5-hexenyl-9-BBN) with a Ziegler–Natta catalyst which was intercalated initially with organically modified montmorillonite (OMMT) and diphenyldimethoxysilane (DDS) as the external donor. The resultant hydroxyl-functionalized PP/MMT nanocomposites exhibit enhanced structural stability against processing conditions as compared with pristine or unfunctionalized PP nanocomposites. In a further attempt by He et al. [44] alkyltriphenylphosphonium-modified montmorillonite (PMMT) was used to prepare PP exfoliated nanocomposites in the presence of a heterogeneous compound catalyst $TiCl_4/MgCl_2$/PMMT. Before initiating the polymerization reaction, the compound catalyst system was pretreated with toluene, followed by $AlEt_3$, dipenyldimethoxylsilane, and, finally, pretreated catalyst slurry or powder were added successively to the monomer to initiate polymerization. Thus, the resulting PP/PMMT nanocomposites were found to be exfoliated (based on XRD and TEM experiments) with improved thermal stability.

3.2 Melt Compounding

This method became popular since its invention by Giannelis et al. in 1993 [13]. It has been the most popular way to prepare PP nanocomposites due to its flexibility with the existing processing techniques, like extrusion, blow, and injection molding. Melt blending involves mixing of the clay or modified clay with PP at above the melting temperature of PP, in an extruder or mixer, preferably under shear. The level of clay dispersion in the nanocomposite depends on the mixing conditions and the compatibility between the polymer and the layer silicates. Favorable enthalpy of mixing between the polymer and layered silicates can be expected when the polymer–clay interactions are more enhanced than the surfactant–clay interactions. In order to improve the polymer–clay interaction in PP nanocomposites, many modifiers or compatibilizers have been reported [16, 21–30, 45–54]. Moad et al. [54] used various copolymers as dispersants/intercalants/exfoliants in PP clay composites without clay modification. Polyethylene oxide (PEO)-based nonionic surfactants and amphiphilic copolymers based on a long chain (meth)acrylate (e.g., octadecyl acrylate) and a more polar comonomer (e.g., *N*-vinylpyrrolidone, methyl methacrylate) were applied during melt mixing. Nanocomposites were prepared by melt mixing in a twin-screw extruder by first forming a master batch having 10–70 wt% clay and other components, which was then melt-blended with PP to obtain a composite of the required clay content. The properties of nanocomposites were found to depend strongly on both the level of dispersant and its overall composition. The tensile modulus increased up to 40 % in comparison to organically modified clays. The additives provide substantially improved clay dispersion and cause partial exfoliation. Cui and Paul [23] prepared a PP-g-NH_3^+ functional compatibilizer from PP-g-MA by reacting with 1,12-diaminododecane and followed nanocomposite preparation by the melt mixing method. In the prepared nanocomposites, the NH_2 or ionic –NH_3^+ groups do not induce a better interaction for a better clay dispersion than the usual PP-g-MA compatibilizer. The PP-g-MA/organoclay and PP/PP-g-MA/organoclay

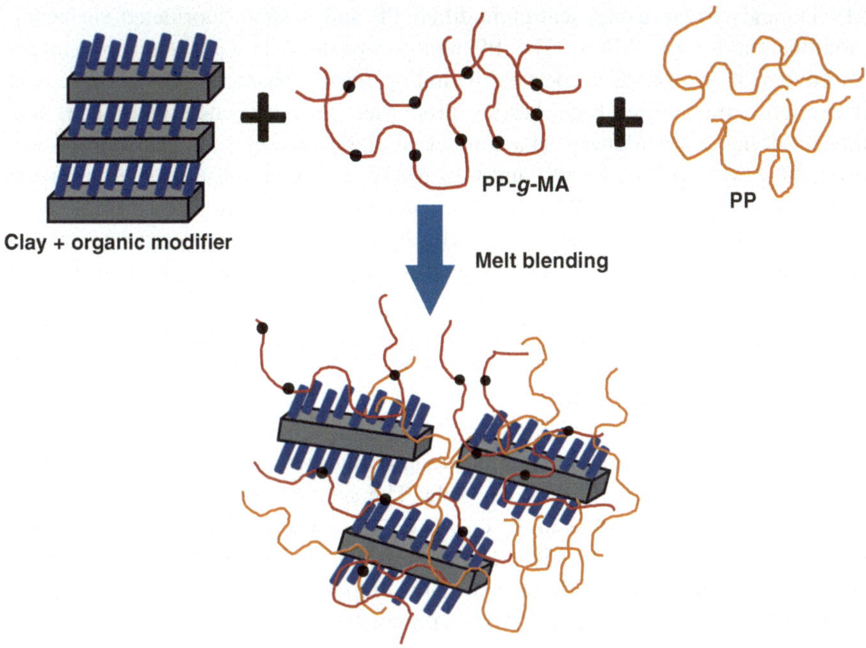

Fig. 8.2 Schematic representation of the melt compounding method used to prepare PPCN in the presence of PP-g-MA compatibilizer

hybrids showed the most substantial increases in mechanical properties and clay dispersion as compared to other polymer clay hybrids. In another attempt, Mulhaupt et al. [55] suggested that octadecylamine reaction with the anhydride group facilitates strong hydrogen bonds between silanol groups of the clay surface. They also discussed the effect of carbon chain length on the clay dispersion.

Among the many compatibilizers, maleic anhydride grafted polypropylene (PP-g-MA) is the most widely used for the development of PP nanocomposites [16, 21–30, 45–53]. This is due to the strong hydrogen bonding formation between the maleic anhydride (or COOH groups generated after hydrolysis) and the oxygen groups of the clay surface. The interaction between PP-g-MA and clay layers help to expand clay galleries; as a result, the interaction between clay layers weakens. In many approaches, first, a master batch with high clay loadings of PP-g-MA is prepared, which is then mixed with PP to form a composite with the required clay content. In this process, when the PP-g-MA intercalated clay contacts with PP under processing, due to a strong shear fields, delamination of the clay layers may be facilitated if there is a great enough interaction strength existing between the PP and PP-g-MA to form a molecularly dispersed clay nanocomposite (Fig. 8.2). Manias et al. [6] reviewed the preparation of PPCNs with the coexistence of exfoliated and intercalated MMT layers by the melt compounding method. In their approach, they applied two steps to prepare the nanocomposites: (i) by introducing functional groups in PP and the use of common alkylammonium

MMTs, and (ii) by using neat/unmodified PP and a semi-fluorinated surfactant modification for the MMT. The PP nanocomposites showed better concurrent improvement in several properties, including tensile characteristics, higher heat deflection temperature, high barrier properties, better scratch resistance, and increased flame retardancy. Okamoto et al. [56] prepared PP nanocomposites using PP-g-MA (0.2 wt% MA) and C18-MMT in a melt extrusion method using a twin-screw extruder at 200 °C. The prepared nanocomposites with 2 wt% clay showed an exfoliated structure; with 4 wt% clay, a disordered intercalated structure; and with 7.5 wt% clay content, the nanocomposite showed an ordered intercalated structure.

3.3 Solution Blending

In this method, the layered silicate or organic modified silicate is first swollen in solvent in which the polymer also soluble. The swollen clay enables the polymer chains to intercalate in between the clay layers by replacing the solvent molecules in the galleries. When the solvent is removed under vacuum, the intercalated structure remains to form intercalated nanocomposites. There are several reports published on the preparation of PP nanocomposites based on the solution blending method [57–61]. Avella et al. [57] prepared nanocomposites by dissolving PP in *O*-dichlorobenzene at 180 °C in a specially made cylindrical flask. The modified organoclay (OMMT) was dispersed in *O*-dichlorobenzene by ultrasonic mixing, followed by mixing of the PP solution. The PP/organoclay nanocomposites were separated by precipitation in cold ethanol, and the solvent was removed in a separating funnel. It was found that, in the nanocomposites with high clay contents (5 %), the clay tactoids remain as clusters in some regions, which cause inhomogeneity, poor adhesion, and, as a result, deterioration of mechanical properties. In another attempt, Chiu and Chu [58] prepared nanocomposites by dissolving PP and clay in 1,2,4-trichlorobenzene (TCB). The solutions were kept at high temperature until clear solutions were formed, and these solutions were cast onto stainless dishes that were kept around 140 °C to evaporate the TCB solvent. XRD and TEM results revealed that the clay layers were dispersed finely (at the nano level), even without using compatibilizer. In another attempt, Oya and Kurokawa [60] demonstrated another somewhat complex procedure to prepare hydrophobic PPCNs. In addition to the solution technique, diacetoneacrylamide was polymerized prior to the addition of PP solution in toluene. However, the expected clay dispersion was not achieved, but, still, the formed nanocomposites had excellent mechanical properties.

3.4 Other Methods

In addition to those common methods to prepare PP nanocomposites, there are several other methods incorporating specially designed instruments and accessories. Nowicki and coworkers [26] introduced radiation-induced modification of

MMT to improve the compatibility between inorganic fillers and the polymer matrix, where the filler was modified with MAH, followed by irradiation with an electron beam. It has been demonstrated that the method is an effective approach in overcoming the organophobic character of inorganic fillers. In polymeric composites, the radiation treatment generates stable carbon-centered free radicals in the organic modifier, which induce filler–matrix linkages. Kato and co-workers [62, 63] developed a direct compounding method for the preparation PP nanocomposites, in which a specially designed twin-screw extruder, equipped with four geometrical sections and a much longer length-to-diameter (L/D) ratio (77:1), is used. In the first section of the extruder, all the materials, comprising PP, PP-g-MA, pristine clay, and alkylammonium salt, are fed and premixed. In the second stage, water is injected into the extruder to disperse the clay into PP in a slurry state, followed by the evaporation of water in the third and fourth sections. With this method, it was found that the clay was dispersed finely in an exfoliated structure. The PP nanocomposites prepared by this method showed better improvement in the properties as compared to composites prepared by the conventional method. Lapshin and Isayev [64] demonstrated an ultrasound-aided extrusion process for the preparation of PPCNs. The nanocomposites were prepared using a co-rotating twin-screw extruder followed by a single-screw extruder equipped with an ultrasonic die. The thus formed nanocomposites showed enhancement in the properties. Wang et al. [65] prepared the PP nanocomposites using a dynamic packing injection molding technique. It was found that, in the nanocomposites prepared by this method, the exfoliation structure increase from the skin to the core.

Marchant and Jayaraman [66] quantitatively studied the role of compatibilizer in the preparation of PPCNs. They found that a significant exfoliation of clay in the composites occurs at an optimum compatibilizer concentration of 10 wt%. Wang and co-workers [67] synthesized end-functionalized PP that contained a functional group (like Cl, OH, COOH, etc.) at the chain end. The ammonium (PP-t-NH_3^+)-terminated PP is directly compounded with clay. The resultant nanocomposites had an exfoliated structure.

4 Structure and Crystallinity of PP Nanocomposites

Semicrystalline polymers are unique among engineering materials, since they are the only common technological materials in which a complicated higher-order structure consists of stacks of lamellar crystals, with amorphous and intermediate regions intercalated between them. For semicrystalline polymers, this structure plays an important role in controlling the mechanical properties of the product. PP has a simple chemical structure ($-[CH_2\text{-}CH(CH_3)]_n-$) but complicated higher-order structure, which may be because it exhibits polymorphism. By adding nanoclay to the PP matrix, the situation becomes more complicated. In this section, the structure and properties exhibited by PP nanocomposites and the mechanism for the formation of such structures will be discussed. That is to say, it is essentially important to reveal the crystal structure, morphology, and orientation of

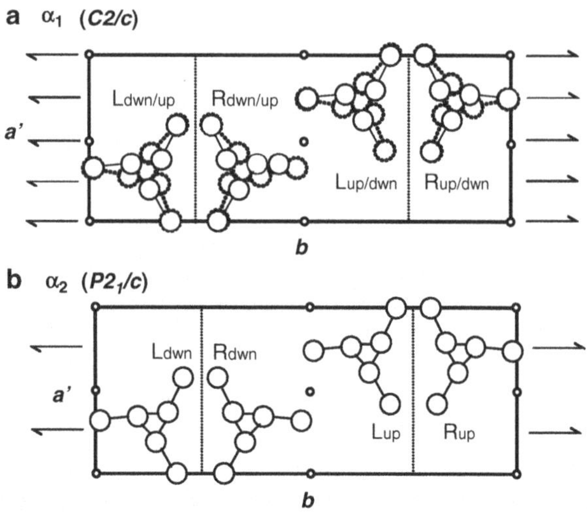

Fig. 8.3 Schematic representation of the crystal structures of iPP. (**a**) α_1 form with statistically random packing of upward and downward chains (space group *C*2/*c*). (**b**) α_2 form with regular chain packing structure (space group $P2_1/c$)

polymer chains developed in the crystalline state, as well as the orientation of silicate layers in the amorphous state, because these microstructures sensitively affect the properties.

4.1 Crystal Structure of PP and Its Effect on Nanofiller Dispersion

The crystalline structure of PP is complex because it crystallizes in several crystalline modifications, depending on the crystallization conditions. It crystallizes in four different crystalline modifications of monoclinic (α_1, α_2), trigonal (β), orthorhombic (γ), and hexagonal (smectic) systems (Fig. 8.3) [68]. In all of the crystal structures, the chains are packed in the crystal lattice as left- or right-handed (or both) (3/1) helix conformation with upward or downward directions of methyl groups. The disordered α_1 is obtained by slow cooling of the melt. The α_1 form exhibits the crystal structure of space group *C*2/*c* or *Cc*, in which the (3/1) helical chains with $(TG)_3$ conformation are packed upward and downward randomly with 50 % probability, where T and G are *trans* and *gauch*e bonds, respectively [68–71]. When the α_1 sample is annealed at higher temperatures, it transforms into the α_2 form with a structure in which the (3/1) helices are packed regularly upward and downward with the space group $P2_1/c$ (Fig. 8.3) [69, 71]. The γ form is obtained relatively easily for the iPP sample polymerized with Ziegler–Natta catalyst under high pressure [72, 73], for the low-molecular-weight compounds at atmospheric pressure [74, 75], and for the ethylene–polypropylene copolymers [76, 77]. The γ form (with space group F_{ddd} and orthorhombic unit cell) can also be obtained from the iPP samples polymerized with the metallocene catalyst system and contains the *regio* and *stereo* defects [77, 78]. The β form (with space groups $P3_1$ or $P3_2$ and trigonal unit cell) can be obtained by adding special nucleating agents [79]. In the actual structure, these crystal forms are created in a mixed fashion at different ratios, depending on the sample preparation conditions.

4.2 Crystalline Morphology and Crystallization Behavior

Zheng and co-workers [80] studied the effect of clay on the polymorphism of PP in PPCNs. They found that, in both extruded PP and PP nanocomposite samples, only α-form crystallites were formed. They also prepared the samples of PP and PPCNs under compression. It was found that, in the pure PP samples which were compressed at 200 °C and crystallized at various temperatures and times, the maximum amount of β crystallites were formed. This indicates that the crystallization temperature has a significant effect on the formation of β-phase crystallites in neat PP. In the compressed films of PPCN, prepared under the same thermal and processing conditions, no β-phase crystals were detected. This reveals that the clay galleries significantly inhibit the formation of β-phase PP crystallites in the nanocomposites (Fig. 8.4). The results also indicated that the clay might support the formation of γ-phase crystallites in PP nanocomposites, as the γ-phase crystallites are nucleated on the α-phase lamellae, which is similar to the case of γ-phase crystal formation in neat PP under pressure. In fact, Nam et al. [56] observed an enhancement in the formation of γ-phase crystals in PP nanocomposites. However, they proposed a different mechanism of γ-phase formation, which is due to the narrow space in the clay galleries of the nanocomposite.

In their report, Medellin-Rodriguez and co-workers [81] studied the effect of nanoclays on the nucleation, crystallization, and melting mechanism of PP. The prepared nanocomposites had tactoid-like morphology because no compatibilizer is used. After the isothermal crystallization of samples, an inverse relationship between nanoclay concentration and the formation of the β-crystalline structure (Fig. 8.5) was observed. These results are not comparable with the results observed by Zheng et al. [80]. It was, therefore, proposed that the exfoliated individual layers, which are present at relatively high amounts at low nanoclay concentrations, must meet the crystallographic requirements for the nucleation and formation of the β-structure of PP in PP nanocomposites. Further increase in the amount of clay in the nanocomposite inhibits the formation of β-crystalline structure. In another attempt, Dong and Bhattacharya [82] studied the effect of clay content and maleated PP content on the PP crystalline structures in nanocomposites. It was found that the presence of clay enhanced the preferred orientation of PP crystals, which is believed to have contributed to the enhancement of tensile and flexural properties of nanocomposites with increase in clay content. It was also observed that MA-g-PP hinders the growth of α-phase crystals, reducing the tensile properties of nanocomposites with high MA-g-PP contents.

Misra and co-workers [83] discussed the effect of silicate layers on the pressure-induced crystallization of PP. It was found that, in neat sample at high pressures exceeding 60 MPa, γ-phase crystal formations are promoted. In the presence of nanoclay, γ-phase is formed, even at a low crystallization pressure of 0.1 MPa, implying the ability of nanoclay to encourage γ-phase nucleation [84]. Maiti and co-workers [49] discussed the effect of crystallization temperature on the morphology of PPCNs. The dispersed clay nanocomposites acted as a nucleating agent for the PP-g-MA matrix. As a result, the spherulite sizes decreased with the increase of clay content known from the light scattering experiments.

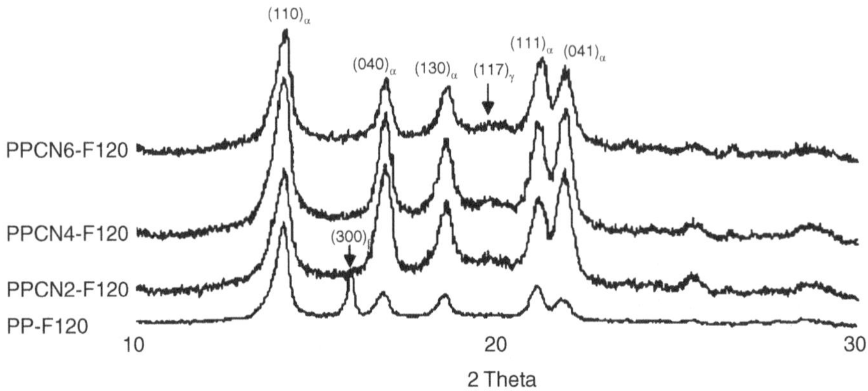

Fig. 8.4 XRD patterns of the neat PP and PPCN film samples prepared by isothermal crystallization at 120 °C on a hot press. Here, PPCN2-F120 identifies the PP sample with 2 wt% clay and annealed at 120 °C on a hot press (Reproduced from Zheng and co-workers [80] with permission)

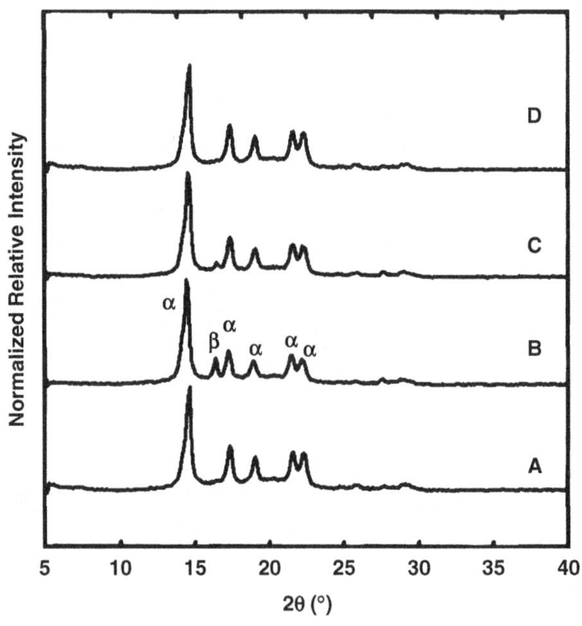

Fig. 8.5 WAXD profiles of the neat PP and PP nanocomposites after isothermal crystallization at (125 °C, 15 min). (*A*) iPP homo polymer. (*B*) iPP + 2 wt%. (*C*) iPP + 4 wt%. (*D*) iPP + 6 wt% nanoclay (Reproduced from Medellin-Rodriguez and co-workers [81] with permission)

The clay particles are well dispersed in the PP composites crystallized at low crystallization temperature (T_c) and the segregation of silicate layers occurs at high T_c. It was also observed that the equilibrium melting point was not affected by the clay content.

The crystallization behavior of PPCNs using differential scanning calorimetry (DSC) in the isothermal crystallization process has been studied by several authors. The crystallization process of pure PP itself is complex, due to the presence of various crystalline modifications in the final product. So, it is very important to reveal the structural evolution process in the presence of nanoclays in the crystallization process from various points of view. It was reported by several authors that the presence of clay layers enhance the crystallization rate of PP in the PP nanocomposites [85, 86]. This is due to the nucleation (heterogeneous nucleation) effect of clays in the crystallization process PP in nanocomposites.

Hegde et al. [87] studied the effect of clay on the crystallization mechanism of PP in PPCNs using a combination of DSC, scanning electron microscopy (SEM), TEM, and polarized optical microscopy (POM). The nanocomposites, prepared with varying clay contents in the range 1–15 wt%, showed intercalated and flocculated morphology for all concentrations. At low wt% (1–5 wt%), well-dispersed clay platelets act as the antinucleating agent (or diluent) and reduced polymer chain mobility, as known from the longer half-time of crystallization or maximum time for crystallization. On the other hand, in the nanocomposite samples with higher (10–15 wt%) clay loadings, due to the high nucleation effect of clay platelets, the polymer chain mobility increased and, as a result, the crystallization rate increased. The phase segregation and phase separation of clay is observed in the inter-spherulite region of samples with a larger amount of clay sample. Solomon et al. [88] studied the flow-induced crystallization phenomenon of PP nanocomposites with intercalated structure using time-resolved small-angle light scattering measurements. It was observed that there was a significant increase in the flow-induced crystallization kinetics for the polymer nanocomposite at applied strain rates. However, a similar flow rate has only a modest effect on pure PP crystallization. Nowacki et al. [89] observed similar results for the crystallization of PP nanocomposites with OMMT and MA-g-PP in isothermal conditions. It was found that the nucleation ability of MMT was enhanced in shear-induced crystallization as compared to the static crystallization conditions.

4.3 Orientation

Wang et al. [90] observed highly oriented structure with both PP and clay platelets along the shear flow direction of PP nanocomposite samples obtained via dynamic packing injection molding. A much higher degree of orientation of PP was observed in the composites as compared with pure PP based on the analysis of the 2D WAXD results. The fibers made from PP and PP nanocomposites (intercalated structure) were studied by Ajji et al. [91]. It was observed that the [001] (normal to the clay platelets plane) plane of clay was oriented perpendicular to the machine direction. The orientation of the *c*-axis of PP in the fibers was slightly higher in the nanocomposite fibers. Preferential orientation of the PP chains is observed in the films of PP nanocomposites (extruded film) based on the XRD studies [92]. The uniaxial drawing in PP nanocomposites induced the silicate surface to align parallel to the sheet surface. The *c* and *a** axes of PP crystals were oriented bimodally to the flow direction (draw direction), and the *b*-axis was oriented to the thickness direction of the specimen [93].

5 Properties of PP Nanocomposites

5.1 Thermal Properties

It is believed that the presence of a small amount of clay can extraordinarily improve the properties of polymers if the clay platelets are dispersed finely. Due to their very high aspect ratio, the silicate layers are capable of dissipating energy rapidly if the silicate layers and the host matrix have interphases at *nano-level* thickness. It was proposed that the thermal resistance of PP increases by filling with clay, which brought about an increase of thermal stability in oxidizing atmosphere [48, 94–105]. The thermo-oxidative degradation of PP nanocomposites was widely investigated using thermogravimetric analysis (TGA). The TGA traces in oxidizing atmosphere showed a shift of the weight loss curves of nanocomposites towards higher temperatures with respect to neat polymer. These results are attributed to the nanoclays bringing about significant improvements in the thermal stability of the PP matrix in oxidizing atmosphere [99]. The shift towards higher temperatures was attributed to the formation of a high-performance carbonaceous silicate char that builds up on the surface, which insulates the underlying material and slows down the escape of volatile products generated during decomposition.

The thermomechanical properties of a polymer material are studied using dynamic mechanical analysis (DMA). The dynamic mechanical behaviors of PP nanocomposites have been studied in order to discover the effect of clay on the viscoelastic behavior of the polymer matrix. It is predicted that, with the increase in the confinement of polymer chains, the viscosity and mechanical properties of the system increases. Maiti et al. [49] measured the dynamic behavior of PP nanocomposites and found that that the storage modulus G' of a nanocomposite with 4 wt% of clay is higher than that for pure PP samples. Also, with an increase of T_c, G' increases, wherein the enhancement is maximal (30 %) for a PP nanocomposite with 4 wt% clay content. For further increases in the clay content, the G' decreases and is found to be 13 %. They attributed that the greater increase of the storage modulus is related to the extent of clay dispersion (amount of intercalation). Kawasumi et al. [50] discussed the temperature dependence of the storage modulus of PPCNs in the presence of PP-g-MA as a compatibilizer.

5.2 Mechanical Properties

The enhancement of the mechanical properties of polymer clay nanocomposites is mainly dependent on the dispersion state of clay, the reinforcing effect, and the interfacial interaction between the clay and the polymer matrix. PPCNs have been investigated in terms of their mechanical properties [106–116] and have been examined by taking into consideration the effect of the extent of clay exfoliation on nanocomposite properties [109]. PP nanocomposites often exhibit improved strength and stiffness by incorporating clay platelets into the PP matrix. To investigate the role of silicate modification and compatibilizer addition on the mechanical properties, clays were synthesized with varying modifier alkyl chain lengths [115]. In order to examine the effect of the alkyl chain length of the silicate modifier on the morphological and mechanical properties, the content of the organophilic layered silicates MEC*n* was varied from 0 to 10 wt% at a MAH-g-PP compatibilizer content of 20 wt%. The significant impact of both MAH-g-PP compatibilizer content and alkyl chain length on the Young's modulus was observed, where MAH-g-PP compatibilizer addition afforded substantially higher stiffness. Silicate modification with C12 amine gave much greater stiffness with respect to that obtained using C8 ammonium-modified silicate MEC8. In the absence of compatibilizer, both the Young's modulus and yield stress were very similar, a result of which improved the exfoliation and stabilization of uniformly dispersed anisotropic nanoparticles. It was concluded that a uniform distribution of clay was the most important for property enhancement. Joshi and Viswanathan [107] observed that the PP clay composite filaments produced by the melt spinning method displayed a substantial improvement in the tensile and dynamic mechanical properties, as well as creep resistance, over neat PP filaments.

The impact strength of the PPCNs increased at low clay (0.5–2 wt%) loadings, which may be due to the exfoliated or intercalated clays hindering crack growth under impact. This effect should increase under more favorable interaction between the clay and the PP matrix or by adding a compatibilizer. By increasing the clay concentration further, a reduction in the impact strength is observed, which is mainly due to poor clay dispersion. Analyses of the fracture and failure mechanisms of PP nanocomposites have been carried out by Bureau et al. [116] through the essential work of fracture (EWF), where tensile testing showed 25–50 % improvements corresponding to the nanoparticle reinforcement effects.

5.3 Other Properties

In general, PP exhibits good water vapor barrier properties, but oxygen, carbon dioxide, and hydrocarbons easily permeate through PP. The presence of the silicate layers must be expected to decrease the permeability due to the more circuitous path for the diffusing molecules that must bypass impenetrable clay platelets. Several authors observed an enhancement in the barrier properties of PPCNs with respect to pristine PP. Sirousazar et al. [46] concluded that adding organoclay into the PP matrix enhances the barrier property of the matrix against O_2, CO_2, and water vapor. A nanoclay master batch has been used to make PP nanocomposite films

Upon the burning of nanocomposites, a carbonaceous char layer is formed, which acts as an excellent insulator to the mass transfer barrier, and, further, increases the flame retardancy of the resulting product [117]. Char formation impedes the movement of volatilized polymer from the interior of a plastic matrix, denying fuel at the air–surface and interface. A flame-retardant formulation containing nanoclay (Nanomer®) has been reported [26–29]. Nanoclays are very effective as a flame-retardant additive as compared to those of the traditional flame retardation chemical systems. Nanoclay also enables the reduction of traditional flame-retardant agents, to reduce toxicity, specific gravity, and provides easy processing for some flame-retardant systems.

6 Degradation of PP Nanocomposites

There are very few reports on the degradation behavior of nanocomposites, in particular, photodegradation [117–121]. The effect of compatibilizers on the photostability of PP nanocomposites has been studied. It was observed that the induction period decreased from 8 to 4 h by using MAH-g-PP as a compatibilizer and a two-phase degradation mechanism was observed based on the infrared spectral changes in the hydroxyl and carbonyl regions. In the first stage (up to 40 h), there was no evidence of the hydroxyl band formation in the infrared spectra, which implied the absence of degradation on the polymer backbone, whereas in the second phase, a dramatic increase in the rate of photooxidation was found. The degradation products were the same in both the composite and the neat polymer [118, 119]. As a whole, PP nanocomposites are less stable against photodegradation as compared to the neat PP. In another report, Lonkar et al. [120] studied the effect of layered double hydroxides (LDH) on the photodegradation of PP nanocomposites under accelerated ultraviolet irradiation. The rates of oxidation are influenced by the nanofiller, depending on the divalent cations of the LDH layers. The LDH phases containing Mg^{2+} have a positive effect on the nanocomposite material, whereas LDH with only Zn^{2+} has no influence on the rate of oxidation of the polymer matrix. Zanetti et al. [121] reported on the thermo-oxidative degradation of PPCNs by using TGA. They found that the nanocomposites were more thermally stable (>50 °C) in comparison to neat polymer. It was speculated that oxygen attacks the carbon radicals within the chain by H abstraction. Around 200–250 °C, hydrogen abstraction becomes more likely, thus, resulting in oxidative dehydrogenation. As the temperature increases, the concentration of chain end radicals increases due to the beta scission of radicals. Direct thermal scission of carbon–carbon bonds was also possible above 300 °C.

7 Applications of PP Nanocomposites

PP and its copolymers are important materials for industrial use because of their combination of properties and light weight as compared to metals, in particular, in the automotive industry. A decade ago, it was General Motors who first introduced commercial PP nanocomposites in automotive applications. The presence of clay enhances not only the mechanical properties but also the heat distortion temperature (HDT). PPCNs are suitable to replace nylon in applications where high HDT is needed. The synergistic enhancements (in terms of the flame-retardant property) of clay nanocomposites for fire safety applications have led to the development of commercial products: a series of polypropylene + organoclay + flame-retardant systems (Maxxam™ FR) produced by PolyOne® [122, 123]. Another important application field is gas barrier materials (packaging materials). Since the addition of clay improves the gas barrier property, their use in the food and drink industry as packaging film to keep foods fresher and for longer should increase tremendously.

8 Future Directions

In this chapter, the author comprehensively summarized reports published on the preparation, structure, crystallization performance, and application of PPCNs. It has been found that the material properties sensitively depend on the dispersion of clay in the matrix. Better clay dispersion in the PP matrix can be achieved through the introduction of functional groups in the main chain or by blending a compatibilizer (PP-g-MA) or by choosing suitable clay modifiers. The addition of nanoclays to the PP matrix not only enhances the crystallization rate and change in the morphology, but it also enhances the mechanical and thermal properties. Although much research has been devoted to develop PP nanocomposites for various applications by achieving fine dispersion of clay platelets, it is still far from addressing various problems such as filler–polymer interactions. So, it is essentially important to understand fundamentally the interaction between polymer–compatibilizer, polymer–filler, or filler–compatibilizer in order to understand structure–properties relationships in PP nanocomposites. Future research essentially requires focusing on better control over the structure and morphology of PP nanocomposites, which is key in achieving the current demands of PP nanocomposites with excellent mechanical properties.

References

1. Usuki A, Kojima Y, Kawasumi M, Okada A, Fukushima Y, Kurauchi T, Kamigaito O (1993) Synthesis of nylon 6–clay hybrid. J Mater Res 8:1179
2. Kojima Y, Usuki A, Kawasumi M, Okada A, Fukushima Y, Kurauchi T, Kamigaito O (1993) Mechanical properties of nylon 6–clay hybrid. J Mater Res 8:1185
3. Ray SS, Okamoto M (2003) Polymer/layered silicate nanocomposites: a review from preparation to processing. Prog Polym Sci 28:1539
4. Usuki A, Hasegawa N, Kato M (2005) Polymer-clay Nanocomposites. Adv Polym Sci 179:135
5. Alexandre M, Dubois P (2000) Polymer-layered silicate nanocomposites: preparation, properties and uses of a new class of materials. Mater Sci Eng 28:1
6. Manias E, Touny A, Wu L, Strawhecker K, Lu B, Chung TC (2001) Polypropylene/montmorillonite nanocomposites. Review of the Synthetic Routes and Materials Properties. Chem Mater 13:3516
7. Okada A, Usuki A (2006) Twenty years of polymer-clay nanocomposites. Macromol Mater Eng 291:1449

8. Biswas M, Ray SS (2001) Recent progress in synthesis and evaluation of polymer-montmorillonite nanocomposites. Adv Polym Sci 155:167
9. Dubinska E, Wiewiora A (1988) Layer silicates in the contact zone between granite and serpentinite. Clay Miner 23(4):459–470
10. Ray SS, Bousmina M (2005) Biodegradable polymers and their latered silicate nanocomposites: in greening the 21st Century material world. Prog Mater Sci 50(8):962–1079
11. Zeng QH, Yu AB, Lu GQ (Max), Paul DR (2005) Clay-based polymer nanocomposites: research and commercial developement. J Nanosci Nanotechnol 5(10):1574
12. Pinnavaia TJ, Beall GW (2000) Polymer-clay nanocomposites. Polymer-clay nanocomposites. Wiley, London
13. Vaia RA, Ishii H, Giannelis EP (1993) Synthesis and properties of 2-dimensional nanostructures by direct intercalation of polymer melts in layered silicates. Chem Mater 5:1694
14. Salamone JC (1996) Polymeric materials encyclopedia. Polymeric materials encyclopedia. CRC Press, Boca Raton
15. Utracki LA (2007) Interphase between nanoparticles and molten polymeric matrix: pressure-volume-temperature measurements. Compos Interfaces 14(3):229
16. Liu HZ, Lim HT, Ahn KH, Lee SJ (2007) Effect of ionomer on clay dispersions in polypropylene-layered silicate nancomposites. J Appl Polym Sci 104(6):4024
17. Cerrada ML, Rodriguez-Amor V, Perez E (2007) Effects of clay nanoparticles and electron irradiation in the crystallization rate of syndiotactic polypropylene. J Polym Sci Part B Polym Phys 45(9):1068
18. Shim JH, Joo JH, Jung SH, Yoon JS (2007) Effect of silane-grafted polypropylene on the morphology of polypropylene and nylon6/clay composites. J Polym Sci Part B Polym Phys 45(5):607
19. Treece MA, Oberhauser JP (2007) Processing of polypropylene-clay composites: Single-screw extrusion with in-line supercritical carbon dioxide feed versus twin-screw extrusion. J Appl Polym Sci 103(2):884
20. Yang KF, Huang YJ, Dong JY (2007) Preparation of polypropylene/montmorillonite nanocomposites by intercalative polymerization: Effect of in situ polymer matrix functionalization on the stability of the nanocomposite structure. Chin Sci Bull 52(2):181
21. Lee SH, Cho ENR, Youn JR (2007) Rheological behavior of polypropylene/layered silicate nanocomposites prepared by melt compounding in shear and elongational flows. J Appl Polym Sci 103(6):3506
22. Lee HM, Park BJ, Chin IJ, Kim HK, Kang WG, Choi HJ (2007) Preperation and characterization PP/organoclay nanocomposites with maleicanhydride. Diffus Defect Data Part B 119:203
23. Cui LL, Paul DR (2007) Evaluation of amine functionalized polypropylenes as compatibilizers for polypropylene nanocomposites. Polymer 48(6):1632
24. Dong Y, Bhattacharyya D, Hunter PJ (2007) Characterization and object-oriented finite element modeling of polypropylene/organoclay nanocomposites. Key Eng Mater 334–335:841
25. Moncada E, Quijada R, Retuert J (2007) Comparative effect of metallocene and Zieglar-Natta polypropylene on the exfoliation of montmorillonite and hectorite clays to obtain nanocomposites. J Appl Polym Sci 103(2):698
26. Nowicki A, Przybytniak G, Kornacka E, Mirkowski K, Zimek Z (2007) Radiation-induced modification of montmorillonite used as filler in PP composite. Radiation Phys Chem 76(5):893
27. Qian G, Lan T, Fay AM, Tomlin AS (2002) Intercalates formed with polypropylene/maleic anhydride-modified polypropylene intercalants. US Patent 6,462,122
28. Qian G, Cho JW, Lan T (2003) Intercalates formed with polypropylene/maleic anhydride-modified polypropylene intercalants. US Patent 6,632,868

29. Liang Y, Qian G, Cho J, Psihogios V, Lan T (2002) Applications of Plastic Nanocomposites. Additives 2002, Clearwater Beach, 24–27 Mar
30. Lee SH, Youn JR (2007) Rheological properties of grafted maleic anhydride based polypropylene composites filled with organoclay and glass fires. e-Polymers 7:411
31. Oriakhi CO (2000) Polymer nanocomposition approach to advanced materials. J Chem Educ 77:1138
32. Biswas M, Ray SS (2001) Recent progress in synthesis and evaluation of polymer-montmorillonite nanocomposites. New polymerization techniques and synthetic methodologies, Advances in Polymer science, vol 155. p 167
33. Bandyopadhyay JR, Ray SS (2010) The quantitative analysis of nano-clay dispersion in polymer nanocomposites by small angle X-ray scattering combined with electron microscopy. Polymer 51:1437
34. Kenneth CC (2008) Use of infrared spectroscopy to characterize clay interaction and exfoliation in polymer nanocomposites. Macromolecules 41:834
35. Jongsomjit B, Panpranot J, Praserthdam P (2007) Effect of nanoscale SiO2 and ZrO2 as the fillers on the microstructure of LLDPE nanocomposites synthesized via in situ polymerization with zirconocene. Mater Lett 61:1376
36. Zapata PA, Quijada R, Benavente R (2011) In situ formation of nanocomposites based on polyethylene and silica nanospheres. Journal of Applied Polymer Science 119:1771
37. Zapata PA, Quijada R, Lieberwirth I, Benavente R (2011) Polyethylene Nanocomposites Obtained by in situ Polymerization via a Metallocene Catalyst Supported on Silica Nanospheres. Macromol. Reaction Engineering 5:294.
38. Zapata PA, Quijada R, Lieberwirth I, Palza H (2011) Synthetic layered and tube-loke silica nanoparticles as novel supports for metallocene catalysts in ethylene polymerization. Applied Catalysis A, 407:181
39. Sun T, Garces JM (2002) High-performance polypropylene–clay nanocomposites by in-situ polymerization with metallocene/clay catalysts. Adv Mater 14:128
40. Tudor J, Willington L, O'Hare D, Royan B (1996) Intercalation of catalytically active metal-complexes in phyllosilicates and their application as propene polymerization catalysts. Chem Commun 2031:89
41. Heinemann J, Reichert P, Thomson R, Mülhaupt R (1999) Polyolefin nanocomposites formed by melt compounding and transition metal catalyzed ethane homo- and copolymerization in the presence of layered silicates. Macromol Rapid Commun 20:423
42. Nishiwaki T (1997) JP 09118792 N Tetsuo. Chem Abstr 127:51586
43. Tudor J, O'Hare D (1997) Stereospecific propene polymerization catalysis using an organometallic modified mesoporous silicate. Chem Commun 603
44. He A, Wang L, Li J, Dong J, Han CC (2006) Preparation of exfoliated isotactic polypropylene/alkyl-triphenylphosphonium-modified montmorillonite nanocomposites via in-situ intercalative polymerization. Polymer 47(6):1767
45. de Paiva LB, Morales AR, Guimaraes TR (2007) Structural and optical properties of polypropylene-montmorillonite nanocomposites. Mater Sci Eng A Struct Mater Prop Microstruct Process 447(1–2):261
46. Sirousazar M, Yari M, Achachlouei BF, Arsalani J, Mansoori Y (2007) Polypropylene/montmorillonite nanocomposites for food packaging. E-Polymers Art No 027
47. Rohlmann CO, Failla MD, Quinzani LM (2006) Linear viscoelasticity and structure of propylene-montmorillonite. Polymer 47(22):7795
48. Wang W, Zeng X, Wang G, Chen J (2006) Preparation and properties of polypropylene filled with organo-montmorillonite nanocomposites. J Appl Polym Sci 100(4):2875
49. Maiti P, Nam PH, Okamoto M (2002) Influence of Crystallization on Intercalation, Morphology, and Mechanical Properties of Polypropylene/Clay Nanocomposites. Macromolecules 35:2042
50. Kawasumi M, Hasegawa N, Kato M, Usuki A, Okada A (1997) Preparation and mechanical properties of polypropylene-clay hybrids. Macromolecules 30:6333

51. Perrin-Sarazin F, Ton-That M-T, Bureau MN, Denault J (2005) Micro-and nano-structure in polypropylene/clay nanocomposites. Polymer 46(25):11624
52. Reddy CS, Das CK, Narkis M (2005) Propylene Ethylene Copolymer Nanocomposites: Epoxy resin grafted nanosilica as reinforcing filler. Polym Compos 26(6):806
53. Utracki LA, Sepehr M, Boccaleri E (2007) Synthetic, layered nanoparticles for polymeric nanocomposites (PNCs). Polym Adv Technol 18(1):1
54. Moad G, Dean K, Edmond L, Kukaleva N, Li GX, Mayadunne RTA, Pfaendner R, Schneider A, Simon G, Wermter H (2006) Novel copolymers as dispersants/intercalants/exfoliants for polypropylene-clay nanocomposites. Macromolecular Symposia. Macromol Symp 233:170
55. Reichertm P, Nitz H, Klinke S, Brandsch R, Thomann R, Mulhaupt R (2000) Polypropylene/organoclay nanocomposite formation: Influence of compatibilizer functionality and organoclay modification. Macromol Mater Eng 275:8
56. Nam PH, Maiti P, Okamoto M, Kotaka T, Hasegawa N, Usuki A (2001) A Hierarchical structure and properties of intercalated polypropylene/clay nanocomposites. Polymer 42:9633
57. Avella M, Cosco S, Volpe GD, Errico ME (2005) Crystallization behavior and properties of exfoliated isotactic Polypropylene/organoclay nanocomposites. Adv Polym Technol 24(2):132
58. Chiu FC, Chu PH (2006) Characterization of Solution-Mixed Polypropylene/Clay Nanocomposites without Compatibilizers. J Polym Res 13(1):73
59. Wang Y, Huang SW (2007) Solution intercalation and relaxation properties of maleated polypropylene/organoclay nanocomposites. Polym Plast Technol Eng 46:1039
60. Oya A, Kurokawa Y (2000) Factors controlling mechanical properties of clay mineral/polypropylene nanocomposites. J Mater Sci 35:1045
61. Kresge CT, Leonowitz ME, Roth WJ, Vartulli JC, Beck JS (1992) Ordered mesoporous molecular sieves synthesized by a liquid-crystal template mechanism. Nature 359:710
62. Hasegawa N, Okamoto M, Kato M, Usuki A, Sato N (2003) Nylon6/Na-montmorillonite nanocomposites prepared by compounding nylon 6 with Na-montmorillonite slurry. Polymer 44:2933
63. Kato M, Matsushita M, Fukumori K (2004) Development of a new production method for a polypropylene-clay nanocomposite. Polym Eng Sci 55:1205
64. Lapshin S, Isayev AI (2007) Ultrasound-aided extrusion process for preparation of polypropylene–clay nanocomposites. J Vinyl Addit Technol 13(1):40
65. Wang ZM, Han H, Chung TC (2005) Synthesis of chain-end functionalized PP and applications in exfoliated PP/clay nanocomposites. Macromol Symp 225:113
66. Marchant D, Jayaraman K (2002) Strategies for Optimizing Polypropylene-Clay Nanocomposite Structure. Ind Eng Chem Res 41(25):6402
67. Wang ZM, Nakajima H, Manias E, Chung TC (2003) Exfoliated PP/Clay Nanocomposites Using Ammonium-Terminated PP as the Organic Modification for Montmorillonite. Macromolecules 36:8919
68. Brückner S, Meille SV, Petraccone V, Pirozzi B (1991) Polymorphism in isotactic polypropylene. Prog Polym Sci 16:361
69. Natta G, Corradini P (1960) Structure and properties of isotactic polypropylene. Nuovo Cimento Suppl 15:40
70. Mencik Z (1972) Crystal Structure of Isotactic Polypropylene. J Macromol Sci Phys B6:101
71. Hikosaka M, Seto T (1973) The Order of the Molecular Chains in Isotactic Polypropylene Crystals. Polym J 5:111
72. Pae KD, Sauer JA, Morrow DR (1966) Interior Morphology of Bulk Polypropylene. Nature 211:514
73. Mezghani K, Phillips PJ (1997) The g-phase of high molecular weight isotactic polypropylene. II: The morphology of the g-form crystallized at 200 MPa. Polymer 38:5725
74. Morrow DR, Newman BA (1968) Crystallization of low-molecular-weight polypropylene fractions. J Appl Phys 39:4944

75. Lotz B, Graff S, Wittman JC (1986) Crystal morphology of the gamma(triclinic) phase of isotactic polypropylene and its relation to the alpha-phase. J Polym Sci, Polym Phys Ed 24:2017
76. Turner-Jones A (1971) Development of the γ-crystal form in random copolymers of propylene and their analysis by DSC and X-ray methods. Polymer 12:487
77. Mezgahni K, Philiphs PJ (1995) γ-Phase in propylene copolymers at atmospheric pressure. Polymer 36:2407
78. Fischer D, Mulhaupt R (1994) The influence of regio- and stereoirregularities on the crystallization behaviour of isotactic poly(propylene)s prepared with homogeneous group IVa metallocene/methylaluminoxane Ziegler-Natta catalysts. Macromol Chem Phys 195:1433
79. Alamo RG, Man-Ho K, Maria JG, Jose RI, Mandelkern L (1999) Structural and Kinetic Factors in the Formation of the Gamma Polymorph of Isotactic Poly(propylene). Macromolecules 32:4050
80. Zheng W, Lu X, Toh CL, Zheng TH, He C (2004) Effect of clay on polymorphism of polypropylene in polypropylene/clay nanocomposites. J Polym Sci, Part B: Polym Phys 42:1810
81. Medellin-Rodriguez FJ, Mata-Padilla JM, Hsiao BS, Waldo-Mendoza MA, Ramirez-Vargas E, Sanchez-Valdes S (2007) The effect of nanoclays on the nucleation, crystallization, and melting mechanisms of isotactic polypropylene. Polym Eng Sci 47:1889
82. Dong Y, Bhattacharya D (2012) Investigation on the competing effects of clay dispersion and matrix plasticisation for Polypropylene/clay nanocomposites. Part II: Crystalline structure and thermo-mechanical behavior. J Mater Sci 47:4127
83. Yuan Q, Rajan VG, Misra RDK (2008) Nanoparticle effects during pressure-induced crystallization of polypropylene. Mater Sci Eng B 153:88
84. Yuan Q, Misra RDK (2006) Polymer nanocomposites: current understanding and issues. Mater Sci Tech 22:742
85. Liu X, Wu Q (2001) PP/clay nanocomposites prepared by grafting-melt intercalation. Polymer 42:10013
86. Xu W, Ge M, He P (2002) Nonisothermal crystallization kinetics of polypropylene/montmorillonite nanocomposites. J Polym Sci, Part B Polym Phys 40:408
87. Hegde RR, Spruiell JE, Bhat GS (2012) Different crystallization mechanisms in polypropylene–nanoclay nanocomposite with different weight percentage of nanoclay additives. J Mater Res 27:1360
88. Somawangthanaroj A, Lee EC, Solomon MJ (2003) Early stage quiescent and flow-induced crystallization of intercalated polypropylene nanocomposites by timeresolved light scattering. Macromolecules 36:2333
89. Nowacki R, Monasse B, Piorkowska E, Galeski A, Haudin JM (2004) Spherulite nucleation in isotactic polypropylene based nanocomposites with montmorillonite under shear. Polymer 45:4877
90. Wang K, Zhao P, Yang H, Liang S, Zhang Q, Du R, Fu Q, Yu Z, Chen E (2006) Unique clay orientation in the injection-molded bar of isotactic polypropylene/ clay nanocomposite. Polymer 47:7103
91. Ajji A, Denault J, Cote D, Bureau M, Trudel-Boucher D (2007) Polypropylene Nanocomposite Fibers: Structure and Some Applications. Int Polym Proc 22(4):368
92. Woods CG. Muzzy JD (2003) 61st Annual Technical Conference-Society of Plastics Engineers, Nashville, TN, May 4-8 (ANTEC) p. 2205
93. Koo CM, Kim JH, Wang KH, Chung IJ (2005) Melt-extensional properties and orientation behaviors of polypropylene-layered silicate nanocomposites. J Polym Sci Part B Polym Phys 43:158
94. Hao JW, Lewin M, Wilkie CA, Wang JQ (2006) Additional Evidence for the Migration of Clay upon Heating of Clay-PP Nanocomposites from X-Ray Photoelectron Spectroscopy (XPS). Polym Degrad Stab 91(10):2482

95. Hu Y, Tang Y, Song L (2006) Poly(propylene)/clay nanocomposites and their application in flame retardancy. Polym Adv Technol 17(4):235
96. Wang K, Liang S, Deng JN, Yang H, Zhang Q, Fu Q, Dong X, Wang DJ, Han CC (2006) The role of clay network on macromolecular chain mobility and relaxation in isotactic polypropylene/organoclay nanocomposites. Polymer 47(20):7131
97. Chen L, Wang K, Toh ML, Kotaki M, He CB (2005) Polypropylene/clay nanocomposites prepared by reactive compounding with an epoxy-based masterbatch. Abstr Papers Am Chem Soc 230:3563, 129-PMSE
98. Valera-Zaragoza A, Ramirez-Vargas E, Medellin-Rodriguez FJ, Huerta-Martinez BM (2006) Thermal stability and flammability properties of heterophasic PP-EP/EVA/organoclay nanocomposites. Polym Degrad Stab 91(6):1319
99. Modesti M, Lorenzetti A, Bon D, Besco S (2006) Thermal behaviour of compatibilised polypropylene nanocomposite: Effect of processing conditions. Polym Degrad Stab 91(4):672
100. Andersen PG (2006) Society of Petroleum Engineers International Conference on Polyolefins 2005, Huston, Texas, USA: The Challenges of Globalization, Vol 1. p.146
101. Ma XY, Lu HJ, Liang GZ, Yan HX (2004) Preparation and properties of intercalated rectorite/polypropylene nanocomposites. ACTA Polym Sin 1:88
102. Hambir S, Bulakh N, Jog JP (2002) Polypropylene/Clay nanocomposites: Effect of compatibilizer on the thermal, crystallization and dynamic mechanical behavior. Polym Eng Sci 42(9):1800
103. Ellis TS, D'Angelo JS (2003) Thermal and mechanical properties of a polypropylene nanocomposite. J Appl Polym Sci 90(6):1639
104. Ding C, Jia D, He H, Guo B, Hong H (2005) How organo-montmorillonite truly affects the structure and properties of polypropylene. Polym Test 24(1):94
105. Parija S, Nayak SK, Verma SK, Tripathy SS (2004) Studies on Physico-Mechanical Properties and Thermal Characteristics of Polypropylene/Layered Silicate Nanocomposites. Polym Compos 25(6):646
106. Chen L, Wang K, Kotaki M, Hu C, He C (2006) Fracture behavior of polypropylene/clay nanocomposites. J Nanosci Nanotechnol 6(12):3969
107. Joshi M, Viswanathan V (2006) High-performance filaments from compatibilized polypropylene/clay nanocomposites. J Appl Polym Sci 102(3):2164
108. Li JM, Ton-That M-T, Tsai SJ (2006) PP-Based Nanocomposites with Various Intercalant Types and Intercalant Coverages. Polym Eng Sci 46(8):1060
109. Peltola P, Valipakka E, Vuorinen J, Syrjala S, Hanhi K (2006) Effect of rotational speed of twin screw extruder on the microstructure and rheological and mechanical properties of nanoclay-reinforced polypropylene nanocomposites. Polym Eng Sci 46(8):995
110. Vladimirov V, Betchev C, Vassiliou A, Papageorgiou G, Bikiaris D (2006) Dynamic mechanical and morphological studies of isotactic polypropylene/fumed silica nanocomposites with enhanced gas barrier properties. Compos Sci Technol 66(15):2935
111. Yuan Q, Awate S, Misra RDK (2006) Nonisothermal crystallization behavior of polypropylene–clay nanocomposites. Eur Polym J 42(9):1994
112. Modesti M, Lorenzetti A, Bon D, Besco S (2006) Effect of processing conditions on morphologyand mechanical properties of compatibilized polypropylene nanocomposite. Polymer 46:10237
113. Ljungberg N, Cavaille JY, Heux L (2006) Nanocomposites of isotactic polypropylene reinforced with rod-like cellulose whiskers. Polymer 47(18):6285
114. Sukhyy KM, Burmistr MV, Shilov VV, Pissis P, Spanoudaki A, Sukha IV, Tomilo VI (2005) Synthesis, structure, thermal and mechanical properties of nanocomposites based on linear polymers and layered silicates modified by polymeric quaternary ammonium salts (ionenes). Polymer 46(26):12226
115. Reichert P, Nitz H, Klinke S, Brandsch R, Thomann R, Mülhaupt R (2000) Polypropylene/organoclay nanocomposite formation: influence of compatibilizer functionality and organoclay modification. Macromol Mater Eng 275:8

116. Bureau MN, Ton-That M-T, Perrin-Sarazin F (2006) Essential work of fracture and failure mechanisms of polypropylene–clay nanocomposites. Eng Fract Mech 73(16):2360
117. Peneva Y, Tashev E, Minkova L (2006) Flammability, microhardness and transparency of nanocomposites based on functionalized polyethylenes. Eur Polym J 42(10):2228
118. Morlat S, Mailhot B, Gonzalez D, Gardett J (2004) Photo-oxidation of polypropylene/montmorillonite nanocomposites. Chem Mater 16:377
119. Mailhot B, Morlat S, Gardett J, Boucard S, Duchet J, Gérard J (2003) Photodegradation of polypropylene nanocomposites. Polym Degrad Stab 82:163
120. Lonkar SP, Therias S, Caperaa N, Leroux F, Gardette JL (2010) Photooxidation of Polypropylene/Layered Double Hydroxide Nanocomposites: Influence of Intralamellar Cations. Eur Polym J 46:1456
121. Zanetti M, Camino G, Reichert P, Mülhaupt R (2002) Thermal behavior of polypropylene layered silicate nanocomposites. Macromol Rapid Commun 22:176
122. Tidjani A (2001) Wilkie. Photooxidation of polymeric-inorganic nanocomposites: chemical, thermal stability and fire retardancy investigations. C A: Polym Degrad Stab 74:33
123. Morgan AB (2006) Flame retarded polymer layered silicate nanocomposites: a review of commercial and open literature systems. Polym Adv Technol 17:206

ABS Based Nanocomposites 9

Michele Modesti, Stefano Besco, and Alessandra Lorenzetti

Contents

Abstract

A discussion about the state of art regarding ABS based nanocomposites has been proposed considering the different aspects regarding their synthesis and production as well as the effects of nanofillers on physical and thermal properties of the polymer itself and of its most common blends. Classic melt-blending process has been compared with low temperature solvent based techniques. The effectiveness of more recent ultrasonic-mixing assisted solution processes has been assessed as well together with the perspective uses of innovative characterization techniques based on confocal microscopy. Regarding thermal stability during processing, the effects of innovative synthetic organic modifiers have been analyzed, with particularly good results reported for imidazolium salts based modifiers. Flammability and fire behavior have been extensively investigated analyzing possible solutions given by the combination of nanofillers with traditional halogen based or alternative

M. Modesti (✉) • S. Besco • A. Lorenzetti
Department of Industrial Engineering, University of Padova, Padova, Italy
e-mail: michele.modesti@unipd.it

J.K. Pandey et al. (eds.), *Handbook of Polymernanocomposites. Processing, Performance and Application – Volume A: Layered Silicates*, DOI 10.1007/978-3-642-38649-7_18,

halogen-free flame retardants. ABS blends have also been considered in order to study the selective localization of OMMT and MWCNTs in complex multicomponent systems.

Keywords
Electrical conductivity • Flame retardancy • Melt blending • Polymer nanocomposites • Processing

1 Introduction

ABS plastics are composed of styrene–acrylonitrile copolymer as the continuous phase and a dispersed phase of butadiene–acrylonitrile rubber or a butadiene-containing rubber onto which styrene–acrylonitrile monomers are grafted.

ABS polymers are a natural outgrowth of the impact polystyrenes which are made from the two monomers (styrene and butadiene). In impact polystyrenes, a significant improvement in toughness over polystyrene is attained through the introduction of rubbery components containing butadiene into the polystyrene matrix. However, with the attainment of toughness, other properties are compromised: these include rigidity, chemical resistance, tensile, and flexural strengths. By introducing acrylonitrile monomer, a significant improvement in all these properties is obtained, as well as outstanding toughness and abuse resistance. Chemical resistance is improved dramatically, hardness level is upgraded significantly, and tensile and flexural strengths and modulus (rigidity) are all improved substantially. Various favorable combinations of properties are possible, thus making these polymers most attractive for a large number of current and newly developed applications. ABS plastics are extremely useful and versatile, since ease of processing and forming allows them to be used for a great number of applications. Due to these peculiar characteristics, ABS has been considered in several researches regarding the possibility to obtain ABS-based polymer nanocomposites in order to evaluate the effects of nanofillers on physical, thermal, and structural properties of the polymer.

In this chapter, a discussion about the state of the art regarding ABS-based nanocomposites will be presented considering at first different issues correlated to synthesis and production of these materials (processing, filler structures, and modifications) and then the effects of nanofillers on physical and thermal properties of polymer. ABS-based nanocomposite blends will also be considered [1].

2 Synthesis of ABS Nanocomposites

2.1 Processing

Several strategies have been considered to prepare polymer–clay nanocomposites. Four main processes are exfoliation–adsorption, in situ polymerization, melt intercalation, and template synthesis. Exfoliation–adsorption method is a solvent-based process in which the polymer matrix and modified clay become in contact with a solution, and after evaporating, the solvent (or precipitating) nanocomposite is produced as a film or bulk material [2]. So far, melt intercalation method was the most commonly used procedure because of its advantages especially from commercial point of view and mass production ability. In this process, molten polymer is mixed with modified clay while applying high shear stresses by using an internal mixer or in small gap between the hot rotating rolls (twin-roll mill). Polymer chains intercalate between the silicate layers of the modified clay and usually a high level of exfoliation would be achieved. High temperatures used in mixing process may lead to some destructive influences on the polymeric molecules of the matrix.

During melt blending (MB), the high temperature, shear stress, and local overheating induced by the shear itself can affect the clay through thermal degradation of the organic modifier, phase separation of the clay, and possible reduction in the aspect ratio of the layered silicate.

Low-temperature processing techniques or OMLS with enhanced thermal stabilities have been developed in order to avoid thermal degradation of clay's surfactant during the preparation of the composite. The thermal degradation of alkyl-ammonium salts, commonly used as surfactant for OMLS, becomes significant above 200 °C; thus, the use of clays in polymers requiring higher processing temperature is challenging.

In their work, Pourabas et al. [3] developed a new approach to the solvent-based preparation method of nanocomposites. A solvent/non-solvent method has been developed in this study for the preparation of ABS/clay nanocomposites. ABS nanocomposite is precipitated after addition of ethanol (non-solvent) containing organically modified montmorillonite from a tertrahydrofurane (THF) solution while it is stirring. The method utilized a special mechanical tool known as homogenizer, which consists of two coaxial cylindrical stator and rotating parts with a small gap between them. The rotating cylinder rotation rate can be increased up to 45,000 rpm. Therefore, the material (precipitating polymer and the OMT) undergoes high shear stresses between the rotor and stator cylinders. In other words, homogenizer operates as a wet twin-roll mill.

Investigations showed that shear stress does not lead to further modification of montmorillonite (MMT) during ion-exchange process. A same interlayer spacing was finally achieved with or without using homogenizer. The same interlayer spacing in a shorter time (2 h against 32 h) was the main advantage of using homogenizer in the modification reaction. In the preparation of the nanocomposites, homogenizer led to an intercalated structure with more uniform interlayer spacing of the clay silicate layers. This uniformity can be detected by X-ray diffraction (XRD) peak shape and transmission electron microscopy (TEM) micrography. From thermal properties point of view, silicate layers work as thermal shield for intercalated polymer chains which in turn lead to an enhanced thermal resistance for the nanocomposites. Isothermal degradation analysis showed that the nanocomposites prepared by using homogenizer must contain higher amount of organically modified montmorillonite (OMMT), which influence the thermal properties of the product totally. As a conclusion, the investigations showed that ion-exchange process and polymer intercalation between the silicate layers are thermodynamically controlled processes. In other words and by taking the results obtained by other researchers into account, it seems that two steps are involved, a preliminary intercalation step and subsequent complete delaminating process. In the first step, an equilibrium state, which can be established between the materials incorporating in the modification or polymer intercalation process, is the final attitude. Applying severe mixing or shear stresses in this step by using, for example, homogenizer has just an accelerator effect for the equilibrium to be reached. After that initial intercalation, further shear stresses via, for example, a twin-roll mill or an internal mixer may lead to the final and complete exfoliated state [4].

By using solvent/non-solvent method, polymers with moderate level of solubility in common solvents can be converted into corresponding nanocomposites. Higher solubility will cause difficulties in the precipitation step so as enormous volume of the non-solvent will be needed. On the other hand, solubility in non-common solvents will not be economically feasible because of the solvent's high cost.

In their work, Modesti et al. [5] choose dimethyl-hexadecyl-imidazolium modified montmorillonite (DMHDIM-MMT) for the preparation of polymer nanocomposites using melt-blending (MB) and solution-intercalation techniques. The aim was to investigate the effectiveness and influence of these two different processes on dispersion, morphology, and properties of the ABS/clay nanocomposites, in order to obtain a material that would combine the excellent mechanical behavior provided by low amounts of inorganic nanoparticles with the versatility and easy processing characteristics of rubber-toughened thermoplastic.

In a previous work of the same authors [6], they studied the effect of the clay surfactant on the properties of the nanocomposites prepared by solution intercalation. The results have confirmed the outstanding stability of DMHDIM-MMT as well as its improved compatibility with ABS matrix with respect to traditional ammonium salt-based surfactants.

Morphology, mechanical properties, and thermal stability of the composites have been assessed by X-ray diffraction (XRD), transmission electron microscopy (TEM), dynamic-mechanical analyses (DMA), thermal gravimetric analyses (TGA), and fluorescence spectroscopy (FS) of optical probes. The last one is an innovative technique for the rapid evaluation of intercalation and exfoliation in polymer–clay nanocomposites [7]. Preliminary findings are reported on probe fluorescence in polymer nanocomposites prepared from organically modified layered silicates (OMLS) with polystyrene (PS) and polyamide 6 (PA6). In particular, Nile Blue A has been found to be a sensitive probe of the local nano-environment and hence a useful fluorescence probe when co-exchanged into clay with traditional quaternary ammonium treatments or high-temperature-stable trialkyl-imidazolium-based surfactants [8–10].

OMLS were dispersed in ABS by ultrasonic solution blending (US) or MB. US samples were prepared by first dissolving and stirring ABS in refluxing acetone in a three-neck reflux flask. A dispersion of OMLS in the same solvent was added to

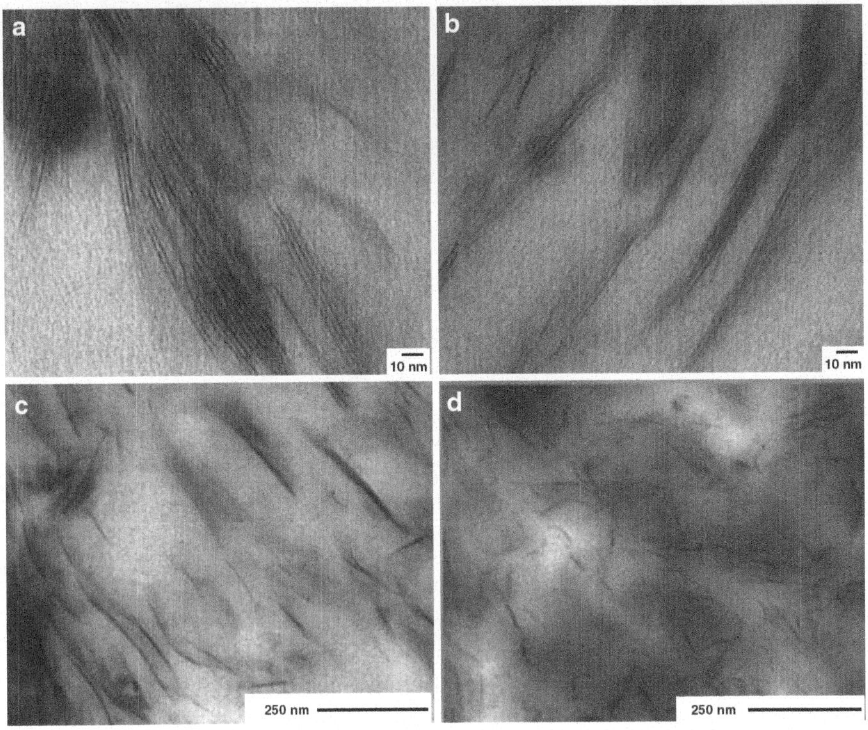

Fig. 9.1 TEM micrographs for ABS/DMHDIM–MMT 5 wt% composites obtained with melt-blending (a, c) and ultrasonic solution blending (b, d) at different magnifications [5]

the solution to obtain 5 wt% concentration of OMLS in the polymer/clay composite after solvent removal (ABS/DMHDIM-MMT 5 wt% – US). Ultrasonic mixing (Bransons 1510, maximum power 70 W at 72 kHz) was applied for 6 h to disperse the OMLS in acetone and the polymer solution after addition of the OMLS dispersion. The solvent was then evaporated under vacuum at 50 °C, and the composite was dried under vacuum at 100 °C for 4 h to remove the solvent residue from the solid.

MB composites (ABS/DMHDIM-MMT 5 wt% – MB) were produced by mixing the molten polymer with 5 wt% of OMLS in a 50 cm^3 Brabender apparatus working at 190 °C and 4.8 rad/s with a residence time of 5 min. The obtained composites were then compression molded with a Collin P200 press at 200 °C and 30 bar during a 600 s cycle (with a cooling rate of 0.5 °C/s) to obtain specimens for morphological and mechanical characterization.

Representative TEM micrographs in Fig. 9.1 show that both samples have a mixed intercalated/exfoliated morphology. At low magnification (Fig. 9.1a and b), the sample ABS/DMHDIM-MMT 5 wt% – MB – shows tactoids containing a larger number of lamellae and, at high magnification (Fig. 9.1c and d), clay layers with higher planar dimension.

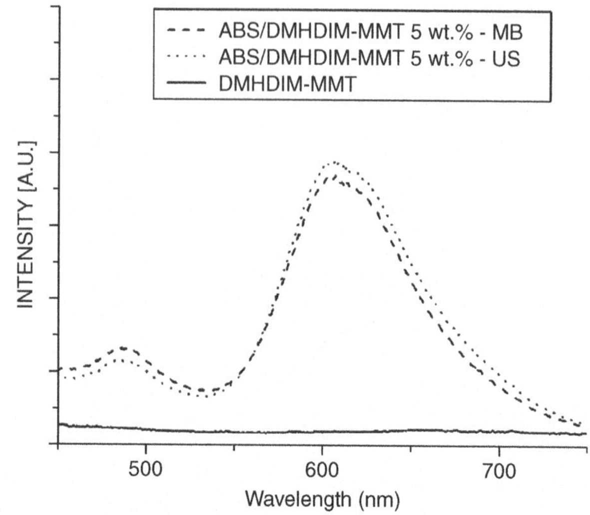

Fig. 9.2 Fluorescence spectra for ABS/DMHDIM–NB–MMT 5 wt% composites obtained with melt-blending (MB) and ultrasonic solution blending (US). Solid line indicatespristine OMLS [5]

This suggests that ultrasonication, as compared to MB, promotes not only a higher extent of exfoliation but also a more severe reduction in the size of clay platelets due to fragile fracture [4, 11] The previous considerations are deduced from a limited number of TEM micrographs, where only a minuscule volume is illuminated, and therefore, it is not assured that this description represents the bulk of the samples' morphology. Spectroscopy data (Fig. 9.2) reveal the presence of fluorescence (denoted by an intense peak at 610 nm and a weak one at 480 nm) in the nanocomposites but no fluorescence in the OMLS itself due to quenching effects between the dye molecules [12]. Fluorescence has been used for monitoring the intercalation/exfoliation of the clay [10]. It has been shown that for PA6/DMHDIM-MMT/Nile Blue nanocomposites, the spectroscopic emission at about 560 nm can be related to intercalation, while fluorescence effects at 610 nm are indicative of mixed intercalation/exfoliation structures. The nature of the emission around 495–500 nm is less clear and could be explained with the desorption of Nile Blue into the polymer matrix or with the presence of unquenched higher-order aggregates on the clay [10]. Emission wavelengths depend on the nano-confinement of the optical probe and on the polar character of the local environment; thus, fluorescence spectra can also be used to investigate the preferential localization of clay in one of the phases composing the polymer structure (i.e., styrene–acrylonitrile and butadiene).

Modified dyed clay has been dispersed in three different solvents at a concentration of 5 wt%. Each solvent is chosen to mimic the polarity of a different ABS phase: heptane for butadiene, acetonitrile for acrylonitrile, and toluene for styrene. In Fig. 9.3 the fluorescence spectra for the composite and the clay in the solvents are shown. The fluorescent spectra for the ABS/OMLS composites reveal two peaks at about 600 and 480 nm (Fig. 9.2) which appear to be the combination of

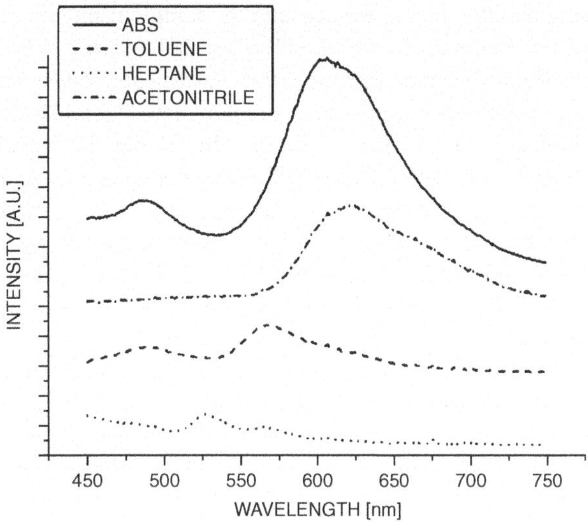

Fig. 9.3 Fluorescence spectra for ABS/DMHDIM–NB–MMT 5 wt% composite obtained with solution-intercalation and for OMLS solutions using representative solvents [5]

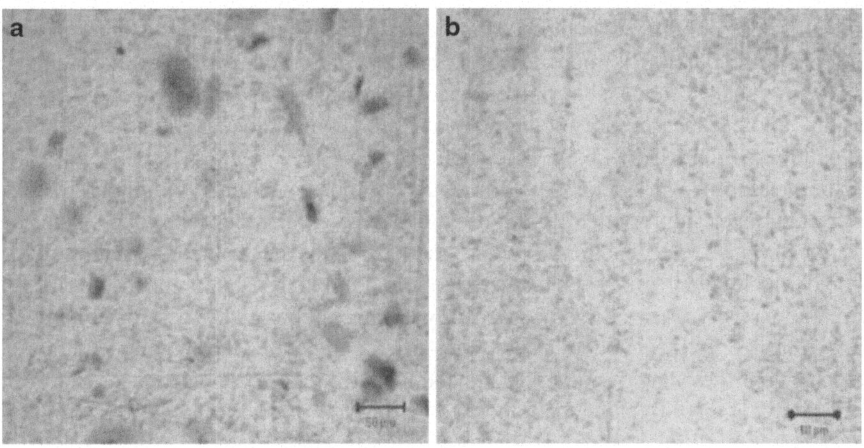

Fig. 9.4 Composite images from the confocal microscope of: ABS/DMHDIM– NB–MMT 5 wt% composites produced by (**a**) melt-blending and (**b**) ultrasonic solution blending [5]

the peaks obtained with toluene and acetonitrile: this observation suggests that clay resides in the styrene–acrylonitrile (SAN) rigid phase, as previously observed by Stretz et al. [7].

Images collected by confocal microscope (Fig. 9.4) show dark areas due to quenching generated by micrometer aggregates of clay tactoids. The clay will not fluoresce until the effective distance between fluorophores is at least 3–5 nm [10]. The dyed molecules in the intercalated tactoids are most likely quenched because, as shown by the XRD, the d-spacing is about 2.9 nm; thus, authors assumed that the fluorescence in the sample is mostly generated by dyed molecules on the external surface of well-dispersed clay tactoids or on exfoliated layers.

In the composite images (Fig. 9.4) (generated by superimposing 20 individual confocal images), aggregates of intercalated tactoids with a maximum dimension of about 50 mm are observed for the MB sample; these aggregates are much smaller in the US composites. This result further supports a more homogeneous dispersion for US sample as observed on the mesoscale and is consistent with the indications discussed above from the TEM data (on the nanoscale). Thus, the combination of the confocal microscopy, XRD, and TEM data indicates that OMLS has a mixed intercalated/exfoliated structure and a nonhomogeneous dispersion with the MB samples exhibiting greater heterogeneity.

Hence, the sonication process reduces the size of these aggregates as compared to MB and improves the degree of dispersion. As expected, the reinforcing action of the nanofiller in terms of elastic modulus measured by DMA increases with the extent of dispersion. A strong reduction in deformation at break has been measured, and it is attributed to the presence of clay aggregates that act as micro defects in the composites, which may initiate the crack propagation. No significant variation in thermal and thermo-oxidative degradation was observed between the nanocomposites prepared by sonication and MB or between the nanocomposites and the neat polymer. All these data clearly show that, for the system studied, solution intercalation is more effective at dispersing and improving mechanical properties than MB and that fluorescence spectroscopy and confocal microscopy using fluorescence-probe modified clay are complimentary characterization techniques when used with WAXS and TEM.

In most of the work done in polymer nanocomposites, organic/inorganic hybrids were obtained simply by co-mixing the polymeric component with alkoxysilanes, followed by a solgel reaction involving the hydrolysis and polycondensation reaction of alkoxysilane [13, 14]. However, colloidal polymer nanocomposites have not been intensively researched so far. A study has been reported by workers at DuPont to synthesize uniform spherical polymer–silica composites in the patent literature [15, 16]. Barthet and co-workers [17] have recently reported the synthesis of novel polymer–silica colloidal nanocomposites by the free-radical (co)polymerization of vinyl monomer in the presence of an aqueous silica sol. More recently, raspberry-like silica/poly(methyl methacrylate) and silica/polystyrene nanocomposites have been recently obtained through emulsion polymerization [18, 19].

Kim et al. [20] recently became interested in utilizing well-dispersed silica sol for the preparation of novel polymer/silica nanocomposites, particularly the incorporation of spherical silica particles into acrylonitrile–butadiene–styrene copolymer (ABS) and the combination of the excellent properties provided by inorganic nanoparticles with the versatility and easy processing characteristics of rubber-toughened thermoplastic. In order to reach a homogeneous dispersion of the particles in the polymer matrix, colloidal silica sol was mechanically mixed with polybutadiene-g-(acrylonitrile–styrene copolymer) (PB-g-SAN) latex, and then ABS/silica nanocomposites were prepared by blending these PB-g-SAN/silica hybrids and SAN resin. The strong interfacial bonding between the SiO_2 and the

ABS might be beneficial for a homogeneous dispersion of the particles in the polymer matrix, even though a completely uniform distribution of the particles cannot be achieved during the compounding process. The interactions between submicrometer PB-g-SAN latex spheres and charged colloidal silica have been studied to determine the conditions for the formation of well-dispersed ABS/SiO_2 hybrid nanocomposites and understand the correlation between the morphology of silica particles and the mechanical properties of the ABS/SiO_2 nanocomposites.

The hybrid nanocomposites with 1.0–1.8 wt% silica particles are found to exhibit the improvement of impact strength by ca. 30 %, without making a sacrifice of elastic modulus, hardness, and tensile strength, which can be explained by promotion of deformation process at the silica particles. The interfacial interaction between the SiO_2 and the grafted ABS phases plays a major role in controlling the microstructures and the properties of the composite materials.

Solid particles such as silica, clay, carbon, hydroxides, metal oxides, basic salts of metals, colloidal silver, and solid organic materials had been used for stabilizing emulsion droplets in oil-in-water (O/W) or water-in-oil (W/O) systems since the early 1900s. Stabilizing emulsion droplets by solid particles involves several mechanisms. Solid particles stabilize emulsion droplets by creating barriers around dispersed droplets, which keep the droplets from cohering [21, 22], and they also build up a three-dimensional network in aqueous phase. The emulsion droplets confined in the three-dimensional network reduce their mobility [23, 24]. Ionic surfactants are commonly used as stabilizers for preventing emulsion droplets from coagulation via electrostatic repulsion between colloidal particles, while block copolymers or nonionic surfactants also act as a stabilizer by steric repulsion after being adsorbed on oil droplets. Most articles on emulsion polymerizations consider electrostatic stabilization, steric stabilization, and electrosteric stabilization by ionic stabilizers, block copolymers, and poly(acrylic) acid [25–33].

Jang and co-workers reported that intercalated ABS sodium montmorillonite nanocomposites were developed with emulsion polymerization [34]. Wang et al. prepared intercalated–exfoliated structure of nano-ABS by melt-blending ABS and an organically modified clay [35].

In their paper Choi et al. [36] elucidate the role of clay on stabilizing ABS latex in emulsion polymerization. Two variables were adopted for the polymerization of ABS clay nanocomposites: (1) surfactant weight was varied at a fixed total weight percentage and ratio of clays during emulsion polymerizations, and (2) sodium montmorillonite/laponite ratio in mixed clays was changed under a fixed surfactant weight during polymerizations. Colloidal properties of ABS clay nanocomposites such as particle sizes, viscosity, and colloidal stability in salt solution were examined and compared with neat ABS latex. Structures and mechanical properties of synthesized ABS clay nanocomposites were also investigated.

From the results, authors summarized that clay works as a colloidal stabilizer due to several factors during ABS emulsion polymerization. Firstly, electrostatic repulsion force between ABS clay nanocomposite latex particles may contribute to stabilize the latex, because of their negative surface charges.

Secondly, laponite layers separate MMT layers and polybutadiene particles, preventing the coagulation. Thirdly, the laponite layers adsorbed on latex particles may contribute to prevent latex from coagulating as a barrier. The clays are believed to provide a steric barrier against ABS latex coalescence so that it prevents coagulation of ABS latex particles. Fourthly, laponite layers, which have a large surface volume, increase the viscosity of latexes, reduce the mobility and the coagulation rate of the particles, and contribute to stabilize the latex. The ABS clay nanocomposite latexes consume less surfactant than neat ABS latex, and their production may need the small amount of surfactant. The storage moduli of ABS clay nanocomposites increase proportionally with MMT content in the mixed clay (MMT/laponite). It explains that a clay with high aspect ratio strongly enhances the mechanical properties of the nanocomposites.

2.2 Filler Modification

Generally, regarding polymer–clay nanocomposites, the compatibility (and thus the quality of the nanodispersion, between the polymer and the clay) has been a subject of much interest and has led to the development of new surfactants for the modification of the clay [37–39].

In their study Chigwada et al. [40] reported that a clay containing a naphthyl substituent was used, and this gives better dispersion in styrenic polymers than that obtained with the commercial clay which contains a single benzene ring. A larger substituent, 4-acetylbiphenyl (BPNC16), was placed on the ammonium cation, and this cation was used to modify the clay, and acrylonitrile–butadiene–styrene-based nanocomposites were prepared by melt blending. The aim of the study was to investigate how different substituents can affect the dispersion of the clay in the polymer. The BPNC16 salt was prepared by the combination of a-bromo-4-phenylacetophenone and N,N-dimethylhexadecylamine. BPNC16 modified clay showed enhanced thermal stability and a larger d-spacing compared to some commercially available clays, and it can be conveniently prepared in a few hours time at room temperature with a minimum amount of solvent, which makes it potentially economical and convenient. The nanocomposites prepared with this clay show improved thermal stability and a significant reduction in the peak heat rate from cone calorimetric measurements.

Modesti et al. [6] investigated the effect of the clay surfactant on the morphology and performance of ABS/modified clay (OMLS) nanocomposites. Several kinds of organically modified layered silicates were used for the preparation of nanocomposites by a solution-intercalation technique. The authors focused their attention on surfactants based on imidazolium and quaternary ammonium salts. The good thermal stability observed for imidazolium modified clays [41] suggested in fact their huge potential in the preparation of nanocomposites suitable for elevated processing and/or operating temperatures. In particular, two commercially available OMLS were used in terms of comparison; the former was derived from

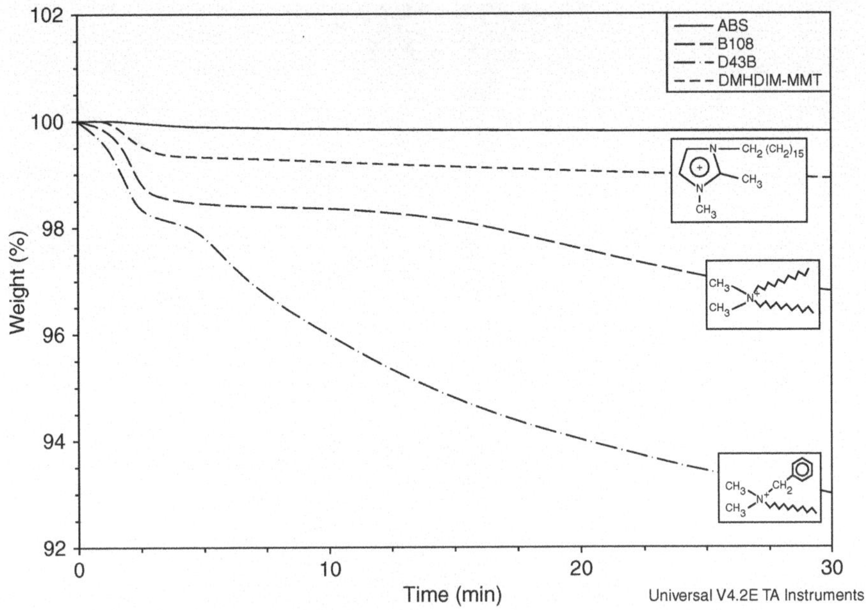

Fig. 9.5 Thermal stability of OMLS and pure polymer (isothermal conditions, 200°C, air) [6]

a natural hectorite modified with alkyl-ammonium salts, while the latter was based on a benzyl-alkyl-ammonium salt modified montmorillonite. Dimethyl-hexadecyl-imidazolium modified montmorillonite was experimentally prepared by a standard cationic exchange procedure [8]. Sodium montmorillonite with an ion-exchange capacity of 92 meq/100 g was ion exchanged with 1,2-dimethyl-3-hexadecylimidazolium (DMHDIM) bromide in water/ethanol (1:1 volume ratio). DMHDIM was prepared and purified as previously reported [41]. The different thermal–oxidative behaviors have been analyzed by means of thermogravimetric analysis (isothermal 200 °C, air) and are shown in Fig. 9.5.

The behavior of alkyl-ammonium (AA-MMT)-, benzyl-alkyl-ammonium (BAA-MMT)-, and alkyl-imidazolium-based OMLS (DMHDIM-MMT) after their dispersion in the polymer was investigated through morphological (XRD, TEM), dynamic-mechanical (DMA), and thermal analyses (TGA). XRD showed the presence of intercalated tactoids characterized by an average interlayer spacing of about 3 nm for all the OMLS, independently from the amount of filler (5 and 15 wt%). As observed by TEM, single delaminated lamellae were also present, and the highest extent of exfoliation among the prepared composites was obtained with ABS/DMHDIM-MMT (Fig. 9.6).

All OMLS exerted a reinforcing action on the polymer matrix in terms of stiffness; this effect was particularly evident for ABS/DMHDIM 5 wt% composite that showed an increase of about 40 % at 25 °C of the storage modulus if compared with the pristine polymer. The superior performance of DMHDIM-MMT

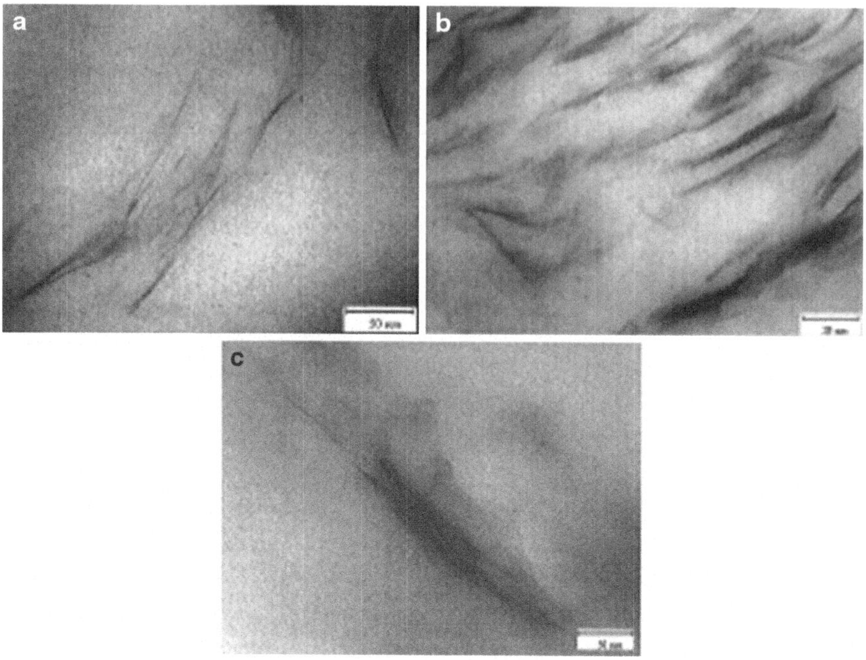

Fig. 9.6 TEM images of ABS/OMLS composites: ABS/DMHDIM-MMT 5 wt.% (**a**), ABS/BAA-MMT 5 wt.% (**b**), ABS/AA-MMT 5 wt.% (**c**) [6]

composite might be due to both the level of dispersion and the thermal stability. In fact, TGA data showed that the processing temperature influences the behavior of the clays modified with quaternary ammonium salts that might degrade and catalyze the decomposition of ABS, whereas in imidazolium-based nanocomposites, no catalytic decomposition is observed (Fig. 9.5).

Sepiolite is another type of clay which also provides an enhancement in the properties of the matrix. Its use as a reinforcing nanoparticle in polymers is now experiencing an increasing interest, but thermal and mechanical properties of sepiolite nanocomposites have not been widely studied yet. It has a laminar structure with tetrahedral and octahedral sheets, similar to montmorillonite (MMT), with an inversion of the sheets in one direction. Due to that, a structure of fibers and channels appears, and this feature is responsible for its high specific surface area (>300 m^2/g) [42]. Inside the structure, there is a relatively low number of inorganic interlaminar cations of Mg and Ca [43], susceptible of being exchanged (cation exchange capacity of sepiolite is around 30 meq/100 g [42] compared to 100 meq/100 g of MMT). Moreover, upon the surface of the fibers, there is a high concentration of silanol groups, giving its hydrophilic character to its surface [44]. Those two morphological characteristics have been employed in a modification process, in order to improve the dispersion of the sepiolite in the matrix. García-López et al. [45] reported that it was necessary to modify sepiolite nanofibers to obtain a higher increment in the final properties. Two different methods of modification have been reported by Tartaglione et al. [46]: using quaternary ammonium salts, as for MMT, or using a silane coupling agents. The first one involves a cation exchange reaction, substituting the inorganic cations inside the structure of fibers by organic cations, such as a quaternary ammonium salt molecule. The second method is based on a grafting reaction between a silane coupling agent and the silanol groups located onto the surface of the fibers. Tartaglione approximately calculated that the amount of organic modifier was around 12–15 % for quaternary ammonium salt-modified sepiolites and around 3–5 % for silane modified ones. But this percentage is very low compared with organically modified MMT.

In their study Basurto et al. [47] compared several modifications of sepiolite by means of thermogravimetric analysis. Clay has been modified with four different surfactant agents, two silane coupling agents, VTMO-2 (vinyl-trimethoxysilane) and HS-0.6 (aminosilane), and two quaternary ammonium salts, BM2TH (benzyl-methyl-dihydrogenated tallow) and 3MTH (trimethyl-hydrogenated tallow). The selection of the modifier used for each grade of ABS has been based on the polarity of the matrix and the agent. Thermogravimetric study has revealed different behaviors of sepiolite depending on the modification process, related to the amount of surfactant or the hydrophobic nature, and the necessity of a modification treatment.

3 ABS Nanocomposites Properties

3.1 Mechanical Behavior

Inorganic particulate fillers have extensively been employed to improve properties and/or lower costs of polymer products effectively. Young's modulus, hardness, heat distortion temperature, thermal expansion (mold shrinkage), etc., of the filled polymers could be improved to various extents. In general, nano-sized fillers are superior to their micron-sized counterparts in improving mechanical and thermal properties of thermoplastics due to the larger interfacial area between the particles and the surrounding polymer matrix [48–53]. Nano-sized fillers could increase modulus and hardness of some polymers but still maintain or even increase their tensile and/or impact strength in a certain filler loading range [54–61]. Nano-sized fillers are also preferred when transparency, surface smoothness, fire retardancy, and barrier property of the composites are the priorities.

Calcium carbonate does not have a layered structure, and therefore, there is no intercalation or exfoliation in nano-sized $CaCO_3$/polymer composites. However, due to the larger interfacial area in nano-sized $CaCO_3$/polymer, its properties are expected to be better than the micron-sized $CaCO_3$/polymer composites. ABS/$CaCO_3$ composites received very limited research interests. Liang et al. studied some rheological properties of the composites such as shear viscosity, extension viscosity, and entry pressure drop by capillary extrusion [62, 63]. Mechanical properties study revealed increase in tensile modulus but decrease in tensile and impact strengths. Particle size and surface treatment were found to have insignificant effect on the properties [64].

In their investigation Jiang et al. [65] compared micron-sized (MCC) and nano-sized (NCC) calcium carbonate/ABS composites. Particle dispersion and mechanical properties were analyzed. In addition, rheological measurements were employed to study their microstructures in the melt state.

Aside from some agglomerations, SEM micrographs demonstrated that NCC particles were distributed in the ABS matrix in much smaller sizes than MCC. MCC increased the modulus of neat ABS but decreased its tensile and impact strengths. NCC was superior to MCC in that it increased the modulus and maintained the impact strength. NCC/ABS property superiority was attributed to its larger interfacial area and cavitation toughening. Rheological tests revealed striking microstructure difference between MCC/ABS and NCC/ABS in the molten state. The addition of MCC simply increased the viscosity of the matrix, while NPCC changed the rheological response of ABS by forming the ordered structure in the matrix. Loss of Newtonian region and yield behavior was among the new rheological phenomena of NCC/ABS.

The effect of clay additions on toughness, particularly toughness as measured by impact methods rather than elongation at break in a tensile test, is rather more uncertain, as often found in other materials [66]. High strain rate testing can change a ductile material to brittle [67] which is one of the reasons why impact testing can give quite contrary results to toughness inferred from tensile tests [66].

There is a tendency for academic researchers to avoid the use of impact tests, perhaps because of their ambiguities: unstable crack growth, a wide range of energy sinks, and ill-defined notch radius. However, for industrial applications, impact strength is rated highly in the list of criteria for materials' selection. Indeed, many otherwise satisfactory plastics have a tendency towards brittle fracture under impact loading [66]. The well-documented property advantages that dispersed smectite clays can offer need to be accompanied by retention of impact strength.

In their work Chen et al. [68] studied the toughness and morphology of polymer–clay nanocomposites focusing also on ABS. The choice was made to assess claims that clay is effective in influencing toughness only if the matrix polymer is above T_g [69]. The argument is that clay platelets are sufficiently small, comparable in dimensions to the radius of gyration of the polymer chain, and sufficiently well attached to the polymer to be able to rotate and reorient during deformation in a way that larger-scale reinforcements, carbon or glass fibers, for example, cannot [70].

In the case of ABS, with an intercalated and exfoliated structure, the results from high-speed impact tests and lower strain rate tensile tests demonstrate an unequivocal and dramatic decrease in toughness after addition of clay, which is supported by field emission SEM (FE-SEM) image on the impact fracture surfaces. This large reduction of toughness in ABS is likely to be because the dispersed clay resists microscopic deformation, preventing the rubbery zones from fulfilling their role in absorbing energy in the crack propagation zone in notched tests and prior to crack initiation in un-notched tests. Parallel experiments indicate that the slight degradation of organoclay that may occur during processing of ABS at 200 °C does not significantly contribute to the decrease in impact strength.

A wider consequence of these results is that increased energy absorption in a tensile test cannot be used to infer increased toughness at impact strain rates. There is insufficient evidence from these results to support the emerging view that nanoclay additions to polymers are effective in maintaining toughness only if the polymer is above Tg and neither is there a correlation between the extent of dispersion as deduced from XRD and TEM and the retention of toughness given that the ABS–clay nanocomposite.

3.2 Fire Behavior and Thermal Stability

One of the main drawbacks of ABS is its inherent flammability [71], and therefore, there is a need to increase its thermal stability and flame-retardant properties. In this way, polymer nanocomposites have been extensively investigated as a possible solution, mainly in combination with traditional or alternative halogen-free flame retardants.

Regarding degradation mechanism, since some changes occur in the degradation of polystyrene nanocomposite compared to virgin polystyrene, Jang et al. [41] designed their study to determine if similar changes are also seen for styrene–acrylonitrile (SAN) and ABS. Because the primary difference between ABS and SAN is the presence of butadiene rubber, the effect of butadiene rubber on the thermal stability and degradation pathway of ABS is also explored in this study. The degradation pathway of SAN and ABS, in general, follows the same degradation pathway as described for polystyrene, where chain scission followed by b-scission (depolymerization).

In the presence of clay, two additional reactions, radical recombination reactions and extensive random scission, become significant as the clay content increases. Since ABS shows a well-dispersed morphology with a separate phase of butadiene rubber in the SAN matrix, the evolved products in the degradation of ABS are not different from those of SAN copolymer. The effect of rubber is similar to that of the clay, but not as effective due to its shorter duration. The difference in the degradation pathways of virgin SAN and virgin PS is in the evolution of dimers and trimers; SAN shows more evolution of these dimers and trimers, implying more radical transfer followed by β-scission. In the presence of clay, SAN nanocomposite produces smaller amount of recombined products compared PS nanocomposite, probably because the tertiary radical of the acrylonitrile unit is less stable than the corresponding styryl radical.

As already underlined, there are various methods of improving fire retardancy of ABS for safety consideration and the flame-retardant ABS is mainly achieved by halogen antimony synergism [71, 72]. However, to some degree, these methods are limited with respect to environmental requirements [73, 74]. Recent studies contain very diverse and efficient strategies for improving ABS fire resistance, such as ABS/clay [36], ABS/halogenated FR [75, 76], ABS/phosphorous FR [77, 78], ABS/intumescent FR [79], nitrogen–phosphorus flame retardants [80], ABS/zinc stannate [81], ABS/ferric chloride [82] synergistic agent [83, 84], and ABS/carbon nanotube [85].

A classic decabromodyphenylether (DB)/antimony trioxide (AO) system was employed by Wang and his colleagues [72], in order to prepare flame-retardant ABS/organically modified MMT (OMMT) nanocomposites, which could pass the rigorous UL94 test. Accordingly, the same type of clay (5 wt%) was well dispersed in ABS along with 15 wt% DB and 3 wt% AO using a twin-roll mill, and the fire properties of the resulting mixture were thoroughly studied. Contrary to the neat ABS and the pertinent nanocomposite, the sample containing both the clay filler and the flame-retardant system managed to get a V-0 rating, presenting also a limiting oxygen index (LOI) value (27.5 vol%) elevated by about 50 % and 28 %, respectively. Furthermore, the performance at cone calorimeter was greatly improved as probed by the 78 % lower peak heat release rate (pHRR) for the flame-retarded nanocomposite relative to that of pure ABS.

The challenge to enhance the thermal stability of ABS nanocomposites along with their resistance to ignite was faced by Ma et al. [86], substituting DB with a brominated epoxy resin (BER). BER is a high molecular weight gas-phase flame retardant with 53 wt% bromine content, designed by ICL Industrial Products (Beersheba, Israel) and used commonly in ABS or PC/ABS blends with AO as synergist. BER molecules, due to their superior polarity than ABS, were found to have higher affinity for clay particles, facilitating the formation exfoliated and, thus, more thermally stable structures. However, the great advantage of the applied approach was that, with small amount of the halogenated compound (12 wt% BER + 4wt%AO), the LOI of ABS containing 2 wt% clay raised from 20.5 to 31.4 vol%, which is far beyond 24 vol%, the LOI value usually required for a material to obtain a V-0 rating [87].

Similarly to the OMMT–DB–AO system, the synergy between OMMT–BER–AO derived from silicates forming barriers that hindered BER pyrolysis and reactions between BER–AO taking place at lower temperatures. Consequently, continuous flame retardancy in the vapor phase could be attained throughout combustion. Moreover, it was suggested that the alkyl-ammonium cations residing in the interlayer decomposed, at around 200 °C, to fragments which could volatilize and expand clay layers promoting silicates dispersion. Exfoliated structures, exhibiting better barrier properties than their intercalated analogues, could delay more efficiently mass and heat transport. On the other hand, the reaction between the surfactant's decomposition products and DB–AO, occurring at high temperatures, resulted in the formation of radical scavengers.

Kim et al. [84] intercalated triphenyl phosphate (TPP) in the galleries of a commercial OMMT (Cloisite 30B, Southern Clay Products, USA) and melt mixed the resulting clay with ABS to formulate nanocomposites. The benefit of this approach is that by "shielding" the normally volatile TPP inside silicates, its evaporation during melt compounding can be suppressed, allowing for more efficient flame retardancy and a wider range of processing conditions. The addition of clay was shown to enhance slightly the thermal stability of ABS owing to the delayed release of TPP; the LOI, yet, remained almost unchanged. A great increase in LOI was achieved when an epoxy novolac system was co-incorporated, and further improvement in thermal stability was observed for the samples containing silane agents also which favored coupling between epoxy and silicates. It is worth noticing that the LOI of 85/9/6 wt% ABS/epoxy/(clay–TPP + silane) formulation raised to 41.2 vol%, while the corresponding value for the sample of 15 wt% TPP was 20.2 vol%, just 9 % above that of the virgin polymer (18.2 vol%). Optical micrographs of the combustion residues revealed that this improvement in flammability was associated with the formation of a more coherent char, lacking of holes and crevices. The synergistic effect of this system was also confirmed when instead of TPP, a tetra-2,6-dimethylphenyl resorcinol diphosphate (DMP–RDP) was employed.

In their work Ma et al. [88] prepared ABS-grafted maleic anhydride (ABS-g-MAH) resin with different grafting degrees, ABS/OMMT, and ABS-g-MAH/OMMT nanocomposites via melt blending. FTIR spectra confirmed that maleic anhydride was successfully grafted onto butadiene chains of the ABS backbone in the molten state using dicumyl peroxide as the initiator and styrene as the comonomer. Electron microscopy images indicated the size of the dispersed domains of ABS-g-MAH increased, and the dispersion was more uniform than that of neat ABS resin. X-ray diffraction (XRD) and transmission electron microscopy (TEM) results showed that intercalated/exfoliated structure formed in ABS-g-MAH/OMMT nanocomposites and rubber phase intercalated into clay layers. TGA results revealed that the intercalated/exfoliated structure of ABS-g-MAH/OMMT nanocomposites has better barrier properties and thermal stability than intercalated ones of ABS/OMMT nanocomposites. The T_g of ABS-g-MAH resin was almost unchanged compared to that of neat ABS, but the addition of clay can improve T_g and the T_g of ABS-g-MAH/OMMT nanocomposites was higher than that of neat ABS/OMMT nanocomposites. The results of cone measurement indicate that grafting MAH onto ABS chains does not show any flame retardancy compared to pure ABS resin. But ABS-g-MAH/OMMT nanocomposites exhibit reduced flammability compared to ABS/OMMT nanocomposites at the same clay content. The chars of ABS-g-MAH/OMMT nanocomposites were tighter, denser, more integrated, and fewer surface microcracks than ABS/OMMT residues. The improvement of flame retardancy of ABS-g-MAH/OMMT nanocomposites can be ascribed to the better dispersion of clay layers and intercalation of clay.

In another paper of the same authors [89], a simple method of melt blending was used to prepare ABS/montmorillonite nanocomposites incorporating the intumescent flame retardant, poly(4,4-diaminodiphenylmethane-spirocyclic-pentaerythritol-bisphosphonate (PDSPB).

The organophilic montmorillonite was prepared by cationic exchange between Na-montmorillonite and octadecyl-trimethyl ammonium bromide (OTAB) in aqueous solution, while PDSPB was synthesized by condensation polymerization [79]. A synergistic effect was found between PDSPB and montmorillonite which improved thermal stability and flame retardancy. The phosphoric acid generated on heating from PDSPB probably reacts with montmorillonite to form silico-aluminophosphate (SAPO). The decomposition of the amine silicate modifier leads to strongly acidic catalytic sites that may further promote the oxidative dehydrogenation cross-linking charring process and increase the char yield. During combustion, the physical process of layer reassembling acts as a protective barrier in addition to the intumescent shield and can limit the oxygen diffusion to the substrate or gives a less disturbing low volatilization rate.

Zinc hydroxyl stannate and zinc stannate have even been used as a highly effective flame retardant. The advantage of zinc hydroxyl stannate and zinc stannate as flame retardant is lower toxicity than antimony compounds [90–93]. Yousefi et al. [94] have been interested in the synthesis of Bi_2S_3 [95], SnS [96], CdS [97–99], and ZnS [100–105] nanostructures using thioglycolic acid, via hydrothermal method.

They prepared SnS nanoflowers by TGA-assisted hydrothermal process at relatively low temperature [94] and studied the influence of inorganic phase on the thermal properties of ABS matrix. The thermal decomposition of the ABS shifted towards higher temperature in the presence of the SnS nanoflowers. Improved thermal stability of composites with respect to the pure ABS can be assigned to partially alter molecular mobility of the polymer chains due to their adsorption on the surface of the filler particles. Also, exfoliated SnS filler particles have significant barrier effect to slow down product volatilization and thermal transport during decomposition of the polymer.

4 ABS-Based Nanocomposite Blends

Polycarbonate (PC)/acrylonitrile–butadiene–styrene (ABS) alloys are well-known commercial polymers. They are widely used in engineering thermoplastics due to an appropriate combination of two components. PC is a resin in which groups of dihydric or polyhydric phenols are linked through carbonate groups. Virtually all general-purpose PCs are based on bisphenol A. PC has high thermal stability and good impact behavior. ABS has easy processability and economic benefits. However, there are few reports about the synthesis of PC/ABS/clay nanocomposites. In their work, Wang et al. [106] synthesized PC/ABS/OMMT nanocomposites through direct melt intercalation and studied the morphology and thermal stability of the nanocomposites compared to that of pure PC/ABS alloys. The morphology of alloy nanocomposites indicates that it is mainly the ABS molecules which are intercalated into the clay layers rather than the PC molecules. The study by TGA shows that the addition of OMMT (5 wt%) can improve the thermal stability of PC/ABS alloys.

Regarding thermal stability studies, Zong et al. [107] synthesized PC/ABS/clay nanocomposites through direct melt intercalation technique and study the thermal stability of the nanocomposite compared to that of pure PC/ABS alloy.

Detailed kinetic analyses of the nanocomposite and the alloy have been performed using thermogravimetric analysis (TGA), to analyze their thermal behavior at different heating rates in the nitrogen atmosphere. There are many methods of evaluation of the non-isothermal kinetic parameters of the thermal degradation of polymers [108], but the Kissinger and Flynn–Wall–Ozawa methods were chosen, which do not require knowledge of the reaction mechanism.

Comparing values for the activation energies for the thermal degradation of the PC/ABS alloy and PC/ABS/OMMT nanocomposite, via both the Kissinger and Ozawa methods, it has been found that there is an obvious trend for the increase of the apparent activation energy of the nanocomposite. This increasing trend coincides with the thermal analysis results that the polymer/clay nanocomposite has a higher thermal stability and lower flammability.

Activation energies for polymer degradation, with or without the OMMT, were determined using the Flynn–Wall–Ozawa method. For this study, conversion values of 30 %, 50 %, and 70 % have been used. Authors clearly observed that the activation energy values (Ea) computed using the Kissinger and Flynn–Wall methods agree on the similar change trend that the activation energies of the thermal degradation for the nanocomposite are more than those of the pure polymer.

The phenomenon indicates an important role of clay in improving the flame retardance of the alloy. This increasing tendency coincides with the thermal analysis results that the polymer/clay nanocomposite has a higher thermal stability and lower flammability. Other studies [109–115] show that the lower flammability of polymer/clay nanocomposite is not due to retention of a large fraction of fuel but, in the form of carbonaceous char, due to the condensed phase. The nano-dispersed lamellae of clay (exfoliation or intercalation) in polymer matrix may change the decomposition process of polymer since the nano-dispersed silicate layers acted as thermal hinder in polymer matrix. The nano-dispersed silicate layers slow the decomposition rate and increase the temperature of degradation (especially as measures at the point of 50 % mass loss) by acting as an excellent thermal insulator and mass transport barrier.

Recently, increasing attention has also been focused on the incorporation of conductive fillers into immiscible/partially miscible polymer blends in order to improve their conductivity at much lower filler content due to the double percolation phenomenon. It refers to the percolation of a conductive filler within one phase in a polymer blend (first percolation), which itself percolates in the mixture (second percolation). The first studies in this direction were presented by Sumita et al. [116] with the analysis of immiscible blends filled with carbon black.

With regard to PC/acrylonitrile–butadiene–styrene/MWCNT (PC/ABS/MWCNT) blends, specific studies by Sun et al. [117] recently focused on the evaluation of the effect of ABS rubber content on the localization of MWCNTs and subsequent composite morphology and electrical properties.

Moreover, Göldel et al. [118] investigated the selective localization and migration of MWCNTs in PC/styrene–acrylonitrile blends (SAN) observing as, regardless the way of introducing nanotubes, they exclusively located within PC phase, resulting in a much lower electrical resistivity than the one of pure PC and SAN. The phenomena have been explained by the differences in interfacial energies of the filler and polymers, which originate from the differing polarities and surface energies.

In their paper Besco et al. [119] examined the correlation between the morphology of PC/ABS/multiwalled carbon nanotube (MWCNT) composites and their electrical properties when a hybrid filler system of MWCNTs and an organically modified synthetic mica (OMLS) are introduced. Considering the cited results from specific literature, the effects of the OMLS on nanotube localization and consequently on the percolation threshold value were investigated using a multiscale approach based on the comparison of electrical and morphological properties that were examined by means of electron microscopy, XRD analyses, and static electrical tests and dielectric analyses.

TEM has been used to investigate blend structure and the extent of MWCNT/OMLS dispersion and distribution in the composite matrix. By way of example, for the PC/ABS/MWCNTs, 2 wt% composite was analyzed to examine the microstructure of the conductive network just above the electrical percolation threshold. Due to the low voltage used (80 kV) for the images shown in Fig. 9.7, the brighter continuous phase represents PC, while the darker grey areas with a size of 0.5–2 μm are ABS. Moreover, as already discussed, ABS has a complicated structure where styrene–acrylonitrile (SAN) phase contains islands of black polybutadiene (PB) rubber (stained by OsO_4) in which white particles of SAN are embedded. Comparing these micrographs, the MWCNTs appear to be homogeneously dispersed within the PC phase and are mainly present as single particles. The results confirmed that when ABS is blended with PC, MWCNTs are selectively localized in the PC phase due to their higher affinity for this polymer with respect to SAN and PB.

Regarding PC/ABS/MWCNT/OMLS nanocomposites, the modified mica was dispersed as small stacks and submicron-size agglomerates, suggesting a predominately intercalated morphology (as evidenced also by XRD analyses) (Fig. 9.8). Exfoliated layers of OMLS were found mainly within the SAN phase at the interface between SAN and PB particles, a behavior observed previously by several authors studying ABS/montmorillonite nanocomposites [7]. The mean dimensions of OMLS intercalated agglomerates are of about 500 μm, and they show a low degree of interaction with the matrix, obvious from the absence of an interface between them and the matrix (holes) in several cases. The same authors attributed this phenomenon to an excessive difference of the surface tensions of the two components which is too large to allow intercalation of the clay and breakup of clay tactoids [7]. Furthermore, the distribution of clay particles within the SAN phase indicates the preferential affinity of the filler with this phase.

The most important information about the influence of OMLS on the distribution of MWCNTs in the polymer matrix arises from the comparison between the images

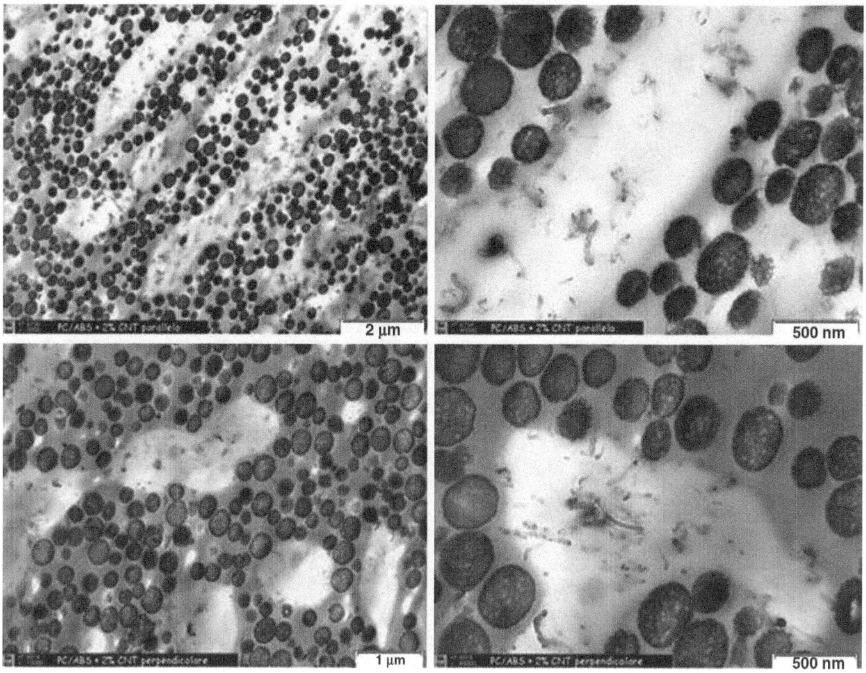

Fig. 9.7 TEM micrographs of PC/ABS-MWCNTs 2 wt % composite [119]

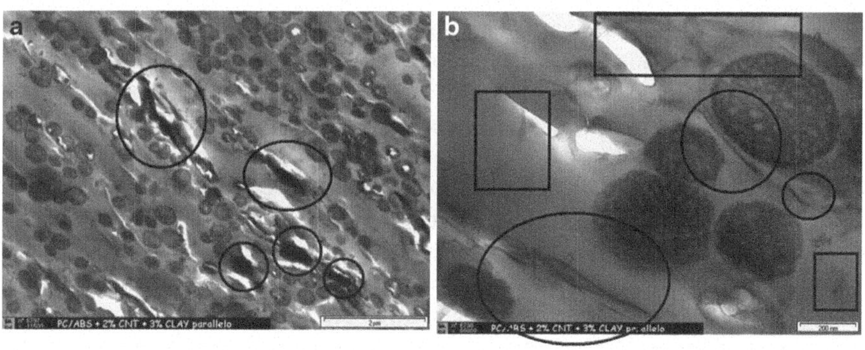

Fig. 9.8 TEM micrographs of PC/ABS-MWCNTs 2 wt % composite with OMLS 3 wt %: particulars of OMLS dispersion. OMLS particles are marked with circles while CNTs are evidenced by rectangles [119]

in Fig. 9.7 with those in Fig. 9.9. When OMLS 3 wt% is present in the blend, large poorly dispersed MWCNT agglomerates can be observed outside brighter PC domains (see Fig. 9.9) suggesting a hindering effect of the clay on nanotube dispersion.

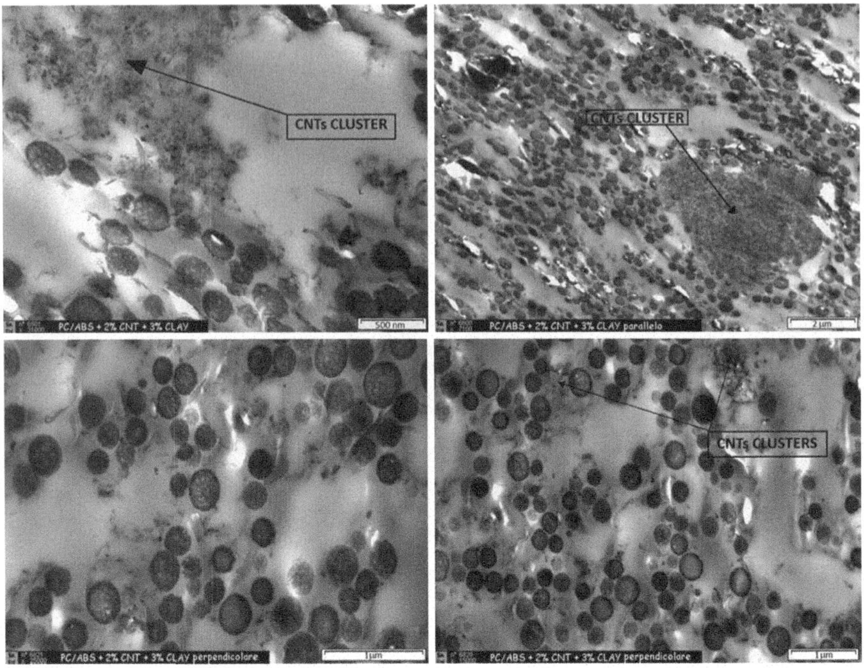

Fig. 9.9 TEM micrographs of PC/ABS-MWCNTs 2 wt % composite with OMLS 3 wt %: particulars of CNTs dispersion [119]

Poor MWCNT dispersion within the continuous PC phase together with the presence of agglomerated nanotubes bundles will result in a less effective nanotube network in particular when CNT content is lower than the one observed (hence below 2 wt%) such that the composites have a relatively high electrical resistivity and higher percolation threshold. This observation is in agreement with electrical and dielectric analyses that showed an increase in electrical percolation threshold when OMLS was added to the composite material. As observed by Göldel et al. [118], the preferential localization of a filler in one of the phases has commonly been explained by the differences in interfacial energies of the filler and the respective polymers, which originates from the differing polarities and surface energies. If, as reported by several papers [120–124] about the role of organoclay in polymer blends, OMLS can act as a compatibilizer modifying interfacial energies, this can be a hypothesis to explain the changes in CNT localization and distribution when clay is present above a certain amount in the considered systems (OMLS 3 wt%).

The effective concentration of MWCNTs within the PC-rich phase results in a remarkable reduction in electrical resistivity and a low percolation threshold. Hence, the electrical conductivity and dielectric properties of PC/ABS/MWCNT composites are strongly affected by MWCNT localization/dispersion, and when MWCNT concentration within PC phase decreases, the conductive network becomes

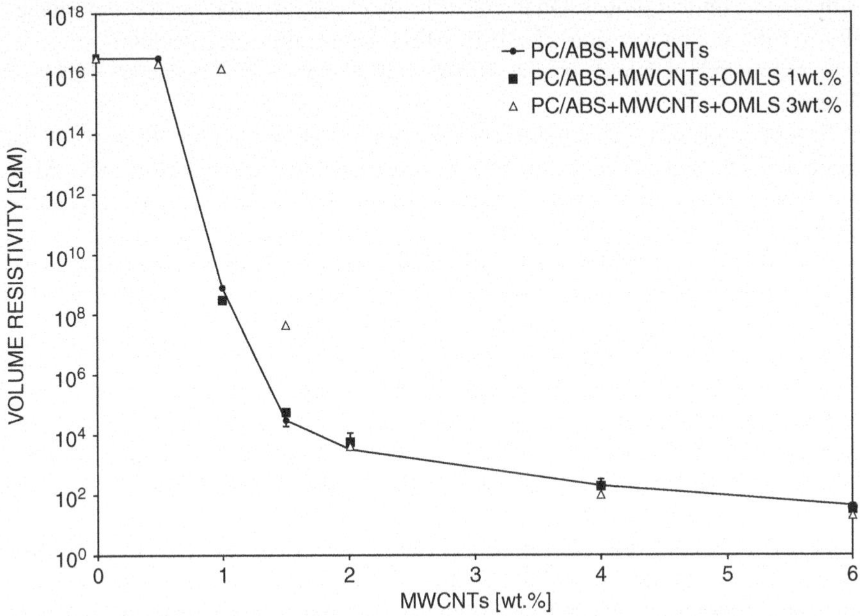

Fig. 9.10 Volume resistivity of PC/ABS based composites varying MWCNTs and OMLS contents [119]

less effective leading to an increase in composite electrical percolation threshold. This is true in particular below a certain CNT content (2 wt%), because above this limit electrical conductivity can be reached due to high CNT overall concentration (Fig. 9.10).

5 Conclusions and Future Perspectives

Acrylonitrile–butadiene–styrene (ABS) is a widely used engineering thermoplastic owing to its desirable properties which include good mechanical properties, chemical resistance, and easy processing characteristics. In this chapter, a discussion about the state of art regarding ABS-based nanocomposites has been presented considering different aspects regarding synthesis and production as well as the effects of nanofillers on physical and thermal properties of the polymer itself and of its most common blends.

Classic melt-blending process has been compared with low-temperature solvent-based techniques that have been proposed in order to prevent the possible thermal degradation of organically modified layered silicates often occurring at temperatures near 200 °C. The effectiveness of ultrasonic-mixing-assisted solution processes has also been assessed, even by means of innovative characterization techniques based on confocal microscopy and experimental nanofillers modified with organic optical probes. On the other hand, classic electron microscopy studies allowed the comprehension of ABS/OMMT nanocomposites peculiar structure, with exfoliated platelets localized mainly at the interface between SAN and PB phases.

Several kinds of organic modifiers have been proposed in order to improve the thermal stability of OMMTs together with the chemical compatibility with ABS, with particularly good results observed for innovative imidazolium salt-based modifiers.

One of the main drawbacks of ABS is its inherent flammability, and polymer nanocomposites have been extensively investigated as a possible solution in combination with traditional or alternative halogen-free flame retardants. Interesting results have been obtained for several formulations regarding the presence of a positive synergism between nanofillers and flame retardants, hence improving ABS performances evaluated with a large series of experimental tests. Thermal and thermo-oxidative degradation mechanisms have also been assessed and compared with the ones typical of parent polymers (even including PC/ABS blend), in order to further clarify the role of nanofillers during these processes.

PC/ABS blends have also been considered in order to study the selective localization of OMMT and MWCNTs in complex multicomponent systems. A peculiar influence of OMMT regarding the preferential localization of MWCNTs has been evidenced, with important consequences on macroscale properties of hybrid nanocomposites.

References

1. Basdekis CH (1964) ABS plastics. Reinhold Pub. Corp, New York, p 147
2. Alexander M, Dubois P (2000) Mater Sci Eng 28:1
3. Pourabas B, Raeesi V (2005) Polymer 46:5533
4. Dennis HR, Hunter DL, Chang D, Kim S, White JL, Cho JW, Paul DR (2001) Polymer 42:9513
5. Modesti M, Besco S, Lorenzetti A, Zammarano M, Causin V, Marega C, Gilman JW, Fox DM, Trulove PC, De Long HC, Maupin PH (2008) Polymer Adv Technol 19:1576
6. Modesti M, Besco S, Lorenzetti A, Causin V, Marega C, Gilman JW, Fox DM, Trulove PC, DeLong HC, Zammarano M (2007) Polym Degrad Stab 92:2206
7. Stretz HA, Paul DR, Cassidy PE (2005) Polymer 46:3818
8. Gilman JW, Awad WH, Davis RD, Shields J, Kashiwagi T, VanderHart DL, Harris RH, Davis C, Morgan AB, Sutto TE, Callahan J, Truelove PC, Delong HC (2002) Chem Mater 14:3776
9. Bottino FA, Fabbri E, Fragala LI, Malandrino G, Orestano A, Pilati F, Pollicino A (2003) Macromol Rapid Commun 24:1079
10. Maupin PH, Gilman JW, Harris RH, Bellayer S, Bur AJ, Roth SC, Murariu M, Morgan AB, Harris JD (2004) Macromol Rapid Commun 25:788
11. Yalcin B, Cakmak M (2004) Polymer 45:6623
12. Baumann R, Ferrante C, Deeg FW, Brauchle C (2001) J Chem Phys 114:5781
13. Huang HH, Wilkes GL (1987) Macromolecules 20:1322
14. Mascia L, Kioul A (1995) Polymer 36:3649

15. Kirkland JJ (1974) US patent no. 3782075
16. Iler RK, McQueston JJ (1977) US patent no. 4010242
17. Barthet C, Hickey AJ, Cairns DB, Armes SP (1999) Adv Mater 11:408
18. Reculusa S, Poncet-Legrand C, Ravaine S, Mingotaud C, Duguet E, Bourgeat-Lami E (2002) Chem Mater 14:2354
19. Chen M, Wu L, Zhou S, You B (2004) Macromolecules 37:9613
20. Kim IJ, Kwon OS, Park JB, Joo H (2006) Curr Appl Phys 6S1:e43
21. Finkle P, Draper HD, Hildebrand JH (1923) J Am Chem Soc 45:2780
22. Levine S, Sanford E (1985) Can J Chem Eng 58:622
23. Abend S, Bonnke N, Gutschner U, Lagaly G (1998) Colloid Polym Sci 276:730
24. Lagaly G, Reese M, Abend S (1999) Appl Clay Sci 14:83
25. Valint PL, Bock J (1988) Macromolecules 21:175
26. Piirma I (1992) Polymer surfactants, Surfactant science series 42. Marcel Dekker, New York, p 127
27. Astafieva I, Zhong XF, Eisenberg A (1993) Macromolecules 26:7339
28. Gilbert RG (1995) Latex polymerization: a mechanistic approach. Academic, London
29. Coen E, Lyons RA, Gilbert RG (1996) Macromolecules 29:5128
30. Müller H, Leube W, Klaus Tauer K, Förster S, Antonietti M (1997) Macromolecules 30:2288
31. Vorwerg L, Gilbert RG (2000) Macromolecules 33:6693
32. Capek I (2002) Adv Colloid Interface Sci 99:77
33. Kukula H, Schlaad H, Tauer K (2002) Macromolecules 35:2538
34. Jang LW, Chul MK, Lee DC (2000) J Polym Sci, Part B: Polym Phys 39:719
35. Wang SF, Hu Y, Song L, Wang ZZ, Chen ZY, Fan WC (2002) Polym Degrad Stab 77:423
36. Choi YS, Xu M, Chung IJ (2005) Polymer 46:531
37. Zheng X, Wilkie CA (2003) Polym Degrad Stab 82:441
38. Su S, Jiang DD, Wilkie CA (2004) Polym Degrad Stab 83:321
39. Zhang J, Wilkie CA (2004) Polym Degrad Stab 83:301
40. Chigwada G, Wang D, Jiang DD, Wilkie C (2006) Polym Deg Stab 91:755
41. Jang BN, Wilkie CA (2005) Polymer 46:9702
42. Alkan M, Tekin G, Mamli H (2005) Microporous Mesoporous Mater 84:75
43. Bokobza L, Burr A, Garnaud G, Perrin MY, Pagnotta S (2004) Polymer Int 53:1060
44. Tartaglione G, Tabuani D, Camino G, Moisio M (2008) Compos Sci Technol 68:451
45. García-López D, Fernández JF, Merino JC, Pastor JM (2010) Compos Sci Technol 70:1429
46. Tartaglione G, Tabuani D, Camino G (2008) Microporous Mesoporous Mater 107:161
47. Basurto FC, García-López D, Villarreal-Bastardo N, Merino JC, Pastora JM (2012) Compos Part B (in press)
48. Lewis TB, Nielsen LE (1970) J Appl Polym Sci 14:1449
49. Sumita M, Shizuma T, Miyasaka K, Ishikawa K (1983) J Macromol Sci Phys B22:601
50. Sumita M, Tsukumo T, Miyasaka K, Ishikawa K (1983) J Mater Sci 18:1758
51. Yano K, Usuki A, Okada A, Kurauchi T, Kamigaito O (1993) J Polym Sci, Part A: Polym Chem 31:2493
52. Messersmith PB, Giannelis EP (1994) Chem Mater 6:1719
53. Messersmith PB, Giannelis EP (1995) J Polym Sci, Part A: Polym Chem 33:1047
54. Kojima Y, Usuki A, Kawasumi M, Okada A, Fukushima Y, Kurauchi T, Kamigaito O (1993) J Mater Res 6:1185
55. Kojima Y, Usuki A, Kawasumi M, Okada A, Kurauchi T, Kamigaito O (1993) J Polym Sci, Part A: Polym Chem 31:983
56. Lee DC, Jang LW (1996) J Appl Polym Sci 61:1117
57. Hasegawa N, Kawasumi M, Kato M, Usuki A, Okada A (1998) J Appl Polym Sci 67:87
58. Wang Z, Pinnavaia TJ (1998) Chem Mater 10:3769
59. Wang SJ, Long CF, Wang XY, Li Q, Qi ZN (1998) J Appl Polym Sci 69:1557
60. Kojima Y, Usuki A, Kawasumi M, Okada A, Kurauchi T (1999) Kamigaito O. J Polym Sci, Part A: Polym Chem 31:1755

61. Liu LM, Qi ZN, Zhu XG (1999) J Appl Polym Sci 71:1133
62. Liang JZ (2002) Polym Int 51:1473
63. Tang CY, Liang JZ (2003) J Mater Process Technol 138:408
64. Tang CY, Chan LC, Liang JZ (2002) J Reinforce Plast Comp 21:1337
65. Jiang L, Lam YC, Tam KC, Chua TH, Sim GW, Ang LS (2005) Polymer 46:243
66. Bucknall CB (1977) Toughened plastics. Applied Science Publishers, London
67. Rothon RN (2003) Particulate-filled polymer composites. RapraTechnology Limited, Shropshire, p 377
68. Chen B, Evans JRG (2008) Polymer 49:5113
69. Shah D, Maiti P, Gunn E, Schmidt DF, Jiang DD, Batt CA (2004) Adv Mater 16:1173
70. Shah D, Maiti P, Jiang DD, Batt CA, Giannelis EP (2005) Adv Mater 17:525
71. Owen SR, Harper JF (1999) Polym Degrad Stab 64:449
72. Wang S, Hu Y, Zong RW, Tang Y, Chen ZY, Fan WC (2004) Appl Clay Sci 25:49
73. Montezin F, Lopez-Cuesta JM, Crespy A, Georlette P (1997) Fire Mater 21:245
74. Lu SY, Hamerton I (2002) Prog Polym Sci 27:1661
75. Bhaskar T, Murai K, Matsui T, Brebu MA, Uddin MA, Muto A, Sakata Y, Murata K (2003) J Anal Appl Pyrolysis 70:369
76. Brebu M, Bhaskar T, Murai K, Muto A, Sakata Y, Uddin MA (2004) Chemosphere 56:433
77. Czégény Z, Blazso M (2008) J Anal Appl Pyrolysis 81:218
78. Hoang D, Wan J (2008) Polym Degrad Stab 93:36
79. Ma H, Tong L, Xu Z, Fang Z, Jin Y, Lu F (2007) Polym Degrad Stab 92:720
80. Nguyen C, Kim J (2008) Polym Degrad Stab 93:1037
81. Petsom A, Roengsumran S, Ariyaphattanakul A, Sangvanich P (2003) Polym Degrad Stab 80:17
82. Cai Y, Hu Y, Song L, Xuan S, Zhang Y, Chen Z, Fan W (2007) Polym Degrad Stab 92:490
83. Lee K, Kim J, Bae J, Yang J, Hong S, Kim HK (2002) Polymer 43:2249
84. Kim J, Lee K, Bae J, Yang J, Hong S (2003) Polym Degrad Stab 79:201
85. Yang S, Castilleja JR, Barrera EV, Lozano K (2004) Polym Degrad Stab 83:383
86. Ma H, Xu Z, Tong L, Gu A, Fang Z (2006) Polym Degrad Stab 91:2951
87. Weil ED, Hirschler MM, Patel NG, Shaki S (1992) Fire Mater 16:159
88. Ma H, Fang Z, Tong L (2006) Polymer Degrad Stabil 91:1972
89. Ma H, Tong L, Xu Z, Fang Z (2008) Appl Clay Sci 42:238
90. Cusack PA (1986) Fire Mater 10:41
91. Andre F, Cusack PA, Monk AW, Seangprasertkij R (1993) Polym Degrad Stab 40:267
92. Cusack PA, Heer MS, Monk AW (1997) Polym Degrad Stab 58:229
93. Cusack PA, Hornsby PR (1999) J Vinyl Add Technol 5:21
94. Yousefi M, Salavati-Niasari M, Gholamian F, Ghanbari D, Aminifazl A (2011) Inorg Chim Acta 371:1
95. Salavati-Niasari M, Ghanbari D, Davar F (2009) J Alloys Compd 488:442
96. Salavati-Niasari M, Ghanbari D, Davar F (2010) J Alloys Compd 492:570
97. Salavati-Niasari M, Loghman-Estarki MR, Davar F (2008) Chem Eng J 145:346
98. Salavati-Niasari M, Loghman-Estarki MR, Davar F (2009) Inorg Chim Acta 362:3677
99. Salavati-Niasari M, Davar F, Loghman-Estarki MR (2009) J Alloys Compd 481:776
100. Marcilla A, Beltrán M (1995) Polym Degrad Stab 50:117
101. Davar F, Salavati-Niasari M, Mazaheri M (2009) Polyhedron 28:3975
102. Salavati-Niasari M, Loghman-Estarki MR, Davar F (2009) J Alloys Compd 475:782
103. Salavati-Niasari M, Davar F, Mazaheri M (2009) Mater Res Bull 44:2246
104. Salavati-Niasari M, Davar F, Mazaheri M (2009) J Alloys Compd 470:502
105. Salavati-Niasari M, Davar F, Loghman-Estarki R (2010) J Alloys Compd 494:199
106. Wang S, Hu Y, Wang Z, Yong T, Chen Z, Fan W (2003) Polym Degrad Stab 80:157–161
107. Zong R, Hu Y, Wang S, Song L (2004) Polym Degrad Stab 83:423
108. Budrugeac P (2003) Polym Degrad Stab 71:185
109. Gilman JW (1999) Appl Clay Sci 15:31

110. Vaia RA, Price G, Ruth PN, Nguyen HT (1999) J Appl Clay Sci 15:67
111. Bourbigot S, Le Bras M, Dabrowski F, Gilman JW, Kashiwagi T (2000) Fire Mater 24:201
112. Gilman JW, Jackson CL, Morgan AB (2000) Chem Mater 12:1866
113. Zhu J, Wilkie CA (2000) Polym Int 49:1158
114. Hu Y, Song L (2001) In: International fire safety conference. Fire Retardant Chemicals Association, 11–14 March, 2001
115. Bourbigot S, Devaux E, Flambard X (2002) Polym Degrad Stab 75:397
116. Sumita M, Sakata K, Asai S, Miyasaka K, Tanemura M (1992) Coll Polym Sci 270:134
117. Sun Y, Guo ZX (2010) Macromol Mater Eng 295:263
118. Göldel A, Kasaliwal G, Pötschke P (2009) Macromol Rapid Commun 30:423
119. Besco S, Lorenzetti A, Donadi S, McNally T, Modesti M (2012) J Appl Pol Sci 124:3617
120. Lipatov YS (2002) Prog Polym Sci 27:1721
121. Voulgaris D, Petridis D (2002) Polymer 43:2213
122. Gelfer MY, Song HH, Liu L, Hsiao BS, Chu B, Rafailovich M, Si M, Zaitsev V (2003) J Polym Sci, Part B: Polym Phys 41:44
123. Khatua BB, Lee DJ, Kim HY, Kim JK (2004) Macromolecules 37:2454
124. Ray SS, Bousmina M, Ray SS, Pouliot S, Bousmina M, Utracki LA (2004) Polymer 45:8403

Polysterene Layered Silicate Nanocomposites

10

Abozar Akbari, Mahsa A. Tehrani, and Hossien Cherghibidsorkhi

Contents

Abstract

Polymer/layered silicate nanocomposites bring about an enhancement of many properties for polymers. They have attached considerable attention since they have the capacity to generate new polymer properties. The performance improvement greatly depended on the distribution and dispersion of layers of the silicates in the polymer matrix. This chapter aims to highlight on the recent developments in preparation and characterization of polystyrene/ layered silicate nanocomposites, and also the effect of different parameters such as type of clay, clay modifiers, and preparation methods on final properties of polystyrene nanocomposites will be reviewed.

A. Akbari (✉) • H. Cherghibidsorkhi
Enhanced Polymer Research Group (EnPRO), Department of Polymer Engineering, Faculty of Chemical Engineering, Universiti Teknologi Malaysia, Johor Bahru, Johor, Malaysia
e-mail: Akbari.Abouzar@gmail.com

M.A. Tehrani
School of Industrial Technology, Universiti Sains Malaysia, Penang, Malaysia

J.K. Pandey et al. (eds.), *Handbook of Polymernanocomposites. Processing, Performance and Application – Volume A: Layered Silicates*, DOI 10.1007/978-3-642-38649-7_16,

Keywords
Clay • Layered silicate • Modification • Nanocomposite • Polystyrene

1 Introduction

Today's one of the most commonly used applied technologies is combination of a polymer matrix with additives on a nanometer scales. These additives must have at least one dimension in the nanometer range. Over the past two decades, polymer/layered silicate nanocomposites have attached considerable attention because of the capability to generate new polymer properties. It is known that the incorporation of just a few percent of such fillers (≤5 wt%) has been found a large in polymer properties such as mechanical properties, barrier properties, thermal, and chemical properties. To achieve these properties, several factors have to be considered: method of producing nanocomposite nature, type and properties of clay, and clay modifier. But the most important factor is the amount of compatibility between clay and polymer domain. The formation of polymer–clay nanocomposites has been carried out by three different methods which are depending on the filler dispersion inside the polymer matrix. In the other words, the association of clay minerals with polymers gives materials with different structures as it is illustrated in Fig. 10.1. The first preparation method involves the formation of aggregate as a microparticle or tactoid from the clay layers in order to create a conventional composite. If in combination between polymer matrix and nanoclay, polymer chains penetrate into the space between clay layers, but do not separate layers thoroughly, as the registry between them still maintained, the nanocomposite is described as intercalated, while if the clay layers are thoroughly detached from each other by the polymer chains while the original registry is lost, the nanocomposite is called exfoliated (Akbari et al. 2011; [1–4]).

In this regard, in recent decades, polystyrene (PS)/clay nanocomposites have been an area of active research in both academic and industry. Actually polystyrene along with polyvinyl chloride, polyethylene, and polypropylene belongs to a group of thermoplastics that are used in extremely wide range of applications. Hence, researchers have been devoting many efforts to the development of PS/clay nanocomposites.

Various methods have been used to prepare polystyrene/clay nanocomposites which include solution blending, melt compounding, in situ polymerization, emulsion/suspension polymerization, and new approach for one-pot preparation of montmorillonite-exfoliated polystyrene (MMT-PS) nanocomposite by ultrasound. It is well-known that for improving thermal stability of PS, the intercalated morphology of nanocomposite is recommended and exfoliated morphology showed better mechanical properties compared to other nanocomposites.

In addition, new developments in PS/clay nanocomposite technology will be reviewed.

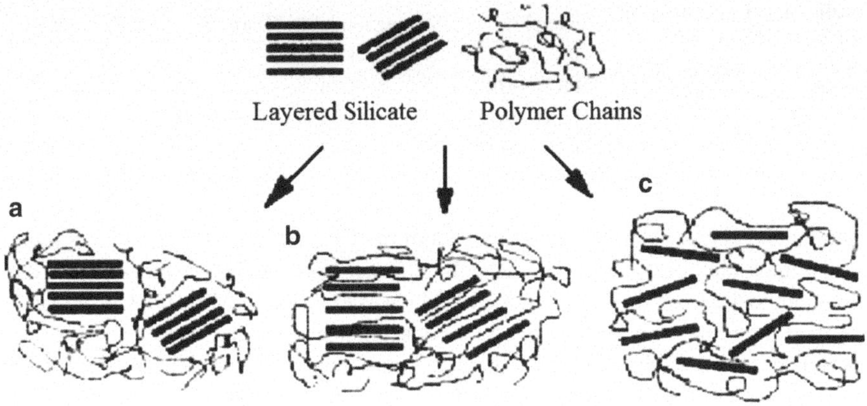

Fig. 10.1 Different structures proposed for clay-polymer materials (**a**) Microcomposite structure. (**b**) Intercalated nanocomposite structure. (**c**) Exfoliated nanocomposite structure [61]

2 Clay Modification

Natural clays are normally hydrophobic and it is important that they be modified in order to increase the miscibility between the clay and the polymer. A cation-exchange process provides the organic modification of the clay by using the inorganic cation, typically sodium, which can be replaced by an organic cation such as ammonium and phosphonium. Moreover, the numbers of alkyl chains attached to the cations are different, and it is obvious that too many alkyl chains cause overcapacity in the gallery space and can lead to the generation of immiscible systems ([5, 6]; Chigwada et al. 2006).

In this context, many researchers have investigated the effect of clay surface coverage on the preparation of PS/clay nanocomposites. For example, Limpanart et al. [4] discovered during melt intercalation of organoclay in PS that the clay with high surface coverage hinders the formation of nanocomposites due to poor wettability of the PS on the clay surface. As a result, a phase separation between the polymer and the organoclay leads to the formation of microcomposites. In addition to method of nanocomposite preparation, types of clay modifiers are also much important. There are several common clay modifiers which are used today in clay modification process.

2.1 Ammonium Salt Modifiers

Yei et al. [5] used two different surfactants for the MMT, the aminopropylisobutyl polyhedral oligomeric silsesquioxane (POSS) and the ammonium salt of cetylpyridinium chloride (CPC), during emulsion polymerization of PS/clay nanocomposite. They revealed that even though the d-spacing of the POSS

Table 10.1 Comparison of effect of different clay modifier on properties of PS/MMT nanocomposite [7]

Sample	Storage modulus (GPa)	T_g (°C)	T_d onset (°C)
PS	1.8	127	408
PS/Na-MMT	1.26	129	419
PS/C-MMT	1.36	130	426
PS/A-MMT	1.65	139	454

intercalated clay is relatively less than that of the CPC-intercalated clay, PS is more easily intercalated and exfoliated by the former than the latter. In addition, the nanocomposite from the POSS-treated MMT was exfoliated, while the CPC-treated one had intercalated morphology. As a consequence, similar to prior surveys, the existence of POSS in the MMT clearly improves the thermal stability of the polystyrene and resultant nanocomposite.

Zwitterion aminoundecanoic acid (AUA) with an amino group ($-NH_2$) and a carboxylic group (−COOH) in the molecule, after protonization in acidic media and creation of cationic $-NH^+{}_3$, could act as a modifier for MMT. It can be said that the amino acid molecules in the acidic medium simply go through the MMT galleries to achieve ion exchange with interlayer cations. The exfoliated PS/A-MMT (AUA-modified MMT) nanocomposite was then prepared via emulsion polymerization, while the long alkyl chain of the amino acid seemed to be helpful to make the clay compatible with styrene. The nanocomposite showed considerable improvements in modulus, Tg, and thermal stability [7]. The results are compared with two different PS composites, incorporated with pristine Na-MMT and organoclay (C-MMT) modified with CTAB, respectively, via emulsion polymerization (Table 10.1).

Clays containing oligomeric units have also been utilized, while a number of alkyl chains attached to the cation could be different. Attaching alkyl groups in the form of long alkyl chains [8] or as bulky groups [9] improves the clay dispersion in the polymeric. Presence of an overly long and bulky substituent on modifier impedes the penetration of the modifier through the clay silicate layers. Moreover, Xie et al. [10] revealed that in final, an alkyl chain with longer possessed a higher glass transition temperature in the PS nanocomposite. In this context, Chigwada et al. [11] prepared PS/clay nanocomposites with intercalated structures via in situ bulk polymerization and direct melt blending method in the presence of pyridine and quinoline-containing ammonium salts surfactants. The experimental results clearly stated that organic modification of clay results in increased compatibility between the clay and the polymer. However, TGA analysis showed that the quinolinium-modified clay provides a higher thermal stability for the nanocomposite compared to pyridinium-modified clay. A parallel work has been done by the same researchers using organically modified clays containing a carbazole or naphthenate unit. They found that in the presence of more than two alkyl chains, the gallery spacing has become too packed which resulted in poor dispersion and lower thermal stability [9]. It can be concluded from their results that quantity, length, and volume of modifiers are very important in the final structure of nanocomposites. Although the presence of long chains in clay modifiers increases d-spacing clay layers and tends to exfoliate structure for nanocomposite, the presence of several bulk substituents with different lengths and sometimes even presence of an overly long chain in modifiers has a negative effect. It can be explained that modifiers with many long and voluminous substituents cannot penetrate into clay layers adequately. In this situation, modifiers not only increase d-spacing between clay layers but also brattle around the clay galleries and are impeded to interact well between clay particles and monomers or polymer chains.

The ammonium salt containing 4-acetylbiphenyl was also used in a similar research study by the above authors, and the high thermal stability for the nanocomposites was obvious along with a reduction in both the peak heat release rate and the mass loss rate of the nanocomposites [11].

In another similar research, Bottino et al. [12] attempted cationic imidazolium surfactant to modify MMT and also to compare the resultant with standard alkylammonium cation-modified MMT in respect to thermal stability improvement. They found that the reactive group in the imidazolium cations on the clay surface could facilitate the interlayer polymerization and enhance the clay dispersion. Additionally, the generation of partially exfoliated PS/OMMT nanocomposites was more probable by incorporation of imidazolium cations due to the presence in their structure of a polymerizable group. Moreover, the improvement in thermal stability of the imidazolium-modified MMT compare to standard alkylammonium cation-modified MMT revealed that the delocalized imidazolium cation had superior thermal stability than the alkylammonium and pyridinium cations.

Elsewhere, three ammonium salts, namely, tetraethylammonium bromide (TEAB), tetrabutylammonium bromide (TBAB), and cetyltrimethylammonium bromide (CTAB), were used in cation-exchange process to prepare organoclay, and then this product was melt blended with PS matrix in order to produce PS/clay nanocomposites [13]. It was shown clearly from XRD and TEM images that the higher clay loading resulted in agglomeration once most of the clay was in thickness range of 30–35 nm. The TBAB-modified clay created the most promoting exfoliated nanocomposite among the others which is attributed to the significant role of ammonium salt's structural stability through the nanocomposite preparation. Moreover, the enhanced separation of clay layers could be resulted from the sonication treatment of clay during modification which led to easier cation exchange.

2.2 Maleic Anhydride Modifiers

Maleic anhydride (MA) or polymer-grafted-MA has been used as a compatibilizer to enhance the compatibility between the polymer and the pristine or organically modified clay [6, 14]. For instance, PS/clay nanocomposites were prepared via melt blending and ion exchange with the oligomer that contains maleic anhydride (MA), styrene (ST), and vinylbenzyltrimethylammonium chloride (VBTACl) terpolymer, in which here it is called MAST. The resulted PS/oligomerically modified MAST showed a mixed immiscible/intercalated/delaminated morphology. However, it is evident that maleic anhydride-modified clay improved the compatibility between the clay and the polystyrene [6].

In another attempt, maleic anhydride-grafted polystyrene (MAHgPS) was utilized as a compatibilizer during the melt compounding of PS with organomontmorillonite (OMMT) which is followed by compression molding, and the resultant nanocomposite was investigated with respect to morphological and flexural properties [15]. The obvious reduction in flexural modulus and slight increase in flexural strength were recognized which the latter could be resulted from developed compatibility between the OMMT and PS. Moreover, FESEM results illustrated the finer particle size of the OMMT by the addition of MAHgPS and also declared that the layered silicates of OMMT were partly intercalated in the PS matrix. This phenomenon could appropriately confirm the decline in interfacial tension between the components in the melt which led to a higher dispersion in the blend.

2.3 Hybrid Modifiers

According to the literature, the layered silicates have been treated with cationic monomers [16] or initiators [17] to obtain monomer–clay and initiator–clay hybrids, respectively. So far, various radical initiator–montmorillonite (MMT) hybrids have been introduced in order to prepare exfoliated polystyrene (PS)/MMT nanocomposites [17, 18]. In consecutive studies, 2,2-azobis{2-methyl-*N*-[2-acetoxy-(2-*N*,*N*,*N*-tributylammoniumbromide)ethyl] propionamide}-MMT (ABTBA-MMT) was found to be an appropriate hybrid component regarding to the interlayer d-spacing and swelling behavior in nonpolar monomers [19]. In the latest one, the above difunctional cationic azo-initiator was applied to prepare exfoliated PS/MMT nanocomposites in different loadings and higher quantities relative to the previous ones. It was depicted that the improvements in various properties, such as storage modulus, Young's modulus, tensile strength, degradation temperature, and the glass transition temperature, were comparable or greater than the prior reported data mostly owing to the good dispersion and exfoliation of the clay in the PS polymeric matrix [20].

Cationic amphiphilic block copolymer of poly(styrene-b-2-hydroxyethyl acrylate) (PS-b-PHEA) used as a clay modifier and the achieved (PS-b-PHEA)-MMT played role of stabilizer in dispersion polymerization of PS. Particle size of the resultant spherical shape PS particles decreased by increasing block polymer loading. It was proved from TEM and SAXS results that the PS/clay nanocomposite has a partially exfoliated morphology at low clay loadings, while at the higher loadings of clay, the intercalated structure is obvious. Furthermore, an improvement in thermal stability, storage modulus, and T_g has been detected for all of PS/clay nanocomposites comparing with the pure PS [21].

Fu et al. [22] used emulsion polymerization to prepare POSS/clay hybrids of polystyrene that were prepared by using two organically modified clays, POSS-NH_2

Table 10.2 Results of thermal and mechanical properties of polystyrene and polystyrene nanocomposites [22]

Sample	T_g (°C)[a]	$T_{0.05}$(°C)[b]	$T_{0.5}$(°C)[c]	Char at 600(°C)%
PS	100 ± 0.5	390 ± 1.7	424 ± 0.8	0
POSS/clay/PS	108 ± 0.6	411 ± 1.3	446 ± 1.2	2.9
C_{20}-POSS/clay/PS	105 ± 0.3	415 ± 1.1	457 ± 0.9	2.4

[a]Glass transition temperature (T_g)
[b]5 % Degradation temperature ($T_{0.05}$)
[c]50 % Degradation temperature ($T_{0.5}$)

and C_{20}-POSS, as intercalated agents. The fine dispersion of exfoliated silicate layers inside the PS matrix caused a remarked improvement in thermal decomposition temperature (T_d) and T_g as well as a great reduction in coefficient of thermal expansion (CTE) of pure PS which leads to the dimensional stability improvement of the matrix. The above trends are mostly contributed to the retardation of PS chain segmental movement during incorporation of organically modified clays (Table 10.2).

3 Effect of Clays on Thermal, Mechanical, Viscoelastic, and Rheological Properties

Since the clay layers in the nanocomposites act as a barrier to heat and mass transfer, incorporation of the clay inside the polystyrene causes the degradation pathway of the composite to change. In addition, the increase of filler surface tension improves the adhesion energy between polymer and clay particle or filler–polymer interaction which follows by increase in the glass transition temperature and also viscoelastic properties of composite. These trends are more obvious for nanocomposites due to the secondary filler aggregation or networking and also the high surface-to-volume ratio of nanofillers, which creates huge immobilized interfacial layer [23, 24]. Furthermore, the organoclay in the nanocomposites tends to play a double role: as a nanofiller to increase the storage modulus and as a plasticizer to decrease the storage modulus [10]. Accordingly, Claytone APA, a commercial organoclay, via in situ polymerization of PS was well dispersed inside the monomer and formed a nanocomposite with superior thermal stability, shear rate viscosity, shear-thinning property, storage and loss modulus, and elastic behavior compared to pure PS [25].

The chemical structure of surfactants used for the modification of clay has also a significant effect on the extent of thermal stability of the polystyrene/clay nanocomposites. When nanocomposites are utilized in high-temperature applications, the low thermal stability of the surfactants, which are usually used to render the layered silicates organophilic, can be a limiting factor [26]. Accordingly, three different surfmer-modified clays and nonreactive surfactant-modified clay which were used to prepare polystyrene/clay nanocomposites confirmed that the surfmer-based nanocomposites had superior thermal stability comparing with nonreactive surfactant-based nanocomposites [26, 27].

Table 10.3 Glass transition temperatures (T_g), initial decomposition temperatures (T_i), and apparent activation energies of degradation (E_a) in static air of PS and studied nanocomposites [28]

Sample	T_g (°C)	T_i (°C)	E_a(kJ/mol)
PS	98	377	128
PS/Na/MMT	99	340	133
PS/ODTA/MMT	98	344	137
PS/C_{12}/MMT	99	343	138
PS/C_{16}/MMT	101	359	145
PS/C_{18}/MMT	102	366	189

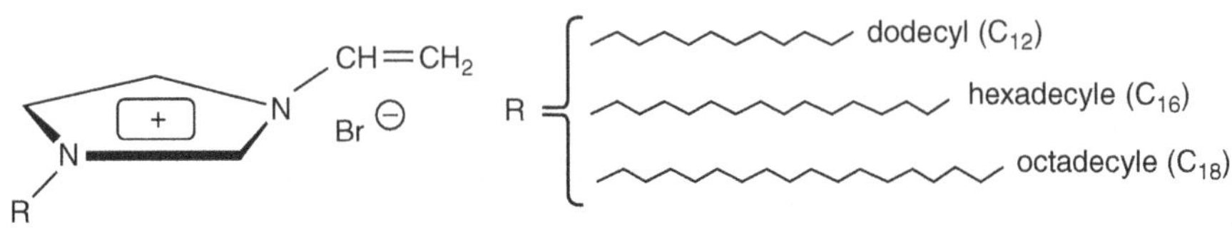

Fig. 10.2 Structure of dodecyl (C_{12}), hexadecyl (C_{16}), octadecyl (C_{12}) surfactants [28]

In terms of UV photooxidative degradation, a study by Bottino et al. [28] deduced that photooxidative instability could be resulted from the degree of exfoliation and also the existence of catalytic active sites on the filler surface. In other words, this effect is more obvious in the highly exfoliated nanocomposites in which the catalytic active sites are more accessible than in the intercalated ones. PS/(OMMT) nanocomposite, which was prepared via bulk polymerization in the presence of long alkyl chain surfactants, exhibited much more faster and stronger photooxidative degradation than the pure PS.

Moreover, thermal study of different PS nanocomposites implied that by increasing the degree of exfoliation, thermal stability and initial decomposition temperature are also increased. A qualitative scale of exfoliation can be recognized by comparing the Ea values of the samples (Table 10.3, Fig. 10.2).

Organo-modified MMT with a mixture of two commercial surfactants (CTAB and ADAB) through the ion-exchange process was dispersed into the styrene monomer during in situ free radical bulk polymerization. These highly exfoliated PS/clay nanocomposites were thermally and mechanically stable [19, 29–31]. From the experimental results, improved dynamic modulus, slower die swelling, better shear thinning, and much greater decomposition temperature were found relative to that of pure PS. Furthermore, the evident yield-like behavior as well as steady amount of storage modulus at low frequencies in the oscillatory shear measurements proves that the enhanced viscoelasticity in these nanocomposites resulted from the formation of a percolating nanoclay network [32].

Three different ammonium salts, TEAB, TBAB, and CTAB, were also incorporated into PS during the melt blending process by Arora et al. [13]. With more investigation in various resultants of nanocomposite researches, it was revealed an improvement in the mechanical properties of nanocomposites relative to the pure PS with the reported order of PS/TBAB system > PS/CTAB system > PS/TEAB system. In addition, the T_g of all the nanocomposites was higher than that of pure PS, while the thermal stability did not show any improvement with clay addition. It can be concluded that the thermal stability of nanocomposites was mostly related to thermal stability of the cationic surfactant; thus, CTAB/clay-PS20 indicated the maximum thermal stability among the other samples.

A polymerizable cationic surfactant, vinylbenzyldimethylethanolammonium chloride (VBDEAC), was synthesized in order to functionalize MMT clay via free radical polymerization of exfoliated PS/clay nanocomposites. Thermal and mechanical investigations demonstrated the remarkable improvements in thermal degradation temperature, T_g, flexural modulus, flexural strength, and impact strength of the nanocomposite compared with pristine PS. These experimental results fairly confirm the better dispensability and increased interfacial adhesion of the clay with PS which follows by retarding the segmental motion of the PS matrix [33].

Another reactive cationic surfactant vinylbenzyldimethyldodecylammonium chloride (VDAC) was synthesized and utilized to produce PS/MMT-exfoliated nanocomposites. It was found that the presence of vinyl benzyl group in the surfactant resulted in exfoliating MMT in PS matrix. This phenomenon leads to improved dynamic modulus of the nanocomposites compared with the neat PS [34].

In another approach, organomontmorillonite (Org-MMT) was intercalated in high-impact polystyrene (HIPS) during melt processing using twin-screw extruder. By increasing Org-MMT loading, die swell ratio and melt elasticity of the composite are going to be decreased due to well orientation of layered silicate under the shear and also hindering the elastic recovery of the confined polymer chains after leaving the capillary die in consequence of unique layered-structure feature of MMT [35].

In another study Greesh [26] revealed that clay modification was not fully effective when sodium MMT clay was premodified using 3-(trimethoxysilyl) propyl methacrylate (MPTMS) in order to prepare exfoliated PS/clay nanocomposites via free radical polymerization in dispersion polymerization. It was exhibited from TEM images that most of the clay platelets were distributed in the dispersing phase; thus, the resultant particles were not fully stable.

4 PS/Layered Double Hydroxide Nanocomposites

Nowadays, the synthesis and properties of polymer/layered double hydroxide (LDH) nanocomposites have caused great interest as a promising class of materials, which can be applied as stabilizers, flame retardants, medical substances, etc. [36–38].

In this section, recent developments in preparation and characterization of different PS/LDH composites will be reviewed. As the first example, it is the investigated results of the research of Manzi-Nshuti et al. [39] who used

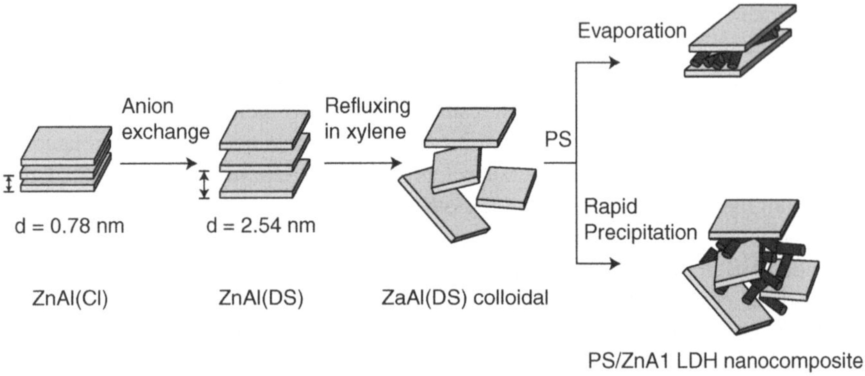

Fig. 10.3 Scheme of the PS/ZnAl LDH nanocomposites prepared by solution intercalation [39]

solution intercalated method to prepare exfoliated PS/ZnAl LDH nanocomposites. As observed in Fig. 10.3, a schematic representation begins with ZnAl(Cl) LDH which contains one molecular layer of Cl- between the LDH sheets with a basal spacing of 0.78 nm. The intercalation of DS anion in ZnAl(Cl) generates the expanded phase surfactant-modified ZnAl(DS) and increases the basal spacing to 2.54 nm. Refluxing ZnAl(DS) in xylene results in a swollen and partially exfoliated LDH colloidal suspension. Concurrently, the LDH layers are broken into small parts in the refluxing process. Since the complete exfoliation of LDH can be achieved during two steps of solvent swelling and layer breaking processes via refluxing, it could be concluded that lower LDH content, shorter refluxing time, and fast precipitation are three valuable factors in order to get accomplished exfoliation [39]. Thermal investigation of PS/ZnAl LDH nanocomposites also demonstrated a remarkable increase in thermal stability and decomposition temperature compare to the neat PS.

In another research study, Ding and Qu [40] produced exfoliated PS/ZnAl LDH nanocomposites via emulsion polymerization (Fig. 10.4). They produced the complete exfoliated PS/LDH nanocomposites even at the 10 wt% LDH content. Moreover, the decomposition temperature of exfoliated PS/LDH with 5 wt% LDH found to be 28 °C higher than that of virgin PS.

Qiu and Qu [41] produced PS/LDH exfoliated nanocomposite with a homogeneous structure by soap-free emulsion polymerization (SFEP) method. This method of polymerization involves an in situ micellization, where the oligomeric radicals generate by means of free radical reaction of an ionic initiator, while the styrene monomers act as surfactant and help to form micelles. In this method, high-molecular-weight PS could exfoliate the regularly structured stacking LDH layers. Since no organic surfactant was used during the process, a considerable increase on the onset decomposition temperature was observed.

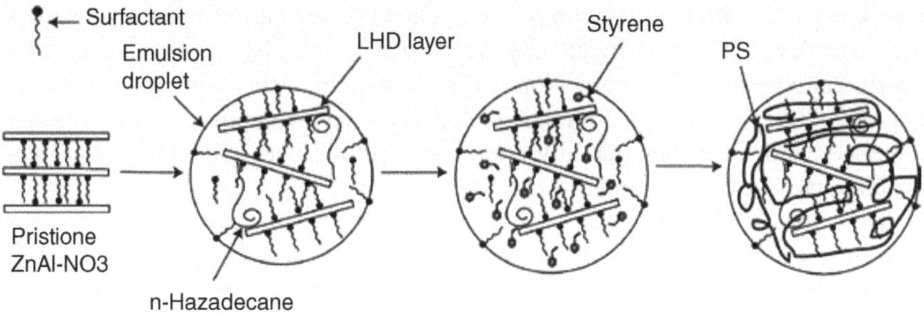

Fig. 10.4 Schematic diagrams of the formation procedure of the exfoliated PS/LDH nanocomposites by emulsion polymerization [40]

The same polymerization method was applied for preparation of core-polystyrene/shell-silica nanocomposite with different silica loading levels by Lee et al. [42]. They revealed that by increasing the addition of silica particles above 10 wt%, an improvement in thermal stability of nanocomposite was obvious which comes from the strong interaction between silica and PS molecules.

In another research, PS/LDH nanocomposite was also prepared by free radical polymerization in presence of an initiator-containing LDH, ZnAl–ACPA [4,40-azobis(4-cyanopentanoate)], and MgAl–ACPA [39]. Morphological studies confirmed the intercalated–exfoliated structure for the composites of ZnAl–ACPA at the same time, MgAl–ACPA leads to form nanocomposite with microstructure morphology.

From the other point of view, glass transition temperature of PS is affected by LDH in different ways; for instance, T_g has been decreased via polymerization of styrene by using MgAl–ACPA, while addition of ZnAl–ACPA has not affected T_g. The above reduction in T_g could be related to the improved polymer dynamics because of the extra free volume at the LDH additive–polymer interface. It was also observed a decline in the onset thermal decomposition temperature of PS/LDH relative to the pure PS which most probably caused from the early decomposition of the LDH [39].

5 Comparison of Polymerization Methods

The preparation methods of the PS/clay nanocomposite have been studied by many researchers. In this context, several polymerization methods have been used such as melt polymerization ([1, 43–45]; Tseng et al. 2001; [46]), bulk polymerization [19, 30, 34, 47–49], solution polymerization [50], emulsion polymerization ([1, 43–45]; Tseng et al. 2001; [19, 34, 47, 48, 51, 52]; Chen et al. 2001; [10, 53–55]), suspension polymerization [10], and living free radical polymerization [36, 50, 56]. Generally, the solution blending yields a better dispersion of the filler compared with the melt compounding, and consequently, superior properties will be achieved [57]. The organically modified clays were incorporated into the PS matrix via both bulk polymerization and melt blending. The experimental results indicated that the bulk polymerization provides nanocomposites with superior dispersion and lower flammability compared with the ones that come from melt blending polymerization [9].

In a comparative study, both of suspension and emulsion polymerizations were performed to prepare PS/clay nanocomposite in the presence of different loadings of ammonium salts as the swelling agent. The emulsion polymerization results in the higher molecular weights and larger d-spacing of PS/clay nanocomposites. The reason behind is that the droplets of suspension polymerization are bigger than the micelles of the emulsion polymerization; therefore, each droplet should have more amount of clay to provide smaller d-spacing. On the other hand, the remaining emulsifier (SDS) can be hardly removed at the end of the emulsion polymerization which probably leads to reduction in thermal stability. The most significant point of suspension polymerization is the easy heat transfer and thus the simply controllable reaction [58].

High-impact polystyrene (HIPS)/MMT nanocomposites were prepared via both bulk polymerization and in situ polymerization of styrene in the presence of polybutadiene, using intercalated cationic radical initiator–MMT hybrid. The bulk polymerization resulted in an incomplete exfoliation of the silicate layers in the HIPS nanocomposites, while the silicate layers were well exfoliated inside the PS matrix by using the solution polymerization. The reason behind this is the low extra-gallery viscosity which leads to the easier diffusion of styrene monomers inside the layers of clay. To put it briefly, the thermal and mechanical properties of the obtained exfoliated HIPS/MMT nanocomposites were improved considerably relative to those of the pure HIPS, mostly due to the full dispersion and exfoliation of clay platelets over the polymer matrix [20].

In terms of living polymerization, different living and controlled/living polymerization methods were used in the preparation of well-dispersed silicate layers, such as nitroxide-mediated polymerization (NMP), reversible addition–fragmentation chain transfer (RAFT), atom transfer radical polymerization (ATRP), living cationic polymerization, living anionic polymerization, ring-opening polymerization (ROP), and ring-opening metathesis polymerization (ROMP) (Fig. 10.5). For instance, in consecutive researches, polystyrene-grafted silica particles dispersed in a polymer matrix during nitroxide-mediated polymerization (NMP). The first study involved the colloidal silica particles, while the fumed silica was used in the second one. For the latter polymerization, two different routes were investigated in order to attach the control agent on the inorganic surface. The benefit of this polymerization system makes it principally smart for the elaboration of polymer layers with controlled structures [58, 59].

Investigation of intercalated PS/clay nanocomposites elsewhere revealed that a typical uncontrolled free radical polymerization takes place without utilization of RAFT agents which leads to produce nanocomposites with high molar

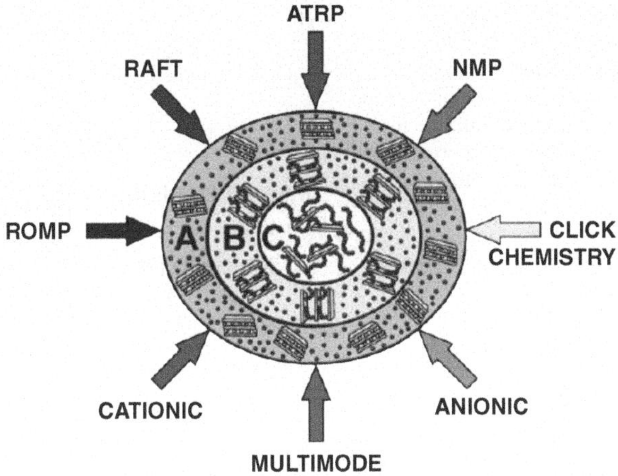

Fig. 10.5 Schematic representation of polymer/clay nanocomposites by various in situ polymerization techniques (*A* monomer immersion, *B* intercalation, *C* exfoliation) [62]

masses and high polydispersity indices. Reversely, polymer chains with narrow polydispersity come from using RAFT agents during the controllable radical polymerization [60].

6 PS/Clay Foams

In recent years, clay nanoparticles with the particularly fine dimensions and large surface area have introduced unique properties to the foaming products compared with the conventional micron-sized filler particles. The appropriate interaction between the filler particles and polymer matrix significantly adjusts the cell growth and cell nucleation which consequently improves the nanocomposite foams properties, provides foam reinforcement, reduces the gas escape rate, and leads to the char formation while the foam is under fire. As a consequence, polymer–clay nanocomposite foam could be a good alternative for applications with the lightweight, high strength, and superior fire resistance requirements.

In this context, the intercalated and exfoliated PS/nanoclay composites were prepared in the presence of CO_2 as the foaming agent by mechanical blending and in situ polymerization, respectively. It was clearly indicated that the addition of small amount of clay decreased the cell size and increased the cell density, in a great extent which is much more evident for exfoliated nanocomposites (Han et al. 2002) [63].

In an interesting attempt, water-expandable PS/clay nanocomposites (WEPSCN) were produced. The preparation process included uniform dispersion of clay particles inside the water which is followed by their addition to styrene monomer by the formation of water-in-oil inverse emulsion (Fig. 10.6). PS polymeric matrix was then heated and expanded to form a cellular structure which was surrounded by a layer of nanoclay. The above layer provided higher water content in the beads and less water loss during the storage.

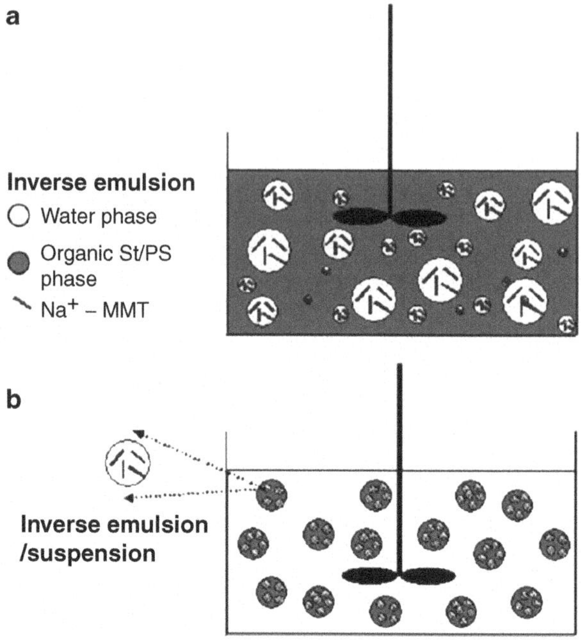

Fig. 10.6 Schematic representation of the preparation process of WEPSCN (**a**) emulsification of water/clay mixture in pre-polymerized styrene/PS, and (**b**) suspension polymerization of styrene/PS droplets containing emulsified water/clay droplets [60]

Furthermore, generation of the bi-model structured foams in the presence of CO_2 co-blowing agent was carried out. During the foaming process, the water cavities developed the cell size extensively and resulted in preparation of foam with ultralow density and low thermal conductivity [60].

7 Conclusion

The intrinsic incompatibility of hydrophilic clay layers with hydrophobic polymer chains causes a poor dispersion of clay nanoparticles through the polymer matrix which follows by generation of weak interfacial interactions. The surface modification of clays could be effective to render the clay layers more compatible with polymer chains. PS/clay nanocomposites with exfoliated structures have been prepared by a variety of polymerization methods. Researcher revealed improvement in mechanical, thermal, and rheological properties as well as flame retardancy and oxygen permeability of exfoliated structure. Among the three frequent methods of polymerization including in situ method, melt blending, and solution casting, nanocomposites with exfoliated structures mostly have been prepared by in situ polymerization process in the presence of appropriate amount of cationic initiators or other reactive cations. A wide variety of novel and applicable techniques are mandatory to develop the nanocomposite manufacturing industry and particularly to accomplish the exfoliation of PS/clay nanocomposites to reach the desired improvement in PS properties.

References

1. Hasegawa N, Okamoto H, Kawasumi M, Usuki A (1999) Preparation and mechanical properties of polystyrene–clay hybrids. J Appl Polym Sci 74:3359
2. Ishida H, Campbell S, Blackwell J (2000) General approach to nanocomposite preparation. Chem Mater 12:1260
3. LeBaron PC, Wang Z, Pinnavaia TJ (1999) Polymer-layered silicate nanocomposites: an overview. Appl Clay Sci 15:11
4. Limpanart S, Khunthon S, Taepaiboon P, Supaphol P, Srikhirin T, Udomkichdecha W, Boontongkong Y (2005) Effect of the surfactant coverage on the preparation of polystyrene–clay nanocomposites prepared by melt intercalation. Mater Lett 59:2292–2295
5. Yei DR, Kuo SW, Su YC, Chang FC (2004) Enhanced thermal properties of PS nanocomposites formed from inorganic POSS-treated montmorillonite. Polymer 45:2633–2640
6. Zheng X, Jiang DD, Wilkie CA (2006) Polystyrene nanocomposites based on an oligomerically-modified clay containing maleic anhydride. Polym Degrad Stab 91:108–113
7. Li H, Yu Y, Yang Y (2005) Synthesis of exfoliated polystyrene/montmorillonite nanocomposite by emulsion polymerization using a zwitterion as the clay modifier. Eur Polym J 41:2016–2022
8. Chigwada G, Wang D, Jiang DD, Wilkie CA (2006) Styrenic nanocomposites prepared using a novel biphenyl-containing modified clay. Polym Degrad Stab 91:755–762
9. Chigwada G, Jiang DD, Wilkie CA (2005) Polystyrene nanocomposites based on carbazole-containing surfactants. Thermochim Acta 436:113–121
10. Xie W, Hwu JM, Jiang GJ, Buthelezi TM, Pan WP (2003) A study of the effect of surfactants on the properties of polystyrene-montmorillonite nanocomposites. Polym Eng Sci 43(1):214
11. Chigwada G, Wang D, Wilkie CA (2006) Polystyrene nanocomposites based on quinolinium and pyridinium surfactants. Polym Degrad Stab 91:848–855
12. Bottino FA, Fabbri E, Fragala IL, Malandrino G, Orestano A, Pilati F, Pollicino A (2003) Polystyrene-clay nanocomposites prepared with polymerizable imidazolium surfactants. Macromol Rapid Commun 24:1079–1084
13. Arora A, Choudhary V, Sharma DK (2011) Effect of clay content and clay/surfactant on the mechanical, thermal and barrier properties of polystyrene/organoclay nanocomposites. J Polym Res 18:843–857
14. Wang D, Wilkie CA (2003) In-situ reactive blending to prepare polystyrene–clay and polypropylene–clay nanocomposites. Polym Degrad Stab 80:171–82
15. Chow WS, Ooi KH (2007) Effects of maleic anhydride grafted polystyrene on the flexural and morphological properties of polystyrene/organo-montmorillonite nanocomposites. Malaysian Polym J 1:1–9
16. Uthirakumar P, Kim CJ, Nahm KS, Hahn YB, Lee YS (2004) Preparation and characterization of new difunctional cationic radical initiator-montmorillonite hybrids. Colloids Surf A Physicochem Eng Asp 247:69
17. Chen TK, Tien YI, Wei KH (1999) Synthesis and characterization of novel segmented polyurethane/clay nanocomposite via poly(ε-caprolactone)/clay. J Polym Sci A Polym Chem 37:2225
18. Park JM, Jana SC (2003) Mechanism of exfoliation of nanoclay particles in epoxy – clay nanocomposites. Macromolecules 36:2758
19. Doh JG, Cho I (1998) Synthesis and properties of polystyrene-organoammonium montmorillonite hybrid. Polym Bull 41(5):511
20. Uthirakumar P, Hahn YB, Nahm KS, Lee YS (2005) Exfoliated high-impact polystyrene/MMT nanocomposites prepared using anchored cationic radical initiator–MMT hybrid. Eur Polym J 41:1582–1588
21. Greesh N, Sanderson R, Hartmann PC (2012) Preparation of polystyrene colloid particles armored by clay platelets via dispersion polymerization. Polymer 53:708–718

22. Fu HK, Kuo SW, Yeh DR, Chang FC (2008) Properties enhancement of PS nanocomposites through the POSS surfactants. J Nanomater
23. Jang BN, Wilkie CA (2005) The thermal degradation of polystyrene nanocomposite. Polymer 46:2933–2942
24. Mortezaei M, Navid Famili MH, Kokabi M (2011) The role of interfacial interactions on the glass-transition and viscoelastic properties of silica/polystyrene nanocomposite. Comp Sci Technol 71:1039–1045
25. Wang GH, Zhang LM (2007) Reinforcement in thermal and viscoelastic properties of polystyrene by in-situ incorporation of organophilic montmorillonite. Appl Clay Sci 38:17–22
26. Nagi Greesh (2011) Preparation of polymer-clay nanocomposites via dispersion polymerization using tailor-made polymeric surface modifiers. Dissertation presented for the Degree of Doctor of philosophy. University of Stellenbosch
27. Jialanella G, Firer E, Piirma I (1992) J Polym Sci A Polym Chem 30:1925–1932
28. Bottino FA, Pasquale GD, Fabbri E, Orestano A, Pollicino A (2009) Influence of montmorillonite nano-dispersion on polystyrene photo-oxidation. Polym Degrad Stab 94:369–374
29. Burnside SD, Giannelis EP (1995) Synthesis and properties of new poly(dimethylsiloxane) nanocomposites. Chem Mater 7(9):1597
30. Fu X, Qutubuddin S (2001) Polymer–clay nanocomposites: exfoliation of organophilic montmorillonite nanolayers in polystyrene. Polymer 42:807
31. Huang X, Brittain WJ (2001) Synthesis and properties of new poly(dimethylsiloxane) nanocomposites. Macromolecules 34(10):3255
32. Zhong Y, Zhu Z, Wang SQ (2005) Synthesis and rheological properties of polystyrene/layered silicate nanocomposite. Polymer 46:3006–3013
33. Tseng CR, Wu JY, Lee HY, Chang FC (2002) Preparation and characterization of polystyrene–clay nanocomposites by free-radical polymerization. J Appl Polym Sci 85:1370–1377
34. Fu X, Qutubuddin S (2000) Synthesis of polystyrene–clay nanocomposites. Mater Lett 42:12–15
35. Dazhu C, Haiyang Y, Pingsheng H, Weian Z (2005) Rheological and extrusion behavior of intercalated high-impact polystyrene/organomontmorillonite nanocomposites. Comp Sci Technol 65:1593–1600
36. Greesh N, Hartmann PC, Sanderson RD (2009) Preparation of polystyrene/clay nanocomposites by free-radical polymerization in dispersion. Macromol Mater Eng 294:787–794
37. Guillot S, Bergaya F, Azevedo C, Warmont F, Tranchant J (2009) Internally structured pickering emulsions stabilized by clay mineral particles. J Colloid Interface Sci 333:563–569
38. Qiu L, Chen W, Qu B (2005) Structural characterisation and thermal properties of exfoliated polystyrene/ZnAl layered double hydroxide nanocomposites prepared via solution intercalation. Polym Degrad Stab 87:433–440
39. Manzi-Nshuti C, Chen D, Su S, Wilkie CA (2009) Structure–property relationships of new polystyrene nanocomposites prepared from initiator-containing layered double hydroxides of zinc aluminum and magnesium aluminium. Polym Degrad Stab 94:1290–1297
40. Ding P, Qu B (2005) Synthesis and characterization of exfoliated polystyrene/ZnAl layered double hydroxide nanocomposite via emulsion polymerization. J Colloid Interface Sci 291:13–18
41. Qiu L, Qu B (2006) Preparation and characterization of surfactant-free polystyrene/layered double hydroxide exfoliated nanocomposite via soap-free emulsion polymerization. J Colloid Interface Sci 301:347–351
42. Lee J, Hong CK, Choe S, Shim SE (2007) Synthesis of polystyrene/silica composite particles by soap-free emulsion polymerization using positively charged colloidal silica. J Colloid Interface Sci 310:112–120
43. Sikka M, Cerini LN, Ghoch SS, Winey JKI (1996) Melt intercalation of polystyrene in layered silicates. Polym Sci B Polym Phys 34:1443

44. Vaia RA, Ishii H, Giannelis EP (1993) Synthesis and properties of two-dimensional nanostructures by direct intercalation of polymer melts in layered silicates. Chem Mater 5:1694
45. Vaia RA, Jandt KD, Kramer EJ, Giannelis EP (1996) Microstructural evolution of melt intercalated polymer – organically modified layered silicates nanocomposites. Chem Mater 8:2628
46. Yoon JT, Jo WH, Lee MS, Ko MB (2001) Effects of comonomers and shear on the melt intercalation of styrenics/clay nanocomposites. Polymer 42:329
47. Laus M, Camerani M, Lelli M, Sparnacci K, Sandrolini F (1998) Hybrid nanocomposites based on polystyrene and a reactive organophilic clay. J Mater Sci 33:2883
48. Okamoto M, Morita S, Taguchi H, Kim YH, Kotaka T, Tateyama H (2000) Synthesis and structure of smectic clay/poly(methyl methacrylate) and clay/polystyrene nanocomposites via in situ intercalative polymerization. Polymer 41:3887
49. Zeng C, Lee LJ (2001) Poly(methyl methacrylate) and polystyrene/clay nanocomposites prepared by in-situ polymerization. Macromol 34:4098–4103
50. Bottcher H, Hallensleben ML, Nuß S, Wurm H, Bauer J, Behrens PJ (2002) Mater Chem 12:1351
51. Chen G, Liu S, Zhang S, Qi Z (2000) Self-assembly in a polystyrene/montmorillonite. Macromol Rapid Commun 21:746
52. Noh MW, Lee DC (1999) Synthesis and characterization of PS-clay nanocomposite by emulsion polymerization. Polym Bull 42:619
53. Chen G, Ma Y, Qi Z (2001) Preparation and morphological study of an exfoliated polystyrene/montmorillonite nanocomposite. Scr Mater 44:125
54. Qutubuddin S, Fu XA, Tajuddin Y (2002) Synthesis of polystyrene-clay nanocomposites via emulsion polymerization using a reactive surfactant. Polym Bull 48:143
55. Voulgrais D, Petridis D (2002) Emulsifying effect of dimethyldioctadecylammonium-hectorite in polystyrene/poly(ethyl methacrylate) blends. Polymer 43:2213
56. Yang WT, Ko TH, Wang SC, Shih PI, Chang MJ, Jiang GJ (2008) Preparation of polystyrene/clay nanocomposite by suspension and emulsion polymerization. Polym Compos 29(4):409–414
57. Schütz MR, Kalo H, Lunkenbein T, Breu J, Wilkie CA (2011) Intumescent-like behavior of polystyrene synthetic clay nanocomposites. Polymer 52:3288–3294
58. Bartholome C, Beyou E, Bourgeat-Lami E, Chaumont P, Zydowicz N (2005) Nitroxide-mediated polymerization of styrene initiated from the surface of fumed silica. Comparison of two synthetic routes. Polymer 46:8502–8510
59. Bartholome C, Beyou E, Bourgeat-Lami E, Cassagnau P, Chaumont P, David L, Zydowicz N (2005) Viscoelastic properties and morphological characterization of silica/polystyrene nanocomposites synthesized by nitroxide-mediated polymerization. Polymer 46:9965–9973
60. Shen J, Cao X, Lee LJ (2006) Synthesis and foaming of water expandable polystyrene clay nanocomposites. Polymer 47:6303–6310
61. Akbari A, Talebanfard S, Hassan A (2011) The effect of the structure of clay and clay modifier on polystyrene-clay nanocomposite morphology. J Polym Plast Technol Eng 49 (14):1433–1444
62. Tasdelen MA, Kreutzer J, Yagci Y (2010) In situ synthesis of polymer/clay nanocomposites by living and controlled/living polymerization. Macromol Chem Phys 211:279–285
63. Han X, Zeng C, Lee LJ, Tomasko DL, Koelling KW (2002) Polymer nanocomposite foams by using supercritical CO_2. J Cell Plast 41(5):487–502
64. Guangming Chen, Suhuai Liu, Shijuan Chen, Zongneng Qi (2001) FTIR spectra, thermal properties, and dispersibility of a polystyrene/montmorillonite nanocomposite. Macromol Chem Phys 202(7):1189–1193
65. Chen-Rui Tseng, Hsin-Yi Lee, Feng-Chih Chang (2001) Crystallization kinetics and crystallization behavior of syndiotactic polystyrene/clay nanocomposites. J Polym Sci Pt B Polym Phys 39(17):2097–2107

Nanoclays as Asphalt-Binder Modifiers 11

Giovanni Polacco, Sara Filippi, Massimo Paci, and Filippo Merusi

Contents

Abstract

Among the solutions to reduce failures of road pavements, polymer-modified asphalts (PMAs) represent the most effective one and have been adopted worldwide in the last decades. PMAs are blends of asphalt and a polymer, where the latter is swollen by the bituminous material, thus forming a network whose structure closely resembles that of gels. Unfortunately, only a limited number of polymers can be used as asphalt modifiers, being the most restrictive requirement of their compatibility with asphalt. Considering that asphalt is a complex organic material often described as a low molecular weight polymer, a recent approach is the addition of clays to both unmodified and polymer-modified asphalts. It is well known that polymer/clay nanocomposites may exhibit improved properties, if sufficient levels of clay exfoliation and silicate layers dispersion are achieved, but recent studies also underlined that clays may be used in polymer blends. In fact, an organoclay, though residing preferentially within the bulk of the most compatible polymer, may localize at the interfacial

G. Polacco (✉) • S. Filippi • M. Paci
Department of Chemical Engineering, University of Pisa, Pisa, Italy
e-mail: g.polacco@diccism.unipi.it

F. Merusi
Department of Civil and Environmental Engineering, University of Parma, Parma, Italy

J.K. Pandey et al. (eds.), *Handbook of Polymernanocomposites. Processing, Performance and Application – Volume A: Layered Silicates*, DOI 10.1007/978-3-642-38649-7_21,

region as well, thus acting as a compatibilizer between the two polymeric phases. This is why the use of ternary systems asphalt/polymer/clay has a double potential advantage: (i) an improvement of polymer and asphalt properties and (ii) a polymer/asphalt compatibilizing effect.

Keywords
Polymer modified asphalt • Compatibility • Phase separation

1 Introduction

For thousands of years, asphalts and other bituminous products have been used to provide waterproofing and protective coverings for roofs and roads [1]. At first, these substances were taken straight from natural bituminous deposits, while today they are mainly derived from crude oil, with most asphalts being the residue of vacuum distillation. Asphalts are mixtures of aliphatic, aromatic, and naphthenic hydrocarbons and small quantities of organic acids, bases, and heterocyclic components containing nitrogen, oxygen, sulfur, and some metal atoms. The chemical composition of asphalts cannot be defined exactly, because asphalt is extremely complex and variable depending on the source of the crude oil. Therefore, the asphalt components are usually grouped in two categories: asphaltenes and maltenes; the latter are further subdivided into saturates, aromatics, and resins. Similarly as composition, also the internal structure is almost unknown, and several models have been suggested by researchers. One of the most used models represents asphalt as a colloid where asphaltene micelles are covered by a stabilizing phase of polar resins, which constitute the interface with the oily maltenic medium where they are dispersed [2]. Pfeiffer and coworkers introduced the concept of sol and gel asphalts [3, 4]. Sol-type asphalts exhibit Newtonian behavior where asphaltenes micelles are fully dispersed and noninteracting, whereas gel asphalts are non-Newtonian and form a structure with fully interconnecting asphaltenes micelles. Most of asphalts belong between these two extremes and exhibit an intermediate behavior where sol-type micelles and a gel structure coexist. Generally speaking, "soft" asphalts have a low asphaltene content and a sol character, while "hard" asphalts have a higher asphaltene content and a gel character. A more detailed description of this aspect is out of the scope of the chapter. Let us stress that such models help to appreciate how important is the chemical composition of asphalts, which affects the internal structure and therefore all the rheological and performance properties of these materials. In conclusion, asphalt is a complex mixture of molecules with different molar mass and polarity that arrange themselves in a more or less organized micellar system whose final architecture depends on the chemico-physical interactions between all the constituents. This is the picture of a pseudo-equilibrium, which strongly depends on temperature and stress and that can be easily altered when additives or modifiers are added to change the material performances.

2 Polymer-Modified Asphalts

During the last decades, the number of vehicles per citizen has increased tremendously. Traffic speed and loads have similarly increased, and most roads are subjected to an overloading which reduces the quality and durability of the pavements. This is why new solutions have been searched for in order to improve pavement performance, and PMAs were developed mainly for this reason. PMAs are blends of asphalt and one or more polymers, usually added in percentages ranging from 3 % to 7 % by weight, with respect to the asphalt. The longer life and better quality of PMA pavements are directly related to both economical and safety requirements, which help to overcome the problem of initial costs that are almost double with respect to that of unmodified base asphalt (BA). However, there are some restrictions and only a limited number of polymers can be used as asphalt modifiers. For example, the high-temperature viscosity of the binder should not change too much after polymer addition so that the existing road-building processes and apparatus can still be used, but this is not the main point; the most important and restrictive requirement is the "compatibility" of polymer with asphalt. This compatibility, when achieved, is the result of interactions within the polymer and the four asphalt components. Polymers preferably interact with maltenes, thus altering the original colloidal equilibrium and leading to a final morphology where a polymer-rich phase and an asphaltene-rich phase coexist in a microscale metastable equilibrium. The polymer-rich phase can be seen as a gel mainly swollen by the lighter asphalt phase occupies a volumetric fraction far greater than that of the polymer mass in the blend and there is a critical polymer content at which the so-called phase inversion occurs and the polymeric phase becomes a continuum where the asphaltene-rich phase is dispersed. The phase inversion coincides with a dramatic change of the binder properties, which markedly reflect those of the polymer. From a thermodynamic point of view, these two phases will always tend to separate. Whether they do or do not is mainly a kinetic condition that assumes critical importance during prolonged storage at high temperatures in the absence of high-shear mixing. Once the binder is cooled down to ambient temperature, the viscosity is high enough to freeze the morphology in and prevent macroscopic separation. In some cases, like in roofing membranes, this freezing immediately follows the mixing phase and the storage at room temperature, thus overcoming the compatibility problem. It is important to underline that this partial solubility and compatibility is necessary in order to obtain asphalt with significantly modified properties, because it allows the polymer to maintain its basic morphological structure and, therefore, its physical characteristics. A perfectly compatible polymer would be completely dispersed in the base asphalt, thus giving a homogeneous mix without problems of phase separation. However, in this case the presence of the polymer corresponds to a simple increase in the average molecular weight with small variations of the basic chemical-physical properties. Therefore, the goal of mixing is to obtain a polymer swollen by the asphaltic material, thus forming a network whose structure "remembers" that of the original polymer. This explains why the compatibility is always the critical point, which strongly limits the usable polymers. In the scientific literature, a great variety of polymers have been tested as modifiers, including thermoplastic elastomers [5–9], plastomers [10, 11], and reactive polymers (which contain functional groups capable of forming chemical bonds with some asphalt molecules) [12, 13]. However, only thermoplastic block copolymers and ethylene-vinyl acetate (EVA) copolymers found extensive application in the industrial practice.

Thermoplastic block copolymers are by far the most frequently used, and among them poly(styrene-*b*-butadiene-*b*-styrene) (SBS) is the preferred one. As it is well known, the styrene and butadiene blocks are not compatible, and below the glass transition of styrene, the morphology of SBS shows glassy styrenic domains dispersed in a butadienic flexible matrix [14]. The rigid domains, interconnected through the flexible chains, constitute the nodes of a physical elastomeric network, whose structure must be preserved after mixing with asphalt. Modification consists in a simple mechanical dispersion of the block copolymer in the molten asphalt under high shear. After cooling below their glass transition temperature, the styrenic blocks reaggregate to form the rigid domains that are dispersed in an olefinic phase swelled by the asphalt molecules (mainly the maltenes). In this way, the physical network is restored, and the elastic properties are transferred to the whole mass of the PMA.

With regard to EVA, its use follows from the hope of using polyolefinic plastomers, which are available in large quantities, at low costs and with a wide range of characteristics. Polyethylene and polypropylene are the two main representatives of the category, but due to their nonpolar nature, they are almost completely immiscible with asphalt and can be used only for the production of impermeable membranes where phase separation does not take place for kinetic reasons. Therefore, for asphalt modification, it is necessary to use polyolefins modified through the insertion of polar groups like in EVA random copolymers (or similar copolymers, i.e., ethylene-butyl acrylate, EBA). Similarly to thermoplastic elastomers, EVA forms a physical network that is swelled by the asphalt components. In this case, the nodes of the network are crystallites that form, thanks to the presence of ethylene segments in the chain. The higher the vinyl acetate content, the lower is the crystallinity and the higher is the polarity of the polymer and its compatibility with asphalt. The degree of crystallinity must be well balanced: if too high, it leads to the formation of large rigid domains interconnected by short and moderately flexible chains not prone to swelling; if the crystallinity is too low, the network is disrupted once the crystallites are disturbed by the insertion of asphalt molecules. Common grades used in asphalt modification are in the range of 18–28 % (by weight) of vinyl acetate.

Even if SBS and EVA are currently used and found application worldwide in the paving industry, the problem of compatibility of these two polymers with asphalt is not completely solved, and manufacturers must keep in mind the risk of phase separation and adapt the mixing process to the characteristics of both source oil and polymer. Therefore, to enhance the performance of already used polymers and also to expand the list of usable polymers, there is an active research on polymer/asphalt compatibilizer. As an example, the chemical interaction between asphalt and polymer can be favored by addition of low molecular weight compounds such as sulfur [15] or mineral acids [16]. Nevertheless, the "magic" additive has still to be found, and organomodified clays recently showed to be a possible candidate for such a purpose. Therefore, there is an increasing interest in modification of asphalts and PMAs with clays.

In fact such compatibilizing effect of clays is not completely unknown to polymer scientists. As many practical applications require blending of polymers with different properties, several studies have been aimed at improving the properties of polymer blends through clay inclusion [17–26]. However, the preparation of such polymer-blend nanocomposites may be quite difficult. Of course, given that the blend constituents are mutually incompatible, it is clearly impossible to find an organoclay that interacts equally well with both of them. Thus, if the goal is that of a mechanical reinforcement of, e.g., the polymer matrix, then the organoclay with optimum affinity for that polymer will be chosen. This strategy was adopted, for example, in several studies dealing with toughening of polyamide/clay nanocomposites. However, both theoretical predictions [27] and experimental results have shown that the organoclay, though residing preferentially within the bulk of the better compatible polymer, may become localized at the interfacial region as well. As a result, the interfacial tension is reduced, the minor phase particles loose the typical rounded shape they possess in the pure polymer blends [42], the droplets coalescence is inhibited, and a finer dispersion is obtained. It is therefore reasonable to suppose that a similar mechanism could operate also in the case of asphalt/polymer mixing.

As it is well known, nanocomposites may be prepared by several routes, such as in situ polymerization, melt compounding, and solution intercalation [28, 29]. In the presence of a melted asphaltic phase, the mixing with clay can be somehow regarded as intermediate between melt compounding and solution intercalation. It is a melt compounding by definition, but we also must take into account the low molecular weight of the asphalt components. In solution intercalation, a three-component system is prepared where the low molecular weight solvent dissolves the polymer and delaminates the clay. Then, the nanocomposite formation takes place in a second step where the solvent is removed from the system. In asphalt/polymer/clay (APC) ternary systems, the asphalt, due to its polarity and relatively low molecular weight, may act like a solvent, with the substantial difference that it remains as the main part of the blend and is not removed. Thus, it is possible to prepare nanocomposite by using the conventional procedures for PMA preparation, without the need of new equipment or procedures.

However, as it often happens in the scientific research, clays were not incorporated into asphalt with this idea in mind. The main idea was to join the concepts of PMAs and nanocomposites, thus preparing asphalt modified by a polymer whose properties are improved by the presence of a nano-dispersed clay. It was mainly expected that such ternary APC binders will have a better performance, and in fact, some improvement has been found. However, the most important advantage observed was the better asphalt/polymer compatibility, and this was probably the main result of the research on clay addition into asphalt binders.

Nevertheless, from a chronological point of view, the interactions between asphalts and clays have been studied well before such ideas and started with the use of the so-called oil sands which represent a potential source of a huge amount of crude oil. It is quite curious to observe that in this case the main goal of the research was exactly the contrary, because it is necessary to remove the clays from the native bitumen. Therefore, depending on the application, clays can be regarded as a useful additive or as an undesired host. The latter case is not of our interest but merits a few words. One of the most known reservoirs of oil sands is the Athabasca in Northern Alberta, which has nearly three-quarters of Canada's oil sands reserves. The Athabasca oil sand deposits typically contain 6–14 wt% bitumzen and 80–85 wt% mineral solids (mainly quartz and clays such as kaolinite, illite, and some montmorillonite [30]), balanced by water. Since about one-tenth of the formation lies within 50 m of the surface, bitumen can be extracted economically by conventional surface mining techniques [31] and produce an oil supply for more than 30 % overall Canadian oil demand [32]. In the extraction process, the oil sands are conditioned with hot water, steam, and caustic to release the bitumen from the sand and other solids. Gravimetric separation and air floatation then produce bitumen froths, which contain significant amounts of water and solids that must be removed before bitumen upgrading. The froth is therefore diluted with partly aromatic naphtha and centrifuged, and then the naphtha diluent and other volatile components are removed by distillation at elevated temperature. In this way, there is a significant reduction of the solid content; however, ultrafine aluminosilicate clay crystallites remain within the bitumen. The reason is that the surfaces of the solid particles are asphaltene-like due to the adsorption of polar, aromatic, toluene-insoluble organics. Moreover, thanks to this, the solids are likely to occur in association with water which form a stable colloidal dispersion in the maltene component of bitumen. Of course, the hydrophobicity of the bitumen surface and the size of bitumen droplets are critical, and they are related to the physicochemical conditions of bitumen extraction systems, such as slurry pH and content of metal ions and solids in the mined ore and in the recycled process water. During the following phases of bitumen upgrading, some of these "fine" solids are carried over to other process units and cause several problems like corrosion (due to the high chloride content of the salt residues), fouling, and coke formation [33]. Therefore, there are several studies dedicated (i) to the interactions between bitumen and clays and (ii) to remove these solids from the bitumen before they enter the upgrading plant [34–36]. The first point was evaluated in several ways, for example, by using zeta potential distribution measurement [32, 37, 38], atomic force microscopy [32], and model systems like particles coated with asphaltenes [39, 40] or model water-in-hydrocarbon emulsions [41]. With regard to clay removal, the simplest approach is to remove the solids together with asphaltenes by treating the bitumen with solvents less polar in nature than the naphtha in current use. Asphaltene coprecipitates with the inorganic particles and water to form a "rag layer," thus leaving bitumen of excellent quality in terms of solids and water content. This is a good "academic" solution, which unfortunately has the important drawback that also a high quantity of polar bitumen component are lost so that the overall extraction yield is unacceptable relative to the conventional froth treatment approach using centrifugation.

3 Asphalt/Clay Binary Systems

In asphalt mixtures, the asphalt is used as binder and mixed together with mineral aggregates which are primarily responsible for the load supporting capacity of the pavement. The aggregates are added in amount of about 92–97 % by weight and 75–85 % by volume and have a grading curve ranging from that of a small stone to that of a powder. In the so-called "dense" graded mixtures, the mineral aggregates are dispersed in a continuous asphalt matrix, while in the "open" graded ones, the binder forms a thin film of about 10–20 μm over the aggregates. In both cases it is quite obvious that the mechanical properties of the binder are directly related to the in-service performances, and it is well known that this properties are strongly influenced by the finest aggregates, or mineral fillers, which can be somehow considered as "modifiers." Considering that asphalt can be seen as a low molecular weight polymer, the introduction of clay dispersed at a nanoscale closely resembles the substitution of conventional fillers into polymeric materials to prepare nanocomposites. As already mentioned, asphalt is composed of several compounds with different molecular weights and structure. The presence of asphaltenes containing functional groups with quite high polarity surely favors the interactions with clay. However, asphaltenes are dispersed in the maltenic phase; since as resins they are located at the interface between the maltenic and asphaltenic phase, the resulting material is hydrophobic. This means that either because shielded by resins or because present in too small quantities, the asphaltenes alone are not expected to promote enough interactions with a native clay. Therefore, to form asphalt-based nanocomposites analogous to those prepared with polymeric materials, the first problem to be solved is related to the compatibility between asphalt and clay, and similarly as with polymers, organomodified clays are the best candidates. This is confirmed by some works where the addition to asphalt of both modified and unmodified clays are evaluated. Yu et al. [42]

compared the modification of a base asphalt with a montmorillonite (MMT) with Na^+ interlayer cation and cation exchange capacity of 90 meq/100 g and an organomodified montmorillonite (OMMT) exchanged with octadecylammonium ions and having a cation exchange capacity of 130 meq/100 g. The X-ray analysis shows that the MMT-modified asphalt formed an intercalated structure, while using the OMMT, there was an exfoliated structure.

Unfortunately, the authors did not specify if this is true for all the investigated concentrations of clay (ranging from 0 % to 10 % by weight). However, the difference between the structures of asphalt modified with MMT or OMMT is clearly reflected in the binder properties. Among the reported data, it is interesting to observe the plot of viscosity versus clay content (Fig. 11.1), where the MMT and OMMT curves diverge until 5 % clay content. Then, the OMMT curve has a sharp change in slope, and the viscosities of the asphalts modified with the two clays seem to converge to a common value. It may be supposed that 5 % corresponds to a limit in the clay content, above which further addition of OMMT does not lead to exfoliation.

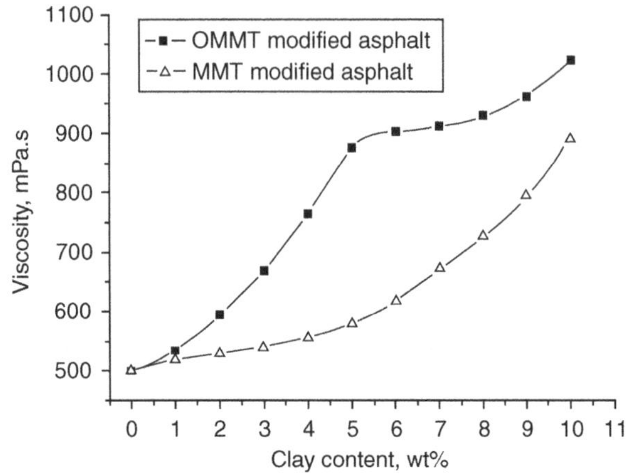

Fig. 11.1 Viscosities of MMT- and OMMT-modified asphalts versus clay contents at 135 °C (From [52])

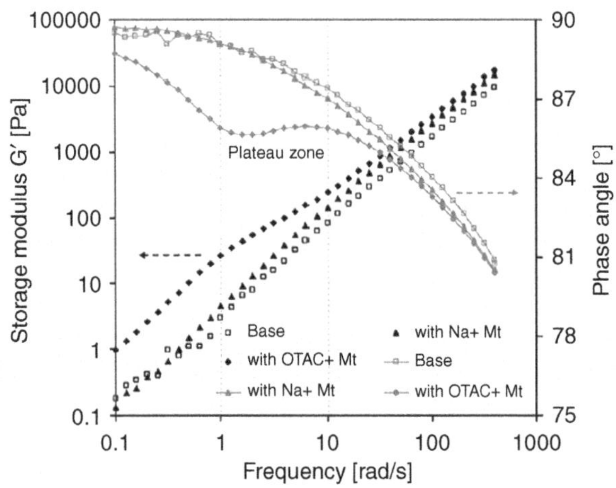

Fig. 11.2 Storage modulus and phase angle at 60 °C. "OTAC + Mt" is the organomodified montmorillonite; Na + Mt is the native Montmorillonite (From [43])

In a subsequent paper, the same research group [43] studied asphalt mixed with MMT (4 % by weight) and reported an intercalated structure for OMMT, while MMT resulted in dispersion only on a micrometer scale. Again, the modified clay has a stronger influence on the binder properties when compared to MMT. As an example, in a frequency sweep test performed at 60 °C, the storage modulus of the original binder and of the MMT-modified binder are very similar to each other and show the classic monotonic trend of asphalts, where the presence of the clay is detectable only, thanks to a modest shift in the modulus magnitude. On the contrary, in the presence of OMMT, the appearance of a plateau zone for the phase angle testifies that the clay has the ability of modifying the internal structure of the binder (Fig. 11.2).

Analogous observation with respect to clay dispersion is reported by Zare-Shahabadi et al. who worked on a sodium bentonite with cation exchange capacity of 110 meq/100 g and the same clay modified with octadecylammonium salt [44]. The sodium bentonite gave intercalation while the other one was able to exfoliate. The authors also show a TEM image of the two modified asphalts, but the micrographies are not easy to interpret.

In conclusion, it is quite clear that the organomodification of clays improves their compatibility with asphalt, but at the same time, there are reported cases where also native clays may at least intercalate in the asphaltic matrix. Other papers can be mentioned. Zhang et al. reported intercalation when using a sodium MMT [45]; Jahromi et al. [46] observed exfoliation when using two different OMMT (one modified with methyl, tallow, bis-2-hydroxyethyl, quaternary ammonium and the second one modified with a not specified "long-chain hydrocarbon"). The authors also characterized the binder properties and those of concretes prepared by using these binders [47–49]. Similarly, complete exfoliation was found by You et al. [50] when using OMMT modified with a quaternary ammonium salt. Hiu et al. compared an MMT and an OMMT, modified with a polysiloxane. In this case no conclusions can be drawn with regard to intercalation or exfoliation because the degree of dispersion was analyzed only through scanning electron microscopy which revealed the presence of large agglomerates ranging from 15 to 50 μm for MMT and with average size of 4 μm for OMMT. However, also a few extremely large agglomerates (about 80 μm) were observed with OMMT, thus revealing that in neither case a really good dispersion was obtained [51]. Nevertheless, significant differences in the binder properties were observed when using the two clays as modifier.

It is interesting to conclude this section underlying that only Yu et al. verified the stability of the binary asphalt/clay systems both stable using either MMT or OMMT [52]. As it was easily predictable, the stability was slightly reduced with increasing clay content and a little bit better when using OMMT compared to MMT. For the readers not used to asphalt technology, it may be useful to shortly recall the procedure to evaluate storage stability. This is a quite old procedure, also called "tube test," where a toothpaste-like tube is filled with the asphaltic binder and then sealed and stored vertically in an oven at high temperature, thus simulating storage without mixing. Temperature and duration of the storage may vary depending on the specification (it may be 163 °C or 180 °C for 2 or 3 days), and during this period, the binder may be subjected to phase separation. In the case of PMAs, if asphalt and polymer are not compatible, due to its lower density, the polymer-rich phase tends to migrate to the top of the tube. On the contrary, an additive or modifier with higher density would migrate to the bottom of the tube. At the end of the storage, the tube containing the modified asphalt is first cooled to room temperature and then frozen and cut horizontally into three equal sections. At this point, the comparison of the top and bottom sections allows to verify a possible phase separation. Specifications usually define a stable binder if the difference in softening points between the top and the bottom sections of the tube is lower than a preestablished value (usually 5 °C). However, independently on the definition of stability, the differences between top and bottom can be quantitatively and/or qualitatively evaluated in several ways, for example, by morphology or by any mechanical or rheological property. This is an important point because to define a "stable" or "not stable" binder (based on the tube test procedure) can be correct according to the technical specification, but it is quite reductive when trying a scientific approach to the problem. In fact, a binder slightly below the allowed limit shows at least a tendency to separation, which may be a better characterization than using only the softening point. This leads to the question of storage stability in the APC ternary systems.

4 Asphalt/Polymer/Clay Ternary Systems

The first research group that modified asphalt with clay and polymer started with block copolymers and used (i) SBS premixed with a kaolinite [53], (ii) SBS premixed with an organobentonite (a silane coupling agent was also added to the blend) [54], and (iii) poly(styrene-*b*-ethylene-*b*-butene-*b*-styrene) (SEBS) premixed with a kaolinite [55]. In all cases, several polymer/clay ratios were used, and the APCs were characterized by storage stability, blend morphology, and mechanical and rheological properties. Therefore, in this series of studies, two types of block copolymers and both native and organomodified clays were used. In all cases, the main finding was that the clay had a positive impact on storage stability. As an example, Table 11.1 shows viscosity (η) and softening point (S) of the top and bottom tube test sections in the case of SEBS and kaolinite. As already mentioned, the migration of the polymer-rich phase toward the top of the tube is driven by a difference in density. In this case SEBS and

Table 11.1 Effect of kaolinite (KC) content on the high-temperature storage stabilities of SEBS/KC-modified asphalts

Formulation		η at 135 °C (Pa s)			S (°C)		
SEBS % (w/w)	SEBS/KC (w/w)	Top (η_t)	Bottom (η_b)	η_t/η_b	Top (S_t)	Bottom (S_b)	S_t–S_b
3	100/0	0.825	0.600	1.38	53.0	50.0	3.0
	100/50	0.642	0.658	0.98	52.5	52.8	−0.3
4	100/0	1.058	0.742	1.43	57.0	53.8	3.2
	100/10	1.008	0.751	1.34	56.5	50.8	5.7
	100/30	0.975	0.800	1.22	55.0	52.0	3.0
	100/50	0.790	0.825	0.96	55.5	55.8	−0.3
	100/50[a]	1.552	1.210	1.28	59.0	52.0	7.0
	100/70	0.875	0.883	0.99	52	52.5	−0.5
5	100/0	2.083	0.992	2.10	70.5	58.0	12.5
	100/50	1.117	1.192	0.94	57.0	58.0	−1.0
6	100/0	2.783	1.258	2.21	85.0	67.5	17.5
	100/50	1.558	1.550	1.01	59	60.5	−0.5

[a]SEBS and KC added directly

asphalt had a density (at room temperature) equal to 0.93 and 1.03 g/cm^3, respectively. Moreover, during swelling the polymer incorporates the lighter oily fraction of asphalt, thus increasing the density difference between polymer-rich and asphaltene-rich phases. Based on this consideration, the authors considered the stabilizing effect due mostly to a variation of the density of the polymer-rich phase, induced by the presence of the clay (its density was 2.57 g/cm^3), mainly bonded with the polymer. This is a "physical" interpretation which is also supported by the data reported in Table 11.1.

Considering the case of 4 % SEBS and different clay contents, it can be seen that increasing the clay content, the difference between top and bottom softening point (S_t–S_b) decreases until it reaches negative values. Since the polymer determines an increase in softening point (with respect to BA), these data seem to suggest that the clay changes the density of the polymer-rich phase. The magnitude of this effect is directly related to the clay content, and therefore, there is an optimal clay content, corresponding to similar densities of the two phases. At high clay loads, the polymeric phases become the one that goes on the bottom of the tube. However, it is quite obvious that the improvement of the high-temperature storage stability may not originate only from variations in densities of the polymeric phase, but originates also from an effect of kaolinite on the compatibility between polymer and asphalt. As it will be shown later, the fact that clay alters the interactions between polymer and asphalt components is shown by its effect on the morphology of the mixes. Shape and dimension of the dispersed phase strongly depend on the clay and inevitably influence at least the kinetics of phase separation.

Therefore, there are at least two effects that improve storage stability: the first one merely related to the density of clay (which is then supposed to be prevalent in the polymer-rich phase) and the second one related to a possible asphalt/polymer compatibilizing action exerted by the clay. This second point introduces another important aspect that must be considered in such kind of ternary systems and is related to the order and procedure of addition of the ingredients. All the mixes but one reported in Table 11.1 were obtained by firstly incorporating kaolinite into SEBS and then adding this mix to asphalt. The ternary mix prepared by directly adding SEBS and kaolinite to asphalt showed to be unstable during high-temperature storage (Table 11.1). Thus, the storage stabilization is obtained only with premixed polymer and clay and not if they are added separately to the asphalt. This means that, during the preliminary mixing of polymer and clay, the two components interact and bond one to each other in a way that is at least partially maintained after the asphalt modification. The same research group found similar results using a styrene-butadiene rubber (SBR) and both a palygorskite and an organomodified palygorskite clays [56]. Unfortunately, in these papers the authors do not investigate if there was a nanocomposite nature of the blends. This aspect has been approached by Polacco et al. in another series of studies. In the first one, BA was modified with SBS and an OMMT (Cloisite 20A, prepared from sodium MMT by treatment with dimethyl dihydrogenated tallow ammonium chloride and subsequently referred to as 20A) by melt mixing [57]. After a preliminary investigation on the binary asphalt/clay and polymer/clay blends, the ternary blends were

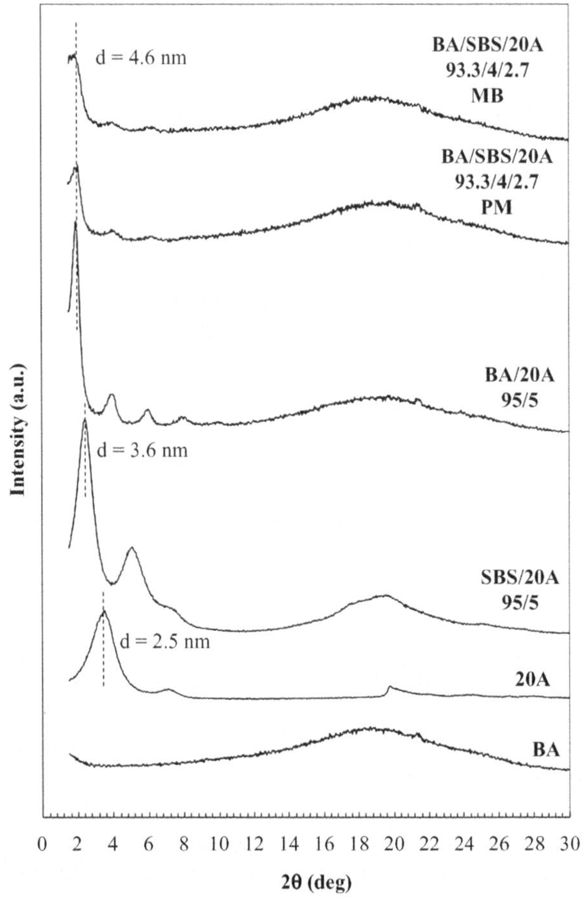

Fig. 11.3 XRD patterns of BA, 20A, SBS/20A = 95/5, BA/20A = 95/5, BA/SBS/20A = 93.3/4/2.7 PM, and BA/SBS/20A = 93.3/4/2.7 MB. All ratios in w/w (From [57])

prepared by adding the clay and polymer to the asphalt, either separately or in the form of a premixed master batch (MB). Thus, three sets of blends were prepared by melt compounding: (1) SBS/20A, (2) BA/20A, and (3) BA/SBS/20A. The two sets of binary blends were characterized by X-ray analysis, in order to find out if, and what type of, nanocomposites were formed when blending BA and SBS separately with the clay. The obtained spectra are reported in Fig. 11.3.

BA shows the typical halo of amorphous materials, while the clay has a main peak at $2\theta = 3.5°$, which corresponds to an interlayer spacing, d_{001}, of 2.5 nm. When the asphalt was mixed with the clay, irrespective of the clay load, the interlayer spacing shifted to a lower angle (corresponding to $d_{001} = 4.6$ nm), thus indicating that the asphalt was able to intercalate the organoclay and probably also to exfoliate part of the stacks. However, none of the reported curves indicated a complete exfoliation of the clay in the asphalt matrix. The storage stability tests of these binary mixes confirm that asphalt and clay were somehow bonded so that no phase separations occurred, even if the two materials had specific densities quite

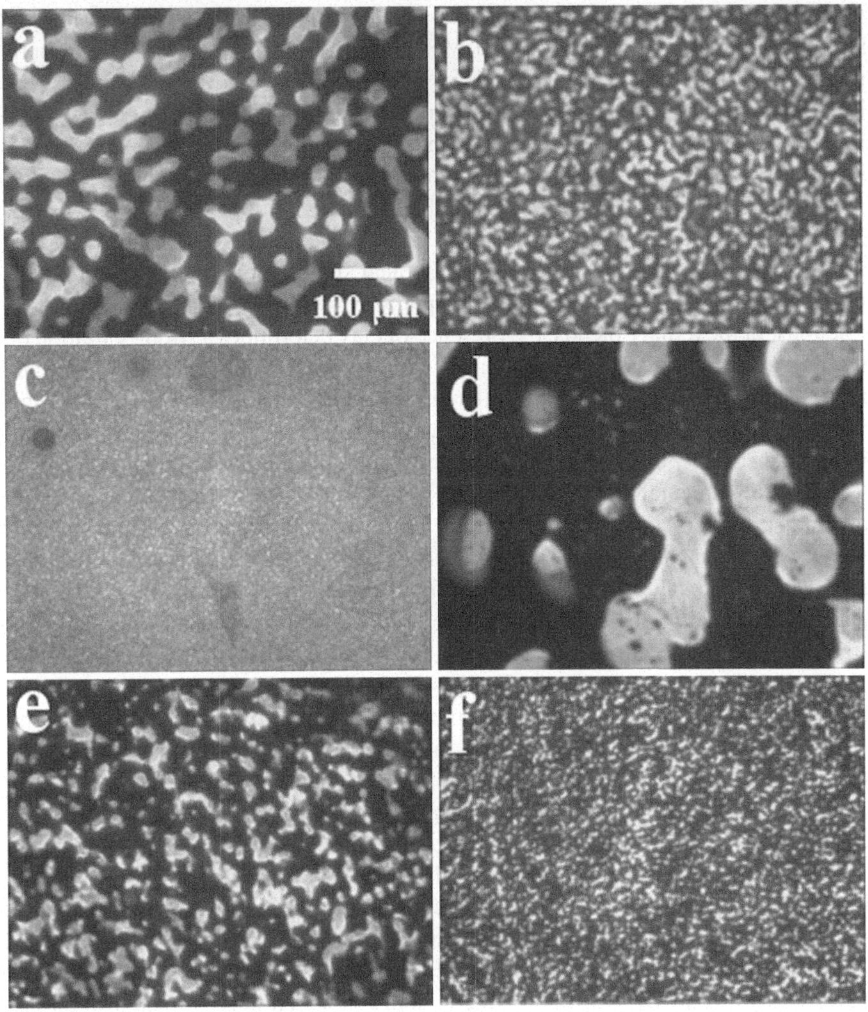

Fig. 11.4 Fluorescence microscopy of (**a**) BA1/SBS (**b**) BA1/SBS/20A PM, (**c**) BA1/SBS/20A MB, (**d**) BA2/SBS (**e**) BA2/SBS/20A PM, (**f**) BA2/SBS/20A MB

different one from each other. Also the SBS/20A binary blends were prepared with different clay loads, and in all cases intercalated nanocomposites characterized by a d_{001} spacing of about 3.6 nm were obtained, irrespective of the clay loading. Therefore, in both binary blends, there was intercalation with interlayer spacing not affected by the clay concentration, and the d_{001} values indicated that the asphalt had greater compatibility with the clay than with the polymer. Finally, two ternary mixes of BA/20A/SBS were prepared: the first one by PM and the second one by MB. The XRD patterns indicated that, in both cases, the mixes resulted in intercalation with an interlayer spacing equal to that obtained by mixing asphalt and clay without the polymer, and this value did not seem to be affected by the mixing sequence. Nevertheless, the PM and MB blends were not equivalent from both stability and rheological points of view. With regard to the first point, it is very interesting to observe the pictures obtained by fluorescence microscopy, which reveal three different morphologies depending on the presence of clay and mixing procedure. In Fig. 11.4, the mixes obtained with two different base asphalts, indicated as BA1 and BA2, are reported (the mixes with BA1 are those whose XRD patterns are reported in Fig. 11.3). In the case of SBS alone and BA1, the aspect is clearly biphasic and reflects the typical situation of a thermodynamically unstable blend in which the polymer is partially swollen by the lighter bitumen fractions. In the ternary blends, the presence of clay determines a strong variation of the internal structure. The PM pattern is similar to the binary BA/SBS blend but looks similarly only at lower enlargement. The smaller dimension of the microdomains indicates a larger interfacial area, thereby denoting higher compatibility and stability of the system. This evolution in the morphological arrangement assigns a fundamental role to the presence of clay and confirms that silicates behave as compatibilizers. This feature became more evident when observing the MB blend, where the canonical aspect of a perfectly compatible PMA was found. The polymer and asphaltene-rich phases are indistinguishable, and the overall image can be described as the so-called orange skin, which corresponds to the highest stability and to a macroscopic behavior that emulates that of the polymer. This dissimilarity, not predicted from the X-ray analysis, justifies the abovementioned differences found in the rheological properties of the two blends. In the case of BA2, the binary mix clearly denotes a lower asphalt/polymer compatibility, with respect to BA1. Nevertheless, the improvement in morphology due to clay and the differences between PM and MB procedures are confirmed also in this case.

Then, a similar procedure was followed by using EVA instead of SBS and two different kinds of organoclays [58, 59]. The first was again Cloisite 20A, and the second was Dellite 43B (subsequently referred to as 43B), derived from naturally occurring montmorillonite and purified and modified by dimethyl dibenzyl hydrogenated from tallow ammonium. Several blends were prepared and it was found that the blend properties depend on the quantity of added polymer and also on the type of nanoclay and preparation method. Also in this case, the clay had a compatibilizing effect as it can be immediately seen from the morphological analysis (Fig. 11.5).

The binary mixture BA/EVA showed a two-phase morphology, with well-defined and quite big dark and white-grey areas, thus denoting a modest affinity of EVA and BA. As it was for SBS, also for EVA the PM ternary mixes with 20A have morphology qualitatively similar to that of the binary mix, but with different dimensions of the patterns. Moreover, a homogeneous morphology was obtained for the MB mixture. In the case of 43B clay, the MB procedure has still a compatibilizing effect better than the PM one, being this effect less pronounced than for 20A.

These results are confirmed by the XRD analysis. In Figs. 11.6 and 11.7, the XRD traces of 20A and 43B clays and their EVA/clay 60/40 (w/w) master batches are reported.

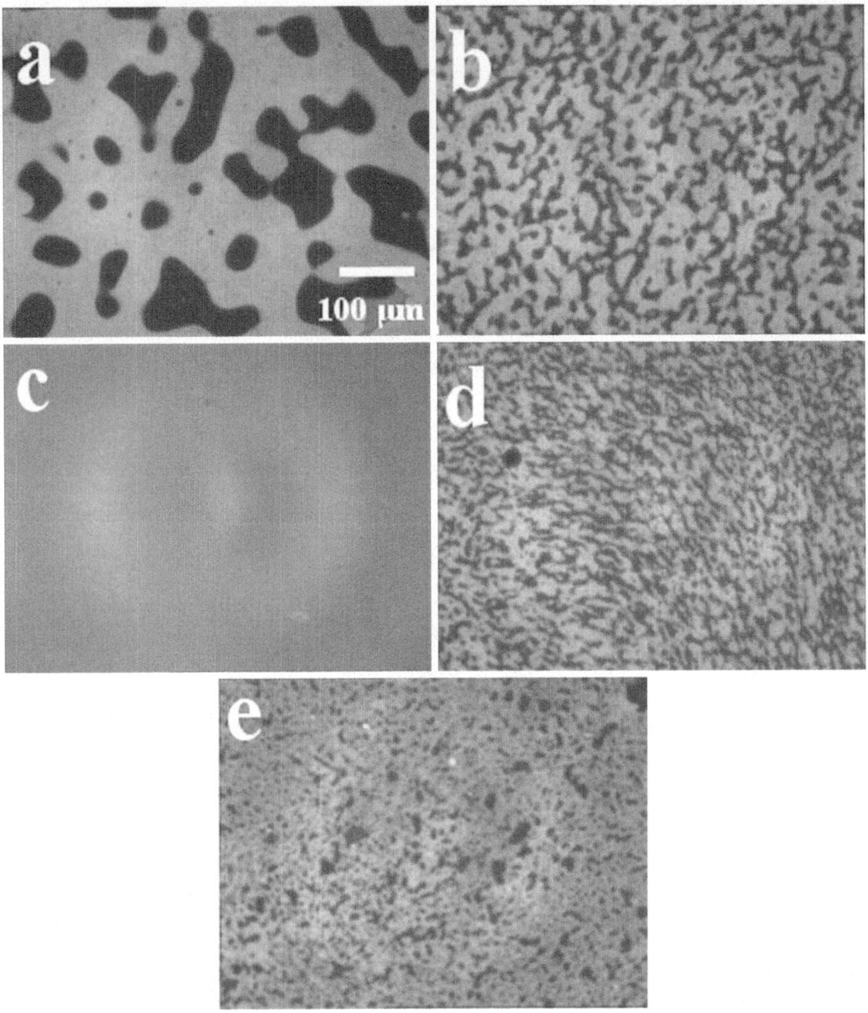

Fig. 11.5 Fluorescence microscopy of (**a**) BA/EVA, (**b**) BA/EVA/20A PM, (**c**) BA/EVA/20A MB, (**d**) BA/EVA/43B PM, and (**e**) BA/EVA/43B MB (From [58])

It is known from the literature [60, 61] that EVA copolymers give, by melt blending with 20A, intercalated/exfoliated nanocomposites, where the higher the vinyl acetate content, the higher is the degree of exfoliation.

According to these studies, in the EVA/20A 60/40 pattern, the interlayer spacing was shifted from 2.5 nm to about 3.9 nm. Therefore, all the tactoids were intercalated by EVA chains even at very high organosilicate concentration. Moreover, the clearly visible d_{002} and d_{003} orders suggest the formation of intercalated tactoids with highly regular and coherent structure. On the contrary, in the EVA/43B 60/40 XRD pattern, the original peak of 43B is still present and a new peak trace at lower

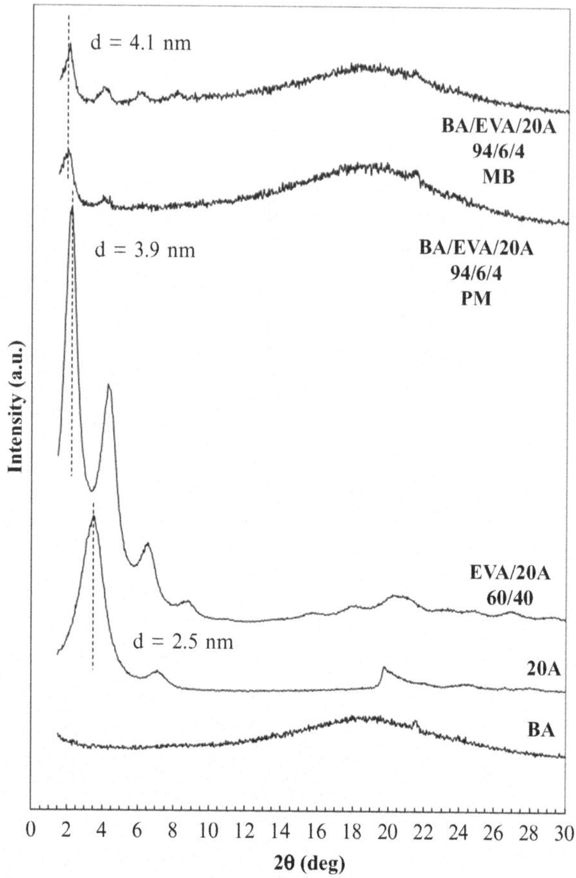

Fig. 11.6 XRD patterns of BA, 20A, EVA/20A = 60/40, BA/20A = 95/5, BA/EVA/20A = 94/6/4 PM, and BA/EVA/20A = 94/6/4 MB. All ratios in w/w (From [58])

angle indicates a partially intercalated hybrid structure. This may be due to the high clay content; however, Li and Ha [62] found similar results using only 3 wt% of Cloisite 10A, which is modified with the same cation as 43B. Moreover, similarly to Olabisi et al. [63], the basal peak of the clay in the hybrid had a shoulder at higher angles, thus indicating that part of the original tactoids collapsed, perhaps due to degradation of the organomodifier. Therefore, in the binary mixes, the 20A clay has more homogeneous clay dispersion when compared with 43B.

With regard to the BA/EVA/20A ternary mixtures, the first observation is that for the PM mixtures, the data were highly scattered, while for the MB mixtures, they were consistent and reproducible. Moreover, only for the mixtures prepared by MB, the $d_{002}$and d_{003} peaks were clearly visible as in the polymer nanocomposites. This indicates a more homogeneous dispersion of the clay using the MB procedure, and it is reasonable to assume that the polymer chains remain in the gallery during mixing with asphalt. Nevertheless, for both MB and PM mixtures, an interlayer spacing of about 4.1 nm was found; this value is quite similar to those obtained for the EVA/20A binary blend. Therefore, in both cases, the clay acts as surfactant for

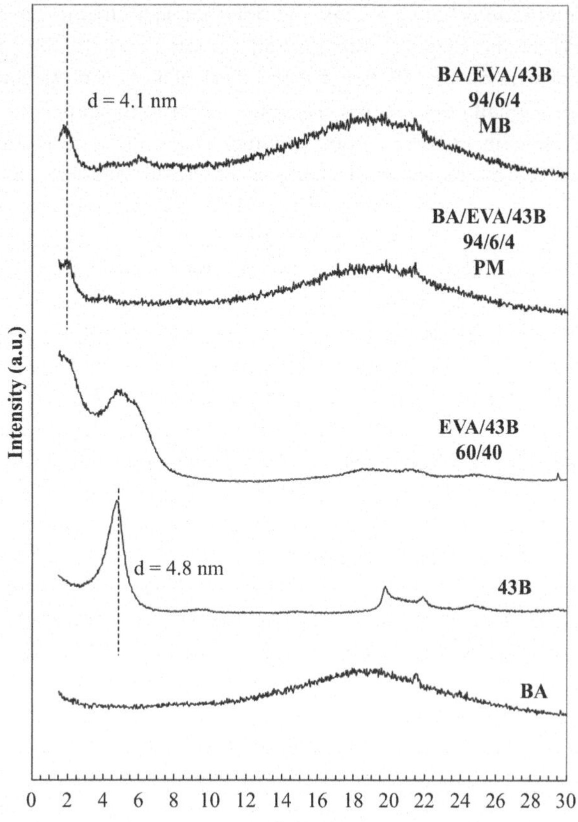

Fig. 11.7 XRD patterns of BA, 43B, EVA/43B = 60/40, BA/20A = 95/5, BA/EVA/43B = 94/6/4 PM, and BA/EVA/43B = 94/6/4 MB. All ratios in w/w (From [58])

the polymer in the asphalt matrix, but when clay and polymer are added separately, the polymer is less efficient in intercalating the clay galleries because it competes with asphalt components. In the case of PM, the surfactant action of the clay is therefore reduced and fewer homogeneous tactoids are obtained.

The XRD patterns of BA/EVA/43B ternary mixtures prepared by PM or by MB showed that for all the blends the original basal peak of the clay disappeared and shifts to a lower angle. Thus, asphalt intercalates almost all the organosilicate galleries (a weak peak at higher angle corresponding to the collapsed tactoids formed during the EVA/43B MB preparation is still visible). Again, the samples prepared by MB are more homogeneous with respect to those prepared by PM, and the surfactant effect of 43B resulted lower than that of 20A because this clay has less affinity for EVA and part of the polymer did not intercalate the clay.

In conclusion, there are significant differences between the mixtures with analogous composition but prepared by MB or PM. Moreover, from the binary mixes with SBS, it is known that clays have a higher affinity to asphalt than to the polymer and this result is easily generalized to EVA. Therefore, mixing by PM, the clays are mainly intercalated by the asphalt molecules, while by MB, the preformed interactions between clay and polymer seem to survive during mixing with asphalt. When the clay is only partially intercalated by the polymer chains during the preparation of MB (as is it the case of Dellite), the intercalation is completed by the asphalt components. Therefore, even if the clay is more compatible with asphalt than with polymer, once polymer/clay interactions are established, they are not altered by the asphalt molecules added subsequently, and the higher polarity of the clay enhances the compatibility between polymer and asphalt. In other words, the clay acts as a surfactant that increases the interactions between the incompatible phases, and the higher the clay/polymer affinity, the higher is this effect. Of course, the different internal structure of the blends influences the macroscopic behavior. Rheological properties studied in small amplitude oscillations, start-up of shear flow, and repeated creep and recovery tests showed improved thermomechanical properties of the ternary blend prepared by MB [59].

In the abovementioned studies, the physical mixing was done by adding polymer and clay simultaneously to the melted asphalt. In three recent papers, asphalt was modified with SBS and OMMT by PM, and the order of mixing was found not to have appreciable effect on the characteristics of the final samples [64–66]. Therefore, the chosen procedure was to add OMMT into SBS-modified asphalt, and the authors claim that an exfoliated structure was obtained. However, the reported XRD spectra do not seem to clearly confirm such affirmation. Nevertheless, improved storage stability due to the presence of the clay is found also in this case. The same mixing procedure was used by Tang et al. [67] who added a MMT to SBS-modified asphalt and characterized the material with XRD. However, also in this case, the reported graph does not clarify the internal structure of the ternary blend, and the authors describe it as a "nanocomposite." The storage stability is not evaluated, and the effect of the clay is mostly evaluated in terms of toughness, strength, and antiaging properties. With regard to the latter point, the authors suggest that the lamellar structure of MMT could prevent oxygen from diffusing and permeating in the asphaltic matrix thus reducing its thermo-oxidative aging. The same authors then evaluated the properties of concretes prepared with this binder [68].

Another quite accurate investigation of the microstructure of OMMT-/SBS-modified asphalts was done recently by Zhang et al. [69]. The mix was prepared by adding first SBS to the melted asphalt and then, after about 30 min of shearing, the OMMT was added, thus obtaining an intercalated structure. The paper also presents some interesting AFM images of BA, SBS-modified asphalt, and OMMT-/SBS-modified asphalt which confirm the dramatic effect of clay on asphalt/polymer interactions and thus on the blend morphology (Fig. 11.8). The BA images show the presence of domains with the typical "bee-like" structure that have been interpreted as asphaltene micelles [70, 71] dispersed in the continuous maltenic matrix. Upon polymer addition, those domains remain visible, but the overall aspect is completely changed and clearly shows the polymer-rich domains. Moreover, compared with BA, the phase contrasts in Fig. 11.8a, b are inverted. This indicates a change in the tip-sample interactions, which is affected by the sample stiffness. After OMMT

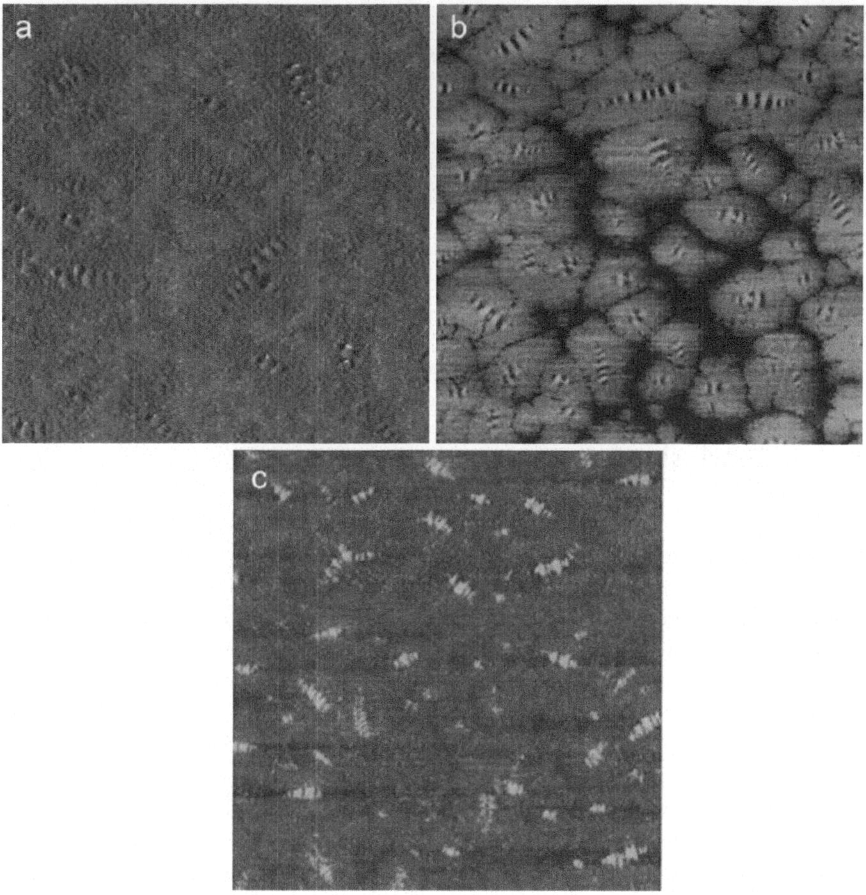

Fig. 11.8 Phase images of (**a**) BA, (**b**) BA/SBS, and (**c**) BA/SBS/OMMT on a scale of 15 × 15 μm (From [69])

modification, the contrast between the dispersed domains and the matrix decreases, and the overall aspect resembles the one of BA, thus demonstrating that the material has a quite homogeneous structure.

Another study related to the same topic but involving other polymers is worth of mentioning. Fang et al. [72] used an OMMT and a polyethylene (PE) packaging waste (from fresh milk bags) to modify an asphalt by PM. The XRD analysis indicated that an exfoliated structure was obtained for the ternary blends, and fluorescence microscopy showed that the OMMT addition is beneficial for the dispersion of WPE in asphalt (Fig. 11.9); the polymer was dispersed in asphalt as small particles with homogeneous distribution, far better than that obtained without clay. Nevertheless, the spherical shape of such particles means that the dispersion is still far from that needed to obtain storage stability.

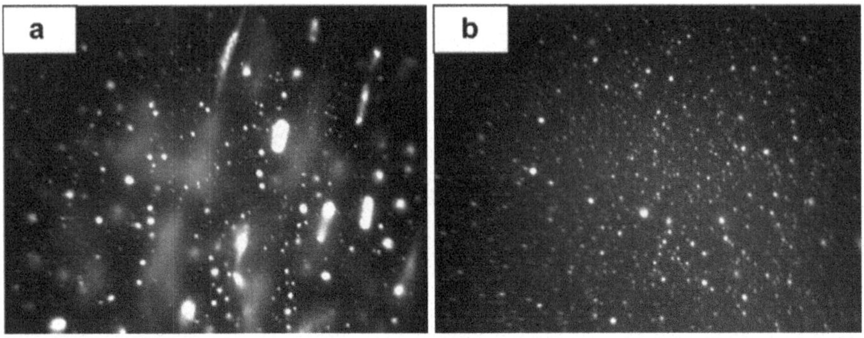

Fig. 11.9 Fluorescence microscopy of (**a**) BA/PE and (**b**) BA/PE/OMMT (From [72])

5 Conclusions

As it is well known, polymeric nanocomposites were invented in 1985 at Toyota Central R&D Labs, Inc. when they introduced the Nylon-6/MMT system and generated great interest in the academic and industrial communities, thus rapidly leading to the production of a great volume of scientific and patent literature. Similarly, polymer-modified asphalts appeared in the scientific literature in the 1980s, and at the moment, their use is a well-established and common practice in the paving industry. However, these two somehow related fields met only recently; the first publication is dated 2005 [53]. The idea was to join the advantages of clay addition to polymers and polymer addition to asphalts. In fact this aspect gave some interesting results, but at the same time, another important gain that has been found in the APC systems is the one related to the enhancement of polymer and asphalt compatibility, which is the main problem of PMAs. Since the use of conventional techniques and apparatus for the production of asphalt pavements is a prerequisite, the APC have always been prepared by melt blending. If we consider the asphalt as a low molecular weight polymer, the ternary APC system can be paralleled to those systems where clay is added to a polymeric blend and acts as a compatibilizer by partially localizing at the interfacial region. At the same time, due to the quite low molecular weight of asphalts components, there are some similarities with solution intercalation where the asphalt plays the role of the solvent, which dissolves the polymer and delaminates the clay, thus helping the polymer/clay interactions. However, in this case, the final step of solvent removal is missing and asphalt remains in the blend. This probably explains the differences that have been found between the PM and MB preparation procedures. As already mentioned, in polymer blends, the clay resides preferentially within the bulk of the most compatible polymer but may also partially localize at the interfacial region, thus acting as a compatibilizer between the two polymeric phases. In the case of asphalt and polymer, the interfacial region can be identified as the one between the polymer-rich and the asphaltene-rich phases, whose morphologies are directly linked with storage stability and thus with asphalt/polymer compatibility. Among asphalt and polymer, the more compatible one is the asphalt, mainly favored by its low molecular weight. Moreover, asphalt usually constitutes 94–96 % by weight of the mixture so that its interaction with the clay is also favored from a statistical point of view. Therefore, in PM, only a minor part of the clay acts as a surfactant and enhances the asphalt/polymer interactions. In comparison, when polymer and clay are premixed to form a classic polymeric nanocomposite and subsequently added to the asphalt, the compatibilization effect is strongly enhanced. This suggests that the preformed interactions between clay and polymer may survive the phase of mixing with asphalt even if the clay is more compatible with asphalt than with polymer. Hence, the introduction of the more compatible part in a second step helps the clay to remain bonded with the polymer and amplifies the surfactant effect. This has to be considered as just an interpretation of the available results. Deep understanding of the mechanisms involved during APC formation is still too far to be reached, and it is particularly difficult due to extremely complex and almost unknown structure of asphalts and polymer-modified asphalts.

6 Future Perspectives

Even if the compatibilizing effect of clays, of course organomodified, and the importance of the preparation technique are quite well established and appear to be promising, to our knowledge a practical application is still missing. With regard to performances of APC systems, compared to classic PMAs, the reader can found several results in the literature cited above. Such aspect has been only marginally exploited in this chapter which, being part of a book on nanocomposites, is mainly targeted to polymer scientists. Nevertheless, the properties and performances of APC binders and aggregates prepared with APC binders must be investigated further, together with economic and practical aspects. With regard to this last point, one of the main drawbacks of APC can be related to the increase of binder viscosity, due to the presence of clay. As a consequence, higher temperatures may be required during paving, and this is in contrast with the recent tendency to warm mix asphalts (WMA). WMAs were introduced over the last years to reduce temperatures traditionally required for production, laying, and compaction of hot mix asphalt without affecting their resistance to traffic and climate. Temperature reduction in the manufacture of asphalt mixtures is highly desirable for many aspects like a decrease in cost, energy consumptions, and atmospheric emissions. Moreover, in addition to the immediate environmental and economic advantages, the development of the WMA technology can induce improvements of health and safety conditions, other than further technical benefits like longer haul distances and construction season. Therefore, the next step for APC, as well as for PMAs, will probably be to join the WMA technology, thus going to a four-component formulation where a “warm” additive is added to asphalt, polymer, and clay.

References

1. Read J, Whiteoak D (2003) The shell bitumen handbook, 5th edn. Thomas Telford, London
2. Lesueur D (2009) Adv Coll Int Sci 145:42
3. Pfeiffer JP, Saal RNJ (1940) J Phys Chem 44:139
4. Saal RNJ (1950) Physical properties of asphaltic bitumen. 1. Rheological properties. In: Pfeiffer JP (ed) The properties of asphaltic bitumen. Elsevier, Amsterdam, p 50
5. Becker IM, Muller AJ, Rodriguez Y (2003) J Appl Polym Sci 90:1772
6. Airey GD (2004) J Mater Sci 39:951
7. Fawcett AH, McNelly T (2001) Polym Eng Sci 417:1251
8. Fawcett AH, McNelly T (2003) Colloid Polym Sci 281:203
9. Lu X, Isacsson U (2000) J Appl Polym Sci 7612:1811
10. Fawcett AH, McNally T (2000) Polymer 4114:5315
11. Airey GD (2002) Constr Build Mat 16:473
12. Polacco G, Stastna J, Biondi D, Antonelli F, Vlachovicova Z, Zanzotto L (2004) J Colloid Interface Sci 2802:366
13. Selvavathi V, Sekar VA, Sriram V, Sairam B (2002) Petrol Sci Technol 20:535
14. Hsiue GH (1996) Styrene-butadiene-styrene triblock copolymer. In: Salamone C (ed) Polymeric materials encyclopedia, vol 19. CRC Press, Boca Raton
15. Wen G, Zhang Y, Zhang Y, Sun K, Fan Y (2002) Polym Test 212:295
16. Trakarnpruk W, Chanathup R (2005) J Metals Mater Miner 15:79
17. Liu X, Wu Q, Berglund LA, Fan J, Qi Z (2001) Polymer 42:8235
18. Tjong SC, Bao SP (2005) J Polym Sci Part B Polym Phys 43:585
19. Gonza´lez I, Eguiza´bal JI, Naza´bal J (2006) Eur Polym J 42:2905
20. Wang H, Zeng C, Elkovitch M, Lee LJ, Koelling KW (2001) Polym Eng Sci 41:2036
21. Yurekli K, Karim A, Amis EJ, Krishnamoorti R (2003) Macromolecules 36:7256
22. Si M, Araki T, Ade H, Kilcoyne ALD, Fisher R, Solokov JC, Rafailovich MH (2006) Macromolecules 39:4793
23. Vo LT, Giannelis EP (2007) Macromolecules 40:8271
24. Chow WS, MohdIshak ZA, Karger-Kocsis J (2005) Macromol Mater Eng 290:122
25. Fang Z, Xu Y, Tong L (2006) J Appl Polym Sci 102:2463
26. Filippi S, Dintcheva NT, Scaffaro R, La Mantia FP, Polacco G, Magagnini P (2009) Polym Eng Sci 49:1187
27. Lipatov YS (2002) Prog Polym Sci 27:1721
28. Okada A, Fukushima A, Kawasumi M, Inagaki S, Usuki A, Sugiyama S et al (1987) Compos Mater Prep 11:22
29. Jordan JW (1949) J Phys Colloid Chem 53:294
30. Oladipo E, Omotoso R, Mikula J (2004) Appl Clay Sci 25:37
31. Kotlyar LS, Sparks BD, Woods JR (1999) Energy Fuel 13:346
32. Liu J, Xu Z, Masliyah J (2004) AIChE J 50:1917
33. Rahimi P, Gentzis T, Fairbridge C (1999) Energy Fuel 13:817
34. Zhou ZA, Xu Z, Masliyah JH, Czarnecki J (1999) Colloids Surf A Physicochem Eng Asp 148:199
35. Hooshiar A, Uhlik P, Liu Q, Etsell TH, Ivey DG (2012) Fuel Process Technol 94:80
36. Li Y (2011) Transport Porous Med 90:333
37. Tshitende K, Zhiang Z, Zhenghe X, Masliyah J (2000) Can J Chem Eng 78:674
38. Liu J, Zhou Z, Xu Z, Masliyah J (2002) J Colloid Interface Sci 252:409
39. Dudasova D, Flaaten GR, Sjoeblom J, Oeye G (2009) Colloids Surf A Physicochem Eng Asp 335:62
40. Dudasova D, Simon S, Hemmingsen PV, Sjoeblom J (2008) Colloids Surf A Physicochem Eng Asp 317:1
41. Sztukowski DM, Yarranton HW (2005) J Colloid Interface Sci 285:821
42. Yu J, Zeng X, Wu S, Wang L, Liu G (2007) Mater Sci Eng A 447:233

43. Liu G, Wu S, De Ven MV, Yu J, Molenaar A (2010) Appl Clay Sci 49:69
44. Zare-Shahabadi A, Shokuhfar A, Ebrahimi-Nejad S (2010) Diffusion Defect Data Solid State Data Pt A Defect Diffusion Forum 297(Pt 1 Diffusion in Solids and Liquids V):579
45. Zhang Z, Wen Y, Pei J, Chen S (2011) Green power, materials and manufacturing technology and applications. Appl Mech Mater 84(Pt 2):662
46. Jahromi SG, Khodaii A (2009) Construct Build Mater 23:2894
47. Jahromi SG, Khodaii A (2009) Amirkabir Int J Model Ident Simul Control 41:49
48. Jahromi SG, Andalibizade B, Khodaii A (2010) J Test Eval 38:549
49. Jahromi SG, Andalibizade B, Vossough S (2010) Arab J Sci Eng Sect B: Eng 35:89
50. You Z, Mills-Beale J, Foley JM, Roy S, Odegard GM, Dai Q, Wei Goh S (2011) Construct Build Mater 25:1072
51. Yao H, You Z, Li L, Shi X, Wei Goh S, Mills-Beale J, Wingard D (2012) Construct Build Mater 35:159
52. Yu J, Zeng X, Wu S, Wang L, Liu G (2007) Mater Sci Eng A 447:233
53. Ouyang C, Wang S, Zhang Y, Zhang Y (2005) Polym Degrad Stab 87:309
54. Wang YP, Liua DJ, Li YF, Wang YP, Gao JM (2006) Polymer Polymer Compos 14:403
55. Ouyang C, Wang S, Zhang Y, Zhang Y (2006) Eur Polym J 42:446
56. Zhang J, Wang J, Wu Y, Sun W, Wang Y (2009) J Appl Polym Sci 113:2524
57. Polacco G, Kříž P, Filippi S, Stastna J, Biondi D, Zanzotto L (2008) Eur Polym J 44:3512
58. Sureshkumar MS, Filippi S, Polacco G, Kazatchkov I, Stastna J, Zanzotto L (2010) Eur Polym J 46:621
59. Sureshkumar MS, Stastna J, Polacco G, Filippi S, Kazatchkov I, Zanzotto L (2010) J Appl Polym Sci 118:557
60. Lili C, Xiaoyan M, Paul DR (2007) Polymer 48:6325
61. Peeterbroeck S, Alexandre M, Jerome R, Dubois P (2005) Polym Degrad Stab 90:288
62. Li X, Ha CS (2003) J Appl Polym Sci 87:1901
63. Olabisi O, Robeson L, Shaw MT (1979) Polymer–polymer miscibility. In: Polymer–polymer compatibility. Academic, New York
64. Galooyak SS, Dabir B, Nazarbeygi AE, Moeini A (2010) Construct Build Mater 24:300
65. Galooyak SS, Sadeghpour S, Dabir B, Nazarbeygi AE, Moeini A, Berahman B (2011) Petrol Sci Technol 29:850
66. Golestani B, Nejad FM, Galooyak SS (2012) Construct Build Mater 35:197
67. Tang X, Kong X, He Z, Li J (2011) Mater Sci Forum 688:175
68. Tang X, Kong X, Huang F, Li J (2011) Mater Sci Forum 688:191
69. Zhang H, Yu J, Wang H, Xue L (2011) Mater Chem Phys 129:769
70. Loeber L, Sutton O, Morel J, Valleton JM, Muller GJ (1996) Microscopy 182:32
71. Wu SP, Pang L, Mo LT, Chen YC, Zhu GJ (2009) Construct Build Mater 23:1005
72. Fang C, Yu R, Zhang Y, Hu J, Zhang M, Mi X (2012) Polymer Test 31:276

Crystallization and Polymorphic Behavior of Nylon-Clay Nanocomposites

12

E. Bhoje Gowd and C. Ramesh

Contents

Abstract

Nylons are very successful semicrystalline polymers and are commercially exploited extensively. The incorporation of nano-sized layered silicates into nylons has a significant influence on the crystallization behavior, polymorphism, and hence on the properties of the polymers. In this chapter, we highlight the preparation of nylon nanocomposites by various methods and the effect of layered silicates on the crystallization, crystallization kinetics, crystal structure, and polymorphic transitions in nylon–clay nanocomposites.

E.B. Gowd (✉)
Materials Science and Technology Division, CSIR-National Institute for Interdisciplinary Science and Technology (CSIR-NIIST), Thiruvananthapuram, India
e-mail: bhojegowd@niist.res.in

C. Ramesh
Polymer Science and Engineering Division, CSIR-National Chemical Laboratory (CSIR-NCL), Pune, India
e-mail: c.ramesh@ncl.res.in

J.K. Pandey et al. (eds.), *Handbook of Polymernanocomposites. Processing, Performance and Application – Volume A: Layered Silicates*, DOI 10.1007/978-3-642-38649-7_12,

Keywords
Crystallization • Layered silicates • Nanocomposites • Nylon • Phase transition • Polymorphism

1 Introduction

Nylon, chemically known as polyamide, was the first commercially successful polymer and the first synthetic fiber to be made. The basic chemical structure consists of methylene sequences, interrupted at regular intervals by intermolecularly hydrogen-bonded amide groups. The interchain interactions that arise from the hydrogen bonding between the amide groups on adjacent chains as well as dipole–dipole interactions play a prominent role in generating the unique properties in nylons. By changing the amide density, one can modify such properties like melting point, moduli, low-temperature impact strength, moisture absorption, and chemical resistance to salts and acids. Nylons are used in a great many applications, including fabrics, bridal veils, carpets, rope, musical strings, and also used for mechanical parts such as gears, coating of metal objects, cable jacketing, and tubing extrusion.

Polymer/layered silicate nanocomposites have attracted great interest because they often exhibit superior properties as compared to those of virgin polymers [1–6]. Discovery of simple and commercially viable preparation methods of these nanocomposites using conventional processing equipments accelerated the research in this field [6]. Since then almost all types of polymers were used to prepare the nanocomposites. Among them, nylon/layered silicate nanocomposites attracted great interest, both in industry and in academia [7–9]. Moreover, it is relatively easy to obtain significant levels of intercalation and/or exfoliation in nylons by virtue of their polar molecular structure. The technological and academic interest of nylon/layered silicate nanocomposites is testified by the number of papers published in the literature and number of patents issued over the last 20 years. One of the main limitations in the commercialization of nylon nanocomposites arises from the difficulty in controlling the morphology developed during process, which, in turn, affects the ultimate material properties. It is known that the physical properties of nylons are greatly dependent on their crystal structures, and hence, it is necessary to understand the role of dispersed layered silicates on the crystallization, crystal structure, and nanoscale morphology of nylon/layered silicate nanocomposites [10]. The final properties of the nanocomposites will depend not only on the distribution of layered silicates but also on the influence of polymer–silicate interaction on polymer crystallization and crystallite morphology.

In this chapter, recent progress made in understanding the crystallization and polymorphism in nylon–clay nanocomposites is described. The chapter is mainly structured into two parts. In the first part, preparation procedures of various nylon–clay nanocomposites are discussed. Second part mainly focuses on the effect of nano-clays on various aspects of crystallization such as crystallization kinetics, crystal structure, and polymorphic transitions.

2 Preparation of Nylon/Layered Silicate Nanocomposites

Montmorillonite, a member of the smectic family is the most widely used layered silicate for making of the polymer nanocomposites. Smectic clays having two tetrahedral sheets around the central octahedral sheet with a relation are known as 2:1 phyllosilicates. This has a platy structure with a thickness of about 1 nm, and the length and the width of these clays can be over a micrometer. The layered structure of clay is hydrophilic, and it is not compatible with most of the nonpolar polymers. The compatibility between the layered silicates and polymer has to be improved and achieved by surface modification of the layered silicates. The most widely used surface modification involves the exchange of hydrated cations of the interlayer with the cationic surfactants such as alkyl ammonium (quaternary ammonium cations), typically with chain lengths longer than eight carbon atoms [3]. The surface modification reduces the attractive forces between the layered silicate sheets and therefore enhances the compatibility. In general, the following methods are used to prepare the polymer nanocomposites:

1. In situ polymerization
2. Melt intercalation
3. Solution intercalation

This chapter mainly focuses on the different processes available to synthesize nylon/layered silicate nanocomposites.

2.1 In Situ Polymerization

In this method, first layered silicates are organically modified using suitable procedure in order to intercalate the monomer into the galleries of silicate layers. Such intercalated monomers can be polymerized either by heat or radiation to obtain the polymer nanocomposites. The growing polymer chains can push the layers apart leading to intercalated or exfoliated structures. However, from an industrial point of view, this may not be a commercially viable method. The presence of layered silicates in the system leads to complicated reaction conditions, and this method requires major changes in the existing production facilities to suit the reaction conditions. Such reasons limit the bulk production of the nanocomposites by this method [11].

Toyota central research and development laboratories successfully produced the nylon/layered silicate nanocomposites for the first time using in situ polymerization method [8, 9]. Layered silicates containing sodium ions between the layers are first modified with α,ω-amino acids ($H_2N\text{-}(CH_2)_{n-1}\text{-}COOH$, with $n = 2,3,4,5,6,8,11,12,18$). This bifunctional organic modifier anchors to the layered silicate through ionic bonding at one end, and the carboxyl group at the other end reacts with the ε-caprolactam, which is a nylon-6 monomer. Organically modified layered silicates are subsequently swollen by ε-caprolactam at 100 °C. The ring opening polymerization of ε-caprolactam is then initiated to

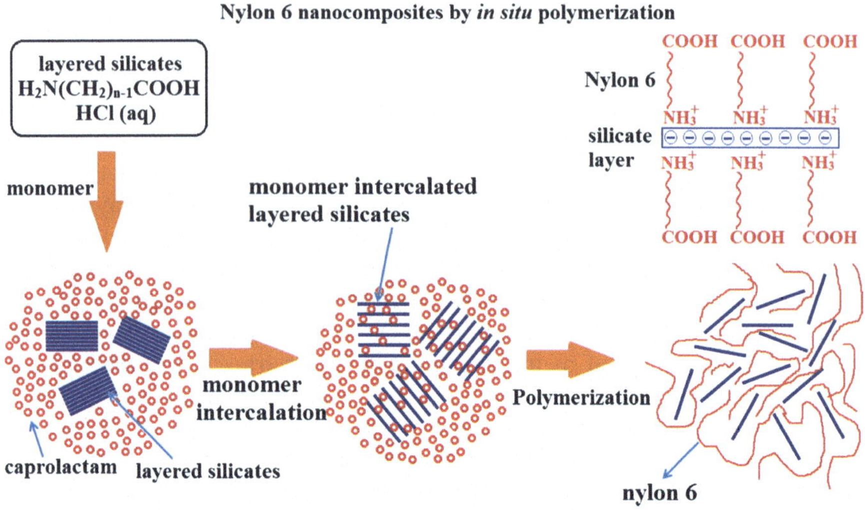

Fig. 12.1 Formation of nylon-6 nanocomposite by in situ polymerization method

obtain nylon-6 nanocomposites. The schematic of such in situ polymerization process is shown in Fig. 12.1. X-ray diffraction and transmission electron micrography revealed that the nanocomposites obtained are completely exfoliated for less than 15 wt% of the filler, whereas intercalated nanocomposites are obtained for filler amounts in the range from 15 to 70 wt% [7]. Nylon-12 nanocomposites are also prepared by in situ polymerization method by using 12-aminolauric acid as the organic modifier and monomer. In this case, partially exfoliated nanocomposites are obtained [12]. In situ polymerization method is also used to prepare exfoliated nylon-10 12/layered silicate nanocomposites by polycondensation polymerization [13].

2.2 Solution Blending

In this method, the organically modified silicate is dispersed in a solvent in which the polymer is also soluble [14–16]. The solvent overcomes the van der Waals forces holding the silicate layers together and helps in dispersing the layered silicate platelets in the polymer solution. When the solvent is evaporated, the polymer chains diffuse and absorb on the layered silicate platelets to form the intercalated or exfoliated nanocomposite structures. Dispersability of the layered silicates determines the nanocomposite structure. The schematic of this process is shown in Fig. 12.2. This method is preferred for polymers that require high processing temperature at which the organo modifier may degrade. However, this preparation method is not suitable for nylon kind of polymers as these polymers are insoluble or poorly soluble in most of the common organic solvents at

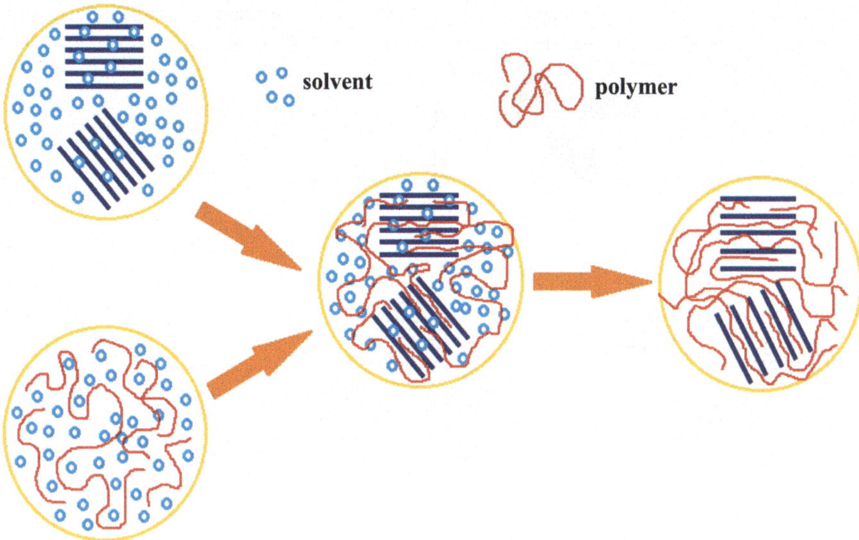

Fig. 12.2 Schematic representation of solution-blending method used to prepare intercalated polymer/layered silicate nanocomposites (not to scale)

room temperature. The use of bulk quantities of organic solvents, which are environmentally unfriendly and high cost, restricts the commercial viability of this method.

2.3 Melt Blending

Giannelis et al. first demonstrated the possibility of melt intercalation method to prepare the nanocomposites [6]. Since then, this method is widely used to prepare polymer nanocomposites due to its compatibility with the existing processing techniques such as extrusion and injection molding. Added advantage is that this method is environmentally benign, as it does not require any organic solvents. In this method, organically modified layered silicates are directly blended with the polymer above the softening temperature of the polymer. At melt conditions, the polymer chains diffuse into the galleries of the silicate layers [6, 17–21]. The level of exfoliation depends on mixing conditions and the compatibility of the polymer with layered silicates.

Many researchers used this method for the preparation of nylon nanocomposites and found that the degree of exfoliation depends on (i) organic modification of layered silicates, (ii) matrix viscosity and polymer/layered silicate affinity, (iii) an optimum balance between the residence time and level of shear in the extruder, and (iv) the processing temperatures [6, 21, 22]. Figure 12.3 shows the schematic representation of the possible polymer/layered silicate hybrid structures by melt blending.

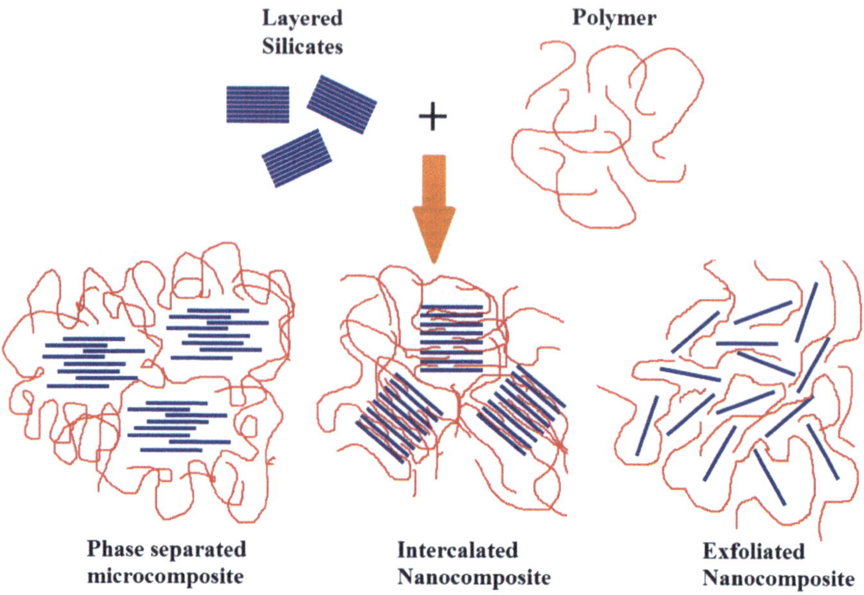

Fig. 12.3 Schematic representations of different nanocomposite structures obtained by melt blending of polymer with layered silicates (not to scale)

3 Crystal Structures of Nylons

Hydrogen bonding dominates the crystal structure of nylons, and hydrogen-bonded sheets are the main feature of the nylon crystal structure [23–46]. The balance between intermolecular hydrogen bonding and van der Waals interactions among methylene units belonging to adjacent chains in the crystal determines the polymorphic form of nylons [25]. Generally, nylons exhibit two well-characterized crystalline forms, namely, α and γ forms. The α form shows two strong reflections (*d*-spacings of these reflections at 0.37 and 0.44 nm) that correspond to the projected interchain distance within the hydrogen-bonded sheets and the intersheet distance between the sheets, respectively [26]. The chains in the α form are in fully extended trans-planar conformation [27]. The γ form is determined by the skew conformation of the CH_2 units adjacent to the amide group resulting in a shorter chain-axis repeat [28]. The twisted chains allow hydrogen bonds to be formed between parallel chains [28]. The characteristic of the γ form is that it exhibits only one strong reflection and the *d*-spacing is 0.42 nm. The γ form is considered to be a less ordered than that of the α form.

In *n*-nylons (nylon-*n*, *n* even), the stability of the α form and the γ form depends on the number of methylene units in the polymer chain. The α form is more stable for n $\leq$6, while the γ form is more stable for n $\geq$8 [29, 30]. The case of nylon-6 is peculiar one, where both the crystal forms can be easily produced due to the small

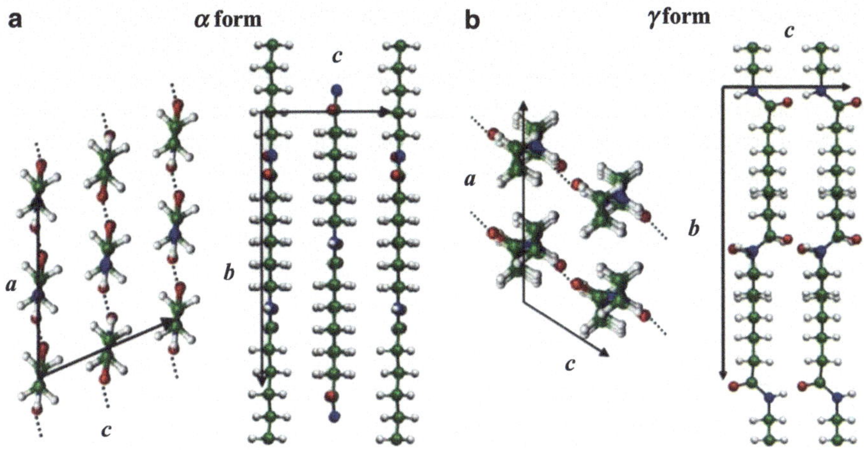

Fig. 12.4 Sketches of the crystalline structures of α and γ forms of nylon-6 (Reproduced from [37] with permission)

difference in energy. Crystal structures of nylon-6 showing different conformations of the polymer chains are shown in Fig. 12.4. The unit cell of the α form is determined to be monoclinic with the dimensions $a = 0.956$ nm, $b = 1.724$ nm (fiber axis), $c = 0.801$ nm, and $\beta = 67.58°$. [27] The calculated density of the α form 1.23 g/cm^3 is in good agreement with the experimental density 1.23 g/cm^3 [28]. On the other hand, the γ form crystallizes into the monoclinic unit cell with the dimensions $a = 0.933$ nm, $b = 1.688$ nm, $c = 0.478$ nm, and $\beta = 121°$ [28]. Although the γ form of nylon-6 was believed to be monoclinic, the chain packing resembles that of a hexagonal structure (pseudohexagonal) where the unit cell dimensions $a = 0.472$ nm, $b = c = 1.688$ nm, and $\gamma = 120°$ calculated based on the monoclinic parameters [31]. Other than these two (α and γ) forms, nylon-6 also exhibits α′ and β forms [32–36]. The α′ is the high-temperature form [32, 33] of α, and β is the disordered α form [34–36].

In *m*,*n*-nylons, the stability of the polymorph depends upon chain conformation, chain directionality, hydrogen bonds, and chain-folding orientation. The stereochemistry of a particular nylon determines the alignment of the hydrogen bonds between adjacent chains, and the hydrogen-bonded sheets do not always stack in the same manner. In most cases the sheets stack with a triclinic unit cell (progressive mode, e.g., nylon-6,6; nylon-6,10; and nylon-6,12). However, in the case of nylon-4,6, the sheets stack with a monoclinic unit cell (alternating up and down mode) [38]. In even–even nylons, the α form is the stable structure at room temperature [26]. In the case of nylon-6,6, the sheets are characterized by a parallel alignment of the adjacent extended molecules, which are spaced with a perpendicular chain-to-chain distance of 0.42 nm and which are successively displaced in chain direction by a distance corresponding to one chain atom. The stacking of the hydrogen-bonded sheets entails a perpendicular sheet-to-sheet distance of 0.36 nm and a displacement of each successive sheet of about 0.5 nm in the chain direction. Many even–even nylons exhibit the pseudohexagonal phase at elevated temperatures and show a single peak with a *d*-spacing of 0.42 nm, but it is different from the γ phase [39–43]. Because of the similarity in the *d*-spacing, the names pseudohexagonal form and γ form are used interchangeably. Most of the even–odd, odd–even, and odd–odd nylons crystallize primarily in the γ form [44, 45]. In this crystalline form, linear hydrogen bonds between adjacent chains cannot be established when an extended conformation is considered and the NH and CO groups are permitted to tilt to satisfy the hydrogen bond requirements [44, 45].

4 Crystallization and Polymorphic Transitions in Nylon/Layered Silicate Nanocomposites

Crystallization of nylons has been thoroughly studied over the past 50 years and understood the complicated nature and mechanisms of crystallization in various nylons [37–44]. Addition of layered silicates into nylons dramatically influenced the crystallization behavior and hence the mechanical and various other properties of the polymer/layered silicate nanocomposites [50–70]. The interaction of the layered silicate with the polymers, the volume percent of the interface, and surface properties of the layered silicate are found to alter the crystallization kinetics, degree of crystallinity, and the crystalline morphology of the polymer matrix. Apart from the crystallization, the presence of layered silicates is also found to modify the crystal structure of nylons. In this section, we mainly focus on the crystallization and polymorphism behavior of various nylon/layered silicate nanocomposites.

4.1 Even Nylons

Among various even nylon/layered silicate nanocomposites, nylon-6-based nanocomposites have attracted great deal of interest from the researchers because of their commercial success. In this section, we will restrict our discussion to nylon-6-based nanocomposites. These nanocomposites are prepared by either in situ polymerization or melt-blending technique and are probably the most extensively studied nanocomposite system [8, 21, 22, 47–53]. The crystallization kinetics of nylon-6 is proved to be strongly affected by the presence of layered silicates. Several authors studied the crystallization kinetics of nylon-6/layered silicate nanocomposites using differential scanning calorimetry, simultaneous small- and wide-angle X-ray scattering, light scattering, and Fourier transform infrared spectroscopy [54–58]. Maiti et al. [59] reported the enhancement of crystallization rate of nylon-6 in presence of layered silicates using light scattering experiments. Fornes and Paul [58] have studied the effect of clay concentration and degree of exfoliation on the crystallization kinetics of the melt-processed nylon-6/layered silicate nanocomposites. They have shown that crystallization kinetics can be significantly increased at low level of filler loadings, while higher loadings retard the overall crystallization process. Moreover, it has been shown that higher levels of exfoliation resulted in the higher crystallization rates. Lincoln et al. [56] have shown that the incorporation of layered silicates in nylon-6 significantly enhances the isothermal crystallization rates and lowers the crystallization half time compared to that of pristine nylon-6. They also found that the strong interaction between the polymer and layered silicates alters the crystallization process, creating much weaker temperature dependence for the crystallization kinetics. Liu and Wu [54] studied the non-isothermal crystallization of nylon-6 nanocomposites and reported that layered silicates enhance the crystallization during the cooling down process from the melt. In all the above studies, the increase in crystallization kinetics has been observed due to the so-called nucleation effect of the layered silicates.

In the case of extruded nylon-6/layered silicate nanocomposites besides the nucleating effect, the silicate layers are also found to retard the crystallization and crystal growth process in nylon-6. At higher concentrations, the silicate layers behave as heterogeneities that retard the nylon-6 crystal growth. This phenomenon is hardly observed in nylon-6 nanocomposites prepared by in situ polymerization methods as discussed above. Lincoln et al. [60] have shown that the presence of well-dispersed silicate layers in nanocomposites reduced the crystallization rate of nylon-6 matrix due to the disruption of lamella growth. Homminga et al. [61] systematically investigated the influence of type and concentration of layered silicates on the crystallization behavior of nylon-6/layered silicate nanocomposites and found that silicate layers have a strong retarding effect on nylon-6 crystal growth. They suggest that the silicate layers hinder the diffusion of polymer chains to the crystal growth front and impurity migration away from the growth front takes place. They have shown that such retarding effect is absent with poorly exfoliated or intercalated nanomorphology. Layered silicates not only change the crystallization rate and crystallization kinetics but also change the crystal structure of nylon-6, which will be discussed in detail in the following section.

4.1.1 Polymorphism in Nylon-6/Layered Silicate Nanocomposites

Nylon-6 exhibits two crystalline modifications, namely, the α form and the γ form [27]. The α form has monoclinic structure, and the hydrogen bonds are formed between antiparallel chains [27]. The γ form also has a monoclinic structure, but the twisted chains allow hydrogen bonds to be formed between parallel chains [28]. The α form is most commonly observed at room temperature and can be transformed into the γ form by treating in aqueous potassium iodide-iodine solution [28, 62–64]. High-speed spinning also yields the γ form. In nylon-6 fibers, both forms can coexist, and the content of the phases depends on the spinning and drawing conditions [30]. The γ form is stable and may be converted into the α form by melting and recrystallization [32]. The characteristic peaks of the α form of nylon-6 at room temperature are at $2\theta = 21°$ and $24°$ and are indexed as 200 and 002/202 reflections, respectively [27, 29]. In the γ form, these peaks are indexed as 001 and $200/\overline{2}01$ and appear at 22° and 23°, respectively. Also, the $200/\overline{2}01$

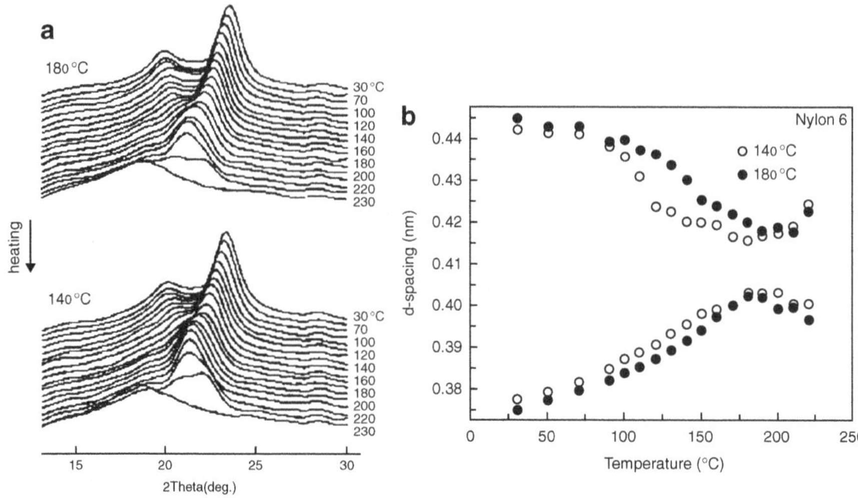

Fig. 12.5 The behavior of (**a**) X-ray patterns and (**b**) *d*-spacing on heating nylon-6, which has been crystallized at 140 °C and 180 °C (Reproduced from [32] with permission)

reflections appear as a shoulder to the 001 reflection [29, 65]. Another observation is that the 020 reflection is weak in the α form, while as it is intense in the γ form. It was reported that rapid crystallization or crystallization at low crystallization temperatures (melt) favors the γ form and slow crystallization or crystallization at high crystallization temperatures (melt) favors the α form [29].

Murthy et al. first reported the presence of α to α' transition in nylon-6 on heating the α form using variable temperature X-ray diffraction (XRD) and NMR measurements [33]. Ramesh and Gowd studied the crystallization of nylon-6 from the melt using variable temperature X-ray diffraction [32]. Nylon-6 crystallized into the high-temperature α' form if crystallized in a narrow temperature range between 200 °C and melting temperature (T_m). On cooling from the crystallization temperature (T_c) to room temperature, the structure transformed into the low-temperature α form at ~180 °C. However, samples crystallized from the melt at crystallization temperatures 140 °C and 180 °C showed the α form at room temperature, but on heating the α form first transformed into a pseudohexagonal form, and before melting the pseudohexagonal form further transformed into the α' form as shown in Fig. 12.5.

Addition of layered silicates into nylon-6 influences the polymorphic behavior. Several authors have reported that the addition of silicate layers into nylon-6 induced the formation of the γ form in the nanocomposites [66]. Mathias et al. investigated the nylon-6/layered silicate nanocomposites using ^{15}N nuclear magnetic resonance and found that the surface of the silicate layers kinetically favored the formation of the γ form. They speculated that the amine end groups which are tightly bound to the clay surface stabilize the γ form by default [67]. Miltner et al. reported that the interaction between the nylon-6 and layered

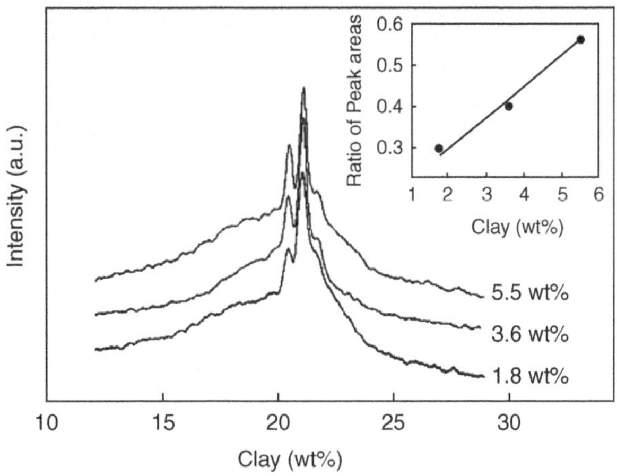

Fig. 12.6 Powder X-ray diffraction patterns of various nylon-6 nanocomposites crystallized at 210 °C. *Inset* shows the ratio of 200 reflections of the metastable and the γ forms at 210 °C (Reproduced from [70] with permission)

silicate reduces the mobility of polymer chains at the interface of the layered silicate which favored the formation of the γ form [68]. Wu et al. showed that the polymorphic behavior in nylon-6/layered silicate nanocomposites is dependent on the layered silicate content and cooling rate from the melt [69]. Higher cooling rate promotes the crystallization of the γ form. They also reported the higher amount of the γ form is obtained when the samples are annealed at higher temperature close to the melting temperature. Fornes et al. used wide-angle X-ray diffraction and DSC to characterize the morphology of the injection molded nanocomposites [58]. They found that the γ form predominates at the skin, while the core contains both forms, which is attributed to the different cooling rates of the skin and core regions. They also reported that the degree of crystallinity of the skin region is higher than that of the core region in nanocomposites, which is opposite in the case of pure nylon-6 [58]. Nair et al. investigated the crystallization behavior of nylon-6 in the presence of layered silicates using variable temperature XRD and Fourier transform infrared spectroscopy [70]. They studied the structure development during isothermal crystallization of nylon-6 layered silicate nanocomposites at 210 °C and subsequent cooling to room temperature. They found the existence of the metastable phase along with the γ form in the nanocomposites crystallized at 210 °C as shown in Fig. 12.6. It was also found that the amount of the metastable phase depends on the amount of layered silicates present in the nanocomposite (see the inset of Fig. 12.6) and also crystallization temperature.

The γ form obtained by the addition of layered silicates helped in improving the mechanical properties and heat distortion temperature of nylon-6/layered silicate nanocomposites. Maiti et al. measured the solid-state storage modulus (G′) of the nylon-6 and their nanocomposites, which are crystallized at different temperatures from the melt (170 °C and 210 °C) [59]. As seen in Fig. 12.7, G′ of the nanocomposites is always higher than that of the pristine polymer, but the percentage of the increment in the modulus with increasing crystallization temperature is higher for pristine polymer than the nanocomposites.

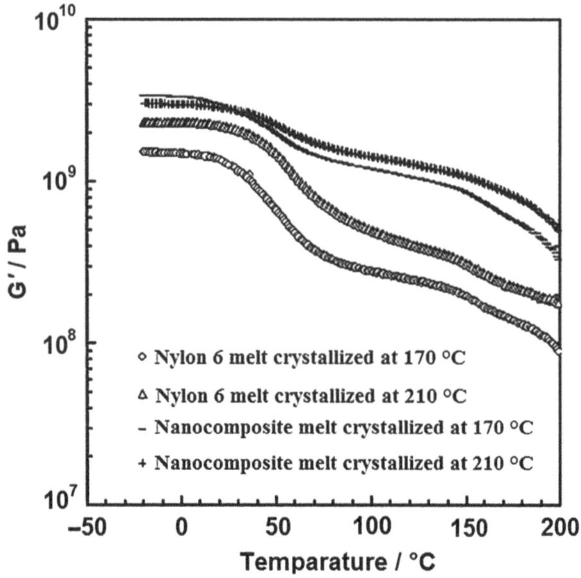

Fig. 12.7 Storage modulus (G′) of nylon-6 and nylon-6 nanocomposite with 3.7 % of layered silicates crystallized at 170 °C and 210 °C as a function of temperature [59]

4.2 Odd Nylons

Among the odd nylons, nylon-11 received considerable interest in recent years due to its bio-based character. Besides nylon-11 is an important commercial polymer and known for its excellent piezoelectric behavior [71]. Nylon-11 exhibits polymorphic behavior, which is greatly dependent on the thermal history and closely related to the piezoelectric and ferroelectric responses [71, 72]. It exhibits at least five crystalline forms, α, α′, δ, δ′, and γ, which are mainly dependent on the thermal history and processing conditions [30, 73–79]. The α form is obtained by annealing the amorphous polymer or solution casting from *m*-cresol, and the α′ form is obtained from melt crystallization [75, 80–82]. These two forms have a triclinic structure. Other three forms crystallize into hexagonal or pseudohexagonal forms with different crystalline lattices [76–78]. The δ form obtained above the room temperature by heating the α form, smectic (δ′) form obtained by melt quenching, and the γ form obtained by solution casting from triflouroacetic acid [30, 73, 74, 76, 78, 83, 84]. Zhang et al. have studied the crystalline transitions of nylon-11 at different drawing ratios [83]. The crystal structure of the α form obtained by casting from *m*-cresol [30] and lamellar crystals grown from glycerine [74] are triclinic; the amide groups lie in a plane tilted to the chain axis. A monoclinic cell is also suggested for nylon-11 grown from the solution in water containing 5 % formic acid at 160 °C [75]. Ramesh and co-workers [85] studied the crystalline transitions in nylon-11 using variable temperature wide-angle X-ray diffraction and variable temperature FTIR. They showed that the α form obtained by precipitating nylon-11 in 1,4-butanediol started transforming to the pseudohexagonal form on heating. They also found that the melt-crystallized sample at 175 °C recrystallized into the

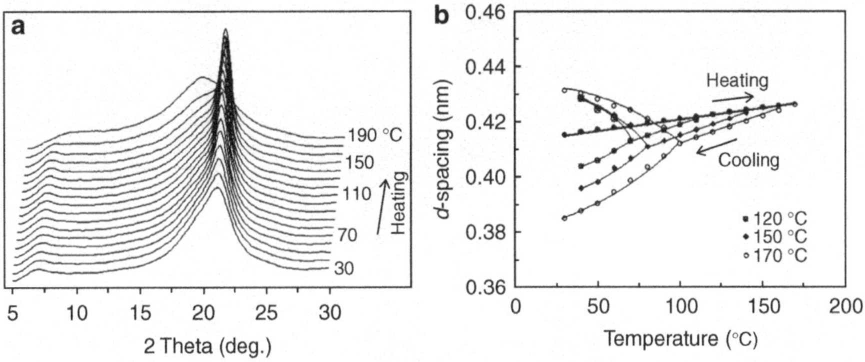

Fig. 12.8 (**a**) Temperature dependence of the powder X-ray diffraction patterns of the quenched nylon-11 from room temperature to melt (**b**) temperature dependence of the *d*-spacings of the melt-quenched sample on heating to 170 °C and on cooling from the crystallization temperatures 170 °C, 150 °C, and 120 °C to room temperature (Reproduced from [85] with permission)

pseudohexagonal form, and on cooling from the crystallization temperature to room temperature, it transformed into the α′ form at about 100 °C. In another case, the melt-quenched sample showed the pseudohexagonal form, which did not change during heating (Fig. 12.8a), but on cooling, it transformed into the α′ form. As shown in Fig. 12.8b, the transition temperature from the pseudohexagonal form to α′ form during cooling is dependent on the crystallization temperature. In this way, complicated phase transitions are reported in the case of nylon-11.

With the aim of improving the performance of nylon-11, many researchers prepared nylon-11-based nanocomposites using layered silicates as nanofillers [86–89]. Nylon-11 nanocomposites can be obtained by either in situ methods or by melt intercalation methods. Zhang et al. prepared exfoliated and intercalated nylon-11/layered silicate nanocomposites by means of in situ polymerization [87]. They showed that thermal properties and crystallization behavior were enhanced in the case of nanocomposites. Liu et al. prepared wide range of nylon-11/layered silicate nanocomposites containing 0, 1, 2, 4, and 8 wt%, and organically modified layered silicate was prepared via melt-compounding method at 220 °C [90]. They showed that the exfoliated nanocomposites are formed at low clay loadings, and a mixture of exfoliated and intercalated nanocomposites is obtained at higher clay contents. The storage modulus increase by 100 % when the clay content is up to 8 wt% in comparison to neat nylon-11, while the glass transition temperature (T_g) steadily decreases as a function of clay loading, probably due to the plasticization effect from the presence of organic modifiers within the layered silicates. Fornes et al. studied the effect of nylon structure on the morphology and mechanical properties of nanocomposites by comparing the nanocomposites of nylon-6, nylon-11, and nylon-12 prepared by melt-processing method at 240 °C [88]. They reported that the polarity of the repeat structure plays an important role in determining the level of interaction of the polymer with the organically modified layered silicates and showed that the extent of layered silicate exfoliation is better

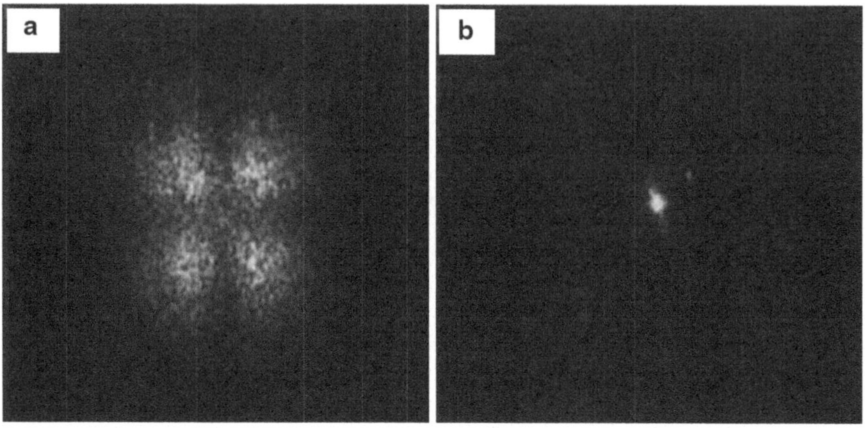

Fig. 12.9 Small-angle light scattering patterns of (**a**) nylon-11 and (**b**) nylon-11 nanocomposites with 1 wt% of layered silicate [87]

in nylon-6 than the more aliphatic containing nylons such as nylon-11 and nylon-12.

Fu and co-workers investigated the crystallization kinetics of nylon-11/layered silicate nanocomposites using polarized optical microscope, small-angle light scattering (SALS) and differential scanning calorimetry [87]. They reported that the addition of the layered silicates to the nylon-11 matrix significantly reduces the size of the spherulites, indeed no spherulites are observed. Figure 12.9 shows the SALS patterns of nylon-11 and its nanocomposite with 1 wt% of layered silicate. As seen in the figure, nylon-11 showed four-leaf clover pattern characteristic of spherulites, and no such pattern is observed in the nanocomposite indicating an absence of spherulitic structure or the crystallites formed are too small to obtain any regular scattering patterns. The decrease in Avrami exponent with the addition of layered silicates in isothermal crystallization indicates that layered silicates acts as a nucleating agent and alters the nucleation mechanism in the nanocomposites.

Zhang et al. investigated the polymorphism in nylon-11/layered silicate nanocomposites by WAXS and variable temperature FTIR [89]. They showed that the addition of layered silicates retards the formation of hydrogen-bonding sheets by forcing the amide groups of nylon out of the hydrogen-bonding sheet, and it resulted in the stabilization of the γ form.

4.3 Even–Even Nylons

Nylon-6,6 is the commercially most successful polymer in this category. For the first time, a reversible crystal-to-crystal transition, from the triclinic to the pseudohexagonal structure, was reported in nylon-6,6 on heating, known as Brill transition [33, 39, 40, 42, 43, 91–95]. Figure 12.10 shows the X-ray diffraction patterns of nylon-6,6 on heating from room temperature to melt and the variation in

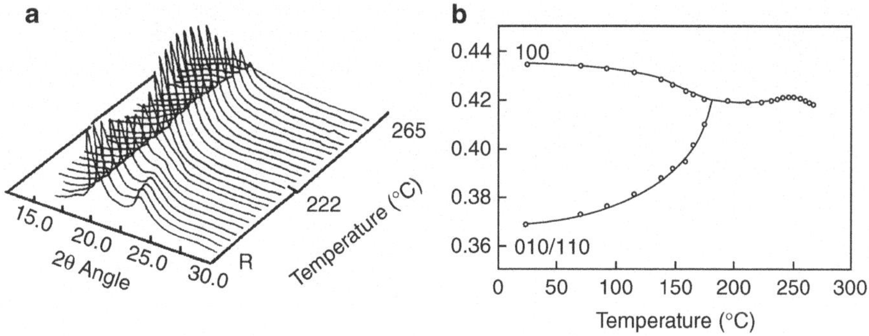

Fig. 12.10 (**a**) Three-dimensional view of the X-ray diffraction patterns of nylon-66 on heating from room temperature to melt and (**b**) variation in the *d*-spacings calculated from (**a**) [41]

the d-spacings with temperature. On heating, the two strong reflections with the spacings of 0.44 and 0.37 nm (100 and 010/110) converge and merge typically at a spacing of 0.42 nm at Brill transition [33, 39, 41, 94–96]. At this transition temperature, the projected interchain distance within the hydrogen-bonded sheets equals to the intersheet spacing. Ramesh et al. investigated the crystallization of nylon-6,6 from the melt and found that the Brill transition temperature depends strongly on the crystallization conditions. The higher the melt crystallization temperature, the higher the Brill transition temperature for nylon-6,6 [41]. Over the years, the Brill transition has been observed in other even–even nylons [96–100].

A little attention has been devoted towards the understanding of the influence of layered silicates on the crystallization and phase transition behavior of nylon-6,6 nanocomposites. These nanocomposites are prepared by either in situ polymerization or melt-blending technique [101–104]. Zhang et al. [101] prepared fully exfoliated nylon-6,6 nanocomposites by melt-compounding method using organically modified layered silicates. In this work, layered silicates are modified in two steps, first with chlorosilane derivative to decrease its cation-exchange capacity and then reacted with dioctodecyl-dimethyl ammonium chloride. The influence of modified layered silicates on the crystallization behavior and phase transition behavior has been studied using DSC, polarized optical microscope, and variable temperature X-ray diffraction. It is found that the melt crystallization temperature increased about 5 °C compared to the pristine polymer because of the heterogeneous nucleation induced by the layered silicates. A number of typical spherulites are observed in the case of nylon-6,6, and upon incorporating the layered silicates into the nylon-6,6 matrix, however, the spherulite texture of the matrix is greatly blurred. The spherulite size decreases significantly, and the well-defined spherulites are absent at higher loadings due to heterogeneous nucleation density from layered silicates. The powder X-ray diffraction pattern of nylon-6,6/layered silicate nanocomposites show two distinct crystalline peaks of nylon-6,6 at 2θ = 20.2 and 23.5°, which are typical for the α-form (triclinic) similar to the pristine

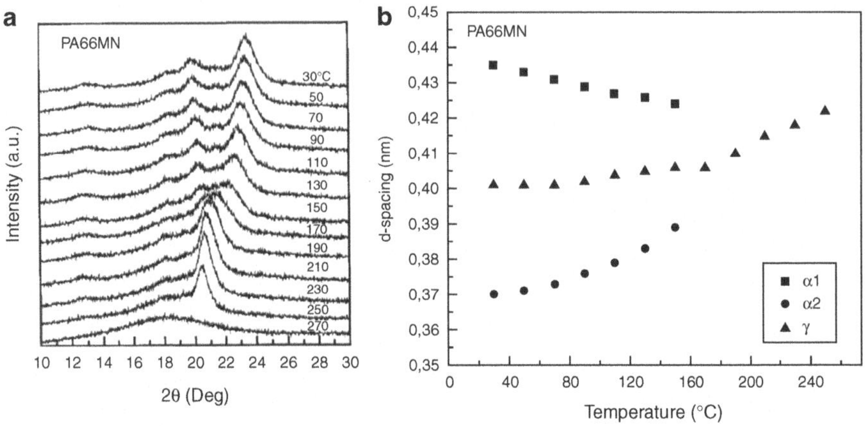

Fig. 12.11 (**a**) Powder X-ray diffraction patterns of nylon-66 nanocomposites (with 5 wt% of layered silicate) which are annealed at different temperatures and (**b**) changes in the *d*-spacings as a function of temperature [103]

polymer. Unlike nylon-6, the addition of layered silicate does not influence the crystal structure of nylon-6,6 in their nanocomposites. Variable temperature X-ray diffraction is performed to understand the influence of layered silicates on the Brill transition, and it is found that addition of layered silicates increased the transition temperature by 10 °C. The higher Brill transition temperature of nanocomposites was attributed to the strong interaction between the nylon-66 and organically modified layered silicates.

Liu et al. [104] studied the polymorphism in nylon-6,6/layered silicate nanocomposites using a new kind of organophilic clay, which was obtained through co-intercalation of epoxy resin and alkyl ammonium into Na-montmorillonite. Melt-compounding method is used to prepare nearly exfoliated nylon-6,6/layered silicate nanocomposites due to the strong interactions between epoxy and nylon-6,6. In this case, they reported a small fraction of pseudohexagonal form along with the α form (triclinic) at room temperature in both slowly cooled samples and quenched samples unlike the Zhang et al. [101] work where they reported only the α form in nylon-6,6 nanocomposites. They also observed that fractions of the pseudohexagonal form increase in nanocomposites with increasing silicate loadings. Phase transitions in these nanocomposites before melting is investigated by variable temperature XRD.

Figure 12.11 shows a series of powder patterns of nylon-6,6/layered silicate nanocomposites (with 5 wt% of layered silicate) at various temperatures and the corresponding changes in the *d*-spacings as a function of temperature. At room temperature, in addition to the characteristic reflections of the α form (triclinic), a weak reflection appears at around $2\theta = 21.97°$ corresponding to the pseudohexagonal form. On heating, a clear phase transition occurred at around 170 °C, where two reflections corresponding to the α form disappeared completely and the reflection corresponding to the pseudohexagonal form become dominant.

The so-called Brill transition temperature in nanocomposites was 20 °C lower than that in nylon-6,6, and it was attributed to the pseudohexagonal form favorable environment caused by the organically modified layered silicates [103]. This observation is in contrast with the reported work of nylon-6,6 nanocomposites by Zhang et al., where they reported the higher Brill transition temperature in nanocomposites compared to the pristine nylon-6,6. This anomalous behavior may be due to the nature of organic modifier used in the preparation of the nanocomposites [101].

As discussed above, a large number of studies have been reported on crystallization and polymorphic behavior of nylon-6 nanocomposites and few studies on even–even and odd nylon nanocomposites. However, it should be pointed out that a limited work or no work has been carried out on the crystallization and polymorphic behavior of other nylon nanocomposite systems which include even–odd, odd–even, and odd–odd nylons.

5 Summary and Outlook

In this chapter, we have attempted to summarize the studies performed on crystallization and polymorphic behavior of nylon/layered silicate nanocomposites. Based on the results of these investigations, it can be concluded that the incorporation of nano-sized layered silicates into nylons significantly influence the crystallization behavior, polymorphism, and hence on the properties of the nanocomposites. In common, it is observed that the presence of silicate layers enhances the crystallization rate, changes the spherulitic morphology, and enhances the nucleation density on the polymer matrix. It is also observed that dispersion of the nano-sized layered silicates and their surface modifications have significant influence on the crystallization and crystal structure of nylons. Crystallization and polymorphic behavior of other nylon nanocomposite systems which include even–odd, odd–even, and odd–odd nylons has not yet been explored fully and offers scope for further work.

Acknowledgments EBG is grateful to the Department of Science and Technology (Government of India) for the award of a Ramanujan Fellowship.

References

1. Ray SS, Okamoto M (2003) Prog Polym Sci 28:1539
2. Usuki A, Hasegawa N, Kato M (2005) Adv Polym Sci 179:135
3. Alexandre M, Dubois P (2000) Mater Sci Eng 28:1
4. Manias E, Touny A, Wu L, Strawhecker K, Lu B, Chung TC (2001) Chem Mater 13:3516
5. Okada A, Usuki A (2006) Macromol Mater Eng 291:1449
6. Vaia RA, Ishii H, Giannelis EP (1993) Chem Mater 5:1694
7. Kojima Y, Usuki A, Kawasumi M, Okada A, Kurauchi T, Kamigaito O (1993) J Polym Sci A Polym Chem 31:983

8. Usuki A, Kojima Y, Kawasumi M, Okada A, Fukushima Y, Kurauchi T, Kamigaito O (1993) J Mater Res 8:1179
9. Kojima Y, Usuki A, Kawasumi M, Okada A, Fukushima Y, Kurauchi T, Kamigaito O (1993) J Mater Res 8:1185
10. Dasgupta S, Hammond WB, Goddard WA III (1996) J Am Chem Soc 118:12291
11. Giannelis EP (1996) Adv Mater 8:29
12. Reichert P, Kressler J, Thomann R, Mulhaupt R, Stoppelmann G (1998) Acta Polym 49:116
13. Wu Z, Zhou C, Qi R, Zhang H (2002) J Appl Polym Sci 83:2403
14. Wu J, Lerner MM (1993) Chem Mater 5:835
15. Ogata N, Jimenez G, Kawai H, Ogihara T (1997) J Polym Sci Polym Phys Ed 35:389
16. Ogata N, Kawakage S, Ogihara T (1997) J Appl Polym Sci 66:573
17. Vaia RA, Jandt KD, Kramer EJ, Giannelis EP (1995) Macromolecules 28:8080
18. Vaia RA, Giannelis EP (1997) Macromolecules 30:7990
19. Vaia RA, Giannelis EP (1997) Macromolecules 30:8000
20. Paul DR, Robeson LM (2008) Polymer 49:3187
21. Cho JW, Paul DR (2001) Polymer 42:1083
22. Liu L, Qi Z, Zhu X (1999) J Appl Polym Sci 71:1133
23. Botta A, Candia FD, Palumbo R (1985) J Appl Polym Sci 30:1669
24. Cannon CG (1960) Spectrochim Acta 16:302
25. Aleman C, Casanovas J (2004) Colloid Polym Sci 282:535
26. Bunn CW, Garner EV (1947) Proc R Soc (Lond) A 189:3279
27. Holmes DR, Bunn CW, Smith DJ (1955) J Polym Sci 17:159
28. Arimoto H, Ishibashi M, Hirai M, Chatani Y (1965) J Polym Sci A 3:317
29. Aharoni SM (1997) *n*-Nylons, their synthesis, structure, and properties. Wiley, New York
30. Kinoshita Y (1959) Makromol Chem 33:1
31. Kohen MI (ed) (1995) Nylon plastics hand book. Hanser, New York
32. Ramesh C, Gowd EB (2001) Macromolecules 34:3308
33. Murthy NS, Curran SA, Aharoni SM, Minor H (1991) Macromolecules 24:3215
34. Auriemma F, Petraccone V, Parravicini L, Corradini P (1997) Macromolecules 30:7554
35. Leon S, Aleman C, Munoz-Guerra S (2000) Macromolecules 33:5754
36. Avramova N, Fakirov S (1984) Polym Commun 25:27
37. Quarti C, Milani A, Civalleri B, Orlando R, Castiglioni C (2012) J Phys Chem B 116:8299
38. Atkins EDT, Hill M, Hong SK, Keller A, Organ S (1992) Macromolecules 25:917
39. Brill R (1942) J Prakt Chem 161:49
40. Hirschinger J, Miura K, Gardner KH, English AD (1990) Macromolecules 23:2153
41. Ramesh C, Keller A, Eltink SJEA (1994) Polymer 35:2483
42. Vasanthan N, Murthy NS, Bray RG (1998) Macromolecules 31:8433
43. Starkweather HW, Jones JA (1981) J Polym Sci Polym Phys Ed 19:467
44. Navarro E, Aleman C, Subirana JA, Puiggali J (1996) Macromolecules 29:5406
45. Bunn CW, Garner EV (1947) Proc R Soc Lond Ser A 189:39
46. Murthy NS (2006) J Polym Sci, Part B: Polym Phys 44:1763
47. Fujiwara S, Sakamoto T (1976) Japan Patent JP-A-51-109998
48. Okada A, Fukushima Y, Kawasumi M, Inagaki S, Usuki A, Sugiyami S, Kurauchi T, Kamigaito O (1988) US Patent 4739007
49. Christiani BR, Maxfield M (1998) US Patent 5747560
50. Fornes TD, Yoon PJ, Hunter DL, Keskkula H, Paul DR (2002) Polymer 43:5915
51. Lincoln DM, Vaia RA, Wang Z-G, Hsiao BS (2001) Polymer 42:1621
52. Dennis HR, Hunter DL, Chang D, Kim S, White JL, Cho JW, Paul DR (2001) Polymer 42:9513
53. Krishnamoorti R, Giannelis EP (1997) Macromolecules 30:4097
54. Liu X, Wu Q (2002) Eur Polym J 38:1383
55. Devaux E, Bourbigot S, Achari AE (2002) J Appl Polym Sci 86:2416
56. Lincoln DM, Vaia RA, Krishnamurti R (2004) Macromolecules 37:4554

57. Ma CCM, Kuo CT, Kuan HC, Chaing CL (2003) J Appl Polym Sci 88:1686
58. Fornes TD, Paul DR (2003) Polymer 44:3945
59. Maiti P, Okamoto M (2003) Macromol Mater Eng 288:440
60. Lincoln DM, Vaia RA, Wang ZG, Hsiao BS, Krishnamurti R (2001) Polymer 42:9975
61. Homminga DS, Goderis B, Mathot VBF, Groeninckx G (2006) Polymer 47:1630
62. Stepaniak RF, Garton A, Carlsson DJ, Wiles DMJ (1979) Polym Sci Polym Phys Ed 17:987
63. Miyasaka K, Makishima K (1967) J Polym Sci A 5:3017
64. Murthy NS (1991) Polym Commun 32:301
65. Vogelsong DC (1963) J Polym Sci A 1:1055
66. Kojima Y, Matsuoka T, Takahashi H, Kurauchi T (1994) J Appl Polym Sci 51:683
67. Mathias LJ, Davis RD, Jarrett WL (1999) Macromolecules 32:7958
68. Miltner HE, Van Assche G, Pozsgay A, Pukanszky B, Van Mele B (2006) Polymer 47:826
69. Wu TM, Liao CS (2000) Macromol Chem Phys 201:2820
70. Nair SS, Ramesh C (2005) Macromolecules 38:454
71. Takase Y, Lee JW, Scheinbeim JI, Newman BA (1991) Macromolecules 24:6644
72. Scheinbeim JI, Lee JW, Newman BA (1992) Macromolecules 25:3729
73. Slichter WP (1959) J Polym Sci 36:259
74. Kawaguchi A, Ikawa T, Fujiwara Y, Tabuchi M, Konobe K (1981) J Macromol Sci, Phys B 21:1
75. Kim KG, Newman BA, Scheinbeim JI (1985) J Polym Sci, Polym Phys Ed 23:2477
76. Newman BA, Sham TP, Pae KD (1977) J Appl Phys 48:4092
77. Schmidt GF, Stuart HAZ (1958) Naturforsch 13:222
78. Sasaki T (1965) J Polym Sci, Part B: Polym Lett 3:557
79. Chen PK, Newman BA, Scheinbeim JI, Pae KD (1985) J Mater Sci 20:1753
80. Rhee S, White JL (2002) J Polym Sci, Part B: Polym Phys 40:1189
81. Rhee S, White JL (2002) J Polym Sci, Part B: Polym Phys 40:2624
82. Rhee S, White JL (2002) Polym Eng Sci 42:134
83. Zhang Q, Mo Z, Zhang H, Liu S, Cheng SZD (2001) Polymer 42:5543
84. Yu HH (1998) Mater Chem Phys 56:289
85. Nair SS, Ramesh C, Tashiro K (2006) Macromolecules 39:2841
86. Yu M, Zhang Q, Fu Q (2004) Chinese J Polym Sci 22:43
87. Zhang Q, Yu M, Fu Q (2004) Polym Int 53:1941
88. Fornes TD, Paul D (2004) Macromolecules 37:7698
89. Zhang G, Li Y, Yan D (2004) J Polym Sci, Part B: Polym Phys 42
90. Liu T, Lim KP, Tjiu WC, Pramoda KP, Chen ZK (2003) Polymer 44:3529
91. Biangardi HJ (1990) J Macromol Sci, Phys B 29:139
92. Radusch MJ, Stolp M, Androsch A (1994) Polymer 35:3568
93. Xenopoulos A, Wunderlich B (1991) Colloid Polym Sci 269:375
94. Itoh T (1976) Jpn J Appl Phys 15:2295
95. Colclough ML, Baker RJ (1978) J Mater Sci 13:2531
96. Ramesh C (1999) Macromolecules 32:3721
97. Tashiro K, Yoshioka Y (2004) Polymer 45:6349
98. Biangardi HJ (1990) J Macromol Sci-Phys B 29:139
99. Jones NA, Atkins EDT, Hill MJ, Cooper SJ, Franco L (1997) Polymer 38:2689
100. Jones NA, Cooper SJ, Atkins EDT, Hill MJ, Franco L (1997) J Polym Sci Ed 35:675
101. Zhang QX, Yu ZZ, Yang M, Ma J, Mai YW (2003) J Polym Sci, Part B: Polym Phys 41:2861
102. Liu X, Wu Q (2002) Macromol Mater Eng 287:180
103. Liu X, Wu Q, Zhang Q, Mo Z (2003) J Polym Sci, Part B: Polym Phys 41:63
104. Liu X, Wu Q, Berglund LA (2002) Polymer 43:4967

Preparation and Characterization of Poly(trimethylene terephthalate) Nanocomposites

13

Jin-Hae Chang

Contents

Abstract

Intercalated nanocomposites comprising poly(trimethylene terephthalate) (PTT) synthesized from 1,3-propanediol (PDO) and terephthalic acid (TPA) incorporated between clay layers were generated by means of in situ interlayer polymerization. The thermal stabilities, mechanical properties, and morphologies of PTT hybrids made with two different organoclays were compared. 1,2-Dimethylhexadecylimidazolium-montmorillonite (IMD-MMT) and dodecyltriphenyl-phosphonium-montmorillonite (C_{12}PPh-MMT) were used as

J.-H. Chang
School of Energy and Integrated Materials Engineering, Kumoh National Institute of Technology, Gumi, Gyeongbuk, South Korea
e-mail: changjinhae@hanmail.net

J.K. Pandey et al. (eds.), *Handbook of Polymernanocomposites. Processing, Performance and Application – Volume A: Layered Silicates*, DOI 10.1007/978-3-642-38649-7_9,

reinforcing organoclay fillers in the fabrication of the PTT hybrid fibers. The variations in the properties of the nanocomposites were discussed with respect to the organoclay content in the polymer matrix and the draw ratio (DR). The thermomechanical properties and morphologies of the PTT hybrid fibers were characterized using differential scanning calorimetry, thermogravimetric analysis, wide-angle X-ray diffraction, electron microscopy, and mechanical tensile properties analysis. Observation of the nanostructures of the hybrid fibers via scanning and transmission electron microscopy demonstrated that the clay layers were well dispersed in the matrix polymer, although some clusters or agglomerated particles were also detected. The addition of only a small amount of organoclay was sufficient for improving the thermal stability and mechanical properties of the PTT hybrid fibers.

Keywords
Fibers • Nanocomposites • Poly(trimethylene terephthalate) • Thermally stable organoclay

1 Introduction

In 1941, the Calico Printers Ass Company [1, 2] successfully fabricated poly (trimethylene terephthalate) (PTT) via polycondensation of 1,3-propanediol (PDO) and terephthalic acid (TPA); nevertheless, PTT could not be obtained on a large scale until the 1990s because of the high cost of PDO. The successful synthesis, using different techniques, of low-cost PDO was reported by many companies such as Shell [24], Dupont [25], and Degussa [15] in the 1990s. These techniques may make it possible to commercialize PTT products, allowing PTT to find its place in the polyester family with counterparts such as poly(ethylene terephthalate) (PET) and poly(butylene terephthalate) (PBT). Ward and Wilding [3, 4] found that PTT exhibited better elastic recovery than PET and PBT. Because of the excellent comprehensive properties of PTT as a fiber material, PTT research has recently become an active pursuit [49, 52, 54–57].

PTT has been widely applied in fibers, injection moldings, and in film production and exhibits good tensile behavior, outstanding elastic recovery, and dyeability. Moreover, PTT fibers have the resilience and softness of nylon fibers, as well as the chemical stability and stain resistance of their PET counterparts, making them ideal candidates for use in applications such as carpets and other textile fiber applications. The low moisture absorption and fiber modulus of PTT endow carpets with a desirable dry and soft feel [4, 5, 29, 51].

Recently, the development of nanotechniques has resulted in blossoming interest in the field of nanocomposites and their special properties. Polymer/clay hybrids are one of the most important classes of synthetic engineering materials [31, 42, 45, 46], given that they can be transformed into new materials possessing the advantages of both organic materials, such as light weight, flexibility, and good moldability, and of inorganic materials, such as high strength, heat stability, and chemical resistance. Materials incorporating polymer/clay hybrids can attain appropriate degrees of stiffness, strength, and gas barrier properties with the introduction of far fewer inorganic additives than are used in conventionally filled polymer composites [13, 16–18]. The resulting materials are therefore lighter in weight than their conventional counterparts and also exhibit two-dimensional stability [20, 22]. A higher degree of delamination results in greater enhancement of these properties in polymer/clay hybrids [7, 11, 12, 23].

Nanoscale composites comprising polymers and clays or organoclays have become the focus of interest, and both intercalated and exfoliated nanocomposites have been prepared using several preparation methods [32, 33, 45]. The three most commonly used methods are polymer solution intercalation, polymer melt intercalation, and monomer intercalation in situ polymerization in which monomers are intercalated into the clay galleries followed by polymerization. The last production method relies on swelling of the organoclay by the monomer, followed by in situ polymerization initiated thermally or by the addition of a suitable compound. The chain growth in the clay galleries accelerates clay exfoliation and nanocomposite formation. This technique of in situ interlayer polymerization is also particularly attractive because of its versatility and compatibility with reactive monomers and is beginning to find commercial application [27, 41, 44, 47, 48].

The thermal stability of the organoclay component is of major importance, as many polymer composites are either melt-blended or intercalated at high temperatures to yield the corresponding nanoscale-sized composites. In the preparation of PTT nanocomposites, elevated temperatures greater than or equal to 265 °C are required for successful in situ intercalation and bulk processing. If the processing temperature is higher than the thermal stability of the organoclay, then decomposition occurs, and the interface between the filler and the matrix polymer is effectively altered. Moreover, the clay layers become hydrophilic, and their ability to affect the overall physical properties of the final nanocomposite may be reduced. For this reason, much research has been directed towards the preparation of organoclays that are thermally stable at high temperatures [34, 35, 37, 40].

The present study evaluates the effects of varying the amount of organoclay and the draw ratios (DRs) on the properties of PTT/clay hybrid fibers. To obtain nanocomposites while avoiding thermal degradation during processing, two thermally stable organoclays were used: 1,2-dimethylhexadecylimidazolium-montmorillonite (IMD-MMT) and dodecyltriphenylphosphonium-MMT (C_{12}PPh-MMT). MMT is a clay mineral consisting of stacked silicate sheets of thickness 1 nm and length ca. 200 nm [47]. MMT possesses a high aspect ratio (ca. 200) and is characterized by a plate-like morphology. MMT also exhibits a high swelling capacity, which is essential for efficient intercalation of the polymer, and is composed of stacked silicate sheets that provide high thermal, tensile, and molecular barrier properties, which may be conferred on the polymeric hybrid materials.

The effects of these organoclays on the thermal stabilities, tensile properties, and morphologies of the resulting hybrid fibers were also examined. In this article, we describe the method used to fabricate PTT nanocomposites, namely, in situ intercalation polymerization. The properties of IMD-MMT/PTT hybrid fibers were compared with those of fibers prepared from C_{12}PPh-MMT/PTT hybrids, using differential scanning calorimetry (DSC), thermogravimetric analysis (TGA), wide-angle X-ray diffraction (XRD), scanning electron microscopy (SEM), transmission electron microscopy (TEM), and tensile-properties testing.

2 Experimental

2.1 Materials

The source clay, Kunipia-F (Na^+-MMT), was obtained from Kunimine and was screened with a 325-mesh sieve to obtain clay with a cationic-exchange capacity of 119 meq/100 g. With the exception of PDO (Shell Co., Ltd., USA), all reagents were purchased from TCI (Tokyo, Japan) and Aldrich Chemical Co. (Seoul, Korea). Commercially available solvents were purified by distillation, and common reagents were used without further purification.

2.2 Thermally Stable Organoclays

Generally, the synthesis of PTT is performed under high-temperature and high-vacuum conditions and in the presence of an effective catalyst. In the case of preparation of PTT nanocomposites, elevated temperatures reaching at least above 265 °C are required for in situ intercalation and bulk processing. The use of processing temperatures higher than the thermal stability of the organoclay results in decomposition, which alters the interface between the filler and the matrix polymer. In real processes with organophilic polymers, the interlayer cations are replaced with alkylammonium cations to enhance the dispersibility. Because the thermal stabilities of these kinds of organoclays containing alkylammonium cations have been problematic, i.e., thermal degradation occurs during processing of PTT above 265 °C, much research has been directed toward the preparation of organoclays that are thermally stable at high temperatures [34, 35, 37, 40].

The chemical structures of some organoclays that are thermally stable at high temperature are presented in Fig. 13.1. The TGA curves shown in Fig. 13.2 indicate that the organoclay samples undergo no weight loss below 300 °C. The TGA curves of the various organoclays are almost identical up to a temperature of approximately 300 °C, above which the initial thermal degradation temperature ($T_D{}^i$) varies with the various chemical structures. The $T_D{}^i$ values of the organoclays (at 2 % weight loss) are listed in Table 13.1. The $T_D{}^i$ values of the various organoclays range between 304 °C and 344 °C, depending on the chemical structures. Similarly, the weight of the residue at 600 °C ($wt_R{}^{600}$) varied with the structure, ranging from 62 % to 79 %, as indicated in Table 13.1.

The XRD patterns of the organoclays and Na^+-MMT shown in Fig. 13.3 provide evidence that both ion exchange and nanocomposite formation are accompanied by increased interlayer distance. In the case of the organoclays, the interlayer spacing

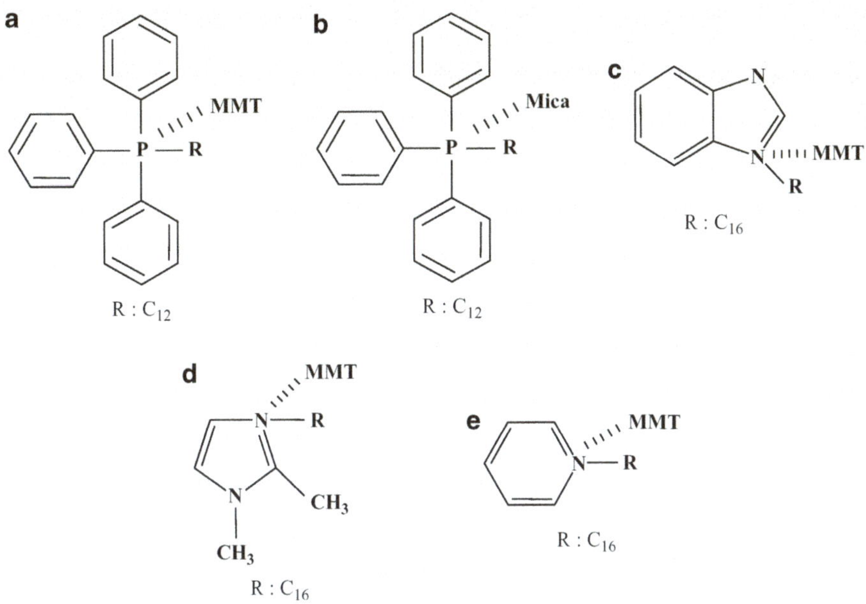

Fig. 13.1 Chemical structures of thermally stable organoclays

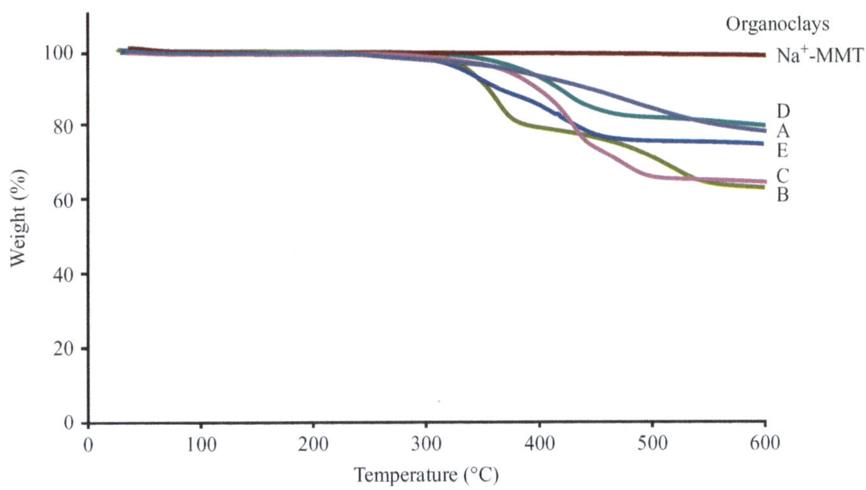

Fig. 13.2 TGA thermograms of thermally stable organoclays. Notations are defined in Fig. 13.1

derived from the characteristic peak was observed to be longer than that of Na^+-MMT ($d = 11.99$ Å). It was thus deduced that the longer lattice spacing corresponds to the presence of the alkyl group (see Table 13.1). The interlayer spacing increased with increasing primary aliphatic-amine length. In general, a larger spacing is

Table 13.1 General properties of thermally stable organoclays for polyester nanocomposite fibers

Organoclay types	$T_D{}^{ia}$ (°C)	$wt_R{}^{600b}$ (%)	2θ (degree)	d in XRD (Å)
A	304	79	2.86	36.08
B	305	62	3.68	27.63
C	326	65	2.72	32.44
D	344	79	4.10	21.55
E	311	75	4.87	18.12
MMT[c]	–	–	8.60	11.99
Mica	–	–	9.23	9.57

[a]Initial weight-loss onset temperature
[b]Weight percent of residue at 600 °C
[c]Montmorillonite

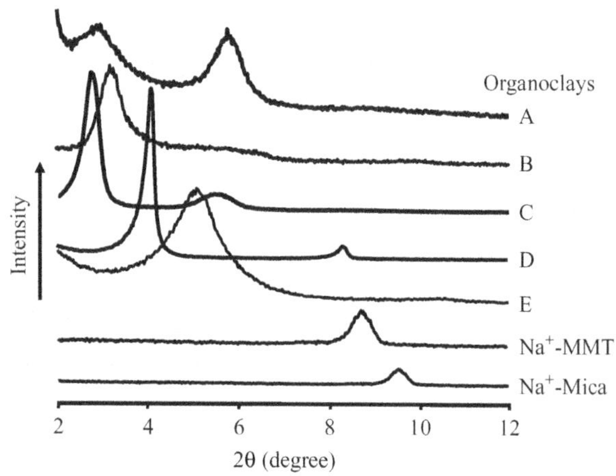

Fig. 13.3 XRD patterns of thermally stable organoclays. Notations are defined in Fig. 13.1

advantageous for polymer intercalation and would also lead to easy dissociation of the clay, resulting in hybrids with better clay dispersion [35].

The two types of organically modified thermally stable clays, C_{12}PPh-MMT and IMD-MMT, used in this study were synthesized using ion-exchange reactions between Na^+-MMT and IMD-Cl^- and between Na^+-MMT and C_{12}PPh-Cl^-, respectively. The two organophilic clays were obtained through a multistep process, as outlined below.

2.3 Preparation of the Organoclay: C_{12}PPh-MMT

The organically modified montmorillonite C_{12}PPh-MMT (type A in Fig. 13.1) used in this study was synthesized using an ion-exchange reaction between Na^+-montmorillonite (Na^+-MMT) and dodecyltriphenylphosphonium chloride (C_{12}PPh-Cl^-) [48]. In the synthesis, 5.0 g (0.02 mol) of bromododecane, 15.8 g (0.06 mol) of triphenylphosphine, and 300 mL of tetrahydrofuran were placed into a 500 mL beaker. This solution was heated at 80 °C for 24 h, after which,

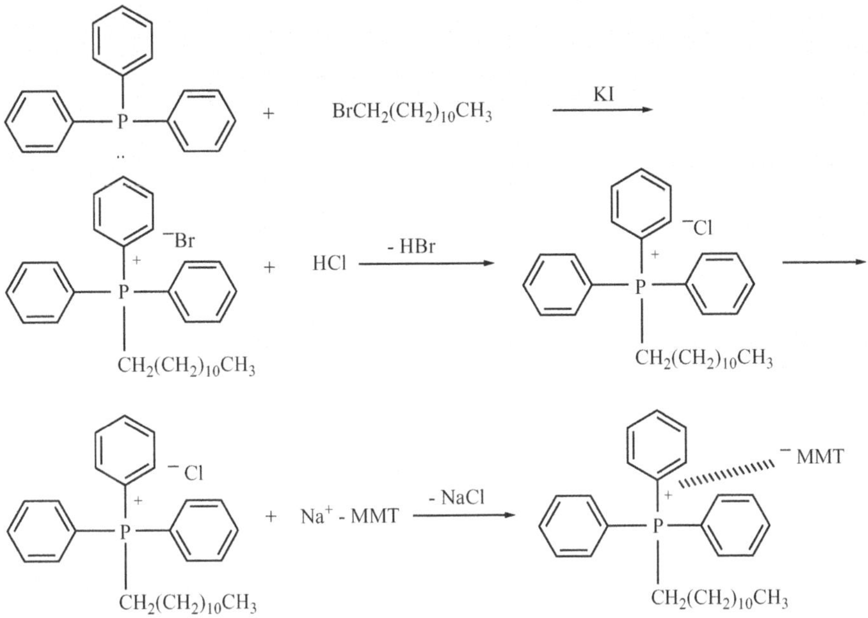

Scheme 13.1 Synthetic route for production of C_{12}PPh-MMT from Na^+-montmorillonite (Na^+-MMT) and dodecyltriphenylphosphonium chloride (C_{12}PPh-Cl^-)

dodecyltriphenylphosphonium bromide, 4.1 g (0.008 mol), was dispersed in a mixture solvent containing 120 mL of deionized water, 30 mL of ethanol, and 3.0 mL of concentrated HCl at 80 °C. A dispersion of 6.0 g of Na^+-MMT in 120 mL of water was added to the C_{12}PPh-Cl^- solution, and the mixture was stirred vigorously for 1 h. The precipitate was isolated by filtration, placed in a 500 mL beaker with 300 mL of water and ethanol, and stirred for 1 h. The product was then filtered and freeze dried. The chemical structure of C_{12}PPh-MMT is shown in Scheme 13.1.

2.4 Preparation of the Organoclay: IMD-MMT

The organically modified montmorillonite IMD-MMT (type D in Fig. 13.1) used in this study was synthesized using an ion-exchange reaction between Na^+-montmorillonite (Na^+-MMT) and 1,2-dimethylhexadecylimidazolium chloride (IMD-Cl^-) [53]. In the synthesis, 9.13 g (0.095 mol) of 1,2-dimethylimidazole, 30.53 g (0.1 mol) of hexadecyl bromide, and 50 mL of acetonitrile were combined in a 500 mL beaker equipped with a reflux condenser. This mixture was heated to 80 °C and refluxed for 10 days. The imidazolium salt was precipitated by addition of a large excess of ethyl acetate. The resulting solid was filtered and washed three times with ethyl acetate to remove the 1,2-dimethylimidazole. The product was dried in a vacuum oven at 80 °C for 24 h.

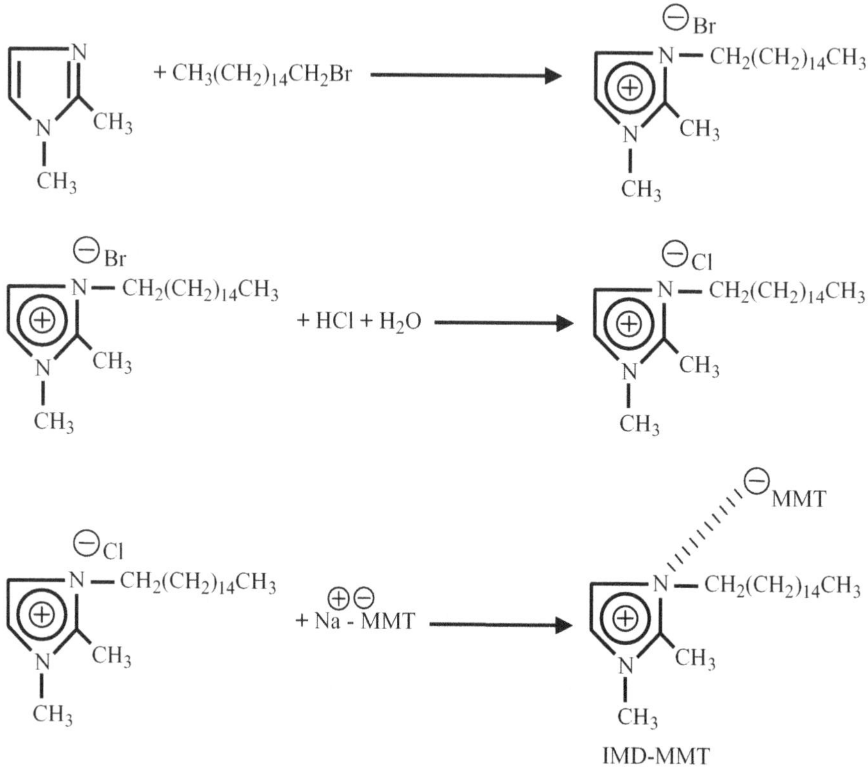

Scheme 13.2 Synthetic route for production of IMD-MMT from Na^+-montmorillonite (Na^+-MMT) and 1,2-dimethylhexadecylimidazolium chloride (IMD^+ Cl^-)

Then, 30.0 g (7.47×10^{-2} mol) of 1,2-dimethylhexadecylimidazolium bromide (IMD^+ ^-Br) was dispersed in a mixture of 500 mL of deionized water and 1.5 mL of concentrated HCl at 80 °C. A dispersion of 46.6 g of Na^+-MMT in 1,500 mL of deionized water was added to the IMD^+Br^- solution, and the mixture was stirred vigorously at 80 °C for 6 h. The precipitate was isolated by filtration and washed with a mixture of water and ethanol. The product was dried in a vacuum oven at 80 °C for 24 h. A schematic illustration of the IMD-MMT synthesis is shown in Scheme 13.2.

2.5 Syntheses of the PTT Nanocomposites

Because the procedure used to synthesize IMD-MMT/PTT was very similar to that used for C_{12}PPh-MMT/PTT, only the preparation of the PTT nanocomposite containing 2 wt% IMD-MMT is described here. PDO (76.0 g, 1.0 mol) and 2.08 g of IMD-MMT were placed in a polymerization tube and stirred for 30 min at room temperature. In a separate tube, dimethyl terephthalate (DMT) (97.0 g, 0.5 mol) and a few drops of isopropyl titanate (1.2×10^{-4} mol) were combined and then added to

Table 13.2 Thermal parameters of PTT hybrid fibers

Organoclay	IMD-MMT					C$_{12}$PPh-MMT			
(wt%)	DR[a]	I.V.[b]	T_m (°C)	T_D^{ic} (°C)	wt_R^{600d} (%)	I.V.	T_m (°C)	T_D^{i} (°C)	wt_R^{600} (%)
0 (pure PTT)	1	0.84	228	362	1	0.84	228	362	1
1	1	0.81	232	367	8				
2	1	0.80	232	370	10	0.86	227	371	10
3	1	0.85	232	372	11	0.83	228	370	11
	3		233	371	11				
	7		232	372	10				
	9		232	371	10				
4	1					0.81	227	371	12
	3						228	370	12
	7						227	370	13
	9						228	371	13

[a]Draw ratio
[b]Inherent viscosities were measured at 30 °C using 0.1 g/dL solutions in a phenol/1,1,2,2-tetrachloroethane (w/w = 50/50) mixture
[c]Initial weight-loss onset temperature
[d]Weight percent of residue at 600 °C

the organoclay–PDO system with vigorous stirring to obtain a homogeneously dispersed system. The resulting mixture was heated at 190 °C for 1 h, after which the temperature was raised to 230 °C and the mixture stirred for a further 2 h under a steady stream of N_2 gas. Methanol was generated continuously during this period. In the final step, the mixture was heated at 265 °C for a further 2 h at a pressure of 1 Torr. The crude solid was washed repeatedly with water and then dried under a vacuum at 70 °C for 24 h to produce the IMD-MMT/PTT nanocomposites.

Attempts were made to synthesize PTT hybrids containing more than 3 wt% organoclay using the in situ intercalation approach. However, repeated attempts to polymerize a 4 wt% IMD-MMT hybrid all failed as a result of the production of bubbles in the polymerization reactor during the transesterification of DMT and PDO. The issue of how to produce high-molecular-weight polymer hybrids with a high organoclay content without the formation of bubbles remains unresolved.

The solution viscosity of the organoclay/PTT nanocomposites was studied in 50/50 (w/w) phenol-1,1,2,2-tetrachloroethane. The inherent viscosity numbers ranged from 0.80 to 0.85 (see Table 13.2). Considering that these viscosity values were obtained from pure PTT derived from the PTT hybrids after removal of the clay contents, these numbers can be regarded as being constant.

2.6 Extrusion

The composites were pressed at 250 °C under a pressure of 2,500 kg/cm^2 for 2–3 min on a hot press. The ~0.5-mm-thick films obtained were dried in a vacuum oven at 110 °C for 24 h, after which they were extruded through the die of a capillary rheometer. The hot extrudates were stretched through the die of the capillary rheometer (INSTRON 5460) at 250 °C and were immediately drawn at the constant speed of the take-up machine to form fibers with different DRs. The pure PTT and PTT hybrids were each extruded through a capillary die into fibers with varying DRs, and the thermal properties and the tensile mechanical properties of the extrudates were examined. The standard die diameter (DR = 1) was 0.75 mm. When the organoclay content of the hybrids was increased from 0 to 3 wt%, all the fibers obtained from the capillary rheometer were bright yellow. The DR was calculated from the ratio of the velocity of extrusion to the take-up speed. The mean residence time in the capillary rheometer was ~3–4 min.

2.7 Characterization

The thermal behavior of the composites was studied at a heating rate of 20 °C/min under a N_2 flow using a Du Pont model 910 differential scanning calorimeter and thermogravimetric analyzer. Wide-angle XRD measurements were performed at room temperature using a Rigaku (D/Max-IIIB) X-ray diffractometer with Ni-filtered Co-Kα radiation. The scanning rate was 2°/min over the range $2\theta = 2$–10°.

The tensile properties of the fibers were determined using an Instron mechanical tester (Model 5564) at a crosshead speed of 20 mm/min, at room temperature. The experimental uncertainties in the tensile strength and the modulus were ±1 MPa and ±0.05 GPa, respectively. An average of at least ten individual determinations was obtained.

The morphologies of the fractured surfaces of the extruded fibers were investigated using a Hitachi S-2400 scanning electron microscope. An SPI sputter coater was used to sputter coat the fractured surfaces with gold for enhanced conductivity. The samples for TEM were prepared by placing PTT hybrid fibers in epoxy capsules and then curing the epoxy at 70 °C for 24 h in a vacuum. Subsequently, the cured epoxies containing the PTT hybrids were microtomed into 90-nm-thick slices, and a layer of carbon, about 3 nm thick, was deposited on each slice on a 200 copper net mesh. TEM photographs of ultrathin sections of the polymer/organoclay hybrid samples were taken on an EM 912 OMEGA TEM using an acceleration voltage of 120 kV.

3 Results and Discussion

3.1 Wide-Angle X-ray Diffraction

XRD patterns of the C_{12}PPh-MMT/PTT nanocomposites are shown in Fig. 13.4. The d_{001} reflection of Na^+-MMT was found at $2\theta = 8.60°$, which corresponds to an interlayer distance (d) of 11.99 Å. The corresponding XRD peak for the surface-modified clay, IMD-MMT, was found at $2\theta = 5.42°$, corresponding to an interlayer distance of 18.94 Å (Fig. 13.4a). The clays modified with organic compounds exhibited improved compatibility with the polymer, enabling the clay galleries to be easily intercalated with the polymer. As expected, ion exchange between the clay (Na^+-MMT) and IMD^+-Cl^- resulted in an increase in the basal interlayer spacing with respect to that of pristine Na^+-MMT as well as a large shift of the diffraction peak toward lower 2θ values. For the PTT hybrid with an organoclay content of 1 wt%, only a low-intensity peak at $2\theta = 3.87°$ ($d = 26.52$ Å) was evident in the XRD pattern of the fiber sample. As shown in Fig. 13.4a, the peak intensity increased with increasing organoclay loading up to 3 wt%.

In the case of C_{12}PPh-MMT (Fig. 13.4b), ion exchange between Na^+-MMT and C_{12}PPh-Cl^- also caused a large shift of the diffraction peak toward lower 2θ values. The interlayer spacing of C_{12}PPh-MMT ($d = 36.08$ Å) was larger than that of IMD-MMT ($d = 18.94$ Å). A second XRD peak was observed at $2\theta = 5.65°$ ($d = 18.20$ Å), which corresponds to the (002) plane of the C_{12}PPh-MMT layers.

The XRD patterns of pure PTT and the PTT hybrid fibers with C_{12}PPh-MMT loadings of 2–4 wt% are also shown in Fig. 13.4b. In the case of the 2 wt% PTT hybrid, no characteristic organoclay peaks are observed in the 2θ range 5–7°; that is, the peak corresponding to the basal spacing disappeared. In the cases of the hybrids with 3 and 4 wt% organoclay loadings, however, a peak was observed at $2\theta = 5.76°$ ($d = 17.81$ Å), the peak intensity of which increased with increasing organoclay loading up to 4 wt%, indicating that agglomeration of a small part of the clay occurred in the PTT matrix. This increase also suggests that dispersion is more efficient at lower organoclay loadings. For all PTT hybrids, the disappearance of the main peak at $2\theta = 2.86°$ is thought to be the result of the presence of the swollen organoclay inserted into the polymer chains exhibiting a diffraction peak lower than 2°.

After intercalation with the organoclays, the *d* value of the polymer increased from 11.99 to 18.94 Å ($2\theta = 5.42°$) for IMD-MMT and from 11.99 to 36.08 Å ($2\theta = 2.86°$) in the case of C_{12}PPh-MMT. This increase in the basal spacing is indicative of intercalation of the polymer chains into the clay galleries [28, 30]. Increasing the amount of organoclay in the PTT matrices resulted in the appearance of sharp peaks in this region; the intensities of these peaks increased linearly with the clay content (see Fig. 13.4a, b). This suggests that a certain portion of the organoclay underwent aggregation and that the dispersion was better at a lower clay content than at a higher clay content. However, the presence of the organoclay had no effect on the locations of the peaks, which indicates that perfect exfoliation of the clay layers in the organoclay does not occur in PTT [26]. For the PTT hybrid fibers, the disappearance of the main peaks (at $2\theta = 5.42°$ for IMD-MMT and at $2\theta = 2.86°$ for C_{12}PPh-MMT) was attributed to the insertion of the swollen organoclay, causing the polymer chains to exhibit a diffraction peak lower than 2° [43].

Figures 13.5a, b show the XRD patterns of the PTT hybrid fibers containing 3 wt% IMD-MMT and 4 wt% C_{12}PPh-MMT, respectively, for various DR values. When DR = 1, peaks were present for both extruded fibers. Because stretching of the fibers increases the alignment of clay platelets in a polymer matrix, the XRD peak intensity usually decreases with increasing DR in coil-like polymers [47]. However, no significant decreases in the XRD peak intensities were observed in either system when the DR was increased from 1 to 9, implying that higher

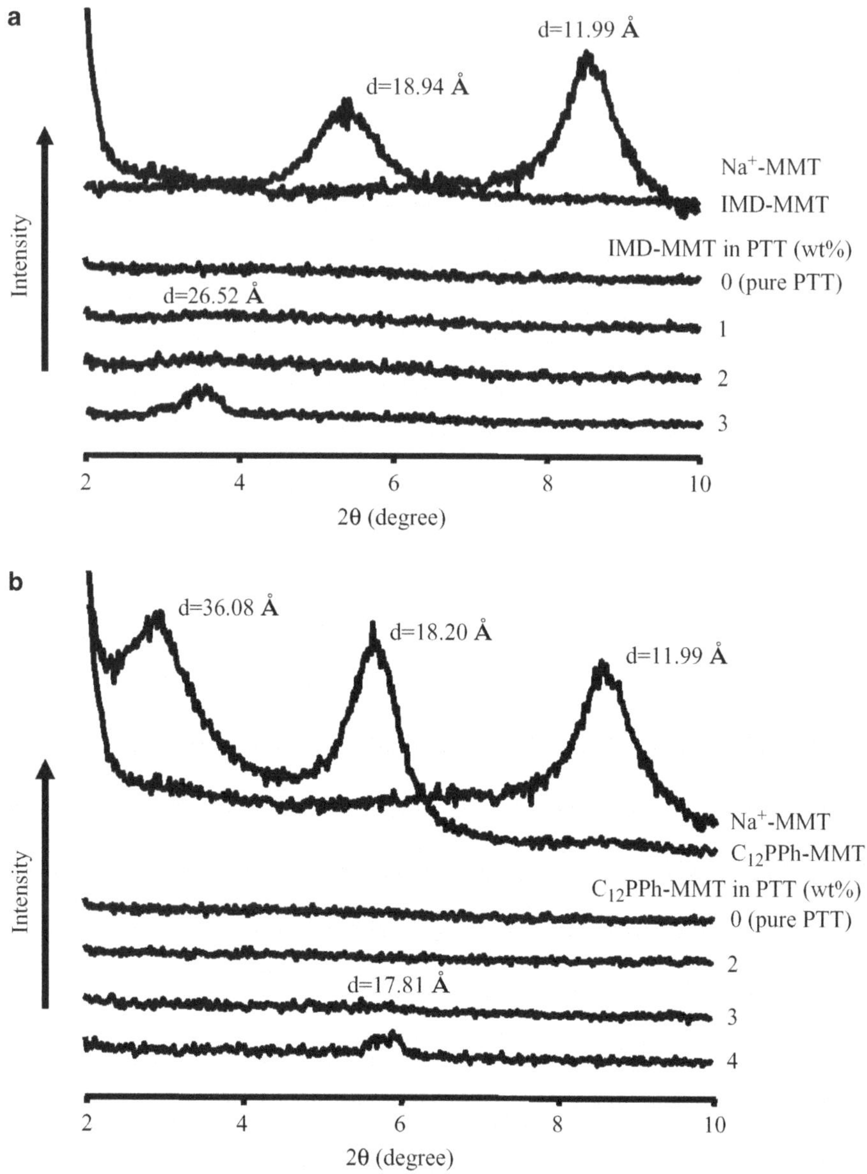

Fig. 13.4 XRD patterns of clay, organoclays, and PTT hybrid fibers with various organoclay contents

stretching of the fiber during extrusion was not effective for good dispersion of the clay in the polymer matrix. The peak position was not significantly affected by the DR values of these systems, which indicates that perfect exfoliation of the organoclay layer structure did not occur in PTT.

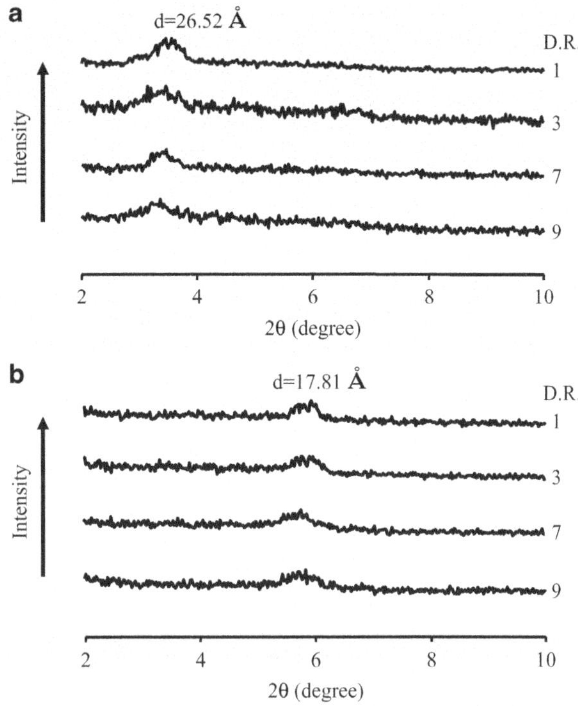

Fig. 13.5 XRD patterns of (**a**) 3 wt% IMD-MMT and (**b**) 4 wt% C_{12}PPh-MMT in PTT hybrid fibers with different draw ratios

3.2 Morphology

The dispersion of clay in the PTT matrix was further evaluated using electron microscopy. Electron microscopy and XRD are complementary techniques and can fill in the gaps in information that other techniques cannot.

The morphologies of the clay aggregates were characterized using SEM. Because of the difference between the scattering densities of clay and PTT, large clay aggregates were readily evident in the SEM images. The SEM images of the fractured surfaces of the PTT hybrid fibers containing the two organoclays are compared in Figs. 13.6–13.9. Figure 13.6 reveals that clay phases were formed within the undrawn PTT hybrid fibers with increasing organoclay content (0–3 wt%). These PTT hybrid fibers consisted of clay domains that were 50–80 nm in size and were dispersed in a continuous PTT phase. Figure 13.7 shows micrographs of the 3 wt% IMD-MMT/PTT hybrid fiber obtained for DRs from 1 to 9. For a DR of 7, the 3 wt% hybrid fiber contained fine clay phases that were 50–60 nm in diameter (see Fig. 13.7c); very similar domain diameters (40–60 nm) were observed for DR = 9 (see Fig. 13.7d). The domain size of the dispersed clay phase and the formed voids were virtually constant, irrespective of DR, which is consistent with the XRD observations shown in Fig. 13.5a.

Figure 13.8 shows the clay phases formed in an undrawn hybrid fiber for organoclay contents ranging from 0 to 4 wt%. The PTT hybrid fibers with 0–4 wt%

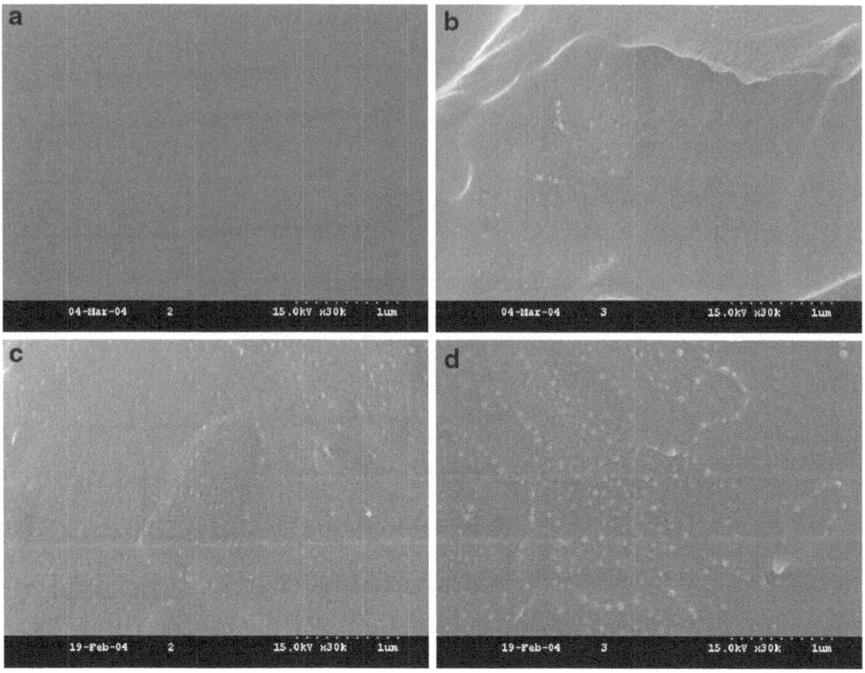

Fig. 13.6 SEM photographs of (**a**) 0 wt% (pure PTT), (**b**) 1 wt%, (**c**) 2 wt%, and (**d**) 3 wt% IMD-MMT in PTT hybrid fibers

C_{12}PPh-MMT consisted of clay domains with sizes ranging from 50 to 70 nm in a continuous PTT phase. The micrographs of the pure PTT (Fig. 13.8a) and the PTT hybrid fibers containing 2 wt% organoclay (Fig. 13.8b) are characterized by smooth surfaces, the latter being the result of good dispersion of the clay particles. Conversely, at 4 wt% organoclay loading (Fig. 13.8c), voids are observed in addition to some deformed regions that may be attributable to the coarseness of the fractured surface. The fractured surfaces exhibited greater deformity when the hybrids contained a higher organoclay content, which is probably a consequence of agglomeration of the clay particles. This is consistent with the XRD data shown in Fig. 13.4b.

Figure 13.9 shows micrographs of the 4 wt% C_{12}PPh-MMT/PTT hybrid fiber for different DRs. The size of the organoclay domain remained virtually unchanged for DRs from 1 to 9. The hybrid fiber with DR = 9 also exhibited fine dispersion with domains that were 30–60 nm in diameter (see Fig. 13.9c). This is also consistent with the XRD data shown in Fig. 13.5b.

TEM studies were used to examine the exact nature of the dispersion of the clay layers in the fiber hybrids. TEM allows a qualitative understanding of the internal structure through direct observation. Typical TEM photographs of the PTT hybrid fibers containing different organoclays are shown in Figs. 13.10–13.13. The dark lines in the photograph are the intersections of the 1-nm-thick clay sheets, and the

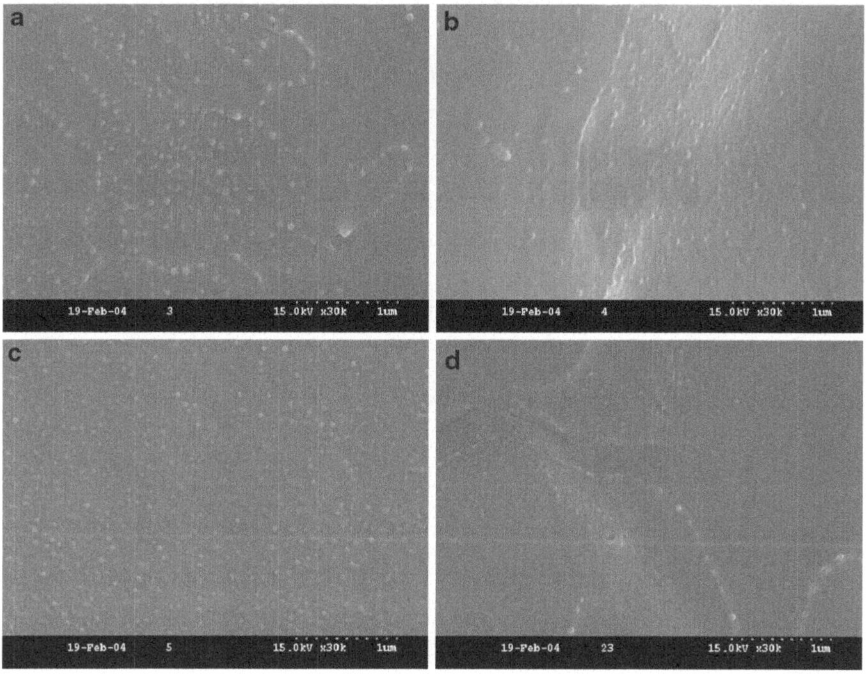

Fig. 13.7 SEM photographs of 3 wt% IMD-MMT in PTT hybrid fibers for draw ratios of (**a**) 1, (**b**) 3, (**c**) 7, and (**d**) 9

spaces between the dark lines are interlayer spaces. The clay layers in Figs. 13.10 and 13.11 exhibit certain regions where there is individual dispersion of delaminated sheets in the matrix, as well as regions where the regular stacking arrangement is maintained with a layer of polymer between the sheets. In the case of the PTT hybrid fibers containing IMD-MMT at 1 and 3 wt% (Figs. 13.10b and 13.11b), some of the clay is well dispersed in the PTT matrix whereas other portions appear as agglomerates of approximately 30 nm.

Typical TEM photographs of the PTT hybrid fibers containing 2 and 4 wt% C_{12}PPh-MMT are shown in Figs. 13.12 and 13.13, respectively. Figure 13.12 shows that the clay was well dispersed in the polymer matrix at all magnification levels, although some particles were agglomerated. For the 4 wt% hybrid fiber (Fig. 13.13a and b), some of the clay was well dispersed within the PTT matrix, whereas the remainder appeared in agglomerations larger than approximately 10 nm. The presence of small peaks in the XRD patterns of these samples can be attributed to these agglomerated layers (see Fig. 13.4b). In contrast with the hybrids containing IMD-MMT, the clay layers of the C_{12}PPh-MMT hybrid fiber (Figs. 13.12 and 13.13) were better dispersed in the matrix polymer. This difference in the clay dispersion for IMD-MMT and C_{12}PPh-MMT might be attributed to the similarity of the chemical structure of the organoclay and to the interactions between the organic region of the clay and the polymer matrix.

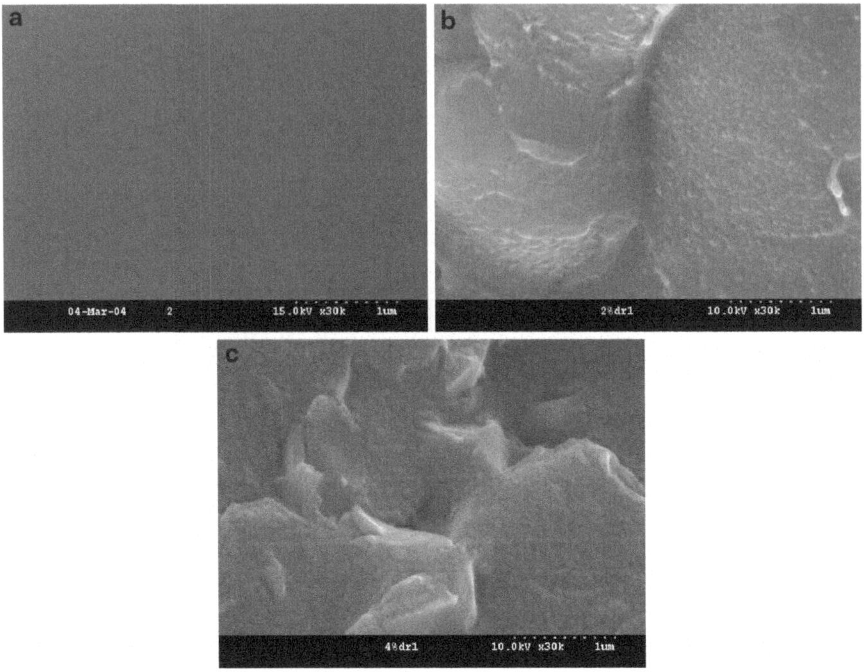

Fig. 13.8 SEM photographs of (**a**) 0 wt% (pure PTT), (**b**) 2 wt%, and (**c**) 4 wt% C_{12}PPh-MMT in PTT hybrid fibers

XRD, SEM, and TEM analyses indicate that for a low clay content, the clay particles were well dispersed throughout the PTT matrix, whereas agglomerated structures were evident at a higher clay content. The unusual thermomechanical properties of these nanocomposites are discussed in the following sections with reference to the homogeneous dispersion of the silicate nanoscale particles.

3.3 Thermal Behavior

Table 13.2 presents the thermal parameters of pure PTT and the PTT hybrid fibers with different DRs. The endothermic peak of pure PTT appeared at 228 °C, which corresponds to the melting transition temperature (T_m). The maximum transition peaks in the DSC thermograms of the IMD-MMT/PTT hybrids shifted from 228 °C to 232 °C as the organoclay content increased to 1 wt% and then remained constant for an organoclay content up to 3 wt% (see Table 13.2). This increase in T_m for the hybrids may be attributable to the thermal insulation effect of the clay layer structure, as well as to interactions between the organoclay and PTT molecular chains [27, 50]. Compared to the results obtained for the IMD-MMT/PTT hybrids, the T_m values of the C_{12}PPh-MMT hybrid fibers remained fairly constant

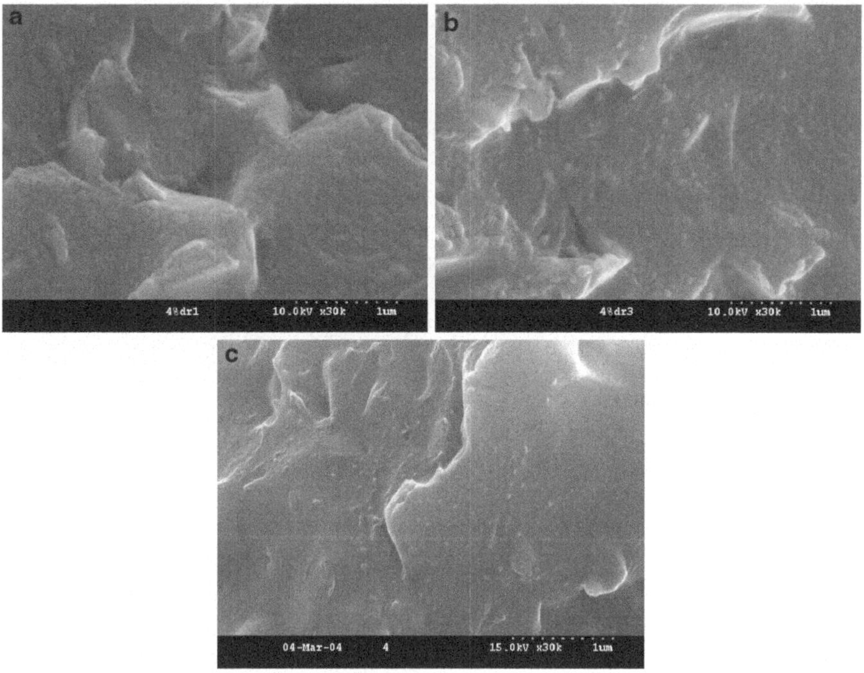

Fig. 13.9 SEM photographs of 4 wt% C_{12}PPh-MMT in PTT hybrid fibers for draw ratios of (**a**) 1, (**b**) 3, and (**c**) 9

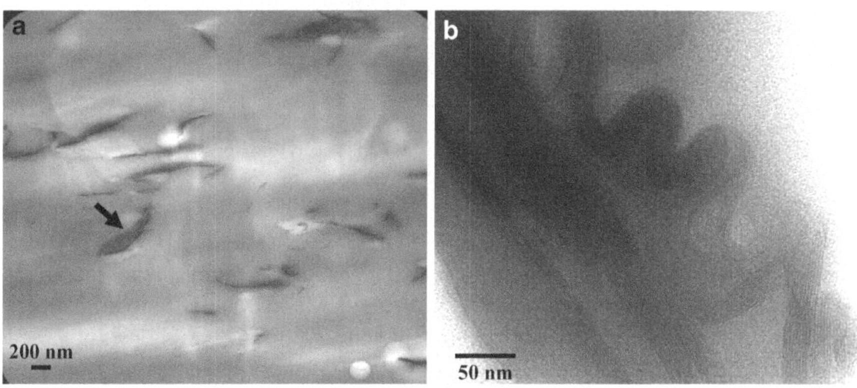

Fig. 13.10 TEM micrographs of 1 wt% IMD-MMT in PTT hybrid fiber at (**a**) low and (**b**) high magnification

(i.e., varying from 227 °C to 228 °C) as the clay content increased from 0 to 4 wt%. This implies that in the latter case, varying the DR of the fiber during the extrusion process did not result in a good insulation effect of the clays in the polymer matrix, as is evident from Table 13.2.

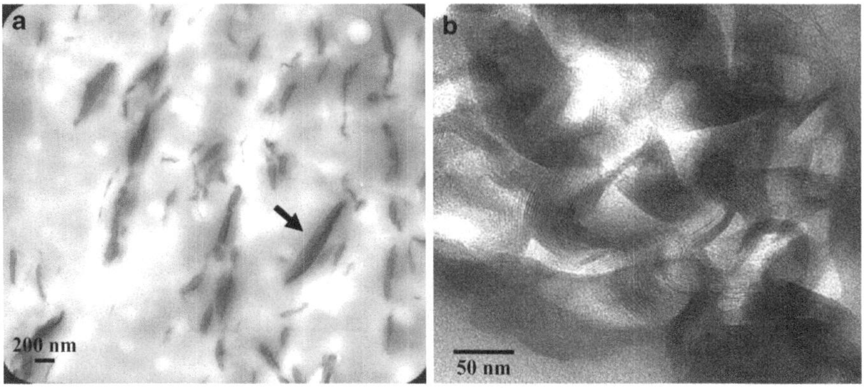

Fig. 13.11 TEM micrographs of 3 wt% IMD-MMT in PTT hybrid fiber at (**a**) low and (**b**) high magnification

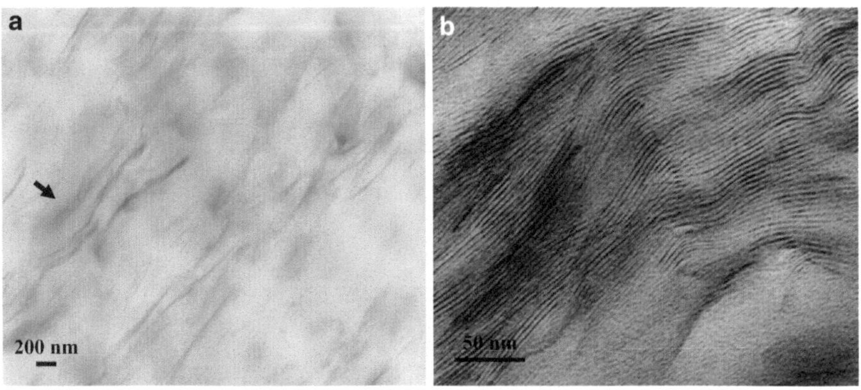

Fig. 13.12 TEM micrographs of 2 wt% C_{12}PPh-MMT in PTT hybrid fiber at (**a**) low and (**b**) high magnification

The TGA data for pure PTT and the PTT hybrids with 0–3 wt% IMD-MMT are compared in Table 13.2. The results presented in Table 13.2 indicate that the initial thermal degradation temperature ($T_D{}^i$) of the IMD-MMT/PTT hybrid fibers increased with increasing organoclay content. The $T_D{}^i$ for 2 % weight loss varied from 362 °C to 372 °C as the clay content in the PTT hybrids increased from 0 to 3 wt%; the largest increase in the initial thermal degradation temperature with respect to that of pure PTT was 6 °C, obtained for 3 wt% IMD-MMT/PTT. This increase in the thermal stability of the hybrid can be attributed to the high thermal stability of clay and to the interactions between the clay particles and the polymer matrix. The weight of the residue obtained at 600 °C increased from 1 % to 11 % as the clay content increased from 0 % to 3 %. This enhancement of char formation with increasing clay content is ascribed to the high heat resistance of the clay [21]. Table 13.2 also presents the thermal stabilities of the C_{12}PPh-MMT hybrid

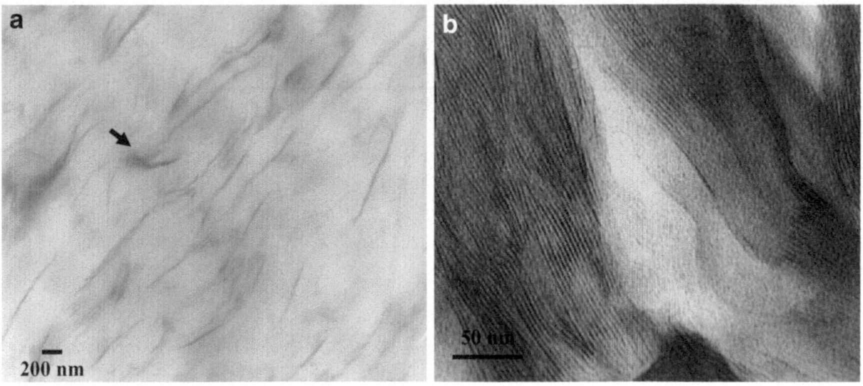

Fig. 13.13 TEM micrographs of 4 wt% C_{12}PPh-MMT in PTT hybrid fiber at (**a**) low and (**b**) high magnification

fibers for various clay contents. $T_D{}^i$ of the C_{12}PPh-MMT hybrid fibers increased from 362 °C to 371 °C for clay contents ranging from 0 to 2 wt% and then remained constant for clay contents up to 4 wt%.

The weight of the residue at 600 °C ($wt_R{}^{600}$) for C_{12}PPh-MMT hybrids increased linearly from 1 % to 12 % as the clay content increased from 0 to 4 wt%. Table 13.2 indicates that the overall thermal properties of the PTT hybrid fibers comprising each of the two organoclays remained virtually constant as the DR increased from 1 to 9. Plots of the thermal analysis of the PTT/organoclay hybrid fibers are presented in Fig. 13.14.

Considering the results presented above, it is perfectly conceivable that the introduction of inorganic components into organic polymers can improve the thermal stability of the latter on the basis of the fact that clays have good thermal stability as a result of the heat insulation effect of the clay layers and because they act as a mass transport barrier to the volatile products generated during decomposition [36, 38].

3.4 Tensile Properties

The mechanical tensile properties of the IMD-MMT hybrid fibers are given in Table 13.3. At DR = 1, the ultimate tensile strength of the IMD-MMT hybrid fibers increased with the amount of clay up to a certain value and then remained constant. For example, the strength of the 1 wt% PTT hybrid fibers was 44 MPa, which is about 40 % higher than that of pure PTT (32 MPa); the ultimate strength was 45 MPa for the 3 wt% organoclay hybrid. This plateauing of the ultimate strength is mainly the result of ineffective dispersion of the clay particles in the polymer matrix when the clay content exceeded a certain threshold, as we have described previously [36, 47]. In contrast with the tensile strength, the initial modulus values increased linearly with increasing organoclay content. For an IMD-MMT content

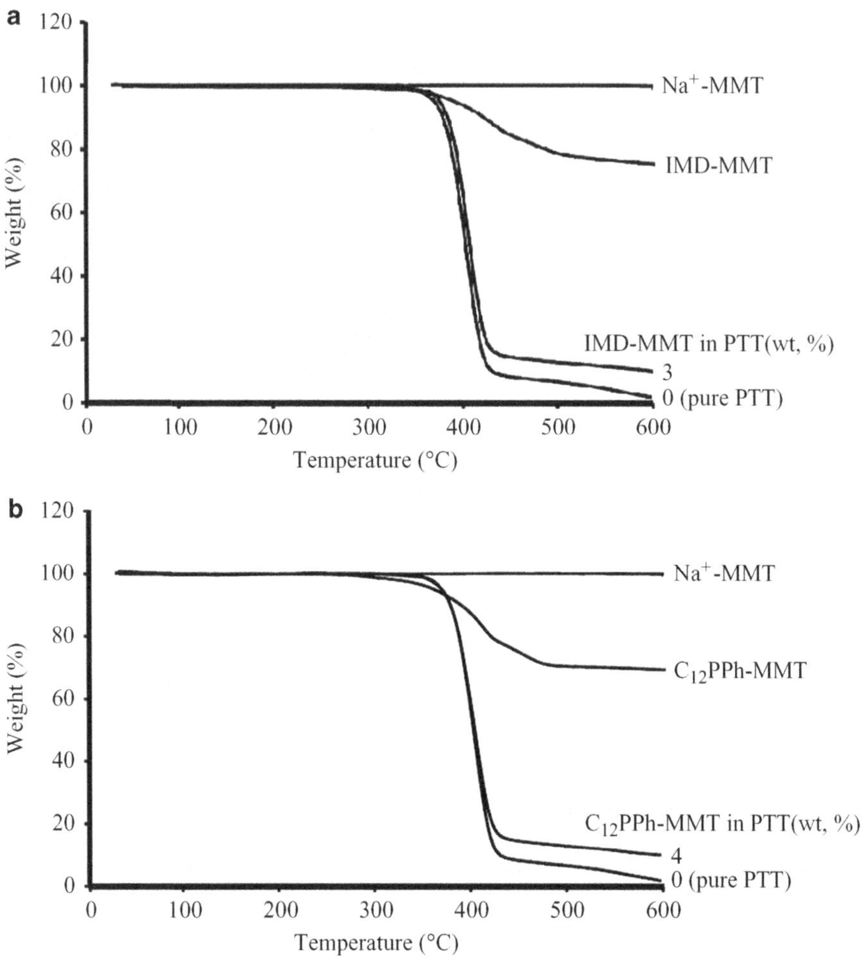

Fig. 13.14 TGA thermograms of PTT hybrid fibers with (**a**) IMD-MMT and (**b**) C_{12}PPh-MMT contents

of 3 wt%, the modulus of the hybrid was 2.54 GPa, which is about 40 % higher than that of pure PTT (1.77 GPa). This enhancement of the modulus is ascribed to the resistance exerted by the clay itself, as well as to the orientation and aspect ratio of the clay layers. Additionally, the enhanced stretching resistance of the oriented backbone of the polymer chain in the gallery also contributes to the enhancement of the modulus.

The tensile properties of the C_{12}PPh-MMT hybrid fibers are also given in Table 13.3. These properties improved with increasing incorporation of organoclay at DR = 1; increasing the organoclay content in the hybrids from 0 to 4 wt% produced a linear increase in the ultimate tensile strength from 32 to 48 MPa.

Table 13.3 Tensile properties of PTT hybrid fibers

Organoclay (wt%)	DR[a]	IMD-MMT			C_{12}PPh-MMT		
		Ult. Str. (MPa)	Ini.Mod. (GPa)	E.B.[b] (%)	Ult. Str. (MPa)	Ini.Mod. (GPa)	E.B. (%)
0 (pure PTT)	1	32	1.77	2	32	1.77	2
	3	35	1.85	2	35	1.85	2
	7	35	1.94	2	35	1.94	2
	9	34	2.02	3	34	2.02	3
1	1	44	2.43	2			
	3	48	2.57	2			
	7	46	2.60	2			
	9	47	2.65	2			
2	1	45	2.46	2	43	2.61	2
	3	45	2.54	2	40	2.58	2
	7	48	2.59	3	41	2.54	2
	9	46	2.70	2	41	2.57	2
3	1	45	2.54	2	45	2.76	3
	3	46	2.65	3	45	2.74	2
	7	47	2.84	2	38	2.75	2
	9	47	2.83	3	38	2.74	2
4	1				48	3.09	2
	3				46	3.10	3
	7				40	3.09	3
	9				41	3.04	2

[a]Draw ratio
[b]Elongation percentage at break

For example, the ultimate strength of the hybrid fiber containing 4 wt% C_{12}PPh-MMT was 50 % higher than that of a pure PTT fiber.

The initial moduli of the hybrids followed a similar trend. For example, the initial tensile modulus of the fiber with 3 wt% organoclay was 2.76 GPa, which is about 50 % higher than the modulus of pure PTT. For an organoclay content of 4 wt% in PTT, the modulus was about 70 % higher (3.09 GPa) than that of pure PTT. This large improvement in the tensile properties can be attributed to the interactions between the PTT molecules and the layered organoclays, as well as to the rigid nature of the clay layers. Because clay is much more rigid than PTT molecules, the tensile properties of pure PTT are improved by hybridization as a result of intercalation and dispersion of the organoclay layers in the PTT matrix. It has been repeatedly reported that nanocomposites exhibit significant improvements in ultimate strength and initial modulus as the organoclay content increases [14, 23, 39].

The ultimate strengths of the IMD-MMT hybrid fibers containing 1–3 wt% organoclay were independent of DR in the DR range from 1 to 9. However, the initial modulus of the IMD-MMT hybrid fibers increased linearly with increasing DR. For example, increasing the DR from 1 to 9 produced an increase in the initial tensile modulus of a hybrid fiber with an IMD-MMT content of 1 wt% from

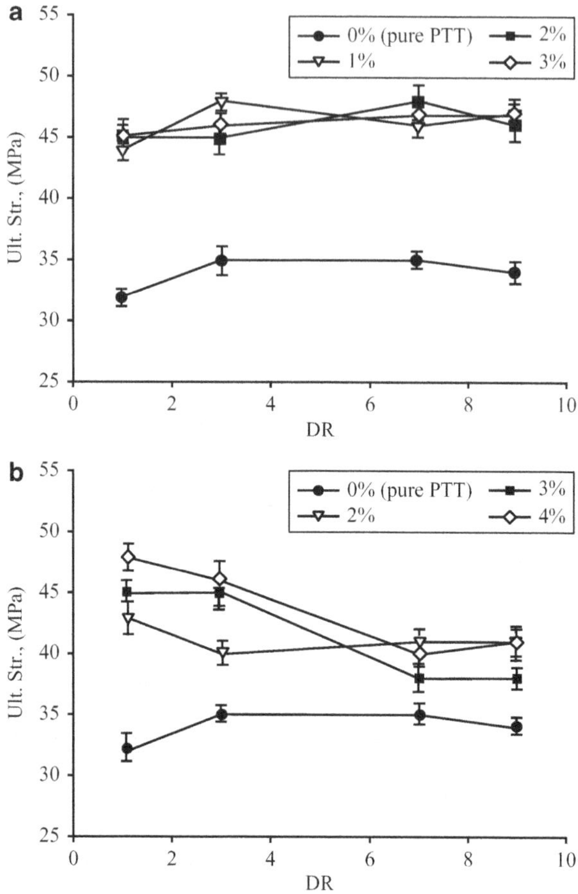

Fig. 13.15 Effects of draw ratio on the ultimate tensile strength of (**a**) IMD-MMT and (**b**) C_{12}PPh-MMT composites

2.43 to 2.65 GPa. The modulus of the PTT hybrid with an organoclay content of 3 wt% was about 40 % higher (2.83 GPa) than that of pure PTT (2.02 GPa) at DR = 9.

In the case of the C_{12}PPh-MMT hybrid fibers, both the ultimate strength and the initial modulus of the hybrid fibers decreased slightly with increasing DR, as is evident from Table 13.3. For example, for a hybrid fiber with a C_{12}PPh-MMT content of 2 wt%, increasing the DR from 1 to 9 was accompanied by respective decreases in the tensile strength from 43 to 41 MPa and in the initial modulus from 2.61 to 2.57 GPa. Similar trends were observed for hybrid fibers with C_{12}PPh-MMT contents of 3 and 4 wt%. The variations of the ultimate strengths and initial moduli of the PTT/organoclay hybrids are plotted against the DR in Figs. 13.15 and 13.16.

An increase in the tensile strength with increasing DR is very common in engineering plastics and is usually observed for flexible coil-like polymers [9, 19]. However, the C_{12}PPh-MMT hybrid system did not follow this trend. The observed decline in the tensile properties with increasing DR seems to be the result

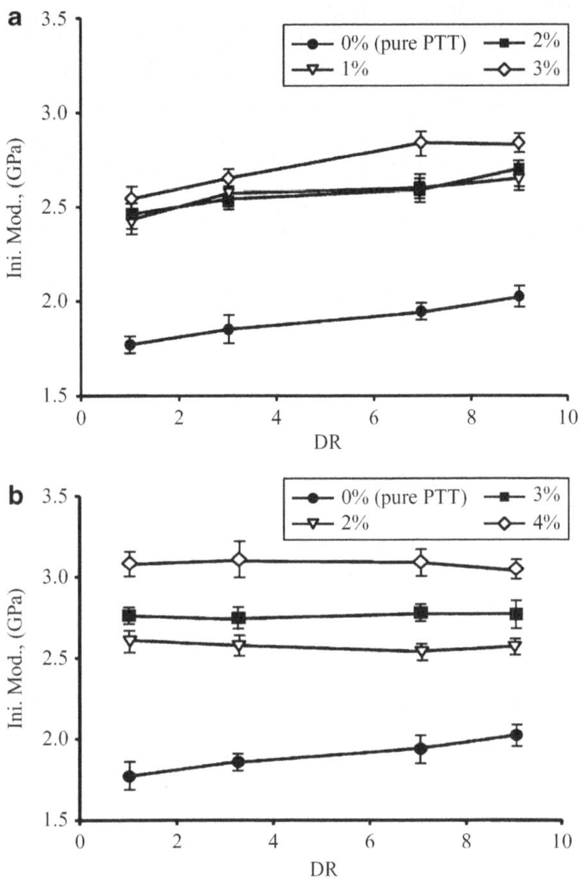

Fig. 13.16 Effects of draw ratio on the initial tensile modulus of (**a**) IMD-MMT and (**b**) C_{12}PPh-MMT composites

of debonding between the organoclay and the matrix polymer and of the presence of many nanosized voids caused by excessive stretching of the fibers. Debonding occurs at the interface because of imperfect bonding, giving rise to a constant interfacial shear stress when strain is applied to the composite. The debonding at the polymer/clay interface at the point of fracture appears to promote some closure forces around the wake of the crack, which produces the observed crack growth resistance. Many studies have shown that an imperfect incursion/matrix interface cannot sustain the large interfacial shear stress that results from an applied strain [6, 8, 10]. This indicates that hydrostatic elongation during extrusion and compression molding operations results in debonding in the polymer chain, as well as in the formation of voids around the polymer/clay interfaces.

The elongations at breakage of the IMD-MMT and C_{12}PPh-MMT hybrids were essentially independent of DR, with only a 1 % change (from 2 % to 3 %) as the DR was increased from 1 to 9. This result is characteristic of materials reinforced with stiff inorganic materials and is indicative of an intercalated morphology.

3.5 Future Prospectives

PTT is one member of the terephthalate polyesters, the most common substances used in engineering thermoplastics. More recently, PTTs have drawn attention for their applications in textile industry due to great reduction in the manufacturing cost of 1,3-propanediol, the monomer used for PTT synthesis. Molecular and crystalline structures of PTT were reported in 1970s. Since then, many studies on PTT have been published. Most of aspects concerning PTT, such as synthetic technique, spinning technology, morphological structure, and properties of fiber, as well as nanoscale composite technology, have been widely investigated. An in-depth survey of recent research and developments in the area of PTT nanocomposites gives an evidence of the growing tendency toward manufacturing and commercialization of such novel materials.

4 Conclusions

Hybrid fibers consisting of PTT and two different organoclays, IMD-MMT and C_{12}PPh-MMT, were synthesized with the objectives of determining the dispersibility of organoclays in the PTT matrix and improving the various properties of PTT hybrid fibers. Hybrid fibers with organoclay contents ranging from 0 to 4 wt% were synthesized via the in situ interlayer polymerization method.

The thermomechanical properties were found to be dependent on both the type and amount of organoclay in the polymer matrix. XRD, SEM, and TEM analyses revealed that the morphologies of the hybrids with a low organoclay content were characterized by a mixture of intercalated and partially exfoliated clay species. Each type of clay was well dispersed in the PTT matrix in individual clay layers, although some agglomerated species with sizes larger than approximately 20 nm were present. Clay dispersion in the polymer matrix was more effective at lower organoclay contents.

The ultimate strength and initial modulus of the IMD-MMT hybrid fibers increased slightly as the DR increased from 1 to 9, whereas the corresponding values for the C_{12}PPh-MMT hybrid fibers decreased slightly with increasing DR. The observed decline in the tensile properties appears to be the result of debonding between the organoclay and the matrix polymer, and also to the presence of many nanosized voids caused by excessive stretching of the fibers. In summary, the addition of small amounts of organoclays to PTT was found to affect the thermal behavior and tensile mechanical properties of the polymer/clay hybrid fibers.

References

1. Whinfield JR, Dickson JT (1941) British Patent 578,079
2. Whinfield JR, Dickson JT (1949) US 2,465,319
3. Ward IM, Wilding MA (1976) J Polym Sci, Part B: Polym Phys 14:263

4. Ward IM, Wilding MA (1977) Polymer 18:327
5. Desborough IJ, Hall JH, Neisser JZ (1979) Polymer 20:545
6. Chawla KK (1987) Composite materials science & engineering. Springer, New York
7. Jaynes WF (1987) Clay Clay Miner 35:440
8. Shia D, Hui Y, Burnside SD, Giannelis EP (1987) Polym Eng Sci 27:887
9. La Mantia FP, Valenza A, Paci M, Magagnini PL (1989) J Appl Polym Sci 38:583
10. Curtin WA (1991) J Am Ceram Soc 74:2837
11. Calvert PD (1992) Mater Res Soc Bull 17:37
12. Usuki A, Kojima Y, Kawasumi M, Okada A, Fukushima Y, Kurauchi T, Kamigaito O (1993) J Mater Res 8:1179
13. Yano K, Usuki A, Okada A, Kurauchi T, Kamigaito O (1993) J Polym Sci, Part A: Polym Chem 31:2493
14. Lan T, Pinnavaia TJ (1994) Chem Mater 6:2216
15. Thomas H, Georg B, Dietrich A (1994) US Patent 5,284,979
16. Lan T, Kaviratna PD, Pinnavaia TJ (1995) Chem Mater 7:2145
17. Usuki A, Koiwai A, Kojima Y, Kawasumi M, Okada A, Kurauchi T, Kamigaito O (1995) J Appl Polym Sci 55:119
18. Vaia RA, Jandt KD, Kramer EJ, Giannelis EP (1995) Macromolecules 28:8080
19. Chang J-H, Jo BW (1996) J Appl Polym Sci 60:939
20. Chuah HH (1996) Chem Fibers Int 46:424
21. Wen J, Wikes GL (1996) Chem Mater 8:1667
22. Brown HS, Chuah HH (1997) Chem Fibers Int 47:72
23. Yano K, Usuki A, Okada A (1997) J Polym Sci, Part A: Polym Chem 35:2289
24. Brown PJ, Blake MS, Richard WP (1998) US Patent 5,770,776
25. Haas T, Yu DH, Sauer J (1998) WO Patent 9,857,913
26. Dagani R (1999) Chem Eng News 77:25
27. LeBaron PC, Wang Z, Pinnavaia TJ (1999) Appl Clay Sci 12:11
28. Zilig C, Mulhaupt R, Finter J (1999) Macromol Chem Phys 200:661
29. Huang JM, Chang FC (2000) J Polym Sci, Part B: Polym Phys 38:934
30. Ke Y, Lu J, Yi X, Zhao J, Qi Z (2000) J Appl Polym Sci 78:808
31. Bharadwaj RK (2001) Macromolecules 34:9189
32. Ishida H, Campbell S, Blackwell J (2000) Chem Mater 12:1260
33. Itagaki T, Komori Y, Kuroda K (2001) J Mater Chem 11:3291
34. Zhu J, Morgan AB, Lamelas FJ, Wilkie CA (2001) Chem Mater 13:3774
35. Zhu J, Uhl FM, Morgan AB, Wilkie CA (2001) Chem Mater 13:4649
36. Chang J-H, Seo BS, Hwang DH (2002) Polymer 43:2969
37. Davis CH, Mathias LJ, Gilman JW, Schiraldi DA, Shields JR, Trulove P, Sutto TE, Delong HC (2002) J Polym Sci, Part B: Polym Phys 40:2661
38. Fornes TD, Yoon PJ, Hunter DL, Keskkula H, Paul DR (2002) Polymer 43:5915
39. Masenelli-Varlot K, Reynaud E, Vigier G, Varlet J (2002) J Polym Sci, Part B: Polym Phys 40:272
40. Saujanya C, Imai Y, Tateyama H (2002) Polym Bull 49:69
41. Wang D, Zhu J, Yao Q, Wilkie CA (2002) Chem Mater 14:3837
42. Chang J-H, An YU, Sur GS (2003) J Polym Sci, Part B: Polym Phys 41:94
43. Danumah C, Bousmina M, Kaliaguine S (2003) Macromolecules 36:8208
44. Hwang SH, Paeng SW, Kim JY, Huh W (2003) Polym Bull 49:329
45. Liang Z-M, Yin J (2003) J Appl Polym Sci 90:1857
46. Osman MA, Mittal V, Morbidelli M, Suter UW (2003) Macromolecules 36:9851
47. Chang J-H, Kim SJ, Joo YL, Im S (2004) Polymer 45:919
48. Chang J-H, Kim SJ, Joo YL, Im S (2004) Polymer 45:5171
49. Frisk S, Ikeda RM, Chase DB (2004) Macromolecules 37:6027
50. Hussain M, Varley RJ, Mathys Z, Cheng YB, Simon GP (2004) J Appl Polym Sci 91:1233

51. Chen K, Shen J, Tang X (2005) J Appl Polym Sci 97:705
52. Jeong YG, Bae WH, Jo WH (2005) Polymer 46:8297
53. Chang J-H, Ju CH, Kim SH (2006) J Polym Sci, Part B: Polym Phys 44:387
54. Kalakkunnath S, Kalika DS (2006) Polymer 47:7085
55. Jia SY, Ren YR, Liu LM (2007) Chin Chem Lett 18:827
56. Luo WA, Chen YJ, Chen XD (2008) Macromolecules 41:3912
57. Yamen M, Ozkaya S, Vasanthan N (2008) J Polym Sci, Part B: Polym Phys 46:1497

Recent Developments in Poly(butylene terephthalate) Nanocomposites

14

Jin-Hae Chang

Contents

Abstract

By comparing PBT hybrids synthesized with varying amounts of thermally stable organoclays, this study aims to investigate the intercalation of polymer chains with organoclays and improve the thermomechanical properties of poly(butylene terephthalate) (PBT) hybrids. Dodecyltriphenylphosphonium-montmorillonite (C_{12}PPh-MMT), dodecyltriphenylphosphonium-mica (C_{12}PPh-Mica), and alkyl ammonium-montmorillonite (NCT-MMT) organoclays were used in the fabrication of PBT hybrids through in situ intercalation polymerization. The variation of their properties with organoclay

J.-H. Chang
School of Energy and Integrated Materials Engineering, Kumoh National Institute of Technology, Gumi, Gyeongbuk, South Korea
e-mail: changjinhae@hanmail.net

J.K. Pandey et al. (eds.), *Handbook of Polymernanocomposites. Processing, Performance and Application – Volume A: Layered Silicates*, DOI 10.1007/978-3-642-38649-7_4,

content of the polymer matrix and the draw ratio (DR) is discussed. The thermomechanical properties and morphologies of the PBT hybrid fibers were characterized using differential scanning calorimetry, thermogravimetric analysis, wide-angle X-ray diffraction, electron microscopy, and mechanical tensile properties analysis. The nanostructures of the hybrid fibers, which were determined using both scanning and transmission electron microscopy, showed that some of the clay layers were uniformly dispersed in the matrix polymer although clusters or agglomerated particles were also detected. The thermal properties of the hybrid fibers were found to be superior to those of pure PBT fibers at a DR of 1. The tensile mechanical properties of the C_{12}PPh-MMT and NCT-MMT hybrid fibers deteriorated with increasing DR. However, the initial moduli of the C_{12}PPh-Mica hybrid fibers slightly increased with increasing DR from 1 to 18.

Keywords
Fibers • Nanocomposites • Poly(butylene terephthalate) • Thermally stable organoclay

1 Introduction

Although Blumstein [1] demonstrated the polymerization of vinyl monomers intercalated into montmorillonite clay as early as 1961, numerous improved methods for the preparation of polymer/clay nanocomposites have been developed by several groups [2–5]. To overcome the problems of macro- and micro-phase separation between the organic polymers and inorganic clays, organic/inorganic polymer hybrids are generally synthesized via one of three methods: solution intercalation [6, 7], melt intercalation [8, 9], or in situ polymerization intercalation [3, 10, 11]. Other approaches, such as the solgel process [12, 13] and monomer/polymer grafting to clay layers [14, 15], have also been used to produce organic/inorganic polymer hybrids. Among these, melt intercalation and in situ intercalation are most commonly used to prepare polymer/clay nanocomposites.

During melt intercalation [16, 17], layered silicate is mixed with a molten polymer matrix. If the silicate surfaces are sufficiently compatible with the chosen polymer, the polymer can enter the interlayer space and form an intercalated or exfoliated nanocomposite. Ha et al. [18] prepared poly(butylene terephthalate) (PBT) nanocomposites via melt intercalation with organoclays and observed good nanoparticle dispersion of the organoclay in the polymer matrix. In contrast, in situ intercalation polymerization is based on the in situ linear polymerization or cross-linking of one or more monomers and was the first method used to synthesize polymer-layered silicate nanocomposites based on polyamide 6 [10]. In situ intercalation relies on the initial swelling of the organoclay by the monomer followed by in situ polymerization that is initiated either thermally or by the addition of a suitable compound. Chain growth in the clay galleries triggers clay exfoliation and nanocomposite formation. Thus, polymer hybrids made using the in situ method have the advantage of lacking physical or chemical interactions between the organic polymer and the inorganic material [3, 10, 11, 19, 20].

Clay minerals comprise alkali metals contained between silicate sheets and can thus swell in polar solvents such as water. There are many types of clays including kaolinite, montmorillonite, hectorite, saponite, and synthetic mica. Among them, montmorillonite (MMT) and mica have recently received significant attention as reinforcing materials for polymers owing to their high aspect ratio and unique characteristics. These clays consist of stacked silicate sheets that are 1 nm thick and about 218 and 1,230 nm long for montmorillonite and synthetic mica, respectively [21, 22].

The synthesis of polyester nanocomposites has not been as successful as for other polymer nanocomposites. Researchers have prepared polyester nanocomposites with quaternary ammonium salt-modified clay using in situ intercalation polymerization [23, 24], but did not achieve good clay dispersion in the nanocomposites or improved mechanical properties. A more commercial method that uses conventional polymer processing approaches is melt intercalation of the polyester with the organoclay [9]. However, this approach usually leads to poorly dispersed clay particles, which may be due to the low decomposition temperature of the organic modifier bound to the clay surface [25]. To obtain nanocomposites from processing at temperatures $\geq$280 °C without thermal degradation, an organoclay that is thermally stable at temperatures higher than the process temperatures must be used [26–30].

PBT is applied worldwide as an engineering plastic and has many useful properties including good thermal stability, excellent flow properties, and good chemical resistance [31–33]. PBT is a thermoplastic and semicrystalline polymer and, as a member of the poly(alkylene terephthalate) family, is derived from a polycondensate of terephthalic acid and 1,4-butanediol (BD).

Many attempts have been made to improve thermomechanical properties of PBT by incorporating various fillers [18, 34–36]. Among these blends, polymer nanocomposites with organo-modified clays are most common. With the addition of only a few percent of clay to pristine polymers, PBT nanocomposites show improved tensile modulus and strength [37], enhanced thermomechanical properties, and elastic modulus [34]. Several experimental studies conducted on PBT nanocomposites can be found in the literature [38–40].

In this chapter, we describe a method for fabricating PBT nanocomposites using in situ intercalation polymerization. The objectives of this study are to evaluate the effects of varying the amount of organoclay and the draw ratio (DR) on the properties of the PBT hybrid fibers. To obtain nanocomposites that do not thermally degrade during processing, we used thermally stable organoclays (i.e., C_{12}PPh-MMT, NCT-MMT, and C_{12}PPh-Mica). This chapter also examines the effects of the addition of these organoclays on the thermal behaviors, tensile mechanical properties, and morphologies of the hybrid fibers. The properties of the fibers from PBT hybrids containing various organoclays were compared with those of PBT fibers.

2 Experimental Materials

The source clay, i.e., Kunipia-F (Na^+-montmorillonite), was obtained from Kunimine Co. By screening this Na^+-montmorillonite (Na^+-MMT) clay through a 325-mesh sieve to remove impurities, we obtained clay with a cationic exchange capacity of 119 meq/100 g. Na^+-type fluorinated synthetic mica (Na^+-Mica), which features fluorine in place of the free –OH group of the mica, was supplied by CO-OP Ltd., Japan. Its cationic exchange capacity was found to be in the 70–80 meq/100 g range.

All reagents were purchased from TCI and Aldrich Chemical Co. The commercially available solvents were purified by distillation. Common reagents were used without further purification.

3 Preparation of Organoclays

Dispersions of Na^+-MMT and Na^+-Mica were added to solutions of the phosphonium salt of dodecyltriphenyl (C_{12}PPh-). The two organophilic clays were obtained through a multistep process described in our previous paper [30, 41], and they are denoted C_{12}PPh-MMT and C_{12}PPh-Mica. The chemical structures of the C_{12}PPh-clays are as follows:

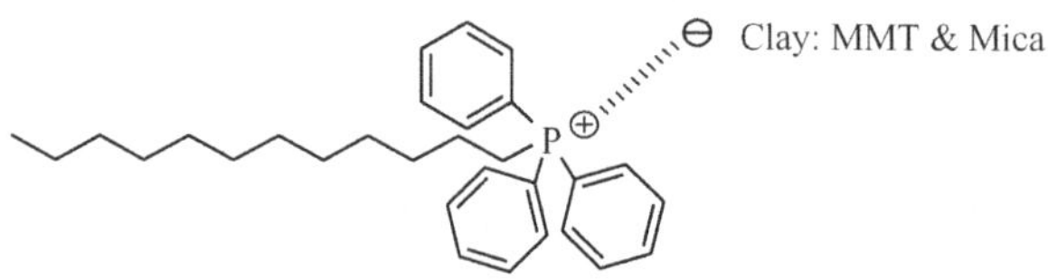

The third organically modified montmorillonite used in this study, i.e., NCT-MMT, was kindly supplied by NC Technology, Co. and was synthesized through an ion-exchange reaction between Na^+-montmorillonite (Na^+-MMT) and alkyl ammonium (NCT) chloride. The chemical structure of NCT-MMT is as follows:

$$\overset{\ominus}{\text{MMT}}\ \overset{\oplus}{\text{N}}\text{H}_3\text{CH}_2\text{CH}_2\text{NHCH}_2\text{CH}_2\text{NH}\overset{\overset{\text{O}}{\|}}{\text{C}}(\text{CH}_2)_{34}\overset{\overset{\text{O}}{\|}}{\text{C}}\text{NHCH}_2\text{CH}_2\text{NHCH}_2\text{CH}_2\text{NH}_2$$

The general properties of the various organoclays used in this work are shown in Table 14.1.

Table 14.1 General properties of organoclays for PBT hybrids

Organoclays	T_D^i [a] (°C)	wt_R^{600} [b] (%)	2θ (degree)	*d* in XRD (Å)
C_{12}PPh-MMT	304	79	2.86	36.08
NCT-MMT	289	81	3.98	25.75
C_{12}PPh-Mica	305	62	3.68	27.63

[a]Initial weight loss onset temperature
[b]Weight percent of residue at 600 °C

4 Preparation of PBT/Organoclay Hybrids

All the samples were prepared via in situ intercalation. Since the synthetic procedures for the hybrids were all very similar, only a representative example, i.e., the procedure for the preparation of the nanocomposite containing 2 wt% C_{12}PPh-MMT, is described here. BD (90.1 g, 1.0 mol) and C_{12}PPh-MMT (2.24 g) were placed in a polymerization tube and stirred for 30 min at room temperature. Dimethyl terephthalate (DMT) (97 g, 0.5 mol) and isopropyl titanate (60 mg, 2.1×10^{-4} mol) were placed in a separate tube to which the organoclay/BD system was then added. Mechanical stirring resulted in a homogeneously dispersed system. This mixture was heated for 1 h at 190 °C under a steady stream of N_2 gas. The temperature of the reaction mixture was then increased to 230 °C and maintained for 2 h under a steady stream of N_2 gas. During this period, continuous generation of methanol was observed. Finally, the mixture was heated for 2 h at 260 °C at a pressure of 1 Torr. The product was then cooled to room temperature and repeatedly washed with water followed by drying under vacuum at 70 °C for 1 day to obtain the hybrid. A schematic of the hybrid synthesis is shown in Scheme 14.1.

The polymers are soluble in mixed solvents. Therefore, phenol/1,1,2,2-tetrachloroethane (w/w = 50/50) was used as the solvent for the

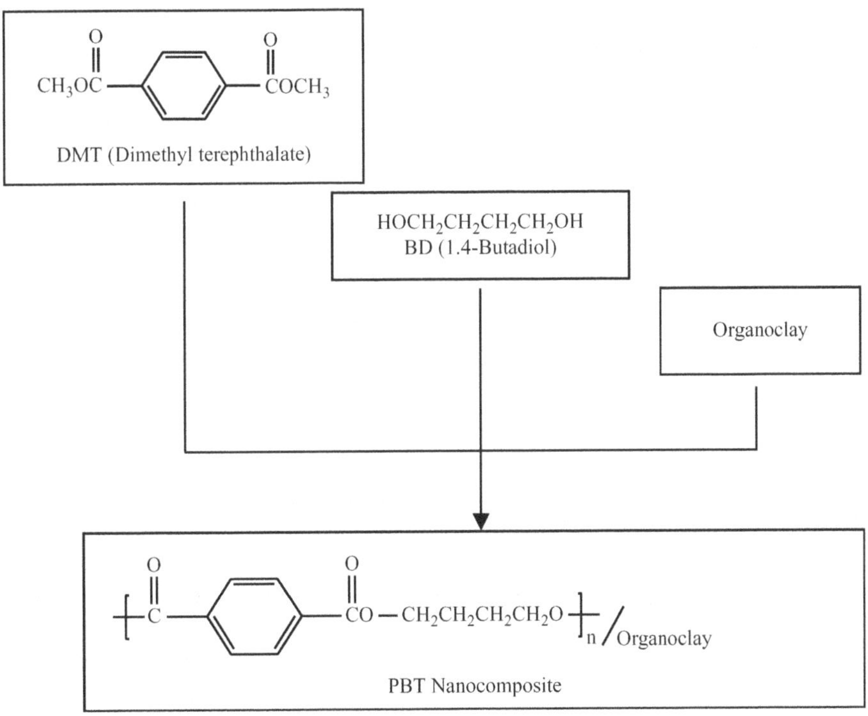

Scheme 14.1 Preparation of PBT nanocomposite

measurement of solution viscosity. The solution viscosity numbers range from 0.80–0.95 for the C_{12}PPh-MMT hybrids, 0.77–1.16 for the NCT-MMT hybrids, and 0.81–0.91 for the C_{12}PPh-Mica hybrids (Table 14.2). Considering that these viscosity values were obtained from pure PBT in PBT hybrids with the clay content removed, these numbers can be regarded as being constant.

We attempted to synthesize PBT hybrids containing more than 2 wt% C_{12}PPh-MMT via in situ intercalation. However, repeated attempts to polymerize the 3 wt% C_{12}PPh-MMT/PBT hybrid failed due to bubbles produced in the polymerization reactor during the transesterification of DMT and BD. The problem of how to produce high-molecular-weight polymer hybrids with high organoclay contents without the formation of bubbles remains unsolved.

5 Extrusion

The composites were pressed at 235 °C and 2,500 kg/cm^2 for 2–3 min on a hot press. The obtained films, which were ~0.5 mm thick, were dried in a vacuum oven at 110 °C for 24 h and then extruded through the die of a capillary rheometer. The hot extrudates were stretched through the die of the capillary rheometer (Instron 5460) at 235 °C and immediately drawn at the constant speed of the take-up machine to form fibers with different DRs. The pure PBT and PBT hybrids were each extruded through a capillary die into fibers with varying DR values; the thermal properties and the tensile mechanical properties of the extrudates were then examined. The standard die diameter (DR = 1) was 0.75 mm. All the hybrid fibers obtained from the capillary rheometer were bright yellow. The DR was calculated from the ratio of the velocity of extrusion to the take-up speed. The mean residence time in the capillary rheometer was ~3–4 min.

6 Characterization

The thermal behavior of the fibers was studied using a DuPont model 910 differential scanning calorimeter (DSC) and a thermogravimetric analyzer (TGA) at a heating rate of 20 °C/min under N_2 flow. Wide-angle X-ray diffraction (XRD) measurements were performed at room temperature using a Rigaku X-ray diffractometer (D/Max-IIIB) with Ni-filtered Cu-Kα radiation. The d-spacing experimental standard uncertainty was ±1.2 Å. The scanning rate was 2°/min over a range of $2\theta = 2$–$10°$.

The tensile properties of the fibers were determined at room temperature using an Instron mechanical tester (Model 5564) at a crosshead speed of 20 mm/min. The experimental uncertainties in the ultimate strength and initial modulus were determined to be ±1 MPa and ±0.05 GPa, respectively, from more than ten independent measurements.

The morphologies of the fractured surfaces of the extruded fibers were investigated using a Hitachi S-2400 scanning electron microscopy (SEM). An SPI sputter

Table 14.2 Thermal properties of PBT hybrid fibers

Organoclay (wt%)	D.R.[a]	C_{12}PPh-MMT I.V.[b]	T_g (°C)	T_m (°C)	T_D^{i}[c] (°C)	wt_R^{600}[d] (%)	NCT-MMT I.V.	T_g (°C)	T_m (°C)	T_D^{i} (°C)	wt_R^{600} (%)	C_{12}PPh-Mica I.V.	T_g (°C)	T_m (°C)	T_D^{i} (°C)	wt_R^{600} (%)
0 (pure PBT)	1	0.85	27	222	371	1	0.84	27	222	371	1	0.91	27	222	366	1
1	1	0.95	36	227	377	6						0.89	33	223	371	2
2	1	0.80	37	227	375	10	1.16	33	230	390	6	0.83	32	222	373	3
	12		36	227	376	10										
	18		36	227	376	10										
3	1						0.77	34	230	388	7	0.81	33	223	374	7
	12												33	222	374	7
	18												34	222	374	7
4	1						0.88	33	229	390	7					
5	1						0.86	33	231	389	9					

[a]Draw ratio
[b]Inherent viscosities were measured at 30 °C using 0.1 g/100 mL solutions in a phenol/1,1,2,2-tetrachloroethane (w/w = 50/50) mixture
[c]Initial weight-loss onset temperature
[d]Weight percent of residue at 600 °C

coater was used to sputter coat the fractured surfaces with gold for enhanced conductivity. The samples were prepared for transmission electron microscopy (TEM) by placing the PBT hybrid fibers into epoxy capsules and curing the epoxy at 70 °C for 24 h in vacuo. The cured epoxies containing the PBT hybrids were then microtomed into 90 nm thick slices, and each slice was deposited onto a mesh 200 copper net. TEM photographs of ultrathin sections of the polymer/organoclay hybrid samples were obtained on a LEO 912 OMEGA (Carl Zeiss) using an acceleration voltage of 120 kV. The dimensions of the morphology were determined using the scale bar in each microphotograph.

7 Results and Discussion

7.1 Dispersion of the Organoclays in PBT

X-ray scattering intensities for the organoclays and PBT/organoclay hybrid fibers with various clay contents are shown in Figs. 14.1–14.3. The organic-modified clays exhibit improved compatibility with the polymers resulting in easy intercalation of the clay galleries with the polymer. As expected, ion exchange between the Na^+-clays (i.e., Na^+-MMT and Na^+-Mica) and dodecyltriphenylphosphonium chloride (C_{12}PPh-Cl^-) resulted in an increase in the basal interlayer spacing over that found in pristine clays. Concurrently, there was a significant shift of the diffraction peak toward lower values of 2θ: the d-spacing increased from 11.99 ($2\theta = 8.56°$) to 36.08 Å ($2\theta = 2.86°$) for C_{12}PPh-MMT, from 11.99 to 25.75 Å ($2\theta = 3.98°$) for NTC-MMT, and from 9.57 ($2\theta = 9.23°$) to 27.63 Å ($2\theta = 3.68°$) for C_{12}PPh-Mica (Figs. 14.1–14.3). The interlayer spacing, d, of C_{12}PPh-MMT was the largest of the three. In general, increased interlayer spacing enhances the intercalation of the polymer chains and leads to easier dissociation of the clay, thereby resulting in hybrids with better clay dispersion [42–44]. A second XRD peak due to C_{12}PPh-MMT is observed in Fig. 14.1 at $2\theta = 5.65°$ ($d = 18.20$ Å) and is associated with the (002) plane of the organoclay layers.

When the C_{12}PPh-MMT/PBT hybrid fibers form, the organoclay peak at $2\theta = 2.86°$ ($d = 36.08$ Å) disappears from the diffraction pattern (Fig. 14.1). This result indicates that the clay layers are exfoliated and dispersed homogeneously in the PBT matrix and provides supplementary evidence that the PBT/C_{12}PPh-MMT hybrid fibers are nanocomposites. XRD offers a convenient way to determine the interlayer spacing due to the periodic arrangement of silicate layers in virgin clay and intercalated polymer-layered hybrids. Although XRD is very useful for measuring the d-spacing of ordered immiscible and ordered intercalated polymer nanocomposites, it may be insufficient for the analysis of disordered and exfoliated materials that do not feature peaks. Therefore, electron microscopy analyses provide the principal evidence for the formation of nanocomposites in our study.

Figure 14.2 illustrates the XRD patterns of an organoclay and its hybrid. As mentioned previously, the characteristic peak for pristine Na^+-MMT clay appears at

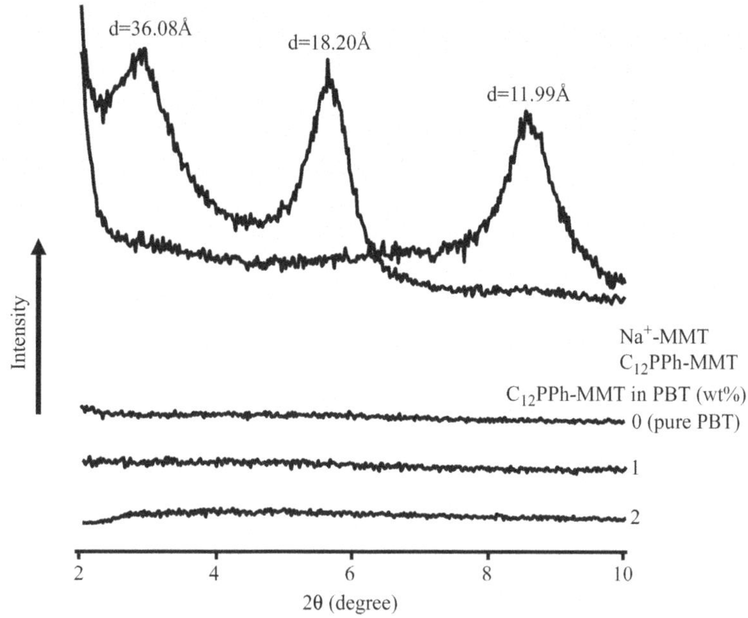

Fig. 14.1 XRD patterns for clay, C_{12}PPh-MMT, and PBT hybrid fibers with various C_{12}PPh-MMT contents

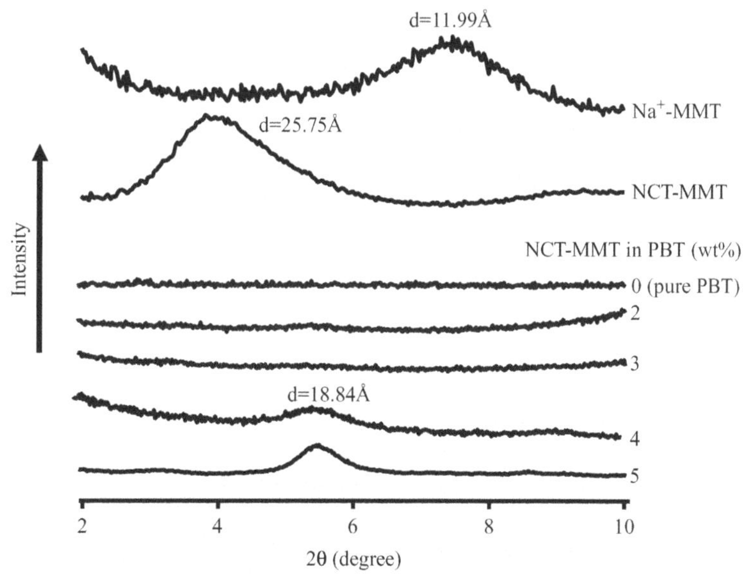

Fig. 14.2 XRD patterns for clay, NCT-MMT, and PBT hybrid fibers with various NCT-MMT contents

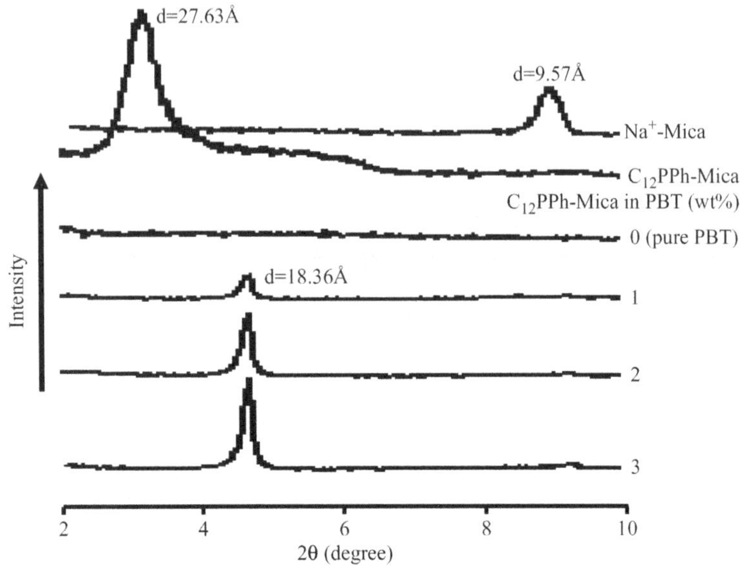

Fig. 14.3 XRD patterns for clay, C_{12}PPh-Mica, and PBT hybrid fibers with various C_{12}PPh-Mica contents

$2\theta = 8.56°$ ($d = 11.99$ Å). For Na^+-MMT reacted with alkylamine, i.e., NCT-MMT, this peak broadens and shifts to $2\theta = 3.98°$ ($d = 25.75$ Å): the difference in *d*-spacing suggests that the clay is swollen. Figure 14.2 also shows the XRD curves of pure PBT and PBT hybrid fibers with 2–5 wt% NCT-MMT loadings. Pure PBT synthesized with an MMT interlayer exhibits its usual XRD peaks. However, the spectra of the 2 and 3 wt% PBT hybrids show no characteristic organoclay peaks in the range of $2\theta = 2$–$8°$: the peak corresponding to the basal spacing has disappeared. However, in the spectra of PBT hybrids with 4 and 5 wt% organoclay loadings, a small peak is observed at $2\theta = 5.44°$ ($d = 18.84$ Å), which indicates that a small part of the clay has agglomerated in the PBT matrix.

As shown in Fig. 14.3, the XRD results for the PBT extrudate fibers with 1 wt% C_{12}PPh-Mica revealed only a small peak at $d = 18.36$ Å ($2\theta = 4.81°$). This indicates that the mica layers of the organoclays are intercalated (not exfoliated) and are not homogeneously dispersed in the PBT matrix. Substantial increases in the intensities of the XRD peaks were observed as the clay loading increased from 1 to 3 wt% suggesting that dispersion is more effective at lower clay loadings than at higher clay loadings. Accordingly, higher clay loadings are expected to result in increased agglomeration of some portion of the clay within the PBT matrix. The disappearance of the main peaks for the PBT hybrid fibers (at $2\theta = 2.86°$ and $2\theta = 3.68°$ for C_{12}PPh-MMT and C_{12}PPh-Mica, respectively) is thought to occur because the diffraction peaks corresponding to the organoclay swollen with polymer chains are present at $2\theta < 2°$ [43, 44].

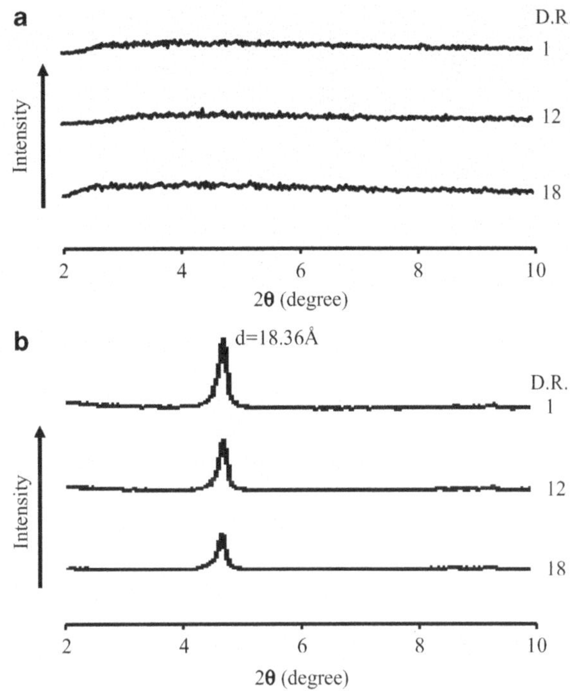

Fig. 14.4 XRD patterns of PBT hybrid fibers with 2 wt% organoclay at different draw ratios: (**a**) C_{12}PPh-MMT and (**b**) C_{12}PPh-Mica

Figure 14.4a shows the XRD results for PBT hybrid fibers containing 2 wt% C_{12}PPh-clay at various DR values. When the DR was increased from 1 to 18, no obvious clay peaks appeared, as shown in Fig. 14.4a. Similar to the case of C_{12}PPh-MMT, no obvious clay peaks were found in PBT hybrid fibers containing 2 wt% NCT-MMT for DR values from 1 to 6 (not shown).

Figure 14.4b also shows the XRD results for PBT hybrid fibers containing 2 wt% C_{12}PPh-Mica at various DR values. For PBT with an organoclay content of 2 wt%, a peak is present at $d = 18.36$ Å ($2\theta = 4.81°$) for a DR of 1. In contrast to the results for the C_{12}PPh-MMT hybrids, substantial decreases in the XRD peak intensities were observed with increasing DRs from 1 to 18, which suggests that an increase in the DR decreases the diameter and clay content in the hybrid fibers leading to decreased XRD peaks.

7.2 Morphology

SEM and TEM were used to visually evaluate the degree of intercalation and amount of aggregation of clay clusters. It is instructive to consider the size and distribution of the clay particles as a prelude to understanding the morphology of the polymer nanocomposites shown in the micrographs.

SEM images of the fractured surfaces of PBT hybrid fibers containing the three different organoclays are shown in Figs. 14.5–14.9. The morphologies of

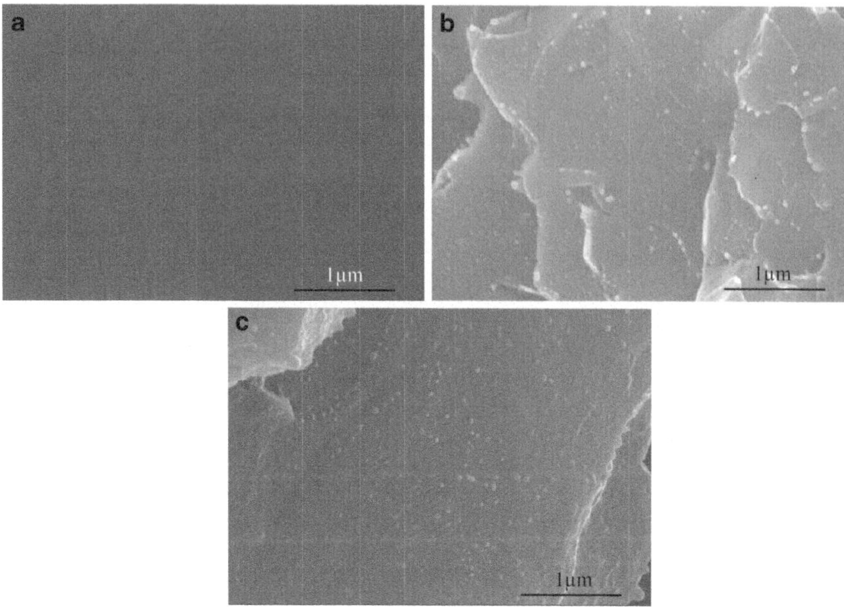

Fig. 14.5 SEM photographs of PBT hybrid fibers with (**a**) 0 wt% (pure PBT), (**b**) 1 wt%, and (**c**) 2 wt% C_{12}PPh-MMT

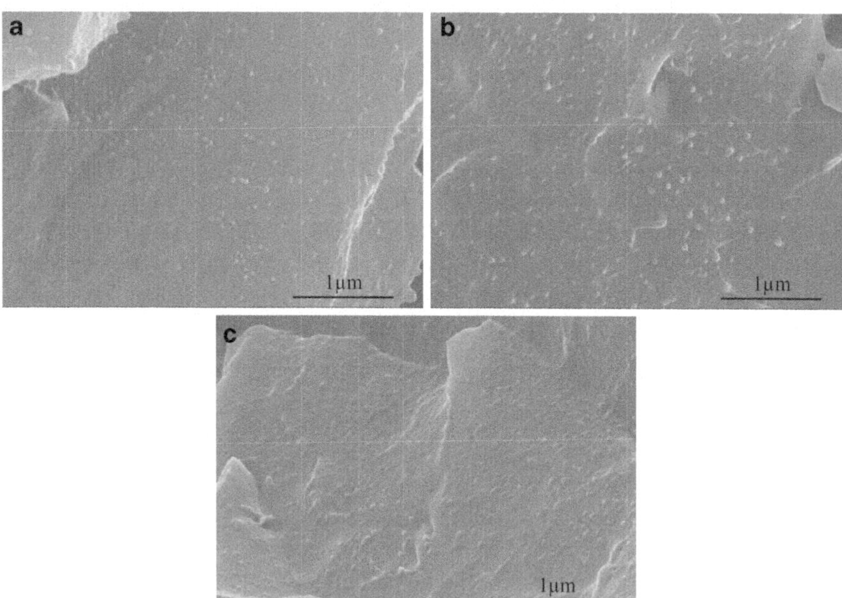

Fig. 14.6 SEM photographs of PBT hybrid fibers with 2 wt% C_{12}PPh-MMT at draw ratios of (**a**) 1, (**b**) 12, and (**c**) 18

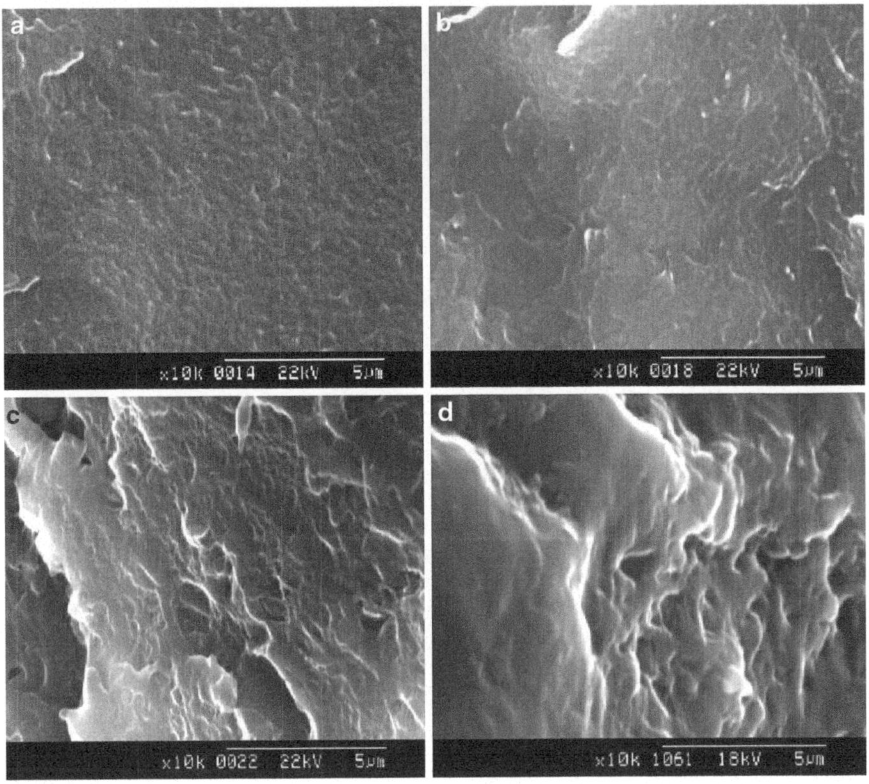

Fig. 14.7 SEM photographs of PBT hybrid fibers with (**a**) 0 wt% (pure PBT), (**b**) 3 wt%, (**c**) 4 wt%, and (**d**) 5 wt% NCT-MMT

the extruded fibers obtained from the PBT hybrid systems with various contents of C_{12}PPh-MMT were determined by performing SEM examinations of their fracture surfaces; the results are shown in Fig. 14.5. Figure 14.5 shows the clay phases that formed in undrawn hybrid fibers with organoclay contents ranging from 0 to 2 wt%; the PBT hybrid fibers with 0–2 wt% C_{12}PPh-MMT have morphologies consisting of 80–100 nm clay domains that are well dispersed in a continuous PBT phase. Figure 14.6 shows micrographs of 2 wt% C_{12}PPh-MMT/PBT hybrid fibers obtained at DRs ranging from 1 to 18. The 2 wt% hybrid fiber with a DR of 12 contains fine clay phases with diameters of 60–110 nm (Fig. 14.6b). Similarly, the hybrid fiber with a DR of 18 exhibits a fine dispersion with domains 30–50 nm in diameter (Fig. 14.6c). The domain size of the dispersed clay phase was found to decrease linearly with increasing DR. The fine dispersion of clay is evidently the result of the stretching of the fiber that occurs when the extrudates pass through the capillary rheometer.

SEM micrographs of the fractured surfaces of NCT-MMT/PBT hybrid fibers prepared with different clay contents are shown in Fig. 14.7. The micrographs of the pure PBT (Fig. 14.7a) and the PBT hybrid fiber containing 3 wt%

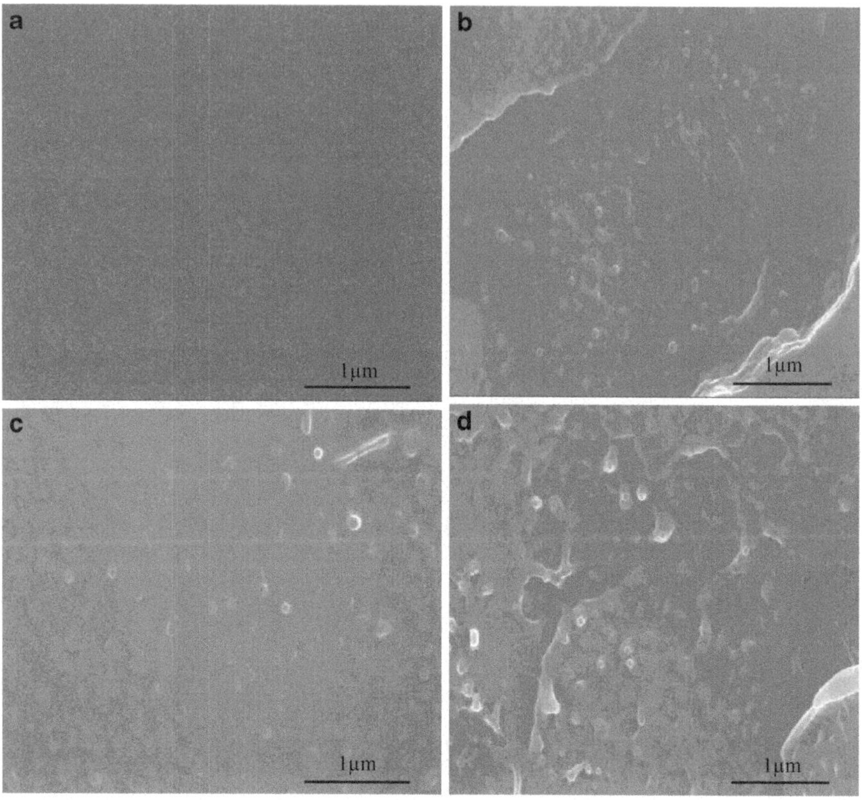

Fig. 14.8 SEM photographs of PBT hybrid fibers with (**a**) 0 wt% (pure PBT), (**b**) 1 wt%, (**c**) 2 wt%, and (**d**) 3 wt% C_{12}PPh-Mica

NCT-MMT (Fig. 14.7b) display smooth surfaces due to well-dispersed clay particles. Conversely, Fig. 14.7c, d shows voids and deformed regions that may result from the coarseness of the fractured surfaces. However, the fractured surfaces of the hybrids featured more deformities when higher contents of organoclay were used. This is probably due to agglomeration of clay particles [45, 46]. These results are consistent with the XRD data shown in Fig. 14.2.

The morphologies of PBT hybrid fibers with 0–3 wt% C_{12}PPh-Mica at a DR of 1 consist of 100–250 nm clay domains (Fig. 14.8). Figure 14.9 shows micrographs of 2 wt% C_{12}PPh-Mica/PBT hybrid fibers obtained at various DRs in the range of 1–18. The 2 wt% hybrid fiber with a DR of 12 contains fine clay phases that are 100–200 nm in diameter (Fig. 14.9b). Similarly, the hybrid fiber with a DR of 18 also contains pullout particles with domain sizes of 100–150 nm (Fig. 14.9c). The average clay particle size was found to decrease slightly with increasing DR, which is consistent with the XRD results shown in Fig. 14.4b.

To examine the precise dispersion of the clay layers in the fiber hybrids, we performed TEM studies. TEM enables a qualitative understanding of the internal

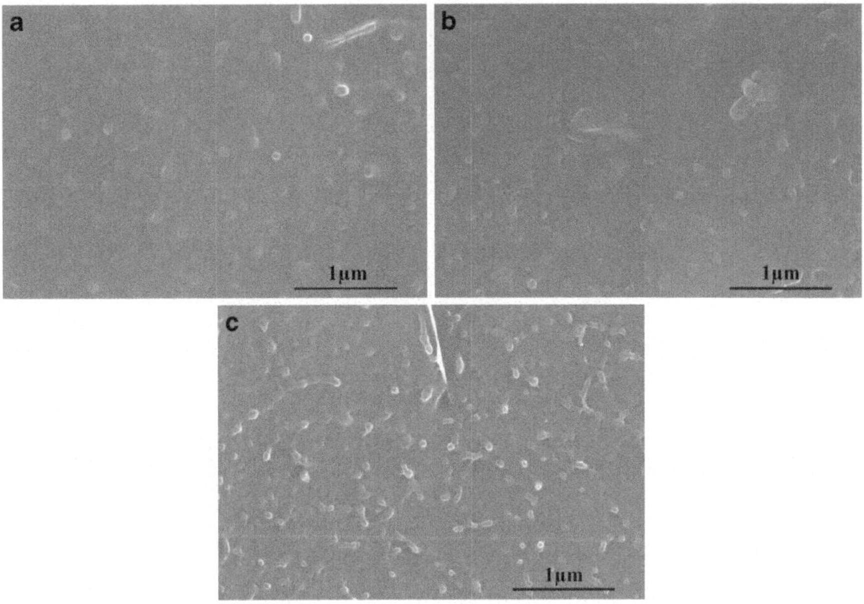

Fig. 14.9 SEM photographs of PBT hybrid fibers with 2 wt% C_{12}PPh-Mica at draw ratios of (**a**) 1, (**b**) 12, and (**c**) 18

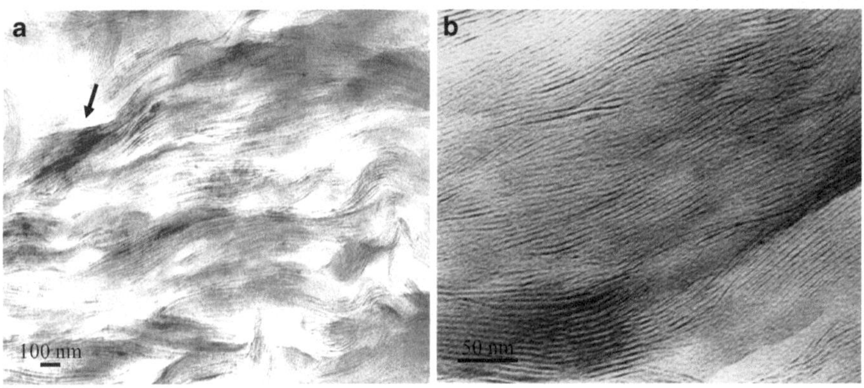

Fig. 14.10 TEM photographs of PBT hybrid fibers with 2 wt% C_{12}PPh-MMT at (DR = 1) increasing magnification levels (from (**a**) to (**b**))

structure through direct observation. Accordingly, direct evidence of the formation of true nanocomposites is provided by the TEM micrographs of the ultramicrotomed sections. The TEM micrographs in Figs. 14.10–14.12 show the structures of the fibers at a DR of 1 at different magnifications. TEM micrographs of C_{12}PPh-MMT/PBT hybrid fibers containing 2 wt% C_{12}PPh-MMT are shown in Fig. 14.10; the dark lines are the intersections of the 1 nm thick sheet layers.

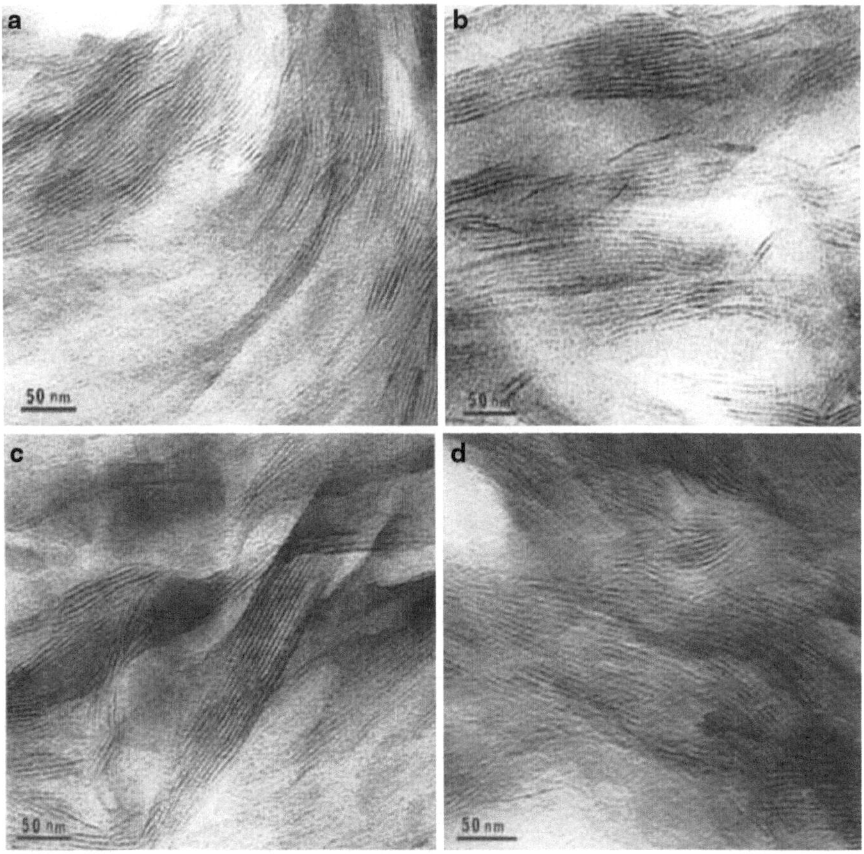

Fig. 14.11 TEM photographs of PBT hybrid fibers with (**a**) 2 wt%, (**b**) 3 wt%, (**c**) 4 wt%, and (**d**) 5 wt% NCT-MMT at a DR of 1

Increasing the magnification in an area occupied by an aggregate reveals that the individual sheets of MMT clay are separated by a layer of polymer, as shown in Fig. 14.10b. Thus, the morphology comprises a mixture of intercalated and exfoliated sheets, i.e., there are regions with a regular stacking arrangement of a layer of polymer between the sheets and regions where completely delaminated sheets are individually dispersed.

TEM micrographs of NCT-MMT/PBT hybrid fibers prepared with different clay contents are shown in Fig. 14.11. Some of the clay layers shown in Fig. 14.11a, b features the dispersion of individual delaminated sheets in the matrix as well as regions with a regular stacking arrangement of a layer of polymer between the sheets. Although face-to-face layer morphology is retained, the layers are irregularly separated by ~4–10 nm of polymer. However, for the 4 and the 5 wt% organoclay-loaded PBT hybrid fibers (Fig. 14.11c, d), some of the clay is well dispersed in the PBT matrix and some forms approximately 4–8 nm agglomerates;

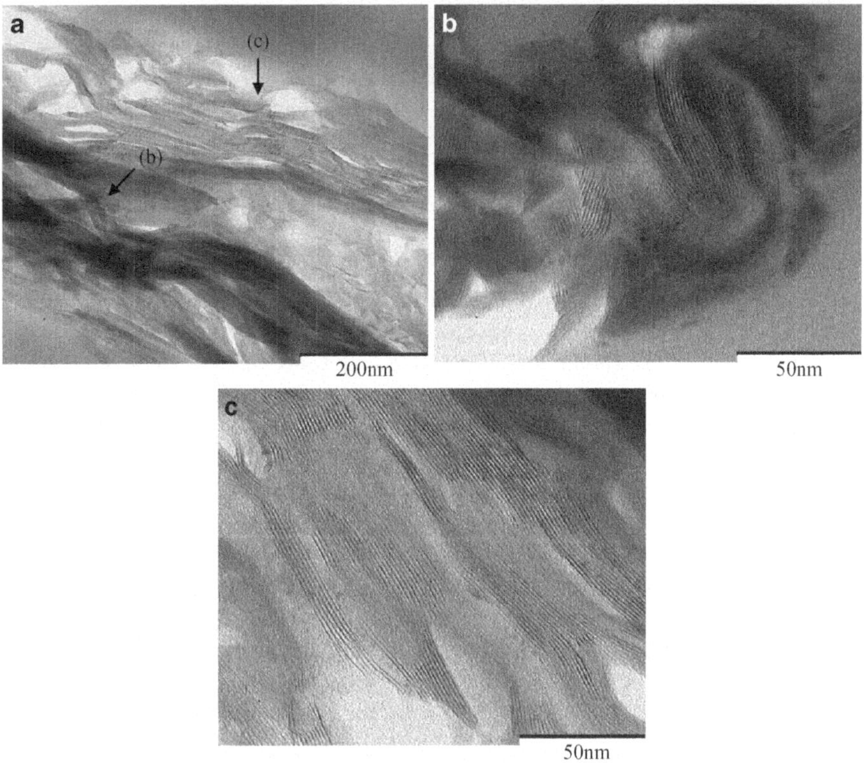

Fig. 14.12 TEM photographs of PBT hybrid fibers with 2 wt% C_{12}PPh-Mica (DR = 1) at increasing magnification levels (from (**a**) to (**c**))

this is consistent with the XRD results shown in Fig. 14.2. From the XRD and electron micrographs results, the morphology at a low organoclay content (<3 wt%) presents a mixture of intercalated and partially exfoliated features. Dispersion is better at a lower organoclay loading than at a high organoclay loading.

Figure 14.12 shows TEM photographs of 2 wt% C_{12}PPh-Mica/PBT hybrid fibers that reveal that the organoclay is well dispersed in the polymer matrix at all magnification levels although some of the clay particles form approximately 20 nm agglomerates. The peaks in the XRD patterns of these samples (Fig. 14.3) are attributed to these agglomerated layers. Compared to the hybrids containing mica, the MMT layers of the C_{12}PPh-MMT and NCT-MMT hybrid fibers (Figs. 14.1, 14.2, 14.10, and 14.11) are more exfoliated within the matrix polymer.

Figures 14.10–14.12 also show that the clay layers are homogeneously dispersed within the matrix polymer, which indicates that nanocomposites formed although some clusters or agglomerated particles were also detected. Therefore, the clays such as MMT and mica are broken down into nanoscale building blocks via in situ intercalative polymerization and dispersed homogeneously in polymer matrices to afford polymer/clay nanocomposites. The unusual thermomechanical properties of these nanocomposites, which are attributed to the homogeneous dispersion of the silicate nanoscale particles, are discussed in the following sections.

7.3 Thermal Behaviors

Table 14.2 presents the thermal behaviors of pure PBT and PBT/organoclay hybrid fibers obtained via in situ intercalation with various DR values. The glass transition temperatures (T_g), melting transition temperatures (T_m), and initial decomposition temperatures at 2 % weight loss ($T_D{}^i$) of the hybrids increase with increasing C_{12}PPh-MMT content up to 1 wt% and then remain constant with further increases in organoclay loading. For example, the T_g, T_m, and $T_D{}^i$ of C_{12}PPh-MMT/PBT hybrid fibers with 1 wt% clay loading are 9, 5, and 6 °C higher, respectively, than those of pure PBT; increasing the clay loading to 2 wt% does not significantly alter these properties. The TGA results for the PBT hybrid fibers with various C_{12}PPh-MMT contents are shown in Fig. 14.13.

The thermal properties of NCT-MMT/PBT hybrids with different organoclay contents are also listed in Table 14.2. The T_g of the PBT hybrids increased from 27 to 33 °C with increased clay loading from 0 to 2 wt% and then remained fairly constant up to 5 wt%. The endothermic peak of pure PBT, which corresponds to the T_m, appears at 222 °C. Similar to the T_g results, the DSC thermograms show that the value of T_m increases from 222 to 230 °C with increasing organoclay content up to 2 wt% and then remains constant for additional organoclay loading up to 5 wt% (Table 14.2). This change in the thermal behavior of the hybrids may result from the heat insulation effect of the clay layer structure, as well as from the interactions between the organoclay and PBT molecular chains [47, 48].

The thermal stabilities determined from TGA analyses of PBT/NCT-MMT are shown in Table 14.2 and Fig. 14.14. In addition to having a higher glass transition temperature and melting temperature, the PBT hybrids are also more thermally stable than pure PBT. The $T_D{}^i$ of the hybrid clearly increases with increasing NCT-MMT content up to 2 wt% and then remains constant regardless of the organoclay loading. At 2 % PBT/NCT-MMT weight loss in the TGA curves, the values of $T_D{}^i$ for the hybrid increased from 388 to 390 °C with increasing clay content from 2 to 5 wt%. The weight loss due to the decomposition of PBT and its hybrids was nearly constant up to 350 °C (Fig. 14.14). Above 350 °C, the $T_D{}^i$ was influenced by the organoclay loading in the hybrids. The addition of clay enhanced the thermal performance by acting as a superior insulator and as a mass-transport barrier to the volatile products generated during decomposition. This improvement in thermal stability has been observed in many hybrid systems [49–52].

Similarly, the T_g and $T_D{}^i$ of the C_{12}PPh-Mica hybrid fibers with 1 wt% clay loading are 6 and 5 °C higher, respectively, than those of pure PBT and do not vary significantly up to 3 wt% clay content. The TGA curves for the clay, organoclay, and PBT hybrid fibers with various C_{12}PPh-Mica contents are shown in Fig. 14.15.

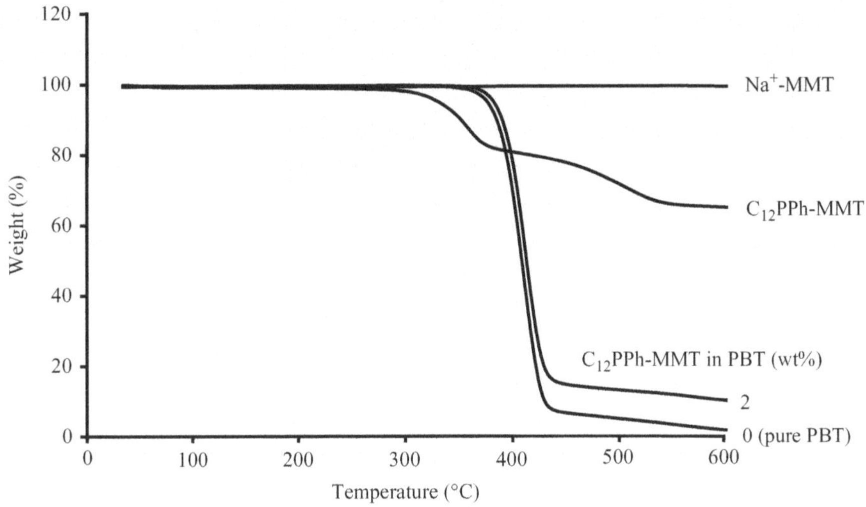

Fig. 14.13 TGA thermograms of clay, C_{12}PPh-MMT, and PBT hybrid fibers with various C_{12}PPh-MMT contents

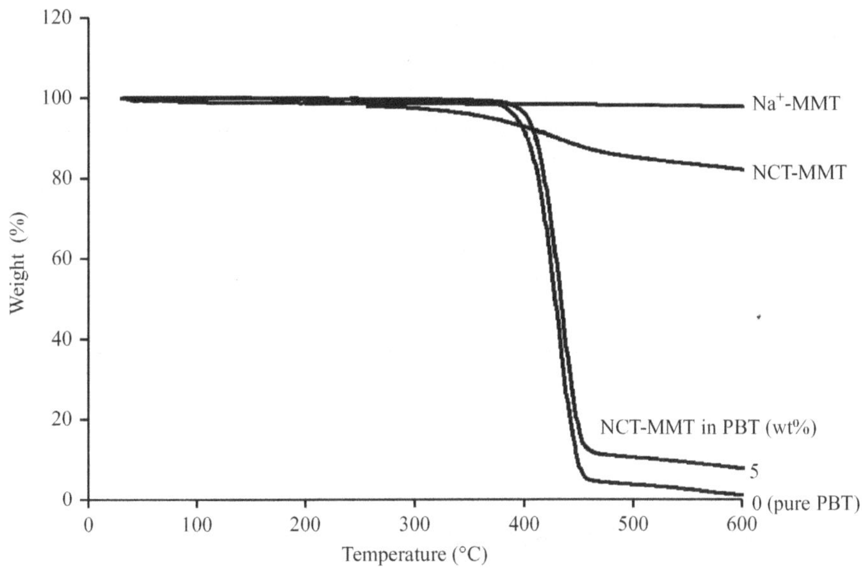

Fig. 14.14 TGA thermograms of clay, NCT-MMT, and PBT hybrid fibers with various NCT-MMT contents

The observed increases in the glass transition temperatures of these hybrids with the addition of the organoclay (Table 14.2) could be the result of several different factors [47, 53]; in particular, the increase in cross-link density and restriction of the segmental relaxation of the chain segments near the clay layers should positively

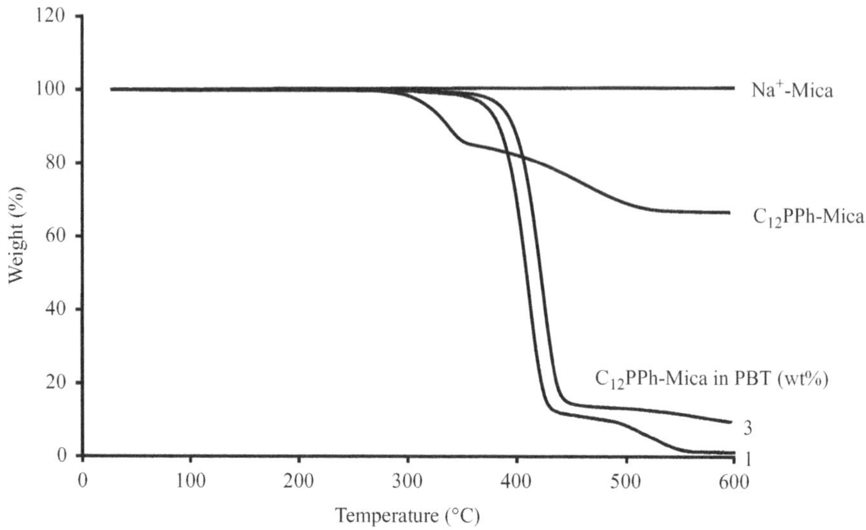

Fig. 14.15 TGA thermograms of clay, C_{12}PPh-Mica, and PBT hybrid fibers with various C_{12}PPh-Mica contents

affect the T_g. Other studies of polymer nanocomposites reported similar results [54, 55]. The increase in T_m with the addition of organoclay may result from the heat-insulating effects of the clay layer structure as well as from the interactions between the organoclay and the PBT molecular chains [56]. The presence of the clay also enhances the initial decomposition temperatures by acting as an insulator and as a mass-transport barrier to the volatile products generated during decomposition [48, 50, 52]. This increase in thermal stability can also be attributed to the high thermal stability of the clay and to the interactions between the clay particles and the polymer matrix.

In contrast to the results for the thermal properties, the weight of the residue at 600 °C increased monotonically with increased clay loading, as shown in Table 14.2. This enhancement in char formation with increasing organoclay content is ascribed to the high heat resistance of the clay [49].

7.4 Mechanical Tensile Properties

The tensile mechanical properties of pure PBT and the PBT hybrids are listed in Table 14.3. Pure PBT and its hybrid fibers were extruded with increasing DRs to examine their tensile mechanical properties. It is evident that the ultimate strengths of the C_{12}PPh-MMT hybrids increase gradually with increasing organoclay content and are highest for a clay content of 2 wt% and a DR of 1. The ultimate tensile strength of the 2 wt% PBT hybrid fibers is 58 MPa, which is about 41 % higher than that of pure PBT (41 MPa).

Table 14.3 Tensile properties of PBT hybrid fibers

Organoclay (wt%)	D.R.[a]	C_{12}PPh-MMT Ult. Str. (MPa)	C_{12}PPh-MMT Ini.Mod. (GPa)	C_{12}PPh-MMT E.B.[b] (%)	NCT-MMT Ult. Str. (MPa)	NCT-MMT Ini.Mod. (GPa)	NCT-MMT E.B.[b] (%)	C_{12}PPh-Mica Ult. Str. (MPa)	C_{12}PPh-Mica Ini.Mod. (GPa)	C_{12}PPh-Mica E.B.[b] (%)
0 (pure PBT)	1	41	1.37	3	41	1.37	3	41	1.37	3
	3				50	1.49	4			
	6				52	1.52	3			
	12	43	1.69	3				43	1.69	3
	18	46	1.71	3				46	1.77	3
1	1	55	2.96	3				48	2.21	4
	12	52	2.57	3				44	2.25	3
	18	48	2.33	2				42	2.28	3
2	1	58	3.33	2	50	1.66	5	50	2.61	3
	12	35	2.46	2				41	2.62	2
	18	28	2.21	2				41	2.67	2
3	1				60	1.76	4	47	2.80	3
	3				35	1.46	5			
	6				29	1.39	4			
	12							45	2.80	2
	18							44	2.87	2
4	1				53	1.80	3			
5	1				49	1.86	3			

[a]Draw ratio
[b]Elongation percent at break

At a DR of 1, the ultimate tensile strength of the PBT/NCT-MMT hybrid fibers increases with the addition of clay up to a critical clay loading then decreases above that critical content. For example, the strength of the 3 wt% PBT hybrid fibers is 60 MPa, which is about 50 % higher than that of pure PBT (41 MPa). However, at 5 wt% organoclay, the strength decreases to 49 MPa. This suggests that the NCT-MMT domains are more agglomerated above 3 wt% organoclay content in the PBT matrix [45, 57, 58]. Evidence for clay agglomeration in the PBT matrix polymer was also provided by XRD, SEM, and TEM, as shown in Figs. 14.2, 14.7, and 14.11.

Similarly, the ultimate tensile strength of the C_{12}PPh-Mica hybrid fibers at a DR of 1 increases with the addition of clay up to a critical clay loading then decreases above that critical content. For example, the strengths of the hybrids were found to increase from 41 to 50 MPa with an increase in the organoclay content to 2 wt%. This improvement arises because the organoclay layers are dispersed and intercalated in the PBT matrix and is consistent with the general observation that the introduction of an organoclay into a matrix polymer increases its tensile strength [57, 59, 60]. However, when the amount of organo-mica in PBT reaches 3 wt%, the strength decreases to 47 MPa. This decrease in ultimate strength is mainly due to the agglomeration of clay particles above the critical clay loading, as we previously reported [41, 61].

The modulus values were found to increase linearly with increasing organoclay content. At a 2 wt% C_{12}PPh-MMT content, the modulus of the hybrid was found to be 3.33 GPa, which is about 2.5 times that of pure PBT (1.37 GPa). The value of the initial modulus increased constantly from 1.37 to 1.86 GPa with increasing NCT-MMT content up to 5 wt%. This enhancement of the modulus is ascribed to the high resistance exerted by the clay. The initial modulus values of the C_{12}PPh-Mica hybrid fibers also increased significantly with increasing organoclay content. For an organoclay content of 3 wt%, the modulus of the hybrid was found to be 2.80 GPa, which is about twice that of pure PBT (1.37 GPa). This increased initial modulus could be the result of two factors: the high resistance of the clay, and the orientation and aspect ratio of the clay layers. Additionally, a higher clay content increases the constraints on polymer chain mobility, which also increases the modulus [62, 63].

On the basis of the aforementioned results, the enhancement of the mechanical properties can be directly attributed to the reinforcement provided by the intercalation of PBT in clay galleries as well as by the fine dispersion of clay particles in the polymer matrix. The increase in the tensile properties caused by the clay layers depends on the interactions between the rigid, rod-shaped polyester molecules and the layered clay as well as the rigidity of the clay layers.

The increases in the tensile strength and initial modulus of pure PBT with increasing DR values were found to be insignificant. For pure PBT, the strength and modulus increased from 41 to 46 MPa and 1.37 to 1.71 GPa, respectively, as the draw ratio increased from 1 to 18 (Table 14.3). Unlike flexible coil-like polymers, the values of the ultimate strength and initial modulus of the hybrid fibers decreased markedly with increasing DR, as shown in Table 14.3. For hybrid fibers with 1 wt% C_{12}PPh-MMT, the tensile strength and initial modulus decreased from 55 to

48 MPa and 2.96 to 2.33 GPa, respectively, as the DR increased from 1 to 18. Similar trends were observed for hybrid fibers with 2 wt% C_{12}PPh-MMT.

To examine the tensile mechanical properties of fibers of the 3 wt% NCT-MMT hybrid, they were also extruded with increasing DRs. As shown in Table 14.3, the strength and initial modulus values for pure PBT fibers were enhanced with increasing DR. However, as expected for a flexible coil polymer, the increases in tensile strength and modulus with increasing DR were insignificant for pure PBT. Table 14.3 shows that the strength and the modulus values of pure PBT increased slightly from 41 to 52 MPa and 1.37 to 1.52 GPa, respectively, with increasing DR from 1 to 6. In contrast, the strength and initial modulus values of the hybrid fibers decreased with increasing DR. For example, the ultimate strength and initial modulus of the hybrid with 3 wt% organoclay decreased from 60 to 29 MPa and from 1.76 to 1.39 GPa, respectively, when the DR increased from 1 to 6. These trends were observed in all our systems that contained 2–5 wt% organoclay.

A similar result was observed for the C_{12}PPh-Mica hybrid fibers. For example, for hybrid fibers with an organoclay content of 1 wt%, increases in DR from 1 to 18 resulted in decreases in the tensile strength from 48 to 42 MPa. Similar trends were observed for hybrid fibers with organoclay contents of 2 and 3 wt%. Increased tensile strength with increased DR is very common in engineering plastics and is usually observed for flexible coil-like polymers. However, our system did not follow this trend.

The decline in tensile properties seems to be the result of debonding of the organoclay and PBT and the presence of multiple micro-sized voids due to excess stretching of the fibers. Similar trends were observed by researchers studying other polymer hybrids [64–66]. Those researchers reported that an imperfect incursion/ matrix interface cannot sustain the large interfacial shear stress that develops as a result of an applied strain; this usually causes yielding and/or debonding to occur at or near the interface. As a result of imperfect bonding, debonding occurs at the interface, which results in constant interfacial shear stress when strain is applied to the composite. The effects of debonding and the presence of voids on the mechanical properties of our system with different DRs will be pursued in future study. The results of this study support two facts: better dispersion of the clay layer in the polymer and a strong interaction between the polymer and clay layer which are essential for achieving improved tensile properties in polymer nanocomposites.

In contrast to the diminished lower tensile strength values, the initial moduli of the C_{12}PPh-Mica hybrid fibers increased slightly with increases in the DR for a given clay content (Table 14.3). This enhancement of the modulus is primarily ascribed to the high resistance of the clay although the stretching resistance of the oriented backbones of the polymer chains in the gallery also contributes.

The elongation at breakage of the C_{12}PPh-MMT and C_{12}PPh-Mica hybrids remained virtually constant despite the variation of the DR, as shown in Table 14.3. For the NCT-MMT hybrid fibers, the percent elongation at the break for all fibers also ranged between 3 and 5 %. These values remained constant for organoclay loadings from 2 to 5 wt%. This result is characteristic of materials reinforced with stiff inorganic materials and is indicative of intercalated morphology.

8 Conclusions

We prepared PBT nanocomposites with organoclay dispersed at the nanometer-scale via in situ interlayer polymerization. PBT hybrid fibers containing various organoclays (i.e., C_{12}PPh-MMT, NCT-MMT, and C_{12}PPh-Mica) were extruded from a capillary rheometer with different DRs in order to investigate the thermomechanical properties and morphology of the hybrids.

Some clay layers were dispersed homogeneously in the matrix polymer although clusters were also detected. The thermal properties (i.e., T_g, T_m, and $T_D{}^i$) of the hybrid fibers were found to be better than those of pure PBT fibers. However, these thermal properties remained unchanged over various DR values. The tensile properties of the hybrid fibers increased gradually with increasing organoclay content at a DR of 1. However, the ultimate strengths and initial moduli of the hybrid fibers decreased markedly with increasing DR.

In this work, the properties of C_{12}PPh-MMT hybrid fibers were found to be better than those of NCT-MMT and C_{12}PPh-Mica hybrid fibers. In summary, the addition of a small amount of organoclay was sufficient to improve the properties of PBT matrix polymers.

References

1. Blumstein A (1961) Bull Chim Soc 899
2. Pinnavaia TJ (1983) Science 220:365
3. Messersmith PB, Giannelis EP (1993) Chem Mater 5:1064
4. Giannelis EP (1996) Adv Mater 8:29
5. Gilman JW (1999) Appl Clay Sci 15:31
6. Greenland DG (1963) J Colloid Sci 18:647
7. Chang JH, Park KM (2001) Polym Eng Sci 41:2226
8. Vaia RA, Jandt KD, Kramer EJ, Giannelis EP (1995) Macromolecules 28:8080
9. Vaia RA, Ishii H, Giannelis EP (1996) Adv Mater 8:29
10. Fukushima Y, Okada A, Kawasumi M, Kurauchi T, Kamigaito O (1988) Clay Miner 23:27
11. Akelah A, Moet A (1996) J Mater Sci 31:3589
12. Mascia L, Tang T (1998) Polymer 39:3045
13. Tamaki R, Chujo Y (1999) Chem Mater 11:1719
14. Lepoittevin B, Pantoustier N, Devalckenaere M, Alexander M, Dana K, Calberg C, Jerome R, Dubois P (2002) Macromolecules 35:8385
15. Tseng CR, Wu JY, Lee HY, Chang FC (2002) J Appl Polym Sci 85:1370
16. Wu J, Lerner MM (1993) Chem Mater 5:835
17. Vaia RA, Jandt KD, Kramer EJ, Giannelis EP (1996) Chem Mater 8:2628
18. Li X, Kang T, Cho W-J, Lee J-K, Ha C-S (2001) Macromol Rapid Commun 22:1306
19. LeBaron PC, Wang Z, Pinnavaia TJ (1999) Appl Clay Sci 12:11
20. Wang D, Zhu J, Yao Q, Wilkie CA (2002) Chem Mater 14:3837
21. Yano K, Usuki A, Okada A (1997) J Polym Sci Part A Polym Chem 35:2289
22. Garcia-Martinez JM, Laguna O, Areso S, Collar EP (2000) J Polym Sci Part B Polym Phys 38:1564
23. Takekoshi T, Khouri FF, Campbell JR, Jordan TC, Dai KH (1996) US 5,530,052
24. Li X, Park H-M, Lee J-O, Ha C-S (2002) Polym Eng Sci 42:2156

25. Matayabas JC Jr, Turner SR (2000) In: Pinnavaia TJ (ed) Polymer clay nanocomposites. Wiley, New York
26. Zhu J, Morgan AB, Lamelas FJ, Wilkie CA (2001) Chem Mater 13:3774
27. Zhu J, Uhl FM, Morgan AB, Wilkie CA (2001) Chem Mater 13:4649
28. Davis CH, Mathias LJ, Gilman JW, Schiraldi DA, Shields JR, Trulove P, Sutto TE, Delong HC (2002) J Polym Sci Part B Polym Phys 40:2661
29. Saujanya C, Imai Y, Tateyama H (2002) Polym Bull 49:69
30. Chang J-H, Kim SJ, Im S (2004) Polymer 45:5171
31. Runt J, Miley DM, Zhangg X, Gallagher KP, McFeaters K, Fishburn J (1992) Macromolecules 25:1929
32. Jeon HK, Kim JK (1998) Polymer 39:6227
33. Sasaki A, White JL (2003) J Appl Polym Sci 90:1839
34. Chow WS (2008) J Appl Polym Sci 110:1642
35. Acierno D, Lavorgna M, Piscitelli F, Russo P, Spena P (2011) Adv Polym Technol 30:41
36. Fabbri P, Bassoli E, Bon SB, Valentini L (2012) Polymer 53:897
37. Wu F, Yang G (2009) Mater Lett 63:1686
38. Acierno D, Scarfato P, Amendola E, Nocerino G, Costa G (2004) Polym Eng Sci 44:1012
39. Xiao J, Hu Y, Wang Z, Tang Y, Chen Z, Fan W (2005) Eur Polym J 41:1030
40. Nirukhe AB, Shertukde VV (2009) J Appl Polym Sci 113:585
41. Chang J-H, Mun MK (2006) J Appl Polym Sci 100:1247
42. Zilig C, Mulhaupt R, Finter J (1999) Macromol Chem Phys 200:661
43. Ke Y, Lu J, Yi X, Zhao X, Qi ZJ (2000) J Appl Polym Sci 78:808
44. Hsiao SH, Liou GS, Chang LM (2001) J Appl Polym Sci 80:2067
45. Chang JH, Park DK, Ihn KJ (2002) J Appl Polym Sci 84:2294
46. Chang JH, An YU, Sur GS (2003) J Polym Sci Part B Polym Phys 41:94
47. Chang JH, Seo BS, Hwang DH (2002) Polymer 43:2969
48. Fornes TD, Yoon PJ, Hunter DL, Keskkula H, Paul DR (2002) Polymer 43:5915
49. Wen J, Wikes GL (1996) Chem Mater 8:1667
50. Fischer HR, Gielgens LH, Koster TPM (1999) Acta Polym 50:122
51. Zhu ZK, Yang Y, Yin J, Wang X, Ke Y, Qi Z (1999) J Appl Polym Sci 3:2063
52. Petrovic XS, Javni L, Waddong A, Banhegyi GJ (2000) J Appl Polym Sci 76:133
53. Yano K, Usuki A, Okada A, Kurauchi T, Kamigaito O (1993) J Polym Sci Part A Polym Chem 31:2493
54. Li F, Ge J, Honigfort P, Fang S, Chen JC, Harris F, Cheng S (1999) Polymer 40:4987
55. Agag A, Takeichi T (2000) Polymer 41:7083
56. Hussain M, Varley RJ, Mathys Z, Cheng YB, Simon GP (2004) J Appl Polym Sci 91:1233
57. Lan T, Pinnavaia TJ (1994) Chem Mater 6:2216
58. Masenelli-Varlot K, Reynaud E, Vigier G, Varlet J (2002) J Polym Sci Part B Polym Phys 40:272
59. Cho JW, Paul DR (2001) Polymer 42:1083
60. Masenelli-Varlot K, Reynaud E, Vigier G, Varlet J (2004) J Polym Sci Part B Polym Phys 40:272
61. Mun M-K, Kim JC, Chang J-H (2006) Polym Bull 57:797
62. Kojima Y, Usuki A, Kawasumi M, Okada A, Fukushima Y, Kurauchi T, Kamigaito O (1993) J Mater Res 8:1185
63. Chen L, Wong SC, Pisharath SJ (2003) J Appl Polym Sci 88:3298
64. Chawla KK (1987) Composite materials science and engineering. Springer, New York
65. Shia D, Hui Y, Burnside SD, Giannelis EP (1987) Polym Eng Sci 27:887
66. Curtin WA (1991) J Am Ceram Soc 74:2837

FRP Esthetic Orthodontic Wire and Development of Matrix Strengthening with Poly(methyl methacrylate)/Montmorillonite Nanocomposite

15

Shuichi Yamagata, Junichiro Iida, and Fumio Watari

Contents

Abstract

Orthodontic treatment using a multibracket appliance is the most popular and precise method for correcting misaligned dentition. We have been working to replace a conventional orthodontic metal wire with a fiber reinforced thermoplastic (FRP) wire and have developed an esthetic transparent FRP wire containing biocompatible glass fibers and poly (methyl methacrylate) (PMMA). However, loss of mechanical properties carried by failure related with glass fibers has not yet been resolved. Meanwhile, we are working on developing a super-fiber (high-strength and high-modulus fibers) reinforced

S. Yamagata (✉) • J. Iida
Department of Orthodontics, Graduate School of Dental Medicine, Hokkaido University, Sapporo, Japan
e-mail: shuic@den.hokudai.ac.jp

F. Watari
Department of Biomedical Materials and Engineering, Graduate School of Dental Medicine, Hokkaido University, Sapporo, Japan

J.K. Pandey et al. (eds.), *Handbook of Polymernanocomposites. Processing, Performance and Application – Volume A: Layered Silicates*, DOI 10.1007/978-3-642-38649-7_19,

thermo-plastic wire; for this we will have to focus on improving the stiffness of PMMA without causing any deterioration in its flexibility. Transparent PMMA/ layered silicate nanocomposites were fabricated by a solution intercalation method. Montmorillonite, organically modified with quaternary alkyl-ammonium ions, was selected as the filler for reinforcement. The nano-composites are transparent enough for their use as esthetic orthodontic wires. The flexural modulus of them was favorably improved. Those results encourage us to make them applicable for other dental use. To the best of our knowledge, there has been little or no research on polymer/layered silicate nanocomposites in the field of dentistry. A novelty and an application potentiality of our challenge are beneficial for evolving dentistry.

1 Introduction

Esthetic preference of dental patients is one of the most important factors necessary for developing dental materials, and orthodontic appliances are no exception. A multibracket appliance is the most popular and precise method for correcting misaligned dentition (Fig. 15.1). Brackets are bonded to teeth and wires are fastened to them, and then, forces exerted from the wire for tooth movement transmit to each tooth through the intervention of brackets. Until recently, both brackets and wires were made of metals. To make these fixed appliances more esthetically acceptable, brackets made from either polymers or ceramics have been introduced. However, wires are still metallic. Consequently, a somewhat unnatural facial appearance caused due to the metal wires discourages many potential patients from opting for this treatment.

As for an overview of the present status of orthodontic material usage, as stated above, commercialized esthetic products exclusively are plastic and ceramic brackets. Major raw materials of the currently available plastic brackets are mostly polycarbonate, and the others are polyurethane, polyoxymethylene, polyamide, or something else [1]. Such preference for bracket materials is probably based on their izod impact strength. Poly(methyl methacrylate) (PMMA), with very low izod impact strength, is not preferred as a bracket material for this reason. However, brackets, if being fabricated only by those raw materials, result in excessive creep deformation when subjected to torsional loads generated by wires. Currently available plastic brackets are improved by reinforcing with ceramic or glass-fiber fillers and/or metallic insertion onto bases of slots for wire setting [2]. Nevertheless, they still have not sufficient strength and wear resistance and have problems due to intraoral plasticization and softening. Moreover, polyoxymethylene has another anxiety about possible formaldehyde release associated with degradation [1]. Meanwhile, ceramic brackets manufactured from polycrystalline alumina, single-crystal alumina, and polycrystalline zirconia provide higher strength and wear resistance than those of plastic brackets. They also have superior esthetics and better color stability. Recently, in addition to less esthetic appearances due to opaque color, no significant advantage of zirconia brackets over polycrystalline alumina brackets

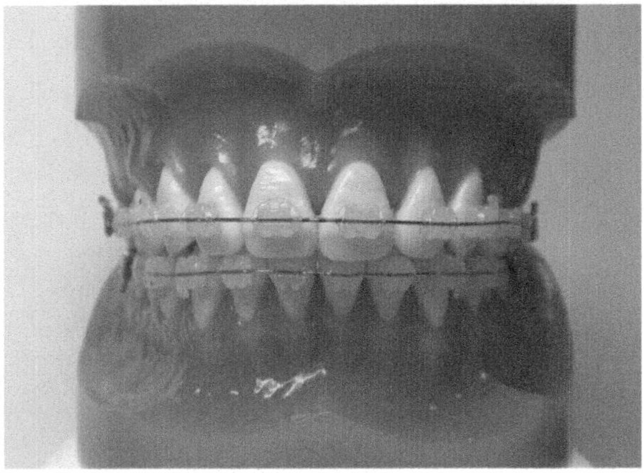

Fig. 15.1 External appearance of the multibracket appliance (frontal view)

in regard to their frictional properties has been reported [2]. Therefore, zirconia brackets have become obsolete. Although ceramic brackets are more attractive than plastic brackets, they have some disadvantages such as low toughness and severe iatrogenic enamel wear of the opposing teeth when they are placed in the lower dentition.

As mentioned above, durability and toughness are emphasized in case of brackets, whereas mechanical properties such as stiffness and elasticity play a much more prominent role in orthodontic wires as well as durability. This is not only due to technical requirements but also related to duration of use. The standard period of time which is normally required to complete an orthodontic treatment using a multibracket appliance is at least 2–3 years. Each bracket is fixed to individual tooth all the while as an intermediate device of orthodontic force generated by wires. In contrast, each wire is ordinarily supposed to be replaced with new one once a month. Thus, PMMA was inevitably selected for our study to develop an esthetic orthodontic wire because of its superior intrinsic flexural modulus, surface hardness, chemical stability, and formability among engineering plastics. Polycarbonate, which has comparable modulus and transparency to PMMA, is a possible candidate material; however, it was not selected by reason of some disadvantages such as inferior formability and a risk of a dioxin emission by incineration.

The intrinsic Young's modulus of PMMA, about 2–3 GPa, is too low to use as an orthodontic wire when it is thermoformed with PMMA alone. We have been working to replace conventional orthodontic metal wires with transparent or translucent wires and have already developed an esthetic transparent glass-fiber reinforced plastic (GFRP) wire containing biocompatible continuous glass fibers oriented in the direction of the long axis and poly(methyl methacrylate) [3–7]. They are 0.5 mm in diameter and their Young's modulus is adjustable in the range from 20 to 40 GPa by changing the volume fraction of glass fibers and is adequate for the middle and latter stages of orthodontic treatment. The form of dentition should have been well arranged to a considerable degree as seen in Fig. 15.2a until

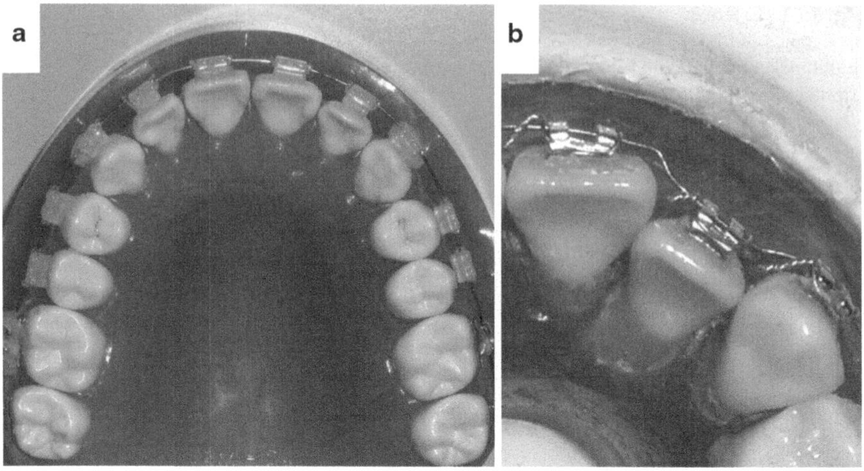

Fig. 15.2 Forms of upper dentition (horizontal angle of view) (**a**) A considerably well-arranged state; (**b**) an example of a misalignment

these stages; any excessive bend is not needed for the wires. In the initial stage of orthodontic treatment, wires should be warped or bent with small radius of curvature due to varying degrees of tooth crowding (Fig. 15.2b). Therefore, if the GFRP wire is used in the initial treatment stage, loss of mechanical properties carried by failure related with glass fibers occurs.

A desired load range to ensure the movement of individual tooth at the initial stage is around 1 N or less. In case of conventional metal wires, a thinner wire can be chosen depending on the degree of teeth misalignment to make it exert adequate load level. Unfortunately, to make our GFRP wires thinner than present products should not be recommended from the aspect of possible breakage risks. For these reasons, we are working on developing a super-fiber (high-strength and high-modulus fibers) reinforced plastic wire; for this we will have to take note of the improvement of the stiffness of PMMA without causing any deterioration in its flexibility. For this idea to work, we focused on nanocomposites. Polymer/layered silicate nanocomposites have attracted considerable interest in the past decade due to their enhanced mechanical and thermal properties as compared to homopolymer or other fillers. However, to the best of our knowledge, there has been little or no research on polymer/layered silicate nanocomposites in the field of dentistry [8–10].

In recent years, montmorillonite (MMT), which is a typical swelling clay mineral, has been the most commonly used reinforcement in polymer nanocomposites [11, 12]. The chemical structure of MMT consists of an edge-shaped octahedral sheet of aluminum hydroxide or magnesium hydroxide sandwiched between two tetrahedral silica sheets. Several studies on PMMA/MMT nanocomposites have been conducted to date [13–16]. However, to the best of our knowledge, there are no previous reports on PMMA/MMT nanocomposites prepared by a solution intercalation method [17–20]. In this chapter, we introduce an outline of the preparation and some characterization of PMMA/organically modified MMT (OMMT) nanocomposites.

2 Preparation and Property Test of the PMMA/OMMT Nanocomposites

Some chemical and mechanical methods, mainly polymerization of an intercalated monomer or melt intercalation, have been utilized to enhance expansion of the *d*-spacing of layered silicates and to achieve the optimum dispersion quality [10, 13–16, 21–23]. However, it is not always possible for some research institutes to conduct such methods needing large-scale equipments. On the other hand, since a simplified and efficient method at low cost is more and more expected, a solution intercalation method is thought to be preferable for experimental stages.

2.1 Preparation of the Nanocomposites

Organically modified montmorillonite (OMMT) was prepared by an ion exchange reaction between MMT and quaternary alkylammonium ion using dimethyldistearylammonium salt. PMMA (ACRYPET™ MD) was purchased from Mitsubishi Rayon (Tokyo, Japan). The grade MD has adequate mechanical properties and a good melt flow rate (6.0 g/10 min at the temperature 503.15 K, 37.7 N) for our purpose.

PMMA not only has desirable physical and mechanical properties but also has superior properties such as esthetic merits, durability, and usability. It is one of the most widely used polymer materials in dental practices, e.g., dental prostheses such as dentures and temporary crowns, adhesives, and removable orthodontic appliances. PMMA is conveniently highly soluble in acetone, an aprotic polar solvent. More importantly, the viscosity of a mixture of dissolved PMMA does not increase easily even if MMT is added. Therefore, the relatively low shear force applied to the mixture was reasonably sufficient to enable good dispersion of OMMT in the solution. The entire fabricating process of the PMMA/OMMT nanocomposites is outlined in Fig. 15.3.

2.2 Property Tests of the Nanocomposite Specimens

Both X-ray diffraction (XRD) analysis and transmission electron microscopy (TEM) observation are essential in the microstructural investigation of the nanocomposites.

Transparency is one of the most important properties for the nanocomposites to be applied as some dental materials. We use a haze meter to measure total light transmittance (Tt), parallel light transmittance (Tp), and haze and usually employ

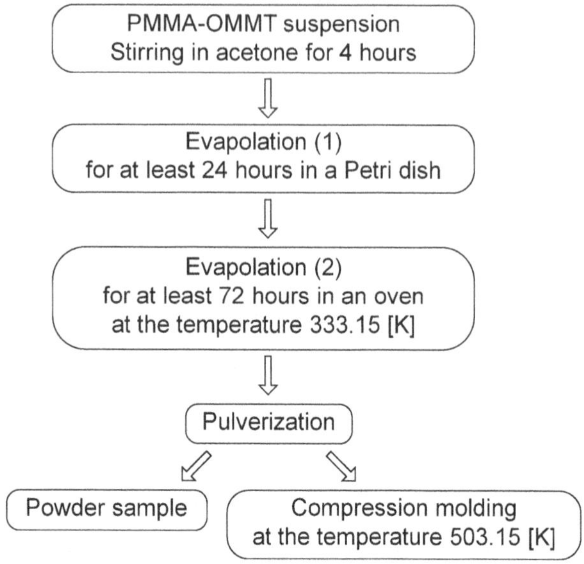

Fig. 15.3 Fabrication process of PMMA/OMMT nanocomposites

an illuminant C as a light source. Haze is defined as the proportion of diffuse transmittance to Tt. Here, the diffusion light is defined as refracting light that deviates more than 2.5° from the parallel light.

Coordination between strength and flexibility of polymer materials is also one of the most important concerns for dental use. Three-point transverse tests (span, 20 mm; crosshead speed, 1 mm/min) are necessary to evaluate the mechanical properties.

3 Characterization of the PMMA/OMMT Nanocomposites

3.1 Intercalation of OMMT with Polymer Molecules

Recently, studies using the solution intercalation method are occasionally reported as a valid method for intercalation. As we know, montmorillonite (MMT), hydrophilic in nature, should be organically modified before being blended with polymer in order to increase its layer spaces and to promote hydrophobicity especially when a solution intercalation method is used. The swelling power of the organically modified MMT (OMMT) we used is 35.0 ml/2 g, which is comparable to that of natural hydrophilic bentonite.

As seen in Fig. 15.4, the XRD pattern of the original OMMT showed a peak at $2\theta = 7.6°$, whereas the PMMA/OMMT samples showed a peak at $2\theta = 5.4°$. The spacing of 1.63 nm, as calculated by Bragg's formula, obtained from the nanocomposite samples is inadequate to conclude that the peak is really (001) *d*-spacing [24]. The shift of 2θ value of the peaks toward lower implies a diffusion of polymer chains into the interlayer regions of OMMT. A small angle scattering should be analyzed after this.

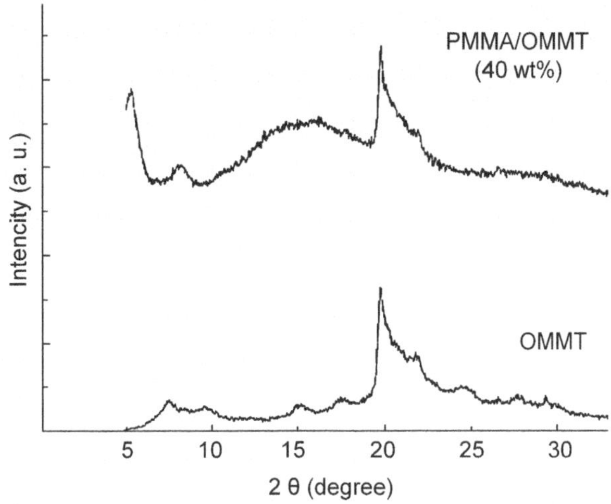

Fig. 15.4 Powder X-ray diffraction patterns of organically modified layered silicates (OMMT) and PMMA/OMMT nanocomposites with the content of 40 wt% OMMT

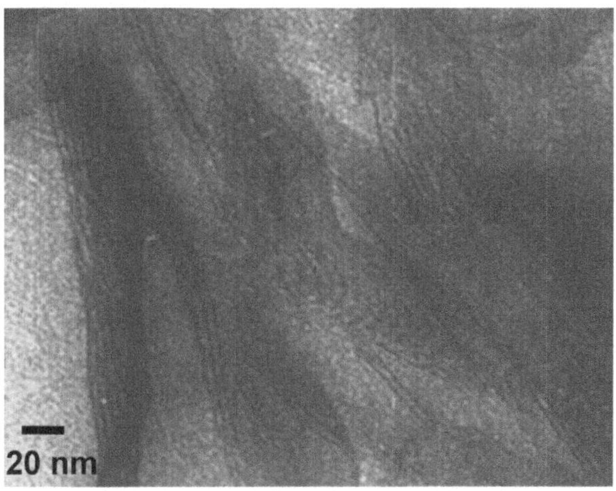

Fig. 15.5 TEM image of a PMMA/OMMT nanocomposite with the content of 2 wt%

The TEM micrograph revealed well-intercalated and partially exfoliated structures of the OMMT (Fig. 15.5). The expansion is probably due to solvation of acetone with long chains or benzene rings of quaternary alkylammonium ion. The stacked silicate layers due to their cohesive force dispersed randomly in the PMMA matrix. These features are mostly affected by both the hydrogen bond as an interaction at edge surfaces and the perpendicular bonding owing to a rearrangement of negatively charged basal plane surfaces and positively charged edge surfaces of silicate layers. Such the inhomogeneous dispersions due to the hydrogen bond and the electrostatic force could be mitigated by silanating the OMMT.

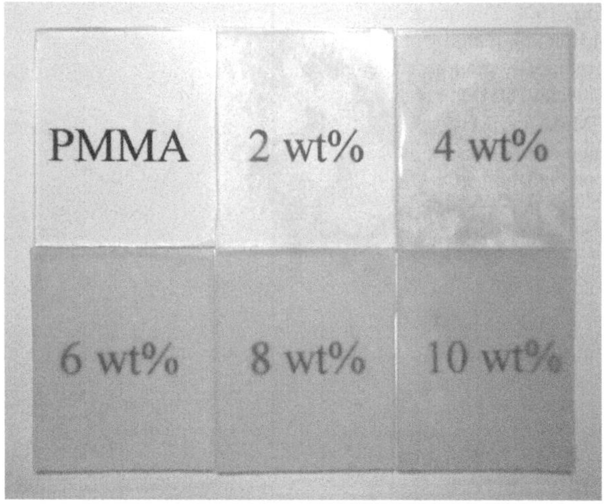

Fig. 15.6 Macroscopic appearance of PMMA/OMMT nanocomposites 1 mm in thickness

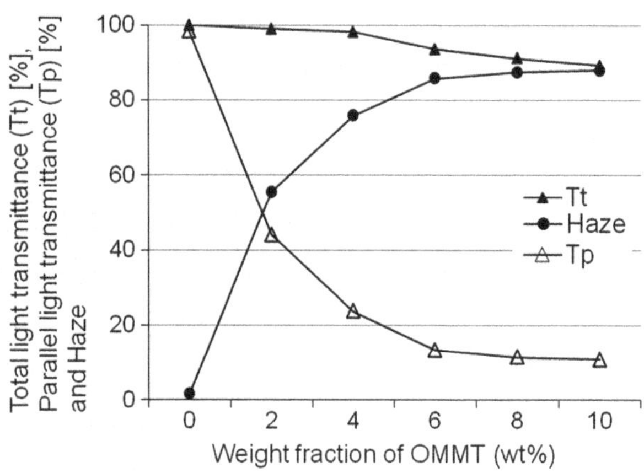

Fig. 15.7 Optical properties of PMMA/OMMT nanocomposites by using illuminant C as a light source

3.2 Retention of Transparency with Varying Contents

Our PMMA/OMMT nanocomposites have favorable macroscopic transparency as shown in Fig. 15.6. Although the translucency increases in shades of brown in accordance with the weight fraction of OMMT, the appearance might be adequate for their use as esthetic orthodontic wires.

The total light transmittance (Tt), parallel light transmittance (Tp), and the haze are unique quantification (Fig. 15.7). Although they all do not directly indicate the degree of transparency and the relative rate of Tp decreases with increasing amount of OMMT, it is considered to be reasonable that the relatively small deterioration that accompanies the increase of OMMT content renders it almost macroscopically equal transparency to that of the original PMMA.

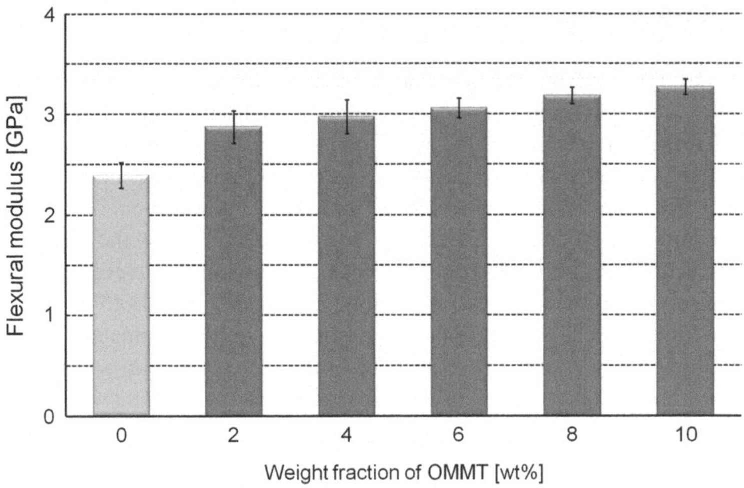

Fig. 15.8 Flexural modulus of the PMMA/OMMT nanocomposites

3.3 Strengthening of PMMA by Intercalated Platelets

The flexural modulus of the PMMA/OMMT nanocomposites increased successfully and almost linearly, relative to that of pristine PMMA, with increasing amount of OMMT content (Fig. 15.8). The improvement in reinforcement may possibly be due to shear deformation and stress transfer to the silicate layers that have exceptionally high tensile modulus. Both an anisotropic morphology of the crystals and an improved wettability by organic modification with PMMA contribute to this phenomenon. Therefore, it is considered that by raising the swelling power of the OMMT or by silanating them, the mechanical properties of the nanocomposites will improve substantially. This would be accompanied by an increase volume of intercalated polymer chains or an increase in the number of exfoliated silicate layers. Further studies will target direct evidence of this.

4 Future Prospects

Introduction of an entirely new concept in development of esthetic orthodontic wires is just beginning. However, from our initial experimental data, PMMA/OMMT nanocomposites show considerable promise as the esthetic wire in the near future. Moreover, PMMA is a preferable biomaterial also in terms of safety such that it does not easily be degraded, and undesirable potential effects of benzene rings are inhibited by the use of it.

Esthetic transparent FRP wires, as Eliades had remarked, will be commercially available as a combination of GFRP structure wires and super-fiber reinforced PMMA/OMMT nanocomposite structure ones during the next several years if the industry finds that introducing them to the market will be profitable.

5 Conclusions

For a technological advancement and an application in polymer nanocomposites of layered silicates, poly(methyl methacrylate)/organically modified montmorillonite (PMMA/OMMT) nanocomposites were prepared using a solution intercalation method. It was found that the polarity, viscosity, and hydrophile-lipophile properties affect the dispersion of the silicate platelets, intercalation of the PMMA molecules into the silicate galleries, and partial exfoliation of the OMMT. It is also noteworthy that the nanocomposites appear to remain stable even though they have been processed by compression molding at temperature 503.15 K. In addition to the good mechanical properties and the optical properties, human body-friendly and environmentally friendly properties of both layered silicates and PMMA are encouraging for their application to esthetic transparent orthodontic wires and many other uses for dental practice. A novelty and an application potentiality of our challenge are beneficial for evolving dentistry.

References

1. Eliades T (2007) Am J Orthod Dentofacial Orthop 131:253
2. Russell JS (2005) J Orthod 32:146
3. Watari F, Kobayashi M, Yamagata S, Nagayama K, Imai T, Nakamura S (1996) Biomechanics 9:469
4. Imai T, Watari F, Yamagata S, Kobayashi M, Nagayama K, Toyoizumi H, Nakamura S (1998) Biomaterials 19:2195
5. Watari F, Yamagata S, Imai T, Kobayashi M, Nakamura S (1998) J Mater Sci 33:5661
6. Imai T, Yamagata S, Watari F, Kobayashi M, Nagayama K, Toyoizumi H, Uga M, Nakamura S (1999) Dent Mater J 18:167
7. Imai T, Watari F, Yamagata S, Kobayashi M, Nagayama K, Nakamura S (1999) Am J Orthod Dentofacial Orthop 116:533
8. Dowling AH, Stamboulis A, Fleming GJP (2006) J Dent 34:802
9. Dowling AH, Fleming GJP (2007) J Dent 35:309
10. Atai M, Solhi L, Nodehi A, Mirabedini SM, Kasraei S, Akbari K, Babanzadeh S (2009) Dent Mater 25:339
11. Fornes TD, Paul DR (2003) Polymer 44:4993
12. Zhang K, Park BJ, Fang FF, Choi HJ (2009) Molecules 14:2095
13. Zeng C, Lee LJ (2001) Macromolecules 34:4098
14. Huang X, Brittain WJ (2001) Macromolecules 34:3255
15. Meneghetti P, Qutubuddin S (2004) Langmuir 20:3424
16. Fan X, Xia C, Advincula RC (2005) Langmuir 21:2537
17. Liu L, Song L, Hu Y, Chen H (2006) Polym Compos 27:660
18. Marras SI, Kladi KP, Tsivintzelis I, Zuburtikudis I, Panayiotou C (2008) Acta Biomater 4:756
19. Yamagata S, Akasaka T, Uo M, Ushijima N, Nodasaka Y, Iida J, Watari F (2009) Nano Biomed 1:151
20. Yamagata S, Hamba Y, Akasaka T, Uo M, Iida J, Watari F (2011) Nano Biomed 3:217
21. Okamoto M, Morita S, Taguchi H, Kim YH, Kotaka T, Tateyama H (2000) Polymer 41:3887
22. Meneghetti P, Qutubuddin S (2005) J Colloid Interface Sci 288:387
23. Hossain MD, Kim WS, Hwang HS, Lim KT (2009) J Colloid Interface Sci 336:443
24. Messersmith PB, Giannelis EP (1994) Chem Mater 6:1719

Development of TGDDM Based Layered Silicate Nanocomposites for High Performance Applications

16

K. Shree Meenakshi, E. Pradeep Jaya Sudhan, S. Ananda Kumar, and M. J. Umapathy

Contents

Abstract

This chapter presents an investigation on TGDDM containing tetraglycidyl epoxy nanocomposites to find its suitability for use in high-performance applications. The synthesis and characterization of the TGDDM epoxy resin denoted as "A" were done. Nanoclay-based layered silicate nanoreinforcement denoted as "N1" was incorporated into the synthesized epoxy resin. Curing was done with diaminodiphenylmethane (DDM) and bis(3-aminophenyl)phenylphosphine oxide (BAPPO) curing agents denoted as X and Y, respectively. The mechanical, thermomechanical, flame retardancy, water absorption, and morphological studies were carried out, and the results are presented.

K.S. Meenakshi (✉) • E.P.J. Sudhan • S. Ananda Kumar • M.J. Umapathy
Department of Chemistry, Anna University, Chennai, TN, India
e-mail: shreemeenakshik@gmail.com

J.K. Pandey et al. (eds.), *Handbook of Polymernanocomposites. Processing, Performance and Application – Volume A: Layered Silicates*, DOI 10.1007/978-3-642-38649-7_15,

1 Introduction

Polymer-layered silicate (PLS) nanocomposites have been studied extensively and have attracted a lot of attention due to their improved properties. They are regarded as new generation materials because they possess unique properties typically not shared by their more conventional micro-composite counterparts and therefore offer new technology and business opportunities. The unique nanometer-size dispersion of the layered silicates with high aspect ratios, high surface areas, and high strengths in the polymer matrix gives nanocomposites improved properties of mechanical performance, thermal stability, barrier performance, and flame retardancy at very low-volume fraction loadings of layered silicates [1–4].

Polymer nanocomposites are two-phase materials in which the polymers are reinforced by nanoscale fillers [5, 6]. The most commonly used filler material is based on the smectite class of aluminum silicate clays, of which the most common representative is montmorillonite (MMT). MMT has been employed in many PLS nanocomposite systems because it has a potentially high aspect ratio and high surface area that could lead to materials which could possibly exhibit great property enhancements. In addition, it is environmentally friendly, naturally occurring, and readily available in large quantities. The layered clays have structural units having two tetrahedral sheets and one edge-shared octahedral sheet of aluminum or magnesium hydroxide between the two. Between these layers, galleries that bond the layers by van der Waals forces exist. Isomorphic substitution of Al^{+3} with Mg^{+2}, Fe^{+2}, Mg^{+2} with Li^{+2}, or Si^{+4} with Al^{+3} generates negative charges balanced with the gallery cations of Na^{+} and Ca^{+2}. The cations in the clay galleries give hydrophilic characteristic to the natural clays, which is a drawback to homogeneously disperse them in organic polymers. It has been demonstrated that this can be overcame by the suitable surface modification of clays. The cations in the clay galleries may be exchanged with the alkyl ammonium cations through an ion-exchange reaction. So, the modified clay becomes organophilic having a lower surface energy and good interactions with the organic polymer. It has been proven in recent years that polymer-based nanocomposites reinforced with a small amount of nanosize clay particles (<5 vol%) significantly improve the mechanical, thermal, and barrier properties of the pure polymer matrix. Clay layers nanocomposites can exhibit lighter weight dimensional stability and a certain degree of strength, stiffness, heat resistance, and barrier properties with far less clay loading than that used in conventional filled polymer composites. Hence, they have attracted considerable attention from both the scientific and practical point of view over the past 15 years. Epoxy-clay nanocomposites are regarded as one of the most promising new materials for industrial applications [7, 8].

In this chapter we have discussed the mechanical, thermomechanical, flame retardancy, water absorption, and morphological studies of TGDDM epoxy-clay nanocomposites in order to find its suitability for high-performance applications.

2 Experimental

2.1 Materials

All chemicals were of reagent grade and were used without further purification. 4, 4'-Diaminodiphenylmethane (DDM) was obtained from Huntsman, USA. Epichlorohydrin and sodium hydroxide were obtained from SD Fine Chemicals, India. Triphenylphosphine oxide was obtained from Alfa Aesar, Germany. Gamma-Aminopropyltrimethoxysilane (γ-APS), stannous chloride, and sulphuric acid were obtained from Merck (Germany). Tetrahydrofuran (THF) and benzene were obtained from Sisco Research Laboratories, India. Nanoclay was obtained from Nanocor, USA. Hydrochloric acid was obtained from Hipure, India.

2.2 Synthesis of *N,N'*-Tetraglycidyl Diaminodiphenylmethane (TGDDM)

The synthesis of TGDDM was carried out as follows. Epichlorohydrin (6.25 mol) was taken in a 1 L three-necked round-bottomed flask provided with a mechanical stirrer, nitrogen atmosphere, and water condenser. This was heated from 50 °C to 86 °C in an oil bath. Then 1.5 mol of DDM was added and stirred continuously for 4 h at about 50–55 °C temperature under nitrogen atmosphere. Chlorohydrin, the intermediate product, was formed and the excess epichlorohydrin used was distilled off under vacuum. Then 2.84 mol of 40 % NaOH solution was added dropwise for 1 h at 54 °C. The reaction was continued at this temperature for a further period of 1 h. The resulting solution was extracted with chloroform. The organic layer was collected and concentrated at a reduced pressure to get the light brown colored liquid epoxy product. The structure of the synthesized *N,N'*-tetraglycidyl diaminodiphenylmethane (TGDDM) is shown below (Scheme 16.1).

2.3 Preparation of Nanocomposites

In order to study the properties of the epoxy resins, neat resin laminates were prepared by curing the synthesized TGDDM (A) by DDM (X) and BAPPO (Y) curing agents as shown in Table 16.1. For the fabrication of nanocomposites, the nanoclay (Nanocor 1.30E) was dried at 24 h at 50 °C under vacuum. The epoxy

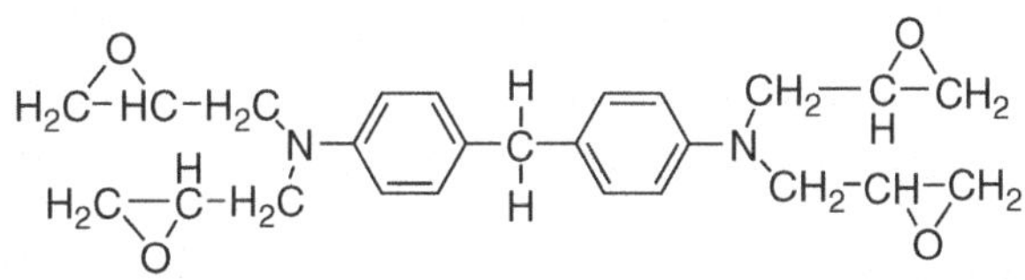

Scheme 16.1 Chemical structure of TGDDM

Table 16.1 Fabrication of resin laminates

Type of epoxy	Matrix name	Nanoreinforcement	Curing agent
TGDDM	AX	–	X
	AXN1	N1	X
	AY	–	Y
	AYN1	N1	Y

resin was mixed mechanically in a reaction vessel with the nanoclay at 50 °C for 3 h. Then it was further mixed in an ultrasonic bath for 30 min to disperse the clay in the resin. Later the mixture was cooled to room temperature in 30 min. The curing agent was then added. After mixing mechanically for 10 min, the mixture was degassed by a vacuum pump to remove the air bubbles and poured into molds. The nanocomposites were cured for 3 h at 120 °C and post-cured for 2 h at 200 °C. After that the resin plaque was cooled to room temperature naturally, and it was cut into specimens of the required dimensions, required for the different testing and evaluation studies.

2.4 Characterization

Various characterization studies like mechanical, thermomechanical, and XRD studies are carried out, and the results are presented.

2.4.1 Mechanical Studies

The mechanical properties such as tensile strength, tensile modulus, flexural strength, flexural modulus, impact strength, and hardness of matrix were studied as per ASTM standards.

2.4.2 Thermomechanical Analysis

Heat Distortion Temperature (HDT)

Determination of heat distortion temperature was carried out as per ASTM D648-72 using heat distortion temperature apparatus. The specimens of size 120 mm in length, 13 mm in thickness, and 5 mm in width were kept in an oil bath under a load of 1.82 MPa. The temperature was raised at a rate of 2 °C per minute, and the temperature was noted when the specimen deflected by 0.25 mm [10].

Dynamic Mechanical Analysis (DMA)

The dynamic mechanical behavior of both unmodified and modified epoxy resins was studied in a Metravib Viscoanalyzer. Each sample of size 10 mm was scanned from 0 °C to 300 °C at a heating rate of 276.15 K/min in an inert atmosphere. During heating the samples were subjected to strain at a frequency of 10 Hz while the storage modulus (E') and the damping factor (tan δ) were recorded. The temperature corresponding to the maximum in tan δ versus temperature plots was recorded as the measurement of glass transition temperature (T_g).

2.4.3 X-ray Diffraction Technique

The X-ray diffraction technique is the most direct and simple method to evaluate the spacing between the clay layers. XRD patterns were obtained using an X-ray diffractometer (X'Pert, Philips) equipped with Cu, Kα radiations and a curved graphite crystal monochromator.

Diffraction data were collected at 0.001°/S steps between 10 °C and 70 °C and were used to determine changes in gallery heights before and after the addition of nanoclay in the polymer matrices.

2.4.4 Flame and Water Absorption Tests

UL 94 vertical burning flame test was carried out to analyze the flammability of the sample [9, 11]. Test specimens prepared as per ASTM D 570 were immersed for 24 h at 30 °C, and the percentage of water absorbed by the specimen was calculated out using equation given below:

$$\%\ \text{water absorption} = (w_2 - w_1) \times 100/w_1$$

where w_1 is the initial weight of the sample and w_2 is weight of the sample after immersion in water for 24 h at 30 °C.

3 Results and Discussion

3.1 X-ray Diffraction

X-ray diffraction patterns were recorded by monitoring the diffraction angle 2θ from 1.5° to 10° on a Rigaku-D/max-γA X-ray diffractometer. The diffractometer was equipped with a Cu Kα (λ = 0.154 nm) radiation source operated at 40 kV and 100 mA. The scanning speed and the step size used were 2°/min and 0.02°, respectively. The XRD patterns of the neat and modified nanocomposites are shown in Figs. 16.1–16.3. The neat and modified resins showed similar diffraction

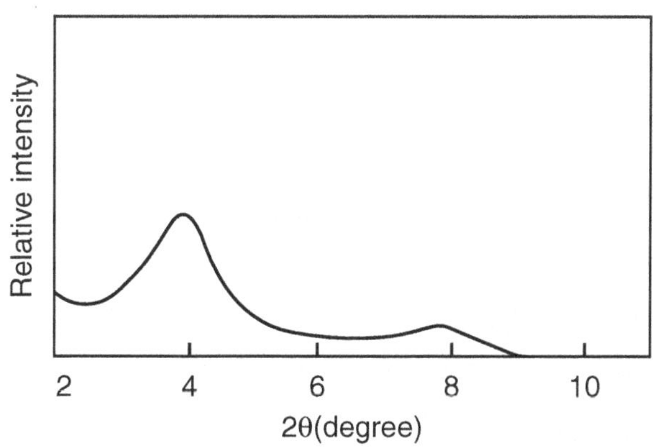

Fig. 16.1 XRD of N1

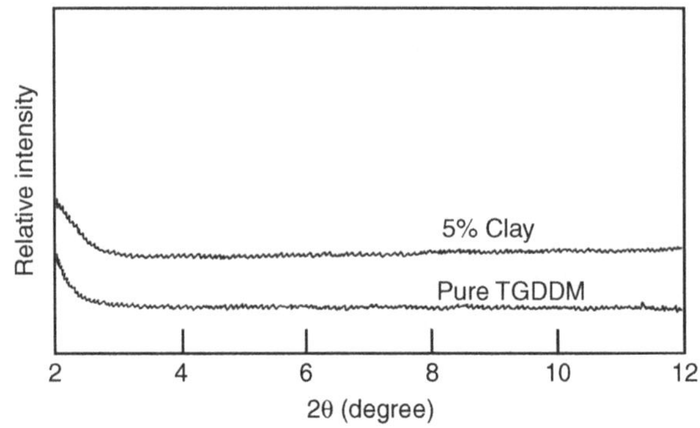

Fig. 16.2 XRD of AX-based systems

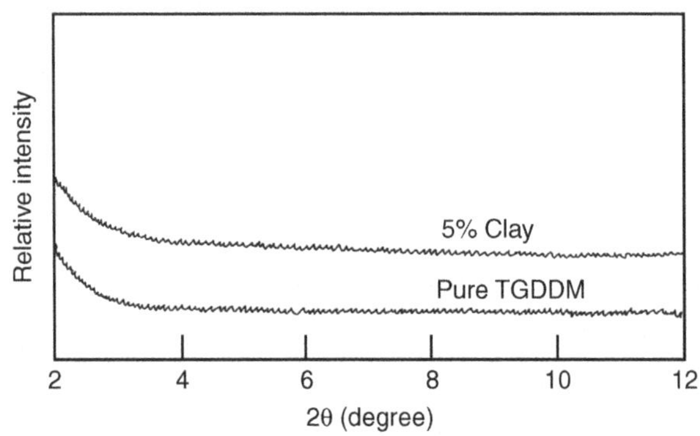

Fig. 16.3 XRD of AY-based systems

pattern. The neat and modified resins cured by X and Y showed similar diffraction pattern. For the nanocomposites cured by both X and Y (Figs. 16.2, 16.3), the (001) peak has completely disappeared suggesting that it has shifted to an angle below 2°, beyond the capacity of the XRD machine. This also means that the d-spacing increased to above 4 nm confirming the formation of nanocomposites [21].

The XRD of organoclay (N1) is shown in Fig. 16.1. The prominent peak corresponding to the basal spacing of nanoclay 1.30E can be seen at 2.37 nm. The XRD pattern of nanocomposites generally gives peaks for intercalated nanocomposites alone. But the XRD pattern of exfoliated nanocomposites does not give any such peaks, indicating the nanoparticulate being exfoliated in the polymer matrix [12]. The XRD patterns of neat and modified epoxy resins and their nanocomposites of our present investigation are depicted in Figs. 16.2, 16.3. From the Figures, it can be clearly understood that the XRD patterns of the neat resin and nanocomposites resemble each other, confirming the formation of completely exfoliated structure with no XRD peaks corresponding to nanocomposites [13]. Similar observation was reported in our previous research studies for epoxy-clay nanocomposites indicating the existence of completely exfoliated structure from XRD studies.

Table 16.2 Results of thermomechanical analysis and HDT

Resin system	Storage modulus (MPa)	Glass transition temperature T_g (°C)	Heat distortion temperature (°C)
AX	4,200	200	202
AY	4,310	207	209
AXN1	4,420	208	210
AYN1	4,530	214	217

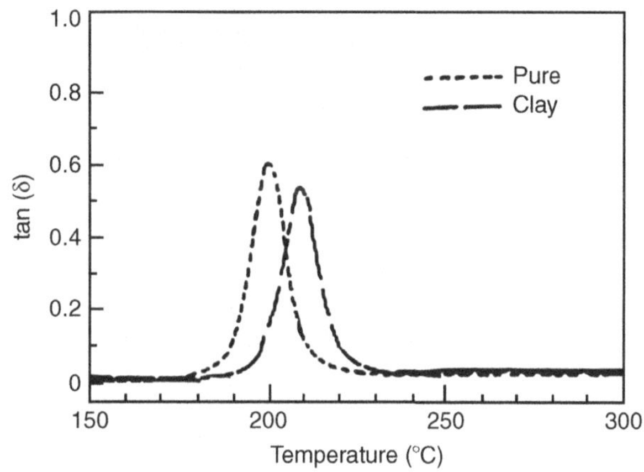

Fig. 16.4 DMA of AX-based systems

3.2 Thermomechanical Analysis

The results of thermomechanical analysis and heat distortion temperature are plotted in Table 16.2. From the table, it is observed that the Tg values and storage modulus of these systems were high. Similar observation was made in the case of the heat distortion temperature of AX systems. This may be due to the enhanced rigidity of these systems which resisted deflection when subjected to heating. In epoxy nanocomposites, a small height of the relation species is observed from the DMA peaks, which corresponds to low cross-linking. The resin system cured with Y showed enhanced dynamic mechanical properties due to increased cross-linking and rigidity. The Tg and storage modulus (as seen from Table 16.2 and Figs. 16.4–16.7) of all the tetrafunctional epoxy resins cured by Y were higher than those of the resins cured by X.

The Tg of AX system was 200 °C, and its storage modulus was 4,200 MPa, whereas for the AY system, the Tg and storage modulus were enhanced to 207 °C and 4,310 MPa, respectively (Table 16.2). Similarly, the HDT value of the AX was 202 °C; while the HDT of the system AY was enhanced to 209 °C. This trend was observed for all the systems. A reduced peak height of tan δ curve was associated with a lower segmental mobility and fewer relaxation species, and therefore, it was indicative of stronger bonding. It was seen that the storage modulus increased as the

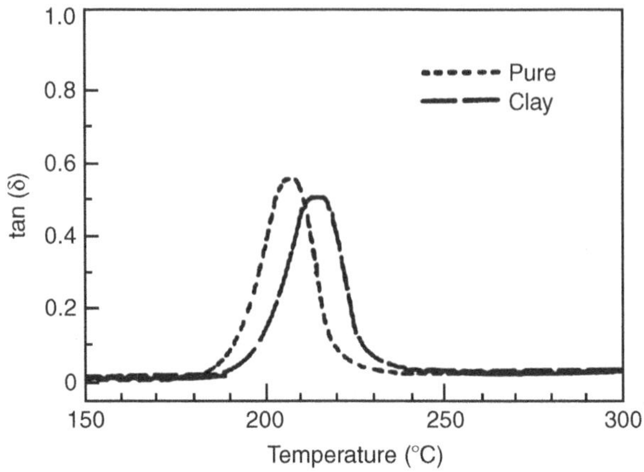

Fig. 16.5 DMA of AY-based systems

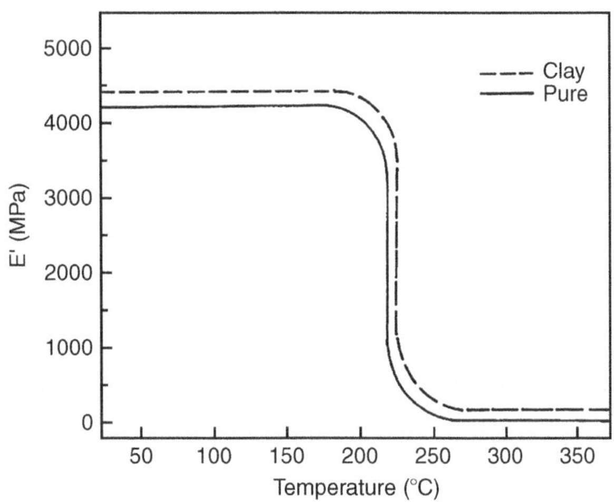

Fig. 16.6 Storage modulus graph of AX-based systems

clay content was increased (Figs. 16.4, 16.5) and can be accounted for an enhancement effect from the addition of rigid inorganic clay particles that produced high stiffness.

The storage modulus for the AX system was 4,200 MPa, and for AXN1 systems, the values were enhanced to 4,420 MPa. As nanoclay content increased, the tan δ peaks of the nanocomposites significantly shifted to higher temperatures (Figs. 16.6–16.7), while the widths of tan δ peaks remarkably broadened. The HDT values were enhanced on the addition of clay into the different tetrafunctional epoxy systems, which might be due to the enhancement in rigidity imparted by the inorganic clay particles which caused steric hindrance to molecular movements. Hence, higher temperature was needed to make molecular chain movement and to obtain a specific deflection under a constant normal load.

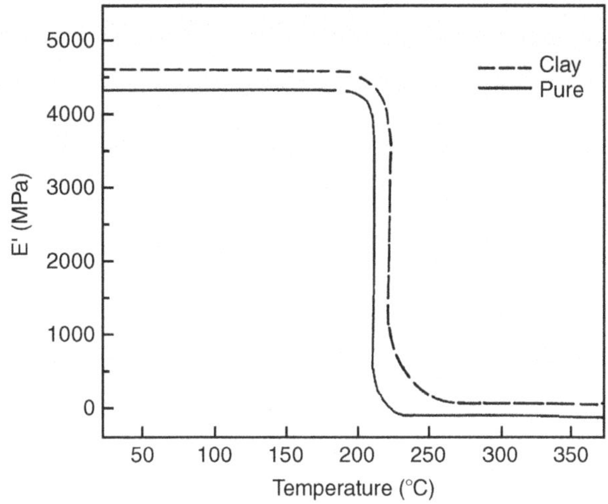

Fig. 16.7 Storage modulus graph of AY-based systems

Table 16.3 Mechanical properties of various systems

Resin system	Tensile strength (MPa)	Tensile modulus (GPa)	Flexural strength (MPa)	Flexural modulus (GPa)	Hardness (D shore)	Impact strength (kJ/m^2)
AX	70	6.12	130	4.21	83	42
AY	76	6.38	138	4.47	84	40
AXN1	77	6.39	137	4.42	85	39
AYN1	83	6.56	144	4.68	88	38

3.3 Mechanical Studies

3.3.1 Tensile, Flexuraly and Impact Strength

The data resulting from the mechanical studies are given in Table 16.3. From the data, it is found that the incorporation of nanoclay (Nanomer 1.30E) into the virgin epoxy results in an increase in tensile and flexural strength. The values of the tensile and flexural modulus are also increased. The reason for the increase in tensile strength and modulus may be due to the uniform dispersion of the nanoparticles of the layered silicates in the matrix of the polymer. The small size of nanoclay restricts the mobility of the polymer chain and thereby increases the modulus, and good interfacial adhesion between the nanoclay and the matrix also contributed to the reinforcement [19, 20]. A lowering of the impact strength was observed; this could be due to the restricted chain mobility due to the increase in cross-linking and reduction in free volume. The impact values are found to be high, for TGDDM nanocomposites, which may be again due to the flexible nature of the Si-O-Si linkage of the nanoreinforcements [14–16].

Table 16.4 Flame retardant studies of TGDDM resin and the nanocomposites

Resin system	UL 94 rating
AX	V2
AY	V1
AXN1	V1
AYN1	V1

Table 16.5 Water absorption studies of TG resin and the nanocomposites

Resin system	% of moisture absorption
AX	.088
AY	.083
AXN1	.076
AYN1	.070

3.4 Flame and Water Absorption Tests

Flame retardant studies -UL −94 V Test (V-0, V-1, V-2) were carried out. The standard vertical ratings for the UL 94 vertical flame test are shown in Table 16.4. The data resulted from the flame retardant studies of the samples of siloxane tetra epoxies and their nanocomposites are shown in Table 16.4. The flame retardancy was better for the BAPPO-cured systems which showed a V0 rating. The flame retardancy enhances on incorporation of clay. A V0 rating was exhibited by the pure and nanoreinforced systems cured by BAPPO. The TGDDM nanocomposites gave a better performance in flame test due to the high thermal stability of Si-O-Si linkage of the nanoreinforcements [17, 22–25]. The data resulting from the water absorption studies (ASTM D570) of the samples of tetra epoxy and the corresponding nanocomposites are shown in Table 16.5. The water absorption tendency decreases on addition of clay which is due to the hydrophobic and partial ionic nature of the Si-O-Si linkage [18, 26–29].

4 Future Prospective

The TGDDM-based layered silicate nanocomposites which have been developed exhibit excellent properties like mechanical strength, thermomechanical properties, and flame retardancy and are hydrophobic. These nanocomposites can be expected to play an important role in the development of high-performance materials for use in automotive, electronics, and advanced aerospace industries. Further research work is being carried out by incorporating suitable reinforcements in the epoxy matrix and studying the improvement in the properties brought about by them.

5 Conclusion

In the current chapter, we have presented an in-depth analysis of TGDDM epoxy nanocomposites. The mechanical studies of the TGDDM epoxy "A" that were carried out indicate that the incorporation of nanoclay (Nanomer 1.30E) into the virgin epoxy results in an increase in tensile and flexural strength. The values of the tensile and flexural modulus are also increased. Thermomechanical (DMA) analysis was carried out on the tetrafunctional epoxies cured by X and Y curing agents in the presence and in the absence of nanoreinforcement to determine their storage modulus and T_g. From the study, it was seen that the Y-cured neat and modified epoxies showed higher storage modulus and T_g values than those of X-cured epoxies.

Morphological studies were carried out by XRD for all tetrafunctional epoxy matrices cured with X and Y. From the XRD studies, it was clearly observed that no diffraction peak in favor of (001) plane was seen indicating an increase in the interlayer distance (d-spacing) of about 7 nm, which confirmed the existence of fully exfoliated nanocomposite structure.

From the data resulted from different studies, it is concluded that these novel organic–inorganic nanohybrid composites developed in the present study, having improved mechanical, thermomechanical, and flame retardant properties and being self-extinguishable and heat resistant at the same time with excellent hydrophobic properties, can very well be expected to find application in automotive, electronics, and advanced aerospace applications for improved performance and longevity than the materials that are currently in use.

References

1. Nielsen LE (1969) Polym Rev 3:1558–3716
2. Kinloch AJ, Shaw SJ, Hunston DL (1983) Polymer 24:1341
3. Yee AF, Pearson RA (1986) J Mater Sci 21:2461
4. Buckingham MR, Lindsay AJ, Stevenson DE, Muller G, Morel E, Costes B, Henry Y (1996) Polym Degrad Stab 54:311
5. Choi J, Harcup J, Yee AF, Zhu Q, Laine RM (2001) J Am Chem Soc 123:11420
6. Li GZ, Wang L, Ni H, Pittman CU Jr (2001) J Inorg Organomet Polymer Mater 11:123
7. Kornmann X, Thomann R, Mulhaupt R, Finter J, Berglund LA (2002) Polym Eng Sci 42:1815
8. Haque A, Shamsuzzoha M, Hussain F, Dean DJ (2003) Compos Mater 37:1821
9. Jose Alcon M, Ribera G, Galia M, Cadiz V (2003) Polymer 44:7291
10. Jagdeesh KS, Shashikiran K (2004) J Appl Polym Sci 93:2790
11. Hussain M, Varley RJ, Mathys Z, Cheng YB, Simon GP (2004) J Appl Polym Sci 91:1233
12. Liu T, Tjiu WC, Tong Y, He C, Goh SS, Chung TS (2004) J Appl Polym Sci 94:1236
13. Ni Y, Zheng S, Nie K (2004) Polymer 45:5557
14. Ahmad S, Gupta AP, Sharmin E, Alam M, Pandey SK (2005) Prog Org Coat 54:248
15. Gacitua W, Ballerini A, Zhang J (2005) Maderas Ciencia y tecnología 7:159
16. Innocenzio P, Kidchob TJ (2005) Sol-Gel Sci Technol 35:225
17. Leszczynska A, Njuguna J, Pielichowski K, Banerjee JR (2005) Thermochim Acta 453:75
18. Liu W, Hoa SV, Pugh M (2005) Comp Sci Tech 65:2364
19. Jia Q, Zheng M, Cheng J, Chen H (2006) Polymer Int 55:1259

20. Lee TH, Kim JH, Bae BS (2006) J Mater Chem 16:1657
21. Araujo EM, Barbosa R, Rodrigues AWB, Melo TJA, Ito EN (2007) Mater Sci Eng A 445–446:141
22. Lu T, Chen T, Liang G (2007) Polymer Eng Sci 47:225
23. Tuncer E, Sauers I, James DR, Ellis AR, Paranthaman MP, Aytug T, Sathyamurthy S, More KL, Li J, Goyal A (2007) Nanotechnology 18:025703
24. Woo RSC, Zhu HG, Leung CKY, Kim JK (2008) Comp Sci Tech 68:2149
25. Huang JM, Huang HJ, Wang YX, Chen WY, Chang FC (2009) J Polym Sci Part B Polym Phys 47:1927
26. Shree Meenakshi K, Pradeep Jaya Sudhan E (2011) Appl Nano Sci 1:109
27. Shree Meenakshi K, Pradeep Jaya Sudhan E, Ananda Kumar S, Umapathy MJ (2011) Prog Org Coat 72:402
28. Shree Meenakshi K, Pradeep Jaya Sudhan E, Ananda Kumar S, Umapathy MJ (2011) Silicon 3:45
29. Shree Meenakshi K, Pradeep Jaya Sudhan E, Prathibha Menon G (2012) Mater Sci Eng A 536:152

Structural and Physical Properties of Polyurethane Nanocomposites and Foams

17

Baiju John

Contents

Abstract

Polyurethanes (PUs) are very versatile polymeric materials with a wide range of physical and chemical properties. PUs also have desirable properties, such as high abrasion resistance, tear strength, shock absorption, flexibility, and elasticity, and

---

B. John
Sealy Centre for Structural Biology and Molecular Biophysics, University of Texas Medical Branch, Galveston, TX, USA

Advanced Polymeric Nanocomposites Materials Laboratory, Toyota Technological Institute, Nagoya, Japan

Graduate School of Science and Technology, Nagasaki University, Nagasaki, Japan
e-mail: jonsbaiju@gmail.com

J.K. Pandey et al. (eds.), *Handbook of Polymernanocomposites. Processing, Performance and Application – Volume A: Layered Silicates*, DOI 10.1007/978-3-642-38649-7_6,

are widely used as elastomers and coatings. PUs have poor thermal stability, mechanical strength, prone to degradation, etc., and can be improved by using treated clay or by polymer-layered silicates. Basic structure of PU is composed of a linear segmented copolymer with a relatively flexible component derived from a macrodiol mainly of polyester, polyether, or polycarbonate chains, called the soft segment, and a relatively hard and stiff component derived from a diisocyanate and a chain extender, called the hard segment. Thermodynamic incompatibility of the so-called soft and hard segments leads to the formation of the microphase separated structure which determines the physical property of the PUs. In the preparation of PU/clay nanocomposites, modifications of clay nanofillers like montmorillonite by using organic modifiers such as alkylammonium improve the compatibility between hydrophilic organomontmorillonite and hydrophobic PUs. These organic modifiers having positively charged ends can be bonded to the surface of the negatively charged silicate layers resulting in an increase of the interlayer gap (exfoliation), is usually associated with large improvements in properties of PU/clay nanocomposites. This entry describes the present state of science and technology of polyurethane (PU) nanocomposites which include nanosilica and organically modified clay (organoclay) with recent trends like PU/clay nanocomposites and its foams.

Keywords
Compressive strength • Exfoliation • Intercalation • Nanofiller • Nanosilica • Organoclay • Phyllosilicates • Polyurethane nanocomposite • Polyurethane nanocomposite foams • Tensile strength

1 Introduction

Polyurethanes (PUs) are widely used as elastomers and coatings because of their excellent mechanical properties like good hardness, adhesive nature, high abrasion, and high chemical resistance [4, 13]. Polyurethane elastomers are composed of linear segmented copolymers with a relatively flexible component derived from a macrodiol mainly of polyester, polyether, or polycarbonate chains, called the soft segment, and a relatively hard and stiff component derived from a diisocyanate and a chain extender, called the hard segment. Thermodynamic incompatibility of the so-called soft and hard segments leads to the formation of the microphase separated structure, which is reflected in a rubber-like matrix containing hard microdomains. The domain structure formed by microscopic phase separation presents similar elastomeric properties to those shown for cross-linked rubber networks [22, 38–42, 44]. The mechanical strength of this structure can be attributed to hard microdomains physically cross linked through hydrogen bonding and dispersion forces, which act as a filler-like reinforcement for the soft segment [28]. The factors that influence microphase separation include composition, chain length of soft and hard segments, symmetry of diisocyanates, thermal history, and so on [1, 2, 21, 30].

Polyurethane nanocomposites consist of bulk polyurethane layered with nanofillers in the matrix region comprising of soft and hard segments [1, 35]. The microstructure of nanocomposites has inhomogeneities in the scale range of nanometers. Fillers have been used for a long time with the goal of enhanced performance of polymers and especially in the polyurethanes. Polyurethane nanocomposites demonstrate often unusual and beneficial for the user properties. Scientific and technical literatures report the improvement or enhancement of properties of polyurethane nanocomposites compared to the pristine polymers. Increase of polymer toughness (J/m^3) is always considered as an improvement [3].

2 Types of Nanofillers

Two types of nanofillers are widely used in the preparation of polymer nanocomposites, namely, nanoparticles like silica and nanoclays like phyllosilicates. The main paradigm is that a valuable nanocomposite is one with the largest possible aspect ratio of nanofiller. Nanoparticles are commercially available from different sources. Sols of nanosilica as colloid solutions in water or in organic solvents are used in preparation of polymer nanocomposites. Fumed silica is available as individual particles (Fig. 17.2a) ranging from 10 to 20 nm to micrometers and can be more or less successfully dispersed in a polymer [25].

Nanoclays belong to the class of layered aluminosilicates (phyllosilicates) clay nanofillers; especially, montmorillonite (bentonite), hectorite, and saponite are widely popular as shown in Table 17.1 [5, 27]. Phyllosilicates have a characteristic distance between nanogalleries with a basal spacing of a gallery of 1 nm. Inorganic cations like Na+ between galleries hold negatively charged galleries together. The replacement of the inorganic cations in the galleries of the native clay by alkylammonium (onium) salts or quaternary amines with long alkyl substituents (surfactants) leads to a better compatibility between the inorganic clay and hydrophobic polymer matrix. The replacement leads to an increase of the space between galleries facilitating the intercalation of polymer molecules into clay [10, 11, 37].

Three main types of polymer nanocomposites are schematically presented in Fig. 17.1.

(a) Intercalated nanocomposites: in this type, the insertion of a polymer matrix into the layered silicate structure occurs in a crystallographically regular fashion, regardless of the clay to polymer ratio. Intercalated nanocomposites are normally interlayer by a few molecular layers of polymer.
(b) Flocculated nanocomposites: conceptually, this is same as intercalated nanocomposites. However, silicate layers are sometimes flocculated due to hydroxylated edge–edge interaction of the silicate layers.
(c) Exfoliated nanocomposites: in an exfoliated nanocomposite, the individual clay layers are separated in a continuous polymer matrix by an average distance that depends on clay loading.

Usually, the clay content of an exfoliated nanocomposite is much lower than that of an intercalated nanocomposite. In most cases exfoliated nanocomposites with

Table 17.1 Chemical formula and characteristics of phyllosilicate nanofillers [20]

Phyllosilicates	Chemical formula	CEC (mequiv/100 g)	Particle length (nm)
Montmorillonite	$M_x(Al_{4-x}Mg_x)Si_8O_{20}(OH)_4$	110	100–150
Hectorite	$M_x(Mg_{6-x}Li_x)Si_8O_{20}(OH)_4$	120	200–300
Saponite	$M_xMg_6(Si_8\text{-}Al_x)Si_8O_{20}(OH)_4$	86.6	50–60

M monovalent cation and *x* degree of isomorphous substitution (between 0.5 and 1.3)

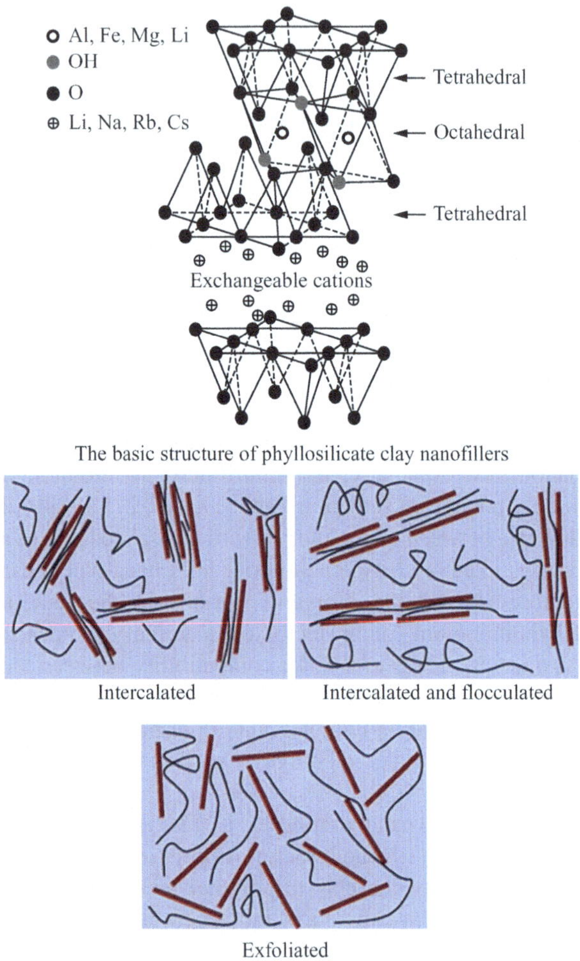

Fig. 17.1 Schematic illustration of three different types of thermodynamically achievable polymer nanocomposites formed by clay nanofillers [43]

a high aspect ratio demonstrate enhanced properties comparable to the same pristine polymers. Usually, the exfoliation of clay nanolayers in a polymer matrix requires polarity match between the clay surface and the prepolymer precursors to allow optimal access to the gallery [23]. There are a number of ways to increase a degree of exfoliation in a nanocomposite, such as in situ polymerization, melt blending, solution blending, sonication, high shear mixing, and melt intercalation [14, 23]. The morphology of nanocomposites is usually studied by X-ray techniques (XRD), transmission electron microscopy (TEM), and scanning electron microscopy (SEM) [23, 43].

3 PU-Nanosilica Composites

Colloidal silica and alumina are widely used for the preparation of PU-nanosilica composites [8, 15]. The colloidal silica/alumina in organic solvent is blended with polyol, and one obtains a sol of nanosilica in polyol. Polyol with silica/alumina sol reacts with diisocyanate forming of PU-nanosilica composite. The loading by nanosilica is in the range 5–15 wt% for desired applications. Silica/alumina sol can be added to monomers at the stage of polyester preparation by polycondensation. Also, nanosilica in PU can be prepared by in situ hydrolysis and condensation of silane-terminated oligomers. The major hurdle in the preparation of polyurethane-nanosilica composites is the agglomeration of nanoparticles [34].

3.1 Morphology

Morphological features of the polyurethane/nanosilica give an idea of the role played by the nanosilica in formation of the composite system. Typical TEM micrographs, of dispersions of colloidal silica particles in silica sol and corresponding polyurethane/nanosilica composites prepared by in situ polymerization and blending methods, are displayed in Fig. 17.2. Basically, homogeneous nanosilica particles appear in polyurethane/nanosilica composite resins prepared by in situ polymerization and corresponding polyurethane/nanosilica composites [Fig. 17.2b, d], whereas some aggregation occurs in polyurethane/nanosilica composite resins prepared by the blending method and corresponding polyurethane/nanosilica composites [Fig. 17.2c, e]. Obviously, this is related to the interaction between nanosilica particles with macromolecular chains. During preparation of polyurethane/nanosilica composite resins by in situ polymerization, some polyol segments were chemically bonded to nanosilica particles. These polyol segments can prevent nanosilica particles from aggregating, whereas by the blending method, very small numbers of polyol segments might be chemically bonded with silica particles, that is, there is less of a protection layer of polyol on the surfaces of nanosilica particles, and, thus, free nanosilica particles with relatively more unreacted hydroxyl groups easily promote aggregation through hydrogen bonding [34].

3.2 Glass Transition Temperature (T_g)

The thermal properties of the polyurethane/nanosilica composites play a significant role in the stability of the same. Figure 17.3 illustrates the effects of silica particle

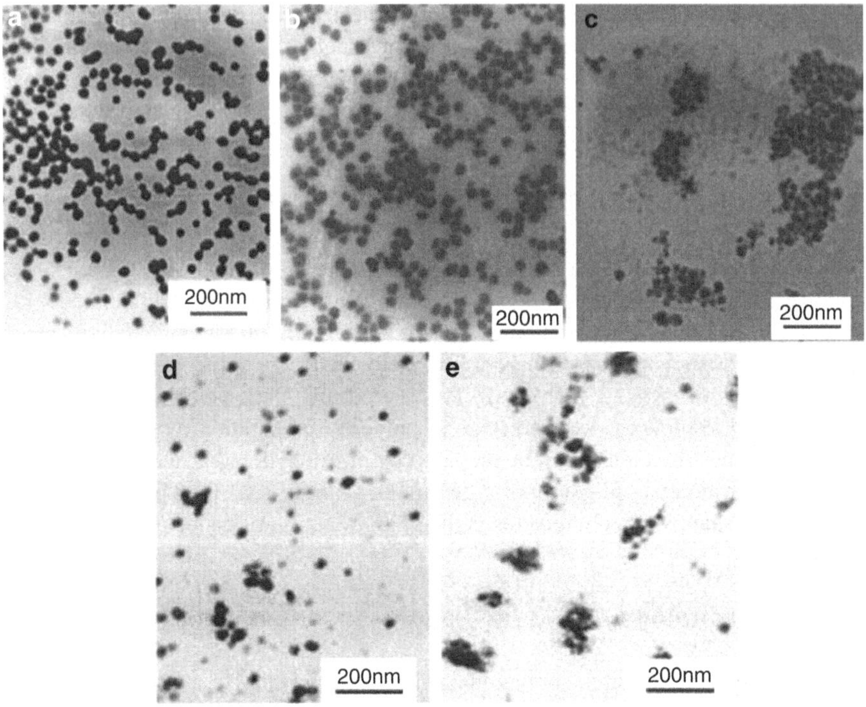

Fig. 17.2 TEM micrographs of the nanosilica particles in (**a**), nanocomposite resins prepared by in situ polymerization (**b**) and blending method (**c**), nanocomposite polyurethanes prepared by in situ polymerization (**d**), and blending method (**e**) [34]

size and preparation method on the T_g values of nanocomposite polyurethane films. As the silica particles are introduced, the T_g values of polyurethane/nanosilica composites clearly increase compared with pure polyurethane, no matter which silica particles or preparation approaches are used; the T_g values of polyurethane/nanosilica composites first increase then decrease as the particle size increases. The maximum T_g values occur at silica particle sizes within the range of 28–66 nm, which is very consistent with the variation of hydroxyl values at the surfaces of silica particles. The nanosilica with sizes of 28–66 nm has the highest –OH values at their surfaces among these particles, which have the strongest interaction with macromolecular chains by hydrogen bonding or chemical action between –OH groups of silanol and –OH or –COOH groups from resin molecules at the same mass level, restricting the segmental motion of amorphous polyol molecular chains. Figure 17.3 also reveals that the PU-nanosilica composites, obtained from in situ polymerization, have much higher T_g values than those of their corresponding composites from the blending method because more polyester segments were chemically bonded to silica particles during in situ polymerization than during the blending method [34].

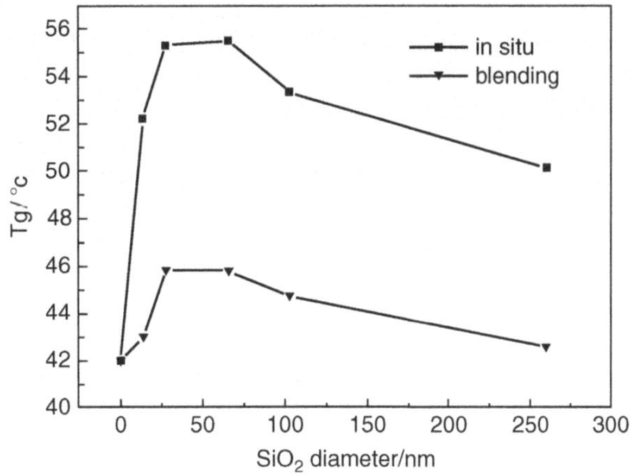

Fig. 17.3 Effect of silica particle size on the T_g of PU/nanosilica composites with 2.25 wt% SiO_2 content [34]

3.3 Tensile Property

The tensile properties decide the practical application of the polyurethane/nanosilica composites. Figure 17.4 demonstrates the tensile strength and elongation at break of the PU-nanosilica composites as a function of silica content. A considerable increase in the tensile strength, for the polyurethane/nanosilica composites even with very low nanosilica content, can be observed regardless of in situ polymerization and blending methods. The nanocomposites, obtained by in situ polymerization, have higher tensile strength than that of those obtained by the blending method at the same mass level because the former preparation method causes more polyester chains to be chemically bonded to nanosilica particles and more homogeneous dispersion of nanosilica in polymer matrix than by the latter method. The elongation at break, of polyurethane/nanosilica composites, decreases as the silica content increases, but the extent of decrease is relatively lower compared with the increasing degree in tensile strength [34].

3.4 Macrohardness

The elastomer property of the polyurethane when blended with the nanosilica is assessed by the macrohardness testing. The influence of particle size of nanosilica on the macrohardness of polyurethane matrix is illustrated in Fig. 17.5. Just as observed in the tensile properties, the macrohardness increases with increasing nanosilica content, regardless of the preparation method. There is also a maximum hardness, at about 28 nm silica, which is consistent with the change of T_g and tensile properties of nanocomposites as a function of particle size of nanosilica. Also, the PU-nanosilica composites, obtained by in situ polymerization, generally have higher hardness than those obtained by the blending method at the same mass level because the former preparation method causes more polyol chains

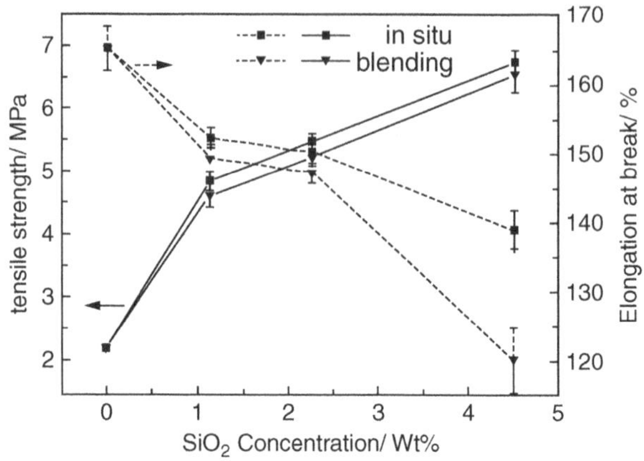

Fig. 17.4 Change of tensile strength and elongation at break of PU/nanosilica composites as a function of silica concentration (silica particle size 66 nm) [34]

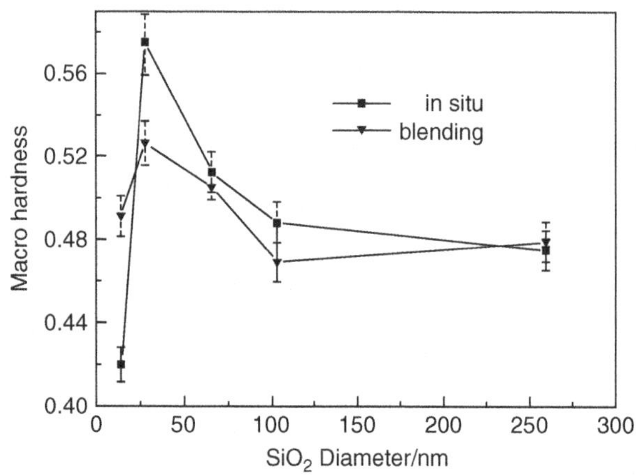

Fig. 17.5 Change of macrohardness of PU/nanosilica composites as a function of silica diameter [34]

to be chemically bonded to nanosilica particles and more homogeneous dispersion of nanosilica in the polymer matrix than by the latter method. However, at relatively high silica concentration, the PU-nanosilica composites, obtained from the blending method, have higher degrees of macrohardness than that of those obtained from in situ polymerization [34].

4 PU-Organoclay Nanocomposites

The modification PU with the organoclays through nanocomposite formation involves ultrafine phase dimensions exhibiting new and improved properties in comparison with their microcomposite or macrocomposite counterparts [9, 12, 19]. For true nanocomposites, the clay nanolayers must be uniformly dispersed in the polymer matrix rather than be aggregated as tactoids. Once nanolayer exfoliation has been achieved, the improvements can manifest themselves as increases in mechanical properties, thermal stability, and gas-barrier properties.

The PU-organoclay composites are prepared by the following procedures: (i) distribution of clay in polyol with a subsequent reaction with diisocyanate, (ii) interaction of PU with clay in organic solvent with a subsequent evaporation of solvent, and (iii) reaction of diisocyanate with hydroxyalkyl groups of organic modifier in the clay with a subsequent reaction with polyol.

4.1 Dispersibility of Organoclays

The dispersibility of clay nanofillers is a significant factor deciding the overall performance of the PU-organoclay nanocomposite. The X-ray diffractograms of the pure PU, organoclays, and PU-organoclay composites are shown in Fig. 17.6. When the clay was modified with an organic compound, it showed improved compatibility with the polymer, so the clay galleries could be easily intercalated with the polymer. The interlayer spacing (d) values of the organoclays are in the following order: C_{16}–MMT ($d = 25.96$ Å) > Cloisite 25A ($d = 19.63$ Å) > DTA–MMT ($d = 16.85$ Å). In the PU-organoclay composites with 1–6 wt% organoclay, obvious clay peaks appeared in their X-ray diffraction (XRD) curves. This indicated that these clay layers of organoclays were intercalated (not exfoliated) and not homogeneously dispersed in the PU matrix. The d values of the hybrid peaks were larger than those of pristine organoclays. For example, after the intercalation by organoclays, the d-spacing increased from 25.96 to 31.22 Å for the C_{16}– MMT, from 16.85 to 27.12 Å for the DTA–MMT, and from 19.63 to 32.60 Å for the Cloisite 25A hybrid. This increase in the basal spacing suggests the intercalating of polymer chains into the clay galleries. A higher loading of clay in PU caused sharp peaks to exist in the region and the peak increased in intensity [20].

4.2 Morphology

The morphology through TEM images is an important tool to analyze the intercalation or exfoliation of the PU chains into the organoclay galleries. Figure 17.7a–c shows the TEM images of PU-organoclay nanocomposite containing different 3 wt% organoclays. The dark lines in the photograph are the intersections of a clay sheet of 1 nm thick, and the spaces between the dark lines are the interlayer spaces. TEM images proved that most clay layers were dispersed homogeneously into the matrix polymer, although some clusters or agglomerated particles were also detected. A similar arrangement of the layers in Fig. 17.7a–c is observed, with the presence of one or several clay layers in the PU matrix. Unlike hybrids containing C16–MMT and DTA–MMT [Fig. 17.7a, b], the clay layers of Fig. 17.7c are more exfoliated and dispersed randomly into the matrix polymer. The nanoparticles possibly exhibit properties different from those of conventional composites in larger particles, such as their optical properties. Because nanoparticles are much

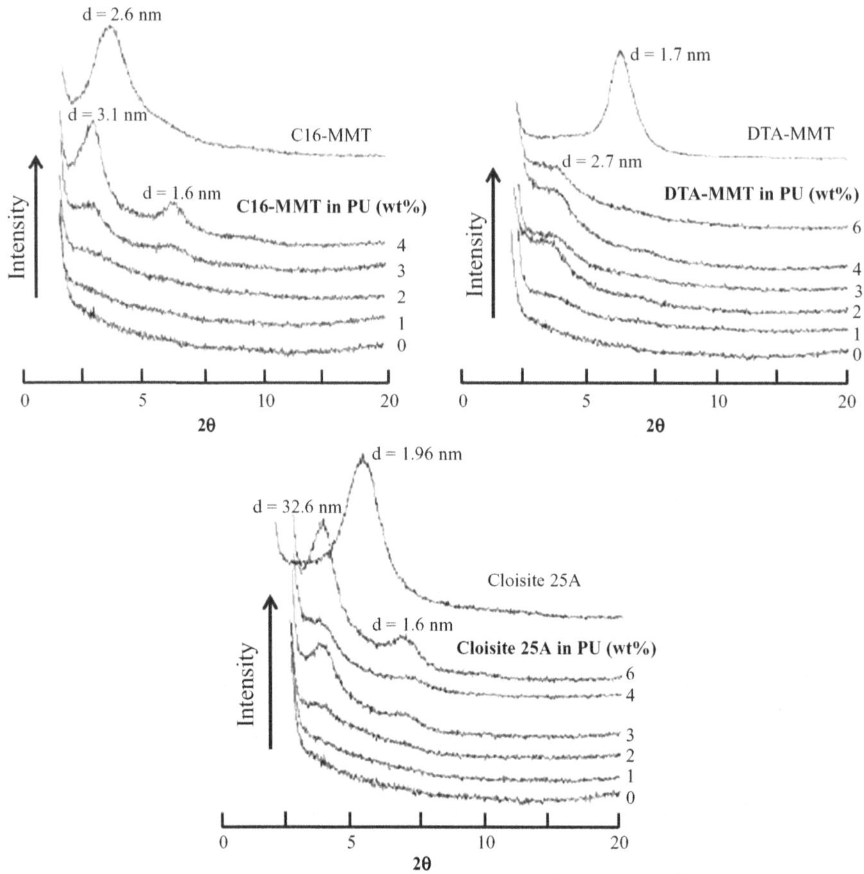

Fig. 17.6 XRD patterns of C16–MMT, DTA–MMT, Cloisite 25A, and their PU/organoclay nanocomposites [20]

smaller than the wavelength of visible light (400–800 nm), the composites may be transparent, although the same matrix with larger particles may not be [20].

4.3 Tensile Properties

The ultimate strength of the PU-organoclay nanocomposite increased remarkably with the content of the organoclay and possessed maximum values at 3 wt% clay content (Fig. 17.8). The tensile strength of the PU-organoclay nanocomposite increased from 45 to 59 MPa for C16–MMT and from 45 to 57 MPa for DTA–MMT, with an increase in organoclay contents up to 3 wt%. For the Cloisite 25A PU-organoclay nanocomposite, when the amount of Cloisite 25A was 3 wt%, the tensile strength increased by about 70 % (76 MPa) compared with that of pure PU (45 MPa).

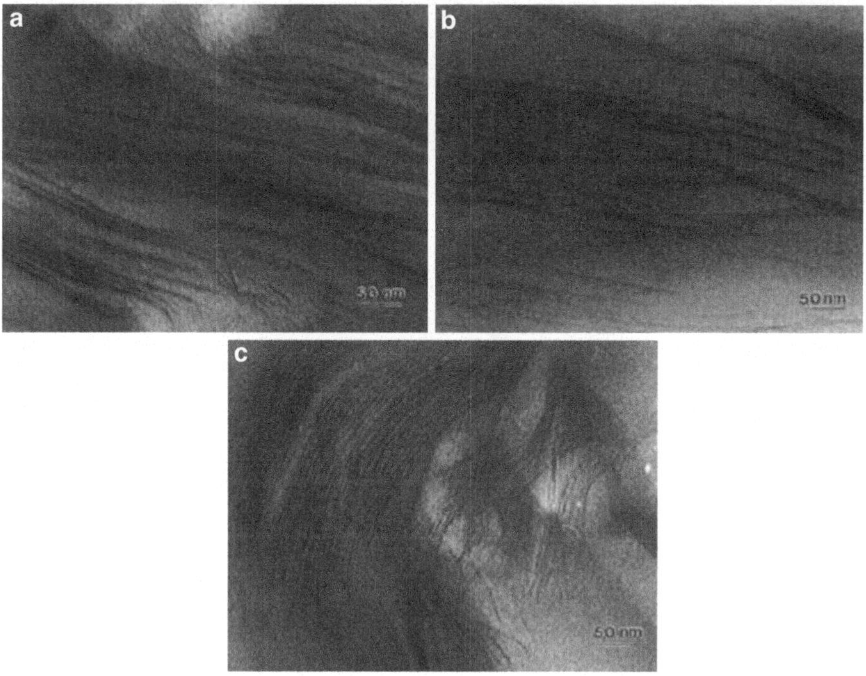

Fig. 17.7 TEM micrographs of PU/organoclay nanocomposites containing 3 wt% organoclay: (**a**) C16–MMT, (**b**) DTA–MMT, and (**c**) Cloisite 25A [20]

The strength values of all hybrids started to decrease above 3 wt% clay content. The modulus values increased linearly with the organoclay content increasing up to 3 wt%. As the amount of Cloisite 25A increased to 3 wt%, the modulus of the hybrid was 13.55 GPa, about two times greater than that of pure PU (7.24 GPa). This enhancement of the modulus is ascribed to the resistance exerted by the clay itself, as well as the orientation and aspect ratio of the clay layers. Additionally, the stretching resistance of the oriented backbone of the polymer chain in the gallery also contributed to the enhancement of the modulus. The elongation at break of PU-organoclay nanocomposite with C16–MMT and DTA–MMT was slightly lower than that of pure PU. This response is characteristic of materials reinforced with stiff inorganic materials and is particularly noteworthy for its intercalated morphology. However, another noteworthy phenomenon is that the Cloisite 25A hybrid had higher elongation than pure PU films, which may be attributed to the good dispersion of Cloisite 25A in the polymer matrix with respect to other organoclays. This was proven by the TEM study shown in Fig. 17.7c. From these results, it appears that there is an optimal amount of organoclay needed in the hybrid to achieve the greatest property improvement. The results also reveal that the tensile properties of Cloisite 25A organoclay nanocomposites were better than those of other hybrids because Cloisite 25A does disperse effectively in intercalation in a PU matrix [20].

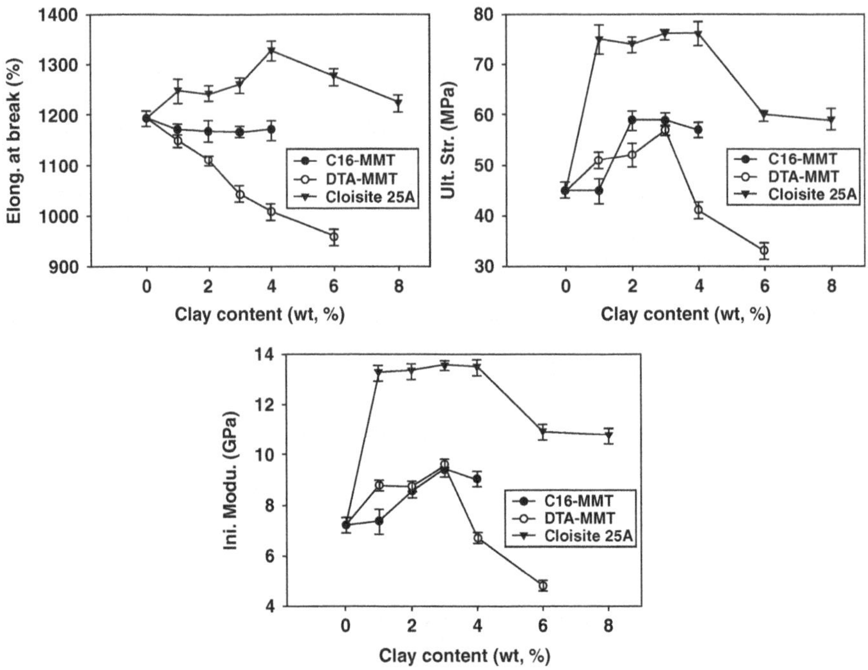

Fig. 17.8 Effects of the clay loading on the ultimate tensile strength, initial modulus, and elongation at break of the PU/organoclay nanocomposites [20]

Table 17.2 Gas permeabilities of PU/organoclay composites [27]

	O_2 (cc/m^2/day)		
Clay (wt%)	C16-MMT	DTA–MMT	Cloisite 25A
0 (pure PU)	7214	7214	7214
2	5447	6953	5817
3	5074	6381	5622
4	3587	6136	4725

4.4 Gas Permeability

In organic polymers filled with inorganic materials, gas permeability can decrease, depending on the polymer–filler adhesion and compatibility [6, 16]. Filler that has good compatibility with the organic polymer matrix usually reduces the permeability, mainly because of the reduction of the transport cross section and the increase in the tortuous paths for gas molecules. Table 17.2 shows gas permeability values for different PU-organoclay nanocomposites. PU-organoclay nanocomposites showed lower permeability values for O_2 than pure PU film, regardless of the difference in organoclays. This behavior could be attributed to the presence of the high aspect ratio and rigid-platelet clay in the polymer matrix. The gas-barrier effect of C16–MMT was higher than that of DTA–MMT or Cloisite 25A. With the addition of 0–4 wt% clay loading, the permeability values decreased from 7,214 to 3,587 cc for C16–MMT, from 7,214 to 6,136 cc for DTA–MMT, and from 7,214 to 4,725 cc for Cloisite 25A. The permeability value in the C16–MMT hybrid was about a half that of the corresponding pure PU. The permeability of O_2 was also influenced by the content of organoclay dispersed in the polymer matrix. With the clay loading increasing from 0 to 4 wt%, regardless of the type of organoclay, the values of the permeability decreased linearly, which is due to the increase in the tortuous paths for gas molecules. It also seems that films containing higher amounts of clay were much more rigid, contributing to a gas permeability decrease [20].

5 PU Nanocomposite Foams

In recent years, PU nanocomposite foams have attracted considerable attention from both a fundamental research and an application point of view due to their remarkable improvement in materials properties [7, 17, 18, 24, 26, 29, 31, 32, 36]. The synthesis of PU nanocomposite foam involves two steps. First is the preparation of PU-nanosilica/organoclay composite through the mixing of nanosilica/organoclay with polyol for 2 h using a high speed mixer. Subsequently, the surfactant, catalyst, crosslinking agent, and distilled water were added to the polyol/nano-organoclay/silica mixture and mixed at 5,200 rpm for 5 min. Finally, diisocyanate was added and stirred for 6 s. The mixture was poured into a mold. The molded PU-nanocomposites are taken for foaming in a foaming chamber through a batch process.

5.1 PU-Nanosilica Composite Foams

Figure 17.9 shows SEM micrographs of the cross section of PU-nanosilica composite foams prepared with three different surfactants. Generally, silicone-based surfactants are used to stabilize gas bubbles for PU foams. Different surfactants produce PU foams with different cell sizes. PU foams prepared with B8629 and L3184 have smaller cell size and higher absorption ratio. Surfactants can change the interfacial energy of the silica-PU interface. It may also act as coupling agent that binds the two phases effectively. The average cell size of the PU-nanosilica foams ranged from 160 to 400 μm. The cell size of the nanocomposite foams was smaller with increasing the nanosilica contents. Tensile strength and Young's modulus of PU-nanosilica foams slightly increase with increasing nanosilica content as shown in Fig. 17.10. The elongations at break of PU/nanosilica composite foams were higher than that of PU, regardless of the nanosilica content present in the system [45].

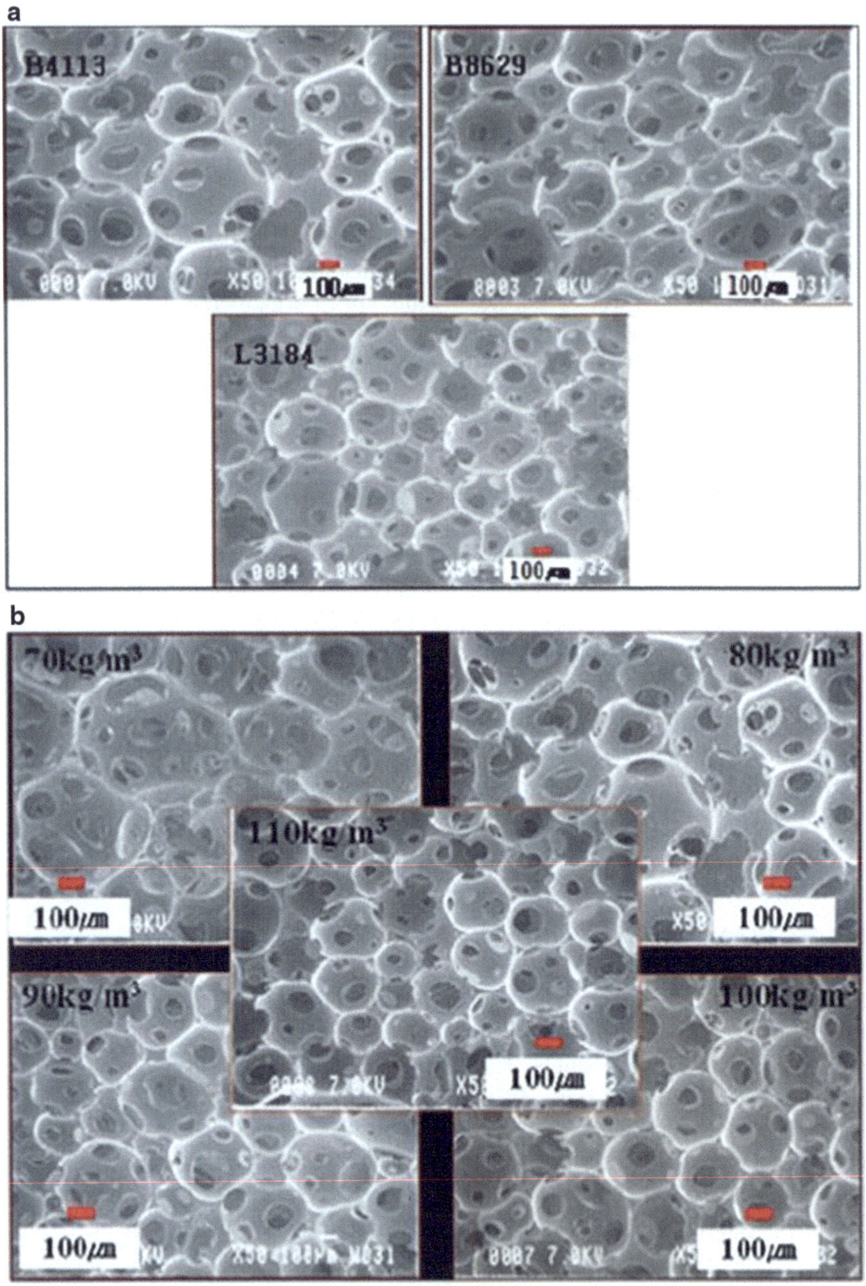

Fig. 17.9 SEM micrographs of the cross section of PU/nanosilica foams (**a**) prepared with three different surfactants and (**b**) prepared with different density [45]

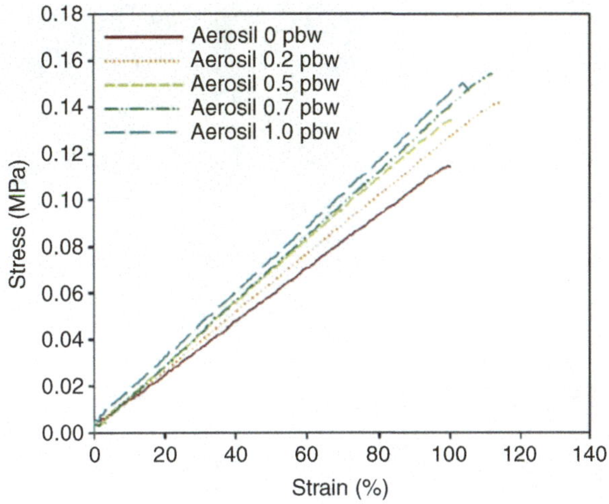

Fig. 17.10 Typical stress–strain curves of PU/nanosilica foams with different nanosilica content [45]

5.2 PU-Organoclay Nanocomposite Foams

Figure 17.11 shows SEM images of the freeze-fractured surface of PU-organoclay nanocomposite foams. All foams exhibited polygon-type closed-cell structures with energetically stable pentagonal and hexagonal faces. The neat PU foam has fewer cells and a larger cell size than PU-organoclay nanocomposites foams with 5 % organoclay. PU foams containing 5 % organoclays have a lower bulk density (ρf) than that of pure PU foam but a higher cell density (N_c) and smaller cell size (d), suggesting that the dispersed clay particles act as heterogeneous nucleation sites during cell formation. However, there is little difference between MMT-Tin/PU (the one used a catalyst during synthesis) and MMT-OH/PU (with no catalyst) nanocomposite foams in terms of cell size and cell density. PU nanocomposite foams were crushed into fine powder, and then, the clay dispersion was determined by XRD spectra of the powder (Fig. 17.12). The clay orientation and dispersion is somewhat affected by the foaming process. The shoulder at a very low angle of the XRD pattern of the MMT-Tin/PU foam is smaller than that of MMT-OH/PU foam.

The PU-organoclay nanocomposite foams exhibit a substantially higher reduced compressive strength and modulus as shown in Fig. 17.13a. The MMT-OH/PU nanocomposite foam showed about a 650 % increase in reduced strength and a 780 % increase in reduced modulus, while the nanocomposites foam with MMT-Tin showed about a 610 % increase in reduced strength and a 760 % increase in reduced modulus as compared to pure PU foams. The glass transition temperatures of the foams shown in Fig. 17.13b indicate that the MMT-Tin/PU foam has the highest T_g and is higher than that of neat PU and MMT-OH/PU foams, respectively.

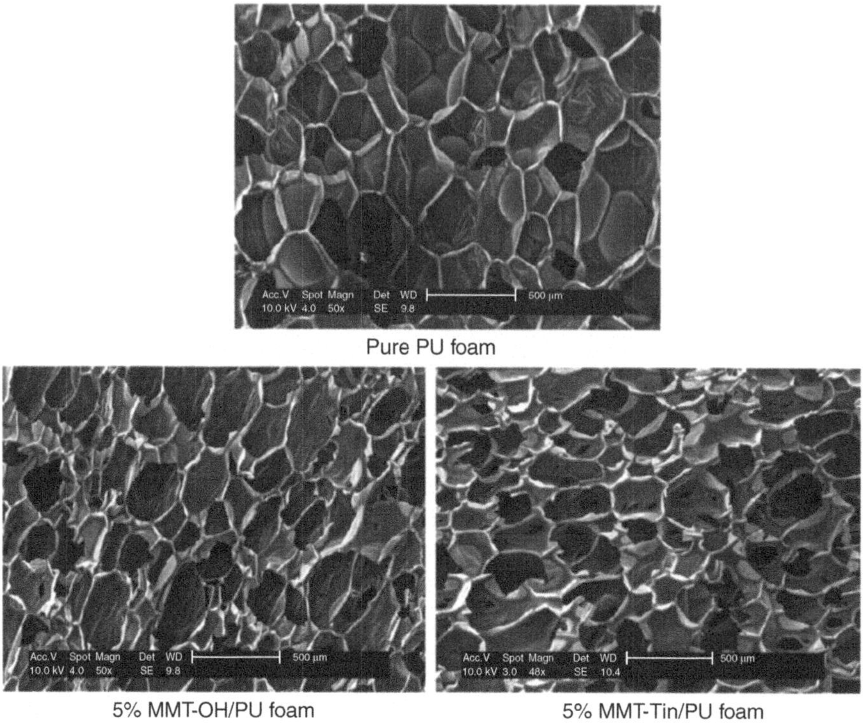

Fig. 17.11 SEM micrographs of PU-organoclay nanocomposite foams at cross sections parallel to foam rising direction [33]

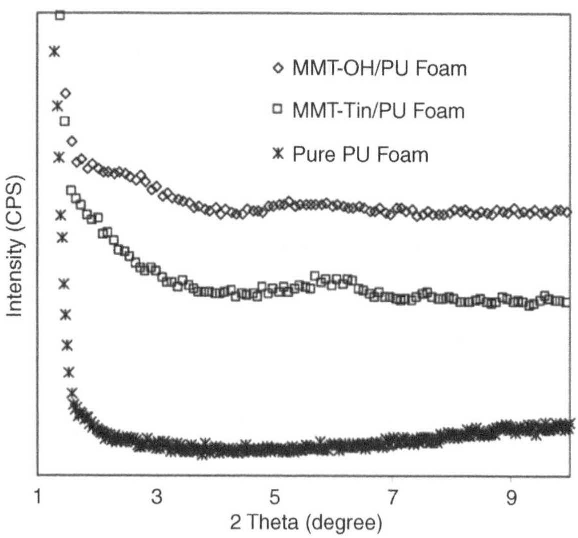

Fig. 17.12 XRD curves of organoclay/PU nanocomposite (5 wt%) foams [33]

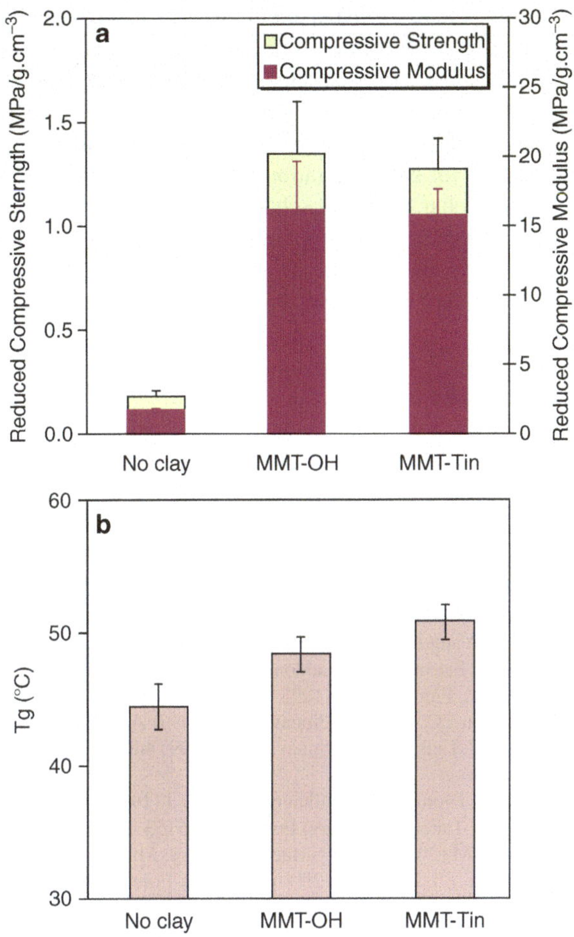

Fig. 17.13 Properties of organoclay/PU nanocomposite (5 wt%) foams (**a**) compressive properties and (**b**) glass transition temperature [33]

The higher Tg of PU nanocomposites foams may result from the confinement of PU molecules by dispersed MMT particles. This effect is stronger in the MMT-Tin/PU foam since the polyol could enhance the intra-gallery reaction [33].

6 Conclusion

The development of the PU nanocomposites is one of the evolutionary steps of the polymer technology, which offer attractive potential for diversification and application of conventional polymeric nanocomposite materials. PU nanocomposites and foams have a great deal of future promise for potential applications as high-performance materials. Colloidal silica and organoclays are the most studied nanofillers widely used in PU nanocomposites. The advantage of nanoscale reinforcement is twofold: (1) when nanoscale fillers are homogeneously dispersed in the matrix, a tremendous surface area is developed that could contribute to polymer chain confinement, which may lead to higher T_g, higher stiffness and tensile strength, increased elongation, and an increase of both flexural and tensile modulus; and (2) nanoscale fillers, especially clays, provide an extraordinary zigzag tortuous diffusion path that lead to enhanced barriers for gas penetration for a gas (dioxygen, others), moisture. The enhanced barrier characteristics, chemical resistance, reduced solvent uptake, and flame retardancy of clay–polymer nanocomposites originate from the hindered pathways through the PU nanocomposite. PU nanocomposites and foams are widely used in automotive, house construction, and coating industry.

References

1. Smith TL (1974) J Polym Sci Polym Phys Ed 12:1825
2. Chang YJ, Wilkes GL (1975) J Polym Sci Phys Ed 13:455
3. Masiulanis B, Zielinski R (1985) J Appl Polym Sci 30:2731
4. Petrovic ZS, Ferguson J (1991) Prog Polym Sci 16:695
5. Giannelis EP (1996) Adv Mater 8:29
6. Yano K, Usuki A, Okada A, Kurauchi T, Kamigaito O (1997) J Polym Sci, Part A: Polym Chem 35:2289
7. Ou Y, Yang F, Yu ZZ (1998) J Polym Sci, Part B: Polym Phys 36:789
8. Javni I, Petrovic ZS, Waddon A (1998) Polyurethanes Expo. Dallas
9. Wang Z, Pinnavaia TJ (1998) Chem Mater 10:3769
10. LeBaron PC, Wang Z, Pinnavaia TJ (1999) Appl Clay Sci 15:11
11. Wang Z, LeBaron PC, Pinnavaia TJ (1999) 8th international conference on polymer additives, an addition
12. Zilg C, Thomann R, Mulhaupt R, Finter J (1999) Adv Mater 11:49
13. Zuo M, Takeichi T (1999) Polymer 40:5153
14. Rosoff M (2000) Nano-surface chemistry. Marcel Dekker, New York
15. Petrovic ZS, Zhang W (2000) Mater Sci Forum 352:171
16. Zoppi RA, Neves SD, Nunes SP (2000) Polymer 41:5461
17. Rong MZ, Zhang MQ, Zheng YX, Zeng HM, Friedrich K (2001) Polymer 42:3301
18. Reynaud E, Jouen T, Gauthier C, Vigier G, Varlet J (2001) Polymer 42:8759
19. Tien YI, Wei KH (2001) Polymer 42:3213
20. Chang JH, An YU (2002) J Polym Sci, Part B: Polym Phys 40:670
21. Ogoshi T, Itoh H, Kim KM, Chujo Y (2002) Macromolecules 35:334
22. Shang XY, Zhu ZK, Yin J, Ma XD (2002) Chem Mater 14:71
23. Triantafillidis CS, LeBaron PC, Pinnavaia TJ (2002) Chem Mater 14:4088
24. Voros G, Pukanszky B (2002) Compos A 33:1317
25. Zhou S, Wu L, Sun J, Shen W (2002) Progr Org Coating 45:33
26. Wu CL, Zhang MQ, Rong MZ, Friedrich K (2002) Compos Sci Technol 62:1327
27. Ray SS, Okamoto M (2003) Prog Polym Sci 28:1539
28. Wang YT, Chang TC, Hong YS, Chen HB (2003) Thermochim Acta 397:219
29. Wetzel B, Haupert F, Zhang MQ (2003) Compos Sci Technol 63:2055
30. Yuan JJ, Zhou SX, Liao JH, Li MW (2003) Chin J Chem Eng 11:483
31. Abbate M, Musto P, Ragosta G, Scarinzi G, Mascia L (2004) Macromol Symp 218:211
32. Garcia M, Van Vliet G, ten Cate MGJ, Chavez F, Norder B, Kooi B, Van Zyl WE, Verweij H, Blank DHA (2004) Polym Adv Technol 15:164
33. Cao X, Lee LJ, Widya T, Macosko C (2005) Polymer 46:775
34. Chen Y, Zhou S, Yang H, Wu L (2005) J Appl Polymer Sci 95:1032

35. Lee HT, Lin LH (2006) Macromolecules 39:6133
36. Kontou E, Anthoulis GJ (2007) Appl Polym Sci 105:1723
37. Paul DR, Robeson LM (2008) Polymer 49:3187
38. Furukawa M, John B (2009) J Rubber Res 12:151
39. John B, Furukawa M (2009) Polym J 41:319
40. John B, Furukawa M (2009) Polym Eng Sci 49:1970
41. John B, Motokucho S, Kojio K, Furukawa M (2009) Sen-I Gakkaishi 65:236
42. John B, Furukawa M (2012) J Fishery Technol (2012) in press
43. Khudyakov IV, Zopf RD, Turro NJ (2009) Des Monomers Polym 12:279
44. Motokucho S, Kojio K, Furukawa M, John B (2009) Japan patent 127154
45. Lee J, Kim GH, Ha CS (2012) J Appl Polym Sci 123:2384

Advanced Electrospun Nanofibers of Layered Silicate Nanocomposites: A Review of Processing, Properties, and Applications

18

Singaravelu Vivekanandhan, Makoto Schreiber, Amar Kumar Mohanty, and Manjusri Misra

Contents

S. Vivekanandhan • A.K. Mohanty • M. Misra (✉)
Bioproducts Discovery and Development Centre, Department of Plant Agriculture, Crop Science Building, University of Guelph, Guelph, ON, Canada

School of Engineering, Thornbrough Building, University of Guelph, Guelph, ON, Canada
e-mail: svivekan@uoguelph.ca; mohanty@uoguelph.ca; mmisra@uoguelph.ca

M. Schreiber
Bioproducts Discovery and Development Centre, Department of Plant Agriculture, Crop Science Building, University of Guelph, Guelph, ON, Canada

Department of Physics, University of Guelph, Guelph, ON, Canada

Department of Chemistry, University of Guelph, Guelph, ON, Canada
e-mail: mschreib@uoguelph.ca

J.K. Pandey et al. (eds.), *Handbook of Polymernanocomposites. Processing, Performance and Application – Volume A: Layered Silicates*, DOI 10.1007/978-3-642-38649-7_26,

Abstract

Nanocomposites fabricated with layered silicates possess improved thermal and mechanical properties along with other functionalities such as accelerated biodegradability, enhanced barrier properties, and flame resistance. These properties can be further enhanced when the composites are made into nanometer-sized fibers due to a high surface area, dimensional confinement of the matrix polymer, and alignment of the layered silicates. The properties of these composite fibers can be tuned by varying the polymer matrix and the type of layered silicate as well as their dispersion within the fiber. In addition, the fiber morphology (size, shape, alignment) plays an important role in determining the properties of the composite. The morphology of the fiber is largely dependent on the fabrication technique that is used. Many methods of nanofiber fabrication have been reported. Among them, electrospinning has proven to be a simple and efficient method for the fabrication of structure-controlled polymer-based fibrous materials with micro to nanometer diameters. The unique properties of electrospun layered silicate nanocomposite fibers allow for potential applications in areas such as scaffolds for tissue regeneration, coatings for food storage containers, and flame-resistant coatings. Because of this, there is an increasing amount of research on electrospun composite nanofibers reinforced with layered silicates. In this book chapter, we review current work on the fabrication and characterization of electrospun layered silicate nanocomposites. The processability of these layered silicate nanocomposite fibers and their properties are summarized for fibers with traditional synthetic polymer matrices and fibers with renewable biopolymer matrices.

Keywords

Barrier properties • Chemical properties • Electrospinning • Exfoliation • Intercalation • Layered silicates • Nanocomposites • Mechanical properties • Nanofibres

1 Introduction

When polymers are fabricated into fibers with nanometer-scale diameters, many wonderful characteristics can emerge which include (i) a higher surface area-to-volume ratio, (ii) enhanced/tuned surface functionalities, and (iii) superior mechanical properties compared to their respective bulk structures [1, 2]. These properties can be tuned precisely, enabling these fibers to be suitable candidates for numerous applications such as environmental filters, reinforcement for composites, biomedical/tissue engineering, and as templates for other methods of material fabrication [3]. A range of techniques have been used for the fabrication of polymers with fibrous structures which include:

Drawing: This process involves the fabrication of long fibers with controlled thickness by mechanically stretching a polymer solution or melt. This process is very similar to industrial dry spinning; however, the challenge is to manipulate the rheological properties of the solution for the fabrication of fibers with desired properties [1, 4].

Template synthesis: The effective utilization of nanoporous architectures (such as membranes) can generate precisely controlled fibril/tubular nanomaterials of polymers, metals, metal oxides, and carbons [5]. However, it is difficult to fabricate continuous fibrous materials using this technique.

Phase separation: This process consists of various steps such as dissolution, gelation, extraction, and freeze-drying for the fabrication of a nanostructured foam archistructure. Though it is a simple process, a major disadvantage of this technique is that each individual step is quite time consuming, making it a very slow process [6].

Self-assembly: The ability of individual molecules to assemble themselves spontaneously can be utilized to form various diversified nanostructures with many desired patterns and functions. While highly controlled architectures can be fabricated through self-assembly, it is very time consuming for the fabrication of polymer nanofibers [7].

Spinning: Spinning processes can be classified into the four major groups of melt spinning, solution electrospinning, melt electrospinning, and force spinning. The advantage of these spinning processes over other existing processes is the ability to fabricate long continuous fibers.

Among the spinning techniques, electrospinning has been extensively utilized for the fabrication of a wide range of polymeric, composite, and ceramic nanofibers with diverse morphologies and functionalities. The history of electrospinning begins in 1934 where it was explored by Formalas, for the fabrication of polymer fibers; resulting in several patents [8–10]. For many years afterwards, the electrospinning process was not significantly researched. However, in 1990, electrospinning began to reemerge along with the large increase in research related to nanoscience and technology. Since then, electrospinning has been utilized for the fabrication of fibrous nanomaterials by many research groups.

The electrospinning technique can be used to fabricate polymer fibers with diameters ranging from 40 to 2,000 nm [11]. These fibers have been utilized for many applications due to their very-high-specific surface area and aspect ratio [11]. Varying the composition of the precursor solution and altering the processing conditions used can enable the fabrication of fibers with the diversified morphologies, physiochemical properties, and multiple functionalities [12]. These types of property enhancements make the electrospun fibers novel candidates for numerous applications in the fields of membrane technologies, tissue engineering, biomedical tools, functional composites, sensors, and electronic/electrochemical devices [13]. Subbiah et al. [14] reviewed research work on the fabrication of electrospun nanofibers from a variety of polymers which possessed different chemical, mechanical, and electrical properties. Their review work summarized that the applied electrostatic forces and the viscoelastic profile of the precursor solution are key factors for the fabrication of nanofibers using the electrospinning process. In addition, various other processing parameters including the polymer injection rate, distance between the spinneret and sample collector, and electrical conductivity influence the morphology of electrospun fibers [14]. The properties of polymeric fibers can be enhanced by incorporating other kinds of nanomaterials such as carbon nanotubes (single wall/multiwall) [15, 16], grapheme [17], cellulose nanostructures [18], metal (silver, gold) [19, 20], metal oxides (TiO_2, ZnO, $Ni_{0.6}Zn_{0.4}Fe_2O_4$) [21–23], metal chalogenides (CdS, CdTe, ZnS) [24, 25, 125], and other ceramics (nano talc, nano clay) as reinforcements. Apart from the enhancement of the fiber properties, these nanostructures provide additional functionalities to the composite structures. Compared to traditional composite materials, nanocomposites can obtain significant property enhancement from much lower levels of reinforcer loadings; typical loadings being <5 %. Among the various types of nanocomposite fibers, polymer/layered silicate nanocomposites have recently attracted many academic and industrial researches due to their enhanced thermal, mechanical, and other functional properties [37, 123].

Layered silicates are composed of tetrahedral silicon and octahedral aluminum/magnesium hydroxide layers [125]. Typically, individual layers are on the order of 1 nm in thickness and can vary in their lateral dimensions from 30 nm to several microns [125]. Polymer nanocomposites reinforced with layered silicates (i.e., clays) exhibit promising physiochemical and various other functional properties compared to their respective neat polymers [37, 123, 125]. These property enhancements can be achieved due to the nanoscopic structure of the individual clay platelets and their superior aspects ratio. The greatest level of reinforcement is achieved when the silicate platelets are exfoliated and dispersed uniformly throughout the matrix. Extensive work has been performed by our research group on the fabrication and characterization of layered silicate reinforced composite structures in both thermoplastic and thermoset platforms [26–29, 30–38, 123]. We found that layered silicates can be used for hybrid reinforcement along with natural fibers [35]. Our research findings indicate that the hybrid reinforcement improved the barrier and thermal properties of the composites. In addition, we have also investigated the effect of surface modifications on layered silicates in order to enhance their compatibility with various matrix polymers [39, 123]. In this book chapter, we will discuss the various aspects of electrospun layered silicate composite nanofibers.

2 Electrospinning of Polymer Nanocomposites

2.1 Instrumentation

The basic electrospinning setup consists of a high-voltage power supply (HVPS), a conductive collector, a syringe pump, and a spinneret as shown in Fig. 18.1 [1, 13]. The output voltage of the HVPS can typically be varied between 0.5 and 50 kilovolts (kV). In most of the cases, the polarity of HVPS is positive but can be changed to a negative polarity depending on the requirements [40, 41]. The syringe pump must be capable of milliliter per hour (ml/h) flow rates which usually range from 0.1 to 10 ml/h. The spinneret is usually a flat-tipped conductive needle

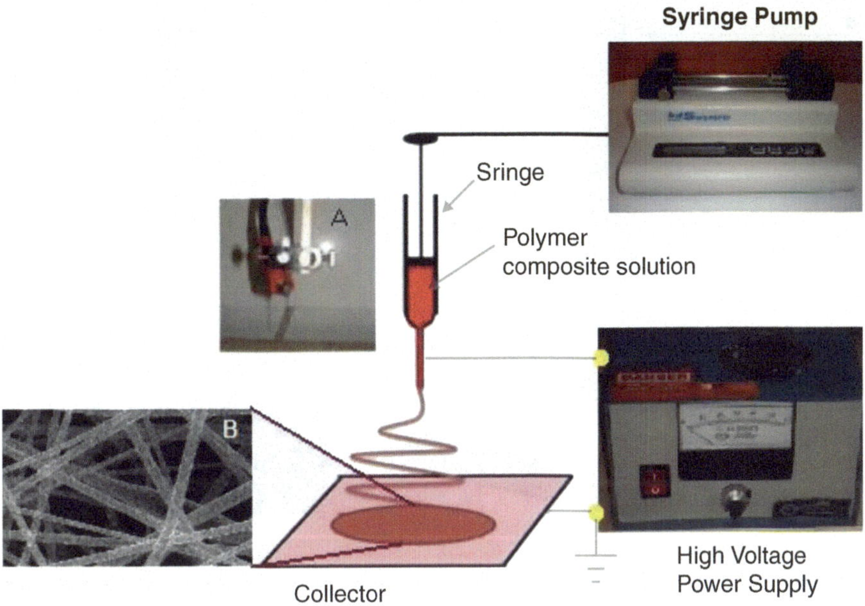

Fig. 18.1 Schematic representation of the electrospinning fabrication process (Reprinted with permission) [2]

connected to the syringe in the syringe pump. The active lead of the HVPS is connected to the electrospinning solution or melt (either directly or through the conductive spinneret), and the grounding lead is attached to the collector [1, 42]. In some cases, a potential opposite to the polarity of the spinneret is applied to the collector in order to increase the electric field strength between the spinneret and collector [12]. Many different types of collectors such as the plate, the drum, and the parallel plate collector have been used for the fabrication of different fiber mat morphologies as shown in Fig. 18.2 [12, 13]. The most commonly used collector is the plate collector. In this collector, the fibers collect in random orientations to produce a nonwoven mat. The drum collector is a rotating conductive cylinder in which the curved portion of the cylinder faces the opening of the spinneret. The rotation speed of the drum must be fast enough to avoid the formation of non-wovens yet slow enough to protect the fibers from stretching. Thus, the rotation speed of the collector must be matched with the deposition speed of the fibers in order to collect continuous aligned fibers [13]. The parallel plate collector consists of two plate collectors placed next to each other on the same plane so that their sides are parallel to each other with a small gap in between them. This effectively splits the electric field which causes the fibers to orient themselves between the gap [13, 43]. With this method, fibers with a high degree of alignment can be obtained, but the lengths of the aligned fibers are limited by the size of the gap between the plates. Other variations on these collectors exist in the literature [12, 13] as do many other unique collectors designed for specific purposes.

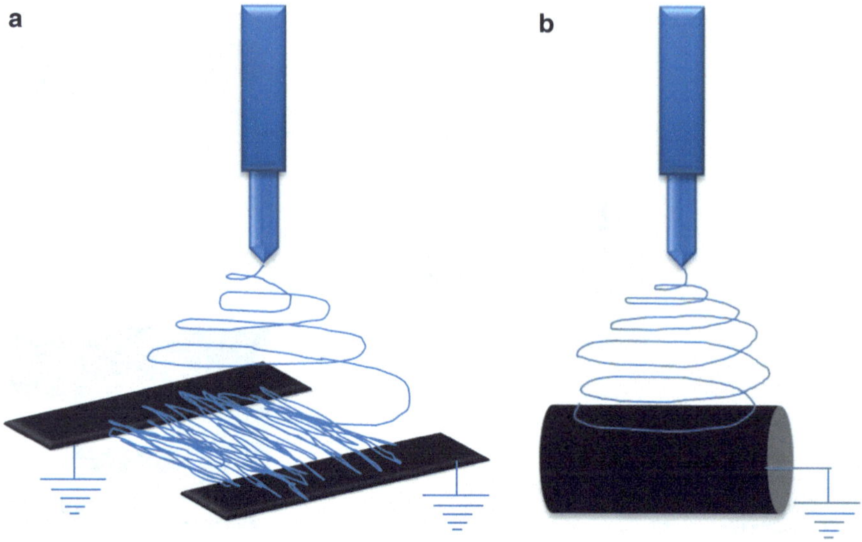

Fig. 18.2 Schematic representation of (**a**) parallel plate and (**b**) rotating drum collector-based electrospinning

Many parameters affect fiber formation through the electrospinning process. These parameters can be organized into three major categories: (i) solution properties, (ii) processing parameters, and (iii) ambient conditions. Solution properties include the viscosity, surface tension, conductivity, and chain entanglement of the polymer solution. These properties can be modified by changing the solvent or the solvent mixture ratio, the polymer concentration or polymer mixture ratio, or by adding other materials to the solution. The processing parameters include the voltage applied by the HVPS, the working distance (distance between the collector and spinneret), size of the spinneret (needle gauge), and flow rate of the solution. These parameters can all be adjusted within the range and resolution of the available equipment. Ambient conditions refer to the surrounding environment of where the electrospinning process takes place. The main ambient factors are temperature, humidity, gas composition, and air flow. These parameters can be controlled by enclosing the electrospinning apparatus within a chamber with environmental controls.

There are two types of electrospinning: solution and melt [1, 42]. Solution electrospinning is more common and involves dissolving the polymer to be electrospun into an appropriate solvent or mixture of solvents. This method allows for more versatility in controlling the solution properties. During the electrospinning process, the solvent evaporates as the jet whips around, gradually eliminating solvent from the fibers. To completely eliminate solvent in the final fibers, the electrospinning jet must be given an adequate flight time so that the fibers that deposit onto the collectors are solid and will not fuse together. Melt electrospinning involves melting a polymer into its liquid phase prior to electrospinning [44]. Typically, polymer melts are nonconductive and have much higher viscosities than solutions [45]. Thus, melt electrospinning cannot produce as fine fibers as solution-based electrospinning. During the electrospinning of a polymer melt, the jet cools and solidifies. Compared to solvent electrospinning, this method is advantageous in that it does not produce any solvent waste [46].

2.2 Challenges and Issues in Electrospinning Polymer Nanocomposites

Nanocomposites, also known as organic–inorganic hybrids due to containing organic/inorganic nanoscale filler materials within the organic polymer matrix [47], are readily becoming materials of interest. Electrospinning can be utilized in order to produce polymer nanocomposite fibers. This can be achieved by mixing an insoluble nanoparticle into a polymer solution followed by electrospinning [48]. The resultant electrospun fiber will have the filler material dispersed throughout the fiber [48]. These materials may possess the general lightweight and flexible properties of the organic polymer as well as the strength and chemical resistance of the inorganic filler [47]. Thus, compared to neat polymers, composite materials can possess improved molecular transport properties, thermal properties, stiffness, strength, and toughness [49]. Nanoscale filler materials, or nanofillers, possess high-specific surface areas and therefore, high levels of interfacing occur within the nanocomposites. This level of interfacing can greatly affect the morphology of the nanocomposite at the molecular scale due to the filler possessing dimensions that are on the same size scale as that of the matrix polymers [49]. In addition, due to the high surface area providing a much higher reinforcement efficiency than traditional filler materials, nanocomposites typically requires a less than 5 wt% nanofiller loading [49].

During the electrospinning process, the polymer solution is drawn into the electrospinning jet from the solution reservoir in a funnel-like manor [50]. In addition, high shearing forces are exerted on the electrospun polymer jets [50]. Due to these two processes, anisotropic fillers such as one- or two-dimensional nanomaterials will tend to have their longer dimensions aligned parallel to the fiber axis in the final deposited fibers [50]. By uniaxially aligning the nanofillers in a nanocomposite, mechanical properties along the alignment direction are greatly increased due to a more efficient stress transfer. For conductive fillers, alignment enhances the conductivity of the composite along the alignment direction. As electrospun fibers are essentially one-dimensional (1D) structures, the most important material properties are those along the fiber axis. Thus, through electrospinning, the reinforcement properties of the nanocomposites can be focused about the fiber axis. In addition, as the diameters of electrospun fibers are very small (on the order of a tens of nanometers to a few micrometers), dimensional confinement of the filler can occur [51]. This causes fillers to pack more densely, thus transferring stress [51] or electric current [52] through the fiber even more effectively. The improved stress transfer efficiency of electrospun fiber nanocomposites can also give rise to different deformation mechanisms from their bulk nanocomposite counterparts such as ductile instead of brittle deformations [51].

Nanocomposite fiber production through electrospinning involves dispersing the inorganic nanoparticle into a polymer solution [53–58]. However, the nanoparticles may agglomerate in the solution, therefore preventing even dispersion and making it difficult to obtain a high concentration of the filler within the composite [15]. Therefore, it is importance to prevent the agglomeration of nanofillers to retain their unique shape and size within the composites [50]. Agglomerated nanoparticles can also form localized charge accumulations while electrospinning; disturbing the spinning process, and leading to poor fiber morphologies [59]. There are several well-established methods of preventing agglomeration and achieving uniform particle distribution in a solution. Capping agents are materials that coat a nanoparticle to stabilize it and prevent self-interactions. Sonication utilizes ultrasonic vibrations to break apart aggregated particles and disperse them throughout a solution. It is important to note that at high frequencies or with prolonged exposure to ultrasonication, the polymer chains may also break down [60]. Additional difficulties can be encountered when preparing an aqueous nanocomposite solution due to the hydrophobic nature of most inorganic materials. However, agglomeration can be prevented, and good dispersion can be achieved by using a surfactant (a type of capping agent with both hydrophilic and hydrophobic ends) or amphiphilic polymers (polymers with both hydrophilic and hydrophobic functional groups) [50]. In addition, the nanofillers can be functionalized to add organic functional groups and make the fillers more compatible with the matrix polymers. The specifics of the dispersion methods used are dependent on the polymer and nanofiller used. The following section goes into the issues specific to the production of electrospun layered silicate nanocomposite fibers.

2.3 Electrospinning of Layered Silicate Nanocomposite Fibers

Layered silicates (also known as phyllosilicates) are the combination of two different types of sheet structures. (i) **The tetrahedral silica sheet** consists of units of silicone atoms tetrahedrally bonded to three oxygen atoms/hydroxyl ions to form a hexagonal planar structure. (ii) **The octahedral metal hydroxide sheet** is usually composed of aluminum or magnesium ions that are surrounded by six hydroxide ions in an octahedral coordination. The hydroxyl ions compose the top and bottom plane of the structure, while the metals form a plane sandwiched between the hydroxyl planes. Depending on the valency of the metal, the structure changes slightly to satisfy the valency. Aluminum hydroxide forms a dioctahedral structure ($Al_2(OH)_6$) meaning that in the middle plane, only two out of every three octahedral centers are occupied by aluminum ions. Magnesium hydroxide forms a trioctahedral structure ($Mg_3(OH)_6$) where all the octahedral sites are occupied.

The octahedral holes can also be substituted with other metal ions such as Li^+, Na^+, Ca^{2+}, Fe^{2+}, or Fe^{3+}. Through condensation reactions, the silicone and metal hydroxide sheets are fused in different ratios to produce different types of layered silicates. The various layered silicates are classified based on the ratio of tetrahedral silica sheets to octahedral metal ion sheets. For example, in 2:1 structures, each layer is composed of two tetrahedral sheets which sandwich an octahedral sheet [61, 62].

In the bulk, layered silicates consist of individual layers separated by an interlayer spacing. In most cases, this interlayer spacing contains adsorbed extraneous cations which balance the overall negative charge of the individual layers. Thus, for layered silicate materials, there are three main measurements of interest: the layer thickness, the interlayer distance, and the basal spacing (the sum of the layer thickness and interlayer spacing). Depending on the clay, the interlayer cations may or may not be exchangeable [62, 63]. In electrospun layered silicate nanocomposites, the most commonly used layered silicate is the natural silicate montmorillonite (MMT) [64]. MMT is a 2:1 phyllosilicate composed of an aluminum (gibbsite) sheet sandwiched by two silica sheets. It is commonly used due to possessing a suitable layer charge density [65] and a high aspect ratio which is between 100 and 200 [66]. It is also highly hygroscopic and prone to swelling as well as miscible with hydrophilic polymers [67]. Other commonly used natural layered silicates include rectorite (REC) which is similar in structure and characteristics to MMT [68, 124]; kaolinite, a 1:1 phyllosilicate of gibbsite and silica [67]; and illite, a 2:1 phyllosilicate with poorly hydrated cations in the interlayer space [67]. Synthetic layered silicates are also used and include Laponite™ [66] and fluorohectorite (FH) [69].

The first report of an electrospun layered silicate nanocomposite fiber was by Fong et al. in 2002 [70]. In this work, a nylon 6-MMT nanocomposite previously produced through a melt extrusion technique was dissolved and electrospun. The intent was to combine a nanostructured material with a nanoscale fabrication technique to achieve high levels of morphological control. By combining the techniques, further understanding could be gained on the effect of processing and silicate layer alignment on the polymer crystallization and nucleation behavior [70]. Layered silicates are incredibly useful nanofillers due to their planar structure which provides them with a high aspect ratio and ability to reinforce in two rather than one dimension [64]. Unlike most inorganic fillers, layered silicates are hydrophilic, allowing them to delaminate in water as well as be compatible with water soluble and polar polymers [59]. The lower density of layered silicates compared to other inorganic fillers allows for composites with high strength and stiffness to be fabricated with low weight [64]. The incorporation of layered silicates into an electrospinning solution can change the solution's surface tension properties, conductivity, and shear viscosity [71, 72]. In addition, it has been found that the incorporation of layered silicates can improve the electrospinnability of a solution due to decreasing the rate of capillary thinning and improving strain hardening [69]. During the electrospinning process, high shear forces act on the polymer jet. It is thus expected that the shearing forces would promote clay layer alignment along the fiber axis as well as promote delamination of the clay layers.

Layer alignment has been readily observed, but delamination seems to be hindered at higher clay loadings [73–75].

Electrospinning of layered silicate nanocomposites can be performed in two ways: polymer templating and direct electrospinning. The polymer templating method involves electrospinning a polymer solution into fibers and then immersing the resultant fibers into a clay dispersion so that silicate layers adsorb onto the surface of the fibers [75]. Direct electrospinning is by far the more common method and involves dispersing the clay particles in a polymer solution that is subsequently electrospun so that the resulting polymer fibers contain dispersed clay particles. This process is favored due to its simplicity and the greater interfacial adhesion that can occur between the polymer matrix and silicate layers [75]. There are many methods of preparing a nanocomposite solution for the direct electrospinning process. Many of the early works involved dissolving a prefabricated nanocomposite into a solvent [70]. This process has the benefit of simplicity but relies on the prefabricated nanocomposites to have well-dispersed silicates and is less efficient due to the additional processing to produce a material that will ultimately be dissolved again. Another method involves first dispersing the clay particles within the solvent then mixing in the matrix polymers. As was studied by Ristolainen et al. [59], it is important to disperse the clay within the solvent first followed by dissolving of the polymer as this allows the clay to swell, increasing its specific surface area and causing more interactions between the clay and the polymer chains. Through this method, it is also possible to incorporate other fillers in addition to the layered silicates into the composites such as silver nanoparticles [76]. The swelling of the clay and orientation of the polymer chains with respect to the clay particles is an important interaction which determines the properties and morphologies of the resulting fibers and is further explored in the next section.

2.4 Interactions Between Layered Silicates and the Polymer Matrix

When layered silicates are dispersed within a polymer matrix, there are three types of structures that can occur [49, 64]. The first type, phase separation, occurs when the polymer and clay are immiscible and the clay retains its original layered structure, with no penetration of the polymer chains into the interlayer spacing (gallery) of the clay. This structure has the minimum amount of interaction between the clay and polymer and is thus undesirable. The second structure is known as an intercalated structure or intercalated tactoid. In this structure, the polymer chains penetrate into the clay galleries, forming multilayers of alternating organic/inorganic phases. The third structure is known as the exfoliated or delaminated structure. While the intercalated structure maintains some degree of interaction between the clay layers, the layers of the exfoliated structure are no longer bound to each other and are uniformly dispersed within the matrix. This is the most desired structure as it exposes the greatest amount of clay surface area and thus high levels of reinforcement are obtained. The electrospinning process can promote intercalation or exfoliation due to using a solvent-based mixing process. When clays are dispersed in polar solvents, the solvent molecules adsorb onto the clay surfaces. During electrospinning of the clay–polymer solutions, large quantities of solvent are evaporated and desorbed from the clay. The increase in entropy by desorption of the solvent molecules acts as a driving force for polymer intercalation as it compensates for the decreased conformational entropy of the intercalated structure [62].

In many cases, the hydrophilic surface chemistry of the silicate layers will not allow it to have good interfacial adhesion with the matrix polymers as hydrophobic polymers are much more common [62]. Thus, before use in composites, the clays are often functionalized with organic modifiers in order to mediate the interactions between the hydrophilic and hydrophobic surfaces [67]. The organic modifiers used can be chosen based on their functionality, length, and packing density to optimize the clay-matrix compatibility [67]. Organophilic layered silicates can by created by exploiting the cation exchange capacity of many clays to exchange their inorganic surface cations with organic surface cations (e.g., amino acids, silanes, alkylammonium ions) [65, 77]. In addition to making the clays organophilic, the organic cations can also change the shape of the layered silicate. Due to the increased size and length of many organic cations compared to inorganic cations, the interlayer spacing can be changed. As was shown by Lan et al. [78], longer alkylammonium chains separate clay layers further and promote the formation of exfoliated clay composite structures. Though clay–polymer interactions can be modified by careful selection of the organic modifier, many organically modified clays are already commercially available.

The large surface area of clay layers serves to reduce the polymer chain mobility through direct interference which hinders the chain mobility as well as from adsorption of the polymer chains onto the surfaces of the clay [79, 80]. Through these mechanisms, the overall polymer rigidity and stiffness of the material are increased [79, 80]. Depending on the matrix polymer, the negative surface charge present on most clays can also play an important role. Depending on the clay and the matrix polymer, the negative charge of the clay can form ionic interactions with the polymer chains, causing an increase in polymer crystallization in the vicinity of the clays and improving the mechanical properties of the composite [81]. Extensive work has been published in the literature on a wide variety of electrospun clay–polymer combinations and is summarized in the following section.

3 Electrospun Layered Silicate Nanocomposite Fibers

Compared with micron-level inorganic fillers, the nanoscale layered silicates exhibit a superior surface area and aspect ratio which brings out various synergistic properties with minimum filler loadings. The addition of layered silicates into biopolymer-based fibers can be beneficial due to increasing the mechanical strength, enhancing the antimicrobial activity, and speeding up biodegradation of the material. In addition, the addition of layered silicates can give the composites barrier properties against gasses and vapors, heat resistance, UV resistance, and improved stiffness without reducing the toughness or optical properties of the composites [82–84]. As the demand for layered silicate nanocomposite fibers grows for various applications in recent years, a wide range of composite fibers have been fabricated through electrospinning and reported in the literature. The fibers can be classified based on the matrix polymer used. Polymers are usually classified based on their source as well as their biodegradable/compostable nature. In this section, we have classified the layered silicate nanocomposite fibers into two major categories based on the source of the polymer matrix: (i) ***petroleum resource based***, which include polymers such as nylon, polyurethane (PU), polyacrylonitrile (PAN), polycaprolactone (PCL), and polyvinyl alcohol (PVA) which are synthetically prepared polymers, and (ii) ***renewable resource based***, which includes polymers such as polylactide (PLA), polyhydroxyl butyrate (PHB), and starch which are either biodegradable or nonbiodegradable.

3.1 Petroleum Resource-Based Layered Silicate Nanocomposite Fibers

3.1.1 Nylon-LS Nanocomposite Fibers

Electrospun nanofibers of various nylon-layered silicate composite have been successfully fabricated and investigated for their thermal, mechanical, and barrier properties. Li et al. [57] reported the fabrication of nylon-LS nanocomposite fibers with a diameter of 100 nm from nylon-6 and the organically modified montmorillonites. They fabricated the composite fiber in two stages: (i) fabrication of bulk composites by melt processing and (ii) electrospinning of nanofibers from a solution of the melt fabricated composites dissolved in aqueous formic acid. They identified that increasing the filler content in the starting solution above 15 % caused a reduction in the mechanical properties. Cai et al. [126] fabricated electrospun fibers of polyamide 6 nanocomposites reinforced with organic-modified Fe-montmorillonite (Fe-OMMT) and investigated their thermal stability. Fabrication of the electrospinning solution began with preparing the Fe-OMMT dispersion in DMF and the PA6 solution in formic acid separately. The two solutions were then mixed together. The authors concluded that the reinforcement of Fe-OMMT into polyamide 6 enhanced the overall thermal stability. Li and his research team [85] reported the fabrication of polyamide 6/montmorillonite composite nanofibers fabricated through the electrospinning process. They aimed to investigate the dynamic water adsorption behavior of the fibrous composites in addition to structural characteristics of the fibers. Their microscopic investigation revealed that the montmorillonite layers were exfoliated and aligned along the axis of fibers and enhanced the water absorption properties of the fibers. Nylon 6 with different molecular weights has been explored for the fabrication of layered silicate nanocomposites in order to understand their molecular mobility as well as the interaction between silicate layers during electrospinning and their influence on the crystallization behavior. Cai and

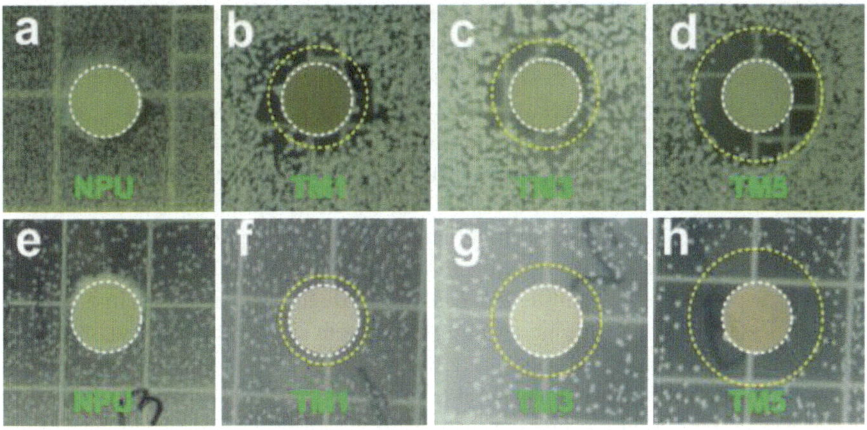

Fig. 18.3 Antimicrobial activity of electrospun polyurethane (*PU*)/tourmaline (*TM*) composite fiber mats; zone inhibition tests against Gram-negative *E. coli* (**a–d**) and Gram-positive streptococci (**e–h**) bacteria (NPU, Neat PU; TM1, 1 wt% TM/PU; TM3, 3 wt% TM/PU; and TM5, 5 wt% TM/PU) (Reprinted with permission) [87]

coworkers [86] reported the fabrication of electrospun polyamide 6/O-MMT composite fibers which were functionalized with Fe_2O_3 by sputter coating. They found that the sputter-coated polyamide 6/O-MMT composite fibers with Fe_2O_3 exhibited enhanced thermal behavior.

3.1.2 Polyurethane (PU)-LS Nanocomposite Fibers

Polyurethane, which exhibits high elastomeric behavior, has been widely used for coatings, adhesives, and fiber fabrication. Enhancement of their overall performance (including thermal, mechanical, and barrier properties) has been reported through reinforcement with various types of nanosized layered silicates. Recently, electrospun PU-LS nanocomposite fibers have received extensive attention due to their superior properties. Hong et al. [73] reported the fabrication of PU/O-MMT nanocomposites by electrospinning and correlated the filler content with the obtained fiber diameter. They found that the addition of O-MMT inhibits microphase separation in the PU phase. The electrospun composite fiber diameters showed a decreasing trend with increasing MMT content in the composites. Tijing et al. [87] fabricated electrospun tourmaline reinforced polyurethane composite fibers and investigated their antibacterial properties along with their superhydrophilic behavior. The antimicrobial behavior of tourmaline nanoparticles arise from their unique self-radiating behavior (far-infrared region) and the ion (negative) releasing mechanism. In this study, they successfully demonstrated their antibacterial and superhydrophilic properties along with the uniform dispersion of tourmaline nanostructure throughout the PU matrix. Figure 18.3 shows the zone inhibition results obtained for the electrospun nonwovens of the virgin and composite PU fibers.

3.1.3 Polyacrylonitrile (PAN)-LS Nanocomposite Fibers

Polyacrylonitrile (PAN) has been widely investigated due to its technological importance, especially for the fabrication of carbon fibers. The cross-linking behavior of PAN makes it resistant to various commonly used organic solvents [88]. Carbon nanofibers can be obtained by carbonizing PAN fibers produced through electrospinning. Thus, an extensive research effort has been made for the fabrication of PAN-LS nanocomposite fibers by electrospinning. Qiao et al. [89] investigated the effect of organic-modified Fe-montmorillonite reinforcement into PAN for the fabrication of electrospun composite fibers. They confirmed the uniform dispersion of clay in the PAN matrix through XRD analysis. In addition, they also identified that the increasing clay content decreases the fiber diameter significantly. Further, they concluded that the mechanical properties of PAN nanocomposite fibers were enhanced due to the reinforcement of Fe-OMMT. Cai et al. [72] also reported the fabrication of PAN/Fe-OMMT electrospun composite fibers for carbon fiber applications. The fabricated carbon fibers were pre-oxidized at 250 °C followed by carbonization at 600 °C. Their SEM and TEM analysis revealed that the Fe-OMMT layers were uniformly dispersed in the composite fibers and the fiber diameter decreased with increasing filler content. Reinforcement of Na-MMT into PAN for the fabrication of electrospun nanofibers was reported by Shami et al. [75]. Their studies also indicated the reduction of fiber diameter with increasing Na-MMT content. Enhanced thermal stability and residual mass were identified for the composite fibers compared to the virgin fibers. Yu et al. [90] investigated the fabrication of Polyacrylonitrile/kaolinite composite nanofibers. The obtained composite fibers were effectively used for the removal of dissolved heavy metal pollutants from water. Reinforcing a small amount of layered silicates into PAN allowed for reduced pre-oxidation and carbonization temperatures to be used which enhanced the yield and mechanical strength of the carbon fibers [89].

3.1.4 Polycaprolactone (PCL)-LS Nanocomposite Fibers

Polycaprolactone (PCL) is a petroleum resource-based synthetic polymer that exhibits a slow biodegradation rate, making it a suitable polymer for long-term implants [91, 92]. Marras et al. [80] added MMT to PCL fibers to investigate its effects on the fiber morphology, mechanical properties, and thermomechanical response of the fibers. It was found that with increasing MMT content, the fiber diameter distributions became narrower and less undesirable bead defects were observed, likely due to increases in the solution conductivity and viscosity. Compared to prepared cast film composites, the MMT in the fiber composites possessed larger interlayer distances due to the great interfacial adhesion between MMT layers and PCL chains. Thermal analysis was not carried out on the composite fibers, but for the films, it was found that the presence of MMT delayed the onset of degradation temperatures. Kladi et al. [93] reported the fabrication of fibrous scaffolds by employing the electrospinning technique from PCL reinforced with layered silicates. They found that the addition of layered silicate into PCL helped for the fabrication of fibers with a uniform morphology. In addition, they also found that layered silicate influenced the thermal behavior and enhanced the stability of the fiber. Recently, Nitya et al. [94] fabricated electrospun nanocomposite fibers from PCL/nanoclay to be used as a scaffold for bone tissue engineering. The obtained composite fibers exhibited enhanced mechanical properties such as tensile strength and modulus. They also found that the composite fibers possessed excellent protein absorption properties with better cell compatibility.

3.1.5 Polyvinyl Alcohol (PVA)-LS Nanocomposite Fibers

Poly(vinyl alcohol)/layered silicate composite nanofibers have been attractive due to the enhanced composite performance with low concentration of fillers particles. Lee et al. [95] first reported the fabrication of electrospun PVA/MMT composite nanofiber mats and optimized the solution concentration, operating voltages, and working distance operating parameters. They also found that the addition of MMT into PVA enhanced the mechanical strength and thermal behavior of the fibers. In addition to the work by Lee et al., Qin et al. [96] fabricated PVA composite nanofibers reinforced with layered double hydroxides (LDH). They found that the addition of LDH to PVA caused a reduction in fiber diameter and enhanced the spinnability of the solution. Park et al. [97] reported the hybrid reinforcement of MMT and silver nanoparticles into PVA. Figure 18.4 shows the TEM images of electrospun PVA, PVA/MMT, and PVA/MM/AgNP fiber morphologies. Their research suggests that the addition of MMT and silver nanoparticles into PVA as hybrid reinforcement materials enhanced the spinnability of the solution and the thermal stability of the fibers. In this hybrid composite, the silver nanoparticles played a key role in exhibiting significant antimicrobial properties. A composite of PVA, chitosan, and REC was produced by Deng et al. [98]. In this composite, chitosan is positively charged, REC negatively charged, and PVA neutrally charged, leaving the possibility for unique interactions between the components. It was found that with an increasing REC concentration, the solution conductivity increased and the resultant fibers had significantly less beading and smaller diameters [98]. While all chitosan containing fibers exhibited antibacterial activity, the addition of REC was found to enhance the bacterial inhibition properties [98]. In some cases, the addition of a clay may reduce properties of a composite. A polyvinyl alcohol (PVA) and Laponite^{TM} composite fiber was studied by Yacoob et al. [66]. As electrospun PVA fibers were found to possess weaker mechanical properties than their cast film counterparts, Laponite^{TM} was added in an attempt to improve the properties [66]. However, it was found that the tensile strength and strain at maximum load were decreased with Laponite^{TM} addition. The authors attribute this to the lower aspect ratio of Laponite^{TM} (25) compared to other clays such as MMT (100–200) leading to more agglomeration and less alignment [66]. However, the flame retardation properties increased with Laponite^{TM} addition due to it acting as a char promoter and creating a protective layer on the fibers [66].

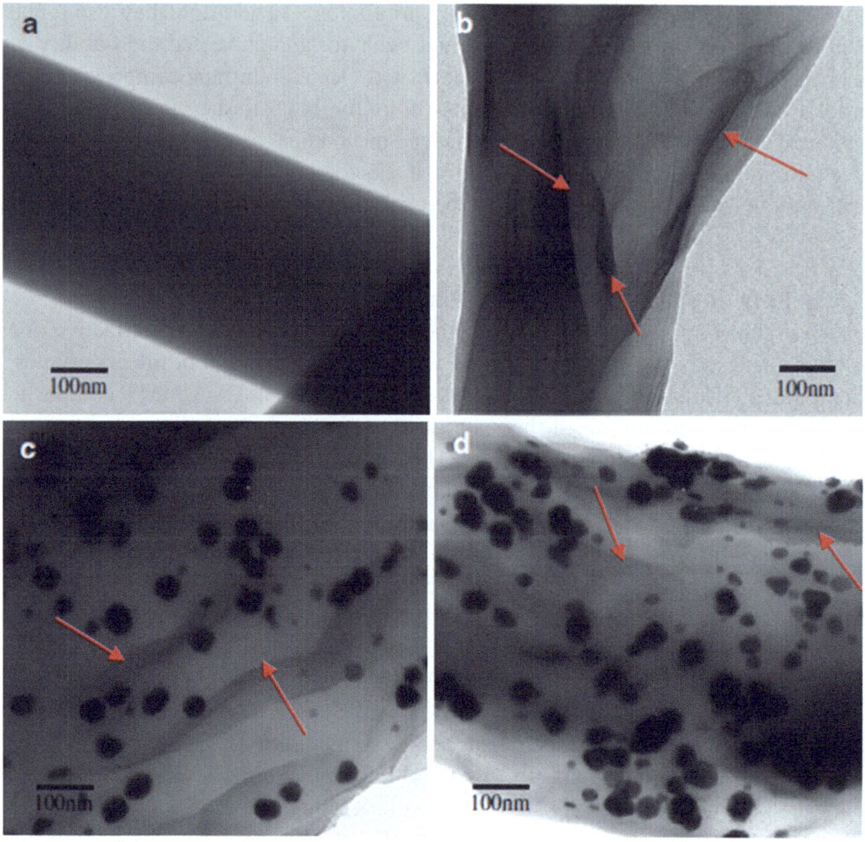

Fig. 18.4 TEM microscope images of (**a**) PVA, (**b**) PVA/MMT, (**c**) PVA/MMT with 1 wt% Ag nanoparticles, and (**d**) PVA/MMT with 3 wt% Ag nanoparticles (Reprinted with permission) [97]

3.1.6 Poly(butylene succinate), Poly(butylene adipate)-LS Nanocomposite Fibers

Poly(butylene succinate), poly(butylene adipate), and their copolymer poly [(butylene succinate)-co-(butylene adipate)] (PBSA) are very useful polymers due to their thermal and chemical resistance and biodegradability. Tsimpliaraki et al. [99] performed a statistical analysis on PBSA/MMT electrospun composites. It was found that the solution concentration and clay loadings had the greatest effect on the fiber morphologies, and thus those parameters were optimized [99]. Additionally, it was found that for the PBSA/MMT composites, a 3 % clay loading gave the most thermally stable fibers, while any clay loading above that would decrease the thermal stability. The fibers with high clay loadings also possessed a secondary peak not observed in the film composite counterparts [99]. This revealed that the increased degradation was likely due to surfactants on the organically modified MMT that was used and the high surface area of the fibers [99]. However, the fiber structure was retained for the MMT loaded fibers after exposure to high temperatures, while the fibrous structure was not retained in the neat PBSA fibers [99].

3.2 Biopolymer-Based Layered Silicate Nanocomposite Fibers

Biopolymers are a special type of polymeric materials that are naturally produced by living organisms or derived from the monomers obtained from renewable resources. As such, they can potentially be used as alternative renewable materials to replace traditional synthetic polymers in a variety of applications [26]. However, they posses various challenges in the processing aspect compared to traditional polymers due to wider variations in molecular weight, crystallinity, purity, and functional group distribution [100]. This section describes the various aspects of electrospun nanocomposite fibers from biopolymers reinforced with layered silicates.

3.2.1 Chitosan-Layered Silicate Nanocomposite Fibers

Chitosan is a biopolymer derived from chitin, a natural polysaccharide produced by many organisms. Chitosan is a nontoxic biodegradable polymer, which also exhibits unique antimicrobial activity [101]. A recent advance in chitosan technology is their fabrication into electrospun nanofibers for various technological applications. These chitosan-based fibers have also been reinforced with various layered silicate nanostructure to produce composite fibers with improved properties. Li et al. [102] reported the fabrication of electrospun chitosan-layered silicate composite nanofibers and investigated their potential for protein delivery. Intercalation of the layered silicate by chitosan was confirmed by the small angle X-ray diffraction. They concluded that the addition of layered silicate into chitosan significantly enhanced the protein release efficiency. The enhanced performance of chitosan-layered silicate composites can be further explored for various other advanced applications including sensors and tissue engineering. Similarly, quaternized chitosan-layered silicate composite nanofibers were fabricated by Deng et al. [103], and the obtained nanofibrous mats were investigated for their antibacterial activity. They have compared the antimicrobial properties of chitosan-layered silicate composite fibers with chitosan-PVA blend fibers and found that the chitosan–silicate composites showed superior antimicrobial properties. Huang et al. [104] reported a layer-by-layer technique for the immobilization of lysozyme–chitosan–organic rectorite composites on electrospun cellulose acetate fibers. Reinforcing organic rectorite into chitosan results in enhancement of their antimicrobial properties. Many of the chitosan-layered silicate composites exhibit unique antimicrobial properties, and hence they are being researched for antimicrobial wound dressing applications. Xin et al. [105] fabricated carboxymethyl chitin/organic rectorite composites and investigated their cell compatibility performance. The results indicated slight toxic effects on the cells.

3.2.2 Polylactide (PLA)-Layered Silicate Nanocomposite Fibers

One major use of biopolymer fibers is in tissue regeneration as a scaffold material. Lee et al. produced electrospun composites of poly (L-lactic acid) (PLLA) and MMT to produce a porous scaffold [74, 106]. PLLA is a very good material for cell culturing due to its biocompatibility but has weak mechanical properties. During cell growth, tissue engineering scaffolds experience many stresses and loadings and undergo a biodegradation processes. Therefore, to be effective scaffolds, especially for hard tissue regeneration, the scaffolds must be sufficiently stiff and robust. Lee et al. [74] found that by varying the amount of MMT in the composite, the biodegradation rate and tensile modulus (stiffness) could be tailored. In addition, during electrospinning, composites with higher MMT loadings suppressed solvent evaporation so that explosive solvent evaporation occurred and the resulting fibers contained craters [106]. The resulting fibers could be transformed into a robust three-dimensional dual porosity scaffold through a salt-leaching/gas-foaming technique [74]. The composite scaffolds maintained their structural integrity after 6 weeks of hydrolysis, while the neat PLLA scaffolds did not [74]. Thus, the addition of MMT greatly improved the applicability of PLLA electrospun fibers for tissue engineering applications.

3.2.3 Zein-Layered Silicate Nanocomposite Fibers

Zein is a corn protein material which is hydrophobic in nature and is highly soluble in alcohol [107]. This gives it a lower toxicity and biodegradability, making it suitable for various plastics, food-packaging, biotechnological, pharmaceutical, textile, and cosmetic applications [67, 107]. Zein-based electrospun fibers must overcome their inherent lower thermal and mechanical performance to improve their applicability [108, 109]. Torres-Giner and Lagaron studied the electrospinning of zein prolamine, a storage protein of corn, with the various phyllosilicates kaolinite, MMT, and mica [67]. It was found that most of the silicates used produced similar results. Fiber diameters decreased with increasing clay contents but at high clay loadings (past 10 %), clay aggregates formed. Between 1 % and 10 % clay loadings, interesting "rolled" clay structures were observed which were aligned along the fiber axis. These rolled structures were likely due to horizontal shifting of the silicates which did not completely delaminate them but allowed for the shifted sheets to curl into rolls [67]. Various morphologies of electrospun zein/modified mica composite fibers are shown in Fig. 18.5. Thermal stability of the fibers increased with up to 10 % clay loadings then began to decrease. Park et al. [107] also electrospun zein/MMT composites and found the presence of MMT increased the thermal stability of the fibers. In addition, MMT was found to influence the crystallization, melting point, and water contact angle of the fibers [107].

Layered silicate composite fibers can be produced using many different biopolymers. In general, the addition of clay is used in order to improve the mechanical properties of biopolymer-based fibers to make them more robust and durable. Clays are also used to compliment the intrinsic properties of the biopolymers to improve their antimicrobial properties and thermal resistance. However, it is important to not load in too much clay as certain properties may start to decrease.

As seen in conventional and biopolymer-based layered silicate nanocomposite fibers, many unique properties are exhibited. The following section summarizes and goes into detail on how the addition of layered silicates gives rise to those properties.

3.3 Unique Properties of Layered Silicate Nanocomposite Fibers

From the literature on electrospun layered silicate composite fiber, a few unique properties have been described and explained. The addition of layered silicates to a polymer matrix has been found to increase the thermal stability, biodegradability, and mechanical properties of the fibers. In addition, the presence of the clays can change the deformation mechanism of the fibers as well as the matrix's crystal conformation.

The enhanced thermal stability and flame resistance of the fibers lies in the ability of the silicate particles to hinder the diffusion of volatile decomposition products from within the nanocomposite and to act as a char promoter [66, 110]. The burning process follows a cycle involving oxidation, heating, heat radiating into the fiber, decomposition of molecular chains within the fiber, volatilization of decomposition products to the fiber surface, and then restarting at oxidation of the surface [66, 111]. Interruption of any of these processes causes the burning cycle to be halted [66, 111]. Thus, through the mechanisms of char formation, a protective layer is formed around the fibers [112], trapping volatile decomposition products, decreasing the burning rate, and increasing the thermal stability of the fibers [110, 112].

The deformation mechanisms of electrospun layered silicate composites can be quite different from that of their bulk composite counterparts. Bulk clay composites exhibit brittle deformation [51]. Through in situ tensile testing in a TEM, it was observed that deformation of a bulk composite initiates as crazing in the region of the largest intercalated tactoid. Beginning from the point of crazing, a crack rapidly spreads through the composite, leading to the observed brittle deformation [51]. In the case of electrospun composite fibers, a ductile deformation was observed. This can largely be attributed to the alignment, dispersion, and confinement of the clay fillers [51]. Deformation initiates with a stress concentration around the filler particles due to the different elastic properties from the matrix [51]. Uniaxial shear yielding is induced at that point due to confinement of the matrix polymer chains between fillers [51]. Once a critical strain is reached, the fiber begins to neck due to overlapping stress fields. During the necking process, the yield strength lowers and the induced shear yielding increases until the fiber fails [51].

Water contact angle measurements indicate the degree of wettability of the material's surface to water. This in turn indicates how strongly the water and solid surface interact with each other. Depending on the application, materials with higher or lower levels of water wettability may be desired. Some research has shown that the addition of clay to a composite fiber can be used to tune the fiber

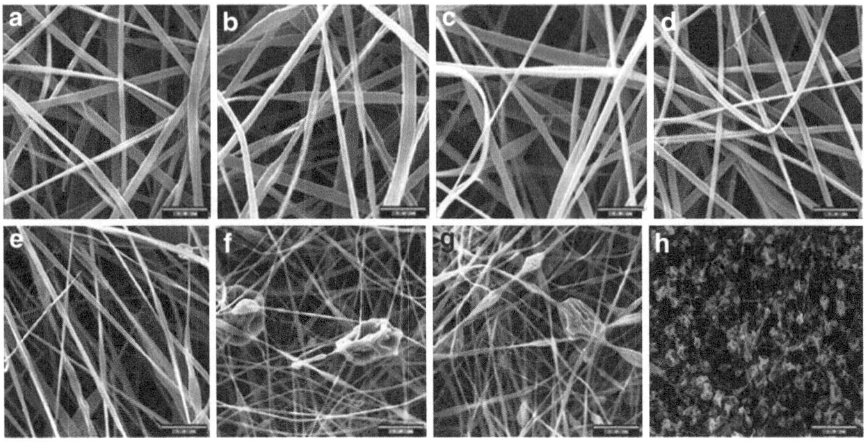

Fig. 18.5 SEM micrographs of electrospun: (**a**) zein, (**b**) zein/MM-99:1, (**c**) zein/MM-95:5, (**d**) zein/MM- 90:10, (**e**) zein/MM-85:15, (**f**) zein/MM-80:20, (**g**) zein/MM-75:25, and (**h**) zein (75 wt%-pure zein solution) (*MM*=modified mica, 99:1 represents w%:w%) (Reprinted with permission) [67]

mat's water contact angle [107]. As unmodified clays are hydrophilic and the water contact angle is related to surface hydrophilicity, addition of MMT has been found to decrease the water contact angle [107]. However, if an organically modified clay is used, the water contact angle will increase due to the organoclay being hydrophobic [59]. As the wetting is also related to surface roughness, nanofiber mats have greater water contact angles than films [59]. However, with higher levels of clay dispersion, the contact angle can be decreased [59]. Thus, through adjustment of the clay, clay dispersion, and fiber mat fiber morphology, layered silicate composite fibers can tune their water contact angles to desired levels.

In some cases, the composition of the clay can play an important role in the functionality of the composite structure. An Fe-MMT was prepared by Cai et al. [72] and incorporated into electrospun polyacrylonitrile (PAN) fibers. The iron ions in the modified MMT acted as a catalyst for the carbonization of the PAN fibers. It was confirmed that the Fe-MMT promoted cross-linking, dehydrogenation, and cyclization in the fibers during thermal treatments [72]. These reactions, along with the presence of the clay layers themselves, acted to hinder degradation of the fibers so that carbonization of the fibers could successfully be performed quickly and at lower temperatures.

The biodegradability of clay nanocomposite fibers was found to increase with increasing clay addition [74, 106]. The surface of the clay layers acts as nucleation sites which promotes the formation of many small crystallites. However, the presence of many small crystallites reduces the overall crystallinity of the material [113, 114]. Thus, the composites possess more amorphous regions which are easier for microbes to break down in the biodegradation process.

Clay composites also exhibit improved antibacterial properties which can largely be attributed to the ability of the clay to promote bacterial adhesion. Layered silicates can selectively adhere to bacteria due to the layer charge density of the clay layers [115]. While the layers are usually of a negative charge, the surface charge of the clays can be made positive through surface modifications and intercalation with select polymers [115]. In addition to the charge interactions, there may also occur hydrophobic interactions between hydrophobic portions of the clay modifiers and lipophilic components of the bacteria [116]. Thus, the bacteria can become immobilized by the clay composite, and the small fibrous structure of the electrospun composites may also penetrate into the cell [115]. Thus, through the enhanced bacterial adsorption by the clay and the large surface area of the fibers, the other antibacterial constituents of the composite can effectively act on the bacteria, improving the overall antibacterial properties of the composite.

Additional effects on the matrix crystallinity by the layered silicate were found by Kim et al. [117] using polyamide 6 (PA6). In bulk composites, the presence of MMT promoted γ-type crystal formation while reducing α-type crystal formation due to a reduced chain mobility. In electrospun PA6-MMT composite fibers, the high shear rate and evaporation rate exerted on the polymer through the Electrospinning process was found to cause γ-crystal formation much more than the presence of MMT. The most interesting effect of MMT on the PA6 crystallization occurs upon heating and cooling the fibers. After going through a heating and cooling cycle, the γ-crystals in neat PA6 fibers would return to α-form crystals, but the PA6-MMT composite fibers retained their γ-form crystals. α-crystals are the preferred crystal structures in PA6. Thus, this shows that clays can provide a molecular memory effect in their electrospun fibers. The mixture of layered silicates and polymers into a fibrous structure leads to a range of unique or improved properties that may be utilized and applied to a variety of applications.

4 Applications

The electrospun nanocomposite fibers made from various types of polymers and layered silicates exhibit very high surface-to-volume ratios with unique structure-controlled morphologies. The inherent properties of these nanostructured fibrous materials combined with the added functionalities by the incorporation of clays make them suitable candidate for various advanced applications including biomedical applications, carbon fiber fabrication, food packaging, filtration, and sensing which are summarized in this section.

Biomedical: One of the major application areas of electrospun fibers is in the biomedical field, which includes tissue engineering, wound dressing, and antimicrobial surfaces. Tissue engineering is a multidisciplinary research field which involves the use of living cells coupled with fibrous scaffolds aiming to repair, sustain, or enhance the biological activity of a particular tissue or organ [118]. Electrospun fibrous nanocomposites can be useful for the fabrication of biocompatible fibrous scaffolds with precisely controlled morphologies. Layered silicate reinforced nanocomposite fibers can offer unique antimicrobial properties as well as protein release mechanisms, making them suitable for wound dressing applications. In the literature, it has also been found that chitosan- or PVA-based hybrid clay nanocomposites with silver nanoparticles showed enhanced antimicrobial activity [97]. These attractive properties can be effectively utilized for various antimicrobial products. Figure 18.6 shows scanning electron microscopy images of electrospun nanofibers with cultured cell after 3 days.

Carbon nanofibers: In recent years carbon nanofibers receive imminence attention due to their demand in many fields including polymer composites, energy storage and conversion, and filters. The electrospinning technique has been widely explored for the fabrication of nanosized polyacrylonitrile fibers which can be converted into carbon nanofibers through a carbonization processes [88]. The addition of layered silicates as a reinforcement into polyacrylonitrile fibers provides many advantages such as reduced pre-oxidation and carbonization temperatures, higher carbon yields, enhanced strength, and controllable fiber morphologies [89].

Food packaging: Chitosan is well known to be antibacterial and nontoxic and is therefore, a desirable material for food-packaging applications. Thus, nanocomposite fibers of silicates with chitosan could be used for applications such as packaging and coatings as the incorporation of clay can improve the thermal, mechanical, and barrier properties of the fibrous coating. In addition, the layered silicates can be used with both conventional plastics and bioplastics without hindering the biodegradability or optical properties of the material. It is also possible to control the release of actives or bioactives in smart food packaging for pharmaceutical and biomedical applications [67].

Filters: Filters have been widely used for the purification of air and water in both domestic as well as industrial environments. They also play a major role in personal protective respirators. Nonwoven mats produced using the electrospinning process have very small pore sizes and high surface areas, making them excellent candidates for filter media. Electrospun polymeric fibers with layered silicates exhibit improved fiber morphologies, durability, and strength. Thus, with the addition of clay, electrospun fibers can be used in many more filtration applications that require tough and durable materials. Recently, it was found that the reinforced layered silicates in electrospun polymer composites can act as absorbing agent for metal toxins in polluted water [90]. By tailoring the fiber morphology and chemistry, the fibers can be used for the absorption or neutralization of a variety of pollutants and toxins.

Sensors: The demands for sensor materials have been increasing for various monitoring purposes. Electrospun materials have been looked at for applications in sensing due to the high surface area-to-volume ratio which enhances their sensor performance. Electrospun nanocomposite fibers reinforced with layered silicates offer excellent mechanical strength and thermal stability, which can be effectively utilized for the fabrication of various sensor

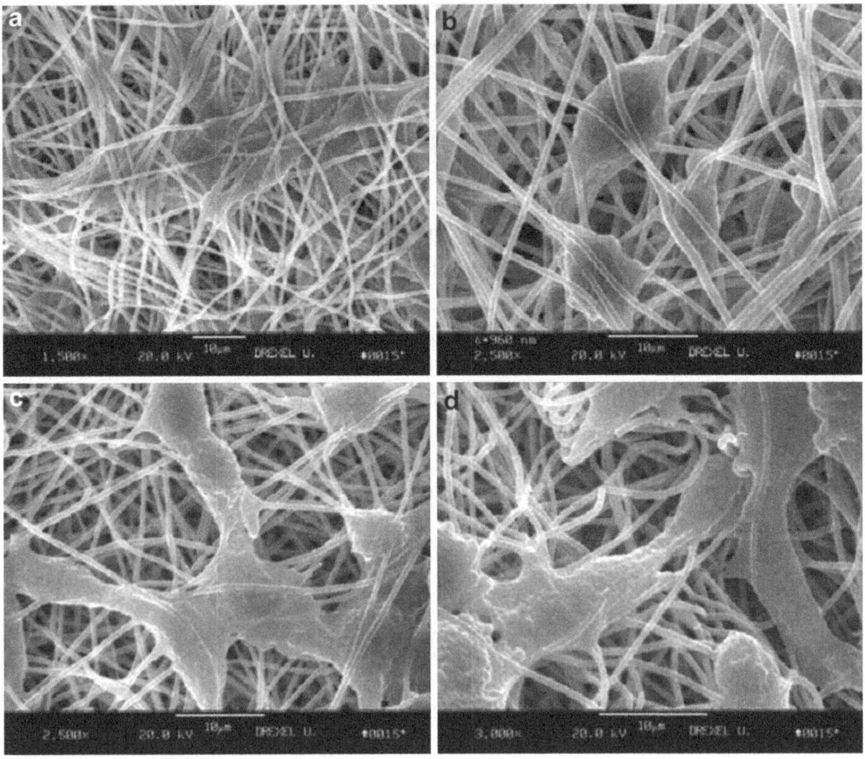

Fig. 18.6 SEM micrographs of cell cultures grown on electrospun nanofibers after 3 days at different magnifications. (**a**) ×1,500; (**b**) ×2,500; (**c**) ×2,500; (**d**) ×3,000 (Reprinted with permission) [118]

materials with wide operational ranges. Possible hybridization of reinforcement materials can create additional functionalities which can be tuned for various sensor applications. However, these processes still need further investigations and optimization before they can be successfully employed in sensing devices.

5 Future Prospects of Layered Silicate Nanocomposite Fibers

Layered silicates can be functionalized with various metals. These functionalized fillers can offer additional functionalities to the nanocomposites. Patakfalvi et al. [119, 120] effectively functionalized the kaolinite structure with silver nanoparticles through a chemical reduction process. Silver nanoparticles are well known for their efficient antimicrobial properties. These silver functionalized layered silicates can be used as an effective filler for the fabrication of improved antimicrobial composite fibres and should be further explored in the future. Electrospinning also allows for the surface modification of the fibers to provide further enhancements in properties. The electrospinning procedure also allows the fabrication of nanofibers with hollow, core-sheath, or porous structures [121]. So far, little work has been done to fabricate layered silicate nanofibers with hollow or core-sheath morphologies which could further enhance the applicability and properties of these materials. As layered silicate composites were identified as effective protein absorbants, numerous research opportunities are ahead in the area of controlled drug release. Recently, there is an increasing interest in developing carbon nanofibers from renewable resources such as lignin. Successive reports are available for the effective fabrication of carbon nanofibers from PAN with the help of various layered silicates. This can be further expanded for the fabrication of renewable resource-based carbon fibers with enhanced morphologies and thermomechanical properties. It is also essential to perform a detailed rheological investigation of the precursor solutions and to correlate the obtained results with the resultant fiber morphology. This will provide a greater understanding for the optimization of the processing conditions.

6 Conclusions

Electrospinning is an efficient tool for the fabrication of structure-controlled nanofibers and has been investigated for the generation of layered silicate reinforced nanocomposite fibers. Structural, morphological, thermal, mechanical, and other functional properties can be precisely controlled by varying the electrospinning processing parameters and the composition of the polymer-layered silicate electrospinning solution. Various types of layered silicates have been utilized as nano-reinforcement for the fabrication of composite fibers via electrospinning. Addition of layered silicates enhances the spinnability of fibers to give better morphological control. Further, the clay composite fibers exhibited improved mechanical, thermal, barrier, and various other functional properties such as antimicrobial activity and protein delivery. These properties can be effectively utilized for various advanced applications. The major challenges in fabricating electrospun nanocomposite fibers with layered silicates are achieving a uniform dispersion of the clay in the polymeric solution and obtaining compatibility between the clay and polymer matrix. Various experimental techniques have been adopted in order to overcome these issues and have been summarized in this book chapter. These electrospun layered silicate composite fibers find potential applications in areas such as tissue engineering. Although extensive research work has been performed on various aspects of electrospinning these materials, much research is still required to further understand the relationship between the clay–polymer interactions and the exhibited properties. Finally, there are still many research areas specific to electrospun polymer-layered silicate composites which have not yet been explored which may act to further improve the properties of these unique materials.

Acknowledgements The authors gratefully recognize the financial support provided by the Natural Sciences and Engineering Research Council (NSERC) Canada, Discovery grant, NSERC CRD grant, NSERC NCE AUTO21 program, Ontario Ministry of Agriculture, Food and Rural Affairs (OMAFRA) – Alternative Renewable Fuels Plus Research Program,; OMAFRA-University of Guelph, Bioeconomy-Industrial Uses Research program, Grain Farmers of Ontario (GFO) and the Ministry of Economic Development and Innovation (MEDI), Ontario Research Fund – Research Excellence Round 4 program. The Department of Physical and Engineering Science is acknowledged for the use of the SEM facilities.

References

1. Huang ZM, Zhang YZ, Kotaki M, Ramakrishna S (2003) Compos Sci Technol 63(15):2223–2253
2. Khan N (2012) Stud Undergrad Res Guelph 5(2):63–73
3. Bognitzki M, Czado W, Frese T, Schaper A, Hellwig M, Steinhart M, Greiner A, Wendorff JH (2001) Adv Mater 13(1):70–72
4. Fu X, Li F, Liu W, Stefanini C, Dario P (2011) Microelectron Eng 88(8):2653–2656
5. Huczko A (2000) Appl Phys A 70(4):365–376
6. Qian L, Zhang H (2011) J Chem Technol Biotechnol 86(2):172–184
7. Smith DK, Hirst AR, Love CS, Hardy JG, Brignell SV, Huang B (2005) Prog Polym Sci 30(3):220–293
8. Formhals A (1939) Method and apparatus for spinning. US Patent 2160962
9. Formhals A (1940) Artificial thread and method of producing same. US Patent 2187306
10. Formhals A (1943) Production of artificial fibers. US Patent 2323025
11. Reneker DH, Chun I (1999) Nanotechnology 7(3):216
12. Teo WE, Ramakrishna S (2006) Nanotechnology 17(14):R89–R106
13. Ramesh KP, Khan N, Vivekanandhan S, Satyanarayana N, Mohanty AK, Misra M (2012) J Nanosci Nanotechnol 12(1):1–25
14. Subbiah T, Bhat G, Tock R, Parameswaran S, Ramkumar S (2005) J Appl Polym Sci 96(2):557–569
15. Ge JJ, Hou H, Li Q, Graham MJ, Greiner A, Reneker DH, Harris FW, Cheng SZD (2004) J Am Chem Soc 126(48):15754–15761
16. Kannan P, Eichhorn SJ, Young RJ (2007) Nanotechnology 18(23):235707
17. Mack JJ, Viculis LM, Ali A, Luoh R, Yang G, Hahn HT, Ko FK, Kaner RB (2005) Adv Mater 17(1):77–80
18. Peresin MS, Habibi Y, Zoppe JO, Pawlak JJ, Rojas OJ (2010) Biomacromolecules 11(3):674–681
19. Hong KH (2006) Polym Eng Sci 47(1):43–49
20. Kim JK, Ahn H (2008) Macromol Res 16(2):163–168
21. Sui X, Shao C, Liu Y (2007) Polymer 48(6):1459–1463
22. Khan W, Asmatulu R, Lin Y, Chen Y, Ho J (2012) J Nanotechnol. doi:10.1155/2012/138438. Article ID 138438, 5P
23. Xie Y, Sun L, Wang C, Wang S, Xu G, Zhang J, Yan E (2012) J Appl Polym Sci 126:1061–1068
24. Yu G, Li X, Cai X, Cui W, Zhou S, Weng J (2008) Acta Mater 56(19):5775–5782
25. Cho K, Kim M, Choi J, Kim K, Kim S (2010) Synth Met 160(9):888–891

26. Mohanty AK, Misra M, Drzal LT (2002) J Polym Environ 10(1/2):19–26
27. Miyagawa H, Mohanty A, Drzal LT, Misra M (2004) Ind Eng Chem Res 43(22):7001–7009
28. Park HM, Liang X, Mohanty AK, Misra M, Drzal LT (2004) Macromolecules 37(24):9076–9082
29. Park HM, Misra M, Drzal LT, Mohanty AK (2004) Biomacromolecules 5(6):2281–2288
30. Miyagawa H, Mohanty AK, Burgueno R, Drzal LT, Misra M (2006) J Nanosci Nanotechnol 6(2):464–471
31. Park HM, Mohanty AK, Drzal LT, Lee E, Mielewski DF, Misra M (2006) J Polym Environ 14(1):27–35
32. Ray D, Sengupta S, Sengupta S, Mohanty AK, Misra M (2006) Macromol Mater Eng 291(12):1513–1520
33. Drown E, Mohanty A, Parulekar Y, Hasija D, Harte B, Misra M, Kurian J (2007) Compos Sci Technol 67(15):3168–3175
34. Mohanty AK, Drzal LT, Park H, Misra M, Wibowo AC (2005) Compositions of cellulose esters and layered silicates and process for the preparation thereof. WIPO Patent 2005111184, 25 Nov 2005
35. Haq M, Burgueño R, Mohanty AK, Misra M (2008) Compos Sci Technol 68(15):3344–3351
36. Haq M, Burgueño R, Mohanty AK, Misra M (2009) Comp Part A 40(4):394–403
37. Haq M, Burgueño R, Mohanty AK, Misra A (2009) Comp Part A 40(4):540–547
38. Das K, Ray D, Banerjee I, Bandyopadhyay N, Sengupta S, Mohanty AK, Misra M (2010) J Appl Polym Sci 118(1):143–151
39. Huda M, Drzal L, Mohanty A, Misra M (2007) Compos Part B 38(3):367–379
40. Tong H-W, Wang M (2010) Biomed Mater 5(5):054110
41. Tong HW, Wang M (2011) Soc Plastics Eng Plastics Res Online 3p. doi:10.2417/spepro.003635
42. Frenot A, Chronakis IS (2003) Curr Opin Colloid Interface Sci 8(1):64–75
43. Li D, Wang Y, Xia Y (2003) Nano Lett 3(8):1167–1171
44. Larrondo L, St John Manley R (1981) J Polym Sci Part B Polym Phys 19:909–920
45. Hutmacher DW, Dalton PD (2010) Chem Asian J 6(1):44–56
46. Góra A, Sahay R, Thavasi V, Ramakrishna S (2011) Polym Rev 51(3):265–287
47. Shao C, Kim HY, Gong J, Ding B, Lee DR, Park SJ (2003) Mater Lett 57(9–10):1579–1584
48. Lu X, Wang C, Wei Y (2009) Small 5(21):2349–2370
49. Kim GM, Lee DH, Hoffmann B, Kressler J, Stöppelmann G (2001) Polymer 42(3):1095–1100
50. Dror Y, Salalha W, Khalfin RL, Cohen Y, Yarin AL, Zussman E (2003) Langmuir 19(17):7012–7020
51. Kim GM, Lach R, Michler GH, Pötschke P, Albrecht K (2006) Nanotechnology 17(4):963–972
52. Hwang J, Muth J, Ghosh T (2007) J Appl Polym Sci 104(4):2410–2417
53. Yang QB, Li DM, Hong YL, Li ZY, Wang C, Qiu SL, Wei Y (2003) Synth Met 137(1–3):973–974
54. Demir MM, Gulgun MA, Menceloglu YZ, Erman B, Abramchuk SS, Makhaeva EE, Khokhlov AR, Matveeva VG, Sulman MG (2004) Macromolecules 37(5):1787–1792
55. Kim GM, Wutzler A, Radusch HJ, Michler GH, Simon P, Sperling RA, Parak WJ (2005) Chem Mater 17(20):4949–4957
56. Tan ST, Wendorff JH, Pietzonka C, Jia ZH, Wang GQ (2005) Chemphyschem 6(8):1461–1465
57. Li L, Bellan LM, Craighead HG, Frey MW (2006) Polymer 47(17):6208–6217
58. Li M, Zhang J, Zhang H, Liu Y, Wang C, Xu X, Tang Y, Yang B (2007) Adv Funct Mater 17(17):3650–3656
59. Ristolainen N, Heikkilä P, Harlin A, Seppälä J (2006) Macromol Mater Eng 291(2):114–122
60. Paulusse JMJ, Sijbesma RP (2006) J Polym Sci Part A Polym Chem 44(19):5445–5453

61. Theng BKG (1974) The chemistry of clay-organic reactions. Wiley, New York
62. Theng BKG (1979) Formation and properties of clay-polymer complexes. Elsevier, Amsterdam
63. Giesse RF (1978) Clay Clay Miner 26(1):51–57
64. Giannelis E, Krishnamoorti R, Manias E (1999) Adv Polym Sci 138:107–147
65. Kornmann X (1999) Synthesis and characterisation of thermoset-clay nanocomposites. Luleå University of Technology, Luleå. http://atmsp.whut.edu.cn/resource/pdf/3802.pdf
66. Yacoob C, Liu W, Adanur S (2010) J Ind Text 40(1):33–48
67. Torres-Giner S, Lagaron JM (2010) J Appl Polym Sci 118(2):778–789
68. Wang X, Du Y, Luo J, Lin B, Kennedy JF (2007) Carbohydr Polym 69(1):41–49
69. Wang M, Hsieh A, Rutledge G (2005) Polymer 46(10):3407–3418
70. Fong H, Liu W, Wang CS, Vaia RA (2002) Polymer 43(3):775–780
71. Daga VK, Helgeson ME, Wagner NJ (2006) J Polym Sci Part B Polym Phys 44(11):1608–1617
72. Cai Y, Xu X, Wei Q, Yao X, Qiao H, Song L, Hu Y, Huang F, Gao W (2011) Int J Polym Anal Charact 16(1):24–35
73. Hong JH, Jeong EH, Lee HS, Baik DH, Seo SW, Youk JH (2005) J Polym Sci Part B Polym Phys 43(22):3171–3177
74. Lee YH, Lee JH, An IG, Kim C, Lee DS, Lee YK, Nam JD (2005) Biomaterials 26(16):3165–3172
75. Shami Z, Sharifi-Sanjani N (2010) Fibers Polym 11(5):695–699
76. Yeum JH, Park JH, Kim IK, Cheong IW (2011) Electrospinning fabrication and characterization of water soluble polymer/montmorillonite/silver nanocomposite nanofibers out of aqueous solution. In: Reddy B (Ed) Advances in nanocomposites- synthesis, characterization and industrial applications. Intech Publishers, Rijeka, pp 483–502
77. Laura RD, Cloos P (1975) Clay Clay Miner 23(5):343–348
78. Lan T, Kaviratna PD, Pinnavaia TJ (1995) Chem Mater 7(11):2144–2150
79. Shi H, Lan T, Pinnavaia TJ (1996) Chem Mater 8(8):1584–1587
80. Marras SI, Kladi KP, Tsivintzelis I, Zuburtikudis I, Panayiotou C (2008) Acta Biomater 4(3):756–765
81. Usuki A, Koiwai A, Kojima Y, Kawasumi M, Okada A, Kurauchi T, Kamigaito O (1995) J Appl Polym Sci 55(1):119–123
82. Sanchez-Garcia MD, Gimenez E, Lagaron JM (2007) J Plast Film Sheet 23(2):133–148
83. Sanchez-Garcia MD, Gimenez E, Lagaron JM (2008) J Appl Polym Sci 108(5):2787–2801
84. Sanchez-Garcia MD, Ocio MJ, Gimenez E, Lagaron JM (2008) J Plast Film Sheet 24(3–4):239–251
85. Li Q, Wei Q, Wu N, Cai Y, Gao W (2008) J Appl Polym Sci 107(6):3535–3540
86. Cai Y, Huang F, Wei Q, Wu E, Gao W (2008) Appl Surf Sci 254(17):5501–5505
87. Tijing LD, Ruelo MTG, Amarjargal A, Pant HR, Park CH, Kim DW, Kim CS (2012) Chem Eng J 197:41–48
88. Nataraj S, Yang K, Aminabhavi T (2012) Prog Polym Sci 37(3):487–513
89. Qiao H, Cai Y, Chen F, Wei Q, Weng F, Huang F, Song L, Hu Y, Gao W (2009) Fibers Polym 10(6):750–755
90. Yu DG, Zhang XF, Shen XX, Zhu LM, Branford-White C, Yang Y, Welbeck E (2009) Polyacrylonitrile/kaolinite hybrid nanofiber mats aimed for treatment of polluted water. IEEE, pp 1–3
91. Ma PX (2002) Encyclopedia of polymer science and technology. Wiley, Hoboken
92. Taddei P, Tinti A, Reggiani M, Fagnano C (2005) J Mol Struct 744–747:135–143
93. Kladi KP, Marras SI, Philippou J, Tsivintzelis I, Panayiotou C (2005) Fibrous Scaffolds by electrospinning of poly (ε-caprolactone)/layered silicate nanocomposites. In: Paper presented at the proceedings of the 3rd international symposium on nanomanufacturing (ISNM 2005), Limassol, 2005

94. Nitya G, Nair GT, Mony U, Chennazhi KP, Nair SV (2012) J Mater Sci Mater Med 23(7):1749–1761
95. Lee HW, Karim MR, Ji HM, Choi JH, Ghim HD, Park SM, Oh W, Yeum JH (2009) J Appl Polym Sci 113(3):1860–1867
96. Qin Q, Liu Y, Chen SC, Zhai FY, Jing XK, Wang YZ (2012) J Appl Polym Sci 126:1556–1563
97. Park JH, Karim MR, Kim IK, Cheong IW, Kim JW, Bae DG, Cho JW, Yeum JH (2010) Colloid Polym Sci 288(1):115–121
98. Deng H, Li X, Ding B, Du Y, Li G, Yang J, Hu X (2011) Carbohydr Polym 83(2):973–978
99. Tsimpliaraki A, Zuburtikudis I, Marras SI, Panayiotou C (2011) Polym Int 60(5):859–871
100. Schiffman JD, Schauer CL (2008) Polym Rev 48(2):317–352
101. Sun K, Li Z (2011) Expr Polym Lett 5(4):342–361
102. Li W, Xu R, Zheng L, Du J, Zhu Y, Huang R, Deng H (2012) Carbohydr Polym 90(4):1656–1663
103. Deng H, Lin P, Xin S, Huang R, Li W, Du Y, Zhou X, Yang J (2012) Carbohydr Polym 89(2):307–313
104. Huang W, Xu H, Xue Y, Huang R, Deng H, Pan S (2012) Food Res Int 48(2):784–791
105. Xin S, Li Y, Li W, Du J, Huang R, Du Y, Deng H (2012) Carbohydr Polym 90:1069–1074
106. Lee JH, Park TG, Park HS, Lee DS, Lee YK, Yoon SC, Nam JD (2003) Biomaterials 24(16):2773–2778
107. Park JH, Park SM, Kim YH, Oh W, Lee GW, Karim MR, Park JH, Yeum JH (2012) J Compos Mater. doi:10.1177/0021998312439221
108. Kayaci F, Uyar T (2012) Carbohydr Polym 90(1):558–568
109. Yang JM, Zha L, Yu DG, Liu J (2013) Colloid Surf B 102:737–743
110. Gilman JW, Jackson CL, Morgan AB, Harris R, Manias E, Giannelis EP, Wuthenow M, Hilton D, Phillips SH (2000) Chem Mater 12(7):1866–1873
111. Roy R (2008) Structure and properties of polymers. Auburn University
112. Gong F, Feng M, Zhao C, Zhang S, Yang M (2004) Polym Degrad Stab 84(2):289–294
113. Park CS, Lee KJ, Nam JD, Kim SW (2000) J Appl Polym Sci 78(3):576–585
114. Park CS, Lee KJ, Kim SW, Lee YK, Nam JD (2002) J Appl Polym Sci 86(2):478–488
115. Wang X, Du Y, Luo J, Yang J, Wang W, Kennedy JF (2009) Carbohydr Polym 77(3):449–456
116. Zhang ZY (2003) Chem Res 36(6):385–392
117. Kim G-M, Michler GH, Ania F, Calleja FJB (2007) Polymer 48(16):4814–4823
118. Li WJ, Laurencin CT, Caterson EJ, Tuan RS, Ko FK (2002) J Biomed Mater Res 60(4):613–621
119. Patakfalvi R, Oszko A, DeKany I (2003) Colloid surf A 220(1):45–54
120. Patakfalvi R, Dékány I (2004) Appl Clay Sci 25(3):149–159
121. Deitzel J, Kosik W, McKnight S, Beck TN, DeSimone J, Crette S (2002) Polymer 43(3):1025–1029
122. Miyagawa H, Misra M, Drzal LT, Mohanty AK (2005) J Polym Environ 13(2):87–96
123. Miyagawa H, Misra M, Drzal LT, Mohanty AK (2005) Polymer 46(2):445–453
124. Wang X, Du Y, Yang J, Wang X, Shi X, Hu Y (2006) Polymer 47(19):6738–6744
125. Wang H, Lu X, Zhao Y, Wang C (2006) Mater Lett 60(20):2480–2484
126. Cai Y, Li Q, Wei Q, Wu Y, Song L, Hu Y (2008) J Mater Sci 43(18):6132–6138

Flame Retardant Properties of Polymer/Layered Double Hydroxide N Nanocomposites

19

Tianxi X. Liu and Hong Zhu

Contents

Abstract

Layered double hydroxide (LDH) as a promising class of anionic clays consists of mixed metal hydroxides and tunable interlayer anions. The highly tunable nature of LDH facilitates the preparation of high-performance and multifunctional polymer nanocomposites. Polymer/LDH nanocomposites have been proved to possess enhanced mechanical properties, thermal stability, and flame retardancy. This chapter mainly concentrates on the recent progress in synthesis of LDH with well-defined morphology and excellent properties, novel strategies for fabricating exfoliated LDH/polymer nanocomposites, and their applications as flame retardants. Synergetic effects by combining LDH with other additives on the flame-retardant properties of polymer nanocomposites are also presented and discussed.

T.X. Liu (✉) • H. Zhu
State Key Laboratory of Molecular Engineering of Polymers, Department of Macromolecular Science, Fudan University, Shanghai, China
e-mail: txliu@fudan.edu.cn

J.K. Pandey et al. (eds.), *Handbook of Polymernanocomposites. Processing, Performance and Application – Volume A: Layered Silicates*, DOI 10.1007/978-3-642-38649-7_22,

Keywords
Flame retardancy • Layered double hydroxide • Nanocomposites • Polymers

1 Introduction

Polymers, with the development of science and technology, are penetrating into every corner of our daily life replacing traditional facilities such as metals, glass, ceramics, cements, and so on. Now, scientists and engineers are able to develop new synthetic polymeric materials that meet specific requirements. Fibers like Kevlar® are used to protect ourselves in battle; plastics may substitute metals in the automotive industry, while most decorating and packing materials use polymers. The benefit from such substitution is obvious that polymers offer better performance and lower cost. However, not all of these changes make a safer and better world because plastic materials are inherent flammable, unlike the items that have been replaced. The increasing usage of polymeric materials increases fire hazards in our daily life, which provide challenges for scientists to develop flame retardants to limit the combustibility of polymers and suppress smoke generated during combustion. As early as 1600s, it was recorded to use a combination of clay and gypsum in fire retard canvas. Now, modern fire retardants mainly include halogen-containing compounds, organic phosphorus/nitrogen compounds, inorganic fillers, intumescent systems, and so on [1]. The family of halogen-containing flame retardants had been the most widely produced and used flame retardants due to the advantages of low cost, processability, and miscibility. However, it has been found that some halogen-containing flame retardants are toxic and the smoke generated from burning is a threat to both the environment and human bodies. As the awareness of health care and environment protection increase, the use of halogen is diminishing in Europe and the United States. In 2004, two formulations of halogen flame retardants were banned in Europe and North America markets, while a third one was banned in 2008 [2]. The need for other halogen-free, environment-friendly flame retardants has become evident and imperative.

One promising approach for fabricating flame-retardant polymers is to incorporate flame retardants into the materials themselves, e.g., flame-retardant fillers. Traditional inorganic fillers are general nonflammable and anticorrosive and have low toxicity, low cost, and low emission of smoke during processing and flaming, but the flame-retardant efficiency is relatively low. Thus, they must be used at very high loading levels, which usually lead to serious deterioration in physical properties of the final products. The emerging of nanoscience and nanocomposites offers new solutions for fabricating flame-retardant polymers. In polymer/layered silicate nanocomposites, only a small amount of clay nanofillers could significantly improve the flame-retardant properties of the nanocomposites, without sacrificing or even enhancing the mechanical properties. One of the key points for achieving high-performance nanocomposites is the fine dispersion of nanofillers in polymer matrices. Clay systems, mainly montmorillonite (MMT), are the most well-developed and used flame retardants in polymer matrices. Unlike MMT, which is a naturally occurring cationic clay mineral, layered double hydroxides (LDHs) are a class of anionic clays whose structure is based on brucite-like layers and derived primarily by synthetic approaches. LDH materials can be very interesting to industry as they combine the features of conventional metal hydroxide-type fillers with the layered silicate type of nanofillers. The main advantage of LDHs, compared with MMT, is that the morphology and properties of the synthesized LDHs can be easily designed and controlled, while the type and quality of MMT are constrained by nature. According to the research work documented in recent literature, polymer/layered double-hydroxide nanocomposites show enhanced thermal stability and significant reduction in PHRR (peak heat release rate). It is believed that the next-generation LDH flame retardants with enhanced properties such as smoke suppression by synergistic effects with other agents will come up soon for commercial use.

In this review, we will first introduce the structure and the synthesis and modification methods of LDHs. LDHs can be synthesized to meet specific requirements with a variety of compositions and functions by well-defined strategies. In order to fabricate exfoliated LDH/polymer nanocomposites, long-chain organic anions are usually introduced into LDH galleries. With the development of total delamination of LDHs in organic solvents, layer-by-layer (LBL) assembly of LDH nanosheets and polymers using nitrate LDHs as precursor provides a new route to fabricate hybrid nanomaterials. Since LDHs are polar fillers, the dispersion state of LDHs in polar and nonpolar polymer matrices is different. As to the applications of LDHs as flame retardants, we will secondly discuss the effects of organo-modification, loading level, and other issues on flame-retardant properties of polymer/LDH nanocomposites, using both polar and nonpolar polymer matrices as typical examples. Besides, synergistic effect by combining LDH with other additives provides further improved flame-retardant properties. However, we still have to maintain critical attitudes towards future commercial applications of LDHs because organically modified LDHs are easily contaminated by carbon dioxide in air. And in most cases, the mechanical properties of LDH/polymer nanocomposites are usually decreased although flame-retardant properties are enhanced.

2 LDHs: Structure, Synthesis, and Modification

LDHs are also known as hydrotalcite-like materials, represented by the general formula $[M^{II}{}_{1-x}\,M^{III}{}_{x}(OH)_2]^{x+}[A^{n-}{}_{x/n}\cdot yH_2O]^{x-}$, where M^{II} is a divalent metal cation (M^{II}, Mg, Zn, Ni, Co, etc.), M^{III} is a trivalent metal cation (M^{III}, Al, Fe, etc.), and A stands for n valence inorganic or organic acid anions [3]. Generally, the value x is in the range of 0.2–0.33, giving rise to a large class of isostructural materials with varying AEC (anion exchangeable capacity) values. Figure 19.1 presents general layered structure of an LDH and idealized surfaces of 2:1 and 3:1 (M^{II}/M^{III})-type LDHs [4, 5]. The basic layered structure of LDH is based on that of brucite [$Mg(OH)_2$] which is formed by stacked sheets of octahedrons of magnesium hydroxide. In LDHs, a fraction of the divalent cations in brucite lattice is substituted by trivalent cations, so that the

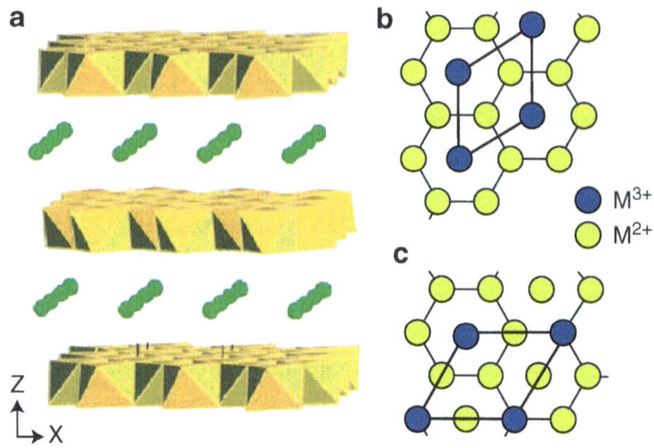

Fig. 19.1 (**a**) General layered structure of an LDH [4]. (**b**) An idealized surface of 2:1 MgAl-LDH. (**c**) An idealized surface of 3:1 MgAl-LDH [5]

layers acquire positive charges, which are balanced by interlayer anions. The interlayer galleries of LDHs contain both interlayer anions and water molecules. Therefore, there is a complex network of hydrogen bonds between layer hydroxyl groups, anions, and water molecules [6]. The bonding between LDH layer and interlayer anions involves a combination of electrostatic effects and hydrogen bonding. In particular, due to the favorable lattice stabilization enthalpy, CO_3^{2-} has greater affinity to LDH layers than other monovalent or multivalent anions [7]. Thus, anion exchange of interlayer carbonate anions is extremely difficult, and the organically modified LDHs that have been stored for a long time usually contain carbonate impurities, which limit the industrial applications of LDHs.

A number of synthetic methods have been successfully employed in preparation of LDHs. The most traditionally used are coprecipitation methods, including coprecipitation in simple alkali solutions [8] and hybrid solution containing surfactants [9] or functional molecules [10]. In order to control the morphology and particle size, methods like coprecipitation in homogeneous conditions [11] and a combination of coprecipitation and subsequent hydrothermal treatment [12] are developed. Recently, LDHs as drug carriers and biological materials require smaller particle size in nanometers and stable dispersion in aqueous systems. An SNAS (separate nucleation and aging steps) method developed by Duan et al. [13] and a sol–gel technique [14, 15] are applied to fabricate colloidal dispersion of LDHs.

The direct coprecipitation method is also known as one-pot synthesis, which enables to scale up in order to produce large quantities of LDHs. In this method, aqueous solutions of mixed M^{II} and M^{III} ions are added dropwise to a dilute alkali solution which contains specified interlayer anions. NaOH and/or $NaHCO_3$, Na_2CO_3, or NH_4OH may be chosen to maintain the pH value in order to ensure simultaneous precipitation. The pH value of the reaction system may vary in the range of 8.0–10.0, depending on the type and concentration of metal ions to be precipitated. The coprecipitation method offers opportunities to synthesize LDHs with unusual cations and rare structures, such as $LiAl_2$- [16], Zn_2Cr-, Cu_2Cr- [17], or even $M^{II}{}_2Zr^{IV}$-LDH

(M^{II} = Mg, Ni, Zn) [18], also with tunable M^{II}/M^{III} molar ratio. Besides, it is often a choice for the preparation of organic–inorganic LDH hybrid materials which are difficult to obtain in other ways. When aqueous soluble organic compounds are introduced into the precipitation system, the preparation should be performed in an inert atmosphere to exclude the carbonate anions. Chemical structures of interlayer anions that are often incorporated into LDHs by coprecipitation are listed in Table 19.1. From small inorganic anions to biopolymers, the diversity of interlayer chemistry attracts a lot of research interests. However, LDHs obtained by this simple coprecipitation are usually associated with low crystallinity and poor morphology, as shown in Fig. 19.2a [19]. There are a number of LDH sheets stacking into large aggregates with secondary particle size of about 1 – 10 μm. The morphology of well-formed and highly crystalline LDHs should be hexagonal platelets, as shown in Fig. 19.2b, which were obtained by urea hydrolysis method. Therefore, a hydrothermal treatment process, after the fast coprecipitation, was developed to increase the crystallinity of amorphous or poorly crystallized materials. Consequently, well-crystallized LDH hexagonal platelets in nanometers were obtained [21], as shown in Fig. 19.2c. It is possible to make stable LDH suspensions in a wide range of compositions. By varying the hydrothermal treatment temperature, time, and LDH concentration, it is also possible to control platelet size in the range of 40 – 300 nm [12].

Another method that enables to control particle size and size distribution of LDH crystals is SNAS developed by Duan's group [13]. In conventional coprecipitation processes, nucleation and aging take place simultaneously during the adding process. It is inevitable that a wide distribution of crystalline sizes is obtained after aging. The key features of this SNAS method are a very rapid mixing and nucleation process in a colloid mill (as shown in Fig. 19.3a) followed by a separate aging process. The deliberate design of colloid mill enables fast separation of slurry of nuclei and salt solutions. Moreover, colloid mill will subject extreme forces to nucleation mixture, thus preventing aggregation and resulting in uniform and well-crystallized LDH platelets, as shown in Fig. 19.3b, c [22]. This method has proved to be very effective in fabricating LDHs in large quantities and is now being commercialized in China.

In polymer composite industry, compared with mono-distribution of LDH crystal sizes, well-formed LDHs with large aspect ratio may be more attractive. It is well known that a fine dispersion of MMT on nanoscale usually leads to enhanced mechanical properties of polymer nanocomposites. Higher aspect ratio of homogeneously dispersed nanofillers will also lead to better physical properties. Urea hydrolysis method, which is also known as "homogeneous precipitation method," is developed to afford large and well-crystallized LDHs of micrometer scale, as shown in Fig. 19.2b. The key point of this method is the use of ammonia-releasing agents, such as urea and hexamethylenetetramine (HMT), which are highly soluble in water and whose hydrolysis rate can be easily controlled by controlling the temperature of mixture. The urea hydrolysis usually undergoes two steps, the formation of ammonium cyanate (NH_4CNO) (Eq. 19.1) and subsequently a fast hydrolysis of cyanate to ammonium and carbonate (Eq. 19.2):

Table 19.1 Various anions being incorporated in LDHs through coprecipitation

Category	Anions	
Simple inorganic anions	Cl^-, NO_3^-, SO_4^{2-}, ClO_4^-, PO_4^{3-}, BO_3^{3-}	
Surfactants	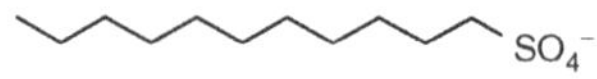	Dodecyl sulfate
	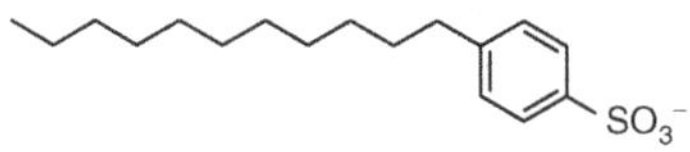	Dodecyl benzenesulfonate
	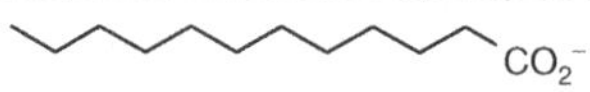	Laurate
Monomers	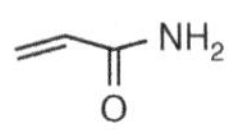 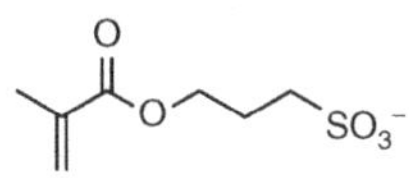	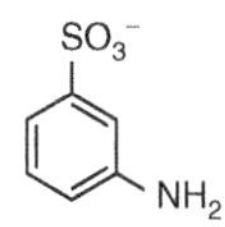 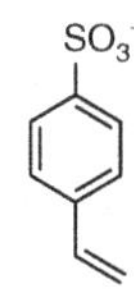
Biopolymers	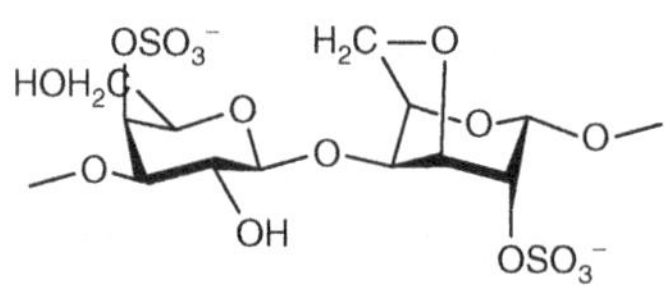	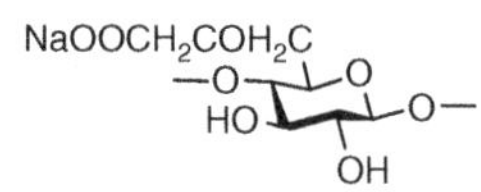
Other functional anions	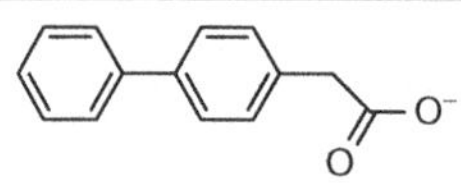	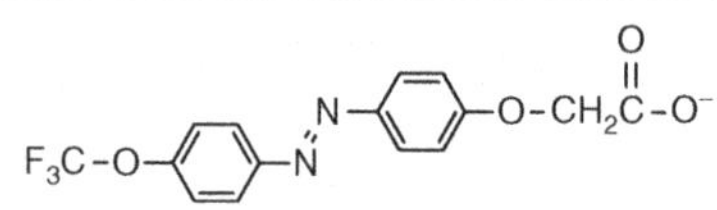
	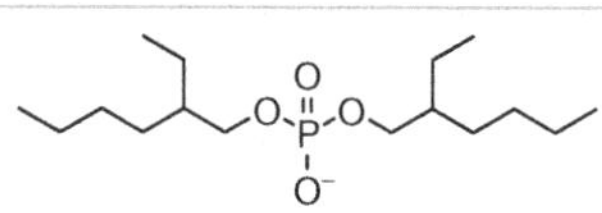	

$$CO(NH_2)_2 \rightarrow NH_4CNO \tag{19.1}$$

$$NH_4CNO + 2H_2O \rightarrow 2NH_4^+ + CO_3^{2-} \tag{19.2}$$

The slow release of ammonium enables a homogeneous solution at a pH value of about 9, which favors a number of metal ions to be precipitated. Costantino [23] and Adachi-Pagano et al. [11] initially prepared MgAl-CO_3-LDH by using this method, while Sasaki and coworkers [24, 25] synthesized highly crystalline transition-metal-bearing (CoAl−, NiAl−, ZnAl−, and FeAl−) LDHs and extremely expanded the chemistry and functions of LDHs. The growth routes and mechanisms

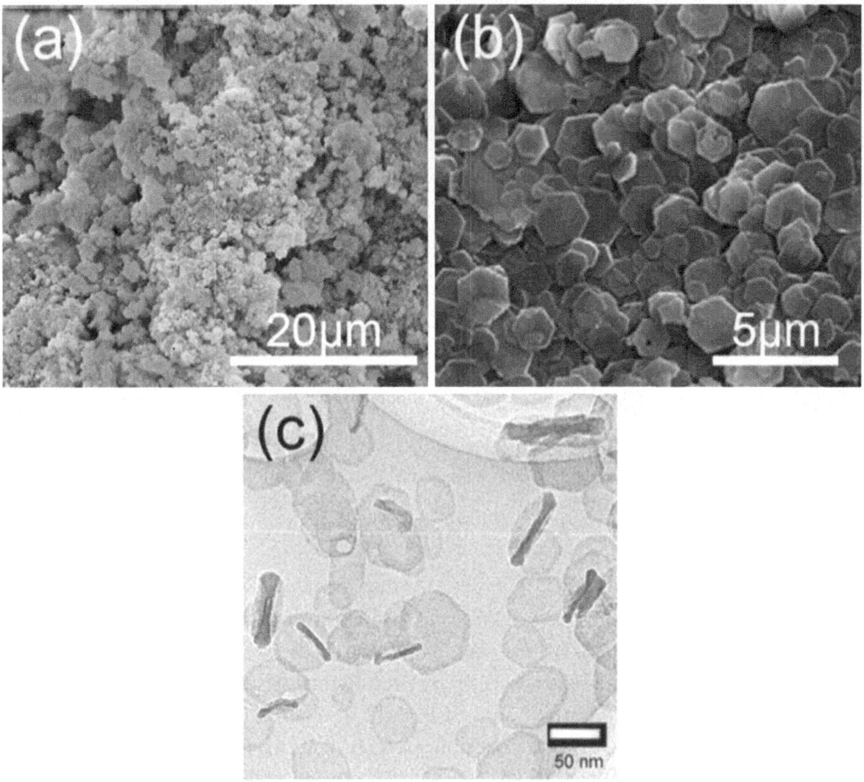

Fig. 19.2 Scanning electron microscopy (*SEM*) images of Mg_2Al-LDH crystals prepared by (**a**) traditional coprecipitation [19] and (**b**) urea hydrolysis [20]. (**c**) Transmission electron microscopy (*TEM*) image of nano-sized Mg_2Al-LDH crystals [12]

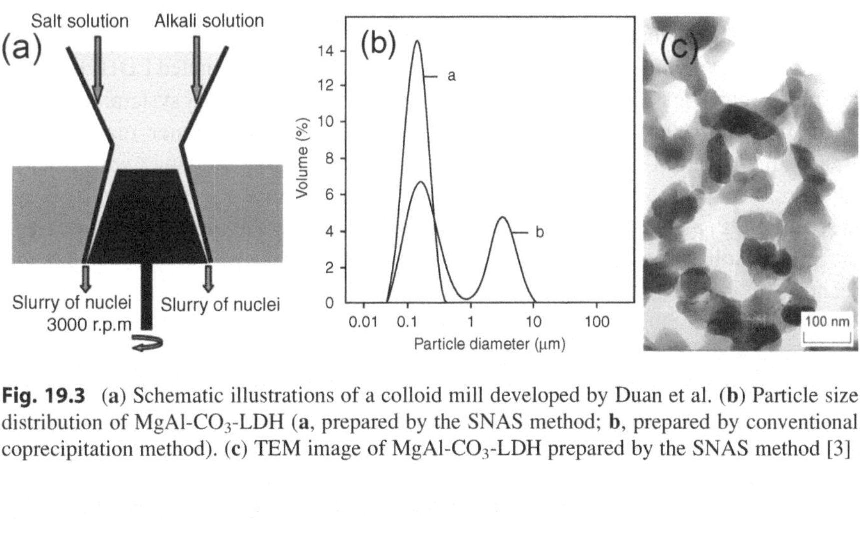

Fig. 19.3 (**a**) Schematic illustrations of a colloid mill developed by Duan et al. (**b**) Particle size distribution of $MgAl\text{-}CO_3$-LDH (**a**, prepared by the SNAS method; **b**, prepared by conventional coprecipitation method). (**c**) TEM image of $MgAl\text{-}CO_3$-LDH prepared by the SNAS method [3]

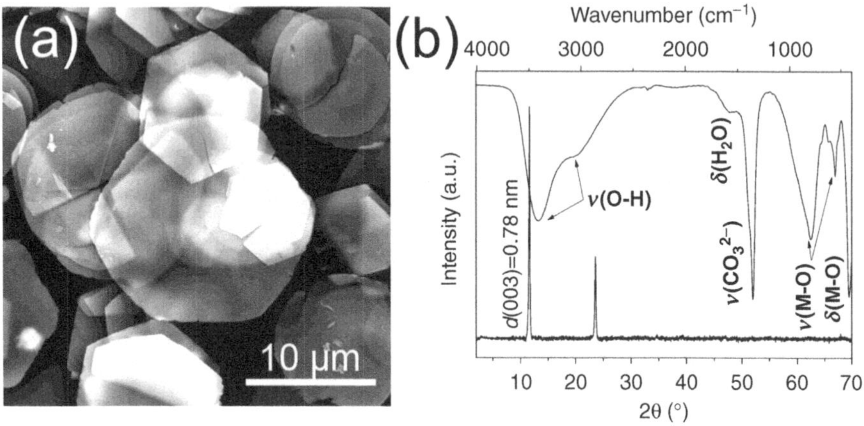

Fig. 19.4 Profiles of $Co_2Al\text{-}CO_3$-LDH with platelet size of micrometers: (**a**) SEM image, (**b**) XRD pattern, and FTIR spectra [26]

of LDHs are also proposed [20]. It is revealed that the size of LDH hexagons tends to be larger at lower concentration of urea. In some extreme cases, CoAl-CO_3-LDHs with lateral size larger than 40 μm (i.e., aspect ratio of 10^3 order) and narrow size distribution can be obtained successfully by urea hydrolysis method [26]. Profiles of a typical $Co_2Al\text{-}CO_3$-LDH are shown in Fig. 19.4. The characteristic *d*-spacing of $Co_2Al\text{-}CO_3$-LDH from X-ray diffraction (XRD) pattern is 0.78 nm. In Fourier transform infrared (FTIR) spectra, the strong absorption peak at 1,354 cm^{-1} belongs to the *v*3 vibration of interlayer carbonate, while the broad peaks and shoulders at 3,420, 3,052, and 1,590 cm^{-1} are associated with stretching and bending modes of hydroxyl groups and water molecules, respectively. The urea hydrolysis reaction may take place in either a glass vessel heating to reflux temperature for 24–36 h or a Teflon-lined autoclave for hydrothermal reactions at desired conditions. Generally, LDHs from hydrothermal reactions have better quality than those from vessel heating. However, most LDHs obtained by this method are carbonated LDHs because carbonates are released along with ammoniums in urea hydrolysis system.

In order to improve the compatibility of LDHs with polymer matrices, an organo-modification treatment is usually necessary. Anion exchange with surfactants or other functional anions is the most common method used for modification. In spite of high affinity of carbonate anions to LDHs, it is shown that by using acid-salt mixed solution, the decarbonation reaction will proceed completely and the decarbonated LDHs thus obtained will show significant anion exchange ability [27–29]. Usually, the anion exchange sequence is CO_3^{2-} (decarbonation), Cl^-, NO_3^-, and then other anions. The affinity of anions for LDH layers depends on their size and charge per area. The solvent medium also affects the anion exchange process, for example, aqueous medium favors the exchange by inorganic anions, while organic solvent favors the exchange by organic anions. Besides, the chemical composition of LDH sheets is another factor which influences the charge density of

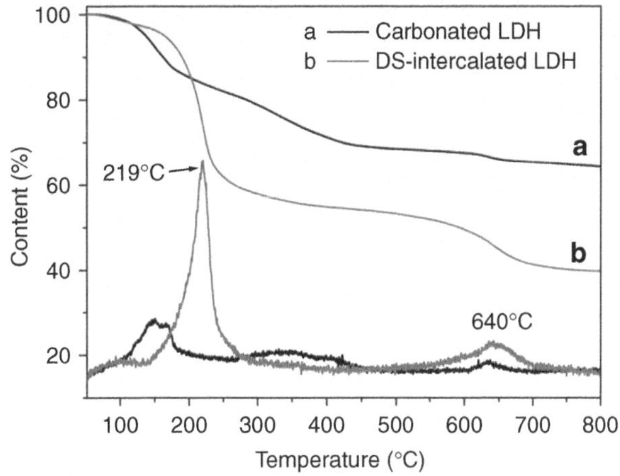

Fig. 19.5 TGA and DTG curves of (**a**) carbonated MgCuAl-LDH and (**b**) DS-intercalated MgCuAl-LDH (prepared through anion exchange reactions)

LDH sheets and the hydration state, thereby affecting the reactions. Figure 19.5 presents a typical thermogravimetric analysis (TGA) profile of dodecyl sulfate (DS)-intercalated MgCuAl-LDH obtained from anion exchange reaction. A sharp weight loss at 219 °C for DS-LDH is related to thermal decomposition of interlayer DS anions, while the weight loss at 640 °C may be due to the decomposition of carbon residues.

Another method for modification is making use of the structural "memory effect" of LDHs. Calcination of LDHs can remove the interlayer water, carbonate, and the hydroxyl groups, resulting in layered mixed metal oxides. As can be seen in Fig. 19.5a, at 450 °C the decarbonation and dehydroxyl processes have completed, and LDO (layered double oxide) is obtained. When the calcined LDO is exposed to water and anions, it is able to regenerate the layered hydroxide structure [30]. Utilizing such a memory effect, organic compounds or surfactants can be incorporated into LDH galleries through this regeneration process.

Recently, the total delamination of LDHs [31–33] opens a new route for organo-modification of LDHs and fabrication of polymer/LDH nanocomposites. Through total delamination of nitrate LDHs in formamide into positively charged nanosheets, self-assembly behavior will take place between LDH nanosheets and organic anions that could be dissolved in formamide. Followed by centrifugation and water washing, the layered structure will be regenerated or restacked with bilayer arrangement in which organic anions are located in the interlayers [34]. The delamination/restacking method will be in favor of fabricating the modified LDHs with high content of organics.

The characterization of organically modified LDHs is usually a combination of several techniques. XRD is primarily used to identify whether the organic compounds have been intercalated into LDH galleries successfully. A shift of the (003) diffraction peak to lower 2θ angle will prove an enlargement of interlayer spaces, due to a successful intercalation of large anions. The morphology of as-prepared LDHs is examined by SEM and TEM. FTIR and TGA are used to examine the chemical structures and the content of interlayer organic anions.

3 New Trends in Fabricating Polymer/LDH Nanocomposites

Polymer/LDH composites have been studied for decades. The conventional fabrication method includes direct intercalation and in situ polymerization. As mentioned above, coprecipitation in an aqueous polymer solution will usually give an intercalated LDH nanocomposite. This method is specially suitable for the fabrication of bionanocomposites [10]. Most polymers are insoluble in water, however, conventional solution mixing or melt blending is thus employed [35–38]. Both direct intercalation and in situ polymerization could be used to fabricate intercalated or exfoliated nanocomposites, mainly depending on the organo-modification of LDHs and their compatibility with polymer matrices. In situ polymerization is carried out by intercalation of monomers in LDH interlayer space, and consequent polymerization will probably enlarge interlayer distance to form intercalated structure or even delamination of LDH platelets. O'Leary et al. [39] reported that DS-intercalated Mg_2Al-LDH was able to be delaminated in acrylate monomers (e.g., 2-hydroxyethyl methacrylate, ethyl methacrylate, methyl methacrylate, ethyl acrylate, methylacrylate) under high-shear stirring, and consequent polymerization did not change the exfoliation state of LDHs. The loading level of exfoliated LDHs could be up to 10 wt% in such polar acrylate monomers.

Along with the successful total delamination of LDHs [32, 33], novel strategies for fabricating exfoliated LDH/polymer nanocomposites have been developed. An early delamination method was introduced by Adachi-Pagano et al. [40] to prepare delaminated Zn_2Al-DS-LDH in butanol under refluxing condition. Until the discovery of formamide as a good solvent for dispersing LDH nanoplatelets, polymer nanocomposites based on delamination/restacking process are rarely studied. Upon the facile delamination of nitrate LDHs in formamide, LBL assembly technique is employed to make nanocomposite films [41–44]. Based on its driving force, LBL assembly can be categorized into several types: electrostatic force, hydrogen bonding, covalent bonding, coordination bonding, and charge transform interaction. Representative strategies for fabricating ultrathin films of polymer/LDH nanocomposites via LBL assembly are presented in Fig. 19.6. Besides LBL assembly, ultrafast facile filtration of exfoliated LDH nanosheets in formamide gives highly oriented and printable LDH/polymer thin films [45]. Utilizing co-assembly of functional anions and positively charged LDH nanosheets, LDH films with tunable functionality and nanostructure can be obtained. The highly oriented biomimetic thin films show marvelous properties and functions. Han et al. [46] developed a new strategy to mimic natural nacre shell with regularly stacked inorganic/organic layered structure by LBL assembly of LDH nano-plates and PVA molecules. After cross-linking by glutaric dialdehyde, the tensile modulus and yield strength significantly increased to 10 and 5 times of that of neat PVA, respectively.

It is interesting to find that when nitrate LDHs take up formamide, they undergo a reversible swollen phase [24]. The swollen structure of LDHs is of long-range ordering with interlayer distance as large as tens of nanometers, whereas a full delamination could be only achieved by shaking or stirring or sonicating in swollen phase. And, when mixing swollen LDHs with water, LDHs with the same

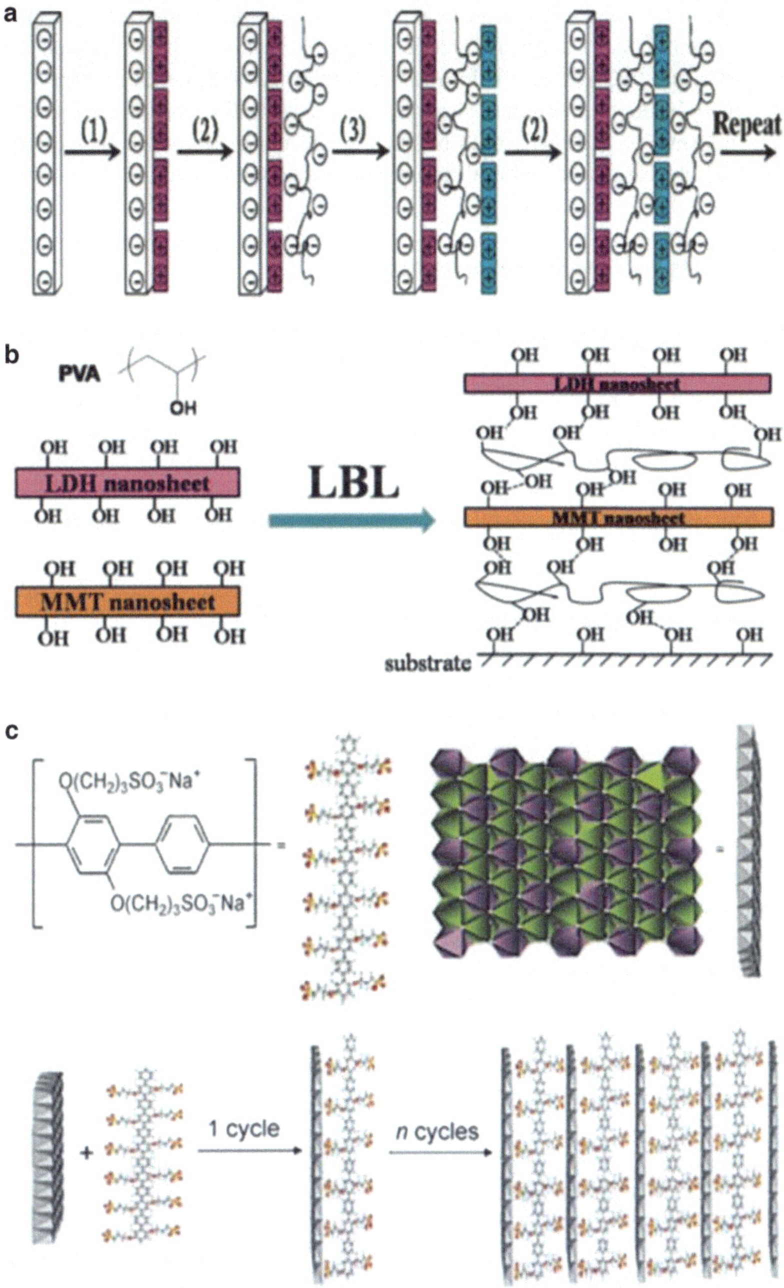

Fig. 19.6 Representative strategies for fabricating polymer/LDH ultrathin functional thin films via LBL assembly: (**a**) conventional (CoAl-LDH/PSS/NiAl-LDH/PSS)$_n$ film [42], (**b**) (LDH/PVA/MMT/PVA)$_n$ film by hydrogen bonding [43], and (**c**) blue luminescent (APPP/LDH)$_n$ film [44]

morphology, size, and crystallinity as those of the original crystals could be restored. Making use of this reversible change, Huang [47] intercalated bulky thiacalix[4]arene (TCAS) anions into LDHs retaining the hexagon morphology. It is a promising route to intercalate various anions with bulky size or polymer chains into the interlayers of LDHs, thus improving the compatibility with matrix and making high-performance and multifunctional nanocomposites.

4 LDHs as Flame Retardants

There is no universal fire retardant for all polymer matrices with high efficiency and no universal mechanism of flame retardancy. Generally speaking, all flame retardants act either in the vapor phase or in the condensed phase to interfere with the combustion process through chemical cross-linking or other interactions such as barrier effect. As to LDHs, the mechanism of flame retardancy is endothermic decomposition into the respective oxides and nonflammable gases such as water and carbon dioxide. The released gases isolate the flame and dilute the flammable gas in the gas phase, while the oxides are high-temperature resistant to prevent flame from spreading [48]. The flame-retardant mechanism of LDHs is just like metal hydroxides, e.g., magnesium hydroxide [$Mg(OH)_2$] and aluminum hydroxide [$Al(OH)_3$]. However, metal hydroxides are difficult to be well dispersed in polymer matrices and extremely difficult to form intercalated or exfoliated nanocomposites. Therefore, metal hydroxides as flame retardants are usually associated with low efficiency and severe deterioration in mechanical properties. On the contrary, the expandable layered structure and proper organo-modification enable LDHs to exhibit better performance in both flame retardancy and enhancement of physical properties than metal hydroxides.

It is figured out that dispersion of fillers in polymer matrices plays a key role in performance reinforcement of the composites. Although Xu [49] reported dodecylbenzene sulfonate had high affinity to LDH layers and the uptake of such surfactant anions could convert the surface of LDH particles from hydrophilic to hydrophobic, the polar nature of LDH layers does not change through modification. The compatibility of LDHs with polar and nonpolar polymers is different, along with different flame-retardant performance in most cases. Therefore, different strategies and formulae should be applied in different flame-retardant systems.

4.1 Polar Polymer Systems

Wilkie et al. [50] prepared the nanocomposites using organically modified LDHs with different polymers, such as polyethylene (PE), polypropylene (PP), polystyrene (PS), and poly(methyl methacrylate) (PMMA), via melt blending approach, and it was found that only PMMA system gave nanocomposites with good dispersion of LDHs. In other words, in spite of being organically modified, the dispersion of LDHs in polar polymers is better than that in nonpolar ones.

PMMA is a transparent thermoplastic in replacement of glass. The thermo-degradation of PMMA is mainly due to depolymerization to monomers, methyl methacrylate, which is highly flammable. Therefore, the flame-retardant strategy for PMMA is usually adding agents that can prevent the depolymerization reactions by cross-linking the active sites. A preliminary study by Li et al. [51] indicated that the 5 % weight loss temperature of 10 wt% LDH/PMMA nanocomposite prepared by exfoliation-adsorption process was significantly increased by 144 °C. Wang [52] compared the thermal stability and fire properties of MgAl-LDH/PMMA nanocomposites with MMT/PMMA and kaolinite/PMMA composites. It is found that both organo-modified MMT and LDHs could form exfoliated/intercalation morphology in PMMA. When the loading level was set at 10 %, the reduction of HRR is 55 % for MMT/PMMA, 45 % for LDH/PMMA, and 23 % for kaolinite/PMMA. MMT gives better flame-retardant results than LDHs. Since the chemical composition of LDHs is tunable, the influence of variable trivalent cations and interlayer anions of LDHs on flame retardancy of PMMA has been systemically studied. MgFe-LDHs were selected to investigate the effect of trivalent metal in LDHs on flame-retardant performance of PMMA composites [53, 54]. It was found that Fe^{III} in LDHs was able to trap radicals at the onset temperature region of decomposition of PMMA, resulting in a significant increase of $T_{0.1}$ (10 % weight loss temperature) by 40 °C. However, the reduction in PHRR for MgFe-LDH/PMMA composites is not that significant as observed in MgAl-LDH composites, because of the poor dispersion of MgFe-LDH in PMMA. As to interlayer organic anions, various interlayer benzyl anions [55] and linear-chain alkyl carboxylates [56] are studied. LDHs intercalated with a range of benzyl anions with different functionalities (carboxylate, sulfonate, and phosphonate) and amino substituent were prepared by coprecipitation and then melt blended with PMMA. Both of the thermal stability and fire performance were enhanced for all the composites, and the largest reduction in PHRR was obtained by 46 % for the composite containing 10 wt% benzoic acid-LDH. For the LDHs intercalated with long-chain linear alkyl carbonates ($CH_3(CH_2)_nCOO^-$ with $n = 8, 10, 12, 14, 16, 20$), however, the effect of LDHs on morphology, thermal stability, and fire properties was different. C10-intercalated LDHs ($n = 8$) gave exfoliated nanocomposites, and C12–16-intercalated LDH/PMMA composites gave a mixed intercalation and exfoliation morphology, while C18, C22-LDHs only gave an intercalated morphology, as shown in Fig. 19.7. As to the enhancement of thermal stability, $T_{0.5}$ (50 % weight loss temperature) of the composites containing LDHs with different loading levels and alkyl chain lengths was almost constant. The largest reduction in PHRR was observed at 10 wt% loading of LDHs no matter what alkyl carbonates were used, as shown in Fig. 19.8. However, it was also found that when the loading level was above 10 wt%, LDHs were no more effective than the mixture of magnesium hydroxide [$Mg(OH)_2$] and alumina trihydrate [$Al(OH)_3$]. Moreover, the tensile strength was significantly reduced compared with MMT/PMMA blends, probably due to lower aspect ratio of LDHs prepared by coprecipitation. Another work [57] reported that the organo-modification by calcination-rehydration method or anion exchange does not have significant difference in fire properties of PMMA

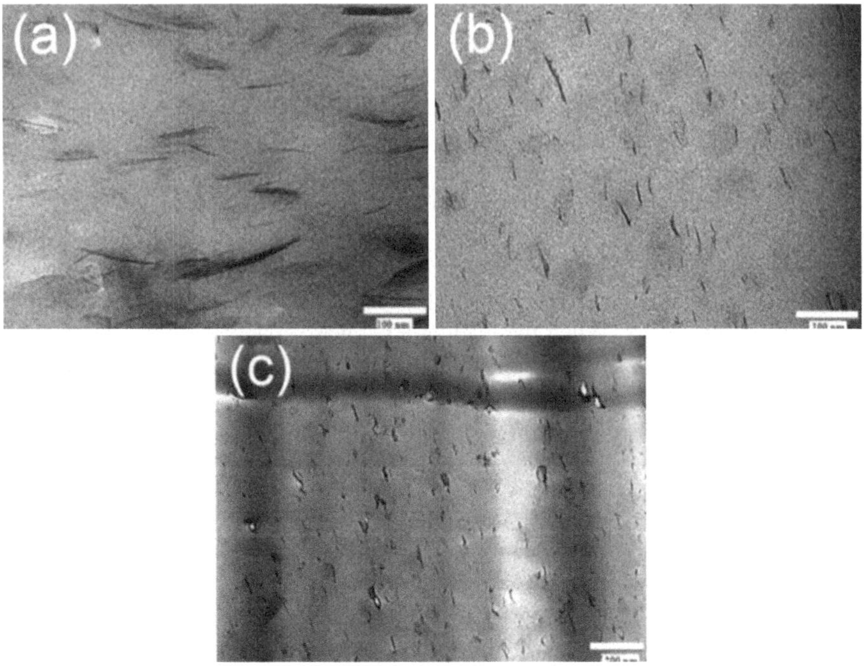

Fig. 19.7 TEM images of (**a**) exfoliated C10-LDH/PMMA nanocomposites, (**b**) C14-LDH/PMMA composite with mixed intercalated and exfoliated morphology, and (**c**) intercalated C22-LDH/PMMA nanocomposites [56]

composites. The synergistic effect of LDHs with other agents on flame retardancy of PMMA composites has also been studied. It was observed that incorporating melamine [58] or phosphate compounds [59] could further improve the flame-retardant performance of PMMA.

Ethylene vinyl acetate (EVA) is the copolymer of ethylene and vinyl acetate. The weight percent of vinyl acetate usually varies from 10 % to 80 % thus giving EVA as thermoplastic, thermoplastic elastomer, or rubber. The fragment of vinyl acetate makes EVA polar and has good compatibility with inorganic fillers, including LDHs. Due to the high polarity of EVA, even carbonated LDHs without modification can be well dispersed in EVA by both melt blending [60] and solution mixing [61]. As indicated by cone calorimeter tests, TGA, limiting oxygen index (LOI), and UL-94 tests, carbonated MgAl-LDHs can greatly improve the fire performance and mechanical properties of EVA composites, compared with $Mg(OH)_2$ and $Al(OH)_3$ [62]. Moreover, incorporation of Fe^{III} into LDH layers can further reduce PHRR due to the radical trapping mechanism of Fe^{III} [63]. Another report by Wang [64] synthesized NiMgAl-LDH via microwave-assisted hydrothermal treatment. Again, Ni cations in LDH layers improved thermal and fire performances of LDH/EVA nanocomposites. Kuila et al. [65, 66] studied the effect of VA content on mechanical and thermal

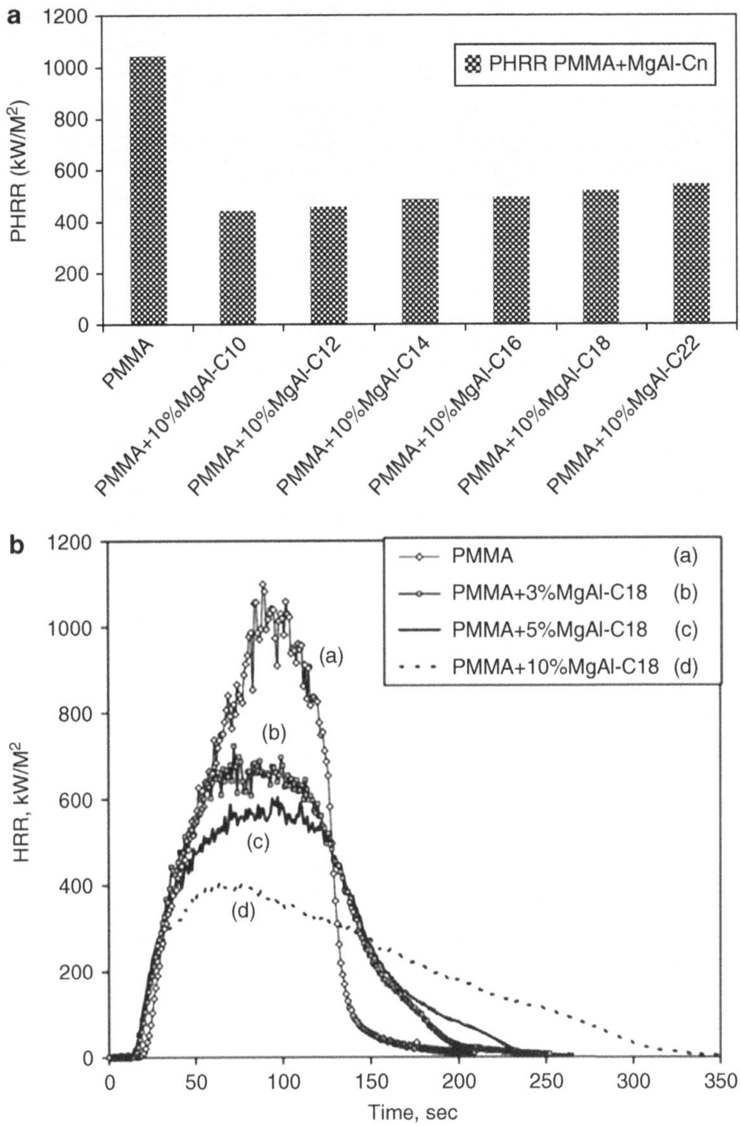

Fig. 19.8 (**a**) PHRR plots of 10 wt% LDH/PMMA composites with alkyl carbonate interlayer anions of different chain lengths; (**b**) heat release rate (HRR) curves for C18-LDH/PMMA nanocomposites [56]

properties of MgAl-DS-LDH/EVA nanocomposites prepared by solution blending. It was shown that more VA content in EVA resulted in better compatibility between LDHs and the EVA matrix and good dispersion of LDHs in the matrix, thus showing higher tensile strength. Wang tried to explain the fire retardancy mechanism of LDH/EVA nanocomposites [67]. The TGA-MS results indicated that LDH plays a key role in initial steps of degradation of EVA. Acetic acid is the main product of side-chain fragmentation of neat EVA, while the addition of LDH catalytically changes the products to acetone. Besides conventional intercalation with DS anions, phosphate [68]-, phosphonate [69]-, and borate [70]-intercalated LDHs were also employed to fabricate halogen-free flame-retardant EVA blends. When phosphorus-containing compounds are introduced into LDH systems, phosphorus will promote the formation of char layers with the P-O-P and P-O-C complexes in the condensed phase. Hence, the thermal stability and flame-retardant properties are improved compared with ordinary LDH/EVA composites [68]. Adding borate-intercalated LDHs into EVA blends, only 3 wt% of LDHs can significantly reduce PHRR by 40 % [70]. The synergistic effects of LDHs with other flame retardants were also reported. It was found that organically modified LDHs can help hyperfine magnesium hydroxide (HFMH) particles to disperse homogeneously in EVA [71]. The synergistic effect of LDHs with HFMH not only makes EVA composites better fire performance but also dramatically decreases the glass transition temperature (T_g) to approximate that of neat EVA, in other words, retaining the low temperature flexibility of EVA nanocomposites. It was also reported by Wilkie et al. [69] that, by blending phosphonate-LDHs with EVA, the time to sustained ignition and average mass loss rate were significantly reduced. The synergistic effect by combining phosphonate and LDHs is quite obvious. Besides, Fe_2O_3 [72], ZnO [73, 74], expandable graphite, and fumed silica [75] were also used to reduce the loading of LDHs while keeping excellent flame-retardant properties due to such a synergistic effect.

Poly(vinyl chloride) (PVC) is an inherent flame-retardant polymer because chlorine builds up 58.6 % of PVC, while the addition of plasticizer during processing can increase the flammability of PVC products dramatically. The degradation of PVC consists of two steps, dehydrochlorination and decomposition of carbon backbones. Large quantities of toxic smoke and gases will be released during degradation which is a big problem for flame retardancy of PVC. The general application of LDHs in PVC is used as thermal stabilizer in replacement of toxic lead salts [76, 77]. Carbonated LDHs have considerable HCl absorption capacity by interlayer anion exchange in the initial stage of decomposition and a further reaction to form metal chlorides [78], while alkyl phosphonate-modified LDHs [79, 80] or toluene-2,4-diisocyanate (TDI)-grafted LDHs [81] were also shown to further improve the thermal stability of PVC. However, the most interesting observation is that incorporation of copper ions into LDHs can significantly reduce the smoke density of PVC/LDH nanocomposites [79]. It was shown that copper compounds might be the most effective class of smoke suppressant for PVC [82]. Therefore, copper-bearing LDHs with the lateral size at micrometer scale were synthesized, and the effect of such LDHs on thermal stability and smoke properties of PVC was studied. At 3 wt% loading level, the optical density of smoke generated by PVC/MgCuAl-LDH composites is 25 % lower than that of PVC/MgAl-LDH and almost 65 % lower than that of neat PVC. Typical smoke behavior of PVC/LDH composites was shown in Fig. 19.9.

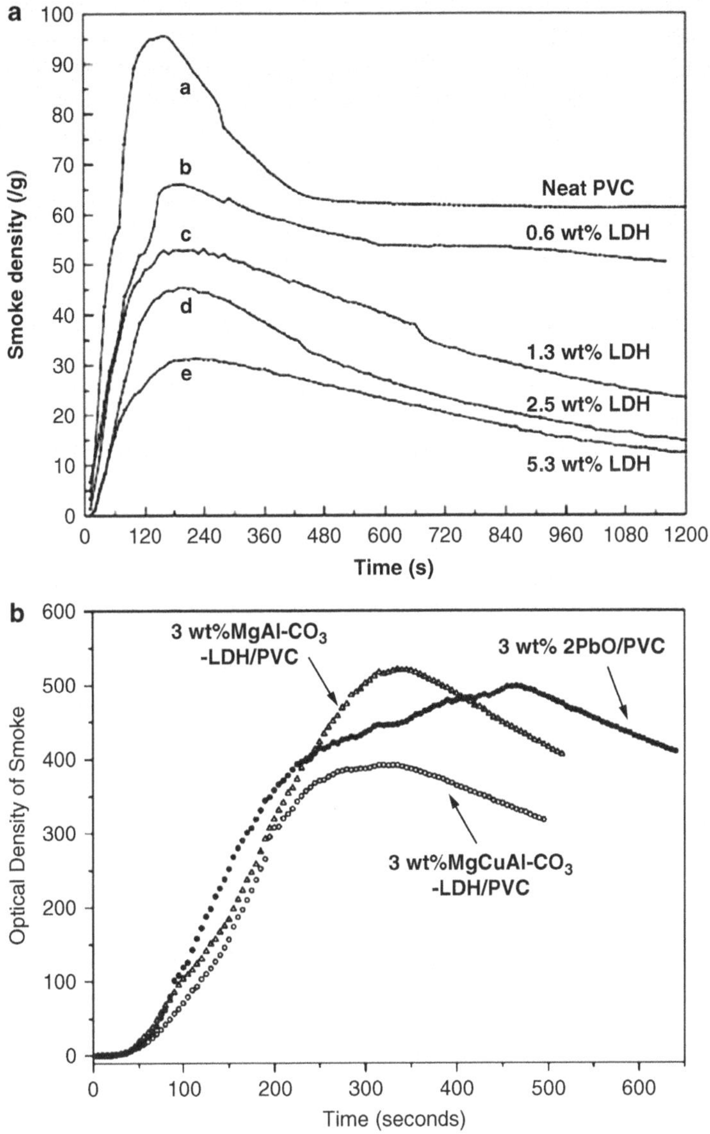

Fig. 19.9 (**a**) Variation of smoke density with time for neat PVC and LDH/PVC nanocomposites [79]. (**b**) Special optical density of smoke generated by 3 wt% lead salts/PVC, 3 wt% MgAl-CO_3-LDH/PVC, and 3 wt% MgCuAl-CO_3-LDH/PVC composites [83]

Polyamide (PA), also known as nylon, is among the most important classes of engineering plastics. The amide bonds link every repeating unit together, endowing polyamide with polarity. By melt blending [84] or in situ polymerization [85], exfoliated LDH/PA6 nanocomposites can be obtained with enhanced mechanical properties while the degradation temperature is decreased. Cone calorimeter tests showed that the PHRR value of 5 wt% MgAl-LDH/PA6 nanocomposite was decreased to 60 % of that of neat PA6 [86]. Huang fabricated LDH/CNTs hybrid nanoparticles via electrostatic assembly of exfoliated LDH nanosheets and modified CNTs. The 3D hybrid nanoparticles were used as reinforcing nanofillers to enhance the mechanical properties of PA6, which increase tensile modulus of 2 wt% hybrid/PA6 nanocomposites to three times of neat PA6 [87]. The unique hybrids which consist of LDHs and CNTs could be potential flame retardants for polymers.

It is also reported that LDHs have considerable flame-retardant effect on poly (vinyl alcohol) (PVA) system [88]. The synergistic effect of addition of ammonium polyphosphate (APP) and LDHs together further improves the fire performance of PVA composites to reach an LOI value up to 33 and UL-94 V-0 rating. Thermal stability and tensile strength of LDH/PVA composites are also enhanced.

Polylactide (PLA) is a well-known biodegradable and biocompatible thermoplastic, which is now widely used as packaging material in food and biomedical applications. PLA and its copolymer are of particular commercial interest as promising replacement of the traditional petroleum-based plastics. However, PLA is still flammable just as common thermoplastics. Therefore, flame retardancy of biodegradable PLA attracts most research interests both in theory and practice. Pan [89] melt mixed PLLA with DS-LDH and found that LDH layers disperse homogenously in PLLA with mixed intercalation and exfoliation morphology. The incorporation of LDH layers has litter effect on crystalline structure of PLLA but increases crystallization rate. Chiang [37] prepared exfoliated LDH/PLLA nanocomposites by solution mixing PLLA with PLA-COOH-modified LDH. The storage modulus of LDH/PLLA nanocomposites is significantly enhanced, but the thermal stability is significantly lower than neat PLLA. The pyrolysis-gas chromatography/mass spectroscopy (Py-GC/MS) characterization of the thermal degradation products indicated that Mg and Al metals in LDH would catalytically depolymerize PLLA to form lactide preferentially [90]. Wang et al. studied the burning behaviors of ZnAl-LDH/PLA nanocomposites [91]. During combustion, HRR and THR of LDH/PLA nanocomposites were significantly decreased. The flame-retardant mechanism is mostly the increasing char formation which may be the result of metal ions in LDHs catalytically cross-linking of degradation fragments of PLA. The synergy effect of LDHs with other flame retardants on PLA is rarely reported.

4.2 Nonpolar Polymer Systems

PE is the most widely used thermoplastic polymer consisting of long chains of ethylene monomer. PE is classified into several different categories mainly based on its density and branching, such as HDPE (high-density polyethylene), LLDPE (linear low-density polyethylene), UHMWPE (ultrahigh molecular weight polyethylene), and so on. The thermal degradation of PE may produce ethylene, pentene, butyl aldehyde, benzene, and other small molecular gases, which are highly flammable. The HRR of PE measured by cone calorimeter is higher than most polymers,

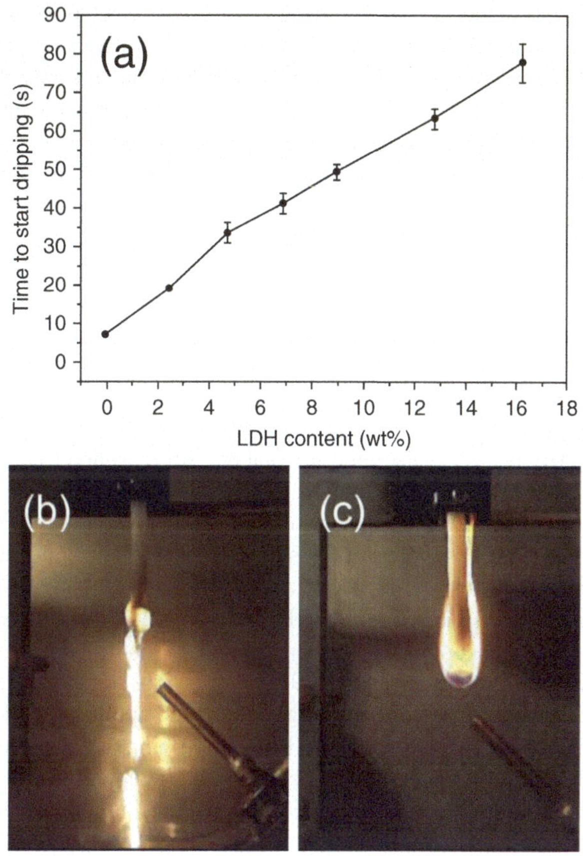

Fig. 19.10 (**a**) Plots of dripping time of LDH/LDPE nanocomposites versus LDH content. (**b**) Dripping behavior of neat LDPE. (**c**) Dripping behavior of 12.75 wt% LDH/LDPE nanocomposites (UL-94 V tests) [84]

and the LOI value is much lower, indicating that PE is highly flammable. It was reported that melt blending of C11-linear alkyl carbonate-modified LDHs at low loadings with PE only gave little improvement on fire performance due to poor compatibility of polar LDH layers with nonpolar PE matrix [50]. When adding 20 wt% of ZnAl-oleate-LDH, PHRR was reduced by 72 %, and the ignition time was also reduced [92]. The ignition problem could be resolved by combining phosphate-containing flame retardants with PE/LDH systems. The studies by Costa [93–95] and Musking [96] showed that using anhydride polyethylene as compatibilizer, a mixed morphology of exfoliation and intercalation could be observed in MgAl-DBS-LDH/LDPE nanocomposites prepared by melt blending. The nanocomposites not only exhibit a reduced total heat release but also lower tendency to fast fire growth. It is well known that PE is easy to drip when burning. The addition of LDHs could also increase the dripping time, as shown in Fig. 19.10 [97]. Qu's group investigated the thermal oxidation mechanisms of polyethylene-grafted-maleic anhydride (PE-g-MA)/MgAl-LDH and LLDPE/ZnAl-LDH [98] and LLDPE/MgAl-LDH system by melt blending [99]. In all the systems, exfoliated

LDHs induce a faster charring process between 210 °C and 370 °C and thus a higher thermal stability above 370 °C. In a comparative study [100], it was shown that the LLDPE nanocomposites with LDHs had much higher thermal stability than those with MMT, and different thermal stability mechanisms for LLDPE/MMT and LLDPE/LDH nanocomposites were proposed. The mechanism of thermal stability improvement for MMT nanocomposites may be due to the formation of protective char layers by MMT, while LDH layers which have high activation energy not only help the formation of char layers but also hinder the dehydrogenation of PE.

PP is another widely used thermoplastic polymer in a variety of applications, including various types of packaging, textiles, stationery, plastic parts, and reusable containers. The flame retardancy of PP is mainly based on flame-retardant inorganic fillers, since chemical modification of PP will induce its degradation. PP/LDH nanocomposites have been proved with enhanced thermal stability, mechanical properties, and lower photooxidation rate than neat PP and PP/MMT composites [101–103]. It is also reported that organo-modified LDH could increase crystallinity of PP [104]. Lonka studied the influence of different cations (Mg or Zn) in LDH layers on the photooxidation of PP and revealed that Mg^{2+} has prodegradant effect on LDH/PP nanocomposites, while Zn^{2+} has no influence on the rate of oxidation [105]. With PP-g-MAH as compatibilizer, homogeneously dispersed LDH nanosheets decrease HRR, PHRR, and THR values of LDH/PP nanocomposites, which means the flame retardancy of PP could be significantly improved with the addition of LDHs [106–109]. Wang synthesized organo-modified CoAl-LDH via a single-step self-assembling method and applied it as flame-retardant nanofiller in PP [106]. Not like to MgAl-LDH improving degradation temperature of PP, CoAl-LDH decreases the onset decomposition temperature although the activation energy calculated based on the Kissinger method is increased. In a further study [108], the kinetic degradation of CoAl-LDH/PP nanocomposites was studied using Friedman and Flynn-Wall-Ozawa methods. The higher activation energy of thermal degradation was observed in nanocomposites with higher LDH content. In intumescent flame retarded PP (IFR/PP) composites, LDHs and MMT play different roles in improving the heat and fire performance of the nanocomposites. LDHs improve thermal stability of IFR/PP nanocomposites in the early stage of heating and combustion, while MMT does in the middle-later stage [35]. If the intumescent flame retardant is initially intercalated into LDH galleries, nanocomposites with excellent optical and mechanical properties will be obtained compared with poorly dispersed microcomposites prepared by conventional co-melt blending [110]. The emerging strategy using LDH nanohybrids is a promising route to further improve flame retardancy of LDH/polymer nanocomposites. Du fabricated LDHs wrapped CNTs through in situ growth of LDHs on surface of CNTs [111]. The combustion experiments revealed that the nanohybrids could confer better flame retardancy on PP that LDHs and CNTs.

PS is an aromatic polymer made from bulk, suspension, or emulsion polymerization of monomer styrene. Due to the ease of polymerization, a variety of strategies for in situ polymerization of PS/LDH nanocomposites have been developed. Leroux et al. [112] used 3-sulfopropyl methacrylate (SPMA)-intercalated LDHs as reactive fillers in PS bulk polymerization system. Being opposite to what was previously thought for reactive fillers, SPMA/LDH at low loadings was not strongly bonded to PS chains, thus no effect was observed on thermal behavior of PS, while for the loadings higher than 10 wt%, SPMA/LDH mainly acted as plasticizer and thus the T_g of PS nanocomposites was decreased. Another unusual PS/LDH nanocomposite was reported by Illaik et al. [113], in which 4-[12-(methacryloylamino)dodecanoylamino] benzenesulfonate acid ($H_2C = C(CH_3)–CO–NH–(–CH_2–)_{11}–CO–NH–C_6H_4–SO_3H$, MADABS) was intercalated into ZnAl-LDH interlayers and then the hybrid material was bulk polymerized with styrene. However, the vinyl part of interlayer MADABS did not participate in polymerization due to strong interdigitation between interlayer anions and LDH layers. It was interesting to observe that from both XRD and TEM characterization, the non-intercalated structure may be dispersed, while the intercalated one may not, indicating that the intercalated structure leads to large-scale aggregation. The bulk polymerization of PS/LDH nanocomposites is hence more complex than the prevailing concepts. Emulsion polymerization of PS/LDH nanocomposites is much straight, and exfoliation morphology is often observed. The emulsion polymerization may take place in the presence of either *N*-lauroyl-glutamate surfactants and *n*-hexadecane [114] or soap-free medium [115] by $S_2O_8^{2-}$ intercalated LDHs and potassium persulfate (KPS) as initiator. The development of living polymerization strategies enables the preparation of PS/LDH nanocomposites by atom transfer radical polymerization (ATRP) [116] and reversible addition-fragmentation chain transfer (RAFT) polymerization using α-bromobutyrate- and 4-(benzodithioyl)-4-cyanopentanoic-intercalated LDHs as initiators, respectively. Compared with the nanocomposites obtained from conventional methods, the thermal stability of the nanocomposites prepared by in situ living polymerization is definitely improved. The flame-retardant properties of PS/LDH composites prepared by melt blending were investigated by Nyambo et al. [117]. It was found that 10-undecenoic acid-intercalated LDHs and APP had strong synergistic effect on the flame-retardant performance. When adding them together, a notable stabilization effect could be observed in the thermo-oxidative degradation stage accompanied with a significant reduction in PHRR.

4.3 Miscellaneous

Epoxy resin (EP) refers to prepolymers that consist of monomers or short-chain polymers with an epoxide group at either end. The cured epoxy is hard and can be used as coatings, adhesives, and composite materials. The LOI value of epoxy varies around 20 %, indicating that epoxy is relatively flammable. The general method for fabricating EP/layered material nanocomposites is to intercalate long-chain curing agents into the interlayer spaces. Zammarano [118] used 3-amino-benzenesul-fonate-, 4-toluene-sulfonate-, and 4-hydroxy-benzenesul-fonate-modified MgAl-LDH in epoxy curing system and found that the epoxy/LDH nanocomposites showed a unique self-extinguishing behavior in UL 94HB tests. The reduction in PHRR for EP/LDH nanocomposites was also more significant than EP/MMT nanocomposites. Chan [119] intercalated long-chain poly(oxypropylene)-amines into MgAl-LDHs as catalyst to initiate the self-polymerization of epoxy. The exfoliated LDH/EP nanocomposites thus formed showed a fast curing and enhanced thermal stability. The LDH-initiated self-polymerization of epoxy provides a new route for preparing EP/layered material nanocomposites.

Elastomer, which is often used interchangeably with the term rubber, generally refers to a polymer with notably low Yong's modulus and high yield strain compared with other materials. Most elastomers only contain carbon and hydrogen atoms in the backbones which are highly flammable. Chemical modification of elastomer molecules is among the most effective flame retardancy of elastomer materials, so is the introduction of flame-retardant fillers. Most elastomer/LDH nanocomposites are prepared via solution mixing. Both ethylene propylene diene terpolymer (EPDM)/LDH [120] and silicon rubber/LDH [121] nanocomposites showed exfoliated morphology with improved mechanical properties and thermal stability. While within a more polar matrix, carboxylated nitrile rubber (XNBR), LDHs could be more efficiently intercalated and exfoliated, and the enhancement of mechanical properties of XNBR/LDH nanocomposites was much more significant [122]. The processing of elastomer compounds usually follows two steps: first, compounding and then cross-linking (vulcanization). It is found that although LDHs exfoliated in XNBR after compounding, a reorganization of LDH nanosheets to more ordered or phase-separated structure would occur in vulcanization step [123]. The tunable nature of LDHs enables the synthesis of Zn-containing LDHs where zinc ions could be accelerators in vulcanization step of rubbers and interlayer stearate ions as activators to promote the vulcanization reaction. Das [124] reported the ZnAl-stearate-LDHs can wholly replace the ZnO and stearate acid in preparation of rubber-based materials. The most significance is that the rubber products vulcanized by ZnAl-stearate-LDHs are transparent, while rubbers vulcanized by other methods usually give opaque products. Polyurethane (PU) is another class of widely used elastomers. Solution blending polyurethane with organo-modified LDHs usually gives exfoliated LDH/PU nanocomposites, which are proved to have enhanced mechanical properties and thermal stability [125, 126]. The dynamic FTIR spectra reveal that LDH/PU nanocomposites are more slowly thermo-oxidized because the barrier effect of LDH nanosheets will decrease the permeability/diffusivity of oxygen [125]. Besides, in polyurethane (PU)/nitrile butadiene rubber (NBR) matrices, nano-dispersed LDHs promoted the formation of compact chars in the condensed phase during combustion [38]. The LOI values of PU/NBR/LDH nanocomposites gradually increased with an increase of filler loading.

5 Summary

LDH materials have been shown to be effective flame retardants for most polymers. The PHRR of polymer/LDH nanocomposites are greatly reduced to almost 50 % of that of neat polymer at 10 wt% LDH loading level. Generally, LDHs have better compatibility with the polar polymer matrices than the nonpolar ones. Through proper modification, LDH materials can be well dispersed in both polar and nonpolar matrices and show significant flame-retardant effects. The general mechanism of flame retardancy for LDHs is the endothermic decomposition to nonflammable gases and heat-resistant oxides. Synergetic effects by combining LDHs with other additives such as APP and phosphorus compounds make the flame-retardant mechanism more complex but more efficient in improving fire performance of polymer composites.

However, flame-retardant polymer/LDH composites with enhanced mechanical properties are still unsolved simultaneously in most reported systems although exfoliated LDH nanocomposites can be obtained. Besides, the effect of large platelet size (e.g., micrometer scale) LDHs, which are potential fillers for mechanical enhancement, on flame retardancy of polymer/LDH nanocomposites is rarely reported. Moreover, easy contamination of organo-modified LDHs by carbon dioxide may be a major barrier to their future commercial production. How to solve this problem is still a big challenge. Despite of these uncertainties, LDH materials as "environment-friendly" flame retardants still attract us to study and develop new methods and materials for flame-retardant polymers and their composites. It is believed that LDH materials will occupy an important place in the future market of flame retardants.

Acknowledgments This work was supported by the National Natural Science Foundation of China (51125011) and "Shu Guang" project (09SG02) supported by Shanghai Municipal Education Commission and Shanghai Education Development Foundation.

References

1. Le Bras M, Wilkie CA, Bourbigot S (2005) Fire retardancy of polymers: new applications of mineral fillers. Springer, New York
2. Betts KS (2008) Environ Health Perspect 116:A210
3. Evans DG, Duan X (2006) Chem Commun 42:485
4. Khan AI, O'Hare D (2002) J Mater Chem 12:3191
5. Leroux F, Taviot-Gueho C (2005) J Mater Chem 15:3628
6. Evans DG, Slade RCT (2006) Structural aspects of layered double hydroxides. In: Layered double hydroxides, vol 119, pp 1–87
7. Dutta PK, Puri M (1989) J Phys Chem 93:376
8. Crepaldi EL, Pavan PC, Valim JB (2000) J Braz Chem Soc 11:64
9. Wang B, Zhang H, Evans DG, Duan X (2005) Mater Chem Phys 92:190
10. Darder M, Lopez-Blanco M, Aranda P, Leroux F, Ruiz-Hitzky E (2005) Chem Mater 17:1969
11. Adachi-Pagano M, Forano C, Besse JP (2003) J Mater Chem 13:1988
12. Xu ZP, Stevenson G, Lu CQ, Lu GQ (2006) J Phys Chem B 110:16923
13. Zhao Y, Li F, Zhang R, Evans DG, Duan X (2002) Chem Mater 14:4286
14. Gursky JA, Blough SD, Luna C, Gomez C, Luevano AN, Gardner EA (2006) J Am Chem Soc 128:8376
15. Prince J, Montoya A, Ferrat G, Valente JS (2009) Chem Mater 21:5826
16. Besserguenev AV, Fogg AM, Francis RJ, Price SJ, Ohare D, Isupov VP, Tolochko BP (1997) Chem Mater 9:241
17. Roussel H, Briois V, Elkaim E, de Roy A, Besse JP (2000) J Phys Chem B 104:5915
18. Tichit D, Das N, Coq B, Durand R (2002) Chem Mater 14:1530

19. Greenwell HC, Holliman PJ, Jones W, Velasco BV (2006) Catal Today 114:397
20. Okamoto K, Iyi N, Sasaki T (2007) Appl Clay Sci 37:23
21. Xu ZP, Stevenson GS, Lu CQ, Lu GQ, Bartlett PF, Gray PP (2006) J Am Chem Soc 128:36–37
22. He J, Wei M, Li B, Kang Y, Evans DG, Duan X (2006) Preparation of layered double hydroxides. In: Layered double hydroxides, vol 119, pp 89–119
23. Costantino U, Marmottini F, Nocchetti M, Vivani R (1998) Eur J Inorg Chem 10:1439
24. Liu ZP, Ma RZ, Osada M, Iyi N, Ebina Y, Takada K, Sasaki T (2006) J Am Chem Soc 128:4872
25. Liu ZP, Ma RZ, Ebina Y, Iyi N, Takada K, Sasaki T (2007) Langmuir 23:861
26. Kayano M, Ogawa M (2006) Clays Clay Miner 54:382
27. Iyi N, Matsumoto T, Kaneko Y, Kitamura K (2004) Chem Mater 16:2926
28. Iyi N, Okamoto K, Kaneko Y, Matsumoto T (2005) Chem Lett 34:932
29. Iyi N, Sasaki T (2008) Appl Clay Sci 42:246
30. Rocha J, del Arco M, Rives V, Ulibarri MA (1999) J Mater Chem 9:2499
31. Hibino T, Jones W (2001) J Mater Chem 11:1321
32. Li L, Ma RZ, Ebina Y, Iyi N, Sasaki T (2005) Chem Mater 17:4386
33. Wu QL, Olafsen A, Vistad OB, Roots J, Norby P (2005) J Mater Chem 15:4695
34. Kang HL, Huang GL, Ma SL, Bai YX, Ma H, Li YL, Yang XJ (2009) J Phys Chem C 113:9157
35. Du BX, Guo ZH, Fang ZP (2009) Polym Degrad Stabil 94:1979
36. Tabuani D, Ceccia S, Camino G (2009) J Polym Sci Part B Polym Phys 47:1935
37. Chiang MF, Wu TM (2010) Compos Sci Technol 70:110
38. Kotal M, Srivastava SK, Bhowmick AK (2010) Polym Int 59:2
39. O'Leary S, O'Hare D, Seeley G (2002) Chem Commun 38:1506
40. Adachi-Pagano M, Forano C, Besse JP (2000) Chem Commun 36:91
41. Chen X, Lei Y, Yang WS (2008) Chem Lett 37:1050
42. Han JB, Lu J, Wei M, Wang ZL, Duan X (2008) Chem Commun 44:5188
43. Huang S, Cen X, Peng HD, Guo SZ, Wang WZ, Liu TX (2009) J Phys Chem B 113:15225
44. Yan DP, Lu J, Wei M, Han JB, Ma J, Li F, Evans DG, Duan X (2009) Angew Chem Int Ed 48:3073
45. Zhu H, Huang S, Yang Z, Liu TX (2011) J Mater Chem 21:2950
46. Han JB, Dou YB, Yan DP, Ma J, Wei M, Evans DG, Duan X (2011) Chem Commun 47:5274
47. Huang GL, Ma SL, Zhao XH, Yang XJ (2009) K. T. Ooi, K.T. Chem Commun 45:331
48. Chen L, Wang YZ (2010) Polym Adv Technol 21:1
49. Xu ZP, Braterman PS (2003) J Mater Chem 13:268
50. Nyambo C, Wang D, Wilkie CA (2009) Polym Adv Technol 20:332
51. Li BG, Hu Y, Liu J, Chen ZY, Fan WC (2003) Colloid Polym Sci 281:998
52. Wang LJ, Xie XL, Su SP, Feng JX, Wilkie CA (2010) Polym Degrad Stabil 95:572
53. Ding YY, Gui Z, Zhu JX, Hu Y, Wang ZZ (2008) Mater Res Bull 43:3212
54. Manzi-Nshuti C, Wang DY, Hossenlopp JM, Wilkie CA (2009) Polym Degrad Stabil 94:705
55. Nyambo C, Chen D, Su SP, Wilkie CA (2009) Polym Degrad Stabil 94:496
56. Nyambo C, Songtipya P, Manias E, Jimenez-Gasco MM, Wilkie CA (2008) J Mater Chem 18:4827
57. Nyambo C, Chen D, Su SP, Wilkie CA (2009) Polym Degrad Stabil 94:1298
58. Manzi-Nshuti C, Hossenlopp JM, Wilkie CA (2008) Polym Degrad Stabil 93:1855
59. Wang LJ, Su SP, Chen D, Wilkie CA (2009) Polym Degrad Stabil 94:1110
60. Du LC, Qu BJ, Xu ZJ (2006) Polym Degrad Stabil 91:995
61. Ramaraj B, Yoon KR (2008) J Appl Polym Sci 108:4090
62. Du LC, Qu BJ (2007) Polym Compos 28:131
63. Jiao CM, Wang ZZ, Chen XL, Hu Y (2008) J Appl Polym Sci 107:2626
64. Wang LL, Li B, Yang MF, Chen CX, Liu YS (2011) J Colloid Interface Sci 356:519
65. Kuila T, Acharya H, Srivastava SK, Bhowmick AK (2007) J Appl Polym Sci 104:1845

66. Kuila T, Acharya H, Srivastava SK, Bhowmick AK (2008) J Appl Polym Sci 108:1329
67. Wang XL, Rathore R, Songtipya P, Jimenez-Gasco MD, Manias E, Wilkie CA (2011) Polym Degrad Stabil 96:301
68. Ye L, Qu BJ (2008) Polym Degrad Stabil 93:918
69. Nyambo C, Kandare E, Wilkie CA (2009) Polym Degrad Stabil 94:513
70. Nyambo C, Wilkie CA (2009) Polym Degrad Stabil 94:506
71. Zhang G, Ding P, Zhang M, Qu BJ (2007) Polym Degrad Stabil 92:1715
72. Jiao CM, Chen XL, Zhang J (2009) J Fire Sci 27:465
73. Jiao CM, Chen XL (2009) J Therm Anal Calor 98:813
74. Jiao CM, Chen XL, Zhang J (2010) J Thermoplast Compos Mater 23:501
75. Du LC, Zhang YC, Yuan XY, Chen JY (2009) Polym-Plast Technol Eng 48:1002
76. Gu Z, Liu WS, Dou W, Tang F (2010) Polym Compos 31:928
77. Liu J, Chen GM, Yang JP, Ding LP (2010) J Appl Polym Sci 116:2058
78. van der Ven L, van Gemert MLM, Batenburg LF, Keern JJ, Gielgens LH, Koster TPM, Fischer HR (2000) Appl Clay Sci 17:25
79. Bao YZ, Huang ZM, Li SX, Weng ZX (2008) Polym Degrad Stabil 93:448
80. Huang NH, Wang JQ (2009) Express Polym Lett 3:595
81. Liu J, Chen GM, Yang JP, Ding LP (2009) Mater Chem Phys 118:405
82. Pike RD, Starnes WH, Jeng JP, Bryant WS, Kourtesis P, Adams CW, Bunge SD, Kang YM, Kim AS, Kim JH, Macko JA, Obrien CP (1997) Macromolecules 30:6957
83. Zhu H, Wang W, Liu TX (2011) J Appl Polym Sci 122:273
84. Zammarano M, Bellayer S, Gilman JW, Franceschi M, Beyer FL, Harris RH, Meriani S (2006) Polymer 47:652
85. Peng HD, Tjiu WC, Shen L, Huang S, He CB, Liu TX (2009) Compos Sci Technol 69:991
86. Du LC, Qu BJ, Zhang M (2007) Polym Degrad Stabil 92:497
87. Huang S, Peng HD, Tjiu WW, Yang Z, Zhu H, Tang T, Liu TX (2010) J Phys Chem B 114:16766
88. Zhao CX, Liu Y, Wang DY, Wang DL, Wang YZ (2008) Polym Degrad Stabil 93:1323
89. Pan PJ, Zhu B, Dong T, Inoue YJ (2008) Polym Sci Part B Polym Phys 46:2222
90. Chiang MF, Chu MZ, Wu TM (2011) Polym Degrad Stabil 96:60
91. Wang DY, Leuteritz A, Wang YZ, Wagenknecht U, Heinrich G (2011) Polym Degrad Stabil 95:2474
92. Manzi-Nshuti C, Hossenlopp JM, Wilkie CA (2009) Polym Degrad Stabil 94:782
93. Costa FR, Abdel-Goad M, Wagenknecht U, Heinrich G (2005) Polymer 46:4447
94. Costa FR, Wagenknecht U, Jehnichen D, Goad MA, Heinrich G (2006) Polymer 47:1649
95. Costa FR, Wagenknecht U, Heinrich G (2007) Polym Degrad Stabil 92:1813
96. Muksing N, Magaraphan R, Coiai S, Passaglia E (2011) Express Polym Lett 5:428
97. Costa FR, Saphiannikova M, Wagenknecht U, Heinrich G (2008) Layered double hydroxide based polymer nanocomposites. In: Wax crystal control: nanocomposites, stimuli-responsive polymers, vol 210, pp 101–168
98. Chen W, Qu BJ (2004) J Mater Chem 14:1705
99. Du LC, Qu BJ (2006) J Mater Chem 16:1549
100. Qiu LZ, Chen W, Qu BJ (2006) Polymer 47:922
101. Ding P, Qu BJ (2006) Polym Eng Sci 46:1153
102. Ardanuy M, Velasco JI, Antunes M, Rodriguez-Perez MA, de Saja JA (2010) Polym Compos 31:870
103. Coiai S, Passaglia E, Hermann A, Augier S, Pratelli D, Streller RC (2010) Polym Compos 31:744
104. Shi YQ, Chen F, Yang JT, Zhong MQ (2010) Appl Clay Sci 50:87
105. Lonkar SP, Therias S, Caperaa N, Leroux F, Gardette JL (2010) Eur Polym J 46:1456
106. Wang DY, Das A, Costa FR, Leuteritz A, Wang YZ, Wagenknecht U, Heinrich G (2010) Langmuir 26:14162
107. Ardanuy M, Velasco JI (2011) Appl Clay Sci 51:341

108. Wang DY, Das A, Leuteritz A, Boldt R, Haussler L, Wagenknecht U, Heinrich G (2011) Polym Degrad Stabil 96:285
109. Wang DY, Leuteritz A, Kutlu B, Landwehr MAD, Jehnichen D, Wagenknecht U, Heinrich G (2011) J Alloys Compounds 509:3497
110. Zhang M, Ding P, Qu BJ, Guan AG (2008) J Mater Process Technol 208:342
111. Du BX, Fang ZP (2010) Nanotechnology 21:6
112. Leroux F, Meddar L, Mailhot B, Morlat-Therias S, Gardette JL (2005) Polymer 46:3571
113. Illaik A, Taviot-Gueho C, Lavis J, Cornmereuc S, Verney V, Leroux F (2008) Chem Mater 20:4854
114. Ding P, Qu BJ (2005) J Colloid Interface Sci 291:13
115. Qiu LZ, Qu BJ (2006) J Colloid Interface Sci 301:347
116. Qiu LZ, Chen W, Qu BJ (2005) Colloid Polym Sci 283:1241
117. Nyambo C, Kandare E, Wang DY, Wilkie CA (2008) Polym Degrad Stabil 93:1656
118. Zammarano M, Franceschi M, Bellayer S, Gilman JW, Meriani S (2005) Polymer 46:9314
119. Chan YN, Juang TY, Liao YL, Dai SA, Lin JJ (2008) Polymer 49:4796
120. Acharya H, Srivastava SK, Bhowmick AK (2007) Compos Sci Technol 67:2807
121. Pradhan B, Srivastava SK, Ananthakrishnan R, Saxena A (2011) J Appl Polym Sci 119:343
122. Pradhan S, Costa FR, Wagenknecht U, Jehnichen D, Bhowmick AK, Heinrich G (2008) Eur Polym J 44:3122
123. Costa FR, Pradhan S, Wagenknecht U, Bhowmick AK, Heinrich G (2010) J Polym Sci Part B Polym Phys 48:2302
124. Das A, Wang DY, Leuteritz A, Subramaniam K, Greenwell HC, Wagenknecht U, Heinrich G (2011) J Mater Chem 21:7194
125. Guo SZ, Zhang C, Peng HD, Wang WZ, Liu TX (2011) Compos Sci Technol 71:791
126. Kotal M, Srivastava SK, Bhowmick AK, Chakraborty SK (2011) Polym Int 60:772

Recent Developments in the Permeability of Polymer Clay Nanocomposites

20

G. Choudalakis and A. D. Gotsis

Contents

Abstract

This chapter focuses on the permeability of gas molecules in nanocomposites. These materials consist of inorganic platelet-shaped nano-sized particles dispersed in polymeric matrices. The dominant mechanisms for the transport of small molecules in polymers and polymer nanocomposites are discussed, and

G. Choudalakis (✉) • A.D. Gotsis
Department of Sciences, Technical University of Crete, Chania, Greece
e-mail: gchoudalakis@isc.tuc.gr

J.K. Pandey et al. (eds.), *Handbook of Polymernanocomposites. Processing, Performance and Application – Volume A: Layered Silicates*, DOI 10.1007/978-3-642-38649-7_11,

some procedures are described for the measurement of the transport coefficients. Two aspects are emphasized: The various models that have been proposed for the prediction of permeability in polymer-clay nanocomposites and the correlation between free volume variations and gas permeation properties. The influence of the characteristics of the anisometric nanoparticles on the free volume and transport properties of the composite membrane are, then, examined using the models and the available experimental data. Some consequences of the methods used to improve the barrier properties of the nanocomposites are examined, and a few applications of these materials as gas barriers are presented.

1 Introduction: Polymer-Clay Nanocomposites

Polymer nanocomposites materials are hybrid systems that consist of a polymeric matrix and dispersed inorganic particles of nanometer scale. The most common inorganic particles belong to the family of layered smectite clay silicates. The layer thickness of each platelet is around 1 nm, and their lateral dimensions may vary from 30 nm to several microns.

The layers are located on top of each other, and van der Waals gaps are created between them, called galleries. The isomorphic substitution of the cations, e.g., the substitution of Al^{3+} with Mg^{2+} or Fe^{2+} with Li^{1+}, generates negative charges that are counterbalanced by alkali and alkaline-earth cations located inside the galleries.

In its initial form, the inorganic phase consists of aggregates or stacks of hundreds of nanoparticles joined together by electrostatic forces. For the nanocomposite formation, the de-agglomeration of the individual nanoparticles or nanolayers (Fig. 20.1) of the primary mineral particles is required. This can be done when the electrostatic forces between the nanoparticles are weakened and if the organic polymer can interact strongly with the inorganic component.

The simple mixing of polymer and layered silicates does not always result in the generation of a nanocomposite, as this usually leads to dispersion of stacked sheets. This failure is due to the weak interactions between the polymer and the inorganic component. If these interactions become stronger, then the inorganic phase can be dispersed in the organic matrix in nanometer scale. An appropriate selection of the primary organic and inorganic components is therefore essential.

The single phyllosilicate follicles present a very high aspect ratio of width/thickness, in the order of 10–1,000. Even for very low concentrations of particles, the total interface between polymer and layered silicates ($\sim$750 m^2/g) is much greater than that in conventional composites.

The nanocomposites have become an area of intensive research activity. The first reason for this development is their high mechanical properties. The complete dispersion of clay nanolayers in a polymer optimizes the number of available reinforcing elements that carry an applied load and deflect the evolving cracks. The coupling between the large surface area of the clay and the polymer matrix facilitates the stress transfer to the reinforcing phase allowing for the improvement of tensile strength and toughness [1].

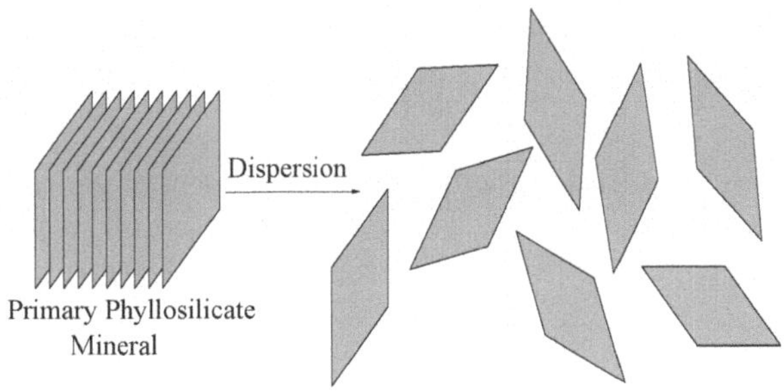

Fig. 20.1 Exfoliation of the primary phyllosilicate mineral into individual dispersed crystalline layers of great aspect ratio

The second major advantage of the nanocomposites is their enhanced barrier properties. The impermeable clay layers force a tortuous pathway for a permeant transversing the nanocomposite. It is reported that gas permeability through polymer films can be reduced further even with small nanoclay loadings. The relevant research on polymer-clay nanocomposite concerns mostly oxygen, carbon dioxide, and nitrogen barrier films, a.o., for packaging food, carbonated drinks, and flexible electronics [2]. Other applications include gas tanks and protective coatings [3].

2 Transport Principles

The transport of gases through a polymer membrane is a complex process which involves the sorption, diffusion, and permeation coefficients. The transport behavior is generally classified into three categories [4]:

- When the diffusion rate is much less than the sorption rate and sorption equilibrium is rapidly established, then the permeation is Fickian. The gas transport, then, follows the solution-diffusion mechanism.
- When the diffusion rate is very rapid compared with the sorption rate, the permeation process is sorption-controlled and is called "anomalous."
- When the diffusion and sorption rates are comparable, then the transport is characterized as non-Fickian. This occurs in the case of liquid penetrants and glassy polymer membranes, and it is the most complicated of the three.

In general, the transport process is described by Fick's first law, in which the gas flux through a polymer membrane, J, is driven by the concentration gradient of absorbed molecules in the polymer matrix:

$$J(\mathbf{r},t) = -D\nabla c(\mathbf{r},t), \tag{20.1}$$

where D is the diffusion coefficient. The above equation in one dimension becomes

$$J(x,t) = -D\frac{\partial c(x,t)}{\partial t}, \tag{20.2}$$

which, in combination with the continuity equation, gives

$$\frac{\partial c(x,t)}{\partial t} = \frac{\partial}{\partial x}\left(D\frac{\partial c(x,t)}{\partial x}\right). \tag{20.3}$$

If the diffusion coefficient is independent of the concentration, then

$$\frac{\partial c(x,t)}{\partial t} = D\frac{\partial^2 c(x,t)}{\partial^2 x}, \tag{20.4}$$

which is often referred to as Fick's second law.

Concentration-dependent diffusion is often observed for many gas penetrants and is usually expressed as [5]

$$D(c) = D_0 e^{\alpha c}, \tag{20.5}$$

where D_0 is the zero-concentration diffusivity and α is the plasticization power which accounts the effectiveness of the penetrant to plasticize the matrix.

2.1 Gas Permeation in Rubbery Polymers

The transport of gases through a rubbery polymer membrane is based on the solution-diffusion mechanism: sorption of gas molecules on the surface of the membrane; diffusion through it; and, finally, desorption of gas from the other surface of the membrane.

The sorption behavior is described by Henry's law:

$$C_D = k_D.p, \tag{20.6}$$

where C_D is the concentration of sorbed gas, k_D is Henry's coefficient or Henry's solubility, and p is the partial pressure of the gas. This represents the gas molecules that exist in the dissolved state (between the polymer chains) and is related to the fluctuating fraction of the free volume of the polymer [6, 7]. The subscript "D" denotes the dissolved state.

Equation 20.4 can be solved analytically for a flat membrane of thickness d under the following initial conditions [8]:

$$c(x;t=0)=0 \quad c(x=0;t)=C_D=S.p \quad c(x=d;t)=0,$$

which results from Henry's law, where S is used instead of k_D for the solubility coefficient and C_D denotes the saturation value for the concentration of the dissolved gas.

The solution of the differential Eq. 20.4 represents the concentration of the gas inside the membrane as a function of depth and time:

$$c(x,t) = Sp\left[1 - \frac{x}{d} - \frac{2}{\pi}\sum_{m=1}^{\infty}\frac{1}{m}\sin\frac{m\pi x}{d}\exp\left(-\frac{m^2\pi^2 Dt}{d^2}\right)\right]. \tag{20.7}$$

The flux of the gas that permeates the membrane is obtained if we set $x = d$ in Eq. 20.7 and then substitute in Eq. 20.2. The result is

$$J(t) = \frac{DSp}{d}\left[1 + \sum_{m=1}^{\infty} 2\cos m\pi \cdot \exp\left(-\frac{m^2\pi^2 Dt}{d^2}\right)\right]. \tag{20.8}$$

2.2 Glassy Polymers

For sorption in a glassy polymer membrane, the sorbed penetrant concentration is the sum of the Henry population (C_D) and the Langmuir population (C_H):

$$C = C_D + C_H. \tag{20.9}$$

The second term represents the adsorption of gas molecules into the free volume holes in the glassy polymer and is given by the relation

$$C_H = \frac{C'_H bp}{1 + bp}. \tag{20.10}$$

Here C'_H is the Langmuir capacity and represents the amount of gas that the glassy polymer can accommodate into its unrelaxed free volume (excess free volume) and is strongly temperature dependent, Fig. 20.2.

$$C'_H = \left(\frac{V_g - V_l}{V_g}\right)\frac{22,400}{V_p}. \tag{20.11}$$

V_g and V_l are the specific volumes in the glassy and the liquid (rubbery) states, Fig. 20.2, and V_p is the molar volume of sorbant gas (in ml) [9]. The parameter b is the hole affinity constant and characterizes the tendency of a given penetrant to sorb into the excess unrelaxed volume in the nonequilibrium matrix [10].

This sorption is characterized by a model which is the combination of two isotherms, a Henry type for matrix absorption and a Langmuir type for site sorption. The solubility coefficient is this case:

$$S = k_d + \frac{C'_H b}{1 + bp}. \tag{20.12}$$

The gas transport behavior in glassy polymers is described by the dual mobility model, which assigns two different diffusivities to the molecules absorbed by different mechanisms:

Fig. 20.2 The three sorption isotherms

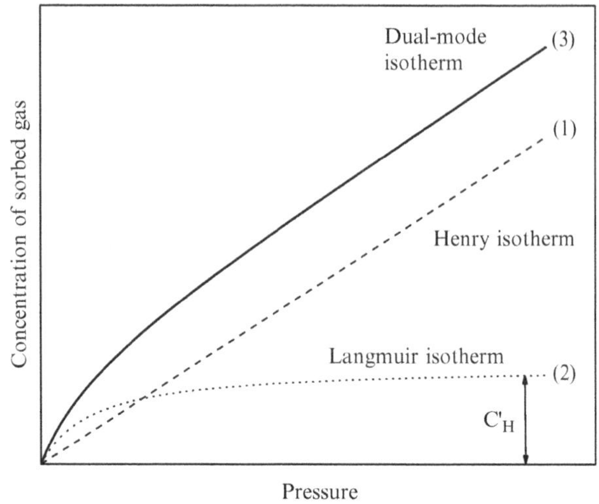

$$J = -D_D \frac{dC_D}{dx} - D_H \frac{dC_H}{dx}, \tag{20.13}$$

where the first term represents the flux due to Henry's law population, while the second is due to Langmuir population. The permeability coefficient can be written in the form of

$$P = k_D D_D + D_H \frac{bC_H^{'}}{1 + bp}. \tag{20.14}$$

Rearranging the terms of Eq. 20.14, this becomes

$$P = k_D D_D + \left(1 + \frac{FK}{1 + bp}\right), \tag{20.15}$$

where

$$F = D_H / D_D \text{ and } K = bC_H^{'} / k_D.\text{.}$$

The ideal selectivity of a membrane between gases A and B is the ratio of their gas permeabilities:

$$\alpha_{A/B} = \left(\frac{D_A}{D_B}\right) \times \left(\frac{S_A}{S_B}\right). \tag{20.16}$$

Solubility selectivity is determined by the relative affinity between the gases and the polymer matrix, while diffusion selectivity depends on the size difference between the molecules of the gases and the minimum average free volume hole size of the polymer.

3 Measurements of the Transport Coefficients

3.1 Permeation Cell Procedure

The transport coefficients can be measured using the permeation cell method or a gravimetric technique in sorption/desorption experiments. In the first case, one can use an arrangement such as shown in Fig. 20.3. A membrane of thickness d and area A separates the cell into two compartments. The gas is introduced in the left compartment (Fig. 20.3) at concentration c_0 under constant pressure p. The concentration of the gas at the other side of the membrane, c_d, is monitored as a function of time.

By integrating Eq. 20.8 from time 0 to t, and since $J(t) = dc/Adt$, with A the area of the membrane, one obtains the concentration, $c_d(t)$ (in mol/m^3):

$$c_d(t) = \frac{ADS_p}{Vd}\left[t - \frac{d^2}{6D} - \frac{2d^2}{\pi^2 D}.\sum_{m=1}^{\infty}(-1)^m \frac{\exp\left(-m^2\pi^2 Dt/d^2\right)}{m^2}\right]. \tag{20.17}$$

For long times, Eq. 20.17 becomes

$$c_d(t) = \frac{ADS_p}{Vd}(t - t_L) \tag{20.18}$$

$$\text{with} \quad t_L = \frac{d^2}{6D}. \tag{20.19}$$

The characteristic time, t_L, is a measure of the time that is required to establish a constant flow through the membrane. The permeability coefficient,

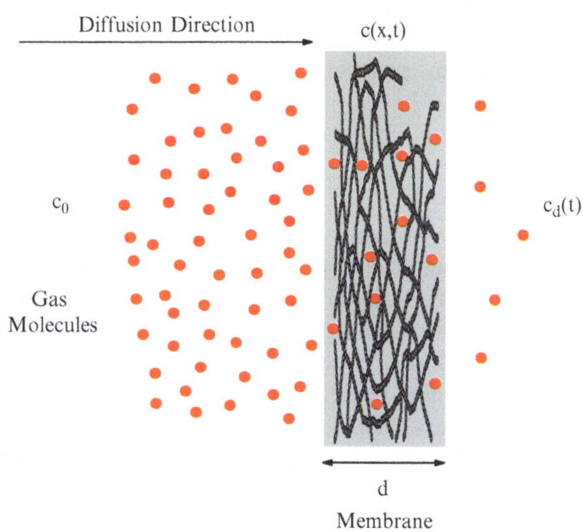

Fig. 20.3 Schematic representation of the permeation cell

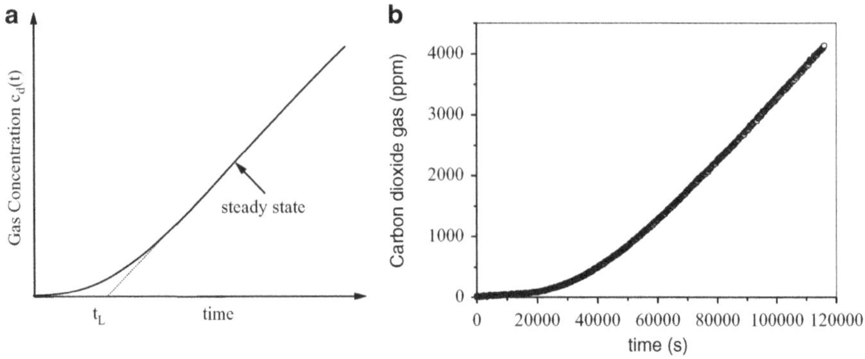

Fig. 20.4 (**a**) The theoretical curve of the concentration of the gas that permeates the membrane as a function of time based on Eq.20.17. The linear part is the steady-state line. (**b**) Experimental curve for a nanocomposite membrane measured by the permeation cell procedure

P (in mol Pa^{-1} m^{-1} s^{-1}), is the product of the diffusion coefficient, D (in m^2s^{-1}), and the sorption coefficient or solubility, S (in mol m^{-3} Pa^{-1}):

$$P = D \cdot S. \tag{20.20}$$

By plotting c_d versus t for sufficiently long times to reach a linear response, we can determine the time lag t_L, as shown in Fig. 20.4, and then the diffusion coefficient D from Eq. 20.19. The slope of the steady-state line gives the product $D \cdot S$, i.e., the permeability coefficient.

The concentration of the gas that permeates the membrane, $c_d(t)$, is recorded as a function of time using a gas sensor or a differential pressure gauge at the right compartment of the cell. In order to prevent external driving forces on the diffusion process, the pressure in both chambers is initially set to be the same.

Equilibrium permeation rates follow the classical Arrhenius temperature dependence [11]:

$$P = P_0 \exp\left(-E_p/RT\right), \tag{20.21}$$

where P_0 is a constant for a given penetrant/polymer pair, E_p is the activation energy ranging generally from 8 to 20 kCal, R is the gas constant, and T is the temperature. Permeation increases by 30–50 % for every 5 °C rise.

Permeation rates are normally given on a per-unit-thickness basis. For all polymers, P varies as t^{-x} where in the ideal case $x = 1$. Extrapolation and prediction of thicker or thinner films than those measured can be erroneous if the permeation versus thickness slope deviates from a value of 1. In most cases, however, x varies between 0.8 and 1.2 [11]. Therefore, a 2 mm film will have a permeability rate anywhere from 40 % to 60 % the value for a 1 mm film.

Possible reasons for such a behavior are [12]:

- Pore flow through defects of the skin layers of membranes can affect the permeation rates. These effects become progressively less important as the thickness of the film is increased.
- If the more permeable phase becomes continuous in a size scale of the membrane thickness, a transmitting channel across the membrane can be formed, which will strongly affect the permeation rate. On the other hand, the length of the diffusion pathways within the membrane can be different in films with different thickness.
- The glassy polymers are not in equilibrium state. Therefore, the rate that they were cooled from the rubbery state may influence their current properties, e.g., the free volume.

The free volume and intrinsic permeation rates are different in thin films from those in the bulk. The density of amorphous polymer films increases when their thickness is reduced. This is caused by the diffusion of excess free volume elements to the film surface during the process of film formation.

At $T, P = constant$, and for long enough measuring times (up to a few days), all changes in the polymer matrix will have taken place, even for the long relaxation times. These relaxations are initiated by the local concentration of the sorbed penetrant molecules. The steady-state quantities P, D, and S may represent the gas transport in the quasi-equilibrium of the steady state.

In contrast, for permeation experiments in glassy polymers, the measuring time is in general too short (some minutes up to few hours) for an equilibration of the polymer matrix after uptake of the penetrant molecules. Because of the reduced chain mobility below T_g, one obtains only an apparent or effective diffusion coefficient D_{eff} in the experiment. This coefficient depends, in general, on concentration and is usually not independent of the measuring time [13].

It should be noted that in the case of a glassy polymer membrane, the permeation cell procedure gives, in fact, the effective transport coefficients, which include the contributions of both Henry and Langmuir populations.

3.2 Sorption/Desorption Experiments

On the other hand, sorption/desorption experiments are used to measure both the diffusion coefficient and the solubility and, therefrom, the permeability (Eq. 20.20). In the simplest form of a sorption experiment, the membrane remains initially under vacuum, and the gas is introduced and maintained at constant pressure. The gas is dissolved and diffuses into the membrane. The weight gain is estimated with a balance. The fractional mass uptake is reported as a function of time. Assuming a flat membrane of thickness d and uniform concentration of the gas in the membrane, there are two possible solutions for Eq. 20.4 [14]:

$$\frac{M_t}{M_{eq}} = 8\sqrt{\frac{Dt}{d^2}}\left[\frac{1}{\sqrt{\pi}} + 2\sum_{m=0}^{\infty}(-1)^m ierfc\left(\frac{md}{4(Dt)^{\frac{1}{2}}}\right)\right], \tag{20.22}$$

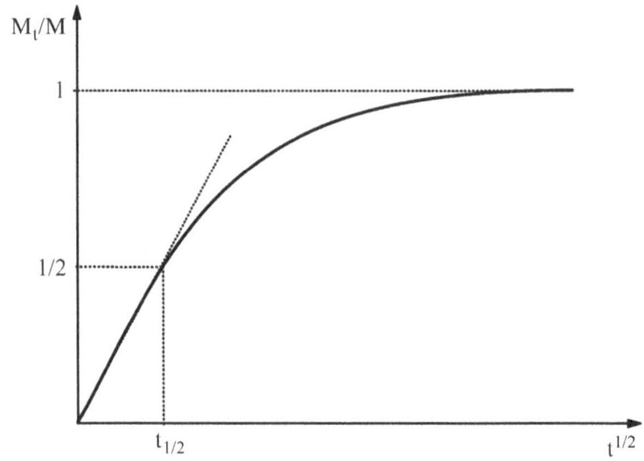

Fig. 20.5 Plot of M_t/M_{eq} as a function of $\sqrt{t}$. The characteristic time $t_{1/2}$ is the half time for sorption and corresponds to the initial linear part of the curve

$$\frac{M_t}{M_{eq}} = 1 - \frac{8}{\pi^2} \sum_{m=0}^{\infty} \frac{1}{(2m+1)^2}.\exp\left[-\frac{D(2m+1)^2\pi^2 t}{d^2}\right], \tag{20.23}$$

where M_t and M_{eq} are the mass uptakes at times t and at equilibrium and $ierfc$ is the error function integral.

Equation 20.22 converges rapidly for short times and can be approximated then by

$$\frac{M_t}{M_{eq}} = \frac{8}{\pi^{\frac{1}{2}}} \left(\frac{Dt}{d^2}\right)^{\frac{1}{2}}. \tag{20.24}$$

By plotting M_t/M_{eq} against $\sqrt{t}/d$, we obtain a straight line until $t_1/_2$, the half-life time when $M_t/M_{eq} = 1/2$ (Fig. 20.6). The diffusion coefficient is calculated from the slope of this line (Fig. 20.5).

Alternatively, Eq. 20.23 converges rapidly for long times. Keeping, then, only the first term and taking the logarithms on both sides, one obtains a straight line in the form

$$\ln\left(1 - \frac{M_t}{M_{eq}}\right) = \ln\left(\frac{8}{\pi^2}\right) - \frac{\pi^2 Dt}{d^2}. \tag{20.25}$$

By plotting the right-hand side against t, we can calculate D from the slope of this line.

Applying these methods, the diffusion coefficient can be determined in isothermal and isobaric sorption/desorption experiments. These methods present, however, an uncertainty in the time interval over which they are valid. A more detailed description of methods that are used to extract the diffusion coefficients from gravimetric data is presented in [15, 16]. Generally, the diffusion mechanism follows equation

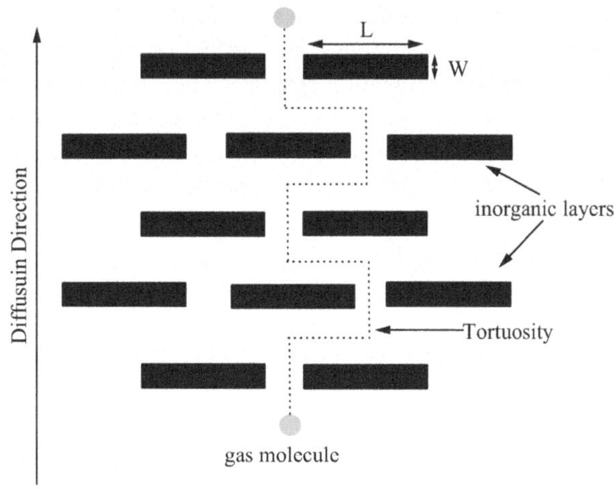

Fig. 20.6 The ideal case of rectangular dispersed nanoplatelets in a periodical arrangement perpendicular to the diffusion direction

$$\frac{M_t}{M_{eq}} = Kt^n, \tag{20.26}$$

where K depends on the structural features of the polymer and n determines the diffusion mechanism [17]. When $n = 1/2$, the transport is Fickian, as it was found previously, while $n = 1$ corresponds to the non-Fickian transport, in which the diffusion rate is rapid compared with the relaxation processes. The case $0.5 < n < 1$ is the anomalous transport behavior and corresponds to the situation where the diffusion rate is comparable with the polymer relaxation. The values of K and n can be determined by plotting log (M_t/M_{eq}) versus log t:

$$\log \frac{M_t}{M_{eq}} = \log\ K + n\ \log\ t. \tag{20.27}$$

The solubility coefficient is found by measuring the concentration of the soluble gas as a function of pressure $C = f(p)$ at constant temperature. By fitting these curves with Eq. 20.9, the parameters k_D, b, C'_H can be extracted. Alternatively, it is possible to deduce the solubility by the relation [18]:

$$S = \frac{M_{eq}}{M_p}, \tag{20.28}$$

where M_{eq} is the mass of the gas sorbed at equilibrium and M_p is the mass of the polymer. The solubility is then expressed as (g of sorbed gas)/(g of polymer).

4 Permeability Models in Polymer-Clay Nanocomposites

In polymer-clay nanocomposite, the dispersed phase consists of inorganic crystalline layers. The gas molecules permeability in these crystals is orders of magnitude lower than that in the amorphous polymer matrix. It is obvious, therefore, that the nanoplatelets would act as barriers to the diffusion of gas molecules through the nanocomposite membrane. This effect is called tortuosity and is the key factor controlling the permeation characteristics of these materials. Thus, in most theoretical treatises, the nanocomposite is considered to consist of a permeable phase (polymer matrix) in which non-permeable nanoplatelets are dispersed. There are three main factors that influence the permeability of a nanocomposite: the volume fraction of the nanoplatelets, their orientation relative to the diffusion direction, and their aspect ratio.

It is generally accepted that the transport mechanism within the polymer matrix follows Fick's law and that the matrix maintains the same properties and characteristics as the neat polymer. Therefore, a decrease of the solubility is expected in the nanocomposite due to the reduced polymer matrix volume as well as a decrease in diffusion due to the more tortuous path for the diffusing molecules.

The reduction of the diffusion coefficient is higher than that of the solubility coefficient. Indeed, the volume fraction of nanoplatelets is low, and, thus, the reduction of the matrix volume is small. The major factor, then, is the tortuosity, which is connected directly to the shape and the degree of dispersion of the nanoplatelets.

The degree of dispersion of the nanoplatelets is determined by the degree of delamination of the clay. The fully delaminated (exfoliated) nanocomposite presents much higher values for the tortuosity factor and the aspect ratio in comparison with the partially delaminated (intercalated) nanocomposite, and it is much more effective to be used in barrier membranes for gases.

The terms *intercalated* and *exfoliated* layered silicate nanocomposites are often used to describe two extreme states of silicate layer organization in the composite morphology. Intercalation implies the insertion of polymer chains in the galleries of the initial-layered tactoids. This leads to a longitudinal expansion of the galleries. Exfoliation implies complete breakage of the initial layer stacking order and homogeneous dispersion of the layers in the polymer matrix.

Several models have been developed in order to describe the mass transfer within the nanocomposites. Most models assume that the platelets have a regular and uniform shape (rectangular, sanidic, or circular) and form a regular array in space. They either are parallel to each other or have a distribution of orientations with the average orientation at an angle to the main direction of diffusion of the gas molecules. Some of these models are outlined in the following.

4.1 Regular Arrangement of Parallel Nanoplatelets

One of the first attempts to describe the permeability of membranes, where a second phase is dispersed in a regular arrangement, was made by Barrer [19]. These authors calculated the diffusion through a regular array of parallelepipeds of a second phase dispersed in a matrix with a different diffusion coefficient. When this approach is applied in the case of dispersed impermeable thin plates, the change of the permeability is found to be proportional to the fractional cross section (slit) that is available for the diffusant to move forward and depends on the tortuosity of the path. A simple permeability model for a regular arrangement of platelets has been proposed by Nielsen [20] and is presented in Fig. 20.6. The nanoparticles are evenly dispersed and considered to be rectangular platelets with finite width, L, and thickness, W. Their orientation is perpendicular to the diffusion direction.

The solubility coefficient of this nanocomposite is

$$S_{\text{composite}} = S_{\text{matrix}}(1 - \phi), \qquad (20.29)$$

where S_{matrix} is the solubility coefficient of the neat polymer and ϕ is the volume fraction of the nanoplatelets that are dispersed in the matrix. In this approximation, the solubility does not depend on the morphological features of the phases.

The platelets act as impermeable barriers to the diffusing molecules, forcing them to follow longer and more tortuous paths in order to diffuse through the nanocomposite. The diffusion coefficient, $D_{\text{composite}}$, is influenced by the tortuosity, τ:

$$D_{\text{composite}} = \frac{D_{\text{matrix}}}{\tau}, \qquad (20.30)$$

where D_{matrix} is the diffusion coefficient of the matrix. The factor τ depends on the aspect ratio, the shape, and the orientation of the nanoplatelets, and it is defined as

$$\tau \equiv \frac{l'}{l}, \qquad (20.31)$$

where l' is the distance that a solute must travel to diffuse through the membrane when nanoplatelets are present and l is the membrane thickness. From Eqs. 20.20, 20.29, and 20.30

$$\frac{P_{\text{composite}}}{P_{\text{matrix}}} = \frac{1 - \phi}{\tau}. \qquad (20.32)$$

The prolonged diffusion length l' is estimated as follows: Each nanoplatelet contributes to the enhancement of the diffusion length by $L/2$ on the average. If $\langle N \rangle$ is the mean number of nanoplatelets that a solute encounters as it diffuses through the membrane, then

$$l' = l + \langle N \rangle \frac{L}{2}. \qquad (20.33)$$

Since $\langle N \rangle = l\phi/W$, the factor τ becomes

$$\tau = 1 + \frac{L}{2W}\phi. \qquad (20.34)$$

Equation 20.32, then, gives

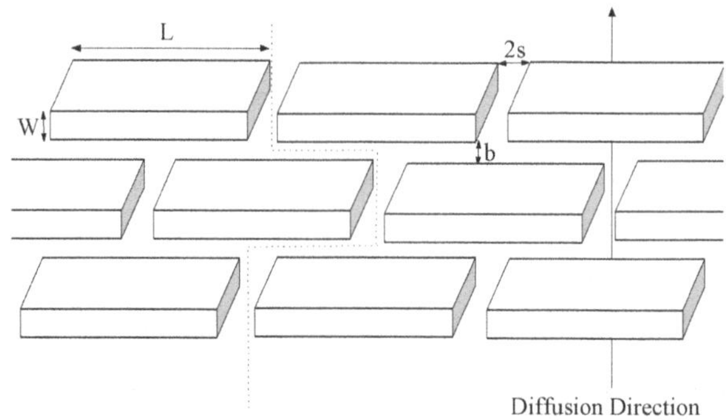

Fig. 20.7 The 2D diffusion case. Rectangular dispersed nanoplatelets in a periodical arrangement perpendicular to the diffusion direction. Some characteristic dimensions are shown

$$\frac{P_{\text{composite}}}{P_{\text{matrix}}} = \frac{1-\phi}{1+\frac{\alpha}{2}\phi}, \tag{20.35}$$

where $\alpha = L/W$ is the aspect ratio of the nanoplatelets. This equation shows that the permeability of the nanocomposite decreases with the increase of ϕ and a. In practice, however, the limit for its validity is $\phi \leq 10\%$, because the particles have a tendency to aggregate which increases with ϕ.

The diffusion through a multi-perforated single laminae was studied by Wakeham and Mason [21]. These authors found that the resistance to diffusion through such a membrane had a contribution from the need of the diffusant to enter the constriction into the pore/slit and a contribution due to the length of the pore. This approach was extended by Cussler et al. [22] to multiple layers of parallel platelets of infinite third dimension (2D diffusion), separated by slits. The result was

$$\frac{P_{\text{composite}}}{P_{\text{matrix}}} = \left(1 + \frac{LW}{2s(W+b)} + \frac{L^2}{4b(W+b)} + \frac{2b}{l}\ln\left(\frac{L}{4s}\right)\right)^{-1}, \tag{20.36}$$

where l is the thickness of the membrane and the other dimensions are shown in Fig. 20.7. In this arrangement, $LW/2[(L/2+s)(W+b)] = \phi$ is the volume fraction of the platelets, $L/2\ W = \alpha$ is the particle aspect ratio, and s/W $= \sigma$ is the slit aspect ratio. Cussler et al. [22] neglected the second term of Eq. 20.36, since it is much smaller, and, for narrow slits ($\sigma \ll 1$), suggested the following simpler expression for the enhancement of the barrier properties of the membrane:

$$\frac{P_{\text{composite}}}{P_{\text{matrix}}} = \left(1 + \frac{\alpha^2\phi^2}{1-\phi}\right)^{-1}. \tag{20.37}$$

This model predicts a faster reduction of the relative permeability at small values of ϕ [22] than the model of Nielsen (Eq. 20.35) [20], which needs either higher volume fraction or higher aspect ratio for the same reduction. This is probably attributed to the different regions over which the above models are applicable.

Two variations of Eq. 20.36 were given by Falla et al. [23] based on Cussler et al. [24] and Wakeham and Mason [21]:

$$\frac{P_{\text{composite}}}{P_{\text{matrix}}} = \left(1 + \frac{\alpha^2\phi^2}{1-\phi} + \frac{\alpha\phi}{\sigma} + \frac{4\alpha\phi}{\pi(1-\phi)}\ln\left[\frac{\pi\alpha^2\phi^2}{\sigma(1-\phi)}\right]\right)^{-1}. \quad (20.38)$$

$$\frac{P_{\text{composite}}}{P_{\text{matrix}}} = \left(1 + \frac{\alpha^2\phi^2}{1-\phi} + \frac{\alpha\phi}{\sigma} + 2(1-\phi)\ln\left[\frac{1-\phi}{2\sigma\phi}\right]\right)^{-1}. \quad (20.39)$$

The second term of these equations, which involves α^2, reflects the contribution of the tortuous path of the diffusant through the crystalline plates. The third term is due to the resistance to diffusion of the slits and depends on their aspect ratio, $\sigma = W/2s$. The last term represents the constriction from the wide space between the plates into the narrow slits, and it should depend on the constriction ratio, α/σ, as it is the case in Eq. 20.38. This term was derived from arguments presented by Aris [24] for Eq. 20.38 and Wakeham and Mason [21] for Eq. 20.39.

Hexagonal flakes arranged in regular parallel arrays were examined by Moggridge et al. [71]. The same arguments as for Eq. 20.37 were used, and the reduction in permeability was found to be

$$\frac{P_{\text{composite}}}{P_{\text{matrix}}} = \left(1 + \frac{2}{27}\frac{\alpha^2\phi^2}{1-\phi}\right)^{-1}. \quad (20.40)$$

The difference due to the specific platelet shape is reflected by the coefficient (2/27), which reduces the effectiveness of the barrier.

4.2 Random Spatial Positioning of Parallel Plates

For the case of slits formed between randomly positioned parallel platelets of infinite third dimension, Cussler et al. [22] suggested the following expression for the permeability:

$$\frac{P_{\text{composite}}}{P_{\text{matrix}}} = \left(1 + \frac{\mu'\alpha^2\phi^2}{1-\phi}\right)^{-1}, \quad (20.41)$$

where μ' is a combined geometrical factor, characterizing also the randomness of the porous media [24].

In a similar case, Lape et al. [26] considered rectangular nanoplatelets of the same aspect ratio arranged randomly but parallel to each other and perpendicular to the diffusion direction. The reduction of the permeability in the nanoplatelet-reinforced membrane is given by the product of the reduced area and the increased path length:

$$\frac{P_{\text{composite}}}{P_{\text{matrix}}} = \left(\frac{A_f}{A_0}\right) \cdot \left(\frac{l}{l'}\right) \tag{20.42}$$

The distance that the solute molecule diffuses through the nanocomposite becomes in this case

$$l' = l + \langle N \rangle \langle d \rangle \tag{20.43}$$

which is the same with Eq. 20.33, except that $L/2$ is replaced by $\langle d \rangle$, the average distance the solute must travel to reach the edge of platelet. Using simple statistical considerations, $\langle d \rangle$ can be estimated considering that the molecule hits the platelet at a random point along its length and that the resistance to mass transfer is proportional to the length of the path traveled. Thus, l' becomes

$$l' = \left(1 + \frac{1}{3}\alpha\phi\right) l. \tag{20.44}$$

Equation 20.44 differs from Eq. 20.34 by a factor 2/3, which is due to the randomness of the positions of the nanoplatelets.

The area available for diffusion, A_f, can be determined by dividing the volume available for diffusion by the distance traveled to cross the membrane:

$$\frac{A_f}{A_0} = \frac{(V_m - V_f)/l'}{V_m/l}, \tag{20.45}$$

where V_m is the total volume of the membrane and V_f is the volume of all platelets. The relative permeability then becomes

$$\frac{P_{\text{composite}}}{P_{\text{matrix}}} = \frac{1 - \phi}{\left(1 + \frac{\alpha}{3}\phi\right)^2}. \tag{20.46}$$

The case where the nanoplatelets are circular disks with radius R, thickness W, and aspect ratio $\alpha = R/W$ (Fig. 20.8) was examined by Fredrickson and Bicerano [27] using multiple scattering theory. When the disks were spaced by an average distance exceeding R (dilute regime), then the diffusion coefficient is given by

$$\frac{D_{\text{composite}}}{D_{\text{matrix}}} = \frac{1}{1 + k\alpha\phi}, \tag{20.47}$$

where $k = \pi/\ln \alpha$. This case corresponds to low values of the volume fraction and the aspect ratio of the nanoplatelets ($\alpha\phi \ll 1$) and presents a form similar to Nielsen's equation (Eq. 20.35). For low values of 0 but high values of α ($\alpha\phi \ll 1$), the disks are overlapping (semi-dilute regime) and the relation becomes

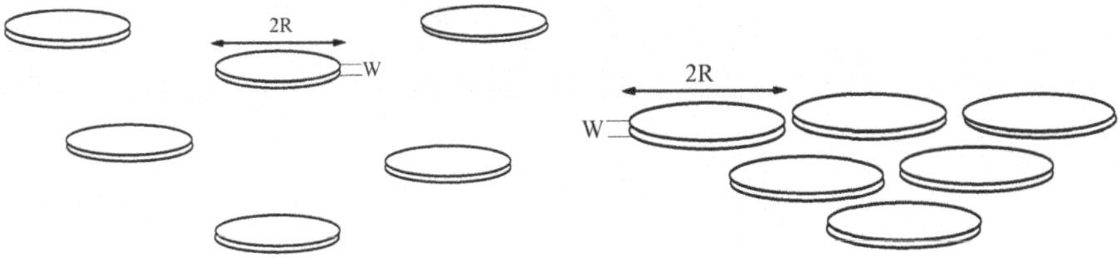

Fig. 20.8 The dilute regime where the distances of the disks exceed R. The semi-dilute regime where the disks are overlapped

$$\frac{D_{\text{composite}}}{D_{\text{matrix}}} = \frac{1}{1 + \mu\alpha^2\phi^2}, \tag{20.48}$$

where $\mu = \pi^2/16 \ln^2\alpha$. This equation has a similar form with the equation of Cussler (Eq. 20.37).

In a different approach, Gusev and Lusti [28] developed a 3D computational model for a random array of parallel circular disks. This model is based on the solution of the Laplace equation for the local chemical potential:

$$\nabla P(r) \cdot \nabla \mu = 0 \tag{20.49}$$

with a space-dependent local permeability coefficient. The latter is assumed to have a value of zero inside the platelets and be equal to P_{matrix} anywhere else. The proposed expression for the global permeability becomes

$$\frac{P_{\text{composite}}}{P_{\text{matrix}}} = \exp\left[-\left(\frac{\alpha\phi}{x_0}\right)^{\beta}\right]. \tag{20.50}$$

The parameters β and x_0 are usually evaluated by fitting experimental data [28]. The values of $\beta = 0.71$ and $x_0 = 3.47$ were used by Picard et al. [29] in order to obtain similar results as with the model of Nielsen (Eq. 20.35) for the diffusion of several gases in nylon-6/montmorillonite composites.

All the models above assume nanoplatelets of uniform size. If, instead, the platelets have a continuous distribution of sizes, $g(R)$, e.g., a Gaussian distribution of average size $\overline{R}$, standard deviation σ'

$$g(R) = \frac{1}{\sqrt{2\pi}\sigma'}\exp\left(-\frac{\left(R-\overline{R}\right)^2}{2\sigma'^2}\right), \tag{20.51}$$

and constant thickness, W, then the mean number of nanoplatelets of size $2R$ encountered by the solute becomes [26]

$$\langle N(R)\rangle = \frac{\phi l}{W}\cdot\frac{g(R)RdR}{\int_0^\infty g(R)RdR}. \tag{20.52}$$

This gives

$$\frac{P_{\text{composite}}}{P_{\text{matrix}}} = \frac{1-\phi}{1+\frac{2}{3}\frac{\phi}{W\overline{R}}\left(\overline{R}^2+\sigma'^2\right)}. \tag{20.53}$$

If one compares this case with the case of nanoplatelets with a uniform size $R=\overline{R}$, then

$$\frac{P_{uni}}{P_{polydisp}} \approx \left[1+\left(\frac{\sigma'}{\overline{R}}\right)^2\right]^2. \tag{20.54}$$

Increasing the polydispersity of the size distribution increases the standard deviation and reduces the permeability. Thus, the barrier properties of the nanoplatelets with polydispersed sizes are better than these of monodispersed sizes of the same mean. Picard et al. [29] modified the model of Lape et al. [26] (Eq. 20.46), taking into account also the polydispersity in the thickness of the particles

$$\frac{P_{\text{composite}}}{P_{\text{matrix}}} = \frac{1-\phi}{\left[1+\frac{1}{3}\phi\frac{\sum n_i\left(\frac{w_i}{t_i}\right)^2}{\sum n_i\frac{w_i}{t_i}}\right]^2}, \tag{20.55}$$

where w_i and t_i are the width and the thicknesses of the fractions i of the platelets. This model [29] is more appropriate for the cases of high loading of the impermeable phase, where there is a distribution in the values of the aspect ratio, due to the presence of agglomerates.

The comparison of the models that have been presented so far is not straightforward because the definition of the aspect ratio of the platelets is not common in all the models. Some authors define this ratio as the half-width-to-thickness ratio, while others as the width-to-thickness ratio. Adopting the latter as the typical definition of the aspect ratio, some of the models should be modified accordingly. In the model proposed by Fredrickson-Bicerano (Eqs. 20.47 and 20.48), where the platelets are considered to be circular disks, the relation between the disk aspect ratio and a rectangular platelet aspect ratio can be derived by comparing the area of the disk (πR^2) to the area of a rectangular platelet (L^2). Thus, $\alpha_{disk}=\alpha/\sqrt{\pi}$ in this model. The coefficients κ and μ should also be modified accordingly.

When all the models are compared, it can be concluded that (relatively) large volume fractions or large aspect ratios are needed for a significant reduction of the permeability. The predictions for the trend of this reduction of the several models differ mainly in the area of low volume fractions and aspect ratios. The best way to compare the predictions of the models, then, is to use the product $\alpha\phi$ as the significant parameter of the membrane.

The models of Nielsen (Eq. 20.35) and of Fredrickson-Bicerano (Eq. 20.47) predict a stronger effectiveness of the platelets in reducing the permeability. In semi-dilute systems and for intermediate aspect ratios ($\alpha \sim 100$), the model of Fredrickson-Bicerano (Eq. 20.48) seems to underestimate the barrier properties, while Cussler's model overestimates them. For higher aspect ratios, Gusev's model predicts almost zero permeability for $\alpha\phi > 30$. As this product goes to higher values (e.g., $\alpha\phi > 60$), the absolute values of the permeability predicted by all the models become very low, in the range of 1–3 % of the value of the unfilled polymer.

From a practical point of view, complete exfoliation is difficult to be achieved at high clay fractions, and the actual aspect ratios are much lower than what is needed for efficient enhancement of the barrier properties. For intermediate loadings, and looking at the comparisons with the available experimental data, it seems that the simple model of Nielsen is adequate to describe the reduction of the relative permeability.

A common feature in all the models that were presented above is the dependence of the permeation properties of the nanocomposite on the volume fraction and the aspect ratio of the reinforcing inorganic phase. In fact, the predictions of all the models are similar. When used in the reverse way, i.e., to predict the aspect ratio from the measured permeation properties, however, these models may give different results [29].

The average aspect ratio of the particles is difficult to determine experimentally, and, therefore, the comparison of the models is not straightforward. It has been observed that during the preparation of the nanocomposites, the obtained aspect ratio depends on the volume fraction. In chlorobutyl/montmorillonite nanocomposites, e.g., the obtained aspect ratio decreases with increasing ϕ, using either modified or nonmodified particles [30]. Obviously, increasing ϕ results in a reduction of the degree of delamination and, consequently, in the reduction of the final particle aspect ratio.

The decrease of the relative permeability is attributed in all models mainly to the inorganic phase. If their aspect ratio and volume fraction are the same, the particles will induce the same relative reduction of the permeability in any matrix they are dispersed, provided that the same degree of delamination is obtained. Further, if there are no special interactions between the solute gas and the particles, the relative enhancement of the barrier properties is not affected by the kind of the gas that is used [31, 32]. In other words, according to the models, the improvement of the barrier properties of the material is determined solely by the dispersed inorganic phase.

Comparisons of the results of Eqs. 20.38 and 20.39 with Monte Carlo calculations for the diffusion through membranes containing impermeable flakes were conducted by Falla et al. [23] and Swannack et al. [33]. Both papers reported rather good agreement between the theoretical and the simulation results, verifying the tortuous paths around the flakes and the effects of the constrictions into the slits and the diffusion through them. Falla et al. [23] showed that Eq. 20.38 is in general more sensitive to changes in the flake arrangement that will influence the diffusion mechanism and can predict the resulting changes in the permeability of the membrane more accurately. Swannack et al. [33] indicated the need for 3D calculations and showed that Eq. 20.38 overpredicts the barrier effect of the membrane in this case, as the transverse slits are omitted.

4.3 Arrangement of the Plates at Angles $\theta \neq 90^{o}$ to the Main Diffusion Direction

The tortuosity, τ, is the highest when the nanoplatelets are aligned perpendicular to the diffusion direction. For other orientation angles, Nielsen's equation should be modified accordingly.

For nonuniform orientation of the platelets, an order parameter, S', is introduced to quantify the degree of their orientation around the diffusion direction [34]:

$$S' = \frac{1}{2}\langle 3\cos^2\theta - 1\rangle, \tag{20.56}$$

where θ is the angle between the diffusion direction and the unit vector normal to the surface of a platelet. The average is taken over all platelets with all possible orientations. When all platelets are parallel to the direction of diffusion ($\theta = 0$), then the order parameter is $S' = -1/2$. When $\theta = \pi/2$, the orientation of the platelets is perpendicular to the diffusion direction and $S' = 1$. For random orientation, $S' = 0$ (Fig. 20.9).

For nonuniform orientation, Nielsen's equation takes the following form [34]:

$$\frac{P_{\text{composite}}}{P_{\text{matrix}}} = \frac{1-\phi}{1+\frac{\alpha}{2}\frac{2}{3}\left(S'+\frac{1}{2}\phi\right)}. \tag{20.57}$$

Obviously, the case $S' = 1$ corresponds to the maximization of the tortuosity factor and, therefore, to the greatest reduction of the permeability. The term $\frac{2}{3}(S'+1/2)$ ($= \langle\cos^{2}\theta\rangle$ from Eq. 20.56) takes the value 1/3 for $S' = 0$. The predictions are then similar to the ones of the models of Lape et al. [26] and Fredrickson and Bicerano [27] for random spatial positioning of clay layers.

An empirical relation for the prediction of the permeability in nanocomposites with randomly (3D) oriented platelets has been proposed by Maksimov et al. [35] in the form of

$$P_{\text{composite}} = \frac{1}{3}\left(P_{\parallel} + 2P_{\text{matrix}}(1-\phi)\right), \tag{20.58}$$

where $P_{\parallel}$ is the permeability of the nanocomposite as given by the model of Nielsen for oriented platelets. The second term corresponds to the permeability of the

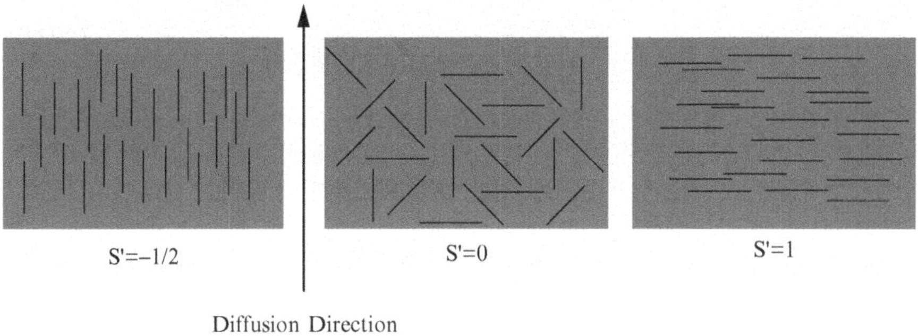

Fig. 20.9 Values of the order parameter S' for three orientations of the particles

composites with platelets oriented parallel to the diffusion direction, and it is assumed to be equal to the permeability of the matrix corrected for its volume fraction.

4.4 Influence of Particle Aggregation

A crucial factor that affects the permeation properties of the nanocomposite is the aggregation of silicate layers. This leads to a decrease of the aspect ratio of nanoparticles. Manninen et al. [6] indicated that the processing method used to prepare the nanocomposites may result in agglomeration of the organoclay layers. These agglomerates may cause the formation of large-scale holes (pores) in the matrix, which can act as low-resistance pathways for gas transport within the nanocomposite. In such a case, the diffusion becomes mostly of the Knudsen flux type.

For high clay contents, it is difficult to keep a high degree of platelet dispersion and avoid the presence of intercalated structures. Nazarenko et al. [36] modified Nielsen's equation to take into account the case of stacks of layers (aggregates) dispersed homogeneously in the polymer matrix and oriented perpendicular to the diffusion direction. The proposed expression for the reduction in permeability is

$$\frac{P_{\text{composite}}}{P_{\text{matrix}}} = \frac{1-\phi}{1+\frac{\alpha}{2N}\phi} \tag{20.59}$$

where N is the number of layers in the layer stack. Obviously, the value $N = 1$ corresponds to complete layer delamination (exfoliation). Again, the larger the degree of stacking, i.e., the higher the value of N, the less efficient is the enhancement of the barrier properties.

One observes that Eqs. 20.57 and 20.59 have a similar form. The former accounts for the orientation of the platelets, while the latter for their degree of stacking. By combining these two equations, we can obtain a single equation that might predict the change in permeability as a function of the volume fraction, the geometry, the orientation, and the degree of stacking of the nanoplatelets:

$$\frac{P_{\text{composite}}}{P_{\text{matrix}}} = \frac{1-\phi}{1+\frac{\alpha}{3N}\left(S' + \frac{1}{2}\right)\phi}. \tag{20.60}$$

In a different approach, Bhattacharya et al. [37] introduced a dimensionless corrective function ψ to accommodate the shape and aggregation effects of the particles instead of aspect ratio and modified Eqs. 20.46 and 20.50 as

$$\frac{P_{\text{composite}}}{P_{\text{matrix}}} = \exp\left[-\left(\frac{\psi\phi}{x_o}\right)^{\beta}\right] \tag{20.61}$$

$$\frac{P_{\text{composite}}}{P_{\text{matrix}}} = \frac{1-\phi}{\left(1+\frac{\psi}{3}\phi\right)^2} \tag{20.62}$$

with $\psi = \rho \cdot \zeta \cdot \xi \cdot \phi$ where ρ is the filler density, ζ is the specific surface area, ξ is the correlation length between the nanoparticles (which can be determined by statistically analyzing TEM micrographs and is at least 100 nm for most nanocomposites), and ϕ is the filler volume fraction.

5 Transport Coefficients and Free Volume

All the above discussions were based on the fact that the incorporation of the particles does not affect the physical characteristic of the matrix. However, this is not always the case since the inorganic phase often alters the free volume properties of the polymer in which it is dispersed. Consequently, the transport properties of the nanocomposite can be modified in such a way depending on the specific impact of the nanolayers on the polymer free volume.

5.1 Free Volume

The term "free volume" in a polymer refers to the volume of the mass that is not actually occupied by the molecules themselves. This is really the room that the parts of the chains have to move around. The concept of the free volume is not uniquely defined. Various definitions are often used, such as hole, configurational, fluctuation, and excess free volume [38]:

Hole free volume: Even if the polymer chains were perfectly aligned, there would be free space between them. This is called "hole free volume."

Configurational free volume: Due to the insufficient chain packing, an additional free space is created.

Fluctuation free volume: The polymer chains or side chains are not fixed but can move (rotate, vibrate, etc.) randomly due to their thermal activation. These motions of the side chains generate transient gaps, which create an extra free volume that is called "fluctuation free volume."

Excess free volume: At temperatures below T_g, the polymer has a specific volume greater than what it would have had if it had been allowed to cool slowly under equilibrium conditions. This difference between the equilibrium free volume and the actual volume under a given quenched condition is called "excess free volume."

As time passes, the polymer molecules will try to rearrange themselves to approach the equilibrium density of the material, thus eliminating the excess free volume.

The sum of the hole free volume and the configurational free volume is often stated as static free volume, while the fluctuation free volume is called dynamic free volume. The excess free volume may be static or dynamic depending on the time scale of the relaxation processes. The ratio of static to dynamic excess free volume is thought to be greater than 1, and it depends on temperature.

The static free volume can, therefore, be defined as

$$V_f = V - V_0, \tag{20.63}$$

where $V = 1/\rho$ is the specific volume of the polymer and V_0 is the volume occupied by the atoms of the polymer chains at 0 K. This volume is often calculated by the empirical relation

$$V_0 = 1.3V_w, \tag{20.64}$$

where the coefficient 1.3 is assumed to apply for all groups and structures and may sometimes result in a faulty interpretation and V_w is the volume calculated using the group contribution method [39] that employs the van der Waals volume of the groups contained in the polymer repeat unit.

$$V_0 = 1.3\sum_{i=1}^{N}(V_w)i, \tag{20.65}$$

where N is total number of groups into which the repeat unit structure of the polymer is divided.

The fractional free volume (FFV) is defined as

$$FFV = \frac{V_f}{V}. \tag{20.66}$$

In general, the polymer free volume can be assumed to consist of holes, the size, shape, and position of which vary with time. Around the glass transition temperature, T_g, the polymer undergoes a dramatic transformation, as the degrees of freedom of the molecular chains and the resulting mobility change considerably and abruptly. From macroscopic point of view, the T_g can be considered as an index of the polymer backbone segmental mobility. At $T < T_g$, the segments motion is almost frozen, and thermal expansion of free volume is almost impossible. In this case, we can assume that the size, shape, and position of the free volume holes remain constant (have very long relaxation times) and, therefore, a definite time-averaged size distribution of holes is established. Assuming spherical in shape holes, the distribution of hole sizes $F(V_h)$ is given by a Gaussian function

$$F(V_h) = \frac{N}{\sigma\sqrt{2\pi}} \exp\left[-\frac{(V_h - V_0)^2}{2\sigma^2}\right], \tag{20.67}$$

where N is total number of the holes and σ is the width of the Gaussian distribution given by

$$\sigma = \sqrt{\frac{\overline{V}_h RT_g}{B}}. \tag{20.68}$$

Here B is the bulk modulus at T_g and $\overline{V}_h$ is the average hole size [40].

It has been found that high T_g polymers (e.g., polymers containing aromatic structures in their backbones) consist of inflexible chains and are limited in their chain mobility, while low T_g polymers have, in general, more flexible chains and present higher segmental mobility. The higher the T_g, the greater the polymer chain stiffness and the broader the distribution of hole sizes, since the available variety of chain conformations is further restricted. A very nice and representative description has been given by Hensema et al. [41]:

> An amorphous matrix of rigid polymer molecules can be compared to a heap of random strands of uncooked spaghetti, requiring substantially more free volume than flexible polymers, comparable to cooked spaghetti, which can pack into a much tighter form. This difference in packing density will result in a different free volume and in its distribution. Flexible polymers will be able to achieve a narrow distribution, while rigid polymers will lack this possibility since these polymers are limited in achieving the same variety of conformations, resulting in a broader free volume distribution, through which the number of larger holes is substantially increased.

Amorphous polymers in the rubbery state have high chain mobility and short relaxation times. Above T_g a liquid-like motion of much longer segments of the backbone chain can take place. This motion is directly related to the increase in free volume caused by the thermal expansion of the system.

$$FFV = (FFV)_{T_g} + (T - T_g)a_{\mathrm{T}} \tag{20.69}$$

where α_T is the thermal expansion coefficient for free volume. According to the WLF theory, $(FFV)_{Tg} = 0.025$ and $\alpha_T = 4.8 \times 10^{-4}\ \mathrm{K}^{-1}$, [42, 43].

Moreover the position of the free volume holes changes with time, owing to the anharmonic oscillation of the polymer chains. That is, the holes can move freely inside the polymer, as no energy change is required for their redistribution [44]. While the activity of main chain segmental motion in the rubbery state results in the enlargement of free volume holes, the question remains on what happens to the number of these holes.

At a given temperature, the number of holes remains unchanged, while their size, shape, and position vary with time. As the temperature increases, the number of holes increases too, which means that additional free volume sites are created.

It is interesting to note that the increase of the number of holes is induced by thermal activation (Brownian segmental motion) and not by an orientational effect. For example, Xie et al. [45] measured the hole size and the number of the holes under mechanical deformation and observed that, while the size of the holes was sensitive to the applied stresses, the number of the holes remained unaffected.

Cross-linking does not contribute to changing the free volume size of the polymer in the glassy state, since the macromolecular motions are frozen (vitrified). However, cross-linking can decrease the size of free volume in the rubbery state, as it confines the mobility of chain segments. Obviously, the more free and mobile the polymer network (e.g., linear polymers), the more effective will be the reduction of free volume caused by cross-linking. On the other hand, in a rigid network structure (e.g., thermosetting polymer), the influence of cross-linking is expected to be smaller.

5.2 Effects of Nanoparticles on Polymer Free Volume

The incorporation of the inorganic nano-fillers into the polymer matrix will inevitably change the morphological features of the organic matrix and, consequently, the free volume properties of the primary polymer. The following effects are expected in the nanocomposite [38]:

Interfacial regions: Because of their large surface area, the inorganic particles create interfacial regions with the organic polymer. These regions are numerous and have finite thickness because of the weak interaction/miscibility between the inorganic reinforcement and the organic polymer and the need for compatibilization. The interfacial adhesion between the two compatibilized components leads to the formation of an interfacial layer, which is helpful for the stress transfer between the nanoparticles and the matrix. These regions, however, may contribute to the enhancement of the overall free volume, especially when the interaction is weak [46, 47].

Interstitial cavities in the filler agglomerates: In the case that the exfoliation is not complete, there will be stacks of nanolayers (intercalated morphology) inside the polymer. The interlayer spacing of these stacks varies from 2 to 5 nm, so it is expected that they contribute to the overall free volume of the composite. If the insertion/penetration of polymer chains in the galleries of these stacks is partial or incomplete, this extra free volume is in layers rather than spherical holes [48].

Insufficient chain packing: The filler nanoparticles can confine the polymer chains and disrupt their packing. The free volume can be enhanced due to this inefficient packing or due to the increase of the distances between polymer segments [25].

Changes of the free volume hole size distribution: The nanoparticles may alter the free volume hole size distribution. Winberg et al. [49] observed broader distribution in the PTMSP/silica nanocomposite than in the unfilled polymer. Similar results have been reported by Choudalakis et al. [50] in acrylic resin/laponite nanocomposites.

Changes of the crystallinity or cross-linking density of the matrix: The incorporation of nanoparticles in the polymeric matrix often reduces the degree of crystallinity or the cross-link density of the matrix polymer. Inevitably, this enhances the free volume. In general, the fillers may act in some cases in different ways depending on their volume fraction, shape, and possible physicochemical interactions with the matrix: They may serve as cross-linker, which decreases the polymer mobility; they may perturb the polymer chain packing and thus reduce the crystallinity in the polymer, increasing, thus, the chain mobility [51]; or they may also act as nucleating agents, increasing the degree of crystallinity and reducing the free volume.

All the above effects of the nanoparticles on the polymer free volume coexist in any nanocomposite system. The question usually is which of them are dominant. This is determined primarily from the degree of interaction between the two components, the volume fraction of the filler and the geometrical features of the particles. Often the dominant factor, which is the interaction, cannot be predicted and usually has to be decided instinctively.

5.3 Solubility and Free Volume

Generally speaking, relations between solubility and free volume are not always obvious, since the way penetrants dissolve in the polymer depends on the total pressure, the affinity between gas and polymer, and the temperature. However, for a specific gas-polymer system, these conditions are identical so only the free volume changes, due to the particles, affect the solubility coefficient. During adsorption, the gas molecules are positioned in the free volume holes of the polymer that are created by Brownian motions of the chains or by thermal perturbations.

Henry's sorption in rubbery polymers cannot be directly associated to the existing free volume hole structure because the penetrants are dissolved in the matrix in a similar mode to the one that occurs in liquids. Thus, Henry's constant depends on the facility of the gas molecule to penetrate in the space between the chains [52]. If we consider the case of holes that are large enough to accommodate gas molecules inside the polymer, then the solubility coefficient should reflect directly the number of the free volume holes and remain unaffected by their sizes. In an opposite case, the value of k_d increases with the increasing hole size as well.

On the other hand, Langmuir's sorption is principally governed by the available free volume. Consequently the variation in the solubility is attributed mostly to C'_H, which is determined by the difference $V_g - V_l$. This, in turn, is influenced by the glass transition temperature. Therefore, higher T_g polymers generally exhibit higher solubility coefficients. It can be said that the FFV is related to the solubility because it corresponds to the free space in the polymer matrix, where the gas molecules can locate [52].

The connection between the solubility coefficient and the number of free volume holes has been verified experimentally by Choudalakis et al. [50] in an acrylic resin/laponite nanocomposite. The free volume in a material can be measured by positron annihilation life time spectroscopy (PALS). In this method, the intensity of the peak corresponding to the lifetime of the ortho-positronium in the material, I_3, corresponds to the total free volume fraction [53]. The value of I_3 was measured as a function of laponite particles loading in acrylic resin nanocomposites and found to have an identical trend with the solubility coefficient, which was measured separately in a permeation cell [50].

5.4 Diffusion and Free Volume

In contrast to the solubility, the diffusion is a kinetic process which depends on the mobility of gas molecules in the polymer matrix. At temperatures below T_g, the polymer backbone is considered to be in a frozen state; segmental chain motions are drastically reduced, as compared to the rubbery state; the number of fee volume holes is fixed; and no hole redistribution is likely. Gas transport is, therefore, assumed to take place via fixed (preexisting) holes [41].

A gas molecule must "find its way" from hole to hole along pathways involving only minor segmental rearrangements. This means that the magnitude of the diffusivity depends largely on the number of holes with an appropriate size, able to accommodate a diffusing gas molecule.

A diffusing molecule can migrate if it can find a nearby hole large enough to jump into it, that is, larger than a critical size v^*. The probability for such a process is proportional to $\exp(-\gamma v^*/v_f)$ where v_f is the mean hole free volume per molecule and γ is an overlap factor between 0 and 1, which accounts for the fact that the same free volume may be available to more than one diffusing molecules. The diffusion coefficient can be written as [54]

$$D = \widetilde{D}_0 \exp(-E^*/kT) \exp\left(-\gamma \xi v^*/v_f\right) \tag{20.70}$$

where $\xi = v_s^*/v^*$ and v_s^* is the critical volume of the penetrant. The activation energy E^* needed for the diffusion of trapped molecules from one hole to another is related to the molecule "jumping length." The proportionality coefficient $\widetilde{D}_0$ is related to the size and shape of the permeant and is given by the relation

$$\widetilde{D}_0 = \frac{RT\sigma^2}{M^{1/2}}, \tag{20.71}$$

where σ is the Lennard-Jones size parameter and M is the molecular weight of the permeant [54].

The amount of free volume alone does not provide information about the possible connectivity of the holes. When free volume elements are added randomly to a system, the percolation threshold will be reached at some point. Then a connected pathway will exist from one side of the system to the other. Since glassy polymers consist of randomly placed free volume elements, this percolation threshold exists and has been determined by Greenfield and Theodorou [55] via molecular simulations. Therefore, as the fractional free volume (FFV) in the polymer increases, the holes become interconnected above the percolation threshold and provide, thus, a flow-through path for the permeating gas. This has been commonly identified with a pore-flow diffusion process.

Diffusivity is better correlated to the accessible (for a given gas) free volume fraction (AFV). The percolation threshold is found to be in the range of 2–4 % *AFV* for conventional glassy polymers. This range is equivalent to 25–31 % *FFV* [56]. This picture adequately describes the diffusion process in some high free volume glassy polymers.

In the rubbery state, the picture of fixed holes needs to be replaced, since the polymer chains are mobile and the free volume holes are governed by a dynamic variation about their size, shape, and position. The gas molecules are able to diffuse within the fluctuating interstitial free volume with greater mobility compared to the glassy state. That is to say, polymers with a stiff backbone have a large jumping step, whereas flexible chains have a smaller one [54]. Equation 20.70 can be written in this case in the following form:

$$D = ART\exp(-B/FFV), \tag{20.72}$$

where the parameter B is equal to $\gamma v_s^*/v^*$. The factor $\exp(-E^*/kT)$ is neglected since no energy for diffusional jumping is required. The parameter A is related to the size and shape of the permeant and replaces the term $\sigma^2/M^{1/2}$ of Eq. 20.70. It seems that the factor B depends not only on the type of gas but also on the polymer since v^* is the minimal free volume hole size required for the penetrant diffusion.

The size of the diffusing molecule has a smaller influence if the polymer is in the rubbery state than in a rigid glassy membrane, where the hole size distribution is frozen [57]. The glassy polymers seem to be more suitable to be used as gas separation membranes.

While a correlation between $\ln D$ and $1/FFV$ of the type of Eq. 20.72 has been verified for various glassy polymers [38, 41, 54], the values of parameters A and B show a significant variance. A linear fit of all the data gives $A = 4 \times 10^{-6}$ cm^2/s and $B = 0.97$. In polycarbonates [58] reported $A = 39 \times 10^{-8}$ cm^2/s, $B = 0.297$ for CO_2. The variance is due to the fact that the parameter B depends on the polymer and not only to the type of gas. Some indicative values of these parameters in rubbery polymers are $A = 7.4 \times 10^{-5}$ cm^2/s, $B = 0.1336$ for CO_2 in polyurethanes [59] and $A = 3.7 \times 10^{-8}$ cm^2/s, $B = 0.035$ for O_2 in NR and XSBR elastomers [60].

In polymer nanocomposite systems, there is an additional feature that strongly affects the gas diffusion process: the tortuosity. In the case of nanocomposites, Eq. 20.72 should be modified to account for the tortuosity effect:

$$D = \frac{A'}{\tau}\exp(-B/FFV). \tag{20.73}$$

Thus, the variation of the diffusion coefficient in a polymer nanocomposite depends on the fractional free volume and tortuosity factor. These two parameters can be either in accordance or competitive. The final increase or decrease of the diffusion coefficient upon adding nanoclay depends on which of the two is dominant.

The curve of ln D versus $1/FFV$ of some PVA-PA6/MMT nanocomposites was linear for $1/FFV < 15$ [61]. That is, relations of the type of Eq. 20.72 are valid in this case only when the fractional free volume is more than 6.7 %. For smaller FFV, the curve is not linear, showing a saturation with increasing $1/FFV$. Similar results were obtained in SBR/rectorite nanocomposite where Eq. 20.72 holds mainly for $FFV > 3.33$ % [47]. This suggests that when the FFV is relatively high, it overcomes the influence of the tortuosity factor, and the diffusion coefficient will be determined by the free volume variations.

There is also the case where the free volume consists of a network of large holes connected by channel-like holes. Diffusing molecules may now move with little friction within the network. Large variations in the size or in the number of larger holes have small effect on the free volume network, whereas a small decrease in the size of smaller holes may destroy the connectivity of the channel network. In this case, the mobility of diffusing molecules may become strongly hindered [44]. This means that a broader distribution of free volume holes can result in higher values of the gas diffusion coefficient than a narrow distribution and the same average size. Therefore, not only the amount of the free volume but also its distribution over the hole sizes are important factors that influence the diffusion coefficient.

The incorporation of nanoparticles in a polymeric matrix can cause redistribution of the free volume hole sizes. Changes in the free volume hole size distribution will affect both rubbery and glassy polymers. Even when the total FFV is identical between two polymers, the differences in their hole sizes distribution may be able to alter their diffusivities [58]. This distribution may strongly affect the diffusion process as very small holes, where the gas molecules cannot enter, do not contribute to the diffusion. The solubility, which is connected to the overall free volume, remains unaffected by the hole size distribution.

This was the case in an acrylic resin/laponite nanocomposite where the particles did not influence the FFV but the diffusion coefficient was increased as a function of laponite loading [50]. This enhancement was ascribed to changes in hole size distribution. Another example is the polymethyl methacrylate (PMMA)/MMT system [6]. While the solubility of CO_2 in the PMMA/MMT is approximately the same as that for the pure PMMA, the diffusion coefficient is much greater than that for pure PMMA. This has been attributed to changes in the free volume size distribution towards a relatively greater number of larger holes.

5.5 Permeability and Free Volume

The permeability coefficient is the product of the solubility coefficient and the diffusion coefficient. Thus, the permeability coefficient of polymer-clay nanocomposites will be equal to

$$P = \frac{A'S}{\tau} \exp\left(-B/FFV\right). \tag{20.74}$$

This equation shows that the permeability depends on the diffusion coefficient, the solubility, and the tortuosity. As described above, the first two depend on the free volume in the system. However, the permeability coefficient seems to be less sensitive to the *FFV* than these two coefficients.

In polymer-clay nanocomposites where the tortuosity factor dominates the fractional free volume variations, the percentage reduction on gas permeability is 20–90 % [38]. Priolo et al. [2] managed to fabricate a super barrier polymer-clay thin film using layer-by-layer assembly technique. This method is based on the alternating exposure of a substrate to aqueous cationic and anionic mixtures. Negatively charged clay nanoparticles seem to form highly structured layer-by-layer films when combined with a positively charged polymer. The oxygen barrier performance of the resulting polymer nanocomposite thin films was reported to be below 10^{-5} cm^3/m^2 day bar at a thickness of just 51 nm.

Recently, Moller et al. [62] developed high-charge coarse-grained Li-hectorites using melt synthesis. The clay was delaminated, spontaneously yielding platelets with aspect ratios typically larger than 1,000. The completely delaminated clay has 800 m^2/g surface, which is transformed into interface when perfectly dispersed into the polymer matrix. The authors managed to disperse the high aspect ratio Li-hectorite particles into a UV-curable polyurethane and to produce a flexible and transparent nanocomposite coating with remarkable oxygen barrier properties.

From the permeability reduction, a value of 410 for the effective aspect ratio of the hectorite filler was deduced. However, since the hectorite precursor had an aspect ratio of 1,000, it is expected that the oxygen transmission rate (OTR) can be reduced by another two orders of magnitude. The reduction of the OTR for the hectorite composite may be impaired either by coating defects or by an increase of the diffusion coefficient through the matrix, which is related to an enlargement of the free volume in the matrix by solvent trapped in the film [62].

5.6 Influence of the Interfacial Regions

In many cases, the permeability of the nanocomposite is affected by the existence of interfacial regions between the matrix and the inorganic particles. Such domains will affect mainly the solubility coefficient, but in some cases, they may also enhance the diffusion coefficient. The interfaces either are caused by the surfactant that is used for the modification of the particles or are due to the formation of voids between the different phases.

The relative diffusion coefficient in such cases has been expressed by [63] as

$$\frac{D_{\text{composite}}}{D_{\text{matrix}}} = \frac{a'}{\tau}, \tag{20.75}$$

where a' is a factor that quantifies the effect of the interfacial regions

$$a' = 1 + \beta' \phi, \tag{20.76}$$

with the parameter β' given by

$$\beta' = \frac{V_s}{V_f} \cdot \frac{D_s}{D_0} - \frac{V_s + V_f}{V_f}. \tag{20.77}$$

D_s, Vs are the diffusion coefficient and the volume of the interfaces, D_0 is the diffusion coefficient for the neat polymer, and V_f is the volume of the nanoplatelets. If V_s/V_f is negligible, then $\beta' = -1$ and $a' = 1 - \phi$. However, the parameters D_s, V_s are not easily measurable. The value of the parameter β' has a strong effect on the diffusivity for all concentrations of the impermeable phase.

The overall transport properties of the nanocomposites are, thus, affected by the presence of interfacial regions. To quantify this effect, Xiao et al. [64] developed a multi-scale hierarchical numerical model and defined a polymer-nanoparticle interaction strength parameter. For high values of this parameter, they showed that the polymer density next to the interface increases, indicating that the enhancement of the barrier properties is due also to the decrease of the free volume there. For large gas molecules, Pryamitsyn et al. [65] showed that the polymer-particle interactions are less important than the tortuosity effect. The mass transport properties, therefore, can be determined by the segmental dynamics of the matrix.

5.7 Improving the Barrier Properties Using Reactive Additives

Passive barriers restrict the rate of mass transport by diffusion or convection but do not chemically interact with the permeating solute. The barrier properties of the nanocomposites may further be improved by the addition to the membrane of chemically reactive components that are incorporated into the polymer. These components could react with the penetrating gas molecules, stopping them from advancing their diffusion. This is used in packaging. Reactive barriers to gas permeation can reduce the effective permeation rates into the package by several orders of magnitude if their reactivity is sufficiently high, at least for a limited time.

The reactive additives are called scavengers of the solute if they are able to remove the solute via an irreversible chemical reaction. The additive must preferably be immobilized inside the membrane and distributed in channels with low resistance to permeation forming a reactive composite. In this way, an increase may be accomplished in the needed time lag, t_0. The process is delayed, thus, while the steady-state properties remain unchanged.

Let us suppose that the reaction between the gas molecules and the additive material is very fast and irreversible. While the additive molecules are immobilized, the permeating gas reacts and consumes them. Merkel at al. [25] assumed that the reaction occurs at a front in depth $l'(t)$, which propagates deeper in the membrane. If the quantity of the additive material is considerable, this front will move slowly.

We assume a quasi-steady state for the diffusion of the gas molecules from the surface to the front:

$$D\frac{d^2z}{dz^2} = 0, \tag{20.78}$$

where z is the distance from the surface of the membrane where the gas molecules enter. This equation is solved for the appropriate boundary conditions to give the penetration depth, $l'(t)$:

$$l' = \sqrt{\frac{2vDHc_{1,0}t}{c_{2,0}}}, \tag{20.79}$$

where $c_{1,0}$ and $c_{2,0}$ are the initial concentrations of the gas and the additive, respectively, v is the stoichiometric coefficient for the reaction between gas and additive, and H is the partition coefficient of the gas.

Setting l' equal to the membrane thickness, we can estimate the time lag

$$t_0 = \frac{l^2}{2vDH} \cdot \frac{c_{2,0}}{c_{1,0}}. \tag{20.80}$$

After this time, the flux comes to steady state. The model fails for $c_{2,0} = 0$, as it gives $t_0 = 0$ then. This is attributed to the approximation of the pseudo-steady state that was used.

If one compares the time lag for the membrane with and without additives, then

$$\frac{t_0}{t_0(\text{without})} = \frac{3c_{2,0}}{vHc_{1,0}}. \tag{20.81}$$

Since H is a property of the polymer, the key parameter that can be adjusted is the concentration of the additive.

The effectiveness of such a reactive barrier can be predicted by a model developed by Solovyov [66]. The model leads to the following expression:

$$\frac{J^P}{J^*} = \frac{\sinh(\tau\phi_0)}{(1-\phi)\phi_0}, \tag{20.82}$$

where J^P is the penetrant flux through the passive barrier and J^* is the penetrant flux through the reactive barrier. ϕ_0 is the barrier reactivity which depends on the reactive particle size, its volume fraction, and the barrier thickness. Loading of 1 wt% of highly reactive 1 μm oxygen scavengers particles into a 0.3 mm-thick PET barrier can potentially result in an initial barrier improvement of the order of 10^5. This improvement is sufficient to make such a reactive PET composite into an ultimate reactive barrier to oxygen permeation until the reactive capacity of the scavenger is exhausted by the reaction [66].

6 Applications

Low-permeability materials are of great interest due to their numerous applications in packaging and coating industries. Among the most frequently used polymers are the polyolefins, such as polyethylene and polypropylene, which cover almost 70 % of the thermoplastic market, with major applications in the automotive and packaging industry [67].

The cost of the packaging materials represents 17 % of the total cost of goods produced. This means that there is a potential for savings through source reduction. For example, there is a general trend to move from glass to lighter plastic bottles, including the replacement of glass also for sensitive beverages such as beer and carbonated drinks, where one tries to keep the gas in the bottle, and fruit juice, where one wants to prevent the diffusion of oxygen. The benefits include cost saving, weight reduction, and reduced breakage [68].

Some of the desired barrier performances can be achieved by multiple layered structures. However, the placing the oxygen barrier films between two water vapor barrier layers is costly. Thus, there is a demand for oxygen barriers that retain their properties in the presence of water.

One additional issue is the increasing demand for transparency where a combination of good barrier properties with (packaged) product visibility is required. Obviously, aluminum and metallized coatings are unable to provide this combination.

Most of the commercial applications of polymer-clay nanocomposites use polar polymers, such as nylon and ethylene vinyl acetate. Highly polar polymers, such as those containing hydroxyl groups, are excellent gas barriers but poor water barriers. Nonpolar hydrocarbon polymers, such as polyethylene, have excellent water barrier properties but poor gas barrier properties. In order to be a truly good barrier material, the polymer must have [11]:

- Some degree of polarity
- High chain stiffness
- Close chain-to-chain packing
- Some bonding or attraction between chains
- High glass transition temperature

In practice it has been found that only liquid crystalline polymers and ethylene vinyl alcohol materials can reach the highest demands for gas barrier properties [69]. Some indicative values for the desirable oxygen permeation rate for food packaging are between 0.1 and 100 $cm^3/(m^2$ day bar), while for gas storage tanks, they are below 0.1 $cm^3/(m^2$ day bar).

Another possible application for polymer nanocomposites is the area of flexible displays. Electronic devices like OLEDs (organic light-emitting devices) demand high barrier properties. For an OLED lifetime $>$ 10,000 h, the requirements are 10^{-6} $g/(m^2day)$ for water vapor transmission rate (WVTR) and 10^{-6} $g/(m^2day)$ for oxygen. Flexible displays are extremely sensitive to water vapor and oxygen which bring about their degradation. The degradation problem can be dealt with by sealing the devices in an inert atmosphere. Therefore, encapsulation of the devices and isolation of the active materials from the atmosphere are very important to prolong the lifetime of flexible devices [70].

Synthetic polymer coatings have been widely used in the treatment of construction materials of historical monuments for conservation of such structures. Among the various properties like optical transparency, high compatibility with the surface, stability against weathering, and UV irradiation, the protective coating must be capable to stop the pollution gases, like sulfide dioxide, and also the water entry [3].

Hybrid polymers are certainly among the most promising challenges in the fields of packaging, flexible displays, and protective coatings in the near future.

References

1. Utracki LA, Sepehr M, Boccaleri E (2007) Synthetic, layered nanoparticles for polymeric nanocomposites. Polym Adv Technol 18:1
2. Priolo MA, Gamboa D, Holder KM, Grunlan JC (2010) Super gas barrier of transparent polymer-clay multilayer ultrathin films. Nano Lett 10:4970
3. Sadat-Shojai M, Ershad-Langroudi A (2009) Polymeric coatings for protection of historic monuments: opportunities and challenges. J Appl Polym Sci 112:2535
4. Chen C, Han B, Li J, Shang T, Zou J, Jiang W (2001) A new model on the diffusion of small molecule penetrants in dense polymer membranes. J Membr Sci 187:109–118
5. Tortora M, Vittoria V, Galli G, Ritrovati S, Chiellini E (2002) Transport properties of modified mont-morillonite-poly(−caprolactone) nanocomposites. Macromol Mater Eng 287:243
6. Manninen A, Naguib H, Nawaby A, Day M (2005) CO_2 sorption and diffusion in polymethyl methacrylate-clay nanocomposites. Polym Eng Sci 45:904
7. Bohning M, Goering H, Hao N, Mach R, Schonhals A (2005) Polycarbonate/SiC nanocomposites-influence of nanoparticle dispersion on molecular mobility and gas transport. Polym Adv Technol 16:262
8. Choudalakis G, Gotsis AD (2009) Permeability of polymer/clay nanocomposites: a review. Eur Polym J 45:967
9. Tsujita Y (2003) Gas sorption and permeation of glassy polymers with microvoids. Prog Polym Sci 28:1377
10. Fu YJ, Hu CC, Lee KR, Tsai HA, Ruaan RC, Lai JY (2007) The correlation between free volume and gas separation properties in high molecular weight poly(methyl methacrylate) membranes. Eur Polym J 43:959
11. Salame M, Steingiser S (1977) Barrier polymers. Polym Plast Technol Eng 8:155
12. Shishatskii AM, Yampolskii YP, Peinemann KV (1996) Effects of film thickness on density and gas permeation parameters of glassy polymers. J Membr Sci 112:275
13. Heuchel M, Hofmann D, Pullumbi P (2004) Molecular modeling of small-molecule permeation in polyimides and its correlation to free-volume distributions. Macromolecules 37:201
14. Neogi P (1996) Diffusion in polymers. Marcel Dekker, New York
15. Ballik CM (1996) On the extraction of diffusion coefficients from gravimetric data for sorption of small molecules by polymer thin films. Macromolecules 29:3025–3029
16. Wong B, Zhang Z, Handa P (1998) High precision gravimetric technique for determining the solubility and diffusivity of gases in polymers. J Polym Sci B: Polym Phys 36:2025
17. Jacob A, Kurian P, Aprem AS (2008) Transport properties of natural rubber latex layered clay nanocomposites. J Appl Polym Sci 108:2623

18. Thomas PS, Sreekumar PA, Aprem AS, Thomas S (2011) Poly(ethylene-co-vinyl acetate)/calcium phosphate nanocomposites: mechanical, gas permeability, and molecular transport properties. J Appl Polym Sci 120:1974
19. Barrer RM, Petropoulos JH (1961) Diffusion in heterogeneous media: lattice and parallelepipeds in a continuous phase. Br J Appl Phys 12:691–697
20. Nielsen LE (1967) Models for the permeability of filled polymer systems. J Macromol Sci A 1:929
21. Wakeham WA, Mason EA (1979) Diffusion through multi-perforate laminae. Ind Eng Chem Fundam 18(4):301
22. Cussler EL, Hughes SE, Ward WJ, Aris R (1988) Barrier membranes. J Membr Sci 38:161
23. Falla WR, Mulski M, Cussler EL (1996) Estimating diffusion through flake-filled membranes. J Membr Sci 119:129
24. Aris R (1986) On a problem in hindered diffusion. Arch Ration Mech Anal 12:83
25. Merkel TC, Freeman BD, Spontak RJ, He Z, Pinnau I, Meakin P, Hill AJ (2003) Sorption, transport, and structural evidence for enhanced free volume in poly(4-methyl-2-pentyne)/fumed silica nanocomposite membranes. Chem Mater 15:109
26. Lape NK, Nuxoll EE, Cussler EL (2004) Polydisperse flakes in barrier films. J Membr Sci 236:29
27. Fredrickson GH, Bicerano J (1999) Barrier properties of oriented disk composites. J Chem Phys 110:4
28. Gusev AA, Lusti HR (2001) Rational design of nanocomposites for barrier applications. Adv Mater 13:21
29. Picard E, Vermogen A, Gerard J, Espuche E (2007) Barrier properties of nylon-6 montmorillonite nanocomposite membranes prepared by melt blending: influence of the clay content and dispersion state: consequences of modelling. J Membr Sci 292:133
30. Sridhar V, Tripathy DK (2006) Barrier properties of chlorobutyl nanoclay composites. J Appl Polym Sci 101:3630
31. Yeh JM, Huang HY, Chen CL, Su WF, Yu YH (2006) Siloxane modified epoxy resin-clay nanocomposite coatings with advanced anticorrosive properties prepared by a solution dispersion approach. Surf Coat Technol 200:2753
32. Jacquelot E, Espuche E, Gerard JF, Duchet J, Mazabraud P (2006) Morphology and gas barrier properties of polyethylene-based nanocomposites. J Polym Sci Part B: Polym Phys 44:431
33. Swannack C, Cox C, Liakos A, Hirt D (2005) A three dimensional simulation of barrier properties of nanocomposite films. J Membr Sci 263:47
34. Bharadwaj RK (2001) Modeling the barrier properties of polymer-layered silicate nanocomposites. Macromolecules 34:9189
35. Maksimov RD, Gaidukov S, Zicans J, Jansons J (2008) Moisture permeability of a polymer nanocomposite containing unmodified clay. Mech Comp Mat 44(5):505
36. Nazarenko S, Meneghetti P, Julmon B, Qutubuddin S (2007) Gas barrier of polystyrene montmorillonite clay nanocomposites: effect of mineral layer aggregation. J Polym Sci Part B: Polym Phys 45:1733
37. Bhattacharya M, Biswas S, Bhowmick AK (2011) Permeation characteristics and modeling of barrier properties of multifunctional rubber nanocomposites. Polymer 52:1562
38. Choudalakis G, Gotsis AD (2012) Free volume and mass transport in polymer nanocomposites. Curr Opin Coll Inter Sci 17:132
39. Bondi A (1964) Van der waals volumes and radii. J Phys Chem 68:441
40. Arnold JC (2010) A free-volume hole-filling model for the solubility of liquid molecules in glassy polymers 1: model derivation. Eur Polym J 46:1131
41. Hensema ER, Mulder MHV, Smolders CA (1993) On the mechanism of gas transport in rigid polymer membranes. J Appl Polym Sci 49:2081
42. Fujita H, Kishimoto A (1958) Diffusion-controlled stress relaxation in polymers. Stress relaxation in swollen polymers. J Polym Sci XXVIII:547
43. Jean YC (1990) Positron annihilation spectroscopy for chemical analysis: a novel probe for microstructural analysis of polymers. Microchem J 42:72

44. Consolati G, Genco I, Pegoraro M, Zanderighi L (1996) Positron annihilation lifetime (PAL) in poly[1-(trimethyl-silyl)propine](PTMSP): free volume determination and time dependence of permeability. J Polym Sci Part B: Polym Phys 34:357
45. Xie L, Gidley DW, Hristov HA, Yee AF (1995) Evolution of nanometer voids in polycarbonate under mechanical stress and thermal expansion using spectroscopy. J Polym Sci Part B: Polym Phys 33:77
46. Paranchos CM, Soares BG, Machado JC, Windmoller D, Pessan LA (2007) Microstructure and free volume evaluation of poly(vinyl alcohol) nanocomposite hydrogels. Eur Polym J 43:4882
47. Wang ZF, Wang B, Qi N, Zhang HF, Zhang LQ (2005) Influence of fillers on free volume and gas barrier properties in styrene-butadiene rubber studied by positrons. Polymer 46:719
48. Becker O, Cheng YB, Varley RJ, Simon GP (2003) Layered silicate nanocomposites based on various high- functionality epoxy resins: the influence of cure temperature on morphology, mechanical properties and free volume. Macromolecules 36:1616
49. Winberg P, DeSitter K, Dotremont C, Mullens S, Vankelecom IFJ, Maurer FHJ (2005) Free volume and interstitial mesopores in silica filled poly(1-trimethylsilyl-1-propyne) nanocomposites. Macromolecules 38:3776
50. Choudalakis G, Gotsis AD, Schut H, Picken SJ (2011) The free volume in an acrylic resin/laponite nanocomposite coatings. Eur Polym J 47:264
51. Muramatsu M, Okura M, Kuboyama K, Ougizawa T, Yamamoto T, Nishihara Y, Saito Y, Ito K, Hirata K, Kobayashi Y (2003) Oxygen permeability and free volume hole size in ethylene-vinyl alcohol copolymer film: temperature and humidity dependence. Radiat Phys Chem 68:561
52. Garcia A, Iriarte M, Uriarte C, Etxeberria A (2006) Study of the relationship between transport properties and free volume based in polyamide blends. J Membr Sci 284:173
53. Pethrick RA (1997) Positron annihilation – a probe for nanoscale voids and free volume? Prog Polym Sci 22:1
54. Thran A, Faupel GKF (1999) Correlation between fractional free volume and diffusivity of gas molecules in glassy polymers. J Polym Sci Part B: Polym Phys 37:3344
55. Greenfield ML, Theodorou DN (1993) Geometric analysis of diffusion pathways in glassy and melt atactic polypropylene. Macromolecules 26:5461
56. Thornton AW, Nairn KM, Hill AJ, Hill JM (2009) New relation between diffusion and free volume: predicting gas diffusion. J Membr Sci 338:29
57. Michaels AS, Bixler HJ (1961) Flow of gases through polyethylene. J Polym Sci 50:413–439
58. Jean YC, Yuan JP, Liu J, Deng Q, Yang H (1995) Correlations between gas permeation and free-volume hole properties probed by positron annihilation spectroscopy. J Polym Sci Part B: Polym Phys 33:2365
59. Wang ZF, Wang B, Yang YR, Hu CP (2003) Correlations between gas permeation and free-volume hole properties of polyurethane membranes. Eur Polym J 39:2345
60. Stephen R, Ranganathaiah C, Varghese S, Kuruvilla J, Sabu T (2006) Gas transport through nano and micro composites of natural rubber (NR) and their blends with carboxylated styrene butadiene rubber (XSBR) latex membranes. Polymer 47:858
61. Cui L, Yeh JT, Wang K, Tsai FC, Fu Q (2009) Relation of free volume and barrier properties in the miscible blends of poly(vinyl alcohol) and nylon 6-clay nanocomposites film. J Membr Sci 327:226
62. Moller MW, Kunz DA, Lunkenbein T, Sommer S, Nennemann A, Breu J (2012) Uv-cured, flexible, and transparent nanocomposite coating with remarkable oxygen barrier. Adv Mater 24:2142
63. Sorrentino A, Tortora M, Vittoria V (2006) Diffusion behavior in polymer-clay nanocomposites. J Polym Sci Part B: Polym Phys 44:265
64. Xiao J, Huang Y, Manke CW (2010) Computational design of polymer nanocomposite coatings: a multiscale hierarchical approach for barrier property prediction. Ind Eng Chem Res 49:7718

65. Pryamitsyn V, Hanson B, Ganesan V (2011) Coarse-grained simulations of penetrant transport in polymer nanocomposites. Macromolecules 44:9839
66. Solovyov SE (2006) Reactivity of gas barrier membranes filled with reactive particulates. J Phys Chem B 110:17977
67. Esfandiari A, Nazokdast H, Rashidi AS, Yazdan-shenas ME (2008) Review of polymer – organoclay nanocomposites. J Appl Sci 8:545
68. Lange J, Wyser Y (2003) Recent innovations in barrier technologies for plastic packaging – a review. Packag Technol Sci 16:149
69. Hanika M, Langowski HC, Moosheimer U, Peukert W (2003) Inorganic layers on polymeric films – influence of defects and morphology on barrier properties. Chem Eng Technol 26:605
70. Choi MC, Kim Y, Ha CS (2008) Polymers for flexible displays: from material selection to device applications. Prog Polym Sci 33:581
71. Moggridge GD, Lape NK, Yang C, Cussler EL (2003) Barrier films using flakes and reactive additives. Prog Organ Coat 46:231

Recent Developments of Foamed Polymer/Layered Silicates Nanocomposites

21

Krzysztof Pielichowski, James Njuguna, and Sławomir Michałowski

Contents

Abstract

In this chapter, recent developments in the area of foamed polymer/layered silicates nanocomposites have been presented. Polymer nanocomposite foams are a new class of materials for lightweight and high strength applications which show higher heat and fire resistance. In recent years, substantial efforts were undertaken to develop novel nanocomposite foams based on thermoplastic or thermoset polymer matrixes and using layered silicates. For polymer foams, the use of layered silicates may cause an increase in cell nucleation rate, provide

K. Pielichowski • S. Michałowski (✉)
Department of Chemistry and Technology of Polymers, Cracow University of Technology, Kraków, Poland
e-mail: kpielich@pk.edu.pl; spri@indy.chemia.pk.edu.pl

J. Njuguna
Institute for Innovation, Design and Sustainability, School of Engineering, Robert Gordon University, Aberdeen, UK

J.K. Pandey et al. (eds.), *Handbook of Polymernanocomposites. Processing, Performance and Application – Volume A: Layered Silicates*, DOI 10.1007/978-3-642-38649-7_80,

foam reinforcement, and result in char formation during flame action. Clay nanoparticles may help to control foam's structure and morphology, as they serve as nucleation sites to facilitate the bubble formation process. Synthesis of foamed polymers, different foamed polymer/layered silicate nanocomposites, including polyurethane- polyolefin-, polystyrene-, polyamide and poly(lactic acid)-based nanostructured materials with layered silicates, and bubble nucleation effects in foamed polymer nanocomposites are discussed. Special attention was focused on the flammability resistance effect of nanoparticles in foamed matrix which is based on the flame barrier mechanism involving formation of a high-performance carbonaceous silicate char yielding insulating action.

Keywords
Foamed polymer nanocomposites • Layered silicates • Bubble nucleation • Structure • Morphology

1 Introduction

Polymer foams also called cellular plastics are a group of polymeric porous materials that consist of at least two phases – one phase is a solid polymer matrix that may be either inorganic, organic, or organometallic, and the second phase is a gaseous phase derived from a blowing agent. Foams are either flexible or rigid depending on their mechanical properties which can be controlled by chemical composition, cell morphology, degree of cross-linking, and crystallinity degree. Rigid foams are widely used in different applications such as building insulations, refrigeration, transportation, packaging, etc., whereas flexible foams are used in furniture parts, in sport applications, shock and sound attenuations, and for protection in transportation industry [1, 2].

Polymer foams can be classified according to the size of the cells as macrocellular (>100 μm), microcellular (1–100 μm), ultramicrocellular (0.1–1 μm), and nanocellular (<100 nm).

Polymeric foams have unique physical, mechanical, and thermal properties, which can be controlled by the cellular structure of polymer matrix. When a gaseous phase is being dispersed in a spherical form within a polymeric matrix, a composite structure is naturally formed. The properties of obtained composite are determined by its constituents and their distributions. The characteristics of polymeric foams are determined by the following structural parameters: cell density, expansion ratio, cell size distribution, open-cell content, and cell integrity. These cellular structural parameters may be controlled by the foaming technology settings, and the foaming technology often heavily depends on the type of polymer to be foamed.

Cellular plastics may show closed-cell or open-cell morphology. In order to fabricate open-cell foams, cell walls are broken and the cell structure consists of mainly ribs and struts. For closed-cell foams, the cell structures are closed and cells are isolated from each other. Therefore, closed-cell foams are characterized by lower permeability, leading to better insulation properties while open-cell foams have better absorptive properties [2].

In recent years, substantial efforts were undertaken to develop novel nanocomposite foams based on thermoplastic or thermoset polymer matrixes and using various nanoparticles, such as layered silicates, carbon nanotubes, or graphite.

Nanocomposites are a new class of materials in which the particles dispersed in polymer matrix have at least one dimension in the nanometer range, usually less than 100 nm. Depending on the nanoparticle's shape, one can distinguish three types of nanocomposites – the first class has all three dimensions in the range of nanometers, and the example could be isodimensional spherical nanosilica. If two dimensions are in the nanometer scale and the third dimension is larger, they are called nanotubes (e.g., carbon nanotubes) or whiskers. When nanoparticles are characterized by only one dimension in the nanometer range, one can distinguish the third class, comprising, e.g., layered silicates (LS). Polymer nanocomposites are already being commercialized in automotive, aerospace, construction, and electronic sectors due to their advantageous mechanical, thermal, and barrier properties.

Layered silicates have been widely investigated because the clay materials are easily available and their intercalation chemistry has been studied for a long time [3–6].

For polymer foams, the use of nanoclay may cause an increase in cell nucleation rate, provide foam reinforcement and lower gas escape rate, and result in char formation when foam is exposed to fire scenario [1]. Well-dispersed clay nanoparticles in a polymer matrix may serve as nucleation sites to facilitate the bubble formation process. Application of functionalized nanoparticles helps to control foam's structure and morphology, as well as various foaming techniques enable to tune the physical properties of the cellular materials obtained. Therefore, polymer nanocomposite foams can be considered as a new class of materials that are lightweight and show enhanced mechanical characteristics. Other beneficial features include higher heat and fire resistance, as well as reduced gas diffusivity.

2 Synthesis of Foamed Polymers

The fundamental goal in the synthesis of nanocomposite foams is to achieve good intercalation/exfoliation of layered silicates in the polymer matrix. Three main methods to synthesize foamed polymer/layered silicate nanocomposites are (i) intercalation of a suitable monomer and subsequent in situ polymerization, (ii) intercalation of polymer from solution, and (iii) polymer melt intercalation [3–7].

A schematic representation of polymer/layered silicate nanocomposite obtained by an in situ polymerization (when the monomer is used directly as a solubilizing agent for swelling the layered silicate) is shown in Fig. 21.1.

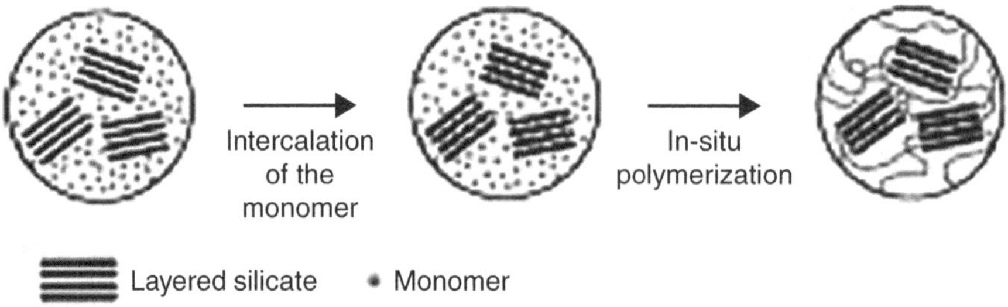

Fig. 21.1 Schematic representation of polymer/layered silicate nanocomposite obtained by in situ polymerization [7]

The second method involves intercalation of polymer from solution; it requires a suitable solvent that can solubilize the polymer. The polymer chains displace then the solvent molecules in the inter-gallery space of the silicate platelets [3–9].

In the polymer melt intercalation method, the polymer chains diffuse under processing conditions in between the silicate platelets. Depending on the degree of penetration, intercalated or exfoliated structures are formed. In the recent years, this method has been widely applied on lab and industrial scale to fabricate polymer/layered silicate nanocomposites [3–12].

It has been found that the mechanical properties, thermal stability, and barrier properties of polymer/layered silicate nanocomposites strongly depend on the possible interactions between organic and inorganic components; therefore, surface modification of nanoparticles is often applied to improve their compatibility with the polymer matrix. Additionally, special dispersing techniques, such as ultrasonication or microwave irradiation, are applied to accomplish good dispersion of nanoparticles.

Polymeric foamed materials can be obtained by using physical or chemical blowing agents [3, 4]. The former are substances that expand under foaming conditions, and the main classes of agents are currently banned chlorofluorocarbons (CFC), volatile hydrocarbons and alcohols, or inert gases such as nitrogen, argon, or carbon dioxide. Chemical blowing agents are usually reactive components that produce gas during the foaming process.

The production of thermoplastic foamed materials (PE, PP, PS, etc.) can be carried out by several methods such as liquid/melt extrusion and injection/compression molding, or in the solid state where gas is brought in the polymer followed by depressurization. Thermoplastic foams can be also obtained by using chemical blowing agents that decompose inside under high pressure and temperature, and within this method, one-step (to produce foams with densities higher than 100 kg/m^3) and two-step foaming processes (<100 kg/m^3) are applied in technology practice. However, most thermoplastic nanocomposite foams are obtained in the latter process [1]. Other, although less common, methods to produce cellular plastics include phase inversion, leaching, or techniques based on supercritical fluids.

The thermoset nanocomposite foams (such as polyurethane/layered silicate foams) can be synthesized by single step reactive foaming. Before the synthesis, nanoparticles should be dispersed in the monomer. Then, a physical blowing agent, e.g., pentene, is mixed with monomer/LS dispersion or chemical foaming agent, such as water that reacts with the polyisocyanate to form urea and carbon dioxide. This approach makes it possible to fabricate porous materials in a wide variety of densities ranging from 10 to over 900 kg/m^3.

3 Layered Silicates

The layered silicates broadly used in polymer nanocomposites belong to the family of 2:1 layered silicates or phyllosilicates. The crystal structure of LS consists of layers that have two tetrahedrally coordinated silicon atoms joined to an edge-shared octahedral sheet of either aluminum or magnesium hydroxide. The layer thickness is around 1 nm, and the lateral dimensions of these layers can differ from 30 nm to several microns or even larger. The most commonly used layered silicates are montmorillonite (MMT), hectorite, and saponite. Layered silicates occur as tetrahedrally or octahedrally substituted structures. For tetrahedrally substituted layered silicates, the negative charge is located on the surface of silicate layers which facilitates the interactions with polymer matrices.

The structure and some characteristic parameters of selected layered silicates are presented in Fig. 21.2 and Table 21.1.

Proper organofilization of LS is the key issue in preparation of "true" polymer/LS nanocomposites. This pretreatment leads to an increase of the inter-gallery spaces and the degree of hydrophobicity of the clay, overcoming thus the immiscibility between the polymer (usually hydrophobic) and the hydrophilic clay.

Other approaches include covalent bond formation through a grafting reaction or direct adsorption of molecules [13, 14].

In technological practice, organo-modification is being performed by using long carbon-chain alkyl or aryl ammonium salts. It has been found that the molecular structure, such as alkyl chain length, number of alkyl chains, and unsaturations, is also the determining factor of the properties (e.g., thermal properties and flammability) of the polymer/MMT nanocomposites. For instance, if the processing temperature is higher than the thermal stability of the organoclay, then decomposition occurs, and the interface between the filler and the matrix polymer is effectively altered [15, 16].

4 Foamed Polymer/Layered Silicate Nanocomposites

Nanocomposite foams can be obtained from thermoplastic or thermoset polymer matrices. They are made of polyurethanes (PU) [17–36], polyolefins

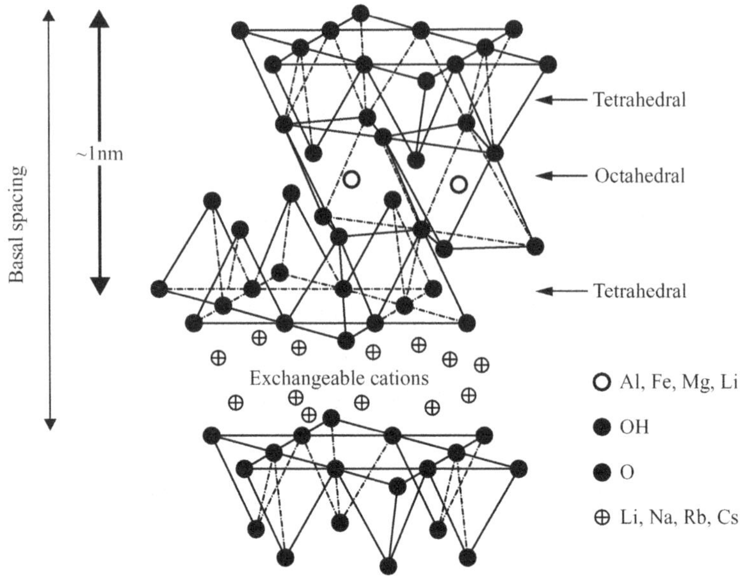

Fig. 21.2 Structure of 2:1 phyllosilicates [4]

Table 21.1 Chemical formula and characteristic parameters of selected 2:1 phyllosilicates [4–6]

2:1 phyllosilicates	Chemical formula	CEC (mequiv/100 g)	Particle length (nm)
Montmorillonite	$Mx(Al_{4-x}Mg_x)Si_8O_{20}(OH)_4$	110	100–150
Hectorite	$Mx(Mg_{6-x}Li_x)Si_8O_{20}(OH)_4$	120	200–300
Saponite	$MxMg_6(Si_{8-x}Alx)Si_8O_{20}(OH)_4$	86.6	50–60
Beidellite	$(Na,Ca_{0.5})_{0.3}$ Al_2 $(Si,Al)_4$ O_{10} $(OH)_2{\cdot}nH_2O$	85	<100
Nontronite	Fe^{3+} $[(OH)_2\vert$ $Al_{0,33}Si_{3,67}O_{10}]{\cdot}Na_{0,33}(H_2O)_4$	125	20–500

(polyethylene (PE) and polypropylene (PP)) [9, 37–57], polystyrene (PS) [58–65], polyamides (PA) [66–73], poly(lactic acid) (PLA) [74–82], and others [83–100].

Depending on the morphology of cell structure and physical properties, the obtained polymeric foams can be categorized as rigid or flexible foams.

4.1 Polyurethane-Based Foams

Polyurethane foams are obtained as rigid or flexible foams that are used in a wide range of applications. Their density, rigidity, cell structure, and foam morphology can be controlled by appropriate formulation adjustments. The rigid foams have found broad use as thermal insulations, in construction, sports, and automotive

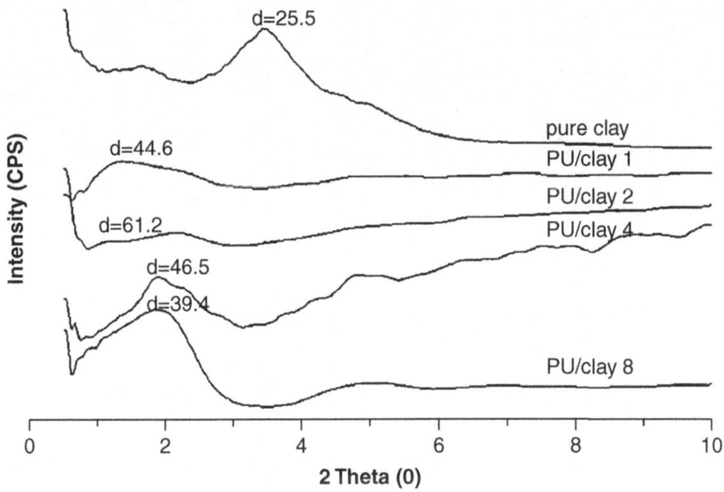

Fig. 21.3 XRD diffraction patterns of the pristine PU foams and rigid PU foam/organoclay nanocomposites [20]

applications. On the other hand, the flexible foams can be applied for seat cushionings, flexible hoses, packagings, in sports, recreation, etc.

Both types of foams can be prepared by one-shot process, *semi*-prepolymer process or prepolymer process. The most commonly applied method is the one-shot process, suitable both for flexible and rigid foams fabrication.

Polyurethane foams are based on the classical reaction of a di-isocyanate with a polyol. The reaction between components is exothermic, and this heat of reaction can be utilized to form a cellular structure by the evaporation of the physical blowing agents, such as chlorofluorocarbons, hydrochlorofluorocarbons, pentanes, and hydrofluorocarbons. Moreover, the foaming process can be carried out using a chemical blowing agent (carbon dioxide) that is generated in the reaction of di-isocyanate with water [12].

The polymer nanocomposite foams with LS are commonly prepared by applying an in situ intercalation/exfoliation process, including two steps. First step is mixing of the monomer or oligomer with the layered silicates followed by the second step – the polyaddition reaction. Intercalation and/or exfoliation can occur at both stages [14].

In a series of experiments, the influence of the layered silicates on the properties of rigid polyurethane foams was investigated. The rigid foams were prepared using a two-step procedure. In the first step, a given amount of layered silicate was dispersed in polyols mixture by means of mechanical stirring or ultrasonication [17–26].

The XRD spectra of the PU/organoclay composites show a characteristic reflection between 1.4 and 2.58 2θ degree, which corresponds to a base spacing of 4.4–6.1 nm due to clay intercalation/exfoliation (Fig. 21.3) [20].

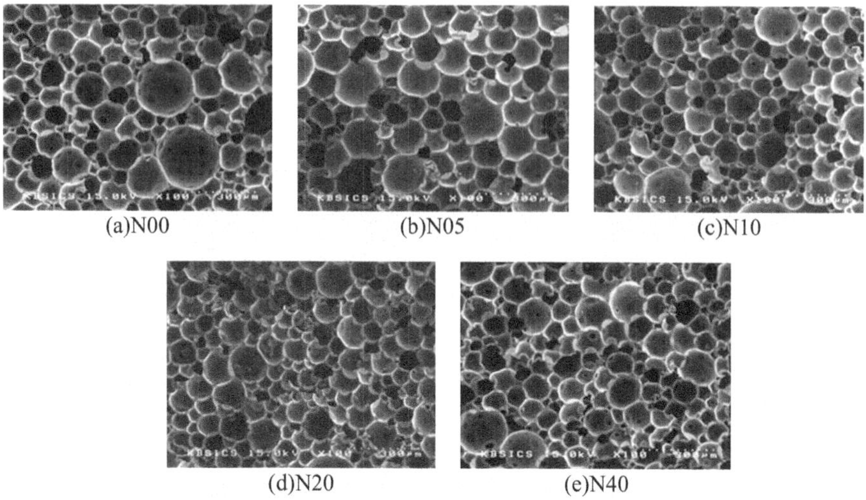

Fig. 21.4 SEM morphology of PUF/clay nanocomposites versus clay content [21]

The morphology of the cell structure of PU foams with different clay content was investigated by SEM technique – Fig. 21.4.

The foams consist of cells with spherical and polyhedral shapes, and the average cell size decreases with an increase of clay content to a minimum of 2 pphp (the formulation is based on 100 parts of the polyol by weight), which suggests that LS may act as a nucleating agent [21].

The mechanical properties of rigid polyurethane foams increase with a small addition of the layered silicate and reach the maximum value at clay content of 2–3 wt % [19, 20, 22]. On the other hand, the thermal conductivity of the composite foams is lower than pristine foam due to the decreased cell size and increased number of cell [21].

In another work, the nanoparticles were dispersed in di-isocyanate component by applying ultrasonication with/without solvent. The rigid polyurethane foams containing LS were then obtained according to the foaming procedure schematically presented in Fig. 21.5 [27].

The application of a solvent (toluene) and ultrasonication enhanced partial exfoliation of the nanoclay in 4,4′-methylenediphenyl di-isocyanate (MDI), as evidenced by shift of the scattering peak. However, the reduction of the peak intensity would indicate that the nanostructure does not have one distinct spacing (Fig. 21.6) [27].

Incorporation of nanoparticles to di-isocyanate caused an increase in the viscosity and a subtle decrease in the mechanical strength of rigid polyurethane foams [27].

The PU foams modified with different amounts of vermiculite were characterized by smaller cells' dimensions and higher uniformity, as compared with unmodified foams [28]. Thermal conductivity of the nanocomposite foam modified

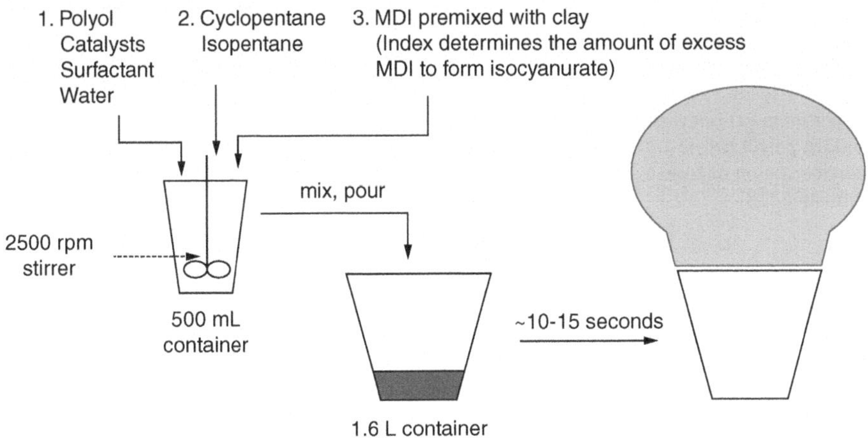

Fig. 21.5 Lab-scale batch foaming procedure [27]

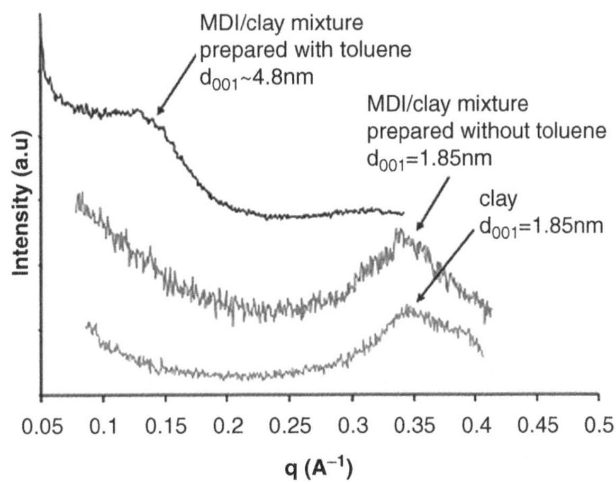

Fig. 21.6 SAXS: intensity versus scattering vector for pure clay and 7.7 wt % clay in MDI prepared with and without toluene [27]

with vermiculite is higher by about 10 %, because vermiculite induces the bubble nucleation, as well as the mechanical properties (strain–stress profiles) are better – Fig. 21.7 [28].

An important issue for technological developments is that ultrasonication-assisted mixing of clay in polymerized MDI (PMDI) causes a reduction of the polymer cell size [12, 28–35].

The flexible polyurethane foams modified by layered silicate were obtained by mechanical or sonification mixing of the nanoparticles with polyols or solvent that were then removed by vacuum distillation. The polyol containing LS, the catalyst, the surfactant, and water are then mixed with the di-isocyanate to produce flexible PU/LS nanocomposite foam.

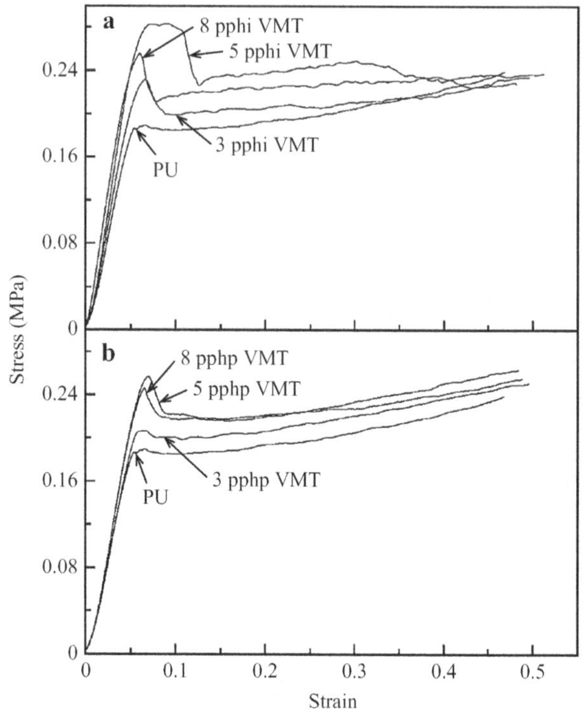

Fig. 21.7 Representative compressive stress versus strain curves of PU and PU-VMT foams. Vermiculite dispersed in (**a**) isocyanate and (**b**) polyol before polymerization or foam formation [28]

However, partial replacement of polyols by LS nanofillers may lead to the strong increase of compression strength and rebound resilience of the foams while not improving the air flow.

Nanoclays disturb the cell's opening mechanism and a decrease of the foams' stability is observed – at the clay content above 3 wt % the foams collapsed [36].

4.2 Expanded Polyolefin-Based Foams

Polyolefin production worldwide gives them the first position among large-scale produced polymers. They are partially utilized for fabrication of foams based on polyethylene, cross-linked polyethylene, and polypropylene.

Khorasani et al. investigated the influence of LS on the microcellular structure, crystalline morphology, and N_2 sorption of HDPE foams. They found that both expansion ratio and cell density can be improved by increasing the PE cell nucleation in the presence of nanoparticles [37]. The mechanical properties of HDPE foams modified by clay showed better tensile properties than the pure HDPE foams. The modulus of elasticity, tensile strength at yield, and elongation at break of nanocomposite foams were also improved, and the most profound effect was found for clay content of 0.5 wt % [38–40].

The addition of a small amount of clay, up to 3 wt %, makes it possible to obtain PE foams with a very high cell density of ca. 1,000 cells/cm^3 with an average cell size considerably smaller than 1 mm. It is thought that the cellular morphology is enhanced due to heterogeneous cell nucleation at the boundaries of the clay particles [41].

In another work, LDPE/EVA/clay nanocomposite foams were prepared by melt mixing, and then a single-stage batch foaming was carried out by using azodicarbonamide (ADC) as a chemical blowing agent and zinc oxide as an ADC activator in a hot-plate press. One observed that the storage modulus and complex/zero shear viscosities increased with addition of nanoclay content. Nanofiller particles acted as nucleation sites, and the reduction of the cells size was observed [42, 43].

LDPE-*g*-MA/clay nanocomposites were obtained by the microcellular foaming of PE with maleic anhydride (MA) and clay. The effect of MA and clay concentration on both PE cell density and cell size is shown in Fig. 21.8.

The morphology of materials obtained changes because the cell's size decreases with an increase of the cells density. However, maleic anhydride enhances exfoliation of clay, as evidenced by lower number of stacks of clay platelets in the nanocomposite [45, 46].

The hectorite nanoplatelets can be also used as nano-additive to polymeric foamed materials. From the research reports published, it seems that the morphology (the average cell size and the cell density) of PE/hectorite foams does not depend on the hectorite content, although it influences the thermal and mechanical properties of foams [9, 47].

The effect of clay dispersion and resultant morphology after the foaming process is shown in Fig. 21.9.

The intercalated structure of clay that is applied then in the polymer foaming process ensures better dispersion and yields higher barrier effect than in case of non-intercalated LS [48].

Zhai and coworkers prepared polypropylene-clay nanocomposite foams using a maleic anhydride as a coupling agent that facilitated the insertion of PP chains into the clay galleries and changed the periodic layered structure of the clay. It has been also found that the diluting of the masterbatch by pristine PP efficiently helps to disperse nanoclay in the polymer matrix [49–51].

It is noteworthy that fine and uniform cellular structures can be produced by appropriate blending of polymeric components, as evidenced for the PP/HDPE blend. By increasing HDPE content, one can eliminate the cell distribution nonuniformity which is due to the distinct interfaces between PP and HDPE phases. It affects the growth of cell diameter and the decrease of cell density.

In another study, the effects of nanoclay and die temperature on polypropylene foam cell structures were investigated. By introducing a small amount of nanoclay to the foam, the cell density and expansion ratio were considerably improved, due to both the enhanced melt strength and the clay nucleation effect [49].

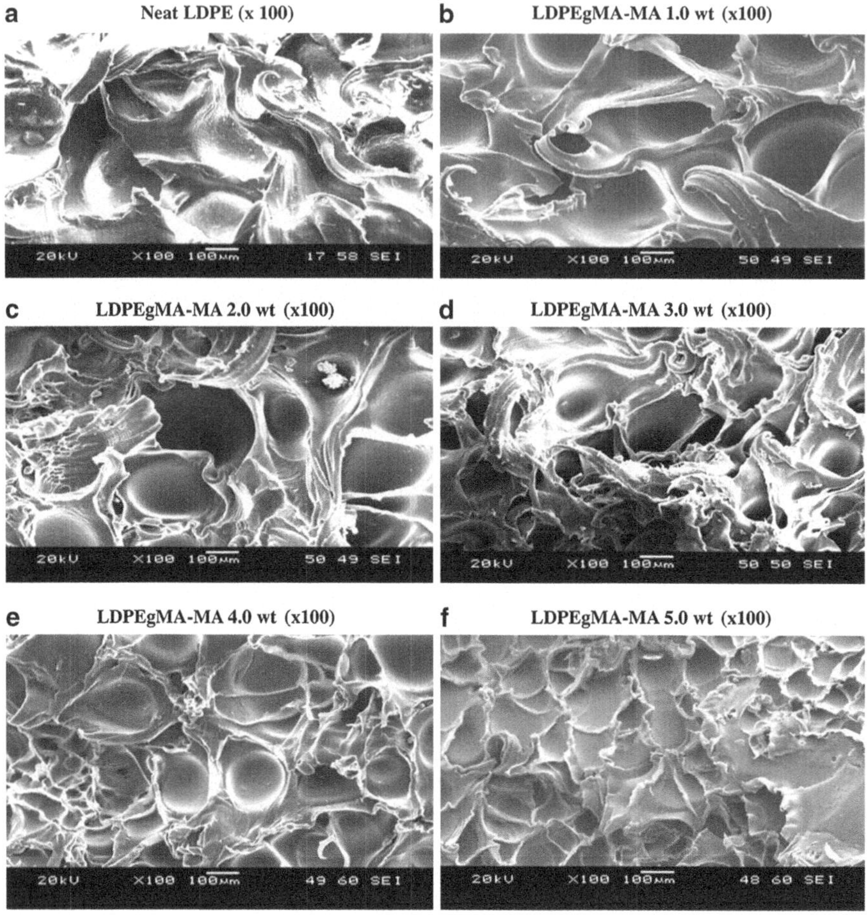

Fig. 21.8 SEM microstructure of LDPE-*g*-MA/clay nanocomposites, 0–5 wt % MA (system A) [44]

The addition of nanoclay of up to 5 wt % to polymeric materials influence the cellular formability due to an increase in cell nucleation sites and a higher melt strength that enhances cell morphology. Die temperature reduction also improves the formability of the polypropylene nanocomposite foams [52].

Polypropylene-clay nanocomposite foams can be obtained by using carbon dioxide as a foaming agent. There is a distinct effect of clay on bubble nucleation and in situ growth rates – the number density of bubbles in composite increases and the bubble growth rate decreases with higher clay content, as the LS platelets decrease the diffusivity of CO_2 [53].

The MMT nanoparticles thermally stabilize the PP foams during expansion, and this effect is even more pronounced with an increase in ADC content [54].

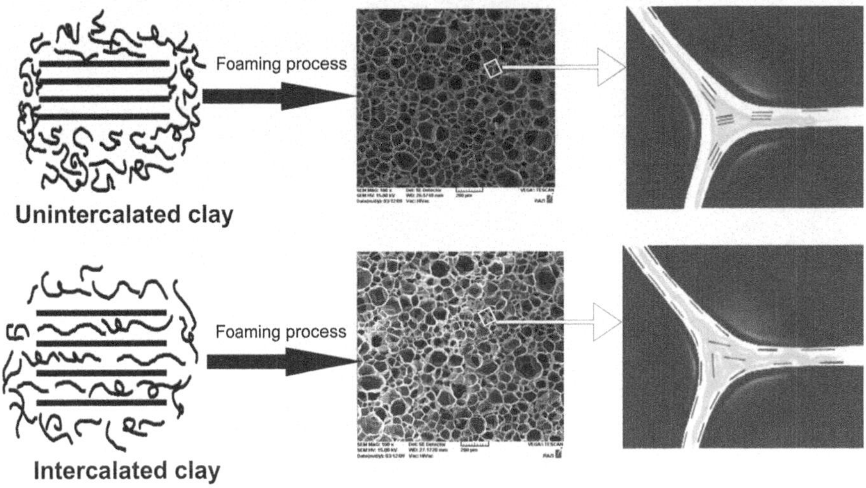

Fig. 21.9 Schematic presentation of the mutual effect of clay dispersion and foaming process [48]

In EPDM/clay foamed nanocomposites modified by maleic anhydride, an increase in mechanical and dynamic mechanical properties, as well as in non-isothermal crystallization rate, has been observed [55].

In the polyolefin foams, efficiency of clay in nucleation is found to decrease with an increase in clay content. This effect can be attributed to the increased interaction potential between the clay particles. The foams' cell density and cell size is followed an inverse relationship. Moreover, with increase in clay content, the cell wall thickness tends to increase [56, 57].

4.3 Expanded Polystyrene-Based Foams

Polystyrene (PS) foams are one of the most intensively developed cellular materials available in two forms as extruded polystyrene foams (XPS) and expanded polystyrene foams (EPS). These foams are light, closed-cell foams with relatively low thermal conductivity and good water resistance.

Layered silicate reinforced polystyrene foams were successfully produced using supercritical CO_2 at temperatures 75–85 °C and pressures 8–12 MPa. Both the concentration and dispersion state of a nanofiller have a strong effect on cell density, shape, and size. As processing temperature is increased, the foam cells are less restricted in growth, and the layered structure is no longer seen.

The nanocomposite orientation around the foam cells promises to yield enhanced physical and mechanical properties of the nanocomposite. Accelerated CO_2 absorption by nanostructured PS-based material can be utilized in fabrication of both micro- and macrocellular foams [58].

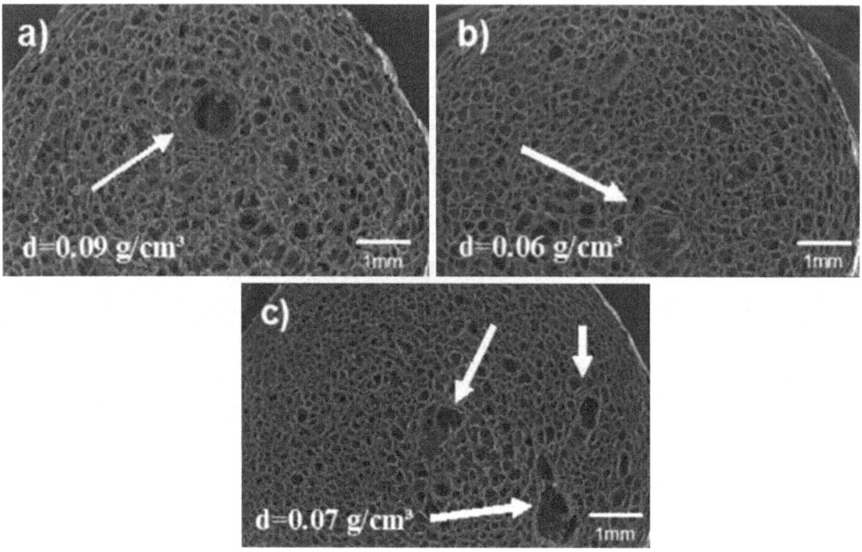

Fig. 21.10 SEM analysis of SAN foams containing 3 wt % of LS (C30B) prepared in one pot in the extruder with CO_2 at different nozzle temperatures: (**a**) 160, (**b**) 155, and (**c**) 150 °C. *Arrows* show some foam defects [64]

Foamed polystyrene/clay nanocomposites can be prepared by styrene polymerization in the presence of LS followed by foaming process using supercritical CO_2. The addition of clay reduces the cells' size and increases the cells' density. However, the morphology of these nanocomposite foams is strongly dependent on the surfactant used [59–63].

Along this line of interest, low-density styrene-acrylonitrile (SAN) foams were successfully produced in a continuous process using supercritical CO_2 as a foaming agent. At higher temperature, these foams are characterized by high density and regular porosity. Incorporation of commercial or pre-exfoliated nanoclays together with the polymer matrix in the foaming extrusion process influences foams' regular morphology, as it can be seen in Fig. 21.10 [64].

Another method to produce PS/clay nanocomposite foams is synthesis via suspension polymerization of water-in-oil inverse emulsion. The obtained materials were characterized by open-cell structure of lower density and high insulation performance [65].

4.4 Polyamide-Based Foams

Polyamides (PA) belong to the most important engineering thermoplastics that can be also produced as microcellular materials. They display good mechanical properties and durability that is required in, e.g., automotive industry.

Yuan et al. investigated the influence of nanoclay (5 and 7.5 wt %) on the cell structure and mechanical properties of injection-molded microcellular PA6 composites. The authors found out that the cell density increases and the cell size decreases for higher content of nanoclay in the microcellular PA6 nanocomposites. For higher content of nanofiller, brittle behavior occurs, and the tensile strength does not improve due to the presence of microcells [66].

For nanocomposites with an optimal amount of nanoclay (5 wt %), the highest crystallization rate was observed, and the dissolved gas did not alter the polymer crystalline structure [67, 68]. In another study, the influence of nanoclay on the microcellular extrusion of PA-6 has been studied. The evenly dispersed clay nanoparticles in the highly crystalline polymer matrix exert a nucleating role, while decreasing the foams cell size to ca. 30 μm and increasing the cell density [69].

Supercritical fluids, such as CO_2, facilitate nanoclay intercalation and exfoliation in the PA-6 microcellular structure formation [70].

PA-6/organoclay nanocomposites were prepared by in situ polymerization and melt-compounding, too. The addition of organoclay to the PA-6 caused an increase in mechanical strength and improved the thermal stability. Moreover, the introduction of organoclay into the PA-6 matrix resulted in a decrease of the cell size [71, 72].

PA-6/clay nanocomposites were also subjected to microcellular injection molding. The microstructure and the mechanical properties of the porous materials obtained were dependent on the process conditions and nanoclay features. The control of the cell size can be achieved by changing the content of nanofiller acting as a nucleating agent. The mechanical properties (impact strength) can be enhanced through formation of microcells at processing conditions [73].

4.5 Poly(lactic acid)-Based Foams

Polylactic acid (PLA) is an important biopolymer derived from lactic acid (2-hydroxy propionic acid) which is widely used as foamed material in packaging, food, and pharmaceutical applications, replacing the commercial plastic foam products from non-renewably raw materials.

The PLA-clay materials were formulated using supercritical CO_2. Incorporation of MMT improves the mechanical properties of PLA matrix and promotes heterogeneous nucleation in PLA/PBS composites [74–76].

Organoclay-reinforced PLA/PVA/NaCl nanocomposites with intercalated structure were processed to produce porous scaffolds for bio-engineering applications. Based on DSC results, incorporation of MMT changed the melting enthalpy and temperature of polymer matrix [76].

The foam processing of pristine polylactide and two different types of PLA/layered silicate nanocomposites, obtained using supercritical carbon dioxide as a foaming agent, was investigated. This method can be used to prepare biodegradable nanocellular polymeric foams with a closed-cell structure [77].

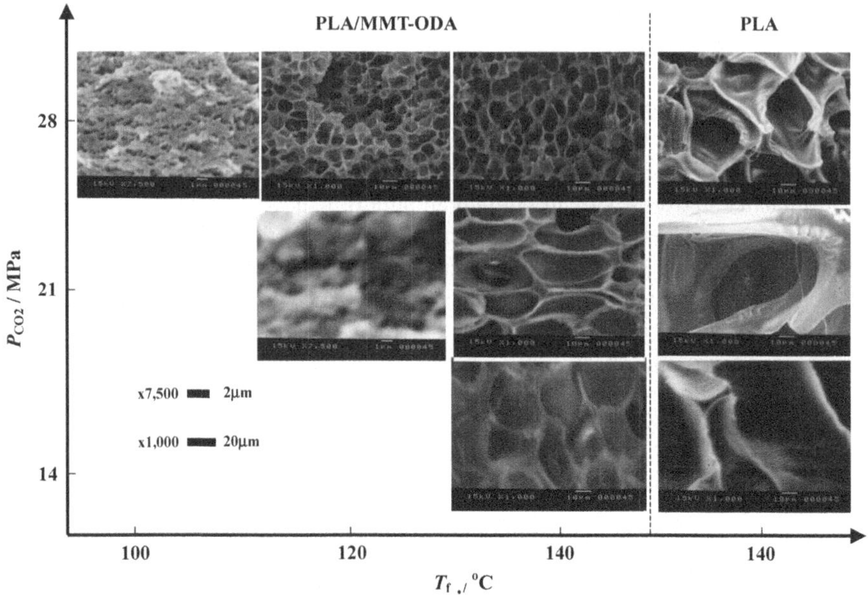

Fig. 21.11 Typical results of SEM images of the fracture surfaces of PLA/MMT–ODA and neat PLA foamed at temperature range of 100–140 °C under different isobaric saturation condition (14, 21 and 28 MPa) [78]

The polylactide/clay nanocomposites were also prepared by melt intercalation method. At low foaming temperature (100–110 °C), the PLA nanocomposite foam showed smaller cell size and larger cell density compared with pure PLA foams. Addition of clay influenced the cells' formation and lowered cell size with increasing clay content (Fig. 21.11) [78].

The PLA-clay foams obtained at subcritical conditions within a short time showed better specific tensile properties compared with the pure PLA foams. These nanostructured microcellular materials can have potentially a broad application spectrum in, e.g., packaging industry since they show high biodegradation rate while maintaining good mechanical properties [79–82].

4.6 Others

Mazzola et al. and Xue et al. investigated solid-state foaming of composite foams based on epoxies. The effect of layered silicates on mechanical properties, foaming efficiency, and cell morphology was evaluated. Since solid-state foaming does not involve processing in the liquid state, the possibility of obtaining an intercalated or exfoliated structures is strongly reduced. It has been found that the composite foams exhibited better mechanical properties when a filler content was in the range of 2.5–3 wt %. Addition of MMT to the polymer matrix during the foaming process reduced the foaming efficiency due to interactions of MMT with polymer matrix [83, 84].

For the PMMA foams containing MMT, the mechanical properties improved, demonstrating the mechanical reinforcement effect of the OMMT nanoparticles. The most profound effect was found for 10 wt % OMMT concentration [85]. In an another development, PMMA/clay nanocomposite foams containing 0.5, 1, and 2 wt % of LS nanofiller were prepared and the mechanical properties were improved at 0.5 wt % clay loading, but further increase in the mass fraction of clay was detrimental to the mechanical properties because of the agglomeration of the clay platelets [86].

Besides, PMMA composites with organoclay were characterized by a decreased value of dielectric constant and a slightly increased thermal conductivity [87].

Porous PMMA/Na^+–MMT membranes are considered as cation-exchange materials for cationic dye removal. An important issue is an optimal clay content to prevent its aggregation inside the membrane which hinders the dye molecules to reach the ion-exchange sites [88].

For poly(ε-caprolactone) (PCL) foams, the role of nanometric particles on the foaming process was investigated. The type, shape, dimension, and surface functionality of the nanoparticles, as well as the kind of blowing agent, considerably affected the foaming process [89].

Highly exfoliated porous PCL/clay nanocomposites were prepared by inserting the blowing agent into the galleries of an organoclay before nanocomposite formation. In this novel approach, the blowing agent carries out a dual role in the foaming process. The first one is formation of bubbles, and the second one is facilitation of clay exfoliation. Two types of compounds were used as the chemical blowing agents – azodicarbonamide (ADC) and sodium bicarbonate (SB). The insertion of the blowing agent into clay galleries before foaming improved the exfoliation degree of clay in poly(ε-caprolactone). The addition of clay influences the nucleation and cell growth, decreasing the pore size by 39–46 % and leading to the formation of more uniform cell structures. The compressive modulus and stress of the porous polymer increased, eliminating the effect of the density changes. These biocompatible porous materials are expected to find applications as biodegradable packagings and carriers of drugs, as well as medical and diagnostic devices [90].

Ethylene vinyl acetate (EVA)-based foamed composites with MMT have been obtained and characterized. EVA nanocomposite foams with and without EtBC-*g*-MAH display lower compression set than EVA/EtBC foams at the same density. Based on the compression set results, OH-MMT (Cloisite 30B) was more effective in improvement of this parameter than DM-MMT (Cloisite 15A) because of -OH-group-induced interactions [91]. Addition of organoclay to EVA matrix resulted in an increase in cell density and mechanical/impact strength, and in a decrease in the cell size, as evidenced by cell morphology measurements (Fig. 21.12).

The rheological results showed a viscosity increase as the organoclay loading increased in the EVA-*g*-MA nanocomposites – Fig. 21.13.

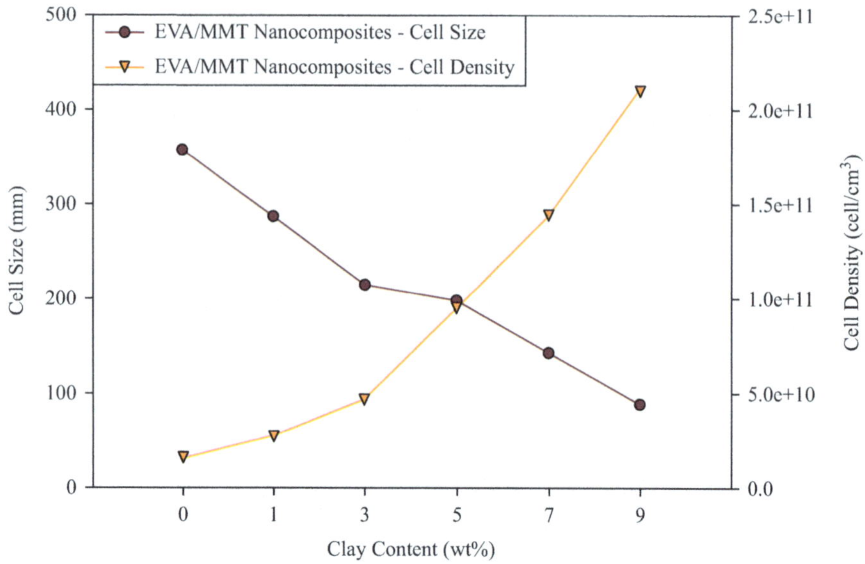

Fig. 21.12 The effect of clay concentration on cell density and cell size in EVA nanocomposite [92]

Another group of cellular nanocomposite materials are hydrogels prepared by in situ free-radical polymerization using inorganic clay as a cross-linker, and $CaCO_3$ as a pore-forming agent. Transmission electron microscopy analysis shows that the clay platelets were partially exfoliated in the nanocomposite hydrogels, and the pore size increased with an increase of the $CaCO_3$ content [93].

Higher water absorption of nanocomposite hydrogels led to a reduction of elasticity modulus due to their soft polymer network in swollen form. These hydrogels can be used in pharmaceutical formulations as controlled-release polymer systems in which the release of solutes is dependent on a diffusion process [94].

Novel porous membranes composed of clay particles dispersed in polysulfone (PSU) matrix were obtained by phase change method and investigated by SEM, TEM, and X-ray diffraction techniques. Among different types of clay applied, Cloisite™ 30B (MMT modified by methyl tallow bis-2-hydroxyethyl quaternary ammonium salt) yielded the highest moduli and the best performance to increase membrane wet ability [95].

In general, these materials reduce the hydrophobic interaction at the surface and require a lower pressure to obtain an acceptable permeation degree [96].

The foam processing of intercalated polycarbonate (PC)/clay nanocomposites was realized as a batch process in an autoclave, having different amounts of clay. As a foaming agent, supercritical CO_2 was used. The incorporation of nanoclay at low loadings induced heterogeneous nucleation because of its lower activation energy barrier compared with homogeneous nucleation, as revealed by the characterization of the interfacial tension between the bubbles and the matrix [97, 98].

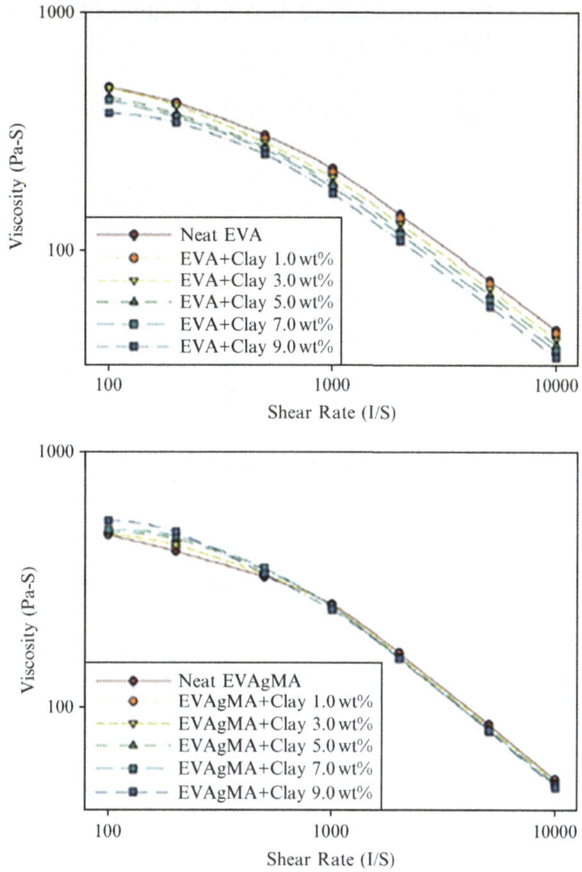

Fig. 21.13 Rheological tests of the EVA nanocomposites [92]

Another material investigated was porous poly(vinyl alcohol) (PVA)/sepiolite in form of scaffolds containing 0–10 wt % of sepiolite nanofibers that were fabricated by a freeze-drying process and cross-linked with a small amount of a low molecular weight poly(acrylic acid) (PAA). The addition of sepiolite changed the degree of anisotropy due to its fibrous structure, and increased the average pore size from 50 (0 wt % sepiolite) to 92 μm (10 wt % sepiolite) and the porosity from 84.6 % to 93.2 %. However, the mechanical properties of the PVA nanocomposite foams were found to worsen with sepiolite additions due to an increase in the porosity and the degree of anisotropy [99].

Styrene-*co*-acrylonitrile (SAN)/clay nanocomposites were prepared in a foaming process using supercritical CO_2 as a blowing agent. Material foaming conditions strongly influence the heterogeneous nucleation efficiency of the nanofiller which affects the final cell density. On the other hand, the foaming with supercritical CO_2 is a quite flexible process which gives access to a large range of foams morphologies from macrocellular to microcellular [100].

5 Bubble Nucleation in Foamed Polymers/Layered Silicate Nanocomposites

The external (nano)particles may serve as heterogeneous nucleating sites during foaming process. However, the broadly applied assumptions in theoretical modeling that those sites are smooth and display planar surfaces are often misleading. On the contrary, the boundaries can contain conical pits due to surface roughness, as shown in Fig. 21.14 [101].

The nucleation agents influence the uniform distribution of cells in a porous material in such a way that there are more cells in the agent-rich area and less cells in agent-deficient areas which leads to a nonuniform cell size distribution. The number and size of the bubbles are determined by the concentration of the foaming agent and by the method used to mix the blowing agents and the polymer matrix.

The role of nanostructured additives in the polymer foaming process is still under investigation. Nanoparticles usually enhance nucleation compared to conventional micro-sized fillers as they decrease substantially the energy barrier for cell nucleation, whereby the change of temperature, gas pressure, and gas concentration in a polymer system influences a secondary nucleation process. This process can result in a broad cell size distribution, because the cells do not generate simultaneously. The shortening of the nucleation time interval yields a narrow cell distribution in microcellular foams, because the cell nucleation rate increases [19, 78, 89, 100, 101].

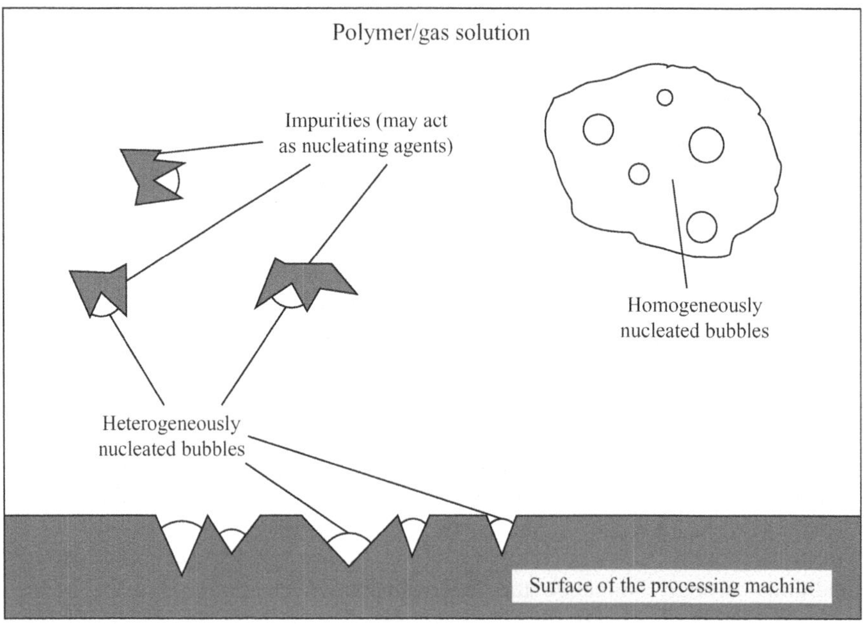

Fig. 21.14 Homogeneous and heterogeneous nucleation in a polymer–gas solution (adopted from [101])

The application of super critical fluids (SCFs) to the preparation of foams leads to reduction of matrix viscosity and/or helps to nucleate nanoscale bubbles. SCFs act as diluent/swelling agents for the polymer phase, solvents for monomers or reactants, and as blowing agents to produce foamed materials.

From technological point of view, the viscosity of the polymer/gas system plays a crucial role during the dispersion of nanoparticles in a polymer matrix, or during the formation of nanofoam bubbles. For instance, when CO_2 is dissolved in polymer matrix, the viscosity is reduced. The rheological behavior becomes more complex due to synergetic effect of nanofillers and CO_2 when the nanometer-sized fillers are dispersed in the polymer matrix. Additionally, SCFs show "gas-like" diffusivity at liquid-like density which are desirable features for nanocomposite polymer foams' production.

6 Conclusions

Development of nanocomposite foams is one of the latest evolutionary technologies of polymeric foams. The nanocomposite foams offer attractive potential for diversification and application of conventional polymeric materials. Some of them are already commercially available and applied as industrial products, opening new promising dimensions for composite foams. However, for a large-scale production, economically viable routes have to be elaborated to yield materials with tailor-made morphology and enhanced properties. Such approaches are being developed and will hopefully lead to commercial products (automotive, aerospace, biomedical, etc.) in the near future [102, 103]. Polymer foams modified with layered silicates show remarkable properties as lightweight and high strength materials. In the most widely applied fabrication method of polymer/layered silicate nanocomposite – polymer melt intercalation – the polymer chains diffuse under processing conditions in between the silicate platelets. Depending on the degree of penetration, intercalated or exfoliated structures are formed. The foaming process can be carried out using physical blowing agents, such as chlorofluorocarbons and pentanes, or chemical blowing agents, such as carbon dioxide that is generated in the reaction of di-isocyanate with water during polyurethane foams synthesis. Up to now, melt intercalation/foaming method has been widely applied on lab and industrial scale to fabricate cellular nanocomposites. However, the crucial issue is to control the cell morphology that is strongly influenced by the presence of nanoadditives. It has been found that by introducing a small amount of nanoclay to the foam, the cell density and expansion ratio were considerably improved, due to both the enhanced melt strength and the clay nucleation effect. In general, nanoparticles usually enhance nucleation compared to conventional micro-sized fillers as they decrease substantially the energy barrier for cell nucleation, whereby the change of temperature, gas pressure, and gas concentration in a polymer system influences a secondary nucleation process. Importantly, the nanocomposite orientation around the foam cells promises to yield enhanced physical and mechanical properties of the nanocomposite, as it was evidenced in both micro- and macrocellular polystyrene foams. When applying processing methods, such as microcellular injection molding, the microstructure and the mechanical properties of the porous materials depend on the process conditions and nanoclay features. Some recent developments refer to the use of supercritical carbon dioxide as a foaming agent to produce polylactide/layered silicate biodegradable nanocellular polymeric foams with a closed-cell structure for, e.g., packaging applications. The application of supercritical fluids to the preparation of foams was found to lead to reduction of matrix viscosity and/or to nucleate nanoscale bubbles. These fluids act as diluent/swelling agents for the polymer phase, solvents for monomers or reactants, and as blowing agents to produce foamed materials. Special attention deserves the flammability resistance effect of nanoparticles in foamed matrix which is based on the flame barrier mechanism involving formation of a high-performance carbonaceous silicate char yielding insulating action. This effect was demonstrated in polyurethane/clay hard nanofoams with superior flame retardancy, commercialized in automobile seats manufacturing. Biomedical applications in the area of tissue engineering scaffolds benefit from the high surface area and rich surface chemistry of layered silicates that can be potentially useful as carriers for desired biofunctionalities. However, this area of applications requires vast toxicological studies to confirm the safe use of nano-modified foams.

Acknowledgments This work was partially funded by the National Science Centre in Poland under contract No. DEC-2011/02/A/ST8/00409.

References

1. Ibeh CC, Bubacz M (2008) Current trends in nanocomposite foams. J Cell Plast 44:493–515
2. Lee LJ, Zeng C, Cao X, Han X, Shen J, Xu G (2005) Polymer nanocomposite foams. Comp Sci Technol 65:2344–2363
3. Hussain F, Hojjati M, Okamoto M, Gorga RE (2006) Polymer-matrix nanocomposites, processing, manufacturing, and application: an overview. J Comp Mater 40:1511–1575
4. Ray SS, Okamoto M (2003) Polymer/layered silicate nanocomposites: a review from preparation to processing. Prog Polym Sci 28:1539–1641
5. Pagacz J, Pielichowski K (2007) Modification of layered silicates for applications in nanotechnology. Tech Trans Chem 1:134–147
6. Alexandre M, Dubois P (2000) Polymer-layered silicate nanocomposites: preparation, properties and uses of a new class of materials. Mater Sci Eng 28:1–63
7. Nguyen QT, Baird DG (2006) Preparation of polymer-clay nanocomposites and their properties. Adv Polym Technol 25:270–285
8. Antunes M, Velasco JI, Realinho V, Solórzano E (2009) Study of the cellular structure heterogeneity and anisotropy of polypropylene and polypropylene nanocomposite foams. Polym Eng Sci 49:2400–2413
9. Velasco JI, Antunes M, Ayyad O, López-Cuesta JM, Gaudon P, Saiz-Arroyo C, Rodríguez-Pérez MA, de Saja JA (2007) Foaming behaviour and cellular structure of LDPE/hectorite nanocomposites. Polymer 48:2098–2108
10. Paul DR, Robeson LM (2008) Polymer nanotechnology: Nanocomposites. Polymer 49:3187–3204. doi:10.1016/j.polymer.2008.04.017

11. Giacomelli M, Pielichowski K, Leszczyńska A (2012) Thermoplastic polymer nanocomposites with montmorillonite-Lab vs industrial scale fabrication. IOP Conf Series: Mater Sci Eng 40:1–6
12. Seo WJ, Sung YT, Kim SB, Lee YB, Choe KH, Choe SH, Sung JY, Kim WN (2006) Effects of ultrasound on the synthesis and properties of polyurethane foam/clay nanocomposites. J Appl Polym Sci 102:3764–3773
13. Zhang J et al (2008) Polymerically modified layered silicates: an effective route to nanocomposites. J Nanosci Nanotechnol 8:1597–1615
14. Indennidate L, Cannoletta D, Lionetto F, Greco A, Maffezzoli A (2010) Nanofilled polyols for viscoelastic polyurethane foams. Polym Int 59:486–491
15. Leszczyńska A, Njuguna J, Pielichowski K, Banerjee JR (2007) Polymer/montmorillonite nanocomposites with improved thermal properties: part I. Factors influencing thermal stability and mechanisms of thermal stability improvement. Thermochim Acta 453:75–96
16. Leszczyńska A, Njuguna J, Pielichowski K, Banerjee JR (2007) Polymer/montmorillonite nanocomposites with improved thermal properties: Part II. Thermal stability of montmorillonite nanocomposites based on different polymeric matrixes. Thermochim Acta 454:1–22
17. Zhu M, Bandyopadhyay-Ghosh S, Khazabi M, Cai H, Correa C, Sain M (2012) Reinforcement of soy polyol-based rigid polyurethane foams by cellulose microfibers and nanoclays. J Appl Polym Sci 124:4702–4710
18. Semenzato S, Lorenzetti A, Modesti M, Ugel E, Hrelja D, Besco S, Michelin RA, Sassi A, Facchin G, Zorzi F, Bertani R (2009) A novel phosphorus polyurethane FOAM/montmorillonite nanocomposite: preparation, characterization and thermal behavior. Appl Clay Sci 44:35–42
19. Kang JW, Kim JM, Kim MS, Kim YH, Kim WN, Jang W, Shin DS (2009) Effects of nucleating agents on the morphological, mechanical and thermal insulating properties of rigid polyurethane foams. Macromol Res 17:856–862
20. Xu Z, Tang X, Gu A, Fang Z (2007) Novel preparation and mechanical properties of rigid polyurethane foam/organoclay nanocomposites. J Appl Polym Sci 106:439–447
21. Kim SH, Lee MC, Kim HD, Park HC, Jeong HM, Yoon KS, Kim BK (2010) Nanoclay reinforced rigid polyurethane foams. J Appl Polym Sci 117:1992–1997
22. Cao X, James Lee L, Widya T, Macosko C (2005) Polyurethane/clay nanocomposites foams: processing, structure and properties. Polymer 46:775–783
23. Zatorski W, Brzozowski ZK, Kolbrecki A (2008) New developments in chemical modification of fire-safe rigid polyurethane foams. Polym Degrad Stab 93:2071–2076
24. Huang G, Gao J, Li Y, Han L, Wang X (2010) Functionalizing nano-montmorillonites by modified with intumescent flame retardant: preparation and application in polyurethane. Polym Degrad Stab 95:245–253
25. Modesti M, Lorenzetti A, Besco S, Hrelja D, Semenzato S, Bertani R, Michelin RA (2008) Synergism between flame retardant and modified layered silicate on thermal stability and fire behaviour of polyurethane nanocomposite foams. Polym Degrad Stab 93:2166–2171
26. Lorenzetti A, Hrelja D, Besco S, Roso M, Modesti M (2010) Improvement of nanoclays dispersion through microwave processing in polyurethane rigid nanocomposite foams. J Appl Polym Sci 115:3667–3674
27. Widya T, Macosko CW (2005) Nanoclay-modified rigid polyurethane foam. J Macromol Sci, Part B: Phys 44:897–908
28. Umasankar Patro T, Harikrishnan G, Misra A, Khakhar DV (2008) Formation and characterization of polyurethane-vermiculite clay nanocomposite foams. Polym Eng Sci 48:1778–1784
29. Kim YH, Choi SJ, Kim JM, Han MS, Kim WN, Bang KT (2007) Effects of organoclay on the thermal insulating properties of rigid polyurethane foams blown by environmentally friendly blowing agents. Macromol Res 15:676–681
30. Harikrishnan G, Umasankar Patro T, Khakhar DV (2007) Reticulated vitreous carbon from polyurethane foam-clay composites. Carbon 45:531–535
31. Mondal P, Khakhar DV (2007) Rigid polyurethane-clay nanocomposite foams: preparation and properties. J Appl Polym Sci 103:2802–2809

32. Han MS, Kim YH, Han SJ, Choi SJ, Kim SB, Kim WN (2008) Effects of a silane coupling agent on the exfoliation of organoclay layers in polyurethane/organoclay nanocomposite foams? J Appl Polym Sci 110:376–386
33. Kim YH, Kang MJ, Park GP, Park SD, Kim SB, Kim WN (2012) Effects of liquid-type silane additives and organoclay on the morphology and thermal conductivity of rigid polyisocyanurate-polyurethane foams. J Appl Polym Sci 124:3117–3123
34. Njuguna J, Michałowski S, Pielichowski K, Kayvantash K, Walton AC (2011) Fabrication, characterization and low-velocity impact testing of hybrid sandwich composites with polyurethane/layered silicate foam cores. Polym Compos 32:6–13
35. Sung CH, Lee KS, Lee KS, Oh SM, Kim JH, Kim MS, Jeong HM (2008) Sound damping of a PU foam nanocomposite. In: Proceedings of IFOST-2008 – 3rd international forum on strategic technologies, vol 4602982, pp 181–185
36. Javni I, Song K, Lin J, Petrovic ZS (2011) Structure and properties of flexible polyurethane foams with nano- and micro-fillers. Journal of Cellular Plastics 47:357–372
37. Khorasani MM, Ghaffarian SR, Babaie A, Mohammadi N (2010) Foaming behavior and cellular structure of microcellular HDPE nanocomposites prepared by a high temperature process. J Cell Plast 46:173–190. doi:10.1177/0021955X09358237
38. Jo C, Naguib HE (2007) Effect of nanoclay and foaming conditions on the mechanical properties of HDPE-clay nanocomposite foams. J Cell Plast 43:111–121
39. Jo C, Naguib HE (2007) Processing, characterization, and modeling of polymer/clay nanocomposite foams. J Phys: Conf Ser 61:861–868
40. Jo C, Naguib HE (2007) Constitutive modeling of HDPE polymer/clay nanocomposite foams. Polymer 48:3349–3360
41. Lee YH, Park CB, Wang KIH, Lee MH (2005) HDPE-clay nanocomposite foams blown with supercritical CO_2. J Cell Plast 41:487–502
42. Jin D-W, Seol S-M, Kim G-H (2009) New compatibilizer for linear low-density polyethylene (LLDPE)/clay nanocomposites. J Appl Polym Sci 114:25–31
43. Riahinezhad M, Ghasemi I, Karrabi M, Azizi H (2010) An investigation on the correlation between rheology and morphology of nanocomposite foams based on low-density polyethylene and ethylene vinyl acetate blends. Polym Compos 31:1808–1816
44. Hwang S-s, Hsu PP, Yeh J-m, Yang J-p, Chang K-c, Lai Y-z (2009) Effect of clay and compatibilizer on the mechanical/thermal properties of microcellular injection molded low density polyethylene nanocomposites. Int Commun Heat Mass Transf 36:471–479
45. Arunvisut S, Phummanee S, Somwangthanaroj A (2007) Effect of clay on mechanical and gas barrier properties of blown film LDPE/clay nanocomposites. J Appl Polym Sci 106:2210–2217
46. Lee YH, Wang KH, Park CB, Sain M (2007) Effects of clay dispersion on the foam morphology of LDPE/clay nanocomposites. J Appl Polym Sci 103:2129–2134
47. Velasco JI, Antunes M, Ayyad O, Saiz-Arroyo C, Rodríguez-Pérez MA, Hidalgo F, De Saja JA (2007) Foams based on low density polyethylene/hectorite nanocomposites: thermal stability and thermomechanical properties. J Appl Polym Sci 105:1658–1667
48. Seraji SM, Razavi Aghjeh MK, Davari M, Salami Hosseini M, Khelgati S (2011) Effect of clay dispersion on the cell structure of LDPE/clay nanocomposite foams. Polym Compos 32:1095–1105
49. Zhai W, Kuboki T, Wang L, Park CB, Lee EK, Naguib HE (2010) Cell structure evolution and the crystallization behavior of polypropylene/clay nanocomposites foams blown in continuous extrusion. Ind Eng Chem Res 49:9834–9845
50. Huang H-X, Wang J-K, Sun X-H (2008) Improving of cell structure of microcellular foams based on polypropylene/high-density polyethylene blends. J Cell Plast 44:69–85
51. Okamoto M, Nam PH, Maiti P, Kotaka T, Nakayama T, Takada M, Ohshima M, Usuki A, Hasegawa N, Okamoto H (2001) Biaxial flow-induced alignment of silicate layers in polypropylene/clay nanocomposite foam. Nano Lett 1:503–505
52. Nofar M, Majithiya K, Kuboki T, Park CB (2012) The foamability of low-melt-strength linear polypropylene with nanoclay and coupling agent. J Cell Plast 48:271–287

53. Taki K, Yanagimoto T, Funami E, Okamoto M, Ohshima M (2004) Visual observation of CO_2 foaming of polypropylene-clay nanocomposites. Polym Eng Sci 44:1004–1011
54. Antunes M, Velasco JI, Realinho V, Solórzano E (2009) Study of the cellular structure heterogeneity and anisotropy of polypropylene and polypropylene nanocomposite foams. Polym Eng Sci 49:2400–2413. doi:10.1002/pen.21488
55. Chang Y-W, Kim S, Kang SC, Bae S-Y (2011) Thermomechanical properties of ethylene-propylene-diene terpolymer/organoclay nanocomposites and foam processing in supercritical carbon dioxide. Korean J Chem Eng 28:1779–1784
56. Bhattacharya S, Gupta RK, Jollands M, Bhattacharya SN (2009) Foaming behavior of high-melt strength polypropylene/clay nanocomposites. Polym Eng Sci 49:2070–2084
57. Nam PH, Maiti P, Okamoto M, Kotaka T, Nakayama T, Takada M, Ohshima M, Usuki A, Hasegawa N, Okamoto H (2002) Foam processing and cellular structure of polypropylene/clay nanocomposites. Polym Eng Sci 42:1907–1918
58. Strauss W, D'Souza NA (2004) Supercritical CO_2 processed polystyrene nanocomposite foams. J Cell Plast 40:229–241
59. Wang Z, Du X, Yu H, Jiang Z, Liu J, Tang T (2009) Mechanism on flame retardancy of polystyrene/clay composites-the effect of surfactants and aggregate state of organoclay. Polymer 50:5794–5802
60. Ngo TTV, Duchet-Rumeau J, Whittaker AK, Gerard J-F (2010) Processing of nanocomposite foams in supercritical carbon dioxide. Part I: Effect of surfactant. Polymer 51:3436–3444
61. Hwang S-s, Hsu PP, Yeh J-m, Hu C-h, Chang K-c (2009) Effect of organoclay on the mechanical/thermal properties of microcellular injection molded polystyrene-clay nanocomposites. Int Commun Heat Mass Transf 36:799–805
62. Han X, Zeng C, Lee LJ, Koelling KW, Tomasko DL (2003) Extrusion of polystyrene nanocomposite foams with supercritical CO_2. Polym Eng Sci 43:1261–1275
63. Zhu B, Zha W, Yang J, Zhang C, Lee LJ (2010) Layered-silicate based polystyrene nanocomposite microcellular foam using supercritical carbon dioxide as blowing agent. Polymer 51:2177–2184
64. Urbanczyk L, Alexandre M, Detrembleur C, Jérôme C, Calberg C (2010) Extrusion foaming of poly(styrene-co-acrylonitrile)/clay nanocomposites using supercritical CO_2. Macromol Mater Eng 295:915–922
65. Shen J, Cao X, James Lee L (2006) Synthesis and foaming of water expandable polystyrene-clay nanocomposites. Polymer 47:6303–6310
66. Yuan M, Winardi A, Gong S, Turng L-S (2005) Effects of nano- and micro-fillers and processing parameters on injection-molded microcellular composites. Polym Eng Sci 45:773–788
67. Yuan M, Turng L-S, Caulfield DF (2006) Crystallization and thermal behavior of microcellular injection-molded polyamide-6 nanocomposites. Polym Eng Sci 46:904–918
68. Yuan M, Turng L-S, Gong S, Winardi A, Caulfield D (2004) Crystallization behavior of polyamide-6 microcellular nanocomposites. J Cell Plast 40:397–409
69. Zheng W, Lee YH, Park CB (2006) The effects of exfoliated nano-clay on the extrusion microcellular foaming of amorphous and crystalline nylon. J Cell Plast 42:271–288
70. Yuan M, Turng L-S (2005) Microstructure and mechanical properties of microcellular injection molded polyamide-6 nanocomposites. Polymer 46:7273–7292
71. Yoon PJ, Fornes TD, Paul DR (2002) Thermal expansion behavior of nylon 6 nanocomposites. Polymer 43:6727–6741
72. Hwang S-S, Liu S-P, Hsu PP, Yeh J-M, Yang J-P, Chang K-C, Chu S-N (2011) Effect of organoclay and preparation methods on the mechanical/thermal properties of microcellular injection molded polyamide 6-clay nanocomposites. Int Commun Heat Mass Transf 38:1219–1225
73. Kharbas H, Nelson P, Yuan M, Gong S, Turng L-S, Spindler R (2003) Effects of nano-fillers and process conditions on the microstructure and mechanical properties of microcellular injection molded polyamide nanocomposites. Polym Compos 24:655–671

74. Ma P, Wang X, Liu B, Li Y, Chen S, Zhang Y, Xu G (2012) Preparation and foaming extrusion behavior of polylactide acid/polybutylene succinate/montmorillonoid nanocomposite. J Cell Plast 48:191–205
75. Sinha Ray S, Okamoto K, Yamada K, Okamoto M (2002) Novel porous ceramic material via burning of polylactide/layered silicate nanocomposite. Nano Lett 2:423–425. doi:10.1021/nl020284g
76. Ozkoc G, Kemaloglu S, Quaedflieg M (2010) Production of poly(lactic acid)/organoclay nanocomposite scaffolds by microcompounding and polymer/particle leaching. Polym Compos 31:674–683
77. Fujimoto Y, Sinha Ray S, Okamoto M, Ogami A, Yamada K, Ueda K (2003) Well-controlled biodegradable nanocomposite foams: from microcellular to nanocellular. Macromol Rapid Commun 24:457–461
78. Ema Y, Ikeya M, Okamoto M (2006) Foam processing and cellular structure of polylactide-based nanocomposites. Polymer 47:5350–5359
79. Liao X, Nawaby AV, Naguib HE (2012) Porous poly(lactic acid) and PLA-nanocomposite structures. J Appl Polym Sci 124:585–594. doi:10.1002/app.34994
80. Tsivintzelis I, Marras SI, Zuburtikudis I, Panayiotou C (2007) Porous poly(L-lactic acid) nanocomposite scaffolds prepared by phase inversion using supercritical CO_2 as antisolvent. Polymer 48:6311–6318
81. Tsimpliaraki A, Tsivintzelis I, Marras SI, Zuburtikudis I, Panayiotou C (2011) The effect of surface chemistry and nanoclay loading on the microcellular structure of porous poly(d, l lactic acid) nanocomposites. J Supercrit Fluids 57:278–287
82. Di Y, Iannace S, Di Maio E, Nicolais L (2005) Poly(lactic acid)/organoclay nanocomposites: thermal, rheological properties and foam processing. J Polym Sci B 43:689–698
83. Mazzola L, Bemporad E, Squeo EA, Trovalusci F, Tagliaferri V (2011) Filler-matrix interaction in solid-state foaming of composite foams. J Cell Plast 47:31–43
84. Xue S, Pinnavaia TJ (2008) Porous synthetic smectic clay for the reinforcement of epoxy polymers. Microporous Mesoporous Mater 107:134–140
85. Realinho V, Antunes M, Martínez AB, Velasco JI (2011) Influence of nanoclay concentration on the CO_2 diffusion and physical properties of PMMA montmorillonite microcellular foams. Ind Eng Chem Res 50:13819–13824
86. Jo C, Fu J, Naguib HE (2006) Constitutive modeling for intercalated PMMA/clay nanocomposite foams. Polym Eng Sci 46:1787–1796
87. Yeh J-M, Chang K-C, Peng C-W, Lai M-C, Hung C-B, Hsu S-C, Hwang S-S, Lin H-R (2009) Effect of dispersion capability of organoclay on cellular structure and physical properties of PMMA/clay nanocomposite foams. Mater Chem Phys 115:744–750
88. Lin R-Y, Chen B-S, Chen G-L, Wu J-Y, Chiu H-C, Suen S-Y (2009) Preparation of porous PMMA/Na^+-montmorillonite cation-exchange membranes for cationic dye adsorption. J Membr Sci 326:117–129
89. Marrazzo C, Di Maio E, Iannace S (2008) Conventional and nanometric nucleating agents in poly(ε-caprolactone) foaming: crystals vs. bubbles nucleation. Polym Eng Sci 48:336–344. doi:10.1002/pen.20937
90. Istrate OM, Chen B (2012) Porous exfoliated poly(ε-caprolactone)/clay nanocomposites: preparation, structure, and properties. J Appl Polym Sci 125:E102–E112. doi:10.1002/app.36336
91. Park K-W, Kim G-H, Chowdhury SR (2008) Improvement of compression set property of ethylene vinyl acetate copolymer/ethylene-1-butene copolymer/organoclay nanocomposite foams. Polym Eng Sci 48:1183–1190
92. Hwang S-S, Liu S-P, Hsu PP, Yeh J-M, Yang J-P, Chen C-L (2012) Morphology, mechanical, and rheological behavior of microcellular injection molded EVA-clay nanocomposites. Int Commun Heat Mass Transf 39:383–389
93. Ma J, Zhang L, Li Z, Liang B (2008) Preparation and characterization of porous poly (*N*-isopropylacrylamide)/clay nanocomposite hydrogels. Polym Bull 61:593–602

94. Guilherme MR, Fajardo AR, Moia TA, Kunita MH, Gonçalves MC, Rubira AF, Tambourgi EB (2010) Porous nanocomposite hydrogel of vinyled montmorillonite-crosslinked maltodextrin-co-dimethylacrylamide as a highly stable polymer carrier for controlled release systems. Eur Polym J 46:1465–1474
95. Monticelli O, Bottino A, Scandale I, Capannelli G, Russo S (2007) Preparation and properties of polysulfone-clay composite membranes. J Appl Polym Sci 103:3637–3644
96. Tran ATT, Patterson DA, James BJ (2012) Investigating the feasibility of using polysulfone-montmorillonite composite membranes for protein adsorption. J Food Eng 112:38–49
97. Ito Y, Yamashita M, Okamoto M (2006) Foam processing and cellular structure of polycarbonate-based nanocomposites. Macromol Mater Eng 291:773–783
98. Mitsunaga M, Ito Y, Ray SS, Okamoto M, Hironaka K (2003) Intercalated polycarbonate/clay nanocomposites: nanostructure control and foam processing. Macromol Mater Eng 288:543–548. doi:10.1002/mame.200300097
99. Killeen D, Frydrych M, Chen B (2012) Porous poly(vinyl alcohol)/sepiolite bone scaffolds: preparation, structure and mechanical properties. Mater Sci Eng C 32:749–757
100. Urbanczyk L, Calberg C, Detrembleur C, Jérôme C, Alexandre M (2010) Batch foaming of SAN/clay nanocomposites with $scCO_2$: a very tunable way of controlling the cellular morphology. Polymer 51:3520–3531
101. Leung SN, Park CB, Li H (2010) Effects of nucleating agents shapes and interfacial properties on cell nucleation. J Cell Plast 46:441–460
102. Okamoto M (2006) Polymer/layered silicate nano-composites advanced polymeric materials engineering, Graduate School of Engineering Toyota Technological Institute, Nagoya, Japan. Int Polym Process XXI:487–496
103. Kurahatti RV et al (2010) Defence applications of polymer nanocomposites. Def Sci J 60:551–563

Polymer-Layered Silicate Nanocomposite Membranes for Fuel Cell Application 22

Ananta Kumar Mishra, Tapas Kuila, Nam Hoon Kim, and Joong Hee Lee

Contents

A.K. Mishra
BIN Fusion Team, Department of Polymer and Nano Science and Technology, Chonbuk National University, Jeonju, Jeonbuk, Republic of Korea

T. Kuila
Department of BIN Fusion Technology, Chonbuk National University, Jeonju, Jeonbuk, Republic of Korea

N.H. Kim
Department of Hydrogen and Fuel Cell Engineering, Chonbuk National University, Jeonju, Jeonbuk, Republic of Korea

J.H. Lee (✉)
Advanced Wind Power Research Center, Department of Polymer and Nano Science and Technology, Chonbuk National University, Jeonju, Jeonbuk, Republic of Korea

Department of BIN Fusion Technology, Chonbuk National University, Jeonju, Jeonbuk, Republic of Korea
e-mail: jhl@jbnu.ac.kr

J.K. Pandey et al. (eds.), *Handbook of Polymernanocomposites. Processing, Performance and Application – Volume A: Layered Silicates*, DOI 10.1007/978-3-642-38649-7_13,

Abstract
Fuel cells have gained increasing interest in this realm due to their promising emission-free energy generation capability. Proton exchange membrane fuel cells (PEMFCs) and direct methanol fuel cells (DMFCs) are the most suitable candidates for this purpose due to their wide range of energy generation capability. The polymer membranes used in PEMFCs and DMFCs play vital role in transporting the protons from anode to cathode. Nafion is the most widely used and commercialized membrane for this application at low-temperature and highly humidified conditions. The drawbacks associated with low-temperature PEMFCs (heat and water management, CO catalyst poisoning, and fuel crossover) can be avoided by increasing the operating temperature of the fuel cells. However, the drastic decrease in the conductivity of Nafion above 80 °C and low humidity has paved the path towards the development of new membranes and technologies. Additions of layered silicates to the polymer membranes have been observed to be beneficial in this regard owing to their high hydrophilicity, low cost, easy availability, and barrier property towards fuel crossover. The resulting composite membranes also infer improved mechanical and thermal properties, along with water uptake of the membranes escorting towards superior performance of the nanocomposite membranes at high temperature compared to the virgin membrane.

Keywords
Cell Performance • Fuel Cell • Membrane • Nanocomposite • Proton Conductivity

1 Introduction

Energy is highly essential for the growth of the modern society due to its requirements by most of the modern technologies. The main source of power in the present circumstances is the nonrenewable fossil fuels. The limited resource and emission of toxic gases from fossil fuels have prompted to the invention of new techniques of power generation. The green techniques adopted for the energy generation are insufficient for this purpose. Fuel cells have emerged as a new emission-free technology to fulfill the power requirements. It involves the direct conversion of chemical energy into electrical energy by avoiding the intermediate steps required by diesel power generators (Fig. 22.1). Hence, fuel cells can minimize the power losses associated with other power generators [1].

Sir William Robert Grove has invented the first fuel cell in 1839 [1]. It was based on the electrochemical conversions of hydrogen and oxygen into electricity and water [2–4]. This has directed towards the development of several fuel cells, namely, proton exchange membrane fuel cell (PEMFC), direct methanol fuel cell (DMFC), solid oxide fuel cell, alkaline fuel cell, molten carbonate fuel cell, etc. [5–14]. Among these fuel cells, PEMFC possesses wide range of energy generation capability, whereas DMFC possesses the merit of longer operational lifetime and the ability to refuel.

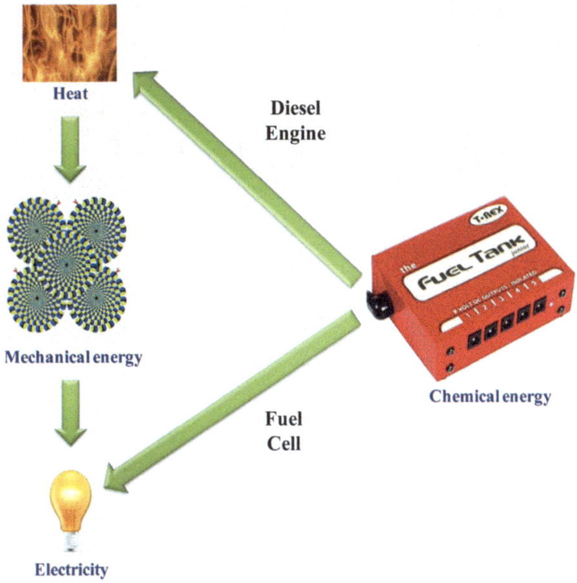

Fig. 22.1 Electricity generation in a diesel engine and fuel cell (see Ref. [1])

In both PEMFC and DMFC, the catalytically oxidized proton at the anode is dragged towards the cathode through a polymer membrane, and simultaneously, the electron generated at the anode moves towards the cathode through an external electrical circuit. The combination of the transmitted protons, electrons, and the reduced oxygen at the cathode surface generate water. However, the electron transmitted through the external electrical circuit is responsible for the generation of electricity [1, 7, 8]. The schematic depiction of PEMFC and DMFC is shown in Figs. 22.2 and 22.3, respectively. The sources of proton in the case of PEMFC and DMFC are hydrogen (H_2) gas and dilute methanol, respectively. The mechanism of proton generation in PEMFCs and DMFCs is shown in Eqs. 22.1 and 22.2, respectively.

$$H_2 \rightarrow H^+ + e^- \tag{22.1}$$

$$2CH_3OH + H_2O \rightarrow 6H^+ + 6e^- + CO_2 \tag{22.2}$$

The role of the membrane in both PEMFC and DMFC is to provide a path for proton conduction from anode to cathode. Hence, limited numbers of membranes are suitable for these applications due to the insulating nature of polymers towards protons. The main requirements of polymer membranes suitable for fuel cell applications include high proton conductivity, good electrical insulation, high thermomechanical and chemical (oxidative and hydrolytic) stability, cost-effectiveness, good barrier property, low swelling stresses, and capability for membrane electrode assembly (MEA) fabrication.

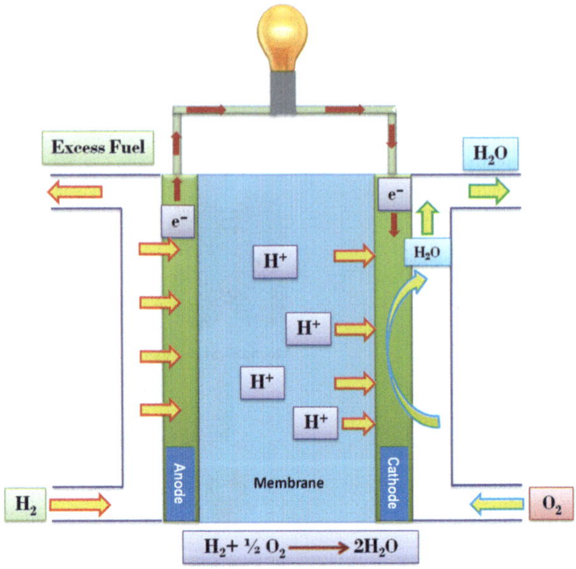

Fig. 22.2 Schematic depiction of PEMFC

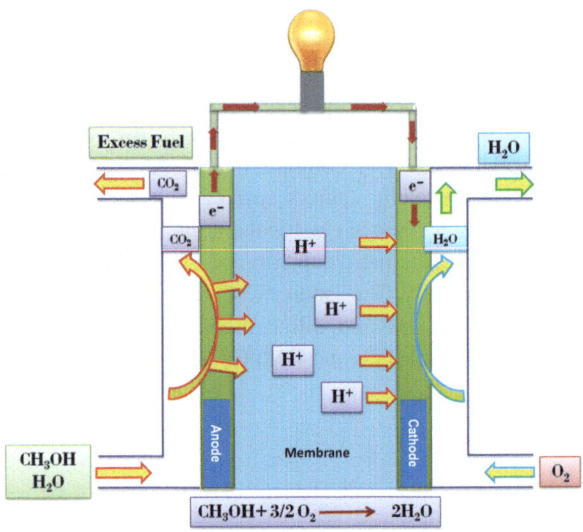

Fig. 22.3 Schematic depiction of DMFC

2 Development of Polymer Membranes for Fuel Cell Applications

The first polymer membrane developed for the Gemini space program by GE was based on the sulfonated polystyrene divinylbenzene copolymer, but high membrane cost and very short life span had limited its application. Nafion® (introduced by Dupont in 1972) is the most suitable and commercialized membrane till date with

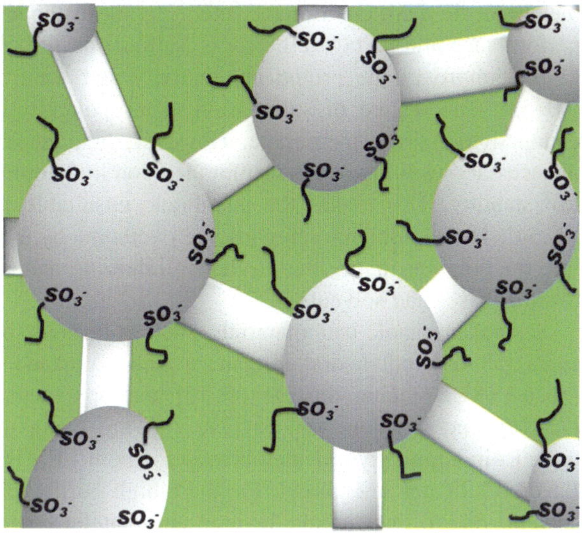

Fig. 22.4 Ion conduction channels present in Nafion under highly humidified conditions

a very high conductivity of 0.7–1.0 S cm^{-1} at room temperature and 100 % relative humidity (RH). The perfluorinated backbone in Nafion® provides strength to the membrane, whereas the sulfonic acid group provides the path for proton conduction. Different models (such as cluster network model [15], core-shell model [16, 17], local-order model [18–20], sandwich model [21], rod model [22], and parallel water channel model [23]) have been proposed based on small-angle X-ray scattering (SAXS), wide-angle X-ray diffraction studies, and solid-state NMR studies to describe the proton conduction mechanism in Nafion. The proposed models are still under investigation; however, all of the models suggest the presence of interconnected ion channels in Nafion® (Fig. 22.4). The $-SO_3H$ groups on the Nafion® backbone self-organize to form a hydrophilic water channel under highly humidified conditions, through which small ions can be easily transported. Under low hydration conditions (at low humidity and high temperature), the ionic clusters are disconnected from each other leading to inferior conductivity of Nafion [1, 7, 13]. PEMFCs operating above 100 °C are preferred over that operating at low temperature to overcome several disadvantages (like CO catalyst poisoning and heat and water management) associated with the later [1, 6, 7, 24]. Inferior proton conductivity of Nafion at high temperature and low humidity has prompted towards the development of new polymer membranes such as polyethersulfone (PES) [25, 26], polyetheretherketone (PEEK) [27–30], polyimide (PI) [31–33], polybenzimidazole (PBI) [34–41], polystyrene-acrylonitrile (SAN) [42, 43], and polyvinylidene fluoride (PVDF) [44–46] to meet the U. S. Department of Energy (DOE) targets [47, 48].

The membrane suitable for DMFC should possess acceptable proton conductivity along with good barrier property to prevent methanol crossover and water transport (by diffusion or electroosmotic drag). Methanol diffusion through the membrane in DMFC from the anode to the cathode imparts reduced fuel cell performance and voltage efficiency. Sulfonation of polymer in both PEMFC and DMFC improves the proton conductivity and water uptake of the membranes [49–52]. The degree of sulfonation is also directly proportional to the proton conductivity of the membrane. However, high degree of sulfonation leads to unnecessary swelling of the membrane upon hydration and decreases the mechanical stability. This makes the membrane unsuitable for MEA fabrication [1, 7, 53, 54]. These problems can partly be overcome by incorporating inorganic fillers into the polymer matrix due to their reinforcing nature and high barrier property towards gases and solvents.

Several inorganic nanomaterials, such as layered silicates (clay) [1, 7, 55–57], silica [58–64], polyhedral oligomeric silsesquioxane [65, 66], titanium dioxide [56, 67, 68], zirconium dioxide [69, 70], heteropolyacids [27, 71], carbon nanotubes [72–76], and graphene oxides [77, 78], are being used for the fabrication of organic–inorganic hybrid membranes for both PEMFC and DMFC applications. Layered silicates are known for their high barrier property towards gases and solvents due to their unique layered and platelet type structure [79]. In addition, they can improve the mechanical properties of their respective nanocomposites due to their reinforcing nature [79–82]. The high hydrophilicity of clay provides additional benefit to improve the proton conductivity of the nanocomposites compared to the virgin membranes [1].

3 Proton Conduction Mechanism in Membrane

The proton conduction through polymer membranes basically follows two types of mechanisms: vehicle and Grotthus mechanism. The proton requires a vehicle or carrier in vehicle mechanism, while it moves from the anode to the cathode through a network of hydrogen bond in Grotthus mechanism. An artistic presentation of the two mechanisms has been proposed by Kreuer (Fig. 22.5) [83]. The membrane

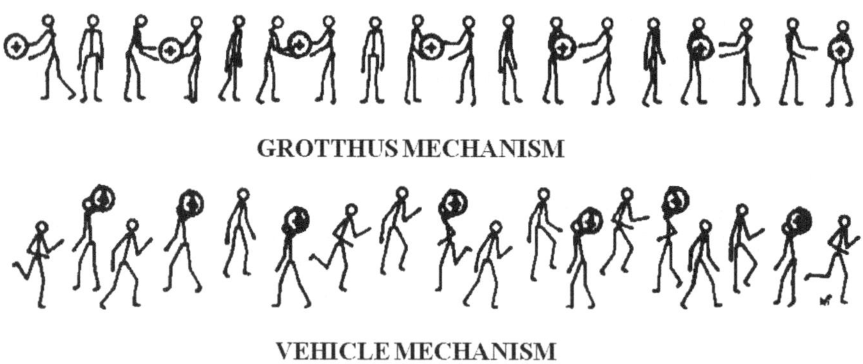

Fig. 22.5 Schematic representation of the phenomenon involved in proton conduction mechanism (see Ref. [83])

conductivity can be further enhanced by the addition of hygroscopic inorganic nanofillers such as silica or clay to the membrane, addition of proton-conducting fillers such as heteropoly acids, modification of the polymer surface with proton-conducting groups like $-SO_3H$ and $-PO_3H$ groups, and doping of the basic polymers (such as polybenzimidazole) with phosphoric acids. Surface modification of inorganic fillers with modifiers end capped with proton-conducting groups can further improve the proton conductivity.

4 Surface Modifications of Nanoclays

Layered silicates are known to provide superior properties to their respective polymer-based nanocomposites in their delaminated (exfoliated) state. In order to increase the compatibility between the organic polymer and inorganic silicates and also to increase the interlayer gallery spacing, surfaces of the silicates are generally modified with organic modifiers [79–82]. Despite the surface modifications, it is difficult to achieve complete exfoliation of clay in the polymer matrix. Generally, a combination of exfoliated and intercalated morphology of layered silicates in polymer matrix is most frequently encountered [79–82].

Inorganic nanoclays are nonconducting in nature towards protons. Hence, nanoclays dispersed in the polymer matrix act as barrier for the proton migration from anode to cathode. Hence, surface modifications of nanoclays with proton-conducting groups can be beneficial for improved proton conductivities of the nanocomposites compared to the virgin polymer and the composites containing unmodified clays [1, 7].

Surface modifications of nanoclays are being carried out in three ways (ionic, covalent, and plasma treatment) [1]. Ionic modifications of nanoclays are performed in two ways (acid activation and conventional ion exchange with alkyl ammonium ions) by the replacement of exchangeable Na^+ ions present in the interlayer gallery spacing of nanoclays. The covalent modification of nanoclays involves the reaction between the –OH groups present on the surface of the nanoclays with alkyl silanes. However, plasma treatment involves the modifiers end capped with vinyl groups. In order to improve the proton conductivity of the nanoclays, surface modifiers containing $-SO_3H$ and $-PO_3H$ groups are being used for the modifications of nanoclays. Different modifiers used for clay modifications via ionic modification technique include sulfanilic acid (SA), dimethyldioctadecylammonium chloride (DMDOC), chitosan, dodecylammonium chloride, and cetyltrimethylammonium chloride (CTAC) [84–89]. Similarly, the modifiers used for the modification of clay by covalent modification technique are 2-acrylamido-2-methyl-1-propanesulfonic acid (AMPS); 3-mercaptopropyltrimethoxysilane (3-MPTMS); 1,3-propane sultone; 1,4-butane sultone; 1,2,2-trifluoro-hydroxy-trifluomethylethane sulfonic acid sultone (FMES); glycidoxypropyl triethoxysilane (GPTES); 3-2-imidazolin-1-yl-propyltrimethoxysilane and amino-propyl trimethoxysilane (APTMS); (3-aminopropyl) triethoxysilane (APS); and imidazolin-1-yl-glycidylpropyltriethoxysilane (IGPTES) [90–92].

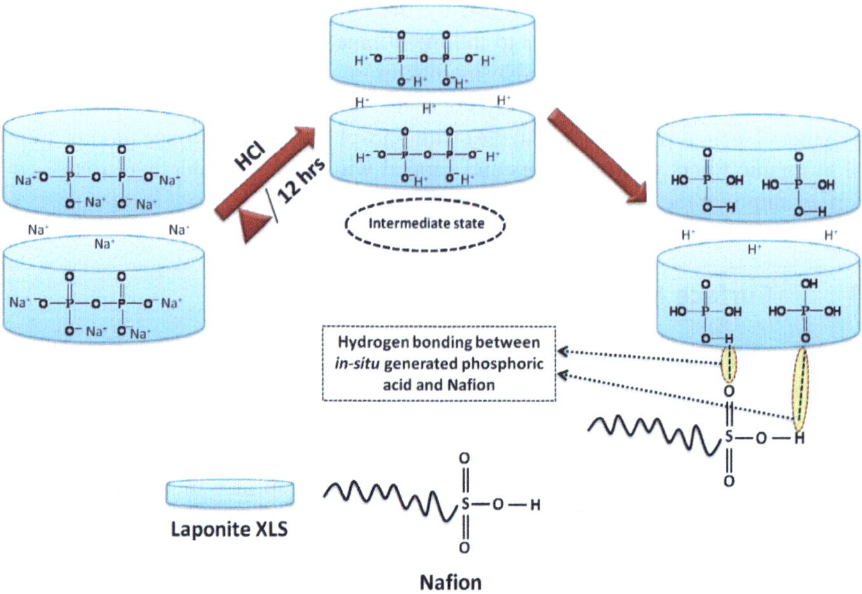

Fig. 22.6 Acid activation of Laponite XLS nanoclay in dilute hydrochloric acid (see Ref. [101])

However, p-styrene sodium sulfonate and SO_2 are being used as modifiers for the modification of clay by plasma treatment technique [93–96].

Acid activation of nanoclay leads to the replacement of Na^+ ions from the interlayer gallery of the nanoclays [97–100]. The ionic mobilities of the H^+ ions are higher than that of the Na^+ ions owing to the lower atomic size of H^+ ions compared to that of the Na^+ ions. Hence, the proton conductivity of the acid-activated clay is expected to be higher than the unmodified clay. Mishra et al. used similar technique for the modification of Laponite XLS (peptized Laponite clay) [101]. Interestingly, in this case, the peptizer ($Na_4P_2O_7$) present on the clay surface hydrolyzed to generate H_3PO_4 (in situ), along with the conventional replacement of Na^+ ions by H^+ ions from the interlayer gallery of the clay (Fig. 22.6). It is worth mentioning here that H_3PO_4 is well known for its high proton conductivity [1, 6].

5 Fabrication of Polymer-Clay Nanocomposite Membranes for Fuel Cell Applications

Different solvents were used for the preparation of Nafion-clay nanocomposite membranes comprised of unmodified, modified, and acid-activated clays [85, 92, 102, 103]. Nafion® 112 solution in ethanol-water (50:50 by weight) was heated at 250 °C in a reactor for 24 h, cooled, and then neutralized by the dropwise addition of 0.1 M NaOH solution. The resulting perfluorosulfonic acid (PFSA) crystals were ball-milled to prepare a white powder. The nanoclay was added to a solution of the PFSA powder in DMF, followed by mixing, sonication, and solvent evaporation to prepare the nanocomposite membrane [56]. Xiuchong et al. mixed modified MMT in 5 % Nafion® dispersion and kept the solution in an autoclave at 150 °C for 4 h with rapid stirring and subsequently evaporated the solvent at 100 °C [104]. Fatyeyeva et al. [105, 106] mixed modified Laponite with polyethylene glycol (PEG) 1500 and water with continuous stirring and ultrasonication. The mixture was then freeze dried to obtain exfoliated clay mixture in PEG. Recast Nafion (from 20 % Nafion® dispersion) was added to the DMF dispersion of the aforementioned clay mixture. The solvent was evaporated in a controlled heating condition to prepare the nanocomposite membrane. However, Bébin et al. added recast Nafion® directly to a 15 % (w/v) DMF-clay suspension to prepare the Nafion®-clay nanocomposite membrane [94].

The polymers like sulfonated polyether ether ketone (SPEEK), polyamide (PAM), sulfonated polysulfone (SPSU), and polystyrene ethylene butylene polystyrene (PSEBS) are soluble in polar solvents like dimethyl acetamide (DMAc), dimethyl formamide (DMF), N-methyl pyrolidone (NMP), and tetrahydrofurane (THF). Hence, the nanoclays and the polymer were dispersed in these solvents to prepare the respective nanocomposites [93, 107–109]. Similarly, water was chosen for the water soluble polymer like poly(vinyl alcohol) (PVOH). The as-prepared PVOH-clay membrane was immersed in HPW solution in water (0.66 wt%) to improve the proton conductivity [110].

6 Effects of Modifications on the Physical Properties of the Hybrid Membranes

As already mentioned earlier, modifications of nanoclays are necessary to improve the compatibility between the organic polymer and the inorganic nanoclays. However, the improvements in the physical properties like thermomechanical, barrier property, water uptake, proton conductivity, and the cell performance of the hybrid membranes mainly depend on the types of modifier to the nanoclay and the degree of dispersion of the nanoclays in the polymer matrix.

6.1 Permeability

Impermeable nanoclay layers in polymer matrix mandate a tortuous zigzag diffusion pathway for solvents and gases to transverse the membrane (Fig. 22.7). Hence, the permeability of the resulting polymer-clay nanocomposites lowers down compared to the virgin polymer. This behavior of the polymer-clay nanocomposite membranes enhances the fuel cell performance by reducing the fuel crossover.

The methanol permeability (defined as the product of diffusivity and solubility) of recast Nafion® was nearly 2.3×10^{-6} cm^2/s compared to that of 1.6×10^{-7} cm^2/s for Nafion®/MMT nanocomposite membranes containing 1 wt% MMT with a nominal

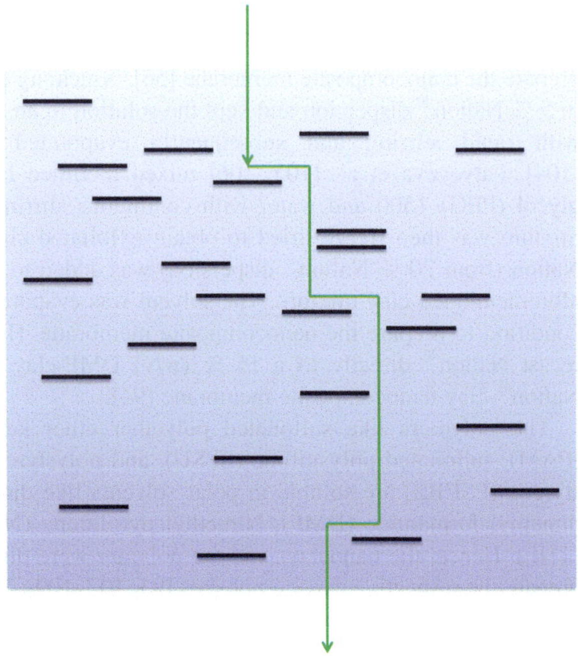

Fig. 22.7 Schematic depiction of the polymer-clay nanocomposite representing the barrier effect of nanoclay

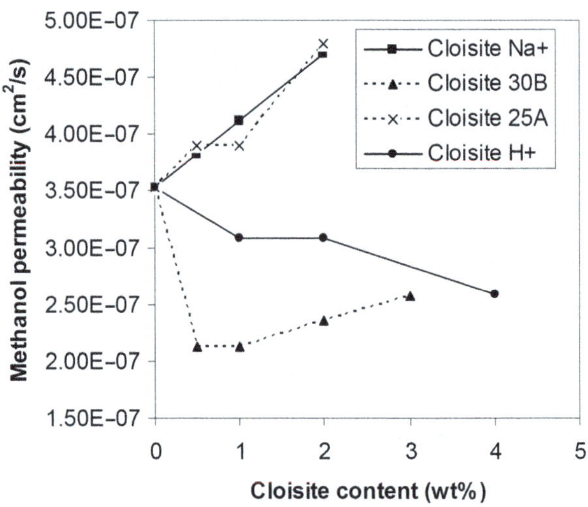

Fig. 22.8 Methanol permeability of Nafion-clay nanocomposites comprised of nanoclays with different counter ions (see Ref. [113])

thickness of 50 μm [111, 112]. Addition of sulfonated MMT and Cloisite 30B to Nafion reduced the methanol and water permeability relative to the virgin Nafion due nm m to their well-dispersed morphology in Nafion [89, 113]. However, poor polymer-filler interactions resulting in the void formation in the polymer-filler interface and inferior dispersion of unmodified MMT in Nafion reduced the barrier

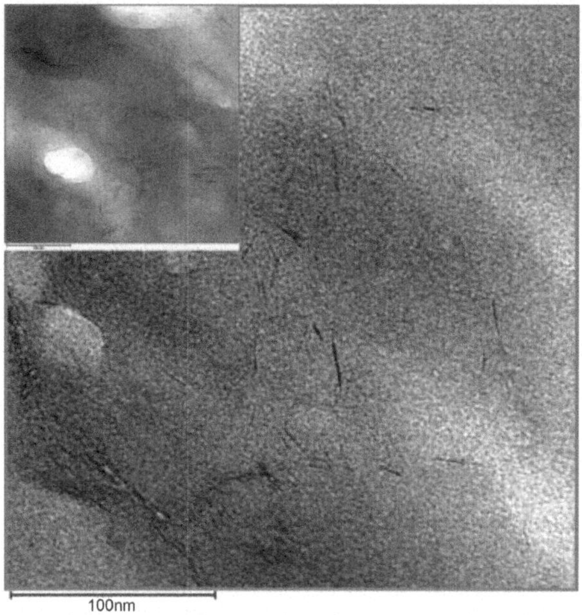

Fig. 22.9 TEM images of SPAS-clay nanocomposite containing 3 wt% Laponite (*inset* shows its low-magnification TEM image, see Ref. [115])

effect of Nafion-based composite towards methanol compared to the virgin Nafion (Fig. 22.8) [113].

The methanol permeability of SPEEK is lower than that of Nafion 115 due to the difference in the microstructure between these two varieties of membranes. Presence of small hydrophilic-hydrophobic interaction, low flexibility of the polymer backbone, and highly branched structure of SPEEK renders narrow proton conduction channels which perplex the methanol transport through the membrane. The methanol permeability further lowers down with the addition of OMMT due to the barrier effect of the nanosized dispersion of OMMT [87]. SPEEK (degree of sulfonation 67 %) possesses the permeabilities of 18 and 127 cm^2/s towards methanol and water, respectively. Addition of unmodified and modified Laponite (modified with imidazoleglycidoxypropyl triethoxysilane) resulted in a decrease in the permeability of the nanocomposites compared to the virgin SPEEK. Methanol and water permeabilities of SPEEK-modified Laponite composite membranes containing 20 wt% clay were reported to be 7 and 74 cm^2/s, respectively, while the corresponding values of the analogous membranes prepared with unmodified Laponite were noted to be 11 and 79 cm^2/s, respectively. This behavior was due to the higher compatibility of the SPEEK with the modified clay compared to that of the unmodified clay [114].

Highly exfoliated structure of Laponite in SPAS matrix (Fig. 22.9) resulted in a decrease in the permeability of the resulting nanocomposite membranes compared to the virgin SPAS. The methanol permeability of the SPAS-clay

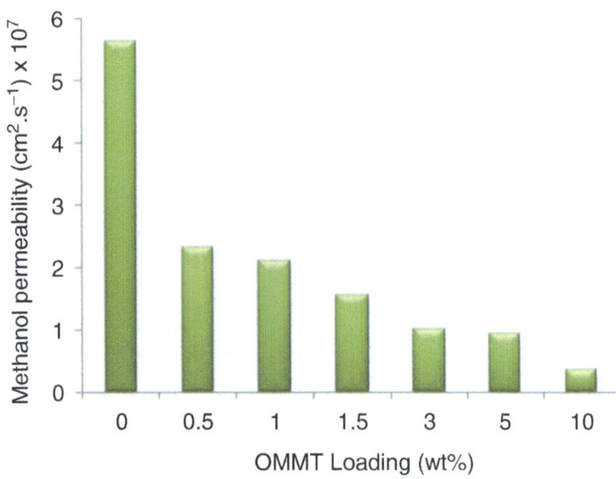

Fig. 22.10 Effect of nanoclay loading on the methanol permeability of the PBI-clay nanocomposite (see Ref. [116])

nanocomposite and SPAS was found to be 1.55×10^{-7} and 3.47×10^{-6} cm^2/s, respectively [115]. A drastic decrease in the methanol permeability of the PBI-MMT nanocomposites was also noted with the increase in the clay content (Fig. 22.10) [116]. However, the methanol permeability of the PVOH-clay nanocomposites was noted to increase with the clay content beyond 7 wt% clay content due to the increased aggregation of clays (or reduced number of individual sheets) [110]. Hence, exfoliated clay morphology and better polymer-clay interaction can reduce the permeability of both water and methanol to a significant extent and vice versa. Types of modifier to the clay do not play a significant role in altering the permeability of the nanocomposites.

6.2 Mechanical Properties

The reinforcing nature of nanoclays enhances the mechanical properties of the nanocomposites compared to the virgin membranes [55, 109]. The tensile strength (TS) and elongation at break (EB) for extruded Nafion® membrane were approximately 30 MPa and 200 %, respectively. Addition of 3 wt% of MMT to Nafion increased the TS and EB by 35 % and by twofold, respectively, compared to the virgin Nafion (Fig. 22.11a,b) [111]. In a similar study, SPAS-clay nanocomposite inferred a significant improvement in the mechanical strength compared to the virgin SPAS [115]. This is possible due to high degree of dispersion of nanoclays in the polymer matrices (Fig. 22.9). The storage modulus (E′) of the virgin PVOH and PVOH-MMT nanocomposite containing 5 and 10 wt% clay was noted to be 1,360, 4,010, and 5,430 MPa, respectively [110]. Increase in the clay content had a positive influence in improving the mechanical properties of the nanocomposites up to a certain amounts of clay content beyond which the mechanical properties started deteriorating [110]. This was ascribed to the inferior clay dispersion in the polymer matrix at higher clay content.

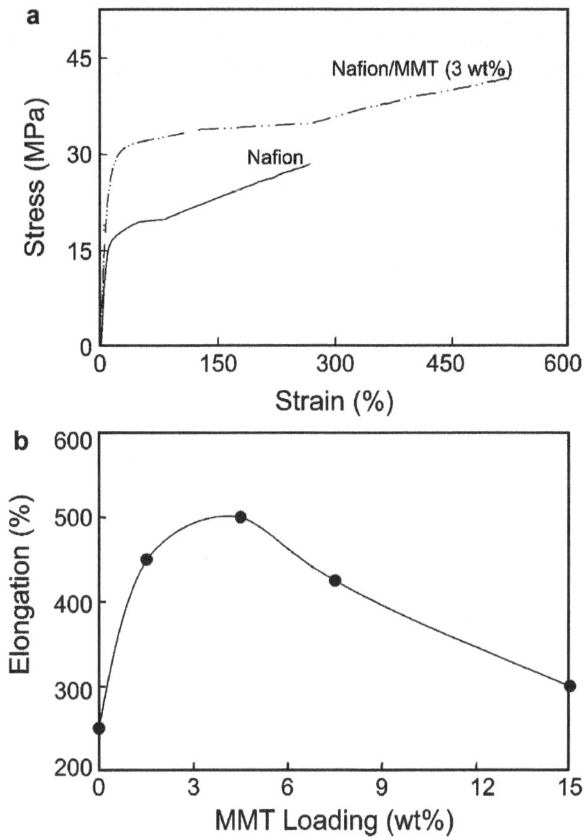

Fig. 22.11 (a) Stress-strain curve and (b) elongation of Nafion-MMT nanocomposite (see Ref. [111])

6.3 Thermal Stability

Three stages of degradation are generally observed for Nafion®. The weight loss up to 300 °C corresponds to the loss of water molecules, whereas the weight loss commencing at 350 °C and between 400 °C and 520 °C corresponds to the decomposition of $-SO_3H$ groups and oxidative degradation of the teflonic polymer backbone, respectively [61]. Addition of modified MMT to Nafion® resulted in an increase in the thermal stability of the nanocomposite membrane compared to that of the virgin Nafion® (Fig. 22.12) [102, 111]. This can be attributed to the strong interfacial bonding between the Nafion and the modified MMT. However, Song et al. reported an increase in the second degradation temperature of the Nafion-MMT nanocomposite despite similar onset degradation temperature to that of the virgin Nafion. The delayed weight loss of the nanocomposites compared to the virgin membranes is due to the fact that the nanodispersed nonconducting clays prevent faster heat transmission through the membrane [112].

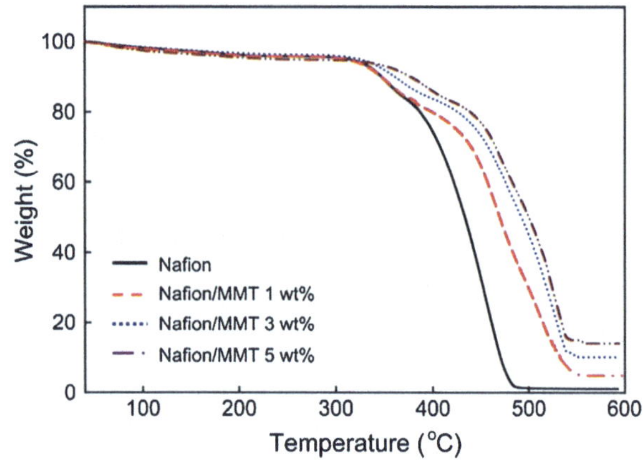

Fig. 22.12 TGA thermograms of Nafion-MMT nanocomposites (see Ref. [111])

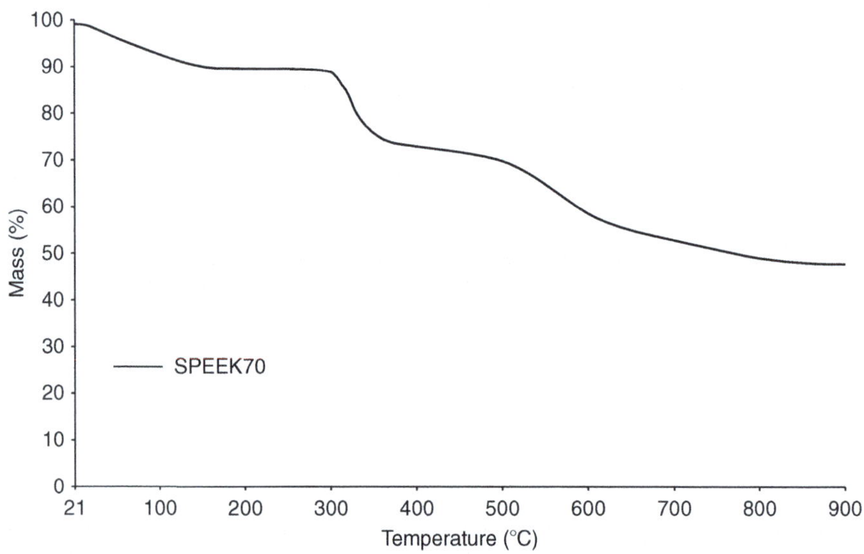

Fig. 22.13 TGA thermogram of SPEEK membrane (see Ref. [117])

A SPEEK membrane produces two degradation peaks at 300–400 °C and 550 °C, corresponding to the loss of $-SO_3H$ groups and main-chain degradation, respectively (Fig. 22.13). Addition of modified clay improved the thermal stability of the nanocomposite compared to that of the virgin SPEEK to a smaller extent [108, 117]. Similarly, the addition of MMT to the SPSEBS and PVOH enhanced the thermal stabilities of the respective nanocomposites compared to the virgin membranes [109, 110]. The increased thermal stabilities of the nanocomposite membranes are due to the high thermal stabilities of the inorganic silicates. However, addition of PWA to PVOH-MMT composites reduced the initial degradation temperature of the nanocomposite membranes due to the catalytic effect on polymer dehydration [110].

6.4 Water Uptake

Water uptake of the membrane is an important parameter due to its assistance in the proton conduction of the membrane. It can be calculated using the following equation:

$$\text{Water uptake} = \frac{W_t - W_d}{W_d} \times 100$$

where W_d is the dry weight of the membrane and W_t is the weight of the membrane after swelling at a particular temperature for 24 h. The water uptakes of various polymer membranes are summarized in Table 22.1.

The room temperature water uptake of Nafion® is ~33 %, and it increases with increased temperature. Addition of nanoclay to Nafion increases the water uptake of the composite membrane due to the hydrophilic nature of the nanoclays. Increasing the nanoclay content in Nafion further increases the water uptake of the composite membranes [92, 104]. Nafion-based composite membranes comprised of sulfonated Laponite RD and unmodified Laponite RD had shown the water uptake of 70 % and 87 %, respectively, compared to 50 % for the virgin Nafion® at 85 °C [94, 106]. In contrast, poor polymer-filler interaction between the Nafion and unmodified MMT lowered the water uptake of the Nafion®-unmodified MMT nanocomposite membrane compared to the virgin Nafion® (Fig. 22.14) [92, 112].

In the case of sulfonated polymers like SPEEK, sulfonated polyarylene sulfone (SPAS), and sulfonated polyether sulfone (SPES), degree of sulfonation (DS) of the polymer determines the water uptake values. SPPEK membranes with DS of 69.4 % and 85 % imparted water uptakes of 42 % and 91 %, respectively [60]. Presence of the hydrophilic groups (like $-SO_3H$ groups) on the modifier to the clay improved the water uptake of the resulting nanocomposite membranes. In contrast, highly hydrophobic modifier (dimethyldioctadecylammonium chloride) to the clay reduced the water uptake of the nanocomposite [112, 117]. Gaowen et al. observed constant water uptake of the nanocomposites irrespective of the temperature due to the matrix stiffening effect of nanoclays on the SPEEK matrix, which restricted the membrane from further swelling at high temperature [87].

Interestingly, the sulfonated PVOH-unmodified MMT nanocomposites imparted low water uptake at low clay content, with a gradual increase of the same at high clay content (Fig. 22.15). This behavior is possibly due to the low affinity of unmodified MMT towards water compared to the sulfonated PVOH along with reduced number of hydrophilic groups (due to the H-bonding between the surface –OH groups of MMT and $-SO_3H$ groups of sulfonated PVOH) [118]. On the contrary, lower water affinity of MMT compared to PWA reduced the water uptake of the PVA-PWA-MMT composite compared to that of the PVA-PWA composite membranes [110].

Table 22.1 Water uptake and cell performance of different membranes

Type of membrane	Water uptake (%)	Operating temperature (°C)	Methanol solution feed	Cell voltage (V)	Current density (mA cm^{-2})	References
Recast Nafion®	50.0	80	–	0.6	600	[94]
Nafion®-10 wt% Laponite	87.0		–		–	
Nafion®-10 wt% sulfonated Laponite	70.0		–		720	
Nafion®	13.5	60	–	60	550	[92]
Nafion®-5 wt% H^+ MMT	13.1		–		–	
Nafion®-5 wt% sulfonated MMT	20.1		–		800	
SPEEK	100	60	–	0.6	80	[108]
SPEEK-10 wt% Laponite	30		–		370	
Nafion	–	40	–	0.2	244	[89]
Nafion-5 % sulfonated MMT	–		–		336	
Nafion-10 % sulfonated MMT	–		–	0.2	286	[89]
Nafion 117	–	–	1 M	0.2	420	[91]
Nafion-3 wt% sulfonated MMT		–			460	
Nafion 117		–	5 M	0.2	210	
Nafion-3 wt% sulfonated MMT		–			390	
SPEEK	–	60	–	0.6	80	[108]
SPEEK-10 wt% Laponite clay	–		–		370	
SPEEK	47.4	80	–	–	–	[100]
SPEEK-5 wt% sulfonated clay	42.2	25	–	–	–	
SPEEK-5 wt% sulfonated clay	58.0	80	–	–	–	

6.5 Proton Conductivity

The proton conductivity of the virgin Nafion® at room temperature and at 100 % RH varies from 0.07 to 0.1 S cm^{-1}. The proton conductivities of the Nafion-clay nanocomposites are highly dependent on the degree of dispersion and types of surface modifier. Improved proton conductivity of the nanocomposites compared to the virgin membrane is mainly due to the high hydrophilicity of the nanoclays which improves the water retention property of both the nanocomposite membranes. This trend remains unchanged even upon UV irradiation of the

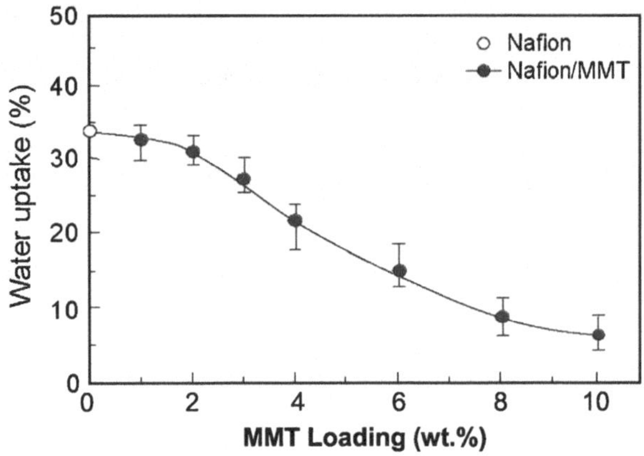

Fig. 22.14 Effect of MMT content on the water uptake behavior of Nafion-based composite membranes (see Ref. [112])

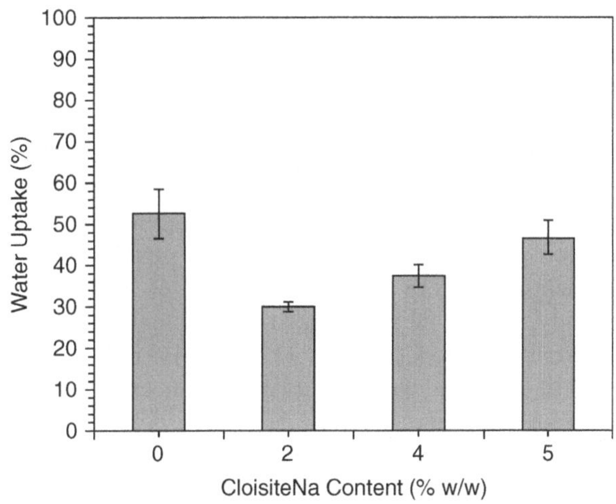

Fig. 22.15 Effect of unmodified MMT content on the water uptake behavior of sulfonated PVA (see Ref. [118])

nanocomposites and the virgin Nafion [104]. A comparative study between the nanocomposites containing MMT with different counter ions (H^+, Na^+, and ammonium counter ions) showed a highest conductivity with the acid-activated MMT (H^+ counter ion) compared to MMT having Na^+ and ammonium counter ions due to the ease in mobility of the smaller counter ions (H^+ ion) (Fig. 22.16) [113]. Sulfonation of the nanoclay surface was also noted to increase the proton conductivity of the resulting membrane compared to the virgin Nafion and Nafion®-unmodified Laponite nanocomposite due to the presence of the highly proton-conducting $–SO_3H$ group on the clay surface [94]. A novel technique was used by Mishra et al. for the preparation of Nafion-Laponite XLS nanocomposite membranes. In this case, the proton conductivity of the nanocomposites was significantly enhanced due to the presence of in situ generated H_3PO_4 (resulting from the acid activation of Laponite XLS). The proton conductivity of the virgin

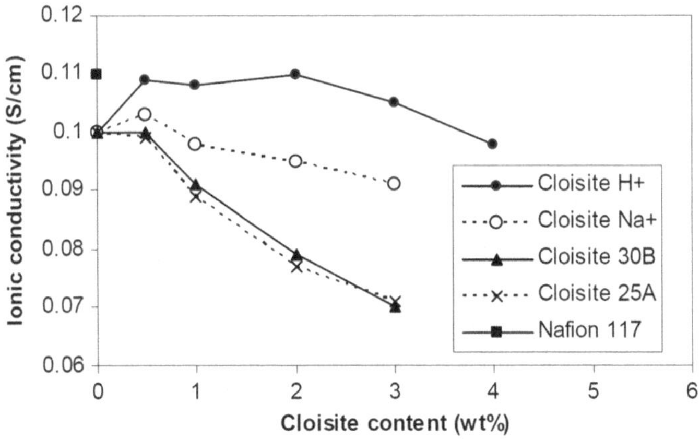

Fig. 22.16 Proton conductivities of Nafion-clay nanocomposites containing clay with different counter ions (see Ref. [113])

Nafion and the nanocomposite membrane containing 3 wt% of acid-activated Laponite were noted to be 0.14 and 0.27 S cm^{-1}, respectively, at 110 °C and 100 % RH [101].

In contrast to the above results, addition of the chitosan-modified MMT to Nafion resulted in a decrease in the proton conductivity with increasing clay contents [85]. The low conductivities of the Nafion-clay nanocomposites arise mainly due to the aggregated morphology of the clay or the presence of the hydrophobic modifier [111, 112]. Inferior clay dispersions in the polymer matrix lead to decreased conductivity of the nanocomposites even upon sulfonation of the clay surface [89].

In a similar study, SPEEK-clay nanocomposites revealed increased proton conductivities up to 10 wt% clay contents. However, further increase in the clay content was detrimental to the conductivity of the nanocomposites due to increased degree of obstacles for the proton mobility [114]. Modifiers to the clay which is indirectly related to the degree of dispersion of clay in the polymer matrix play a vital role in increasing or decreasing the proton conductivity. Hence, despite the presence of the proton-conducting groups like $-SO_3H$, SPEEK-clay nanocomposite based on SA-modified clay was lower than the virgin SPEEK, while the composite based on DMDOC-modified clay was higher than the virgin SPEEK (Fig. 22.17) [117]. The reduction in proton conductivity is mainly due to the blockage in the proton conduction channels created by the nonconducting nature of modifier or the aggregated clays [87, 107, 109, 116].

SPSU membrane with 72 % degree of sulfonation inferred proton conductivities of 0.09 and 0.17 S cm^{-1} at 30 °C and 85 °C, respectively, under 100 % RH [93]. The conductivity of the SPSU-sulfonated Laponite composite was enhanced by 25 % compared to that of the virgin SPSU due to increased sulfonated sites in the composite membranes.

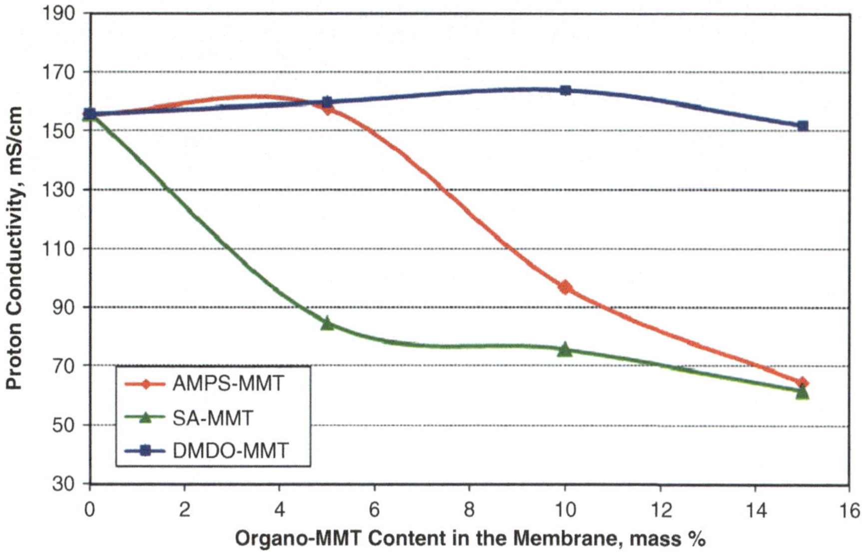

Fig. 22.17 Change in the proton conductivity of SPEEK-clay nanocomposite membranes with varying clay content and different types of clay (see Ref. [117])

Exfoliated morphology of Laponite in the SPAS matrix (Fig. 22.9) enhanced the conductivity of the nanocomposite than that of the virgin SPAS. The conductivity of the SPAS-clay nanocomposite varied from 0.099 to 0.187 S cm^{-1} within a temperature range of 20–70 °C [115]. PVOH membranes possess very low proton conductivity, and hence, it is doped with highly proton-conducting PWAs to improve the proton conductivities of the membranes. However, addition of PWA resulted in inferior mechanical properties of the nanocomposite membranes. PWA being highly conducting in nature compared to the clay, addition of MMT to PVOH-PWA composite resulted in a decrease in the conductivity of the PVOH-PWA-MMT composite compared to the PVOH-PWA composite [110]. In a similar study, addition of MMT (Cloisite 30B) to the highly conductive PAM-PS blend decreased the proton conductivity of the nanocomposite membrane compared to the PAM-PS blend [119]. Hence, despite the high hydrophilicity of nanoclays, degrees of dispersion of nanoclays in the polymer matrix play a vital role in improving the proton conductivity of the nanocomposites. The types of modifiers used for the clay modification can contribute to a very small extent in this regard.

6.6 Cell Performance

Cell performance study of the membranes can determine the suitability of the membrane for its end usage in fuel cell applications. Current densities and power

Table 22.2 Proton conductivities of different membranes

Type of membrane	Operating temperature (°C)	RH (%)	Proton conductivity (S cm^{-1})	References
Nafion®	90	98	0.200	[92]
Nafion®-5 wt% sulfonated MMT			0.160	
Nafion®-5 wt% protonated MMT			0.085	
Nafion®	95	98	0.064	[94]
Nafion®-10 wt% unmodified Laponite			0.065	
Nafion®-10 wt% sulfonated Laponite			0.080	
Nafion	110	100	0.136	[101]
Nafion-3 wt% acid-activated Laponite			0.270	
SPSU	90	100	0.170	[93]
SPSU-5 wt% sulfonated Laponite			0.220	
Nafion	25	100	0.086	[85]
Nafion-1 wt% chitosan-modified MMT			0.083	
Nafion-5 wt% chitosan-modified MMT			0.059	[85]
SPEEK	80	100	0.125	[100]
SPEEK-1 wt% sulfonated clay			0.166	
PVA	70	100	0.043	[120]
PVA-10 wt% MMT			0.032	

densities are the key factors which determine the cell performance of the membranes. The current density of the membrane is highly dependent on the temperature, humidity, and operating voltage. Table 22.2 summarizes the current densities of different clay-based nanocomposites under different experimental conditions.

In the case of PEMFC, hydrogen and oxygen gas are used as the fuel to determine the cell performance. The current density and the power density of the membranes are highly dependent on the thickness of the membrane. Hence, the maximum power densities of Nafion® NRE 212 (thickness 50.8 μm) and Nafion® NRE 211 (thickness 25.4 μm) were noted to be 0.97 and 1.27 W/cm^2, respectively, at 80 °C and 100 % RH (Fig. 22.18a, b) [106]. Addition of Laponite, modified by sultone and p-styrene sulfonic acid to Nafion, enhanced the maximum power densities to 1.36 and 1.41 W cm^{-2}, respectively. The current density of Nafion was noted to be 600 mA cm^{-2} at 0.6 V, 80 °C, and under highly humidified conditions. The current density of the nanocomposite was increased to 720 mA cm^{-2} under similar conditions, due to the incorporation of sulfonated Laponite to Nafion. The current densities of the nanocomposites were also noted to be dependent on the

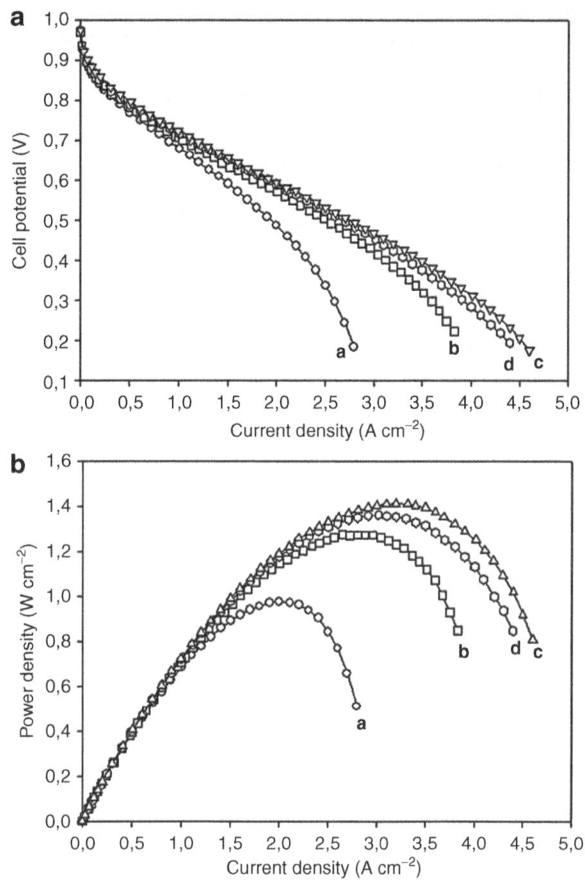

Fig. 22.18 Polarization curves (**a**) and power density curves (**b**) at 80 °C, total pressure of four absolute bars under dried H_2/O_2: *a* Nafion NRE212, *b* Nafion NRE211, *c* Nafion-Laponite-*p*-styrene sulfonic acid, *d* Nafion-Laponite-sultone membranes (see Ref. [95])

clay content. The current densities of Nafion-sulfonated MMT membranes increased with the clay content up to 5 wt% clay content (with a maximum value of 336 mA cm^{-2} at a potential of 0.2 V) and then deteriorated with further increase in the clay content due to increased aggregation of clay beyond 5 wt% clay content. The maximum power density of 67 mW cm^{-2} was also achieved for the composite membrane containing 5 wt% of sulfonated MMT (Fig. 22.19) [89].

In the case of DMFC, dilute methanol and oxygen gas are used as the fuel to determine the cell performance. Hence, the crossover current densities of the membranes are also dependent on the concentration of methanol. Hasani-Sadrabadi et al. reported the crossover current densities for Nafion 117 and the Nafion-clay nanocomposite to be 156 and 123 mA cm^{-2}, respectively, at 1 M methanol concentration and 518 and 320 mA cm^{-2}, respectively, at 5 M methanol concentration. Similarly, the limiting current densities for Nafion 117 and the Nafion-clay composite at the anode side were 530 and 555 mA cm^{-2}, respectively, at 1 M methanol concentration, whereas the values were 260 and 630 mA cm^{-2},

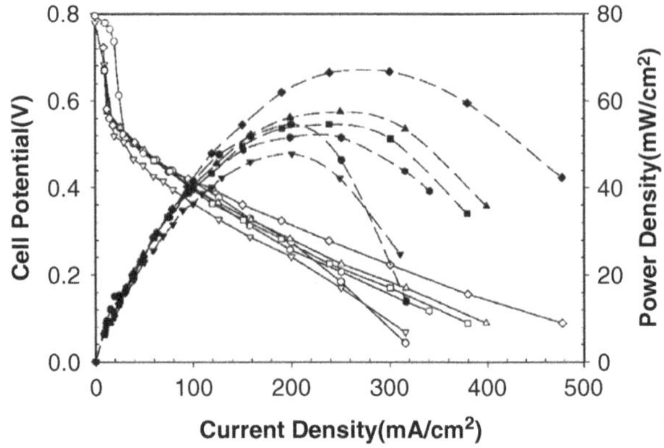

Fig. 22.19 Polarization curves for the MEA made with Nafion115 and composite membranes operated at 40 °C (2 M methanol/air flow rate): ● Nafion 115, ▼ Nafion-3 wt% unmodified MMT, ■ Nafion-3 wt% sulfonated MMT, ♦ Nafion-5 wt% sulfonated MMT, ▲ Nafion-10 wt% HSO3-MMT, ⬢ Nafion-15 wt% sulfonated MMT. The same symbols (but open ones) denote cell potentials for corresponding samples (see Ref. [89])

respectively, at 5 M methanol concentration. The higher open circuit voltage (OCV) values of the nanocomposite compared to the Nafion 117 indicated a drastic reduction in the methanol crossover from the anode to the cathode, which hampers the catalytic activity for oxygen reduction and also leads to the reduction in the fuel efficiency. Hence, addition of nanoclay to Nafion resulted in increased fuel cell efficiency, especially at high methanol concentration. The maximum power density of the Nafion 117 and the composites was noted to be 47 and 171 mW cm^{-2}, respectively, at 5 M methanol feed (Fig. 22.20) [85]. The DMFC cell performance study of SPAS-clay nanocomposite was found to be higher than the Nafion 115 membrane. The power densities of SPAS-clay nanocomposite were obtained to be 110, 145, and 191 mW cm^{-2} at 50 °C, 60 °C, and 70 °C, respectively, compared to those of 77, 119, and 142 mW cm^{-2} at 50 °C, 60 °C, and 70 °C, respectively, for Nafion 115. The increase in the operating temperature had also a positive influence in increasing the power density (Fig. 22.21 a, b) [115].

The current densities of the acid-doped PBI membrane, PBI-clay nanocomposite (3 wt% clay content), and Nafion 117 were noted to be 290, 260, and 351 mA cm^{-2} at 1 M methanol feed and 635, 723, and 420 mA cm^{-2} at 5 M methanol concentration, respectively, at a constant potential of 0.2 V. Similarly, the power densities of the acid-doped PBI membrane, PBI-clay nanocomposite (3 wt% clay content), and Nafion 117 were noted to be 59, 51, and 77 mW cm^{-2}, respectively, at a methanol concentration of 1 M and 130, 145, and 83 mW cm^{-2}, respectively, at a methanol concentration of 5 M and at a constant potential of 0.2 V [116]. The reason behind the decreased current density and power density of PBI-clay nanocomposite compared to the virgin PBI at 1 M methanol concentration and the reverse trend at 5 M methanol concentration is not known, and it has to be investigated further.

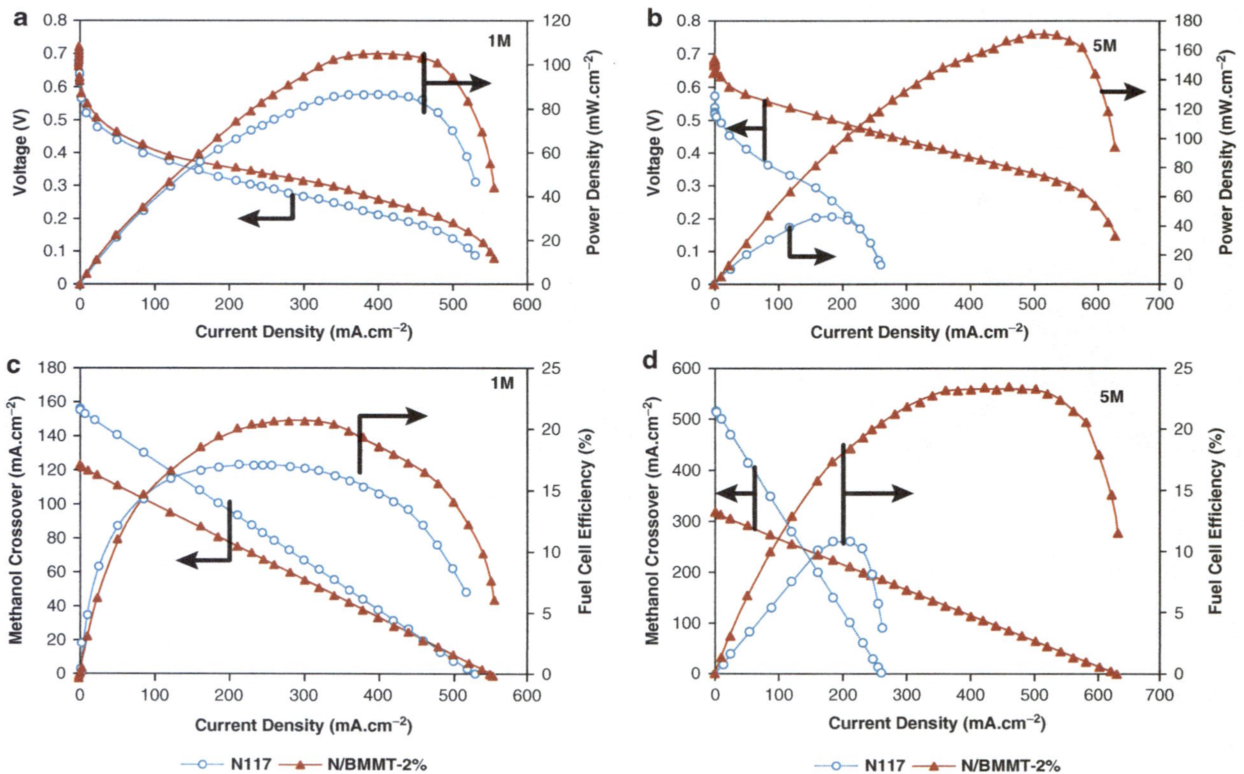

Fig. 22.20 Polarization curves of DMFC cells consisting of Nafion-2 wt% chitosan-modified MMT and Nafion 117, at (**a**) 1 M and (**b**) 5 M methanol solution at 70 °C. Methanol crossover and efficiency of the corresponding fuel cells at (**c**) 1 M and (**d**) 5 M methanol solution (see Ref. [85])

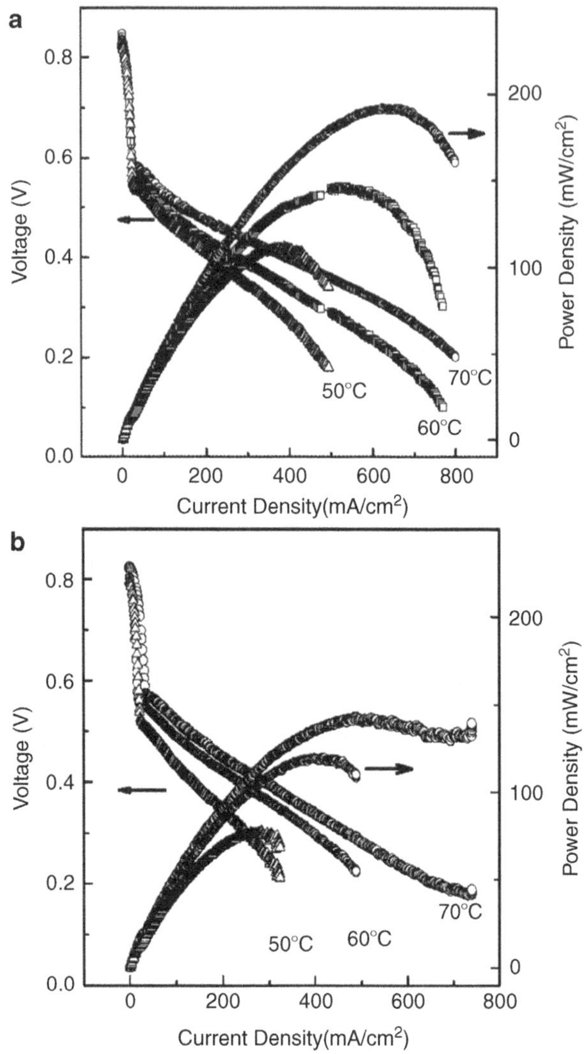

Fig. 22.21 Polarization curves of DMFC single cell test for SPAS-clay nanocomposite containing 3 wt% Laponite (**a**) and Nafion 115 (**b**) operated at different temperatures (see Ref. [115])

Thomassin et al. reported that the fuel cell performance of the membranes depends on the counter ion of the clay. Among three varieties of nanoclays studied, Cloisite 30B (with alkyl ammonium counterion), Cloisite Na^+ (with Na^+ counter ion), and Cloisite H^+ (with H^+ counter ion), Cloisite H^+ and Cloisite 30B provided the best and worst fuel cell performance, respectively. The current densities of Nafion 117 and Nafion-Cloisite H^+ were 100 and 60 mA cm^{-2}, respectively, at 80 °C and at a potential of 0.3 V (Fig. 22.22) [113].

It is worth mentioning here that the decrease in the proton conductivity of the nanocomposite does not confer similar trend in the fuel cell performance as well. Highly aggregated morphology of clay in the polymer matrix leads to the decrease

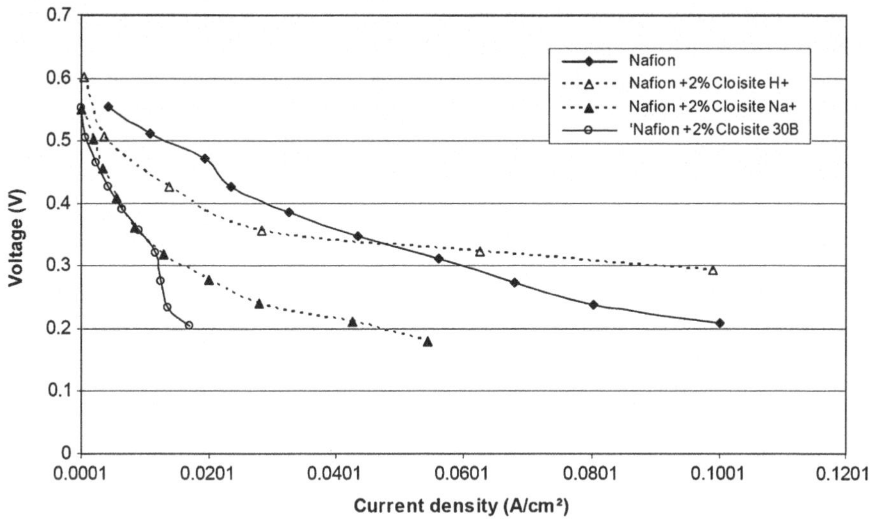

Fig. 22.22 Polarization curves for Nafion-clay nanocomposites at 80 °C (see Ref. [113])

in the water uptake and fuel cell performance [89, 92]. The operating temperature in PEMFC and methanol concentration in DMFC play vital role in improving the fuel cell performance.

7 Summary and Future Direction

The inferior proton conductivity of the Nafion membrane at high temperature and low humidity conditions along with the demerits associated with low-temperature fuel cells lead to the development of new membranes and techniques. Addition of nanoclay to the polymer is one of the techniques adopted to improve the operating temperature. This is due to the high hydrophilicity of the nanoclays and their water retention ability. In addition to that, the mechanical and thermal properties of the nanocomposites also increase with the incorporation of nanoclay to the polymer.

Improvements in all the physical properties of the clay-based nanocomposites are highly dependent on the degree of clay dispersion in the polymer matrix. Surface modifiers to the clay play a little role in improving the proton conductivity and the cell performance of the nanocomposites unless the clay platelets are well dispersed in the polymer matrix. However, in the case of the polymer-clay nanocomposites with well-dispersed clay, presence of $-SO_3H$ and $-PO_3H$ groups on the clay surface provided an additional path for the proton conduction and proved to be beneficial for high proton conductivity of the nanocomposites. High degree of clay dispersion is also responsible for the improvements in the thermomechanical property and water uptake of the nanocomposites along with the cell performance. Addition of clay to the polymer matrix leads to the slight increase in the operating temperature due to the hydrophilic nature of clay. Despite tremendous efforts are being made to replace Nafion with other varieties of polymer membranes, no breakthrough has been achieved so far. Hence, plenty of research still has to be performed before the end usage of the clay-based nanocomposites. Highly conducting nature of ionic liquid can be beneficial for improved proton conductivity of the clay-based nanocomposites at low humidity conditions. The barrier effect of clay can prevent ionic liquid from leaching out of the membrane upon continuous usage. The plasticizing nature of the ionic liquid can also be counterbalanced by the reinforcing nature of the clay.

Acknowledgement This study was supported by the Converging Research Center Program (2013K000404) through the Ministry of Science, ICT & Future Planning and the Basic Science Research Program through the National Research Foundation (NRF) funded by the Ministry of Education of Korea (NRF-2013R1A1A2011608).

References

1. Mishra AK, Bose S, Kuila T, Kim NH, Lee JH (2012). Prog Polym Sci, 37:842–869
2. Grove WR (1839) Philos Mag 15:287–293
3. Grove WR (1839) Philos Mag 14:127–130
4. Grove WR (1842) Philos Mag 21:417–420
5. Couture G, Alaaeddine A, Boschet F, Améduri B (2011) Prog Polym Sci 36:1521–1557
6. Bose S, Kuila T, Nguyen TXH, Kim NH, Lau K-T, Lee JH (2011) Prog Polym Sci 36:813–843
7. Tripathi BP, Shahi VK (2011) Prog Polym Sci 36:945–979
8. Neburchilov V, Martin J, Wang H, Zhang J (2007) J Power Sources 169:221–238
9. Wasmus S, Kűver A (1999) Methanol oxidation and direct methanol fuel cells: a selective review. J Electroanal Chem 461:14–31
10. Couture G, Alaaeddine A, Boschet F, Améduri B (2011) Polymeric materials as anion-exchange membranes for alkaline fuel cells. Prog Polym Sci 36:1521–1557
11. Boudghene Stambouli A, Traversa E (2002) Solid oxide fuel cells (SOFCs): a review of an environmentally clean and efficient source of energy. Renewable Sustainable Energy Rev 6:433–455
12. Yang C, Costamagna P, Srinivasan S, Benziger J, Bocarsly AB (2001) J Power Sources 103:1–9
13. Souzy R, Ameduri B (2005) Prog Polym Sci 30:644–687
14. Rozière J, Jones DJ (2003) Annu Rev Mater Res 33:503–555
15. Hsu WY, Gierke TD (1983) J Membr Sci 13:307–326
16. Fujimura M, Hashimoto T, Kawai H (1982) Macromolecules 15:136–144
17. Fujimura M, Hashimoto T, Kawai H (1981) Macromolecules 14:1309–1315
18. Dreyfus B, Gebel G, Aldebert P, Pineri M, Escoubes M, Thomas M (1990) J Phys France 51:1341–1354
19. Gebel G, Lambard J (1997) Macromolecules 30:7914–7920
20. Gebel G (2000) Macromolecules 33:4850–4855
21. Haubold H-G, Vad T, Jungbluth H, Hiller P (2001) Electrochim Acta 46:1559–1563
22. Rubatat L, Rollet AL, Gebel G, Diat O (2002) Macromolecules 35:4050–4055
23. Schmidt-Rohr K, Chen Q (2008) Nat Mater 7:75–83
24. Li Q, He RH, Jensen JO, Bjerrum NJ (2003) Chem Mater 15:4896–4915
25. Harrison WL, Hickner MA, Kim YS, McGrath JE (2005) Fuel Cells 5:201–212
26. Miyatake K, Chikashige Y, Higuchi E, Watanabe M (2007) J Am Chem Soc 129:3879–3887

27. Zaidi SMJ, Mikhailenko SD, Robertson GP, Guiver MD, Kaliaguine S (2000) J Membr Sci 173:17–34
28. Liu B, Robertson GP, Kim DS, Guiver MD, Hu W, Jiang Z (2007) Macromolecules 40:1934–1944
29. Li L, Zhang J, Wang Y (2003) J Membr Sci 226:159–167
30. Zhang H, Fan X, Zhang J, Zhou Z (2008) Solid State Ion 179:1409–1412
31. Chen S, Yin Y, Kita H, Okamoto K (2007) J Polym Sci, Part A: Polym Chem 45:2797–2811
32. Miyatake K, Yasuda T, Hirai M, Nanasawa M, Watanabe M (2007) J Polym Sci, Part A: Polym Chem 45:157–163
33. Yin Y, Yamada O, Tanaka K, Okamoto K (2006) Polym J 38:197–219
34. Cho Y-H, Kim S-K, Kim T-H, Cho Y-H, Lim JW, Jung N, Yoon W-S, Lee J-C, Sung Y-E (2011) Electrochem Solid-State Lett 14:B38–B40
35. Zhang J, Tang Y, Song C, Zhang J (2007) J Power Sources 172:163–171
36. Ng F, Péron J, Jones DJ, Rozière J (2011) J Polym Sci, Part A: Polym Chem 49:2107–2117
37. Asensio JA, Sánchezab EM, Gómez-Romero P (2010) Chem Soc Rev 39:3210–3239
38. Li Q, He R, Jensen JO, Bjerrum NJ (2004) Fuel Cells 4:147–159
39. Asensio JA, Gómez-Romero P (2005) Fuel Cells 5:336–343
40. Bhadra S, Kim NH, Lee JH (2010) J Membr Sci 349:304–311
41. Bhadra S, Kim NH, Choi JS, Rhee KY, Lee JH (2010) J Power Sources 195:2470–2477
42. Lin B, Cheng S, Qiu L, Yan F, Shang S, Lu J (2010) Chem Mater 22:1807–1813
43. Diao H, Yan F, Qiu L, Lu J, Lu X, Lin B, Li Q, Shang S, Liu W, Liu J (2010) Macromolecules 43:6398–6405
44. Jones DJ, Rozière J (2008) Adv Polym Sci 215:219–264
45. Mustarelli P, Carollo A, Grandi S, Quartarone E, Tomasi C, Leonardi S, Magistris A (2007) Fuel Cells 7:441–446
46. Sel O, Soulès A, Améduri B, Boutevin B, Laberty-Robert C, Gebel G, Sanchez C (2010) Adv Funct Mater 20:1090–1098
47. Garland NL, Kopasz JP (2007) J Power Sources 172:94–99
48. Borup R, Meyers J, Pivovar B, Kim YS, Mukundan R, Garland N, Myers D, Wilson M, Garzon F, Wood D, Zelenay P, More K, Stroh K, Zawodzinski T, Boncella XJ, McGrath JE, Inaba M, Miyatake K, Hori M, Ota K, Ogumi Z, Miyata S, Nishikata A, Siroma Z, Uchimoto Y, Yasuda K, Kimijima K-i, Iwashita N (2007) Chem Rev 107:3904–3951
49. Park CH, Lee CH, Guiver MD, Lee YM (2011) Prog Polym Sci 36:1443–1498
50. Higashihara T, Matsumoto K, Ueda M (2009) Polymer 50:5341–5357
51. Iojoiu C, Maréchal M, Chabert F (2005) Fuel Cells 5:344–354
52. Jones DJ, Rozière J (2001) J Membr Sci 185:41–58
53. Hickner MA, Ghassemi H, Kim YS, Einsla BR, McGrath JE (2004) Chem Rev 104:4587–4612
54. Kerres J, Zhang W, Cui W (1998) J Polym Sci, Part A: Polym Chem 36:1441–1448
55. Xing D, He G, Hou Z, Ming P, Song S (2011) Int J Hydrogen Energy 36:2177–2183
56. Mura F, Silva RF, Pozio A (2007) Electrochim Acta 52:5824–5828
57. Felice C, Qu D (2011) Ind Eng Chem Res 50:721–727
58. Zou H, Wu S, Shen J (2008) Chem Rev 108:3893–3957
59. Durand N, Gaveau P, Silly G, Améduri B, Boutevin B (2011) Macromolecules 44:6249–6257
60. Reinholdt MX, Kaliaguine S (2010) Langmuir 26:11184–11195
61. Gnana Kumar G, Kima AR, Nahma KS, Elizabeth R (2009) Int J Hydrogen Energy 34:9788–9794
62. Jin YG, Qiao SZ, Xu ZP, Yan Z, Huang Y, Diniz da Costa JC, Lu GQ (2009) J Mater Chem 19:2363–2372
63. Jin YG, Qiao SZ, Xu ZP, Diniz da Costa JC, Lu GQ (2009) J Phys Chem C 113:3157–3163
64. Choi Y, Kim Y, Kim HK, Lee JS (2010) J Membr Sci 357:199–205
65. Choi J, Lee KM, Wycisk R, Pintauro PN, Mather PT (2010) J Electrochem Soc 157: B914–B919
66. Chhabra P, Choudhary V (2010) J Appl Polym Sci 118:3013–3023

67. Zhengbang W, Haolin T, Mu P (2011) J Membr Sci 369:250–257
68. Di Vona ML, Sgreccia E, Donnadio A, Casciola M, Chailan JF, Auer G, Knauth P (2011) J Membr Sci 369:536–544
69. Aparicio M, Klein LC (2005) J Electrochem Soc 152:A493–A496
70. Park KT, Jung UH, Choi DW, Chun K, Lee HM, Kim SH (2008) J Power Sources 177:247–253
71. Shao Z-G, Xu H, Li M, Hsing I-M (2006) Solid State Ion 177:779–785
72. Tripathi BP, Schieda M, Shahi VK, Nunes SP (2011) J Power Sources 196:911–919
73. Kannan R, Kakade BA, Pillai VK (2008) Angew Chem Int Ed 47:2653–2656
74. Thomassin J-M, Kollar J, Caldarella G, Germain A, Jérôme R, Detrembleur C (2007) J Membr Sci 303:252–257
75. Ijeri V, Cappelletto L, Bianco S, Tortello M, Spinelli P, Tresso E (2010) J Membr Sci 363:265–270
76. Kannan R, Aher PP, Palaniselvam T, Kurungot S, Kharul UK, Pillai VK (2010) J Phys Chem Lett 1:2109–2113
77. Zarrin H, Higgins D, Jun Y, Chen Z, Fowler M (2011) J Phys Chem C 115:20774–20781
78. Cao Y-C, Xu C, Wu X, Wang X, Xing L, Scott K (2011) J Power Sources 196:8377–8382
79. Ray SS, Okamoto M (2003) Prog Polym Sci 28:1539–1641
80. Mishra AK, Chattopadhyay S, Rajamohanan PR, Nando GB (2011) Polymer 52:1071–1083
81. Mishra AK, Nando GB, Chattopadhyay S (2008) J Polym Sci, Part B: Polym Phys 46:2341–2354
82. Mishra AK, Rajamohanan PR, Nando GB, Chattopadhyay S (2011) Adv Sci Lett 4:64–73
83. Kreuer K-D (1988) J Mol Struct 177:265–276
84. Zhang X (2007) J Electrochem Soc 154:B322–B326
85. Hasani-Sadrabadi MM, Dashtimoghadam E, Majedi FS, Kabiri K, Mokarram N, Solati-Hashjin M, Moaddel H (2010) Chem Commun 46:6500–6502
86. Ramírez-Salgado J (2007) Electrochim Acta 52:3766–3778
87. Gaowen Z, Zhentao Z (2005) J Membr Sci 261:107–113
88. Jung DH, Cho SY, Peck DH, Shin DR, Kim JS (2003) J Power Sources 118:205–211
89. Rhee CH, Kim HK, Chang H, Lee JS (2005) Chem Mater 17:1691–1697
90. Gosalawit R, Chirachanchai S, Shishatskiy S, Nunes SP (2008) J Membr Sci 323:337–346
91. Hasani-Sadrabadi MM, Dashtimoghadam E, Majedi FS, Kabiri K, Solati-Hashjin M, Moaddel H (2010) J Membr Sci 365:286–293
92. Kim Y, Choi Y, Kim HK, Lee JS (2010) J Power Sources 195:4653–4659
93. Buquet CL, Fatyeyeva K, Poncin-Epaillard F, Schaetzel P, Dargent E, Langevin D, Nguyena QT, Marais S (2010) J Membr Sci 351:1–10
94. Bébin P, Caravanier M, Galiano H (2006) J Membr Sci 278:35–42
95. Fatyeyeva K, Bigarré J, Blondel B, Galiano H, Gaud D, Lecardeur M, Poncin-Epaillard F (2011) J Membr Sci 366:33–42
96. Fatyeyeva K, Chappey C, Poncin-Epaillard F, Langevin D, Valleton J-M, Marais S (2011) J Membr Sci 369:155–166
97. Lee W, Kim H, Kim TK, Chang H (2007) J Membr Sci 292:29–34
98. Kim Y, Lee JS, Rhee CH, Kim HK, Chang H (2006) J Power Sources 162:180–185
99. Kim TK, Kang M, Choi YS, Kim HK, Lee W, Chang H, Seung D (2007) J Power Sources 165:1–8
100. Fu T, Cui Z, Zhong S, Shi Y, Zhao C, Zhang G, Shao K, Na H, Xing W (2008) J Power Sources 185:32–39
101. Mishra AK, Kuila T, Kim NH, Lee JH (2012) J Membr Sci 389:316–323
102. Zhang L, Xu J, Hou G, Tang H, Deng F (2007) J Colloid Interface Sci 311:38–44
103. Felice C, Ye S, Qu D (2010) Ind Eng Chem Res 49:1514–1519
104. Xiuchong H, Haolin T, Mu P (2008) J Appl Polym Sci 108:529–534
105. Fatyeyeva K, Chappey C, Poncin-Epaillard F, Langevin D, Valleton J-M, Marais S (2011) J Membr Sci 369:155–166

106. Fatyeyeva K, Bigarré J, Blondel B, Galiano H, Gaud D, Lecardeur M, Poncin-Epaillard F (2011) J Membr Sci 366:33–42
107. Hasani-Sadrabadi MM, Emami SH, Ghaffarian R, Moaddel H (2008) Energy Fuel 22:2539–2542
108. Chang J-H, Park JH, Park G-G, Kim C-S, Park OO (2003) J Power Sources 124:18–25
109. Swaminathan E, Dharmalingam S (2010) Int J Plast Technol 13:150–162
110. Thomassin J-M, Pagnoulle C, Caldarella G, Germain A, Jérôme R (2006) J Membr Sci 270:50–56
111. Song M-K, Park S-B, Kim Y-T, Kim K-H, Min S-K, Rhee H-W (2004) Electrochim Acta 50:639–643
112. Song M-K, Kim Y-M, Kim YT, Rhee H-W, Smirnova A, Sammes NM, Fenton JM (2006) J Electrochem Soc 153:A2239–A2244
113. Thomassin J-M, Pagnoulle C, Bizzari D, Caldarella G, Germain A, Jérôme R (2004) e-Polymers 018:1–13
114. Karthikeyan CS, Nunes SP, Prado LASA, Ponce ML, Silva H, Ruffmann B, Schulte K (2005) J Membr Sci 254:139–146
115. Choi YS, Kim TK, Kim EA, Joo SH, Pak C, Lee YH, Chang H (2008) Adv Mater 20:2341–2344
116. Hasani-Sadrabadi MM, Dorri NM, Ghaffarian SR, Dashtimoghadam E, Sarikhani K, Majedi FS (2010) J Appl Polym Sci 117:1227–1233
117. Doğan H, Inan TY, Koral M, Kaya M (2011) Appl Clay Sci 52:285–294
118. Duangkaew P, Wootthikanokkhan J (2008) J Appl Polym Sci 109:452–458
119. Deyrail Y, Mighri F, Bousmina M, Kaliaguine S (2007) Fuel Cells 07:447–452
120. Yang C-C, Lee Y-J (2009) Thin Solid Films 517:4735–4740

Polymer Nanocomposites: Emerging Growth Driver for the Global Automotive Industry

23

Vivek Patel and Yashwant Mahajan

Contents

1 Introduction

Polymer nanocomposites (PNCs) have become an exciting area of current research and development in the nanotechnology arena and have attracted a lot of interest from both academia and industry during the past two and a half decades. Exfoliated clay-based PNCs have dominated the automotive R&D as well as markets, but there are a large number of other significant areas (carbon nanofibers, carbon nanotubes, graphene, nanometals, and nanometal oxides) of current and emerging interest. They have advantageous and superior properties and can be applied in a wide range of automotive applications. It should be noted that PNCs have smaller filler size,

V. Patel (✉) • Y. Mahajan
Centre for Knowledge Management of Nanoscience & Technology (CKMNT), Secunderabad, AP, India
e-mail: vivepatel@gmail.com

J.K. Pandey et al. (eds.), *Handbook of Polymernanocomposites. Processing, Performance and Application – Volume A: Layered Silicates*, DOI 10.1007/978-3-642-38649-7_23,

which leads to higher surface-to-volume ratio compared to conventional composites reinforced with micro-sized fillers. The absolute alignment and contact of polymer or composites with filler interface is critical for the transfer of mechanical load, electrical conductivity, thermal conductivity, and corrosion resistance. The automotive industry is exploring ways to maximize the use of PNCs in order to tackle fuel economy, cost reduction, product differentiation, and environmental concerns. In this chapter, we aim to analyze the current status and perspectives of PNCs background, advantages, limitations, key drivers and growth drivers, materials, properties, and applications, as well as future directions. This is not meant to be an overview of the field, but rather a comprehensive analysis with particular focus on PNC strengths and potential.

PNCs represent a new class of multiphase engineering materials containing dispersion of nano-sized filler materials such as nanoparticles, nanoclays, nanotubes, and nanofibers within the polymer matrices (thermoplastic or thermoset polymer) that have exhibited high strength-to-weight and modulus-to-weight ratios, even compared with conventional as well as some metallic materials. Owing to their nanoscale size features and very high surface-to-volume ratios, they possess unique combination of multifunctional properties not shared by their more conventional composite counterparts reinforced with micro-sized fillers. The nanoclays-reinforced thermoplastic is the most commonly used material in automotive applications today, but nanoparticle/nanopowder (carbon nanotubes, graphene, metal, and metal oxides)-enabled thermoplastic composites and metal matrix composites also hold potential. It has been estimated that significant use of polymer nanocomposites as structural components could yield a 20–40 % reduction in vehicle weight and gain in fuel economy by up to 10–20 % [1]. To date, nanoclays and nanoparticles reinforcement of polymer composites has been commercialized, but efforts are underway in many laboratories and research institutes/universities to understand how nano-reinforcement actually works with polymer matrix and results in major changes in material properties. The understanding of these mechanisms will extend their scope to the reinforcement of more complicated anisotropic structures and advanced polymeric nanocomposite systems.

PNC materials have been a part of the automotive industry for several years, with the earliest application in 1991 when Toyota Motor Co. first introduced nylon 6/clay nanocomposites in the market to produce timing-belt covers as a part of the engine for their Toyota Camry cars, in collaboration with Ube Industries [2]. These PNCs have been used for power trains application with low production volumes, because of their substantial demand, high cost, and long lead times relative to conventional timing-belt covers fabrication. The major key drivers of the growth of polymer nanocomposites have been the reduced weight; fuel economy; low emissions; superior mechanical properties; outstanding combination of optical, electrical, thermal, magnetic, and other physicochemical properties; and part consolidation opportunities, as well as design flexibility. Although these benefits are well recognized by the automotive industry, use of polymer nanocomposites has been slowed by high material costs, low demand, and high-production lead times. Reduced weight of vehicles offers a potential reduction in fuel consumption and greenhouse gases emission from the automotive sector. It has been estimated that for every 10–20 %

reduction in the weight of the total vehicle, fuel economy improves by 5–10 %, and for every kilogram of vehicle weight reduction, there is the potential to reduce greenhouse gases emissions by 30 kg [3].

Today's average automobile consumes about 10–12 % plastics and composites (i.e., about 150–170 kg/vehicle), a more than 50 % increase from 100 kg in 2005 [4]. Small passenger vehicles generally possess more plastics, while larger vehicles comprise more composites by weight. There is a huge potential for polymer nanocomposites to tap the automotive markets as currently it captures merely 1 % market share of global automotive plastics and composites [3, 4]. Polymer nanocomposites provide a wide range of potential automotive applications such as body panels, exterior panels, bumper systems, power trains, chassis, tires, instrument panels, suspension, steering, air intake manifolds, fuel cells, brakes, leaf springs, drive shafts, sensors, wheel covers, fascias supports, valve covers, antiglare coatings, glazing, filters, cross vehicle beams, compressed natural gas fuel tanks, and other accessories, where the industrial demand varies widely by application and geographical region. The usage of polymer nanocomposites is predominantly in structural applications of the automobile today and is mostly as nanoclays-reinforced thermoplastic polymers.

1.1 Advantages and Limitations of Polymer Nanocomposites

The properties of PNCs are mostly influenced by the composition, blending, filler size (nano-size dimension influences the macroscopic properties), shape, state of compatibility and agglomeration, and degree of matrix filler adhesion. The multifunctional PNCs offer a number of significant advantages over traditional materials such as metals, plastics, and composites. Conventional composites usually require a high loading of fillers (10–50 wt%) to impart the desired mechanical, thermal, and electrical properties. But these properties with PNCs can be achieved with typically 1–6 wt% of the filler loading. In addition to this, other properties of polymer nanocomposites such as optical clarity, corrosion resistance, noise damping, and thermal stability over metallic materials can be achieved by altering the volume fraction of filler particles. In general, PNCs have the following advantages and limitations:

Advantages:

- Improved physico-mechanical properties such as enhanced tensile strength and modulus, fatigue, and corrosion resistance.
- Superior structural and thermal stability (high heat deflection temperature, Enhanced heat deflection temperature, etc.).
- Enhanced electrical conductivity can be achieved with conductive fillers.
- Improved noise, vibration, and harshness (NVH) properties.
- Low density relative to metallic and ceramic materials.
- Ease of production: PNCs can be fabricated through various precise and automated manufacturing techniques such as injection molding, compression molding (SMC/BMC), resin transfer molding, contact molding, and vacuum-bag molding.

Limitations:
- Difficulty in uniform dispersion and compatibility
- Oxidative and thermal instability of nanoclays
- Formation of agglomerates and voids
- High viscosity yields with high amount of filler
- Consistency and reliability in volume production

1.2 Automotive Industry Major Requirements and Commercialization Factors

Since its early commercial application in late 1991, the PNC industry has gone through an industrial revolution. With changing market needs and ever-increasing competition from other materials, the industry is meeting new challenges and developing new materials, technologies, and markets. Still the polymer nanocomposites industry has a long way to go. Its usage in the automotive market is a small fraction as compared to the usage of steel, aluminum, and metals. In order to achieve rapid growth, the PNC industry needs to focus more on high-volume applications and must compete with steel, aluminum, and metals in terms of price (life cycle cost) as well as performance. Automotive industry major requirements and commercialization factors are listed hereunder:
- Light weight, fuel economy, and lower greenhouse gas emissions
- Superior structural and functional properties requirements
- Consistent physical and mechanical properties
- High production rate
- High cost/performance ratio
- Improved colorability and painting
- Class A surface finish for exterior parts and styling feature
- Temperature resistance from -20 °C to 120 °C for body and interior parts
- Rust and Corrosion Proof

PNCs are considered one of the competitive materials for the automotive industry to solve the most common issues, i.e., improving fuel-consumption efficiency (more miles per gallon), reducing greenhouse gas emission, and reducing manufacturing costs.

1.3 Key Drivers, Growth Drivers, and Challenges for the Use of Polymer Nanocomposites in Automotive Industry

High performance at a reduced weight is the key driving force for the selection of polymer nanocomposite materials for automotive applications. Some of the other features, which drive the growth of the polymer nanocomposite industry, are listed below:
- PNCs provide capabilities for part integration, i.e., lesser number of tools and process steps are required. Several metallic components can be replaced by a single PNC component.

- PNCs have high specific stiffness (stiffness-to-density ratio) and offer the stiffness of steel at one-fifth the weight and equal the stiffness of aluminum at one-half the weight.
- The specific strength (strength-to-density ratio) of a PNC material is very high. Due to this, vehicles move faster with better fuel efficiency and low emission of greenhouse gases. It is typically in the range of 3–5 times improvement over steel and aluminum alloys. Due to higher specific stiffness and strength, PNC parts are lighter than their counterparts.
- The fatigue strength (endurance limit) is much higher for PNCs, which leads to excellent dimensional stability and reliable painting.
- PNCs offer a high amount of design flexibility. For example, the coefficient of thermal expansion (CTE) of PNC structures can be made zero by selecting suitable materials and processes.
- Complex parts, appearance, and special contours, which are sometimes not possible with metals, can be fabricated using PNC materials without welding or riveting of separate pieces. This increases reliability and reduces production times. It offers greater manufacturing feasibility.
- PNCs offer better crash-worthy properties because of the multiple failure modes that are present in the polymer composite structures. Crash-worthy characteristics can be achieved by proper design of polymer composite structures. Steering columns, lumber beams, chassis, and A and B pillars of the automobile can be designed in such a way that most of the energy during a collision can be absorbed by these systems without getting transferred to the passenger.
- Noise, vibration, and harshness (NVH) characteristics are better with PNC materials. PNCs dampen vibrations an order of magnitude better than metals.
- PNCs offer resistance to minor impact, i.e., no denting, ease of packing, handling and transportation, and no repairing.
- E-coat compatibility (online paintability) and ESD coating is possible for PNC components (SMC parts).

The automotive industry is one of the major industrial backbones of the world economy and it has a clear roadmap for fuel economy and low-carbon technologies, which will contribute to long-term low fuel consumption and carbon growth. Government, Tier I/II/III, and OEMs should look at the opportunities both on an economic level and an environmental level in supporting the automotive sector. Figure 23.1 deals with growth drivers and challenges related to market dynamics.

2 Materials and Properties

Among nanomaterials, nanoclays are the most commonly used commercial additive for the preparation of nanocomposites, accounting for nearly 80 % of the volume used. Carbon nanofibers, carbon nanotubes (mainly MWCNTs), and polyhedral oligomeric silsesquioxanes (POSS) are also being used commercially in nanocomposites, gaining ground fast with improvements in cost/performance and processability characteristics. The largest demand for nanoclays in nanocomposites

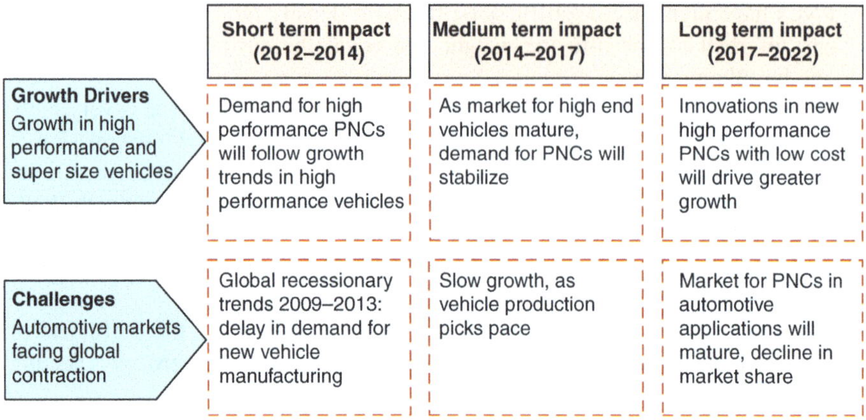

Fig. 23.1 Growth drivers and challenges of PNCs for automotive sector

is primarily driven by the need to meet the customers' (OEMs/Tier I, II, or III) cost expectations and performance of end-use parts, as nanoclays are less expensive ($6–$8/kg) than other nanomaterials and exhibit improved balance of stiffness and toughness, excellent mechanical and barrier properties, enhanced heat deflection temperature without loss in elongation, improved colorability, and improved scratch and mar resistance [5].

Nanoclays also offer a reduction in relative heat release, excellent dispersion and exfoliation, excellent flame-retardant synergy, and reduced weight. Polyolefin is commonly used as host polymer, and thermoplastics such as polyamide (nylon), polyphenylene sulfide (PPS), polyetheretherketone (PEEK), polyethylene terephthalate (PET), polycarbonate, thermosets such as epoxy, conductive polymers, and thermoplastic elastomers such as butadiene-styrene diblock copolymer are also used in more demanding automotive applications. The use of thermoplastics as a matrix material in nanocomposites has been growing steadily, especially in automotive applications, largely due to thermoplastic intrinsic properties such as enhanced mechanical, thermal, electrical, and barrier properties, excellent fracture toughness over thermosets, as well as the ability to be easily joined by mechanical joining and welding techniques [5].

Organoclays are most widely used as nanofiller, while carbon nanotubes are also gaining acceptance. The two major producers are Southern Clay Products Inc. with its Cloisite product line and Nanocor Inc. with its Nanomer products. Carbon nanotubes impart electrical and thermal conductivity, which allows electrostatic painting and when blended with nylon, to protect against static electricity in the fuel system. However, their commercial development is hindered by their high-price tags ($20/g), although they are available in masterbatches (containing 15–20 % nanotubes) for about $50–60/lbs. Green nanocomposites or biodegradable nanocomposites (cellulosic bio-plastic reinforced with clay) are the next generation of materials for automotive applications. They have the potential to replace or substitute existing petroleum derived nonbiodegradable Thermoplastic PolyOlefin

Table 23.1 Comparison of properties of PNCs with conventional composites [3–5]

Properties	Conventional polymer composites	PNCs (2–5 % nanoclay)
Tensile strength	1x	>1.5–2x
Flexural modulus	1x	>4–5x
Notched Izod impact strength	1x	>0.5–1x
HDT	1x	>1.5–2x
LCTE	1x	<1.5–2.5x
Specific gravity	1x	<0.5–1x
Shrinkage and warpage	1x	<0.5–1x

(TPO)-enabled nanocomposites [5]. Table 23.1 shows properties comparison of PNCs with conventional composites while Table 23.2 depicts some examples of commercially available nanocomposites and nanomaterials.

Most commonly used nanoparticles include:
- Montmorillonite organoclays (MMT)
- Polyhedral oligomeric silsesquioxanes (POSS)
- Carbon nanofibers (CNFs)
- Carbon nanotubes
- Graphene
- Nanosilica
- Nano-aluminum oxide (Al_2O_3)
- Nano-titanium oxide (TiO_2)
- Magnetic nanoparticles

Most commonly used as polymer matrices for making PNCs include:
- Polyolefin (mainly PP)
- Polyamides (nylons)
- Polystyrene
- Polyethylene terephthalate (PET)
- Polyurethanes
- Epoxy resins
- Polyimides
- Ethylene-vinyl acetate (EVA) copolymer
- Conducting polymer
- Styrene butadiene rubber (SBR)

3 Applications

PNCs have superior physico-mechanical and chemical properties and therefore possess the potential for very broad range of applications, i.e., structural and functional in many areas of the automotive industries. Polymer nanocomposites

Table 23.2 Raw material/nanointermediate manufacturers [5]

Product	Key benefits	Applications	Manufacturers
Polymers			
Polyolefin nanocomposites	Twofold cheaper than nylon PNCs, do not need drying, stiffer, stronger, less brittle, lighter, more easily recycled, improved flame retardancy, improved temperature resistance and stiffness, very good impact properties	Step assist, heavy-duty electrical enclosure, door frames, seat backs, sail panel, box rail, center bridge, fascias, rocker covers, side trim, grills, hood louvers, instrument panels, vertical and horizontal body and closure panels, fenders	LyondellBasell, Ube Industries Ltd.
Nylon nanocomposites	Improved modulus, strength, heat distortion temperature, barrier properties	Timing-belt cover, engine cover, fuel line, fuel hoses/valves, fuel tanks	RTP Company, Ube Industries Ltd., UNITIKA Ltd., Nylon Corporations of America (NYCOA), SABIC Innovative Plastics Pvt. Ltd.
Elastomeric nanocomposites (butyl, ethylene propylene diene monomer (EPDM), natural, rubber/clay)	Higher durability, reduced weight, reduced rolling resistance, easy rubber processing, high-barrier coatings	Tires	Yokohama Tire Corp., Pirelli SpA, The Goodyear Tire & Rubber Co., Continental AG, InMat Inc.
Nanomaterials			
Nanoclays	Low cost ($6–$8/kg), organophilic, enhanced flexural and tensile modulus, barrier properties and flame retardancy of polymer matrix	Additives and reinforcements	Southern Clay Products Inc., Nanocor Inc., Elementis Specialties Inc., Süd-Chemie Inc., Laviosa Chimica Mineraria, SpA
Carbon nanotubes	High electrical and thermal conductivity and low coefficient of thermal expansion	Additives, reinforcements, electrostatic painting for ESD protection, and EMI shielding	Bayer, Hyperion Catalysis International Inc., Nanocyl, Zyvex Corp.

could reduce production cycle time by up to 20–30 % in injection-molding machines (80–90 % of parts of automotive are being produced) and energy savings of up to 10–20 % as compared to conventional polymer composites [3–5].

PNCs have provided a fillip to automotive research, and this technology has been applied commercially and has received great attention in recent years. The major development in polymer nanocomposites has been carried out over the last two decades. Polymer nanocomposites are attracting increasing funding and investment from governments and industry around the world. At present, total global spend would be around $1 billion, but this is set to rise [3–5].

Table 23.3 World's top 25 vehicle-producing countries [9]

	Total (in thousands)	
Country	2010	2011
China	18,265	18,419
USA	7,763	8,654
Japan	9,629	8,399
Germany	5,906	6,304
South Korea	4,272	4,657
India	3,557	3,936
Brazil	3,382	3,406
Mexico	2,342	2,680
Spain	2,388	2,354
France	2,229	2,295
Canada	2,068	2,135
Russia	1,403	1,988
Iran	1,599	1,649
Thailand	1,645	1,478
United Kingdom	1,393	1,464
Czech Republic	1,076	1,200
Turkey	1,095	1,189
Indonesia	703	838
Poland	869	837
Argentina	717	829
Italy	838	790
Slovakia	562	640
Belgium	555	562
Malaysia	568	540
South Africa	472	533

A market research report, "Nanocomposites-A Global Strategic Business Report" (Electronics.ca Publications), states that world nanocomposites market is forecast to reach 1.3 billion pounds (lbs) by the year 2015, and growth in the nanocomposites market will be driven by robust demand outlook in the emerging application possibilities in automotive market [6]. By 2011, the automotive sector is expected to become the third-largest market for polymer nanocomposite applications, with over 15 % of the market [7]. According to a recent market report released by Frost & Sullivan, it is expected that carbon nanotubes will penetrate about 3.6 % within automotive composites. If 1 % of CNTs are loaded in these materials, it is estimated that CNT market for auto composites would be $ 35.52 million [8]. It is interesting to note that the projected market share in 2015 of automobile tires is 23 % of the total nanotech market in the automobile sector, and it is fully dedicated to the field of nanocomposites [8]. Tables 23.3 and 23.4 list the world's top 25 vehicle-producing countries and world motor vehicle production by countries in 2011.

Table 23.4 World motor vehicle production by countries in 2011 [9]

Country	Cars	Commercial vehicles	Total	% Change
Argentina	577,233	251,538	828,771	15.7 %
Australia	189,503	34,690	224,193	−8.1 %
Austria	130,343	22,162	152,505	45.2 %
Belgium	562,386	0	562,386	1.3 %
Brazil	2,534,534	871,616	3,406,150	0.7 %
Canada	990,483	1,144,410	2,134,893	3.2 %
China	14,485,326	3,933,550	18,418,876	0.8 %
Czech Republic	1,191,968	7,866	1,199,834	11.5 %
Egypt	53,072	28,659	81,731	−30.0 %
Finland	2,540	0	2,540	−61.9 %
France	1,931,030	363,859	2,294,889	2.9 %
Germany	5,871,918	439,400	6,311,318	6.9 %
Hungary	200,000	2,800	202,800	−4.1 %
India	3,053,871	882,577	3,936,448	10.7 %
Indonesia	561,863	276,085	837,948	19.3 %
Iran	1,413,276	235,229	1,648,505	3.1 %
Italy	485,606	304,742	790,348	−5.7 %
Japan	7,158,525	1,240,129	8,398,654	−12.8 %
Malaysia	496,440	43,610	540,050	−4.9 %
Mexico	1,657,080	1,022,957	2,680,037	14.4 %
Netherlands	40,772	32,379	73,151	−22.3 %
Poland	740,000	97,132	837,132	−3.7 %
Portugal	141,779	50,463	192,242	21.1 %
Romania	310,243	24,989	335,232	−4.5 %
Russia	1,738,163	249,873	1,988,036	41.7 %
Serbia	15,050	740	15,790	−12.4 %
Slovakia	639,763	0	639,763	13.9 %
Slovenia	168,955	5,164	174,119	−17.6 %
South Africa	312,265	220,280	532,545	12.8 %
South Korea	4,221,617	435,477	4,657,094	9.0 %
Spain	1,819,453	534,229	2,353,682	−1.4 %
Sweden	188,969	0	188,969	−13.0 %
Taiwan	288,523	54,773	343,296	13.1 %
Thailand	549,770	928,690	1,478,460	−10.1 %
Turkey	639,734	549,397	1,189,131	8.6 %
Ukraine	97,585	7,069	104,654	25.9 %
UK	1,343,810	120,189	1,463,999	5.1 %
USA	2,966,133	5,687,427	8,653,560	11.5 %
Uzbekistan	146,300	33,260	179,560	14.5 %
Others	368,615	127,215	495,830	2.2 %
Total	**59,929,016**	**20,163,824**	**80,092,840**	**3.2** %

Source: International Organization of Motor Vehicle Manufacturers (OICA)

3.1 Structural Applications

The first commercialization of polymer nanocomposites took place in 1991 (Toyota Motor Co.'s timing-belt covers); at about the same period, UNITIKA Co. of Japan introduced nylon 6 nanocomposite for engine covers on Mitsubishi GDI engines [10]. Manufactured by injection molding, the product is said to offer a 20 % weight reduction and excellent surface finish. In 2002, General Motors launched a step-assist automotive component made of polyolefin reinforced with 3 % nanoclays, in collaboration with Basell (now LyondellBasell Industries) for GM's Safari and Chevrolet Astro vans, followed by the application of these nanocomposites in the doors of Chevrolet Impalas [11–13]. The real surge in the commercialization of nanocomposites production has occurred over the last 10 years. In 2009, a one-piece compression-molded rear floor assembly was made by General Motors (GM) for their Pontiac Solace using nano-enhanced sheet molding compounds (SMCs) developed by Molded Fiber Glass Companies (MFG), Ohio. This technology is also in use in GM's Chevrolet Corvette Coupe and Corvette ZO6. The nanofilled SMCs exhibit significantly lower density than conventional SMCs resulting in improved fuel efficiency. The automotive industry can benefit from polymer nanocomposites in several applications such as engines and powertrain, suspension and breaking systems, exhaust systems and catalytic converters, frames and body parts, paints and coatings, lubrication, tires, and electric and electronic equipment.

Although research and development of nanocomposites related to theory, manufacturing, property characterization, and potential applications can be historically traced back to the period before the 1980s, the development of nanocomposites was first reported by Carter et al. [14] in 1950, when they developed organoclays with several organic onium bases to reinforce latex-based elastomers. In 1963, Nahin and Backlund of Union Oil Co. disclosed layered silicate-/polyolefin-based nanocomposites with strong solvent resistance and high tensile strength by irradiation-induced cross-linking [15]. However, they did not succeed in achieving complete exfoliation/intercalation characteristics of the organoclays within the polymer matrix. In 1976, it was Fujiwara and Sakomoto of the Unichika Co. who described the first organoclays hybrid polyamide nanocomposite [16]. In the latter part of the 1980s and the beginning of the 1990s, a research team from Toyota Central Research Development Laboratories (TCRDL) in Japan reported work on a nylon 6/clay nanocomposite and disclosed improved methods for producing nylon 6/clay nanocomposites using in situ polymerization similar to the Unichika process [17–21]. The research findings demonstrated a significant improvement in a wide range of physico-mechanical properties by reinforcing polymers with clay on the nanometer scale [22, 23]. The Toyota research team also reported various other types of clay nanocomposites based on polymers such as polystyrene, acrylic, polyimides, epoxy resin, and elastomers using a similar approach [24–27]. Since then, extensive research in nanocomposites field has been carried out worldwide. Figure 23.2 shows timeline for the commercialization of structural products by automotive players.

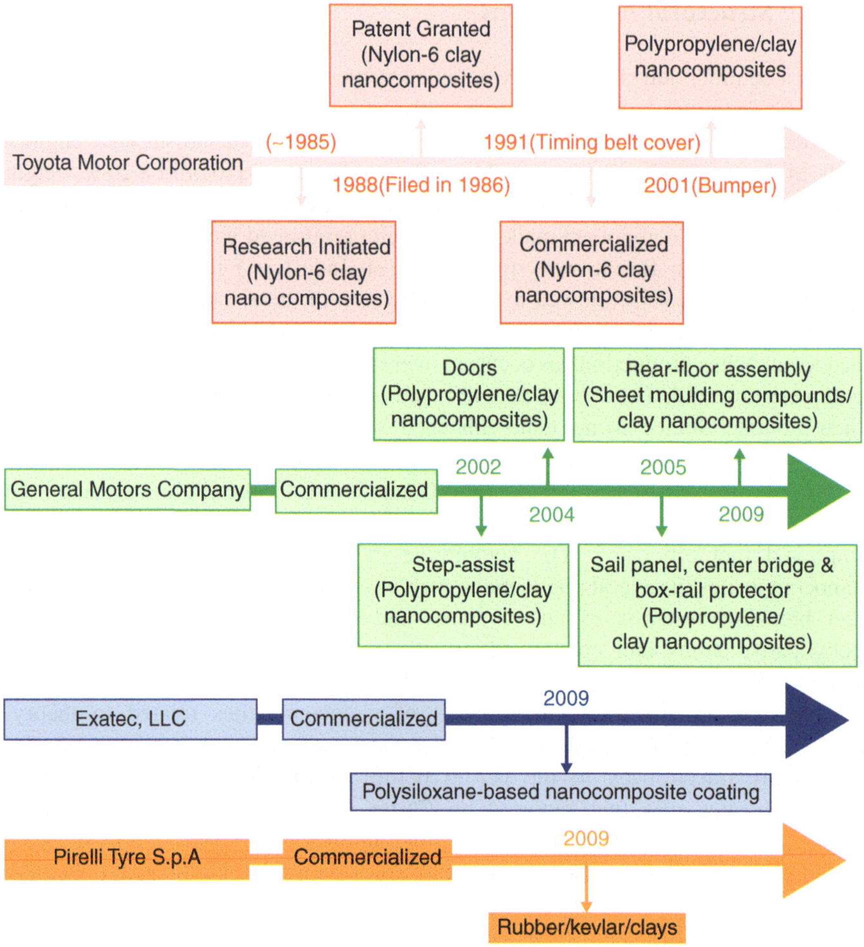

Fig. 23.2 Timeline for the commercialization of structural products by automotive players [5]

Advancements in the automotive industry and commercialization of PNC-enabled components promise to offer improvements in capabilities across a spectrum of structural applications. PNCs show significantly higher mechanical, thermal stability, and chemical properties without much increase in the specific gravity and sometimes retaining the optical clarity to a great extent, and therefore it is attracting increasing attention commercially from OEMs. The incorporation of 2–3 wt% inorganic nanoclay with Thermoplastic PolyOlefin (TPO) exhibits enhanced stiffness and is much lighter than parts with 5–10 times the amount of conventional talc filler. The use of PNCs for stiff and light exterior and interior parts like the step assist, trims, bumpers, fascias, and body side molding has an impact on structural design and applications, primarily by providing a comfort, safer, faster, and eventually low-cost transportation in the near future.

An exterior and interior component plays a key role in a passenger vehicle's appearance, as consumer requirements for product quality such as surface aesthetics, dimensional stability, and low gap tolerances are steadily increased. In addition to consumer requirements, original equipment manufacturers (OEMs) are faced with cost competitiveness issues like lighter and more energy-efficient vehicles and seek ways to reduce product development times and boost productivity while maintaining performance. TPO/clay is the single most widely used nanointermediate material in PNCs for the automotive structural application because of its superior impact and modulus properties while retaining low specific gravity. There are often many additional advantages that TPO/nanoclays could offer, such as low linear coefficient of thermal expansion (LCTE), increased heat distortion temperature (HDT), dimensional stability, controlled shrinkage and warpage, and improved surface appearance.

3.2 Functional

3.2.1 Under-the-Hood Applications

In automotive applications, increasing performance demands, the ability to reduce weight by replacing metal, call for reduced fuel consumption, EURO V and VI legislation, and environmental requirements as well as the rising cost of metals is prompting OEMs and design engineers to consider PNC technology (SMC/BMC/thermoplastic) as a replacement for existing materials. Thanks to their versatility, superior mechanical strength, light weight, corrosion resistance, electrical nonconductance, and exceptional thermal properties (resistance to high temperatures) and moldability, PNCs are an excellent choice for under-the-hood applications and metal conversion, such as from cast steel and aluminum as well. The industry demands for automotive under-the-bonnet applications are changing constantly, for example, the use of smaller engines with higher turbo pressures and exhaust gas recirculation (EGR) has increased. It can be applied to both structural (front-end modules, GORs) and functional applications.

Under-the-hood applications, high temperatures, and pressures tolerance, along with the need for chemical resistance, are industry requirements, which make adoption of PNCs (polyamides) in these areas move much more slowly. The worldwide commercialization of PNCs for under-the-bonnet/hoods/power trains applications could refer to Toyota Motor Co. first when it introduced nylon 6/clay nanocomposites in the market to produce timing-belt covers as a part of the engine for their Toyota Camry cars, in collaboration with Ube Industries in 1991 [5]. At about the same period, UNITIKA Co. of Japan introduced nylon 6 nanocomposite for engine covers on Mitsubishi GDI engines [6] manufactured by injection molding; the product is said to offer a 20 % weight reduction and excellent surface finish. This rapid growth can be attributed to a reduction in the component costs and reduction in overall vehicle weight and government regulations requiring increases in fuel economy. In the engine components, it has competition with glass fiber reinforced polyamides; today, around 50 % of passenger cars are equipped with glass fiber reinforced polyamides and about 80 % in Europe and the USA. Experimental and early test programs in applications for timing-belt covers and engine covers in the 1990s demonstrated PNC's capability for resisting heat and chemicals, paving the way for serious consideration in under-the-hood applications.

Key developments were pioneered in Japan, Germany, and the USA, where smaller cars, higher oil and gas costs, and emission control mandates drove innovation. By the 1990s, automotive researches/scientists were beginning to trust PNCs (nanoclay/nylon) in automotives' key applications. Noted for their ability to reduce weight and cost, while boosting engine performance, PNC use in under-the-hood applications gained a foothold in Japan as early as the 1990s and spread quickly in the late 1990s to Germany and the USA. Use of PNCs (nanoclay/nylon) not only reduced weight and cost, it also enhanced the fuel efficiency and performance of emission control devices.

PNC performance be led to under-the-hood innovations such as air intake manifolds, radiators, brake fluid and power steering fluid reservoirs, end tanks and gas tank caps, engine-cooling flex fans, transmission thrust washers and spring guides, and air cleaner support brackets. Even valve stem oil deflectors, which are required to resist oil and temperatures as high as 150–160 °C, could be converted to PNCs. It is believed that air intake manifolds can significantly reduce carbon dioxide and nitrogen oxide emissions, which are responsible for global warming and acid rain. PNC-enabled air intake manifolds could also lower fuel consumption, EGR system performance, and hot-start performance (because lower thermal conductivity of PNCs provides denser air).

Weight and cost savings are switching factors from metals to PNCs. Along the way, many researchers found additional benefits in PNC's design adaptability, aesthetics, virtually unlimited options for creative parts integration, and manufacturability. The undisputed benefits of using PNCs in under-the-hood applications led many industry experts to predict that at least 30 % of air intake manifold production would be PNCs by the end of 2020 and the average passenger cars will use 3–4 kg of PNCs in under-the-hood applications by 2020. Development programs and funding pertaining to fuel systems and other under-the-hood components are underway with the prospect of a long period of growth ahead.

Bearing retainers made of PNCs could replace metal retainers, because it effectively reduces the high premature failure rates, which is associated with metal retainers. The resistance of PNCs to hydrolysis, chemicals, and temperature opens a new realm of opportunities in air brake and air suspension system valve seats, gears, bushings, inlet and exhaust valves, exhaust stem, thermostat housings, radiator end caps, and overflow tubing.

Under-the-hood applications are adding a whole new dimension of performance capabilities to polyamide-based PNCs. Fuel filler neck assembly is the other potential application for clays/nylon composites, which could substantially reduce cost compared to metals and alloys (steel), while absorbing vibration between the body-mounted fuel tank and fuel neck projecting through the body panel. It also finds new applications in a passenger car cruise control system, fan with flexible blades for engine cooling, fuel vapor emission control canisters, sealing lids, integral mounting tabs and hooks, and power steering reservoirs.

PNCs have the ability to replace steel and save weight. For example, a steel front-end module can weigh 6–8 kg, while a PNC front-end module is about 50 % lighter (3 kg). The metal fuel tanks have a tendency to corrode, thereby requiring frequent flushing to eliminate rust buildup. Because PNCs (nylon/SMC/BMC) do not corrode, it saves annual service maintenance cost and is environment-friendly (recycling is possible).

At present, PNCs have limited applications and niche markets, but in the future it will penetrate the market with a substantial amount of under-the-hood applications, as recent government imposed safety and fuel economy regulations will spur PNC growth. PNC materials reduce weight and cost through parts integration and offer outstanding resistance to chemical, gasoline, oil, battery acid, and salt-spray corrosion compared to steel.

3.2.2 Paints and Coatings

For auto buyers, the optically smooth surface and aesthetic appeal are of utmost importance. Not all paints have excellent adhesion to a variety of substrates and excellent durability. Electrostatic painting can be directly utilized instead of a primer if the components are composed of electrically conductive PNCs. This process provides more uniformity and better-controlled thickness during the painting. During the electrostatic painting, attraction between the paint particles and the electrically conductive parts occurs, which enables strong adhesion and minimal wastage of the paint by drawing the paint droplets to the rear of the component. In addition to this, it also reduces environmental impact by controlling the emission of volatile organic compounds in paint. The use of PNC-enabled electrically conductive coatings may be able to replace the use of primer in exterior automotive applications to reduce cost and weight. In 1997, Ford used electrically conductive PNCs to produce mirror housings for their Taurus and Mercury Sable variants. GE plastics and Hyperion Catalysis International Inc. also produced a multiwalled carbon nanotube (MWCNT)-enabled conductive resin for an electrostatic painting process developed by United Technologies Automotives.

In advanced automotive systems, the rapidly growing volume of onboard electronics and microprocessor-controlled systems requires that the electronic subassemblies (ESA) in the vehicle must meet electromagnetic compatibility (EMC) requirements, i.e., electronics device should not emit levels of electromagnetic energy that cause electromagnetic interference (EMI) in other electronic components in close proximity. The different forms of electromagnetic energy that can cause EMI are electrostatic discharge (ESD), conducted, and radiated. EMI problems with automotive devices can be critical, not only from the technical standpoint but also from the view of public safety and comfort. For example, if airbag, antilock braking, cruise control, and other electronically controlled assemblies are adversely affected by EMI, operation of the vehicle systems could be affected due to nonperformance of an integrated circuit internally.

Today, electronic circuits are becoming smaller and more sensitive to external interference like ESD. Since ESD is a transient overvoltage generated by the friction of the human body, as automotive electronic systems become more portable and the transient susceptibility of semiconductors increases, government regulations are mandatory to maintain a minimum level of performance in all equipment. Europe is so serious about the problem that they require that equipment be certified via testing to meet IEC 61000-4 series ESD test specifications after 1996.

Advancement of automotive electronic systems has led to more and more stringent requirements for EMC and EMI shielding design. Mechanical and electrical design interfaces are challenging, especially for a new product development, in which a critical and early design decision has to be made either assuming EMC can be achieved with good electronic design to obviate the need for an EMI shield or anticipating the inclusion of an EMI shield. Moreover, the EMI shielding design should be optimized to meet the EMC requirements with the cost as low as possible. This also has increased the demand for selecting the correct EMI shielding materials and to develop new materials for EMI shielding applications.

One technique to meet EMC requirements is to shield/block the EMI signals from being emitted and/or penetrating into a defined space. The thin sheet metal sheathing (galvanized steel, Coated steel, etc.) is an effective EMI shielding material in automotive applications. However, these materials are expensive, heavy, and prone to rust, while adding to the complexity and cost of manufacturing processes. Conductive coating represents the most economical and reliable technique for solving EMC problems. Currently, silver paint, silver-coated copper paint and hybrid paint are commercially used as conductive paints and coatings, but they maximize substrate attack on engineering plastics like polycarbonate and polycarbonate blends.

Conductive polymer nanocomposites offer a potentially cost-effective alternative to metals. Conventional conductive fillers such as metal fibers, carbon fibers, and metal flakes are being used, which are dispersed in a polymer matrix to create an electrically conductive network that acts like a Faraday cage. EM radiation is either reflected or absorbed by the shielding polymer composite materials. Recently, conductive polymer nanocomposites have attracted great interest, both in academic and in industry, because of their potential applications in many areas including EMI shielding and electrostatic discharge (ESD). In contrast to conventional conductive fillers, nanofillers such as carbon nanotubes (CNTs) and graphene have at least one dimension in the nanometer range (high-aspect ratio); therefore they form conductive networks much more readily than conventional conductive fillers. Due to high-aspect ratio and larger filler-matrix interface, mechanical and thermal properties may also be improved. Furthermore, conductive polymer nanocomposites are lighter, but processing is a considerable challenge. Because of the strong tendency of conductive nanofillers to agglomerate, uniform dispersion of these fillers in the polymer matrix is a critical issue. There is need for the formation of an effective conductive network at low filler loading. Advancements could be achieved in this area through the use of a proprietary compounding technique.

PNC coating has the ability to achieve excellent electrical conductivity and shielding effectiveness even in a high-volume production environment with less coating. PNC coating is an alternative to traditional paint systems that is more durable and cost-effective. It also offers design possibilities that go beyond color, like brushed metal and chrome finish. Apart from enhanced properties, the use of PNC coating reduces cost by 10–12 %, due to elimination of traditional painting.

In 2012, Carbon Motors Corporation has launched the world's first police car coated with nanocomposite paint films (Fluorex® paint film) on thermoplastic exterior (body panels), which has been developed by Soliant LLC. These coatings demonstrate resistance against chemicals, weathering, UV, scratches, chips, dings, and dents.

3.2.3 Fuel Cells

Applications of fuel cells in automobiles have attracted significant research interest in the last few years due to growing government regulations and public concern about air pollution and consequent environmental problems. Currently, there are at least six generic fuel cell systems in varying stages of development, i.e., phosphoric acid fuel cells (PAFCs), molten carbonate fuel cells (MCFCs), solid oxide fuel cells (SOFCs), polymer electrolyte fuel cells (PEFCs), alkaline fuel cells (AFCs), and direct methanol fuel cells (DMFCs). But for automobiles, PEFC development is receiving the most attention from OEMs because of its low operating temperature, rapid start-up characteristics, and robust solid-state construction. Electrically conductive PNCs could be used as bipolar plates in PEFCs and to eliminate the need for painting. The world's leading vehicle manufacturers undertaking the development of PEFCs are Toyota, DaimlerChrysler, Nissan, Renault, Ford, General Motors, Fiat, Peugeot, Volkswagen, Mitsubishi, Suzuki, Honda, and Hyundai, among others.

In the automotive industry, electrostatic painting and bipolar plates in PEFCs are the two major applications of conducting polymers. The addition of conductive fillers such as carbon black, carbon fibers, carbon nanotubes, and graphene to polymer matrix enables the PNC to be electrically conductive.

3.2.4 Sensors

In automotive systems, sensor is used to collect data from various electromechanical devices working in the vehicle and senses the physical or chemical amount of parts and sends it to the engine control unit for vehicle control, comfort, driver safety, information, and emission control. Sensors in passenger cars are widely used in safety applications as well as engine and transmission applications, where they play a key role in reducing fuel consumption and lowering emissions. Sensors find its application in most of the automotive systems which include power train, chassis, and body, i.e., airbag, tire pressure monitoring systems, anti-blocking system, braking system, exhaust system, fuel supply system, cooling system, lubrication system, ignition system, steering, and suspension system. Today, advanced vehicles use a number of microcontroller-based engine control units, and as such sensor has become a vital component for the automotive system.

It is estimated that a typical vehicle consists more than 30 electrical and electronic systems and over 100 sensors. Although some sensor applications have been dominated by accelerometers and silicon micro-engineering technology and they compete with each other within a given application. There are different types of sensors used in automobile, which work on the various principles like variable reluctance, Hall effect, magnetoresistor, electrooptics, electromagnetic, piezoelectricity, piezo-resistive, and capacitive principle. PNC systems have potential to be used as sensor in all application areas of automotive systems such as power train, drive train, extra vehicular, and comfort and communications systems.

PNCs can be used in the fabrication of giant magnetoresistance (GMR) sensor, where it utilizes ferromagnetic/nonmagnetic layered structures made up of atomically thin films, in the range of 2–5 nm thickness. GMR has found applications as rotational and angle sensors and weak-field measurements in automotive systems and it exhibits sensitivities to variations of applied magnetic field which are up to 20 times greater than those metallic GMR sensors and it responds primarily to field orientation/direction rather than to field strength (use magnetic fields to conduct measurement information between physical value and sensor). GMR sensors offer several key advantages such as excellent rotational measurement, contactless angle measurement up to 180°, wear-free operating principle for angular and linear measurement, large permissible air gap tolerances up to 3.5 mm, withstanding extreme operating conditions (up to 150 °C), full redundancy possible, failsafe design, easy and cost-effective fabrication, flexible integration, and high bandwidth for measurements in time slots of less than 100 ms. However, the problem is associated with the uniform dispersion of nanoparticles throughout the polymer matrix and need for tightly controlled limits on its bias point.

A conductive polymer such as polyaniline is the first choice for GMR sensors, reinforced with magnetite (Fe_3O_4) nanoparticles. Polyurethane and vinyl ester as polymer matrix and magnetic nanoparticles such as iron, cobalt, nickel, and their alloys among them or with others have captured the intense attention for conducting fillers, owing to their unique magnetic and electrical properties. They have not only the excellent magnetic properties but also the admirable conductive properties for the applications in GMR sensors, which are being synthesized using a facile surface-initiated polymerization (SIP) technique. The dielectrical properties of these PNCs are strongly related to the magnetic nanoparticles loadings and their unique negative permittivity.

Existing and future development of the active and passive safety systems needs to be taken into consideration for the fabrication of GMR sensors, which will enable the vehicle systems to sense and interpret its surrounding environment as well as recognize potentially dangerous situations and provide a predetermined level of support to the driver. Parking assistance systems, blind spot detection and monitoring, lane change monitoring, and roundabout shunt prevention are examples of vehicle systems that could be enabled through the deployment of PNC-enabled GMR sensors.

3.2.5 Tires

Polymer-layered silicate (PLS) nanocomposites/clays have generated enormous industrial interests in the field of tires for some enhanced properties such as increased flexibility with excellent tensile strength, tread for lower rolling resistance, increased traction, improved flame-retardant property, outstanding antifatigue properties, superior gas barrier property, and low cost-to-performance ratio. The applications of the PNCs can be found in tire inner tube, tire inner liner, off-the-road (OTR) tire tread, and conveyer belts. Industry experts estimate that the tire industry could contain eight different polymer types, but natural rubber (NR), styrene butadiene rubber (SBR), butadiene rubber (BR), isoprene rubber (IR), and halogenated butyl rubber (HBR) are the main polymers used for tires production. Out of these polymers, SBR is now the most common PNC material being used in tires.

Reinforcement of nanofillers is very important for tire applications; nano-carbon black was first recognized as useful reinforcing filler for rubber in 1904, and from then carbon black reinforced rubber nanocomposites have been widely used in various tire products such as TB Tread, PC Tread, sidewall, and inner liner. In addition to nano-carbon black, nanoclay is also being used as nano-reinforcing filler and is promising for industrialization due to superior cost/performance ratio, excellent exfoliation, and easy preparation process.

Elastomeric nanocomposites are gaining momentum in automotive industry especially for tires application due to their lower rolling resistance, lower weight, and superior performance in terms of fuel saving. The key drivers for these materials are growing demand for fuel efficiency, strict automotive standards for safety, enhanced durability, and noise reduction. For example, elastomeric replacement of traditional inner liners with nanocomposite inner liners reduces permeability by 50 % and thus results in a decrease of approximately 2 kg per truck and also improvement in fuel efficiency (2 %). Major automotive tire producers involved in the development and manufacturing activities pertaining to nanocomposites are Yokohama Tire Corp., Japan; Pirelli SpA, Italy; The Goodyear Tire & Rubber Co., USA; Continental AG, Germany; InMat Inc., USA; etc. Elastomeric nanocomposites-enabled tire models include Goodyear UltraGrip Ice^{+}, Continental Eco Contact 5, Michelin Energy Saver, and Pirelli Cinturato P1 and are into the automotive market. Lanxess AG, Germany; Evonik Degussa GmbH, Germany; Cabot Corp., USA; Nanocor, USA; FCC Inc., China; Elementis Specialities plc, UK; Tokuyam, Japan; and Rhodia, France, are some of the leading producers of nanofillers. China has now become the world's largest manufacturer of automobiles and tires as well as consumer of automobile and latex rubber. The first production line of PNC-enabled tires with a capacity of approximately 10,000 t/year is established in China.

4 Product Outlook

There have been significant research and development activities in PNCs for the automotive industry, with polymer nanocomposites finding commercial applications since 1991 in bumpers, step assists, gas tanks, fuel pumps, interior and underbonnet parts, body panels, electrical parts and appliances, power tool housings, packaging and building components, shock absorbers, coatings, lubricants, and coolants. In addition, PNCs offer a variety of functions in automotive end-use parts such as structural plastic parts that exhibit higher mechanical performance with reduced weight, tires reinforced with nanoparticles for better abrasion resistance and improved gas permeability, fuel-borne catalysts for soot prevention in

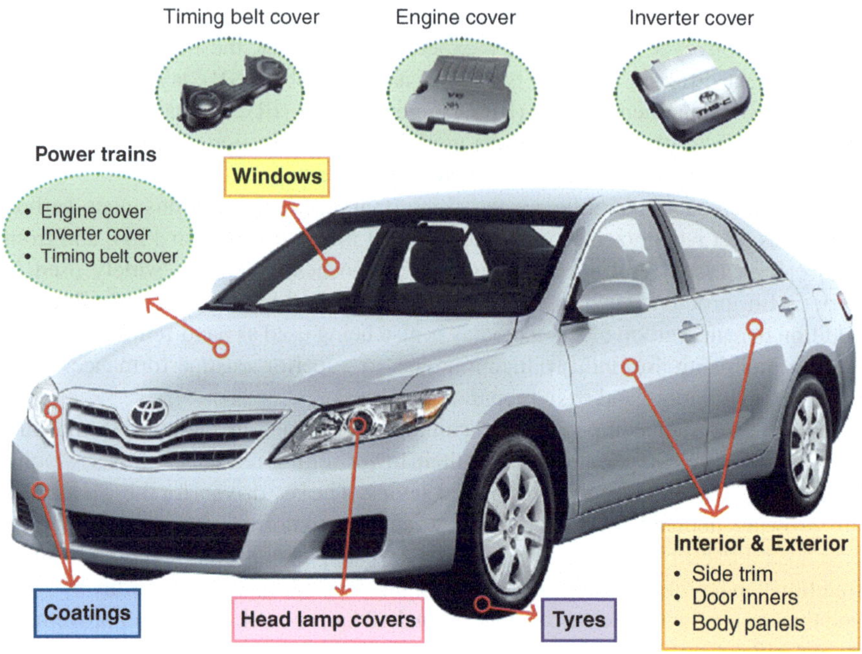

Fig. 23.3 Illustration of the usage of PNC structural parts [5]

particulate filters, car body coatings for greater scratch resistance and improved gloss, and antifog coatings for headlights and windshields.

The performance-to-cost ratio has been a major hurdle for gaining broader market acceptance as nanocomposites should meet the OEMs/molders/customer's cost expectations. Some early commercialized products have lapsed for cost reasons; damages include an automotive timing-belt cover based on nylon 6/clay nanocomposites from Japan's UNITIKA and an automotive mirror housing of conductive polyphenylene oxide (PPO)/nylon blends nanocomposites from GE Plastics. However, tremendous effort has been put forth by OEMs and molders to commercialize more volume of nanocomposites in automotive components.

Today, demand for Thermoplastic PolyOlefin/polypropylene PNC has moved beyond nylon 6/clay nanocomposites, mainly because of their low cost and enhanced physico-mechanical properties. In the past, the automotive industry was more inclined towards using nylon 6/clay nanocomposites for under-the-hood applications, where higher heat deflection temperature, enhanced stiffness, and light weight were the goals. The performance-to-cost ratio was a main constraint which halted the rapid growth of PNCs. However, nylon 6/clay nanocomposites (more costly) are still used for under-the-hood applications, fuel lines, and fuel system components. Considering the plurality of nanocomposites applications in the automotive industry, some major commercialized products are listed below (depicted in Fig. 23.3 and Tables 23.5 and 23.6):

Table 23.5 An overview of PNC's structural and functional application for automotive systems

Application	Components	PNC's advantages	Nanointermediates	Status
Structural	Exterior and interior structural parts, under the hood	Improved strength, greater stiffness, less filler loading, lower weight, good dimensional stability, easier manufacturing, greater design flexibility, use of lighter engine and suspension	Nanoclay/ thermoplastic olefin-based nanocomposites	Commercialized (medium to high volume)
Functional	Engine/power train/under-the-hood components	Part consolidation, excellent chemical resistance, electrochromic	Nanoclay/nylon nanocomposites	Commercialized (low volume)
	Paints and coatings	Improved corrosion resistance, self-cleaning, UV resistance, scratch resistance, antireflective, self-repairing, water and dirt repellent, moisture absorption, ESD, transparency, iridescence, photochromic, permeability	Conducting polymers, carbon nanotubes/ polyoxymethylene, nanostructured quantum dots, silica and soot	Commercialized (low volume)
	Fuel cells and batteries, fuel delivery/ storage	Faster response times, greater fuel and engine efficiency, increased temperature tolerance, reduced emissions and noise dampening, improved safety and lifetimes	Conductive polymers with CNTs, nanoceramics, nanometals	Yet to take off (low volume)
	Suspension/ braking systems	Comfort and flame retardancy	Nanoceramics, nanometals/ nanocomposites	Commercialized (low volume)
	Sensors	Comfort and safety	Conductive polymers and magnetic nanoparticles	Yet to take off (low volume)
	Tires	Wear resistance, lower rolling resistance, lower weight and superior performance in terms of fuel saving improved road comfort, holding pressure retention	Elastomeric nanocomposites (butyl, ethylene propylene diene monomer (EPDM), natural, rubber/clay)	Commercialized (medium to high volume)

Table 23.6 Snapshots of PNC's present, near-term, and long-term applications for automotive systems

	High-volume applications	Medium-volume applications	Limited-volume applications	Growth opportunity
Present	Body panels, fascias, bumpers, step assist, tires, door trim, engine cover armrest and armrest bezels	Timing-belt cover, inverter cover, handles, airbag, spoilers, instrument panels, door panels, head rests grilles, mirror housings, rub strips, wheel trim, cladding	Conducting paints and coatings, diesel fuel tanks, battery cases	Optimistic (10–15 %)
Near term (less than 5 years)	Gear parts, clutches, clutch thrust bearings, door sills, radiator grills, door-mirror housings, wheel covers and parts for electrical fittings such as cable harnesses, straps and connectors, headlamp housings, lamp holders and fuse boxes	Steering wheels, steering column mountings, roller bearing cages and fastening clips, housings and functional parts in electric drives, housings and mountings for various electrical and electronic components, and connectors	Sensor	Moderate (7–10 %)
Long term (beyond five years)	Gas filler caps, tank level sensors, gasoline pump housings and parts, valves, seat-belt release buttons, windshield wiper clips, suspension stabilizer links, levers and rods, sun roof frames, ball sockets, roll-over valves and wash nozzles, air deflectors and spoilers, frames for windows/sunroofs, air intake manifolds, front-end grill, opening panels, battery casings and covers, heat shields (engine transmission), cylinder head (e.g., valve, rocker, cam) covers, pillars (e.g., "a" and "c") and coverings	Interior (instrument panels, seating systems, interior trim, and HVAC); exterior (bumper systems, spoilers, running boards, and trim); under the hood (flexible products, air induction, and fluid management); and fuel systems, fan shrouds battery box/trays, air cleaner/housings, HVAC, canisters	Oil pan, fuel cells	Pessimistic (2–7 %)

- Toyota Motor Corporation was the first automotive manufacturer to produce nylon 6-/clay-based nanocomposites for commercial car parts in 1991, and since then, more nanocomposite-enabled automotive components have been commercially launched with increased interest from other automotive manufacturers. Since 2001, Toyota Motor Corporation has been using nanocomposites in a bumper that makes it 60 % lighter than the conventional plastic bumpers and twice as resistant to denting and scratching.
- In 2002, General Motors Company (GM), one of the world's largest automakers, successfully launched the first commercial auto exterior application (step assist) of Thermoplastic PolyOlefin (TPO) nanocomposites in its Safari and Astro van variants. After the instant success of step assists, two years later (January 2004) GM expanded use of nanocomposites to the doors (body side molding) of Chevrolet Impalas and is now using approximately 350 metric tons (MT) of nanocomposites annually for various automotive exterior parts and body panel applications, which is the highest volume of polyolefin-based nanocomposite material used in the world. GM's Hummer H2 SUT variant (sport utility truck), which was launched in 2005, uses about seven pounds of molded-in-color nanocomposite parts for its sail panel, center bridge, and box-rail protector applications. In 2009, a one-piece compression molded rear floor assembly was made by GM for their Pontiac Solstice (small sports car) using nano-enhanced sheet molding compounds (SMCs) developed by Molded Fiber Glass Companies (MFG), Ohio. This technology is also in use in Chevrolet Corvette Coupe and Corvette ZO6. The nanofilled SMCs show significantly lower density than conventional SMCs resulting in improved fuel efficiency. In addition, GM is also exploring the use of nanocomposite materials for exterior claddings and interior parts and in nonsupport trim and carbon nanotubes with plastics to replace current thermoset structural composites parts for automotive applications. General Motors teamed up with Basell (polyolefin resin supplier), Southern Clay Products Inc. (nanoclays supplier), and Blackhawk Automotive Plastics (molder) to produce a TPO nanocomposite for step-assist components that assist occupants in stepping into and out of the vehicle. The nanocomposite parts save weight, offer cost and performance advantages, are more dent- and scratch-resistant than the conventional materials, and provide superior toughness and heat deflection temperature as they do not change shape when subjected to temperature fluctuations, thus enhancing the overall quality of the vehicle.
- Other car manufacturers commercializing or using nanocomposites include Renault, Maserati, Ford Motor Company, Daimler AG (formerly DaimlerChrysler), and Audi. Renault's Clio and Megane variants are using nanotube-filled polyamide/polyphenylene ether blend for their sport fenders uses. Maserati's Quattroporte (luxury sports saloon) engine bay covers are made of nylon 6 nanocomposites, containing 2 % nanoclays by weight. Daimler AG uses nanocomposites for Smart Forfour's inner door handle. Audi and Putsch GmbH have developed polypropylene/polystyrene/clay nanocomposites for the heater vent applications of the Audi A3 model, as this material

replaces painted acrylonitrile butadiene styrene (ABS) parts and exhibits improved scratch resistance and a luxurious surface feel. Ford uses carbon nanofiber-filled polymer nanocomposites in Taurus sedan's mirror housing.

- Hyperion Catalysis International Inc. has added five thermoplastic families to its commercial masterbatches, including polystyrene (PS), polyetheretherketone (PEEK), polyphenylene sulfide (PPS), polyetherimide (PEI), and nylon, containing FIBRIL MWCNTs. They have vast experience in all aspects of manufacturing Class A body panel and trim parts for the automotive industry.
- Exatec, LLC, a wholly owned subsidiary of SABIC Innovative Plastics, has developed a polysiloxane-based nanocomposite coating used in conjunction with advanced polycarbonate glazing systems for the automotive market. These nanocomposites reduce environmental impact, enhance weatherability and abrasion resistance, enhance styling and aerodynamics, and reduce weight for greatly enhanced fuel efficiency.
- US-based Dow Automotive Systems has developed a reactive extrusion technology to produce nanocomposites based on nanoclays and cyclic butylene terephthalate (CBT®, supplied by Cyclics Corporation).
- Noble Polymers, a Cascade Engineering Family Company, manufactures and supplies Forte™, and Nubrid™ nanocomposites, which are made of halloysite nanotubes or clay/polypropylene, for structural seat backs in Honda's Acura variant.
- The Dow Chemical Company, a leader in specialty chemicals, has developed nanocomposites by using nanoclays as catalyst support for in situ polymerization of polypropylene homopolymer for semi-structural automotive uses.
- Süd-Chemie AG and Putsch Kunststoffe GmbH have developed ELAN XP nanocomposites based on nanoclays and blends of polypropylene and polystyrene for automotive interior applications, which offer a uniform matt surface, high scratch resistance, and good tactile properties.
- BASF of Germany has introduced nanocomposites based on Hyperion's carbon nanotubes and a conductive polyoxymethylene (POM), specifically for automotive fuel systems.
- Degussa's high-performance nanocomposites, which are based on CNTs-reinforced nylon 6, are used as tubes in the flexible portion of the fuel line systems.
- Nylon Corporations of America (NYCOA, US) manufactures nanoclays-/nylon 6-based nanocomposites that offer improved toughness and barrier properties and find application in electrical connectors, fluid reservoirs, molded brackets, wire jacketing, windshield washer tubing, spiral tubing, and convoluted tubing.
- RTP Co., a specialty thermoplastic compounder from the USA, is continuing to expand its nylon 6 nanocomposites product line and manufacturing capacity. The nanocomposite products offer 53 % higher heat deflection temperature (HDT) over unfilled nylon 6, with very little change in specific gravity.
- Elastomeric nanocomposites-enabled tire models include Goodyear UltraGrip Ice^{+}, Continental Eco Contact 5, Michelin Energy Saver, and Pirelli Cinturato P1.

5 Future Development and Directions

It is quite evident from the foregoing discussion that PNCs are finding many applications in the automotive industry, and the market for these materials is on the path of growth and expansion. Global automobile industry has witnessed a double-digit growth in recent years and is expected to grow at the rate of 12–14 % in the next couple of years. The OEMs/Tier I, Tier II, Tier III, raw material/nanointermediate manufacturers, researchers, and technologists are realizing that other than clays, nanomaterials like graphene, CNTs, carbon nanofibers, magnetic nanoparticles, nanofoams, multiscale hybrid reinforcement, biodegradable and natural nanocomposites, and graphene-enabled rubber nanocomposites could drive the market dynamics.

The price and performance advantages of graphene are challenging carbon nanotubes in PNC applications due to its intrinsic properties, and it is predicted that a single, defect-free graphene platelet could have an intrinsic tensile strength higher than that of any other material. In June 2010, a US Patent was granted to The Trustees of Princeton University for functional graphene-rubber nanocomposites [28], which can be produced at a much lower cost than carbon nanotubes and exhibit excellent mechanical strength, superior toughness, higher thermal stability, and electrical conductivity. This graphene-rubber nanocomposite can be employed in all the areas for gas barrier applications including tires and packaging. A similar patent was granted in 2011, for a composite material of nanoscale graphene and an elastomer for vehicle tire application [29]. The multiscale hybrid reinforcement is another potential polymer nanocomposite material for the automotive industry due to its enhanced load transfer at the reinforcement/matrix interface, i.e., by tailoring the interfacial shear strength, which is made of micro-sized carbon-fiber yarns and fabrics coated with carbon nanostructures. The high-performance racing cars and high-end sports cars require excellent properties such as structural stiffness, heat shielding, and impact and compressive strength. The polymer nanocomposite foams could be the right choice of material as they exhibit improved thermal insulation properties, superior peak load-bearing capacity, and higher threshold loads and impact loads.

The list of automotive applications for PNCs is growing steadily because of continued efforts to reduce weight and cost and improve fuel economy. PNCs are moving into ever more exterior, interior, paints and coatings, and under-the-bonnet applications in automotive components. The USA and China use 150 and 130 kg of plastic per passenger vehicle, respectively, while in India, Standard Sedan consumes about 135 kg of plastic (Volkswagen Group India uses), i.e., 10 % of the total weight of the car, and this is primarily used in bumpers, spoilers, front-end module carriers, interior trims, dashboards, seats and airbags, engine components, carpets, and many more, which shows the ample opportunity in global automobile sector for PNCs. In automotive sector, plastics like polypropylene, polyurethanes, polyamides, and thermosetting composites find widespread applications. Whilst most plastic in terms of volume and value remains polypropylene, which also represents the largest end-use market for PNCs. Specialty plastics like polyphenylene sulfide (PPS), polyether ether ketone (PEEK), and high-end polyamides like PA12 and PA46 are also being used extensively for niche applications.

Table 23.7 The top ten PNC R&D countries in the world

	PNC R&D	
Country	Structural applications	Functional applications
Japan	High	Moderate to high
USA	High	Moderate to high
Germany	Moderate to high	High
South Korea	Moderate to high	Moderate
United Kingdom	Moderate	Moderate to high
France	Moderate	High
Italy	Moderate	High
China	Moderate	Moderate
Sweden	Moderate	Moderate
India	Low	Low

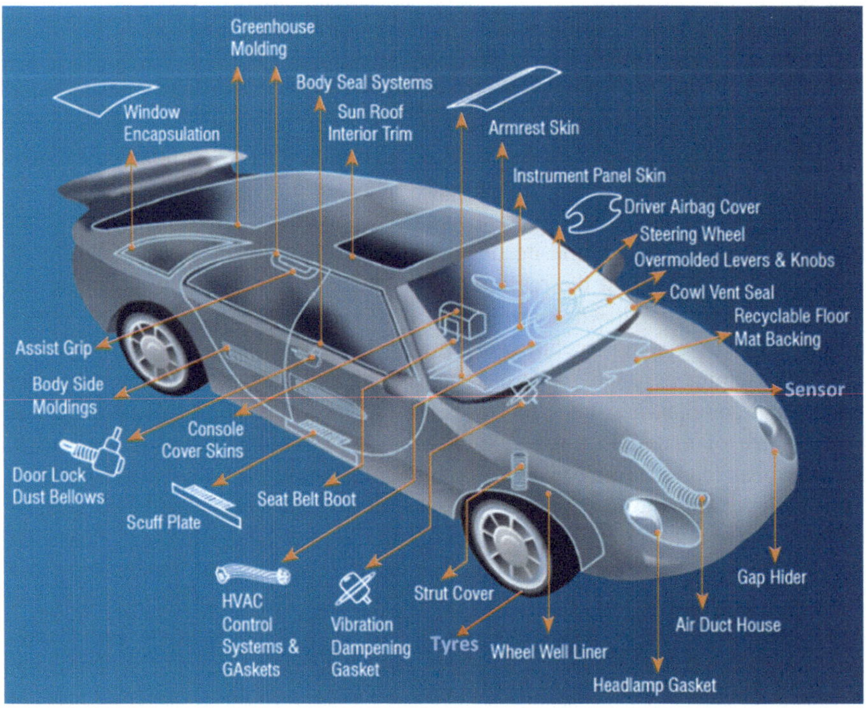

Fig. 23.4 Potential applications of polymer nanocomposites in the automotive

The various stakeholders in the automotive value chain need to take note of polymer nanocomposites technology and development, which has a growing market globally, but the higher cost of the end-use components is a shortcoming which needs to be overcome. Table 23.7 below lists the top ten PNC R&D countries in the world (Fig. 23.4).

6 Conclusion

PNCs undoubtedly are one of the rising stars in the field of automotive today. There are many other potential uses, excluding mentioned applications here, of PNCs because of its intriguing combination of spectacular properties. Polymer nanocomposites offer a number of advantages to the automobile manufacturers who are looking for easy transition from metals to comply with the fuel-efficiency legislations. The commercial success of nano-enabled products for the automotive market has been slow and used currently only in niche applications such as external body parts, interior and under-bonnet parts, and coating and fuel system components but is expected to be a major growth area in the coming era.

Today, in PNCs, most commonly preferred nanofiller is nanoclays because of their low cost, availability in large scales, and enhanced physico-mechanical properties such as light weight, high strength-to-weight ratio, high impact strength, dimensional stability, long-term durability, directional strength as well as low thermal conductivity, low coefficient of thermal expansion, flame retardancy, noise dampening, and corrosion resistance. There have been also huge interests for using magnetic nanoparticles, conductive nanofiller such as CNTs, graphene, carbon nanofibers, and core shell as fillers for PNCs.

The growth in activity surrounding PNCs continue unabated as more R&D funds are poured in by the funding agencies, venture capitalists, and companies as they look to exploit the expanding range of novel properties that are being discovered. To give an example, in Europe, the automotive industry invests over 5 % of its annual turnover in R&D and the major focus is on developing better coatings and paints, and stronger, more durable end-use parts. The use of polymer nanocomposites in the automotive market is thus set to escalate over the next 10 years.

References

1. Patel V (2012) Polym Soc 4:44–47
2. Carter LW, Hendricks JG, Bolley DS (1950) Elastomer reinforced with modified clay. US Patent No. 2,531,396, November 28, 1950 (Filed on March 29, 1947), Assignee: National Lead Co
3. Patel V (2012) Polym Soc 5:39–42
4. Patel V (2012) FRP Today 9:44–47
5. Patel V, Mahajan YR (2011) Nanotech Insights 2:17–24
6. Nanocomposites – a global strategic business report, March 2011, published by Electronics.ca, http://www.electronics.ca/presscenter/articles/1404/1/GlobalNanocomposites-Market-to-Reach-13-Billion-Pounds–by-2015/Page1.html
7. Nanocomposites, nanoparticles, nanoclays, and nanotubes, June 2006, published by BCC Research, http://www.bccresearch.com/report/NAN021C.html
8. Potential market for carbon nanomaterials' applications, February 28, 2011, published by Frost & Sullivan, www.nist.gov/cnst/upload/Valenti-NIST.pdf
9. http://oica.net/category/production-statistics
10. Nahin PG, Backlund PS (1961) Organoclay-polyolefin compositions, US Patent No. 3,084,117, April 2, 1963 (Filed on April 4, 1961), Assignee: Union Oil Co

11. Fujiwara S, Sakamoto T (1976) Method for manufacturing a clay/polyamide composite. Japanese Kokai Patent Application No.109, 998 (1976), Assignee: Unichika K.K., Japan
12. Fukushima Y, Inagaki S (1987) J Incl Phenom 5:473–482
13. Okada A, Fukushima Y, Kawasumi M, Inagaki S, Usuki A, Sugiyama S, Kurauch T, Kamigaito O (1988) Composite material and process for manufacturing same. US Patent No. 4,739,007, April 19, 1988 (Filed on September 19, 1986), Assignee: Kabushiki Kaisha Toyota Chuo Kenkyusho, Japan
14. Kawasumi M, Kohzaki M, Kojima Y, Okada A, Kamigaito O (1989) Process for producing composite material. US Patent No. 4,810,734, March 7, 1989 (Filed on March 15, 1988), Assignee: Kabushiki Kaisha Toyota Chuo Kenkyusho, Japan
15. Kojima Y et al (1993) J Mater Res 8:1185–1189
16. Kojima Y et al (1993) J Appl Polymer Sci 4:1259–1264
17. Usuki A, Kojima Y, Kawasumi M, Okada A, Fukushima Y, Kurauch T, Kamigaito O (1993) J Mater Res 8:1179–1184
18. Usuki A, Mizutani T, Fukushima Y, Fujimoto M, Fukumori K, Kojima Y, Sato N, Kurauch T, Kamigaito O (1989) Composite material containing a layered silicate. US Patent no 4,889,885, December 26, 1989 (Filed on March 4, 1988), Assignee: Kabushiki Kaisha Toyota Chuo Kenkyusho, Japan
19. Okada A, Fukumori K, Usuki A, Kojima Y, Sato N, Kurauchi T, Kamigaito O (1991) ACS Polym Prepr 32:540–541
20. Okada A, Usuki A (1995) Mater Sci Eng 3:109–115
21. Yano K, Usuki K, Okada A, Kurauch AT (1992) Polyimide composite material and process for producing the same. US Patent No. 5, 164, 46, Nov.17, 1992 (Filed on May 30, 1991), Assignee: Kabushiki Kaisha Toyota Chuo Kenkyusho, Japan
22. Yano K, Usuki A, Okada A, Part A (1997) Polym Chem 35:2289–2294
23. Gao F (2004) Mater Today 7:50–55
24. Edser C (2002) Plastic Addit Compd 4:30–33
25. Kurauchi T, Okada A, Nomura T, Nishio T, Saegusa S, Deguchi R (1991) Nylon 6-clay hybrid-synthesis, properties and application to automotive timing belt cover. SAE Technical Paper Ser, 910584
26. Cox H et al (2004) Nanocomposite systems for automotive applications. In: Presented at 4th world congress in nanocomposites. EMC, San Francisco, 1–3 Sept 2004
27. Zhao QZ, Nardelli MB, Bernholc JA (2002) Theoretical study. Phys Rev B 65:144105
28. Prud'homme R, Ozbas B, Aksay I, Register R, Adamson D (2010) Functional graphene–rubber nanocomposites. U.S. Patent 7,745,528, June 29, 2010 (Filed on October 06, 2006), Assignee: The Trustees of Princeton University
29. Zhamu, Aruna, Z. Jang, Bor (2009) Pristine nano graphene-modified tires. US Patent 7,999,027, August 16, 2011 (Filed on August 20, 2009), Assignee: Nanotek Instruments, Inc

The manufacturer's authorised representative in the EU is Springer Nature Customer Service Centre GmbH, Europaplatz 3, 69115 Heidelberg, Germany. If you have any concerns regarding our products, please contact ProductSafety@springernature.com

Printed and bound by CPI Group (UK) Ltd, Croydon, CR0 4YY

23/09/2026

02221529-0005